TUNNELS AND UNDERGROUND CITIES: ENGINEERING AND
INNOVATION MEET ARCHAEOLOGY, ARCHITECTURE AND ART

PROCEEDINGS OF THE WTC2019 ITA-AITES WORLD TUNNEL CONGRESS, NAPLES, ITALY, 3-9 MAY, 2019

Tunnels and Underground Cities: Engineering and Innovation meet Archaeology, Architecture and Art

Volume 5: Innovation in underground engineering, materials and equipment - Part 1

Editors

Daniele Peila
Politecnico di Torino, Italy

Giulia Viggiani
University of Cambridge, UK
Università di Roma "Tor Vergata", Italy

Tarcisio Celestino
University of Sao Paulo, Brasil

CRC Press
Taylor & Francis Group
Boca Raton London New York

CRC Press is an imprint of the
Taylor & Francis Group, an **informa** business

A BALKEMA BOOK

Cover illustration:

View of Naples gulf

CRC Press/Balkema is an imprint of the Taylor & Francis Group, an informa business

© 2020 Taylor & Francis Group, London, UK

Typeset by Integra Software Services Pvt. Ltd., Pondicherry, India

Published by: CRC Press/Balkema
        Schipholweg 107C, 2316XC Leiden, The Netherlands
        e-mail: Pub.NL@taylorandfrancis.com
        www.crcpress.com – www.taylorandfrancis.com

ISBN: 978-0-367-46870-5 (Hbk)
ISBN: 978-1-003-03161-1 (eBook)

Tunnels and Underground Cities: Engineering and Innovation meet Archaeology, Architecture and Art, Volume 5: Innovation in underground engineering, materials and equipment - Part 1 – Peila, Viggiani & Celestino (Eds)
© 2020 Taylor & Francis Group, London, ISBN 978-0-367-46870-5

Table of contents

*Tunnels and Underground Cities: Engineering and Innovation meet Archaeology,
Architecture and Art, Volume 5: Innovation in underground engineering,
materials and equipment - Part 1 – Peila, Viggiani & Celestino (Eds)
© 2020 Taylor & Francis Group, London, ISBN 978-0-367-46870-5*

Preface

The World Tunnel Congress 2019 and the 45th General Assembly of the International Tunnelling and Underground Space Association (ITA), will be held in Naples, Italy next May.

The Italian Tunnelling Society is honored and proud to host this outstanding event of the international tunnelling community.

Hopefully hundreds of experts, engineers, architects, geologists, consultants, contractors, designers, clients, suppliers, manufacturers will come and meet together in Naples to share knowledge, experience and business, enjoying the atmosphere of culture, technology and good living of this historic city, full of marvelous natural, artistic and historical treasures together with new innovative and high standard underground infrastructures.

The city of Naples was the inspirational venue of this conference, starting from the title Tunnels and Underground cities: engineering and innovation meet Archaeology, Architecture and Art.

Naples is a cradle of underground works with an extended network of Greek and Roman tunnels and underground cavities dated to the fourth century BC, but also a vibrant and innovative city boasting a modern and efficient underground transit system, whose stations represent one of the most interesting Italian experiments on the permanent insertion of contemporary artwork in the urban context.

All this has inspired and deeply enriched the scientific contributions received from authors coming from over 50 different countries.

We have entrusted the WTC2019 proceedings to an editorial board of 3 professors skilled in the field of tunneling, engineering, geotechnics and geomechanics of soil and rocks, well known at international level. They have relied on a Scientific Committee made up of 11 Topic Coordinators and more than 100 national and international experts: they have reviewed more than 1.000 abstracts and 750 papers, to end up with the publication of about 670 papers, inserted in this WTC2019 proceedings.

According to the Scientific Board statement we believe these proceedings can be a valuable text in the development of the art and science of engineering and construction of underground works even with reference to the subject matters "Archaeology, Architecture and Art" proposed by the innovative title of the congress, which have "contaminated" and enriched many proceedings' papers.

Andrea Pigorini
SIG President

Renato Casale
Chairman of the Organizing Committee WTC2019

Tunnels and Underground Cities: Engineering and Innovation meet Archaeology,
Architecture and Art, Volume 5: Innovation in underground engineering,
materials and equipment - Part 1 – Peila, Viggiani & Celestino (Eds)
© 2020 Taylor & Francis Group, London, ISBN 978-0-367-46870-5

Acknowledgements

REVIEWERS

The Editors wish to express their gratitude to the eleven Topic Coordinators: Lorenzo Brino, Giovanna Cassani, Alessandra De Cesaris, Pietro Jarre, Donato Ludovici, Vittorio Manassero, Matthias Neuenschwander, Moreno Pescara, Enrico Maria Pizzarotti, Tatiana Rotonda, Alessandra Sciotti and all the Scientific Committee members for their effort and valuable time.

SPONSORS

The WTC2019 Organizing Committee and the Editors wish to express their gratitude to the congress sponsors for their help and support.

Tunnels and Underground Cities: Engineering and Innovation meet Archaeology,
Architecture and Art, Volume 5: Innovation in underground engineering,
materials and equipment - Part 1 – Peila, Viggiani & Celestino (Eds)
© 2020 Taylor & Francis Group, London, ISBN 978-0-367-46870-5

WTC 2019 Congress Organization

HONORARY ADVISORY PANEL

Pietro Lunardi, President WTC2001 Milan
Sebastiano Pelizza, ITA Past President 1996-1998
Bruno Pigorini, President WTC1986 Florence

INTERNATIONAL STEERING COMMITTEE

Giuseppe Lunardi, Italy (Coordinator)
Tarcisio Celestino, Brazil (ITA President)
Soren Eskesen, Denmark (ITA Past President)
Alexandre Gomes, Chile (ITA Vice President)
Ruth Haug, Norway (ITA Vice President)
Eric Leca, France (ITA Vice President)
Jenny Yan, China (ITA Vice President)
Felix Amberg, Switzerland
Lars Barbendererder, Germany
Arnold Dix, Australia
Randall Essex, USA
Pekka Nieminen, Finland
Dr Ooi Teik Aun, Malaysia
Chung-Sik Yoo, Korea
Davorin Kolic, Croatia
Olivier Vion, France
Miguel Fernandez-Bollo, Spain (AETOS)
Yann Leblais, France (AFTES)
Johan Mignon, Belgium (ABTUS)
Xavier Roulet, Switzerland (STS)
Joao Bilé Serra, Portugal (CPT)
Martin Bosshard, Switzerland
Luzi R. Gruber, Switzerland

EXECUTIVE COMMITTEE

Renato Casale (Organizing Committee President)
Andrea Pigorini, (SIG President)
Olivier Vion (ITA Executive Director)
Francesco Bellone
Anna Bortolussi
Massimiliano Bringiotti
Ignazio Carbone
Antonello De Risi
Anna Forciniti
Giuseppe M. Gaspari

Giuseppe Lunardi
Daniele Martinelli
Giuseppe Molisso
Daniele Peila
Enrico Maria Pizzarotti
Marco Ranieri

ORGANIZING COMMITTEE

Enrico Luigi Arini
Joseph Attias
Margherita Bellone
Claude Berenguier
Filippo Bonasso
Massimo Concilia
Matteo d'Aloja
Enrico Dal Negro
Gianluca Dati
Giovanni Giacomin
Aniello A. Giamundo
Mario Giovanni Lampiano
Pompeo Levanto
Mario Lodigiani
Maurizio Marchionni
Davide Mardegan
Paolo Mazzalai
Gian Luca Menchini
Alessandro Micheli
Cesare Salvadori
Stelvio Santarelli
Andrea Sciotti
Alberto Selleri
Patrizio Torta
Daniele Vanni

SCIENTIFIC COMMITTEE

Daniele Peila, Italy (Chair)
Giulia Viggiani, Italy (Chair)
Tarcisio Celestino, Brazil (Chair)
Lorenzo Brino, Italy
Giovanna Cassani, Italy
Alessandra De Cesaris, Italy
Pietro Jarre, Italy
Donato Ludovici, Italy
Vittorio Manassero, Italy
Matthias Neuenschwander, Switzerland
Moreno Pescara, Italy
Enrico Maria Pizzarotti, Italy
Tatiana Rotonda, Italy
Alessandra Sciotti, Italy
Han Admiraal, The Netherlands
Luisa Alfieri, Italy
Georgios Anagnostou, Switzerland

Andre Assis, Brazil
Stefano Aversa, Italy
Jonathan Baber, USA
Monica Barbero, Italy
Carlo Bardani, Italy
Mikhail Belenkiy, Russia
Paolo Berry, Italy
Adam Bezuijen, Belgium
Nhu Bilgin, Turkey
Emilio Bilotta, Italy
Nikolai Bobylev, United Kingdom
Romano Borchiellini, Italy
Martin Bosshard, Switzerland
Francesca Bozzano, Italy
Wout Broere, The Netherlands
Domenico Calcaterra, Italy
Carlo Callari, Italy

Luigi Callisto, Italy

Elena Chiriotti, France

Massimo Coli, Italy

Franco Cucchi, Italy

Paolo Cucino, Italy

Stefano De Caro, Italy

Bart De Pauw, Belgium

Michel Deffayet, France

Nicola Della Valle, Spain

Riccardo Dell'Osso, Italy

Claudio Di Prisco, Italy

Arnold Dix, Australia

Amanda Elioff, USA

Carolina Ercolani, Italy

Adriano Fava, Italy

Sebastiano Foti, Italy

Piergiuseppe Froldi, Italy

Brian Fulcher, USA

Stefano Fuoco, Italy

Robert Galler, Austria

Piergiorgio Grasso, Italy

Alessandro Graziani, Italy

Lamberto Griffini, Italy

Eivind Grov, Norway

Zhu Hehua, China

Georgios Kalamaras, Italy

Iurij Karlovsek, Australia

Donald Lamont, United Kingdom

Albino Lembo Fazio, Italy

Roland Leucker, Germany

Stefano Lo Russo, Italy

Sindre Log, USA

Robert Mair, United Kingdom

Alessandro Mandolini, Italy

Francesco Marchese, Italy

Paul Marinos, Greece

Daniele Martinelli, Italy

Antonello Martino, Italy

Alberto Meda, Italy

Davide Merlini, Switzerland

Alessandro Micheli, Italy

Salvatore Miliziano, Italy

Mike Mooney, USA

Alberto Morino, Italy

Martin Muncke, Austria

Nasri Munfah, USA

Bjørn Nilsen, Norway

Fabio Oliva, Italy

Anna Osello, Italy

Alessandro Pagliaroli, Italy

Mario Patrucco, Italy

Francesco Peduto, Italy

Giorgio Piaggio, Chile

Giovanni Plizzari, Italy

Sebastiano Rampello, Italy

Jan Rohed, Norway

Jamal Rostami, USA

Henry Russell, USA

Giampiero Russo, Italy

Gabriele Scarascia Mugnozza, Italy

Claudio Scavia, Italy

Ken Schotte, Belgium

Gerard Seingre, Switzerland

Alberto Selleri, Italy

Anna Siemińska Lewandowska, Poland

Achille Sorlini, Italy

Ray Sterling, USA

Markus Thewes, Germany

Jean-François Thimus, Belgium

Paolo Tommasi, Italy

Daniele Vanni, Italy

Francesco Venza, Italy

Luca Verrucci, Italy

Mario Virano, Italy

Harald Wagner, Thailand

Bai Yun, China

Jian Zhao, Australia

Raffaele Zurlo, Italy

Innovation in underground engineering, materials and equipment

Tunnels and Underground Cities: Engineering and Innovation meet Archaeology,
Architecture and Art, Volume 5: Innovation in underground engineering,
materials and equipment - Part 1 – Peila, Viggiani & Celestino (Eds)
© 2020 Taylor & Francis Group, London, ISBN 978-0-367-46870-5

Artificial intelligence technique for geomechanical forecasting

M. Allende Valdés, J.P. Merello & P. Cofré
SKAVA Consulting, Santiago, Chile

ABSTRACT: After every blast, rock mass classification is performed. However, no person can see beyond the face. There are methods commonly used for this purpose: core drilling, measurement while drilling and probe holes. In this project, machine learning techniques (Artificial Intelligence), were applied to geotechnical information from probe hole drilling and face mappings, in order to find patterns and infer functions based on data used for training. Information had to be organized to be accessed by the machine learning method. The model learns from training data; it is understood that non-experienced situations cannot be predicted. Because of this, it was assumed that every project would need its own training. Data from a testing tunnel was considered. Once the model was trained, it was used as a forecasting tool during the performance of new Probe Holes. Results show that the model has an accuracy of +85% forecasting rock mass classification.

1 INTRODUCTION

For the construction of every underground facility in rock, the site's geology is the most important condition to consider for every stage of the project, from feasibility studies to detail engineering and construction. Prospecting and studies are performed following the level of certainty required for the project subjected to technical requisites and budget. For tunnels, in the early stages of engineering, core drilling is by far the most used method of prospecting. Depending on the overburden of the tunnel, a large total of meters of core drillings will be needed to reach a single point of the axis of the tunnel. However, there are other forms of prospecting available and used worldwide.

The geology to consider for the design of a project is the result of the interpretation of all prospections, review of references, studies and all available information, generally resulting in a Geotechnical Baseline Report. However, this information gives a general overview and does not cover the drastic variations that geology can have within a matter of meters during the construction of underground facilities resulting in changes in support conditions and in most severe cases, tunnel collapses.

This is the reason why during the construction of tunnels the forecasting of rock mass is always wanted in order to perceive in advance the changes in the geology that could result in construction delays and in worst case scenarios, damage to equipment or loss of life. In other words, a trustworthy forecast of the rock mass to be excavated will result in the reduction of geological risk.

This paper describes some of the current methods for rock mass forecasting and how the application of Artificial Intelligence (AI) improves rock mass forecasting.

2 ROCK MASS FORECASTING METHODS

Currently, rock mass forecasting is performed by several methods. Each method has advantages and disadvantages.

2.1 Core drillings

Core drilling can be precise, but expensive and slow (Chapman 2000). Only specific conditions will justify the investment, because advancing face has to be halted to perform the task. If it is done in parallel, then the advance rate of the face will probably be higher than that of the core drilling, returning useless information. This occurred during the construction of the tunnel detailed in this paper. At the end, a special niche had to be constructed to perform the core drilling without stopping the excavation face advance. Due to the complex logistics and numerous interferences between core drilling and advance teams, the drill & blast cycle had important delays while coexisting with each other. The advance rate of the core drilling was only a couple of meters a week faster than the advance face, so the useful information provided was good only to forecast one or two rounds. Also, a joint system subparallel to the core drill was not detected, producing small deviations in the forecasting information.

At the end, core drilling was a very expensive tool (a special niche had to be built) that contributed with limited forecasting information. Nevertheless, this was high quality information.

2.2 Measurement while drilling

Measurement while drilling (MWD) is a very powerful tool. The interpretation of digital drilling parameters can give useful information about what lies behind the face, but the investment in computerized equipment, and training of personnel imposes a huge constraint (Rivera 2012). Even when the MWD is considered for tunnel construction, the interpretation of the parameters must be done by a geologist with special training in such tools. In most of the cases the geologist is in charge of the drilling equipment while the MWD is in use. This can generate difficulties when the geologists in charge of the definition of the rock mass support belong to an independent third party, because the contractor must allow an external member to use their equipment.

In the case study presented in this paper, the drilling equipment (Jumbo) did not offer the option to include automated MWD tools, so this option was not considered.

2.3 Probe holes

Probe holes, on the other hand, are cheap, fast and easy to perform (Bilgin 2016). The problem is that the data interpretation is very subjective, generally leaving the responsibility in the hands of the operator. The analysis will vary depending on which shift, equipment and operator is drilling. Probe holes are destructive perforations (i.e. no core recovery), so all information is gathered during the drilling process. Also, in old drilling equipment like the one used in the project listed in this paper, maintenance operations can modify the jumbo's performance.

All the information resulting from probe holes will be summarized and interpreted by the operations manager or geologist in charge. This will create a large scattering of information if no unified criteria are applied to the project. Also, as mentioned before, the handling by the drilling operator plays a very important role, because the perforation with old equipment like that used in the study case depends on 2 variables: drilling pressure and rotation speed. So, the operator has to modify these 2 variables to drill most effectively.

2.4 Others

Nowadays, there are other methods available for the forecast of rock mass. Just to name one, Tunnel Seismic Prediction (TSP) uses seismic induced waves and geophysics to interpret properties of the upcoming rock mass and significant changes (i.e. faults, shear zones, etc.).

3 DEFINITION OF THE DATA

In order to work on new forecasting options, a set of data was needed. A project located in Perú was selected for this purpose. The project consists of a river diversion tunnel of 7.7 km in length and with a cross section of 23 m^2. The construction method selected for the project was Drill & Blast.

During the engineering stages of the project, the rock mass support was to be qualified using Barton's Q Index. In total, 6 support types were defined, assigning support type 1 to the best rock mass competency and support type 6 to the weakest rock mass. The following table shows the support classes limits.

The support consisted in combinations of sprayed shotcrete (with and without fiber), passive rock bolts, steel mesh and lattice girders.

The geology in the project area was largely dominated by igneous rocks, mainly granodiorite and riolite. There was an altered zone near the project, which produced a large number of geologic faults with very poor geotechnical conditions. The variation of geotechnical conditions were abrupt in many cases, changing from support type 3 to support type 6 in a matter of meters. These variations had a high impact in the coordination of construction directly impacting the advance rates and stability of the tunnel between blasting and support installation. Also, during rainy seasons, a significant amount of infiltration waters entered the excavation section, producing a decrease in the geotechnical rock mass quality.

3.1 Rock mass face mapping and probe holes

Due to contractual requirements and the site conditions previously described, a geological face mapping had to be performed after every blast during construction by an independent third party (this was SKAVA's role in this project). In such face mapping, a geologist had to evaluate the geological conditions of the rock mass and calculate the Barton's Q Index. Also, in the face mapping a full register of joint systems, shear zones, infiltrations and pictures were included, giving a very complete set of data on the excavated rock mass.

Also due to contractual requirements and site conditions, the contractor had to perform Probe Holes (destructive) 30 meters long with a minimum overlap of 5 meters. The contractor used 3-meter rods for perforation, so each probe hole had a total of 10 perforation rods. The independent third party geologists were responsible for the parameter measurements and interpretation.

While performing the probe holes, the following data were recorded:

- Initial chainage of the probe hole.
- Date and time of the probe hole.
- Orientation of the probe hole.
- Equipment used for the probe hole (jumbo)
- Boring pressure for each rod.
- Advance speed for each rod, in meters per minute
- Advance speed for each rod, in seconds per meter
- Detritus colour
- Water inflow

After the total completion of the probe hole, the third party geologist had to write a report and inform the Owner and the contractor of the project about the findings obtained from the probe hole. Such report was purely factual and not interpretative.

Table 1. Rock mass support types limits (Q index).

Support Type	Lowest Q Value	Highest Q Value
Type 1	10	
Type 2	5	10
Type 3	0.4	5
Type 4	0.1	0.4
Type 5	0.03	0.1
Type 6		0.03

4 FIRST APPROACH OF FORECAST

During the construction of the tunnel an important number of overbreaks resulted after blasting, the drastic changes in the geology being one of the most important root causes. In this scenario, the forecast performed through probe holes was studied in detail by the third party geologist in order to give this database a more useful purpose.

The first action was to gather, standardize, label, organize, and store all the existing information obtained from all historic probe holes of the project. A total of 560 sets of information were analysed, each dataset comprised all the information corresponding to one perforation rod.

Since the focus of this research was to improve the information behind the excavation face, a first examination of the data showed that for this purpose the boring pressure and the advance speed of the rod were the most useful information.

After selecting what information to consider, an historical review of the existing information was performed. As a way to simplify the forecast expected, the first correlations were made looking for support types to be installed, thus all information selected was compared with the actual support type installed. In other words, a linear comparison was performed, aligning information from the probe holes and the rock mass support type effectively installed for the same chainages.

In this line, the following Figures 1, 2 and 3 present, graphically, the history of the database from the probe holes of the project (the aforementioned 560 data sets) and the effectively installed rock mass support type.

Through visual and numerical processing of the information plotted, it was decided that the boring pressure was the most representative information obtained from probe holes to relate to support type. Figure 4 shows how the simple average of the boring pressures relate to the support type for the 560 data sets considered. Also, a simple linear regression was included in this graph to show that the average boring pressure and the support type relation has a R^2 value of 0,692. As expected, the boring pressure increased with the highest rock mass quality. An important fact is that for all the data sets considered, no support type 6 had been installed, thus this support type was not statistically considered in this first approach. As shown in Figure 4, the average boring pressures are within a range of 20 bars, from 60 bars for support type 5, to 80 bars for support type 1.

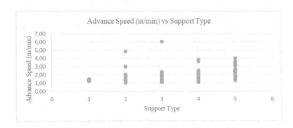

Figure 1. Advance speed (m/min) vs support type.

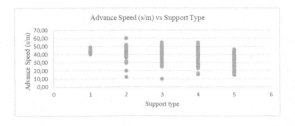

Figure 2. Advance speed (s/m) vs support type.

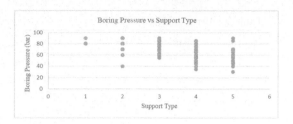

Figure 3.   Boring pressure vs support type.

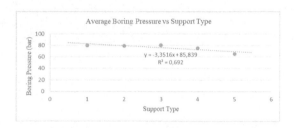

Figure 4.   Boring pressure vs support type.

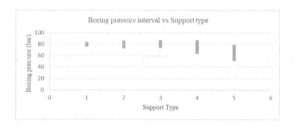

Figure 5.   Boring pressure interval vs support type, "Abacus".

Table 2.   First results using the
"Abacus".

Forecasts	Unit	(%)
Correct	158	74,2%
Incorrect	55	25,8%
Total	213	100,0%

The next step was to generate an simple tool for rock mass forcasting. So, an "abacus" that had a pressure interval for each support type was generated. Each interval consisted of the average boring pressure plus and minus a single standard deviation (Figure 5). The abacus was applied as follows: The geologist supervising the probe hole recorded the average boring pressure for each rod and then through the abacus determined the support types that are contained in such boring pressure. Since the boring pressure intervals have common values, simple boring pressure data could forecast 2 or 3 support types. And so a single boring pressure measurement was used for a single forecast, but such forecast included 2 or 3 support types.

After some time forecasting with the "Abacus" a review of the forecast results was analyzed. A total of 213 forecasts were made using the "Abacus". Table 2 shows the results obtained

Table 3. Error analysis.

Incorrect forecasts	Unit	(%)
Conservative	5	9,1%
Non-conservative	50	90,9%
Total	55	100,0%

with its usage. From all the forecasts, a 74,2% proved to be correct, the forecasted support types with the real mapped support type were a perfect match.

The incorrect forecasts were analyzed to realize whether the errors were conservative or non-conservative. A conservative error was defined as when the forecasted support type having a lower Q value than the real Q value mapped. Table 3 shows the analysis of the errors made.

5 MACHINE LEARNING APPROACH

After the first approach for forecasting the support type using single average values, the idea of including Artificial Intelligence (AI) was implemented.

5.1 Method used and data considered

Supervised Learning is the machine learning task of inferring a function from labelled training data (Mohri et al 2012). The goal of this project was to apply Supervised Learning to geotechnical information, in order to find patterns and infer functions based on training data. The chosen technique was Support Vector Machine (Kotsiantis 2006), and the data from Geotechnical Mapping and Probe Holes drilling was used for training.

Information from both sources had to be standardized, labelled, organized and stored in such a way that it would be easily accessed by the supervised learning method. The model only learns from that training data, and it is understood that non-experienced situations cannot be predicted. Because of this, it was assumed that every project would need its own training process.

As mentioned before, the testing tunnel was located mainly in a Granodiorite region from Superior Cretaceous-Tertiary, with a Rhyolite unit from Cretaceous-Tertiary, and some presence of Dacite. Geomechanical stability was structurally dominated, with a periodical appearance of two fault systems. Water and stresses were not dominant factors.

5.2 First models

After the selection of the AI method to use, a model needed to be developed. This time the model should forecast the Q index value, not the support type. Once the model was complete and ready to forecast, in order to get an idea on how fast the model learned, 4 different models were trained with different sample sizes. Figure 6 shows, in red, the Q value forecasted and in black the real Q value mapped for four different training models with sample size 100, 200, 300 and 500.

As can be seen in Figure 6, more data available (i.e. sample size increases), the more accurate the trained model becomes. For the same models analyzed at this time, a support type forecast was made considering the forecasted Q value that would assign a support type. This was compared with the real support type mapped in the tunnel. The result was found correct within the exact support type or within 2 consecutive support types from the real ones mapped. Table 4 shows the results from this initial 4 models.

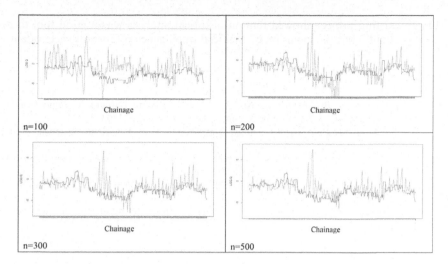

n=100 n=200

n=300 n=500

Figure 6.   Improvement of forecast with Machine Learning.

Table 4.   First Machine Learning results.

Sample Size	Exact support type	2 support types range
100	35%	45%
200	50%	69%
300	51%	71%
500	54%	75%

5.3   Trained model

The final model considered 1692 registers that were used to train it, and 423 registers were used as test set, all from the same tunnel. Once the model was trained, it was used as a forecasting tool during the performance of new Probe Holes. Results show that the model has an accuracy of +85% forecasting rock mass classification behind the face in a way that is suitable for advising a decision maker. Figure 7 shows the results obtained with the final model.

Figure 7.   Final forecast with trained model.

6   CONCLUSIONS

Forecasting the rock to be excavated has been always a crucial tool during tunnel construction. Current methods offer some predictions, but all of them have certain limitations. Data used from an existing project showed that an historical analysis was a good tool to forecast the rock mass quality, but the inclusion of Artificial Intelligence offers more accurate results.

REFERENCES

Bilgin, N. & Ates, U, 2016. Probe Drilling Ahead of Two TBMs in Difficult Ground Conditions in Turkey. *Rock Mechanics Rock Engineering* Volume 49 (Issue 7): pages 2763–2772.

Chapman, R.E., (ed) 2000. Petroleum Geology, Developments in Petroleum Science. Amsterdam: Elsevier.

Kotsiantis, S.B., 2006, Supervised Machine Learning: A Review of Classification Techniques. *Artificial Intelligence Review* Volume 26 (Issue 3): pages 159–190.

Mohri, M, and Rostamizadeh A. & Talwalkar, A, 2012, Foundations of Machine Learning. Cambridge, MA, USA: MIT Press.

Rivera, D, 2012, Aplicación de la tecnología Measurement While Drilling en Túneles. Santiago, Chile: Universidad de Chile Repository.

*Tunnels and Underground Cities: Engineering and Innovation meet Archaeology,
Architecture and Art, Volume 5: Innovation in underground engineering,
materials and equipment - Part 1 – Peila, Viggiani & Celestino (Eds)*
© 2020 Taylor & Francis Group, London, ISBN 978-0-367-46870-5

TBM drive along curved alignments: Model based prognosis of shield movement

A. Alsahly, A. Marwan & G. Meschke
Institute for Structural Mechanics, Ruhr-University Bochum, Bochum, Germany.

ABSTRACT: In current mechanized tunneling practice, the position of the shield machine during TBM-advancement is controlled by the shield driver with the aid of monitoring-based guidance systems. This results in an uneven movement of the machine. This contribution proposes a computational model able to support the TBM-steering during tunnel drives along curved alignments. A computational framework is developed to simulate the advancement and excavation processes, and to enable an efficient and realistic 3D-modeling of the advancement process and the shield-soil interactions during tunneling along arbitrarily curved alignments using the finite element method. The proposed framework combines a newly developed steering algorithm to simulate the TBM-movement during the excavation process. The steering algorithm serves as an artificial guidance system to automatically determine the exact position and the driving direction of the TBM in 3D-space. This modeling approach allows for better prediction of the shield behavior and thrusting forces during the advancement process.

1 INTRODUCTION

In numerical simulations of shield driven tunneling processes, the realistic modeling of both the excavation process and the advancement of the Tunnel Boring Machine (TBM) is a challenge. For a better understanding of these processes during tunnel construction, the interactions between the shield machine and the surrounding soil need to be investigated, yet this excavation process is difficult to model with existing finite element models. In addition, the simulation of the advancement of the machine as an independent body, that interacts with all relevant component of the model, requires a realistic kinematics model of the shield machine which is often not included (Sugimoto & Sramoon 2002). A prototype for a process-oriented three-dimensional finite element model for simulations of shield-driven tunnels in soft, water-saturated soil has been developed and successfully used for systematic numerical studies of interactions in mechanized tunneling (Kasper & Meschke 2004). This model has been reformulated and extended to partially saturated soils and more advanced constitutive models for soils in the context of a integrated design support system for mechanized tunneling (see, e.g. Meschke et al. 2011). Furthermore, several finite element models have been proposed, addressing the difficulties inherent in the simulation of the excavation process. Many of these models account for excavation by removing finite elements from the excavated volume in front of the machine, and then by applying the nodal forces necessary to preserve equilibrium (Bernat & Cambou 1998, Clough et al. 1983). A more realistic representation of the excavation process, based on mesh adaptation, using so-called "excavating elements" in front of the machine has been proposed by (Komiya et al. 1999). In this paper, a steering algorithm is presented in the context of 3D modeling TBM advancement processes. The algorithm serves as an virtual guidance system which automatically determines the exact position and the driving direction of the TBM in three dimensional space. Hereby, the model provides the vertical and horizontal deviations of the machine, its shield orientation, and provides direct input for the jacking cylinders. In addition, a problem specific re-meshing procedure is introduced for the

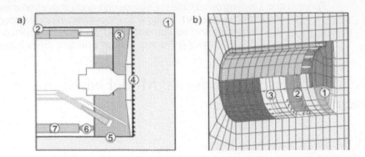

Figure 1.   a) Main components involved in mechanized tunneling: (1) soil, (2) tail gap, (3) pressurized support medium, (4) cutting wheel, (5) shield skin, (6) hydraulic jacks and (7) segmented lining. (b) Modeling of interactions between soil and TBM in the simulation model *ekate*: (1) heading face support, (2) frictional contact between shield skin and soil and (3) grouting of the tail gap) and components of the simulation model: TBM, hydraulic jacks, lining and grouting mortar.

simulation of the excavation process. In this process, a hybrid mesh generation procedure adapts the spatial discretization in the vicinity of the tunnel face, according to the actual position of the TBM, in order to better capture the excavated geometry and predict more realistic shield behavior.

2   COMPUTATIONAL MODEL

The numerical 3D finite element method for shield tunneling used here as a basic framework for the new TBM advancement model has been presented in a number of publications (see, e.g. Alsahly et al. 2016). It has been implemented in the object-oriented finite element framework KRATOS and is denoted as ekate (Enhanced Kratos for Advanced Tunneling Engineering). This finite element model takes into account all relevant components involved in shield tunneling such as the tunnel boring machine (TBM), the hydraulic jacks, the lining structure, the frictional contact between the shield skin and the soil and the supporting measures at the face and the tail gap, respectively, and their interactions.

Figure 1 shows the main components involved in mechanized tunneling (left) and their representation in the finite element model (right). These components are considered as independent sub-models interacting with each other. The interaction between the surrounding soil and the shield machine is accounted for by means of a surface-to-surface frictional contact formulation in the framework of geometrically nonlinear analysis. The hydraulic jacks are represented as CRISFIELD truss elements connected to the surface of tunnel lining element from one side and to the pressure wall of the TBM by tying utility. The simulation of the tunnel advances is performed in a step wise procedure. The soil and the grout are formulated within the framework of the theory of porous media (TPM) as two-phase materials.

3   FINITE ELEMENT MODELING OF THE ADVANCING PROCESS

The modeling of the advancement process in mechanized tunneling is one of the most complex processes, since it requires a good understanding of the real kinematics of the shield machine and its interaction with the surrounding environment (Finno & Clough, 1985). Therefore, a nonlinear kinematic analysis of the shield, based on the action forces imposed on the shield and on the inertial forces due to the shield, is performed. The action forces result from hydraulic jacks pushing against the machine, earth/slurry pressure at the cutting face, friction with surrounding soils, and the fluid flow of processes of the support fluid and grouting mortar, whereas the inertial forces are due to the self-weight of the shield and of the equipment. Furthermore, the taper and the thickness of the shield skin are accounted for in the geometrical representation of the TBM, guaranteeing a realistic distribution of the ground reaction forces in both circumferential and longitudinal directions. The TBM is advanced by hydraulic jacks that are attached to the machine and work as groups in different sectors which push against the previously installed lining ring. From the physical point of view, the minimum pressure exerted by these jacks must overcome the resistance generated by the surrounding soil.

In accordance with tunneling practice, a reliable steering algorithm that provides the numerical model with the required information to keep the TBM on the track is developed. This TBM advancement algorithm serves as an artificial guidance system which automatically determines the exact position and the driving direction of the TBM in 3D space providing the vertical and horizontal deviation, shield orientation and direct input for the jacking cylinders (Figure 2).

The simulation of the advancing process for arbitrary alignments by means of the proposed steering algorithm requires a continuous adaption of the finite element mesh in the vicinity of the tunnel face. Furthermore, the finite element mesh should match the actual motion path of the shield machine resulting from the FE-analysis in each excavation step. For this purpose, a re-meshing algorithm is developed in order to automate the process of mesh generation in a domain in the vicinity of the tunnel face within the advancing process (Figure 3).

Both the steering and the re-meshing algorithm represent the main components of the proposed framework for modeling the excavation process during the advancement of the tunnel

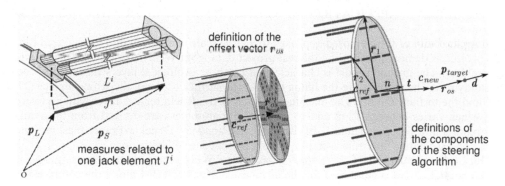

Figure 2. Geometrical quantities involved in the steering algorithm.

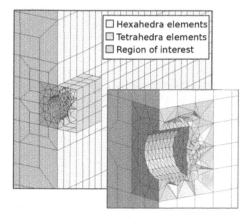

Figure 3. Non-overlapping subdomains and the region of interest (ROI) constituting a compatible hybrid finite element mesh.

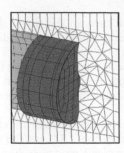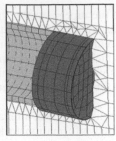

Figure 4. Automatically generated mesh at the tunnel face for two advancement and excavation steps.

boring machine. The advancement of the TBM involves a change in geometry and removal of excavated soil. The kinematic analysis of the shield within the steering procedure provides the exact position and geometry of the new facets as well as the center of the cutting wheel and the updated positions of reference points. In order to avoid over or under excavation it is necessary to adapt the finite element mesh ahead of the shield face. This is accomplished by generating a new finite element mesh ahead of the shield face such that the excavated volume matches the geometric shape and size of the incremental shield advance as provided by the steering algorithm, see Figure 4.

4 NUMERICAL APPLICATION TO WEHRHAHN LINE METRO IN DÜSSELDORF

The applicability of the proposed approach to a large scale simulation of a mechanized tunneling project, is demonstrated using the real project data from Wehrhahn-Line (WHL) metro in Düsseldorf. The ground model is characterized by three geological layers that mainly consist of sandy soil. These layers are the filling layer at the surface (2–3 m), the Rhine terrace (17–29 m), and the tertiary (30 m). The machine is driven in the Rhein terrace layer with an overburden which varies between 12 m and 16 m. The soil parameters are derived from the available soil reports and listed in Table 1. In the numerical analysis, Drucker-Prager plasticity model with non-associative flow rule is employed to represent the soil behavior.

During the simulation of this complete section of WHL, the proposed steering algorithm (Alsahly et al. 2016) is employed for the advancement of the TBM along the complex tunnel path which contains several straight and curved parts, see Figure 5. According to the machine data, the TBM was driven by means of 14 pairs of hydraulic jacks, working in 6 groups: A (2 pairs with 26° cw), B (3 pairs with 90° cw), C (2 pairs with 154°), and D, E and F symmetrical to A, B and C w.r.t. the vertical axis. In order to replicate the real movement of the TBM and predict more realistic distributions of thrusting forces, the steering algorithm is adapted in

Table 1. Material parameters for the finite element model for the selected section of the Wehrhahn-Line.

Soil parameters	Layer 1	Layer 2	Layer 3
Unit weight	17.32	20.38	21.40
Young modulus	21.0	50.0	67.0
Poisson ratio	0.25	0.25	0.3
Friction angle	30.0	35.0	33.0
Cohesion	2.0	0	0
Hardening modulus	5.80	50.0	67.0
Porosity	0.4	0.25	0.3
Permeability	0.01	0.001	10^{-6}

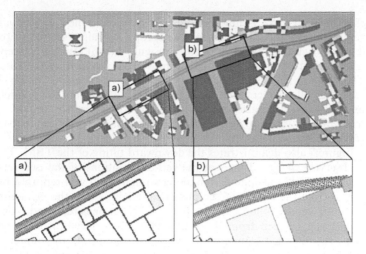

Figure 5. WHL tunnel route, a) section with straight alignment b) section with curved alignment.

Figure 6. Selected results for different advancement steps obtained from the numerical analysis of the selected section of the Wehrhahn-Line in Düsseldorf.

order to account for these Jack's grouping systems during the simulation by enforcing equal elongation for all jacks of the same group. Figure 6 presents selected results for different advancement steps obtained from the parallelized numerical analysis of the selected section of WHL.

In order to investigate the influence of the remeshing procedure during the advancement process, two sections (each of 100 m long) located along different parts of the tunnel route of the WHL project are analysed and observed. These two sections are located at the straight and the curved parts of the tunnel route as shown in Figure 5a and Figure 5b, respectively. As the simulations contain both straight and curved alignment, they suit well to demonstrate the capabilities of the develop simulation model with respect to the steering algorithm and the soil-shield interactions. Furthermore, the predicted shield behaviors and thrusting forces are validated using the real machine data.

4.1 Straight tunnel section

This tunnel section considers a horizontal straight alignment located between Ring-Nr. 1260 and Ring-Nr. 1335 with 75 lining rings. Within the simulation, the TBM (D = 9:49 m) is driven with an overburden of 16 m applying a constant face support pressure equal to 240 [kPa] and grouting pressure equal to 300 [kPa]. The steering algorithm is employed during the simulation of the advancement process in order to push the machine forward and account to the drift-off phenomena. Figure 7 shows a comparison between the real excavated geometry, and the one stemming from the finite element model for the straight section of the tunnel. The real excavated geometry is schematically shown in Figure 7 (top). As can be seen, the machine moves along the straight path with a wriggling movement confirming the so-called "snake-like" motion. This motion results in an uneven excavation boundaries. Similarly, it is noticed from the result of the finite element analysis in Figure 7 (bottom) that the predicted movement of the TBM and the excavation geometry follow a pattern similar to the real ones. It is also apparent that machine drives in a zig-zag pattern along the target alignment which leads to the uneven excavation boundaries obtained from the FE model.

In addition, Figure 7 also demonstrates the downward tilting of the TBM. This can be also deduced from observing the trend lines of the vertical positions of the two reference points. Similar behavior of a shield machine driven along straight path was reported by (Festa et al. 2015). It is noticed from the results of the numerical analysis that the behavior of the TBM resulting from the FE model follows similar patterns to the ones recorded from the machine data. Moreover, it can be seen from the predicted vertical deviation of the two reference points in Figure 7 (as obtained from the steering algorithm during the numerical analysis of the tunneling process) that the tilt of the TBM is trending downward. Hence, the FE model predicts reasonably well the actual shield behavior during the advancement process along straight path.

A further investigation considering TBM-soil interactions is performed employing a similar simplified geometrical approach proposed by (Festa et al. 2015) using machine data. This approach utilizes the reference points (front and rear) to determine shield position and orientation within the obtained theoretical excavated profile at each tunnel advancement step. Assuming a rigid TBM body, the relative distance between the shield periphery and the theoretical excavated soil profile will be quantified. Here, two specific points on the shield tail (top and bottom) are investigated, and the results are presented in Figure 8a. It can be shown from Figure 8a that, upon driving, the top point on the shield tail (point B) is always in contact with the excavated geometry and displacing the soil, whereas the bottom point on the shield tail (point A) originates a gap between the shield and the excavated profile. This is fully

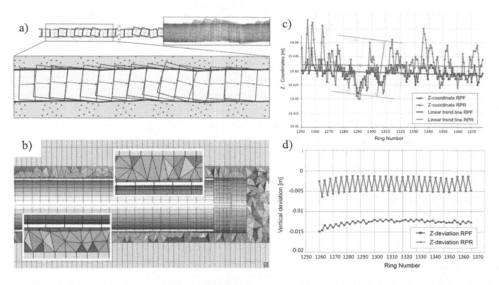

Figure 7.   a) Schematic representation of the real excavated geometry, b) The excavated geometry obtained from FE model, c) Monitored vertical position of two reference points along shield axis, i.e. Reference Point Front (RPF) and Reference Point Rear (RPR), during all advancement steps, d) Computed vertical deviation (position) of two reference points RPF & RPR along shield axis as obtained from FE-analysis employing steering algorithm during the simulation of the advancement process.

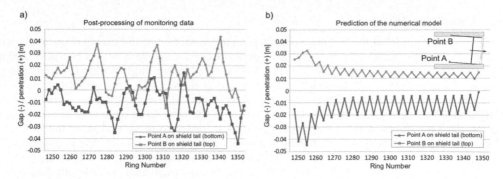

Figure 8. TBM-soil interactions: a) the relative distance between the shield tail and the theoretically excavated soil profile at two selected points (A & B) computed from monitoring data, b) the computed gap distribution from the FE-analysis for the two points during the simulation of the advancement process.

consistent with the reported vertical tendency of the TBM movement along this tunnel section. As for the theoretical, two specific points on the shield tail (B at top and A at bottom) are monitored within the numerical model during the simulation. It can be seen from Figure 8b that the numerical model tends to predict similar pattern of these points which is again consistent with the predicted TBM vertical tendency.

Applying the developed steering algorithm during the simulation of the advancement process, the distribution of jack thrusts exerted by each group, i.e. groups A, B, C, D, E and F, within the thrusting system are obtained. These are visualized and compared with their counterparts value from the machine data in Figure 9. The predicted values are obtained during the advancement of the shield considering a constant face support pressure equal to 240 [kPa] and the cutting forces required to excavate the soil, which are crucial for valid comparison with the measurements. In the numerical model, these cutting forces are considered as a constant static loading on the pressure wall and adopted from the recorded machine data with a mean value of 20 [kPa].

As expected, the thrusting forces of the groups at the invert (group C and group D) are the highest and conversely the thrusting forces of the groups at the crown (group A and group F) are the lowest to prevent the TBM from diving, and to correct the tilting of the machine. From Figure 9, it is evident that the predicted distribution of the jack forces is in reasonably good agreement with the measured thrusting forces. These results demonstrate that enhanced prediction capabilities of the numerical model can be obtained by applying the developed steering algorithm.

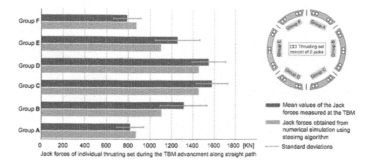

Figure 9. Comparison between the measured thrusting forces of each individual group (obtained from machine data) and the predicted thrusting forces from FE model applying the steering algorithm during the simulation of the advancement process.

4.2 Curved tunnel section

The simulation model of this example considers a curved section of the tunnel route as shown in Figure 5b. This tunnel section simulates rightward curve alignment with a radius of 500m located between Ring-Nr. 1400 and Ring-Nr. 1480 with 80 lining rings. Within the simulation, similar to the straight case, the TBM (D = 9:49 m) is driven with an overburden of 16 m applying a constant face support pressure equal to 240 [kPa] and grouting pressure equal to 300 [kPa]. The steering algorithm in conjunction with the remeshing technique are employed during the simulation of the advancement process in order to drive and steer the machine along the designed curved path by means of 14 pairs of truss elements working in 6 groups A, B, C, D, E and F. This approach allows the TBM to move independent of the soil and, thus, allows for better prediction of the shield behavior within the excavated curved path during the advancement process. Figure 10 shows the generated FE model for the selected curved tunnel section together with the so-called excavated domain or "region of interest".

In this FE model, the TBM is driven along the curved path employing the steering algorithm. Simultaneously, the finite element mesh of the excavation domain is automatically generated according to the actual position of the TBM after each advancement representing the complete curved path of the excavated geometry. Furthermore, the position and orientation of the shield machine within the excavated domain is demonstrated in Figure 10 at three different advancement steps.

An indication of the actual shield behavior during the advancement process is given by the relative positions of the front and rear reference points, defined as RPF and RPR respectively, at each step. In order to understand the actual behavior of the shield upon driving along the given rightward curved path, the horizontal tendency of the TBM-shield is demonstrated using the real machine data in Figure 11 for all advancement steps. The real machine tendency can be deduced by observing the horizontal deviation of the front and rear reference points from the designed path. It can be observed that the TBM is driven with continuous horizontal rightward tilting. A similar behavior of the shield machine driven along curved path was also reported in (Festa et al. 2015).

It is noticed from the results of the numerical analysis that the behavior of the TBM resulting from the FE model (Figure 11-left) follows similar patterns to the ones recorded from the machine data in Figure 11-right. Moreover, it can be seen from the predicted horizontal deviation of the two reference points in Figure 11-left (as obtained from the steering algorithm during the numerical analysis of the tunneling process) that the tilt of the TBM is trending rightward following the curved path. Hence, the FE model reasonably predicts the actual shield behavior during the advancement process along curved path as well. It should, however, be noted that obtained oscillation in the results using FE methods, is due to the application of the steering algorithm, which corresponds to the deviation-correction procedure during the advancement steps. It is evident that the steering algorithm tries to correct the position of the machine during the advancement process in order to follow the defined curved path which explains this oscillation.

Figure 10.   The excavated geometry obtained from FE model employing the developed steering algorithm in conjunction with the automatic remeshing technique and the shield position and orientation within the excavated profile at three different advancement steps. process.

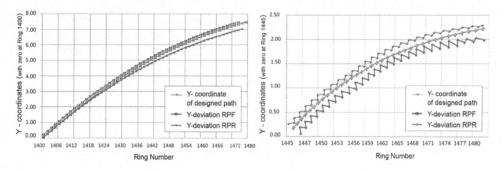

Figure 11. Monitored horizontal position (Y - coordinate) of two reference points (RPF & RPR) along shield axis during all advancement steps as obtained from machine data (left) demonstrating the rightward horizontal tilting and the computed horizontal position (deviation) of the same reference as obtained from FE-analysis (right) employing steering algorithm during the simulation of the advancement process.

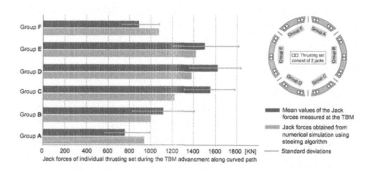

Figure 12. Comparison between the measured thrusting forces of each individual group (obtained from machine data) and the predicted thrusting forces from FE model applying the steering algorithm during the simulation of the advancement process.

Applying the developed steering algorithm during the simulation of the advancement process, the distribution of jack thrusts exerted by each group, i.e. groups A, B, C, D, E and F, within the thrusting system is obtained. These jack thrusts are visualized and compared with their counterparts value from machine data in Figure 12. The predicted thrusting forces of the groups at the invert (group C and group D) are higher than the thrusting forces of the groups at the crown (group A and group F) to prevent the TBM from diving and correct the vertical tilting of the machine due to its self weight. Additionally, in comparison to the obtained thrusting forces for the case of the straight path, higher thrusting forces in group E is predicted, as expected. These results are consistent with the designed curved path where higher thrusting forces on the left side of the shield are required for driving along rightward curvature. From Figure 12, it is evident that the predicted distribution of the jack forces are in reasonably good agreement with the measured thrusting forces.

5 CONCLUSION

In this work, a new computational framework for the simulation of the advancement and the excavation process of tunnel boring machines (TBMs) during mechanized tunneling was proposed using the finite element method. Within this framework, the TBM advancement and excavation processes are simultaneously simulated in such a manner to enable the efficient and realistic three-dimensional modeling of the tunneling process for an arbitrary alignment. The stepwise excavation process is modeled by means of an adaptive spatial discretization strategy used in conjunction with a steering algorithm. This combination of algorithms provides a high resolution of the stress and deformations in the direct vicinity of the TBM and thus also a realistic kinematic description of the movement of the TBM during advancement. The coupled re-meshing– TBM steering procedure embedded in the simulation model takes the actual deformations of the soil and the TBM-soil interactions into account. This method automatically corrects "drift-off" phenomena, as are often observed in TBM tunneling, and continuously keeps the TBM on the desired course during the simulation. The applicability and effectiveness of the proposed computational framework for the advancement and excavation process was demonstrated and verified by 3D numerical simulations of tunnel drives along both straight and curved tunnel paths. The thrust force distribution predicted by the proposed adaptive re-meshing and steering strategy along the shield of the TBM has been shown to be in good agreement with the observed field data for both cases. The machine drives along a zig-zag pattern around the target alignment, which corresponds well with data collected from TBM guidance systems. The applicability of the proposed technique is shown by means of a case-study in which the a TBM was driven along a curved alignment. This study has shown that the proposed framework is able to realistically simulate the motion of the TBM and the TBM-soil interactions along an arbitrary path with minimum effort.

ACKNOWLEDGEMENTS

Financial support was provided by the German Research Foundation (DFG) in the framework of project C1 of the Collaborative Research Center SFB 837 Interaction modeling in mechanized tunneling. This support is gratefully acknowledged.

REFERENCES

Alsahly, A & Stascheit, J. & Meschke, G. 2016. Advanced finite element modeling of excavation and advancement processes in mechanized tunneling. Advances in Engineering Software 100: 198–214.

Festa, D., W. Broere, & J. Bosch. Kinematic behaviour of a Tunnel Boring Machine in soft soil: Theory and observations. Tunnelling and Underground Space Technology 49, 208–217, 2015.

Finno, R. & G. Clough (1985). Evaluation of soil response to EPB shield tunneling. Journal of Geotechnical Engineering 111(2),155–173

G. Clough, B. Sweeney, & R. Finno. Measured soil response to EPB shield tunneling. Journal of Geotechnical Engineering, 109(2):131–149, 1983.

G. Meschke, F. Nagel, & J. Stascheit. Computational simulation of mechanized tunneling as part of an integrated decision support platform. Journal of Geomechanics (ASCE), 11(6):519–528, 2011.

K. Komiya, K. Soga, H. Akagi, T. Hagiwara, and M. Bolton. Finite element modelling of excavation and advancement processes of a shield tunnelling machine. Soils and Foundations, 39(3):37–52, 1999.

M. Sugimoto & A. Sramoon. Theoretical model of shield behaviour during excavation. I: Theory. Journal of Geotechnical and Geoenvironmental Engineering, 128(2):138–155, 2002

S. Bernat & B. Cambou. Soil - structure interaction in shield tunnelling in soft soil. Computers and Geotechnics, 22(3/4):221–242, 1998.

T. Kasper & G. Meschke. A 3D finite element model for TBM tunneling in soft ground. International Journal for Numerical and Analytical Methods in Geomechanics, 28:1441–1460, 2004.

Tunnels and Underground Cities: Engineering and Innovation meet Archaeology,
Architecture and Art, Volume 5: Innovation in underground engineering,
materials and equipment - Part 1 – Peila, Viggiani & Celestino (Eds)
© 2020 Taylor & Francis Group, London, ISBN 978-0-367-46870-5

Soil conditioning adaptation to the heterogeneous volcanic geology of Mt. Etna, Sicily, Italy

A. Alvarez, S. Mancuso & N. Bona
Cooperativa Muratori e Cementisti, Ravenna, Italy

ABSTRACT: In the TBM world, one of the most important key success factors is soil conditioning. While many TBM projects encounter mixed geology, the case of Mt. Etna and the volcanic geology of the Catania Metro is truly singular. Soil conditioning can change the balance of a project, that is economical and feasibility balance. Especially in EPB projects, this factor defines the rules to lead the excavation, it could enhance our performance, decrease our wear drastically and also help us make important savings. In most projects, soil conditioning parameters rarely change. These parameters follow the primary characteristics of the site's geology. In our Catania Metro tunnel, soil conditioning has changed continuously, and we have had to adapt to the changing mixed face. The aim of this article is to show how the heterogeneous geology of Etna has been excavated and the problems found during this challenge.

1 INTRODUCTION

The following article aims to highlight a recent and increasingly common fact, the use of EPB TBMs in mixed excavation faces with high presence of hard rock, and how conditioning is fundamental to bore in these complicated conditions.

The Catania project has given us a rich experience of conditioning in mixed faces, a critical situation for an EPB, mainly designed to work in soft ground. In this scenario, conditioning is even more important than normal, and its continual adaptation to a constantly changing mixed face has been a reality, every meter of excavation.

The use of chemical agents like foams, polymers and natural agents like bentonite can make a radical difference in the excavation parameters and especially in the wear of tools and contact surfaces. For this reason we will describe some aspects encountered during the construction of the tunnel of the metropolitan area called Nesima-Misterbianco lot 1, linking the differences between the lithology forecast by the geological studies and the geology actually encountered.

2 PROJECT GEOLOGY

The Nesima-Misterbianco tunnel is the first tunnel to be bored by an EPB TBM in the soil particular to Etna, but also in an urban context. The tunnel runs with an overburden of between 12 to 20 meters. The geological profile in Figure 1 and the photographs in Figure 2 show the constantly changing lithology characteristic of this project.

The excavation commenced in a geology mainly characterised by a mixed face with non-coherent materials in the upper section and under the ground water table. In particular, the upper part of the tunnel is affected by the presence of alluvial materials, mainly characterised by clayey silts, while the central and lower portions are made up of lava rocks with a significant degree of fracturing and the presence, within them, of terrigenous materials combined with beige silts.

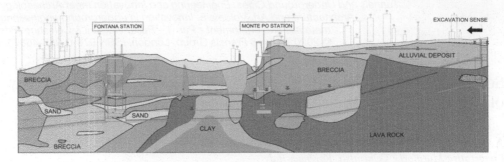

Figure 1. Geological profile updated after excavation.

Figure 2. Different excavation faces: Lava-clay, lava, breccia, rifusa.

As excavation advances, the front presents a predominantly rocky terrain interrupted by sections of breccia and consecutive reappearances of rock near the Monte Po station. In addition, the water table and the tunnel soffit gradually align.

The next tunnel section is affected significantly by the heterogeneous excavation face, with materials having different physical-mechanical behaviors. In particular, in the area immediately after Monte Po station the face is characterised by the presence of lava in the upper section and clayey and alluvial materials in the lower section. Near the Fontana station the tunnel is in turn affected by not very compact breccia. The remaining excavation is characterised by the presence of mixed faces consisting of compact lava-volcanoclastic breccia (lava in the upper section and breccia in the lower section) and breccia alternating with layers of volcanoclastic sands. In particular, the so-called "Rifusa" which was predominant in the final section of tunnel and shows a reddish colour due to thermal factors of contact between lava flows and clays. This mixed face was also characterized for the presence of voids, which creates serious risk to the stability of the excavation, and of course a big risk for the environment over the tunnel.

The article will describe the high degree of fracturing of the rock mass, as well as the abundant and unexpected water inflows which characterised the excavation, this fact caused difficulties due to the incessant application of the adaptation measures necessary to deal with these changes.

3 PREVIOUS STUDIES

3.1 Cutter head

For the construction of the tunnel of it has been foreseen the use of an EPB TBM with an excavation diameter of 10.60 m.

The cutter head tools include 63 cutters with a diameter of 19", including 4 bi-discs, as well as 182 scraper blades and 16 buckets. Its degree of opening is 33–36%.

Figure 3. Cutter head view.

3.2 *Conditioning tests*

After the design of the TBM head, the studies focused on the conditioning of the soil. Numerous tests were carried out in order to reproduce the excavation conditions on a reduced scale. A full range of preliminary tests was performed to define the correct excavation procedure in order to establish the exact balance between the machine parameters and the geology.

However, due to the strong level of subsoil heterogeneity, from the first stage of progress, the conditions represented by the studies were not completely reliable both for the sequence of the different geological formations, and for the configuration of the discontinuities and the percentage of fracturing rock.

For the conditioning tests, the terrains studied were representative of the three geological formations prevalent in the excavation of the Catania tunnel: lava, sand and clayey silt, with different percentages of water added in order to analyse the behaviour of the ground in the different conditions of natural saturation.

In collaboration with the University of Torino, it has been made an in-deep study of the different lithology.

The soil samples were mixed with different foaming agents, using a foam generator that allowed the control of water, foam and airflow, as well as the control of the dosage of the foaming agent. For each lithology an in-depth study using slump testing was carried out with the aim of identifying an optimal conditioning set (foam, added water and polymer) for each combination of foaming agent and type of soil studied. The polymer was also used to evaluate conditioning in the presence of the ground water. The objective was to find the optimum plasticity condition for the fluid to maintain the chamber pressure, and also to be transported by the screw conveyor.

The testing confirmed the feasibility of the conditioning to obtain acceptable levels of plasticity.

These tests were therefore developed as a tool to indicate parameters and levels of reference for conditioning.

However, as already described previously, during excavation the front was rarely composed of homogeneous material, but characterised by constantly changing percentages of lithotypes (mixed faces), making impossible to follow the indications developed by the experimental study. The consistency of paste required for transportation could not be achieved and more importantly, it was truly complicated to maintain pressure in the excavation chamber, especially in the case of rock.

3.3 *Analysis of the wear phenomenon*

Laboratory tests were carried out to evaluate the phenomenon of wear caused by the excavated soil, simulating the interaction between tools and the excavated material, to verify the effect of water content and conditioning.

The tests were carried out using a steel container filled with soil, inside of which a disc made of the same steel used to manufacture the cutter head was placed and rotated, to reproduce at a reduced scale the cutter head of the TBM, measuring rpm and torque and plotting them on a graph.

Before carrying out the test, the weight of the disc was measured. The disc was then rotated inside the steel container for 10 minutes and finally weighed again, to determine the weight loss.

The results obtained were plotted in terms of average weight-loss and torque related to differing water contents, comparing the wear of metal tools in sand and lava rock with the data obtained in a quartz granular soil.

From the results we learnt that fractured rock is more abrasive than volcanic sand and that a higher water content decreases the wear, but only up to a certain value, after which, the abrasiveness is accentuated, especially in the case of sand.

This aspect was confirmed during excavation, in areas where a high water content combined with abrasive volcanic sand caused abnormal wear of the cutting tools.

4 CONDITIONING

Although the guidelines for good conditioning of the ground traversed during the excavation had been defined, it was truly complicated to follow the established procedures.

The high degree of fracturing of the rock, the high presence of water and the frequent lithological changes prevented standard excavation as the values of the machine parameters were never constant due to the incessant interruptions of the production activity.

The reduced excavation speed made the conditioning even more problematic as it was not possible to maintain constant foam flow rates. The low excavation speed meant that the foam flow rates were also low and that created several blockages of the foam lines as the pressure in the excavation chamber was higher than the pressure in the foam lines.

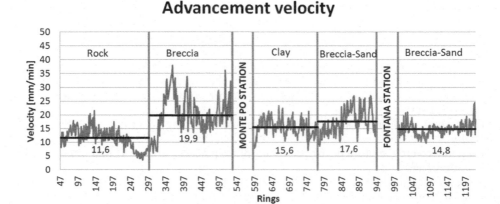

Figure 4. Average boring speed in the different geologies.

In correspondence with the hard rock, the TBM produced a material with an abnormally low presence of fines. That fact, in addition to the large quantity of water affluent in the tunnel, determined the difficulty of correct conditioning of the ground and therefore the difficulty to operate in EPB mode.

In fact, the abundant presence of water separating the fine material from the blocks and thus the difficulty of adequate conditioning forced the operation of the TBM with an open face. Despite the use of foaming agents we rarely obtained the expected fluid material.

Even the use of the polymer was not sufficient to create the pasty mixture typical of good conditioning. However, the use of the polymer was essential to absorb the excessively liquid material and for it to be conveyed by the screw conveyor, avoiding its accumulation at the point of discharge on the belt, facilitating the activity of the material conveyor.

In these conditions, all attempts to excavate the rock under pressure failed. As the excavation chamber filled with rocky material there was a proportional decrease of the torque, and a decrease of the penetration until reaching the complete stop of the TBM as shown in the graphic below, Figure 5.

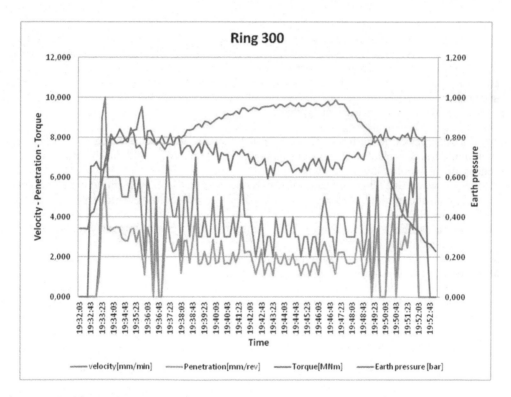

Figure 5. TBM parameters.

As for the tunnel excavation in EPB mode in the lava and with the presence of a large amount of water, the fundamental issue was that of adequate conditioning. Not reaching it easily, we repeated on site the scale studies carried out previously in the laboratory, with a sample of material taken directly from the face and using the foam generator of the TBM. These tests showed the extreme difficulty of conditioning once again.

In the case of clay, conditioning was less problematic even though the sticky nature of this type of material requested a continuous survey of parameters in order to avoid clogging. The effect of conditioning were specially monitored in the TBM parameters, especially in the torque.

Compounding this situation was the absence of suitable excavation tools, which had previously been disassembled in anticipation of boring though rock showed in the geological profile. In fact, the lack of action of the scraping knives did not allow the correct fragmentation of the clayey chips, hindering the action of the conditioning agents.

Also, the irregularity of the excavation and the consequent variation of the abnormal low speed did not allow the correct adaptation of the parameters of the foam system in terms of flow rate and injection pressure to the conditions required by the geology found.

5 ADAPTATION OF THE TBM

As already described, the excavation face was mainly a mix of different materials, and as shown in the photographs, the face was almost always presented with different percentages of these lithotypes (mixed faces).

Although the TBM had all the necessary equipment to excavate all the geological formations present, it was not expected that the changes would be so sudden, and so frequent, rendering very difficult to adequately organise the correct configuration of the cutter head and tools.

As is known, when passing from rock materials to clay materials or vice versa, the replacement of tools involves a forecast and important stoppage of excavation in a stable or consolidated foreseen area. Instead, during our experience, we had to continuously adapt the cutter head configuration to the geological conditions of the moment without ever reaching the optimal configuration for the excavation, and unfortunately having serious delays in our production due to the continual stoppages.

In the beginning of the excavation due at the mix face mainly characterised by fractured lava, the openings of the cutter head has to be reduced also because of the entry into the screw of the boulders, that reached dimensions of up to 60–80 cm, and could have compromised the integrity of the whole conveyor system.

However, when the TBM excavated through loose materials in EPB mode it was necessary to re-establish the original opening ratio of the head, removing the closures to allow the passage of the spoil.

All attempts to limit the breakage of the discs due to the impacts that they received at each turn of the head in the fractured rock, by reducing the rotation speed or reducing the penetration, proved ineffective, as the consequent increase in the excavation time favoured breaking due to continuous non-uniform contact with the face.

Limiting the size of the pieces of rock that could pass through the cutting head was also important to reduce the risk of blockage of the screw conveyor and of the cutter head.

On the other hand, by closing these openings the torque and thrust were increased and the penetration of the machine was reduced, increasing also the wear of tools and cutter head and increasing also the consumption of energy.

Figure 6.   Modification of the cutter head opening.

Figure 7.   a) Bolt deformation; b) knife deformation; c) cover plates.

Many of the stoppages of the production activity were caused by the locking of the cutter head as large blocks interlocked between the openings of the cutter head.

This fact led us to an abundant replacement of tools, always trying to adapt the cutter head to the geological situation found.

Owing to the excessive breaking of the scraping knives due to the continuous impacts against the fractures in the rocky face and boulders in correspondence with the fractured rock formations, we replaced the knives by protection plates. The knives were literally torn from their housing, materialised by the screws suffering an elongation effect.

The objective of the new configuration was not only to avoid the deformation and breakage of the knives and fixing tools, but also to avoid the damage caused by the broken off parts that interfered with the cutter traces, which in turn were also damaged, causing a chain effect that created significant damages and delays.

This new configuration with the protection plates was installed during the TBM stoppage in Monte Po station in parallel to the various maintenance activities. The knives were replaced by cover plates that protected the knife bases, avoiding in this way the loss of knives and the damages that they can generate.

This intervention was carried out in anticipation of a face mainly composed of lava, as represented in the geological profile obtained during the project phase. However, after the restart from the station, subsequently to a very short stretch of rock, the TBM bored through clayey-silty terrain for about 200 meters, as opposed to the geology forecast by the studies.

The cutter head configuration was therefore not suited to the type of soil to be excavated, with the difficulty of replacing tools in a hyperbaric environment or consolidated area.

Without the knives, and with the clay in direct contact with the support protections, the clay was torn out mainly in big blocks causing significant wear of the protection and bases of the knives. As the clay pouring from the screw conveyor onto the conveyor belt, it clogged and obstructed the flow of material at the different discharge hoppers between the belt

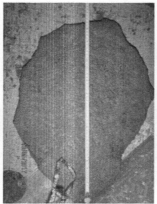

Figure 8.   a) Clay blocks; b,c) boulders.

sections of the entire plant. In addition, the big shapes of the silty sticking masses caused breakdowns in the conveyor system, especially in the vertical conveyor belt.

6 PROBLEMS DURING EXCAVATION

Some of the problems related to the excavation have already been described in the last chapter, as they are related to modifications made to the TBM.

6.1 Excessive water flow

The presence of the water table in correspondence with the lava mass was detected during the preliminary investigation phase.

The excavation experience confirmed that the crushed lava, which remains in suspension in the water present in the chamber, constitutes an abrasive mix that accumulates in the lower part of the face of the excavation chamber, inside which the cutting tools of the head and the screw conveyor must operate.

Also the significant amount of ground water flowing into the excavation chamber in an irregular and discontinuous way through underground fracture systems and the accumulation of water present randomly in the subsoil, led to constant slowdowns in production due to the

Figure 9. Tunnel floods.

impossibility of creating a plastic material leaving the screw. The overflow of the liquid material forced a stoppage at each ring to clear the shields and the dewatering system.

The difficult operating conditions described were not localised, but, as mentioned, they gradually became normal as the excavation progressed through the lava.

In order to reduce the amount of liquid sludge coming out of the conveyor belt it was necessary to modify its inclination and speed.

6.2 *Clogging*

Clogging is another issue that affected the excavation. The mix between clay and rock, and the high temperatures produced by the excavation of hard rock, generated sticky material in the central part of the material chamber.

In fact on many occasions, due to the high temperatures reached in the material chamber, the excavated material, became "cooked" creating a dehydrated crust on the whole metal surface of the cutter head. Therefore, to unlock the head it was necessary to reach the maximal rotation force to obtain values of torque equal to 9000 KN/m.

Different chemical agents were tested in order to avoid this effect, with significant impact. However this still slowed down the excavation rate, and it was continuously necessary to first modify the foam parameters, and when that solution didn't work to directly clean the chamber with high pressure water.

6.3 *Cutter break because of impacts*

Another problem found was the continuous breakage of cutters. It was necessary to work on the cutter head, replacing the excavation tools damaged by the diffused structural discontinuities. The cutters, covering the circular trajectory of the whole excavation section and passing on materials of different compactness, from soft soil to stone, were subjected to impacts that would break the cutting edge, which was often lost inside the excavation chamber and created consequently damage to other cutters and knives.

Even though a significant consumption of cutters was expected, especially near the joints between different geologies throughout the tunnel path, the actual consumption was exponential compared to the previsions.

6.4 *Mechanical breakdown*

The frequent clogging of the cutter head, as well as the presence of rock blocks that did not allow the correct turning of the cutter head, created very high values of torque that caused damage to the main systems of the TBM.

Going beyond the normal levels of the machine parameters led to failures of the pumps that supply the rotation, motors and reducers of the main drive, and also of the screw, with consequential financial impact.

The extreme wear produced by the combination of water and sand already described above, caused damage to the main drive joint seal system. This failure produced contamination of gear oil that has been monitored throughout the excavation of the tunnel.

6.5 *Continuous adaptation of the team*

Interruptions due to mechanical failures produced by the extremely geological conditions together with all the geological stoppages (flooding, etc.), influence the continuity of the excavation.

The recovery of the boring pace, was always long due to the continuous change in the typology of works, cleaning shields for the coming of sand and water, mechanical or electrical repairs, consolidation works, works demolition for head unlocking.

The adaptation of the team and the supply of what was necessary for each task described above was a complicated forecasting mission.

Also the continuous changes to the logistics needed to face the different works were a real challenge for the external supply team that feeds all requested materials as well as the warehouse and procurement department.

All of these facts made it almost impossible to keep a rhythm of excavation that allow us to improve the excavation parameters.

7 CONCLUSION

The **TBM** has been a radical innovation in the excavation method in Catania, allowing to schedule work phases and to predict the end of the excavation activities despite the significant heterogeneity of the ground that characterises the Catanese subsoil.

Moreover, the great change brought about by the first mechanized excavation in Catania concerns the homogeneity of the excavation sections, no longer variable depending on the lithology present but constant throughout the tunnel length.

It is very important to carry out an in-depth study of the geology before the excavation of a tunnel. Drill tests are fundamental in this study, and in the case of a heterogeneous soil like in Catania, it is very important not only to see the material extracted from the drills, but also to appreciate how the drill has been done. Today, with new technology, we can have full parameters of the drilling machine during the excavation, for example torque and penetration that can help us to understand the different joints between lithology and also the empty areas found.

REFERENCES

Peila, D., Oggeri C., AND Borio L. (2009), Using the slump test to assess the behavior of conditioned soil for EPB tunneling, Environmental & Engineering Geoscience, XV (3), 167–174.
Borio, L. & Peila D. 2010. Study of the Permeability of Foam Conditioned Soils with Laboratory Tests, American Journal of Environmental Sciences, Vol. 6, 365–370.
Borio, L. & Peila, D. 2011. Laboratory test for EPB tunnelling assessment: results of test campaign on two different granular soils. Gospodarka Surowcami Mineralnymi, 27(1), 85–100.
Fiore, S. & Neri, S. 2017. The Circumetnea Metropolitan Underground Railway though the great variety of Etna Lava. *Gallerie e grandi opere sotterranee* 124: 21–29.

Tunnels and Underground Cities: Engineering and Innovation meet Archaeology,
Architecture and Art, Volume 5: Innovation in underground engineering,
materials and equipment - Part 1 – Peila, Viggiani & Celestino (Eds)
© 2020 Taylor & Francis Group, London, ISBN 978-0-367-46870-5

Particularities of tunnel primary support modelling in BIM environment

M. Andrejašič, M. Huis, A. Likar, J. Likar & Ž. Likar
Geoportal d.o.o., Ljubljana, Slovenia

ABSTRACT: Karavanke tunnel is considered the most challenging tunneling project in recent time in Slovenian area. Nowadays the project execution design is prepared for the eastern tube and the construction phase will take place in the following years. One of the bigger challenges for tunnel designers was the client requirement for implementation of building information modelling (BIM). The present article discusses the implementation of BIM in tunnel excavation and primary support in the case of Karavanke tunnel project. Because of the uncertainties in tunnel construction, some different approaches had to be taken compared to standard BIM. In fact, lining deformation tolerance is expected to be up to 50 cm, which directly affect the changing of primary lining geometry during the construction phase. The article also compares two different modelling approaches. One approach emphasizes on tunnel support geometry, the other focuses on smart attribute/property attribution with less element modeled in 3D space.

1 INTRODUCTION OF BIM APPROACH IN THE FIELD OF CIVIL ENGINEERING

Building engineering industry has caught up with the digitalization process, taking place in the 21. century. This technological breakthrough was made possible by the engineers that have seen the benefits that the digitalization process is bringing and by the development of the hardware and dedicated BIM software. The process of digitalization in construction industry is known as BIM (Building Information Modelling/Model). As we can deduct from the name, it is a process of modelling the building environment with emphasis on information. If the goals of BIM are successfully followed, the result will be a well-structured database of the building. With BIM the client will get an organized, better defined and well-coordinated project, which will be useful from the construction phase, through the operating phase for the maintenance purposes and thill the demolition phase of the structure.

BIM approach brings many advantages, for example:

- Clear representation of the building elements, which are better understood by the less skilled project participants, who are not able to create a proper 3D structure appearance from the 2D drawings.
- Greater coordination of the project, especially in the case of multi-company project preparation.
- Better defined project quantities and execution schedules.
- Longer lasting digital project information compared to a printed copy
- Faster and easier accessible project information compared to manual searching through the archived printed project versions.

The advantages of BIM have been efficiently used in the field of building construction for some time now. On the other hand, BIM technology has spread into the field of civil engineering just recently. The main problem for BIM implementation in civil engineering projects was

Tunnel segment					Top heading				Bench				Invert		
Seg.	km	Length	BT	%	ST	%	Length	Invert compl.	ST	%	Length	Excav.	ST	%	Length
15	2250.17	349.63	BT3	15%	2P-H-K-6/12,42	50%	26.2 m	200.0 m	2P-H-S-5/7,08	50%	26.2 m	1/1	2P-H-TO-5/4-BB-25	50%	26.2 m
					2P-H-K-6/10,52	50%	26.2 m	200.0 m	2P-H-S-5/8,21	50%	26.2 m	1/1	2P-H-TO-5/4-BB-25	50%	26.2 m
			BT4	80%	2P-H-K-6/14,32	20%	55.9 m	200.0 m	2P-H-S-5/9,51	20%	55.9 m	1/1	2P-H-TO-5/4-BB-30	20%	55.9 m
					2P-H-K-6/16,23	20%	55.9 m	200.0 m	2P-H-S-5/10,74	20%	55.9 m	1/1	2P-H-TO-5/4-BB-30	20%	55.9 m
					2P-H-K-6/18,11	30%	83.9 m	200.0 m	2P-H-S-5/12,16	30%	83.9 m	1/1	2P-H-TO-5/4-BB-30	30%	83.9 m
					2P-H-K-7/16,42	30%	83.9 m	70.0 m	2P-H-S-6/9,73	30%	83.9 m	1/1	2P-H-TO-6/4-BB-30-O3	30%	83.9 m
			BT8	5%	2P-H-K-7/18,69	40%	7.0 m	70.0 m	2P-H-S-6/11,16	40%	7.0 m	1/1	2P-H-TO-6/4-BB-35	40%	7.0 m
					2P-H-K-7/20,89	30%	5.2 m	70.0 m	2P-H-S-6/13,47	30%	5.2 m	1/1	2P-H-TO-6/4-BB-35	30%	5.2 m
					2P-H-K-7/23,44	30%	5.2 m	70.0 m	2P-H-S-6/15,40	30%	5.2 m	1/1	2P-H-TO-6/4-BB-35	30%	5.2 m

Figure 1.   Example of the tunnel supporting measures distribution for the tunnel segment.

lacking of software support for BIM modeling of civil structures. This was mainly cause by the changing project and therefore model boundary conditions represented by the geological strata.

2   PARTICULARITIES OF BIM IN THE FIELD OF TUNNEL CONSTRUCTION

A special segment of civil engineering is tunneling design. It can be divided in two parts based on the designing procedure. The inner-secondary tunnel lining has constant boundary conditions, maintained by the primary lining. On the other hand, primary lining is exposed to changing boundary conditions produced by changes in geological strata. Modelling of the tunnel secondary lining and other inner structures can be somehow compared to standard building modelling, since structure elements precise location and displacements can be defined in the project design phase. We can assume that in the case of inner lining, the challenge is mainly to coordinate concrete structure with electrical and mechanical equipment. These kinds of issues are common in the field of building engineering and are frequently solved with BIM programs.

Design of the primary tunnel support brings a different challenge. In this case the project boundary conditions are represented by the surrounding geological strata. The geological structure is investigated by specialists in the project design phase with geological, hydrological and geomechanical investigations. These investigations though, cannot predict the exact structure and mechanical characteristics of the surrounding ground. Moreover, geologist have a tough job of predicting all the locations and directions of fault zones, through which the planed tunnel would pass. The consequences of these uncertainties, are changing boundary conditions, which can be described with a certain probability of appearing. The result of this is a geological model (boundary condition model) which is defined with a certain probability of appearing on a particular tunnel segment. An example of such of distribution is shown in Figure 1. Every engineering decision is based on this probability geological model. Consequently, the distribution of the tunnel supporting measures is defined with a certain probability of appearance on a tunnel segment.

In essence, the distribution of primary support measures is not finalized in the design phase but is a probabilistic prediction that has to be confirmed in the execution phase. This uncertainty issue represents a significant difference compared to standard building projects. This kind of issues are not common in BIM environment and consequently, no tools are developed, which could make BIM modelling of tunnel support a standard BIM modelling workflow.

3   DESIGN OF BIM KARAVANKE TUNNEL PRIMARY SUPPORT FOR EASTERN TUBE

Nine companies have been involved in the process of designing the Slovene part of east tube of Karavanke tunnel with the length of 7.820 m of which 3.446 m are within Slovenian country borders. The client - DARS the Slovene national highway company, has concluded a contract with the joint venture group - Design group Karavanke. They have agreed that the project will be delivered in standard 2D drawings and in BIM technology. The project consists of a tunnel part and a daylight rode part. BIM should represent all the major building structures: tunnel primary support, inner lining, electrical and mechanical installations, rode structure and road equipment, drainage, bridges and locations of material disposal from tunnel excavation. BIM modelling has been a part of tunnel design throughout all the project phases. With the BIM requirement the client wanted to have a better coordinated and defined project, thus the total cost of the project would be lower, despite the higher initial financial input for a BIM modelling.

The project has been subjected to some particular circumstances, which have caused some issues during design phase:

- Nine companies which were using different software packages had to deliver a harmonized project
- Open BIM principles were adopted in the project
- First BIM project of such of magnitude in the region
- Different coordinate systems were used for the tunnel and daylight road
- The client did not prescribe the exact BIM project specification but has legitimately expected a unified project delivery
- IFC 2x3 format, in which the project is delivered, has no dedicated attributes for tunnel design
- This is a bilateral project between Slovenia and Austria with different national legislation and coordinate systems.

Because of all the mentioned project characteristics, the project resulted to be very complex and difficult to manage. For easier management of large data quantities, more than one hundred partial BIM models have been produced. They were divided into two larger groups. Models of daylight rode and model of tunnel structures. As mentioned before, models were based on different coordinate systems with an exactly defined joining point on the portal area. Furthermore, the tunnel model was divided into primary support models and inner lining models with all the related electrical, mechanical, drainage and road layer models.

This article focuses on the tunnel primary support model. The particularity of the model, because of which it significantly differs from other BIM models, is that the geometry of tunnel lining changes between the installation phase and the final construction phase. This is caused by large surrounding rock mass stresses and the deformation capacity of primary support and surrounding rock mass. In fact, the location in 3D space (x, y, z coordinates) for every structure element of the primary lining change for various centimeters up to 50 cm, during construction works and this represents a collision with one of the basic BIM principles that states that every element in the model has a precisely defined location in actual space.

This discrepancy represents a big challenge for the engineer and the client to find a solution, that will efficiently solve the issue. It has to be practical, accurate and has to produce correct material takeoff quantities.

The Karavanke tunnel project was supposed to be developed based on the LOD description in the internationally adopted specifications »LOD Specification 2015«. Tunnel elements are not separately described in the specifications (2015). Nevertheless, it was written that in case of tunnel structure it is advised to refer to the basic principles based on the development level of the model. For LOD 300, required by the client, it is written: »The Model Element is graphically represented within the model as a specific system, object or assembly in terms of quantity, size, shape, location, and orientation. Non-graphic information may also be attached to the Model Element«. This raises a question about which is the exact location of the tunnel support elements? Is this the location of the installed elements or the final location at the end of construction works or an average of both? Based on the fact that 2D drawings were produced without accounting on construction and deformation tolerances, the decision of producing the BIM model with the same principle was taken. This model somehow represented the final position of structural elements after all the deformation had already taken place.

Since poor rock mass quality was predicted, a great numbers of rock bolts and spills were prescribed to satisfy the standards requirements (Figure 1). While modeling all of these elements, some issues that could be divided in two main groups have appeared:

- Problems with project attributes – managing/checking of attributing consistency
- Problems with element rendering, as a combination of hardware and software limitations.

Despite using smart object filtering it has been a difficult task to check if object had the right parameters assigned, or a human or other kind error had occurred. Problems were detected especially in case or rock bolts and spills which together took around 50% of the whole model size. Rendering of the model due to automatic software hiding of elements was noticeable and different from program to program. Moreover, the loading time of the model exceeded user-friendly time frame. All these issues could rise doubts if the project was delivered as it should have been.

During project modelling we have noticed other issues that were not crucial at that point of the project, but would have affected project modelling in future project processes. The complicity of the project would cause a major problem during generation of 4D and 5D project representations. In essence the large quantity of repeating elements slows down the model, but does not contribute to rising the model value. Furthermore, the client wanted to follow the construction works with BIM technology. That is why, it wished for a simpler model which would be easier to adapt to the project changes in the execution phase.

It was confirmed that the decision of modelling the tunnel geometry on the basis of the tunnel cross section profile without deformation tolerances, was appropriate. The major advantage noted was, it was easier to compare the BIM model quantities with quantities derived from 2D drawings. Moreover, it was easier to check if the geometry of the primary lining model and the inner lining model were well coordinated.

After the revision of the project the client has confirmed the quality of the delivered models. Nevertheless, before the next project phase we (designers and client together) have thought about how to improve the model in the next project phase. Founded on the written conclusions, we have prepared a set of new specifications for BIM modelling of primary lining.

We have decided to divide the model in two parts. One would consist of all the supporting elements modeled in 3D for the length of two excavation steps. The other would account for the project quantities and it would consist of elements of lining sprayed concrete and would run throughout the whole tunnel alignment (Figure 2). This would enable project material quantification with the usage of smart attribute assignments.

Afterwards, another idea about color coding was confirmed for a faster and more intuitive detection of stability of the problematic rock mass zones. Again, we have thought about the geometry of the tunnel cross section, which would be the base for our final design model. Based on the fact that the 2D drawings, in the final design phase, have considered the construction and deformation tolerances, we have decided to model the BIM model on the same principle, considering construction and deformation tolerances. Large tunnel lining deformations were predicted in the final numerical analysis of 0,5 m range. The result was a significant change in the tunnel lining diameter compared to the previous model, which did not consider all the tolerances. Tunnel diameter enlargement resulted in an increased number of rock bolt installed in the case of adopting the tunnel cross section with accounting of tunnel tolerances. We were aware that a possible 3D scan of the tunnel primary lining would differ from the BIM model in the range of the tunnel deformations that have taken place after the support installation.

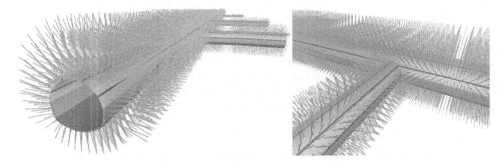

Figure 2. Representation of the BIM model of the primary support.

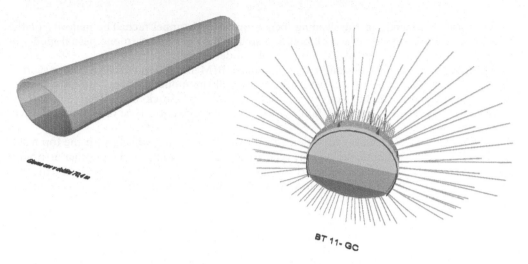

Figure 3.   An example of the model prepared in accordance with the client.

Even though we have optimized the modelling procedure, we still maintained all the advantages that BIM brings – clear representation, precise quantities derivation, the possibility of 4D and 5D construction representation and finally producing a well-structured database of the tunnel. The model has shown to be consistent, manageable and reliable. With the introduced changes we have solved the issues that appeared in the previous design phase.

4   CHARACTERISTICS AND COMPARISION OF BOTH MODELS

4.1   *Model with emphasis on geometry modelling*

As mentioned before, the model that emphasizes on the geometry modelling was modeled with the LOD 300 requirements. We have followed the LOD specification for the year 2015 as it was specified in the contract. In the LOD specifications it was specified to follow and incorporate principles from other described construction elements in case of tunnel structure elements. The result was a model of all the tunnel supporting elements designed for the project. The model was divided in two parts: one would account for tunnel rock mass excavation and the other for tunnel supporting measures (Figure 3). Furthermore, the tunnel supporting measures model was divided in five partial models of the total size of 821 MB. It consisted of rock bolts, spills, rock bolt in tunnel face, deformation elements, reinforcement mesh, steel

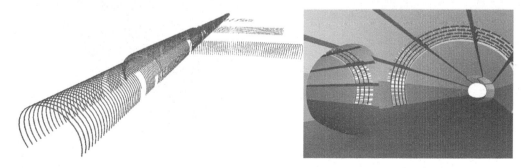

Figure 4.   Renders from the model with the emphasize on tunnel element geometries.

arches, sprayed concrete in top heading, bench invert and tunnel face. The majority of the model size is represented by the rock bolt and spills which are joined in one partial model of 408 MB size, which represent 49,7% of the whole model size.

The model was somehow similar to a standard BIM model in which all the structure elements are modeled. Attributes were attached to all the model elements, describing their properties. Beneath can be seen an example describing a rock bolt. The rock bolt is described by its location in space, failure load, yield load, length, type of bolt and the pay item (Figure 4).

The model of the rock mass excavation was divided into excavation steps for the top heading, bench and invert. Beside the support type attribute of the lining segment, other attributes were added, which derive from the matrix method. These are: length of the line 1a, excavation area, rating area, deformation tolerance and other data linked to the excavation quantity (Figure 5).

4.2 *Final design model based on smart attribute assignment*

The model was prepared based on the proposal that we have prepared together with the client (Figure 2). The resulting model is divided in two parts, one accounts for tunnel rock mass excavation and the other for tunnel supporting measures (Figure 3). The model for tunnel support is divided in two parts one accounts for tunnel support quantities, the other for 3D support representation and elements properties definition. The model for representation of supporting measures is composed of rock bolts, spills, bolts in excavation face, steel arches, deformation elements, reinforcement mesh, sprayed concrete in top heading, bench, invert and tunnel face. The model was enhanced with color coding based on the parameter »behavior type«. The geometry of the model is based on a cross section that considers the displacement and construction tolerances. Based on an additional request from the client, the representations of supporting measures models are positioned in line with the tunnel axis (Figure 6). The whole model was composed of 38 partial models with the total size of 233 MB. From which 35 partial models represents different supporting type measures with the total size of 190 MB. As per agreement with the client, every support type representation should be placed in its own file.

The model that accounts for support elements quantities is built in the way that every support element pay item attribute is prescribed to an element of sprayed concrete besides the quantity of the element/Pay item (Pay_item_N…) (Figure 7). The quantities can be

Property	Value
Definition	TU_OL_BO_DYWIEX9m550kN
Diameter	38 mm
Failure load	550 kN
Ident. št. elementa_ALLright	0010RaE0000002162
Length	9 m
Material	Steel
NOI_UUID	a687d12b-60bb-4c5b-85a9-87bf...
Name	TU_OL_BO_DYWIEX9m550kN
Type	Self-drilling
URL	TU_OL_BO_DYWIEX9m550kN
Unit	m3
Yield load	450 kN

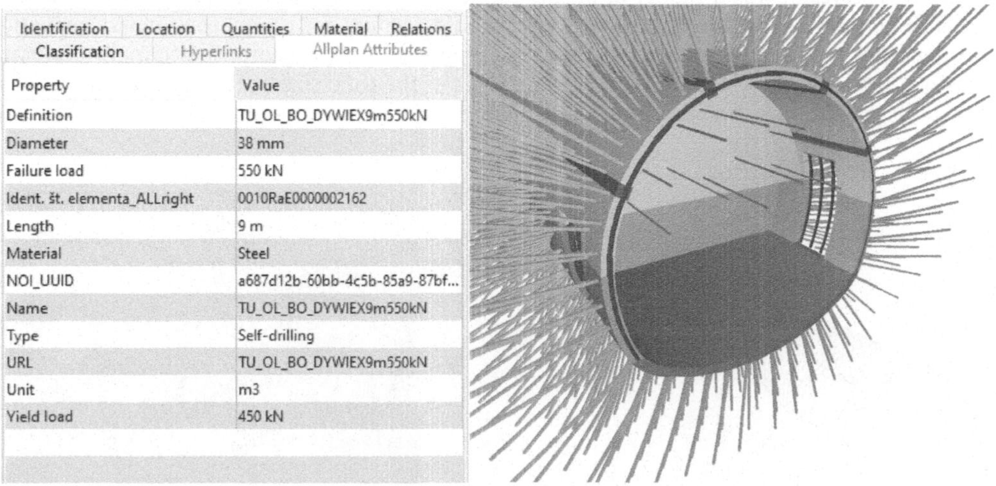

Figure 5. Representation of the attributes used in the model in case of a rock bolt.

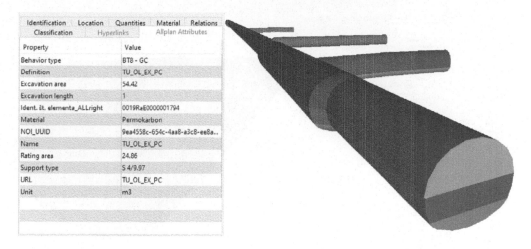

Identification	Location	Quantities	Material	Relations
Classification		Hyperlinks		Allplan Attributes

Property	Value
Behavior type	BT8 - GC
Definition	TU_OL_EX_PC
Excavation area	54.42
Excavation length	1
Ident. št. elementa_ALLright	0019RaE0000001794
Material	Permokarbon
NOI_UUID	9ea4558c-654c-4aa8-a3c8-ee8a...
Name	TU_OL_EX_PC
Rating area	24.86
Support type	S 4/9.97
URL	TU_OL_EX_PC
Unit	m3

Figure 6. Representation of the rock mass excavation model and attributes used to describe one segment of top heading excavation.

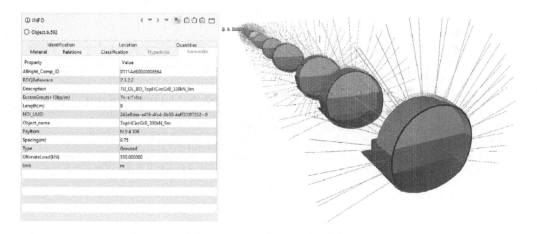

INFO
Object.b.592

Identification		Location		Quantities
Material	Relations	Classification	Hyperlinks	Karavanke

Property	Value
Allright_Comp_ID	0111AdE0000008564
BOQReference	2.3.2.2.
Description	TU_OL_BO_TopHCircGr8_330kN_9m
ExcessGrouti>10ko/m)	True/False
Length(m)	9
NOI_UUID	243e8dee-e416-4fc4-8b00-4aff32087352--0
Object_name	TopHCircGr8_330kN_9m
PayItem	N 9.4.106
Spacing(m)	0.75
Type	Grouted
UltimateLoad(kN)	330.000000
Unit	m

Figure 7. Representation of the supporting measures and attributes for a rock bolt - final design model.

derived from the tunnel support measures model, or transcribed from quantities written in 2D drawings. As can be seen from the figure bellow, an attribute named »Support type« is added to the attribute collection. It is meant as the classification number within the matrix method.

The model of rock mass excavation differs from the previous version mainly by introduction of color coding which is linked to the attribute named »Support type«. The model for rock mass excavation is divided in excavation steps for top heading, bench and invert. Other attributes linked to the matrix method are prescribed as length of line 1a, excavation area, rating area, deformation tolerance and other data connected with the excavation quantity.

The described final design model is well organized, attribution allocation is transparent, while the benefit of 3D support representation is preserved. Furthermore, the model enables a relatively simple way to develop 4D and 5D construction representations and an easy way to input the installed supporting measure during construction phase. The model has shown to be reliable, manageable and useful even when using older, less efficient hardware, always preserving all the benefits of 3D representation and the accuracy in analyzing project material quantities.

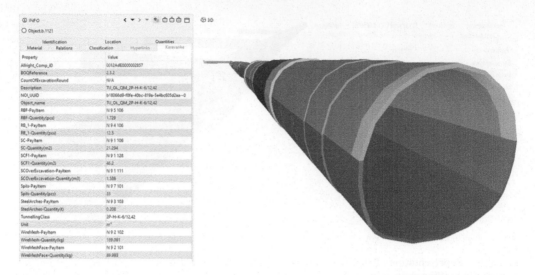

Figure 8. Representation of the model that accounts for support elements quantities – final design model.

4.2.1 *Attribute structure*

Beneath, two out of all quantity model attributes are presented. The pair of attributes is added to one of the sprayed concrete elements (Figure 7) and represents the quantity and type of installed rock bolts:

- RB_1_PayItem – label of the attribute which describes the supporting element type. The label is uniquely defined by the pay item that is written in the quantification schedule. All the pay items can be found in a project database.
- RB_1_Quantity(pcs) – label of attribute that describes the quantity of a support measure element in one excavation step. The attribute label prescribes the unit in which the quantity should be expressed. Quantities are expressed in numbers or text in case of Yes/No input.

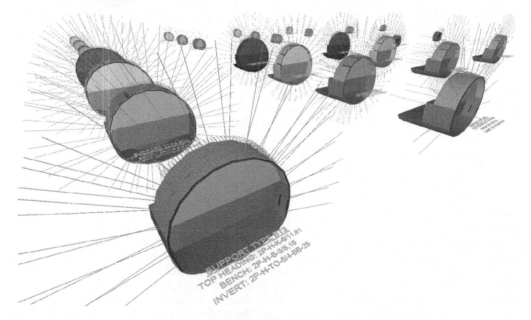

Figure 9. Special model for representation of every supporting measure used in the project.

4.3 *Possibilities of improving the models*

While modelling the primary support system some other ideas came across our heads, that could make the model better, but they were not included in the project because of certain disagreements between engineers or with the client. One of them is producing a model that would include all the support type models, which means moving the support models from the tunnel axes to the sides of the axes (Figure 8). Every support type model has its support type written in 3D letters in front of the model. Doing so we facilitate the visual inspection of the model.

The next idea is about material takeoff procedure. We would not assign every material quantity to a primary lining element. Instead we would prescribe just two essential attributes »Support type« attribute and the volume of the lining element. The quantity of a running meter of supporting measure would be derived from the 3D supporting measure models. Afterwards, the final quantity would be calculated by simply multiplying the support element quantity for a running meter with the length of the tunnel segment, which would be derived from the volume of elements from the tunnel support quantity model.

$$l = V_{SUM}/V_1$$

where l = is the length of the tunnel segment with a specific support type; V_{SUM} = total volume of lining element for a specific support type; V_1 = volume of a running meter of the lining element.

5 CONCLUSION

BIM modelling of the tunnel primary support system is a complex challenge which differs from the standard BIM modelling procedure. Because of the uncertainties in the boundary conditions which are a consequence of the surrounding ground behavior and the probabilistic approach to describe it, it is not possible to model the tunnel excavation and support in the BIM environment in the standard way. While modelling the Karavanke tunnel project we have used two different principles to produce the BIM model. With the first approach we emphasized on the geometry of supporting measures, with the second on the smart attribute assignment. Both approaches resulted to be feasible and both reached the desired goals of accurate 3D representation, project coordination feasibility and precise quantity takeoff derivation. On the other hand, the geometry emphasized approach resulted to be more size consuming and hardware demanding. The model based on smart attribute assignment resulted to be more versatile in case of project changes and more manageable from the point of element quantity and information that they carry. Nevertheless, we believe there is still room for improvement in the modelling approach, as explained in the previous chapter. In our opinion the client has made the right decision when requiring BIM approach in the project. It was proven several times, that project inconsistencies were discovered when modelling or analyzing the BIM model. These might have remained unresolved if BIM approach would not have been used in the project. Cases of discordances were wrong element naming, errors in element property description, conflicts between references in 2D drawing and material takeoff reference, discrepancies between various cross section of the same supporting type, errors in calculating the volume and area of elements and adoption of a wrong principle of tunnel face rock bolt quantity calculation. Based on these examples we can conclude that BIM has contributed to a more accurate project material takeoff. Furthermore, the 3D representation of the support structure with the belonging attributes has shown to be a well usable database of the project structures. Despite the noticeable differences between the common buildings BIM approach modelling and BIM modelling of tunnel excavation and tunnel supporting measures the client observed all the benefits of a standard BIM project. At the end they truly did get a better coordinated and defined project with a more precisely defined material takeoff. By adopting BIM technology to the project, we have produced a useful base for future investments in maintenance of the tunnel structure and have made a step forward in empowering good relations between client and construction companies in charge of excavating and constructing the Karavanke Tunnel project.

REFERENCE

BIM Forum. 2015. Level of Development Specification. Specification BIM Forum.

Tunnels and Underground Cities: Engineering and Innovation meet Archaeology,
Architecture and Art, Volume 5: Innovation in underground engineering,
materials and equipment - Part 1 – Peila, Viggiani & Celestino (Eds)
© 2020 Taylor & Francis Group, London, ISBN 978-0-367-46870-5

Design and construction of cast in-situ steel fibre reinforced concrete headrace tunnels for the Neelum Jhelum Hydroelectric project

B. Ashcroft, G. Peach & J. Mierzejewski
Multiconsult Norge AS, Oslo, Norway

B. De Rivaz & M. Yerlikaya
Bekaert, Kortrijk, Belgium

ABSTRACT: Twin 10 km long parallel headrace tunnels were excavated as part of the headrace tunnel system for the Neelum Jhelum Hydroelectric Project, using two Tunnel Boring Machines (TBMs). The permanent support normally comprised a shotcrete lining applied over initial support elements. However, concrete lining was required in certain areas of extremely poor ground conditions. Once placement of conventionally reinforced concrete commenced, it was found to be taking longer than anticipated, prompting a search for alternative solutions. Steel Fibre Reinforced Concrete (SFRC) proved to be the most viable option for accelerating the concrete lining programme. This paper briefly outlines the requirement for a concrete lining in the aforementioned areas. It includes an assessment of the suitability of a conventionally reinforced concrete lining versus an SFRC lining, the design basis, and the actual design itself. A comparison of the costs and durations to install the two types of lining is also presented.

1 PROJECT DESCRIPTION AND OVERVIEW

The Neelum Jhelum hydroelectric project is located in the Muzaffarabad district of Azad Jammu & Kashmir (AJK), in northeastern Pakistan. Geographically, the area consists of rugged terrain between 500 and 3 200 m in elevation within the Himalayan foothill zone known as the Sub-Himalayan Range.

The project is a run-of-river one, employing 28.6 km long headrace and 3.6 km long tailrace tunnels to cut off a major loop in the river system, transferring the waters of the Neelum River into the Jhelum River, for a total head gain of 420 m (Figure 1). The headrace tunnels comprise both twin (69 %) and single (31 %) tunnels, while the tailrace tunnel consists of a single tunnel. Design capacity of the waterway system is 283 cumecs.

The project, which was completed in 2018, has an installed capacity of 969 MW, generated by four Francis-type turbines located in an underground powerhouse.

At commencement of construction in 2008, all tunnels were to be excavated using conventional drill & blast techniques. However, it soon became apparent that with the equipment being employed, a 13.5 km long section of the headrace twin tunnels underlain by high terrain that precludes construction of additional access adits, would take too long to excavate.

Consequently, the construction contract was amended to allow the operation of two 8.5 m diameter open gripper hard rock TBMs to each excavate some 10 km of the twin headrace tunnels (Figure 1), with an initial centre-to-centre lateral spacing of 33 m, later increased to 55.5 m.

The tunnel excavation diameter was 8.5 m diameter giving a total face area of 56.75 m^2. Excavation direction was upstream to promote drainage, with a typical gradient of 0.8 %.

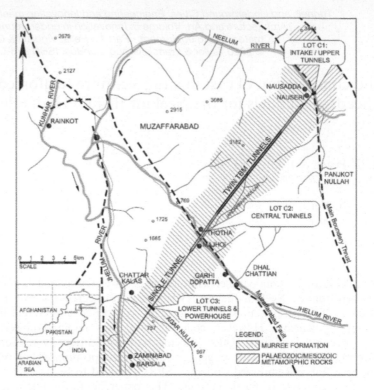

Figure 1. Neelum Jhelum project layout showing TBM Twin tunnels (in bold), major faults (dashed) and alignment geology.

The open gripper design was selected to give the most flexibility for the expected conditions – possible squeezing ground given the relatively weak rock mass and overburden up to 1 870 m, and the potential for rockbursts in the stronger beds. Excavation of the TBM tunnels commenced in January 2013 and was completed in May 2017.

The initial design for the entire project had employed a shotcrete lining throughout, but with short sections of full concrete lining in zones of poor ground, estimated during the feasibility study to add up to about 10 % of the total length.

However, shortly after commencement of the project, the change was made to a full concrete lining for all headrace tunnels, with the exception of the TBM portion, which retained a shotcrete lining for a number of reasons. Specifically, it was judged that the TBM excavation method had several advantages over the Drill & Blast method that resulted in a significantly less-disturbed rock mass, a much smoother tunnel profile, a circular rather than horseshoe shape, and the ability to spray a much more uniform shotcrete layer.

Nevertheless, it was recognized that some sections of concrete lining placed over the shotcrete lining would be required in zones of poor ground, albeit with a shorter aggregate length than the original 10 % estimate, judging by conditions in the early tunnels. It must also be stated that the importance of such local reinforcements had been unavoidably highlighted by the issues encountered on the Glendoe project in Scotland a few years earlier.

It is these sections of concrete lining, which of necessity had to be mostly completed before tunnel excavation had finished, that are the subject of this paper.

1.1 Geological setting

The entire project was excavated in the molasse-type sedimentary rocks of the Murree Formation, which is of Eocene to Miocene age. The succession comprises intercalated beds of sandstone, siltstone and mudstone that have been tightly folded and tectonized, with generally steep bedding dips and a northwesterly regional bedding strike, rarely far from perpendicular to the tunnel azimuth. Weakness zones and local faults were commonly observed, and were invariably oriented parallel to the regional bedding strike.

1.2 TBM configuration and rock support installation

The two TBMs were conventional in their layout, and were based on the successful Gotthard Base TBMs. Nearly all rock support elements, including rock bolts, mesh, channel sections and TH ring beams, but with shotcrete application limited to the maximum extent possible, were installed in the so-called L1 zone immediately behind the shield.

The shotcreted tunnel invert was installed between the L1 and L2 zones, while the majority of shotcrete was sprayed in the L2 zone, some 60 m behind the face, using robots installed outside a cylindrical shield that kept workers and equipment free of overspray and rebound.

2 TBM EXCAVATION

2.1 Progress

Both TBMs started headrace tunnel excavation in early 2013 with completion by the first TBM in October 2016 and the second TBM in May 2017. The lengths of the left and right tunnels, respectively (looking downstream), were 10.428 km and 9.893 km, giving an average daily excavation rate of 8.02 m and 6.37 m. For simplicity, the chainages used in this report start at zero where TBM excavation commenced, and increase in the direction of advance. (In practice, construction records reflect actual chainages, which decreased with upstream advance, and which took into account a section of drill & blast tunnel upstream.) Over most of the excavation programme, the left TBM was generally the lead TBM.

2.2 Encountered ground conditions

Overall, encountered ground conditions were better than expected, in that the squeezing conditions anticipated in the weaker mudrocks were never encountered, despite an overburden of up to 1 870 m. The most likely explanation is that the closely intercalated nature of weak and strong lithologies meant that there was always a 'skeleton' of stronger sandstones and siltstones that provided support for the excavation. Also of benefit to the excavation was the almost complete absence of groundwater ingress.

Furthermore, with the exception of a single major fault, encountered some 1.5 km into the drives, and described below, no other major faults were encountered. Small-scale faulting was common, as were shear zones along beds of weak mudstone, but the strike of most of these features followed the regional bedding strike, which the alignment usually beneficially intersected close to perpendicularly (with the notable exception of the location of the largest rockburst).

On the negative side, however, the incidence of rockbursts was significantly higher than had been anticipated, primarily it is thought because of the existence of elevated horizontal stresses that were unanticipated. It is these rockbursts that primarily contributed to the overall low daily production rates. The left tunnel experienced 937 documented rockbursts, ranging from small to major violent events, while the (usually trailing) right tunnel experienced 590 rockbursts. The zone of the most intense rockbursts persisted for approximately 3 km, before diminishing relatively abruptly.

The two longest sections that required concrete lining were the major fault encountered in both drives, and the location of the largest rockburst experienced on the project, which was named '5/31' after the day on which it occurred, May 31st, 2015. Both features, which are described in more detail below, necessitated construction of concrete lining in both tunnels for over 120 m.

The junction with an access adit, A2, required 48 m of concrete lining in the left tunnel to ensure stability and improve hydraulics. Additional, shorter sections of concrete lining were required in the left tunnel only. Two of these sections were in sheared mudstone that was judged to require additional support, and one was in an area of badly delaminated shotcrete lining. The sections of concrete lining that were placed in the TBM tunnels are shown in Table 1. They amount to 612 m of tunnel, or 3.0 %, significantly lower than the approximately 10 % estimated during the feasibility study.

2.3 Encountered geological features requiring extensive concrete lining

Two geological features required concrete lining in both tunnels in excess of 100 m length. The first is a 95-110 m wide zone of highly sheared mudstone that unusually was associated with groundwater inflows of up to 10 L/min. During the initial encounter by the left TBM, a cavity formed above and ahead of the cutterhead, and the TBM subsequently became jammed at Ch. 1+430 m by the pressure of the collapsed fault gouge on the shield. Stabilizing the excavation and freeing the TBM required installation of a pipe roof canopy, followed by excavation of a top-heading above the cutterhead, and extensive chemical and cement grouting to consolidate the ground ahead of the TBM.

The second feature was the 5/31 rockburst that occurred in the right TBM tunnel, which at the time was trailing the left TBM by 180 m, separated by a 24.5 m wide pillar. The event had a calculated energy release equivalent to a Richter magnitude 2.4 earthquake, causing extensive damage to the TBM, ancillary equipment and rock support over a 60 m section of tunnel as well as significant damage to the tunnel lining of the already-excavated neighbouring TBM tunnel. It disabled the TBM for 7.5 months.

2.4 Requirement for rapid construction of concrete lining

Given the obstacles posed by everyday rockbursts, not to mention the catastrophic 5/31 event, it is perhaps not surprising that the tunnel excavation programme slipped significantly behind schedule, and that as the commissioning date approached, it had become one of the project structures (but by no means the only one) on the critical path.

To meet project deadlines, as much of the concrete lining as possible had to be placed within each tunnel while the TBMs were still operational. Failing that, remaining installation had to proceed after completion of the excavation, but while parts of the TBMs were being removed and other tunnel finishing works, such as grouting, were being completed.

Throughout the lining installation, full tunnel logistical access by way of the rail track had to be maintained. These factors significantly influenced the methodology selected. Principally, the solution was to install the additional concrete lining in two stages. Stage one consisted of the installation of the upper 270° of the concrete lining, while maintaining full tunnel access

Table 1. Summary of TBM tunnel sections with concrete lining installed.

Tunnel	Chainages	Length (m)	Lining Type	Reason for Lining
Left	6+022 to 5+854	168	Conventional RC	5/31 rockburst
Right	5+818 to 5+698	120	Conventional RC	
Left	1+649 to 1+553	96	Conventional RC	Major Fault
Right	1+512 to 1+404	108	Conventional RC	
Left (U/S)	1+710 to 1+698	12	SFRC	Adit A2 Hydraulic Improvements
Left (D/S)	1+698 to 1+662	36	SFRC	
Left	1+327 to 1+303	24	SFRC	Sheared mudstone 1
Left	1+188 to 1+164	24	SFRC	Sheared mudstone 2
Left	0+466 to 0+442	24	SFRC	Delaminated shotcrete
TOTAL		612		

for other tunnel related work. Stage two consisted of the installation of the lower 90°, after the tunnel rail track had been removed and was no longer required.

2.5 *Locations of additional concrete lining*

Concrete lining works commenced at the upstream end of the tunnel and worked backwards downstream. However, it gradually became apparent that the logistical complexity of the operation, with the various concrete lining sections distributed in discrete zones often hundreds of metres apart, all the while maintaining rail access, was delaying completion more than anticipated. One way of streamlining the process was to prioritise construction of the remaining lining sections further downstream initially. This became the driver for assessing and developing alternative concrete lining methodologies.

A major task in conventionally reinforced concrete lining construction involves the careful installation of reinforcing bars at their required locations. It was realized that considerable time could be saved by the implementation of a SFRC lining, and this technique rapidly became the preferred acceleration option.

A comprehensive design review was initiated for the remaining concrete lining locations, and the computations showed that SFRC met all the required specifications, so much so, that its use for most of the required concrete lining sections would have been possible from the start.

The design methodology used for the SFRC lining is presented in detail below.

3 DESIGN

3.1 *Introduction*

Although the majority of the reinforced concrete tunnel linings on the Neelum Jhelum project had been done so with conventional steel bars, steel fibres were considered as a suitable alternative for certain sections within the TBM tunnels since:

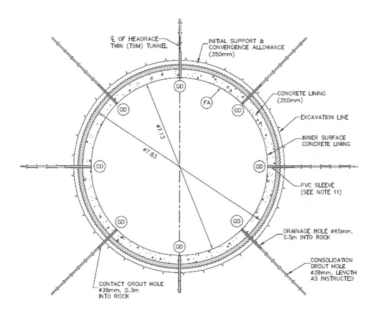

Figure 2. Cross-section of concrete lining.

- The lining shape was circular (as opposed to the horseshoe shape of the portion of the tunnels excavated by Drill & Blast)
- High compressive axial forces would act on the lining with low bending moments
- Steel fibres work to prevent the formation and widening of cracks

3.2 Geometry

As previously mentioned, the excavated diameter of the TBM tunnels was 8.53 m. Due to the severity of some of the ground conditions encountered, large quantities of initial support had been installed in sections where a permanent concrete lining was required, e.g. heavy steel ring beams at 700 mm centres longitudinally along the tunnel and thick layers of shotcrete. A generous allowance of 350 mm had therefore been stipulated for the initial support and any convergence. The minimum design concrete lining thickness was specified as 350 mm. This thickness was a compromise between finding a constructible solution that achieved the design intent, whilst reducing the cross-sectional area of the waterway as little as possible in order to minimize the head loss and thereby the hydraulic penalty.

3.3 Design loads and tunnel design cases

The design loads and tunnel design cases applied in the design are summarised in Table 2.

Table 2. Design loads (with their factors) assumed to act on the lining.

Tunnel Design Cases	Construction Case	Filling Case	Operational Static Case	Transient Water Hammer Case	Dewatering Case	Faulted Ground Special Case
Design Loads:						
Dead Load	1.1	1.1	1.3	1.1	1.1	1.1
Contact Grouting Pressure	1.4					1.4
External Water Pressure	1.4	1.4			1.4	1.4
Transient Water Pressure				1.1		
Wedge Failure Of Rock	1.2	1.4	1.4		1.4	
Faulted Ground						1.2

3.4 Structural analysis of the lining and design assumptions

Elastic continuum, closed form analysis was used to determine the stresses acting on the lining. The analysis is based on excavation and lining of a hole in a stressed isotropic and homogeneous elastic medium.

Once bending moment and ring thrust in a lining have been determined, or a lining distortion estimated based on rock-structure interaction, the lining must be designed to achieve acceptable performance. Since the lining is subjected to combined normal force and bending, the analysis is carried out using the moment-axial force capacity curve, (U.S. Army Corps of Engineers, 1997).

Due to local areas of overbreak or variations in the convergence and thickness of initial support, the concrete lining may in reality deviate slightly from the actual design value. As well as obviously affecting the centroid radius value of the tunnel, the thickness of the lining affects the second moment of area, I, of the lining. The second moment of area is also affected by the reduced stiffness effect of having joints in the lining (the upper 270° is placed first and then the lower 90° is cast below the advancing formwork). In this analysis, as well as calculating the standard value of I based on the thickness of the lining, a sensitivity analysis with different scenarios was also performed where upper and lower bound values were calculated by taking a percentage variation of the 'standard' value.

3.5 Design of conventionally reinforced and fibre reinforced lining

3.5.1 Material properties
The material properties employed in the design are presented in Table 3:

Table 3. Material properties used for design.

Material	Property	Value
Concrete	Specified compressive strength of concrete (MPa)	30
	Modulus of elasticity of concrete (GPa)	25.74
	Poisson's ratio	0.15
Steel reinforcement bars	Specified yield strength of reinforcement (MPa)	400
	Tensile strength of reinforcement (MPa)	620
	Elastic Modulus of steel (GPa)	200
Steel fibres	Fibre length (mm)	35
	Fibre diameter (mm)	0.55
	Aspect ratio l/d	65
	Tensile strength (MPa)	1345
	Young's modulus (GPa)	210
	Minimum Dosage (kg/m^3)	40
	CMOD 0.5mm, $f_{R1,m}$ (MPa)	3.7
	CMOD 1.5mm, $f_{R2,m}$ (MPa)	3.9
	CMOD 2.5mm, $f_{R3,m}$ (MPa)	3.6
	CMOD 3.5mm, $f_{R4,m}$ (MPa)	3.2

3.5.2 Serviceability Limit State (SLS) design

3.5.2.1 SERVICEABILITY LIMIT STATE (SLS) DESIGN FOR CONVENTIONALLY REINFORCED LINING

The minimum required cover according to the applicable standards was 75 mm for the structure. #8 (25 mm diameter) and #6 (19 mm diameter) reinforcement bars were specified at centres of 175 mm in the circumferential and longitudinal directions respectively. This reinforcement configuration satisfied the minimum reinforcement requirements and other durability considerations stipulated in the relevant standards such as:

- Minimum reinforcement
- Requirements for flexural crack control
- Requirements for temperature and shrinkage reinforcement

3.5.2.2 SERVICEABILITY LIMIT STATE (SLS) DESIGN FOR STEEL FIBRE REINFORCED LINING

Crack control is one of the main benefits provided by steel fibres to structural elements. If these cracks do not exceed a certain width, they are neither harmful to a structure nor to its serviceability. The limitation of crack width means that steel fibres provide a post crack strength to the concrete.

Fibres may have been originally introduced for strengthening of the matrix, without distinguishing the difference between material strength and material toughness. (Toughness is used for describing the post-peak response of structural members that quantifies the energy absorption characteristics.) The most significant effect of fibre addition to the brittle cementitious matrix is the enhancement of toughness.

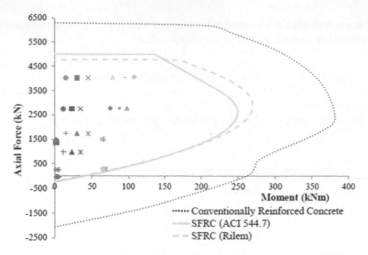

Figure 3. Moment - Axial Force Interaction Diagram - Circumferential Direction from Elastic Continuum Method.

One of the greatest benefits to be gained by using steel fibre reinforcement is improved long-term serviceability of a structure. SFRC in a tunnel lining offers: a) Excellent ductility, b) Reduction in the shrinkage of concrete, c) Elimination of mistakes in conventional reinforcing, d) Shorter construction periods compared to traditional ones, e) Increased tensile and flexure strengths that are equal in all directions, f) Easy crack control and high absorbed energy after matrix failure.

The distance between steel fibres is much smaller than typical spacing between traditional bars. Unlike reinforced concrete, fibres are distributed throughout the whole section. Hence, there is no concrete cover without reinforcement. Furthermore, stresses in the root of a crack can be picked up more quickly. This is why crack propagation and crack patterns change when compared to plain or even reinforced concrete (Vitt, 2005).

3.5.3 *Ultimate Limit State (ULS) design*

The ULS axial loads and bending moments were determined for each type of scenario for each tunnel design case. The moment – axial force capacity curve for the conventionally reinforced section was plotted assuming that there was an equivalent area of circumferential reinforcing steel of 2805 mm^2/m in each face of a column under combined axial load and bending. The moment – axial force capacity curves for the SFRC lining were generated using both the Rilem Method and the method detailed in Appendix A of ACI 544.7 These curves are plotted in Figure 3. The singular points on the diagram are the results from the elastic continuum analysis for different scenarios for each of the tunnel design cases listed in Table 2. Figure 3 shows that the points fall within the capacity curves.

The shear was checked using the equation: $V_{max} = [(M_{max} - M_{min})/R]$

The shear capacity of the concrete alone (without separate shear reinforcement or taking the beneficial effect of the fibres into account) was found to be easily adequate, even for the maximum factored shear force of 58 kN.

4 DURABILITY CONSIDERATIONS WHEN USING SFRC

SFRC is often used for concrete structures subject to severe exposure conditions e.g. bored tunnels exposed to saline ground water containing high levels of sulphates (Edvardsen, 2018). Two phenomena need to be examined when analyzing the durability of SFRC:

- The reinforcement has to provide good "tightness" against infiltration of water by controlling in situ crack opening
- Corrosion of the reinforcement should not cause a notable reduction in the bearing capacity of the lining

Two different configurations must be taken into account when analyzing the corrosion of steel fibres and its consequences:

1. The fibre does not cross a crack emerging on the surface
2. The fibre crosses a fracture crack on the surface

In the first case, apart from some stains which affect the appearance of the structures, corrosion of the fibres does not lead to any serious problems for the durability or the bearing capacity of these structures in SFRC.

In the second case, the bearing capacity of the SFRC is not reduced significantly with crack openings of 250 μm or less.

The corrosion resistance of SFRC is governed by the same factors that influence the corrosion resistance of conventionally reinforced concrete. Processes such as carbonation, penetration of chloride ions and sulphate attack are in direct proportion to the permeability of the cement matrix.

As long as the matrix retains its inherent alkalinity and remains intact, deterioration of SFRC is not likely to occur. It has been found that good quality SFRC, when exposed to conditions conducive to reduced alkalinity, will only carbonate to a depth of a couple of millimeters over a period of many years (Kern & Schorn 1991, Hannant & Edgington 1975).

5 COMPARISON OF COSTS AND DURATIONS

5.1 Comparison of time and cost of different methodologies

The requirement to meet a completion date necessitated the implementation of a detailed project management system to record all aspects of the concrete lining. This data was recorded by both design and supervision teams, and the key points are presented in Table 4.

The sections numbered 1 to 4 refer to additional concrete lining sections constructed using concrete with steel reinforcement bars, or what may be considered the conventional methodology. Sections numbered 5 to 9 refer to the sections of lining constructed using SFRC.

The sections numbered 1, 2, 3, 7, 8 and 9 were installed whilst maintaining full logistical access for other tunnel works by way of the rail track. Sections 4, 5 and 6 were the last sections to be installed when there was no requirements to maintain tunnel access for other works. Furthermore, they were located directly adjacent to access Adit A2, affording uninterrupted access to surface facilities. (It should be noted that while section 4 was one of the last to be placed, it was one of the first to be designed and issued to the contractor, hence the use of conventionally reinforced concrete rather than SFRC.)

5.2 Comparison by time

Table 4 presents three columns relating to time required for lining installation. "Time per Section" provides the total time for the concreting works required for that section length. "Time per Lm" is the average time taken to install one linear metre of tunnel lining. "Delays during concrete placement" presents the extent of production delays for each section and is expressed as a percentage of the total time.

5.3 Findings from time data

The data from Table 4 shows that lining installation using conventionally reinforced concrete ranged from 43 to 175 minutes per linear metre. However, section 4, as mentioned previously,

Table 4. Comparison by Time and cost of Lining Types per Unit Length.

No.	Reason for Lining	Length (m)	Lining Type	Time per Section (hours)	Time per Lm (minutes)	Placement Delays	Cost per Lm
1	5/31 Rockburst (Left)	168	Conventional RC	465.25	166[*]	35%	Special
2	5/31 Rockburst (Right)	120	Conventional RC	350	175[*]	28%	Cases
3	Major Fault (Left)	96	Conventional RC	214	134	26%	100%
4	Major Fault (Right)	108	Conventional RC	76.5	43	9%	82%
5	Adit A2 Strengthening and	12 (u/s)	SFRC	19.5	98	9%	35%
6	Infill for hydraulic improvement (Left)	36 (d/s)	SFRC	45	75	2%	35%
7	Sheared Mudstone 1 (Left)	24	SFRC	24.75	62	3%	41%
8	Sheared Mudstone 2 (Left)	24	SFRC	69	173	13%	41%
9	Delaminated lining (Left)	24	SFRC	34	85	Nil	33%
	Total:	612					

* *Values have been adjusted to account for the fact that the volumes of concrete placed per linear metre for these sections were larger than the other sections due to the enlarged tunnel profile that resulted from the 5/31 rockburst event. The adjustment allows for a just comparison of the values.*

enjoyed direct access to concrete trucks via Adit A2, resulting in a far more favourable result. Consequently, section 4 should be discounted, and the results from sections 1, 2 and 3 (i.e. 134 to 175 minutes per linear metre) are adopted as a more typical time per linear metre for conventionally reinforced concrete.

The associated time delays when using the conventionally reinforced concrete varied from 9% to 35%. However, once again the result from section 4 should be ignored, and the results from sections 1, 2 and 3 (i.e. 26 to 35%) should be regarded as more representative.

The data from Table 4 shows one outlier, at 173 minutes per linear metre for SFRC. This anomaly was a result of some of the batches of SFRC being rejected and extensive stoppages due to the blockages of pipes during placement and is therefore only shown here for the sake of completeness. Excluding this outlier (section No. 8), the time taken to install the SFRC lining ranges from 62 to 98 minutes per linear metre giving an average of 77 minutes per linear metre, offering a significant time saving over conventionally reinforced lining installation.

The associated typical time delays using the SFRC lining type range from 0 – 9% (discounting the outlier).

5.4 *Findings from cost data*

The details of the costs for each section are presented in the last column of Table 4. Information for sections 1 and 2 has been excluded and noted as 'special cases' because these sections required lining as a result of the 5/31 severe rockburst, which caused near-complete destruction of the existing shotcrete lining and surrounding strata. A far more extensive effort was therefore required to repair these two sections, making a comparison with other sections misleading.

The first section of concrete lining to be completed was section 3, associated with the major fault. This has been taken as the reference point for the costs analysis, i.e. this cost is expressed as 100% and all other costs are compared to this bench mark.

The data from Table 4 shows that the lining installed using conventionally reinforced concrete ranged from 82 to 100%.

The data from Table 4 shows that the lining installed using fibre reinforced concrete ranged from 33 to 41% of the reference cost of section number 3, a very significant saving.

6 CONCLUSIONS

In summary, the adoption of SFRC over conventional reinforcement proved to be a notable success. Not only did SFRC lining meet the same design criteria as a conventionally reinforced lining, it offered the following advantages:

- Saving cost over the actual quantity of steel employed
- Saving time by being quicker to install
- Producing a lining with smaller crack widths and improved durability over the life of the structure
- Producing a lining that is more efficient at resisting stresses due to groundwater loads and ground loads than conventionally reinforced concrete

REFERENCES

American Concrete Institute. 2016. *544.7R-16 Report on Design and Construction of Fiber-Reinforced Precast Concrete Tunnel Segments*. Farmington Hills: American Concrete Institute.

Department of the Army, U.S. Army Corps of Engineers. 1997. *Engineer Manual EM 1110-2-2901, Tunnels And Shafts In Rock*. Washington: US Army.

Edvardsen, C. 2018. Consultant's view of durable and sustainable concrete tunnel constructions in the Middle East. *World Tunnel Congress*: Dubai.

Hannant, D. & Edgington, J. 1975. Durability of steel fibre concrete. *Proceeding, RILEM Symposium on Fibre reinforced cement and concrete, Vol 1*, September 1975: pp 159–169.

Kern, B. & Schorn, H. 1991. 23 Jahre alter Stahlfaserbeton, *Beton and Stahlbetonbau*, V 86, September 1991: pp 205–208.

Rilem TC 162-TDF. 2003. *Test and design methods for steel fibre reinforced concrete: σ ε design method*. Mater. Struct. 36 (262): pp 560–567.

Vitt, G. 2005. Crack control with combined reinforcement: from theory into practice. *The 1st Central European Congress on Concrete Engineering: Fibre Reinforced Concrete in Practice. Graz, 8–9 September 2005*. Berlin: Ernst & Sohn.

Tunnels and Underground Cities: Engineering and Innovation meet Archaeology,
Architecture and Art, Volume 5: Innovation in underground engineering,
materials and equipment - Part 1 – Peila, Viggiani & Celestino (Eds)
© 2020 Taylor & Francis Group, London, ISBN 978-0-367-46870-5

Analysis of two-component clay sand backfill injection in Japan

P.J. Ashton, S. Takigawa, Y. Kano & Y. Masuoka
TAC Corporation, Okayama, Japan

ABSTRACT: Ground subsidence is one of the main concerns contractors and engineers have when undertaking tunnel excavation projects. Particularly vulnerable from subsidence are older as well as landmark buildings, which form major focal points in cities worldwide. It is with a view to protecting these structures during tunneling that a two-component injectable non-hardening Clay Sand system was envisaged 30 years ago in Japan. This paper will analyze the journey of this "Made in Japan" method from its inception, to industry acceptance and finally it's widespread application in major tunneling projects. We will seek to show, via the analysis of real-world data, the multipurpose nature of this method for enabling excellent thrust stability during sharp curve excavation and for applications with EPB and Slurry TBMs alike. We will also show how this injectable Clay Sand material is evolving to meet the needs of bigger and more complex tunneling sites in the future.

1 INTRODUCTION

1.1 *Early problems with tunneling in Japan.*

With the success of the first Shield Machine excavated tunnel in Japan in 1922, the almost 100-year long evolution of tunneling technology to the present day had begun. TAC Corporation's own journey within the shield construction industry started in 1976. As the Japanese tunnel market grew, so did TAC's successes with continuous improvements of technologies and materials specifically developed to prevent subsidence at every new opportunity, earning the company the reputation and motto of "Pursuing Zero Ground Subsidence in Shield Construction". In Japan, it was with the evolution of the Shield machine into the two main modern types, Slurry and EPB; and the invention by TAC of simultaneously injected two-component air entrained backfill grout, that the quite regular occurrence of 30–40cm subsidence caused by the deficiencies of Blind shield tunneling in the predominantly ultra-soft clay ground commonly found in Japan, was finally reduced to around 5cm, after the tail of the shield machine had passed, which at the time, was an all-time low. However, it was through an unexpected discovery that subsidence during excavation would be brought under control to virtually zero. Unlike the rapidly hardening two-component cement based backfill grout system that TAC was famous for at the time, this final reduction in subsidence would be achieved by implementing and modifying the use of a non-hardening thixotropic gel material initially invented for its friction reducing properties.

1.2 *Solutions for reducing subsidence.*

In the 1970's, TAC was busy developing a quick setting, air entrained, high flow two-component grout, while the rest of the industry worldwide was generally utilizing single-component, relatively slow setting, sand or gravel-based grouts. This new Shield Thixotropic-gel Grout System, was very well accepted in Japan because of its ability to reduce ground settlement in ultra-soft clayey-sand type ground conditions. Two-component grout injection, became the cornerstone of the foundation for the technical skills to finely control the modulation of A:B

Liquid injection ratios that would eventually be required to tame Clay Sand into a workable solution for further reducing subsidence in the future.

Backfill grout used to be almost exclusively be applied via injection through the segment to stabilize the tunnel segments during excavation. A shift in industry practice in the 1980s saw some machine makers start to inject grout via the tail skin using a simple tube in the skin of the machine which opened into the tail void. However, these systems were prone to blockages, even when properly maintained. In 1982, TAC developed a new kind of cleanable injection pipe. This reliable, easily maintainable injection pipe further decreased the rate of subsidence occurring in tunnel construction, however, it was insufficient to totally eliminate uncontrolled settlement occurring in tunnel construction. Eventually, the inspiration for the development of a two-component Clay Sand solution injection as a subsidence prevention material came from a phenomenon noticed post-project completion.

2 PAST APPLICATIONS

2.1 Subsidence control as a side effect.

In the early 1983, Nishimatsu Construction Company was working on the construction of a 1.3 km section of the Osaka City, Midosuji Subway Line with a 6.98m Kawasaki Heavy Industries slurry shield machine. During the design phase, concerns were raised as to perceived problems arising from the combination of driving through a curve, while also ending the drive, since the final part of the drive included a R=160m curve. Nishimatsu Construction Corporation requested TAC to create a solution to mitigate the risks posed by the overcut in such a vulnerable position. Two solutions were presented; The first was to apply a high air content backfill grout mixture containing 80% air which was to be sacrificed, once hardened, as the machine moved around the curve; however, this was discarded due the fact that once the machine began to drive through the curve, the sacrificed grout on one side of the machine would be replaced by a new void on the opposite side of the machine, thus negating the effect of the injected material in the first place. The second, to use a combination of clay-sand, fly ash and thickener to create a non-hardening material with high viscosity, and low friction resistance to lubricate the machine through the curve was adopted. This option would flow freely around the whole of the machine during the drive under the pressure applied from the movement of the machine. The material called "Clay-Shock", was adopted in conjunction with a segment stabilization system called the "Mini-Packer Method", and the difficult curve was driven through, completing the tunnel as planned (Figures 1–2).

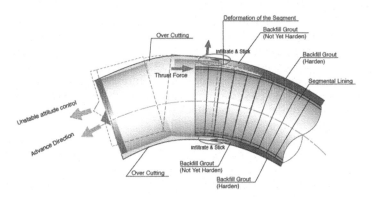

Figure 1. The problem with sharp curve construction. Ground surrounding the TBM during a curve must be filled quickly with grout. The segmental lining must be fixed properly to the ground to ensure stability of TBM attitude. Due to unsatisfactory ability to fill tail void with grout sufficiently, the segmental lining may become unstable due to the thrust of the TBM.

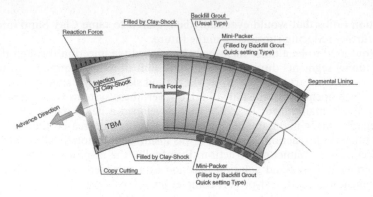

Figure 2. Sharp curve construction using Clay Shock and the Mini-Packer Method. Clay Shock is injected into the overcut to ease the frictional resistance around the TBM, while also preventing the collapse of the overcut wall. Thrust needed to drive the TBM through the curve is reduced. Mini-Packer bags are inflated with quick setting, high strength grout to transfer the thrust force directly into the ground, without affecting the segmental lining. TBM attitude control is improved.

It was only some weeks after the tunnel had been completed, and the final data was analyzed by Osaka City office that a noticeable variation in settlement between the shield machine during a straight drive and a curved drive was discovered. During the straight drives there had been a measurable settlement in places of 3–5mm, and during the curved drives it was expected to be even more. However, the settlement during the curved drive had been reduced to zero. The only difference was the injection of Clay-Shock into the overcut during the curved drives. Based on that result, a new non-hardening process to limit settlement during tunneling had been discovered. This discovery was to become the foundation of a new branch of material research for TAC, which would guide the next 30 years of material development for the company.

2.2 *Clay Shock acceptance and uptake.*

Consensus for the application of Clay Shock in the Japanese tunnel market was not immediately forthcoming, even after quantifiable laboratory and field results had been presented. Clay Shock was utilized effectively for curved drives and as a subsidence reducing material at 5 different sites between 1983 and 1989 leading to the hypothesis that instead of being used exclusively to surround the machine, this new material could possibly be used as an additive for EPB shield machines.

In 1989, Maeda Construction Company was constructing a drainage tunnel in Kakogawa City, Hyogo Prefecture, Japan using a 2.28m EPB machine. The 1.37 km tunnel was to be driven through a mostly coarse gravel and cobble layer with very low silt and clay content, and what was predicted from boring data to be low water content. In reality, the machine's progress was hindered severely by high water content during the initial drive, so much so that tunnel operations were halted. TAC was contacted, and the research which had been undertaken into the properties of the new Clay Shock product started to yield results. Rigorous lab testing of various concentrations and combinations of A & B liquids had produced reliable data showing Clay Shock to be able to produce consistent results, shown in Table 1 and Figure 3. It had also been found that when Clay Shock when prepared to the ratio shown in Table 2, could support up to a 1kg weight on its surface (Figure 4)

Based on the research, the general contractor chose to apply Clay Shock as an additive. Due to the gelling properties of the mixture, it was decided to adopt a similar backfill grout two-component injection method via the tail skin for Clay Shock, using a mixing nozzle arrangement to thoroughly mix the A and B liquids. The forward-facing nozzles on the Shield machine were also repurposed and Clay Shock was injected around the side of the machine

Table 1. Clay Shock β and βII Viscosity Test Results.

Clay Shock β							Clay Shock βII						
A Liquid (1m³ batch)			B Liquid %				A Liquid (1m³ batch)			B Liquid %			
Clay Sand	Water	Viscosity dPa·s					Clay Sand	Water	Viscosity dPa·s				
		0%	1%	3%	5%	7%			0%	1%	3%	5%	7%
419kg	839L	0.4	45	150	170	100	308kg	881L	3	80	150	170	120
454kg	826L	0.6	55	250	300	200	347kg	867L	5	90	270	350	250
488kg	813L	0.8	70	300	400	300	384kg	853L	8	130	400	470	380
520kg	800L	1.0	100	400	500	400	419kg	839L	15	200	550	700	600
552kg	788L	1.5	150	500	750	650	454kg	826L	30	300	700	900	850
582kg	777L	10	250	800	1200	1050							

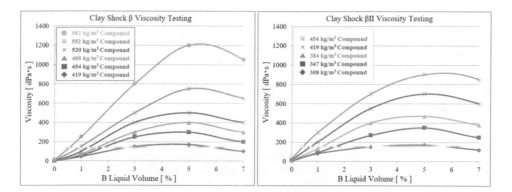

Figure 3. Graph of Clay Shock tests with concentrations and viscosities of various A:B liquid combinations.

Table 2. Mix design for 1m³ of Clay Shock used for Kakogawa City drainage tunnel project.

A Liquid		B Liquid	Viscosity
Clay Sand	Water	Sodium Silicate	
520kg	800L	50L	500 dPa·s

Figure 4. Mixing and testing of Clay Shock compound until viscous enough to support a 1kg weight on its surface.

through to the cutterhead. This facilitated the penetration of the chamber, and upon machine restart, the coarse gravel and cobble layer was transported through the machine (Figure 5). Clay Shock's secondary property of water resistance (Figure 6), also contributed to the success of this project. Owing to the flow of Clay Shock around the tail skin and through into the cutterhead, a percentage of the material ended up filling the cracks and crevices in the ground in front of the excavation face. Previously unachievable on the project, this flowability, combined with resistance to wash out and dilution with water, limited water ingress significantly, which enabled the stable extraction of spoil from the screw conveyor.

It had now been proved on site that Clay Shock was a versatile material solution, with many possible applications. In the 7 years between its development in 1983 and 1990, Clay Shock had been used at a total of 7 sites across Japan mainly as a friction reduction material but also as an additive. Between 1990 and 2000, widespread acceptance of Clay Shock as not just a product, but as a distinct tunneling method, had led to its use at 40 sites within a ten-year time-span. Furthermore, due in part to the official registration of Clay Shock Method with the Japanese Ministry of Land, Infrastructure, Transport and Tourism's New Technology Information System (KT-16002-A), the uptake of Clay Shock between 2000 to 2010 had risen to over 140 sites across Japan with applications as wide and varied as preventing water ingress during machine launch and breakthrough at arrival shaft; settlement control; ground stabilization when passing though fluvial sand layer; as an additive when mining in mixed ground; mining bedrock under high water pressure; to prevent leakage of slurry via fractures in bedrock; preventing collapse of the excavation face in unstable ground; to maintain pressure during cutter head interventions of shield machines; use in pipe jacking sites in fractured ground and Clay Shock had even been used onsite to course correct jammed TBMs in severe nose down scenarios.

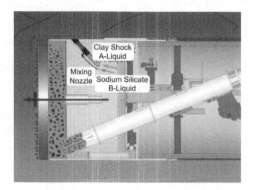

Figure 5.   Injection of Clay Shock through the Tail Skin, to the front of the EPB machine.

Figure 6.   0h, 6h and 12h dilution testing of Clay Shock.

3 RECENT APPLICATIONS

3.1 Breaking new ground.

In 2010 the Hanshin Highway Corporation began construction of a brand-new highway bisecting South Osaka connecting the two ends of the Osaka Loop Line Highway. Called the Yamatogawa Route, the new highway would stretch 9.7 km under one of the most densely populated cities in the world, below 11 separate over ground railway lines, and to within 2.2m beneath one of the busiest subway lines in Osaka, the famous Midosuji Line. This 6.8 m diameter tunnel holds the main Osaka Metro North-South line, travelling 24.5 km through the center of Osaka, and carries 1.2 million people daily. (Figure 7).

The two main contractors, Daitetsu JV and Obayashi JV, shared the task of excavating the eastbound and westbound Yamatogawa Route tunnels respectively, with the segments between the two tunnels having a minimum separation in places of only 1m. The tunnels would be excavated with a single IHI made (now JIM Technology Corporation), 12.54m EPB Shield machine, with TAC providing the Backfill Grout above ground batching plants and underground support machinery, Backfill Grout Injection Pipes, Additive and Foam generation equipment and control panels. The construction would also incorporate TAC's Clay Shock Method to ensure the perfect passing of the EPB Shield Machine under the Midosuji Line.

While passing beneath the Midosuji Line, the EPM would be driven twice, (once for the north tunnel, and once for the south tunnel), through a combination of Ds (Diluvial Sandy Soil) and Dc (Diluvial Clay Soil) layers, (Figure 8). The upper half of the cutterhead would pass though the diluvial sandy soil layer, which has a minimum Coefficient of Uniformity of 8.2, and a minimum fine grain content of 2.4%. Through experience of past tunnel construction in the area, it is known that this kind of soil is very prone to disintegration upon excavation. (Shima, Minamikawa, Nishiki, Nishimori & Miyake, 2015). This would be further compounded by the double passing of the TBM through the high-risk zone under the Midosuji subway tunnel, creating possibly high stresses close to the calculated limits of the segment structure itself.

3.2 Understanding subsidence caused by TBMs and Shield Machines

The key to the success of the Yamatogawa Tunnel was the accuracy of pressure management of the EPB and backing that up with subsidence prevention of the earth surrounding the shield machine. With the Shield Machine within 100m of the Midosuji underpass section, it was decided to test the actual settlement in the ground without using Clay Shock. A settlement meter was placed 2 m above the crown of the cutter head, and a normal drive was commenced,

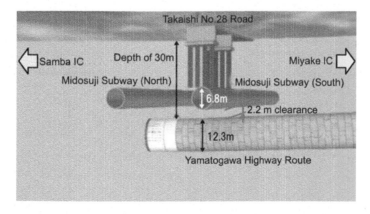

Figure 7.   Yamatogawa Route Tunnel passing under the Midosuji Subway Line Tunnel.

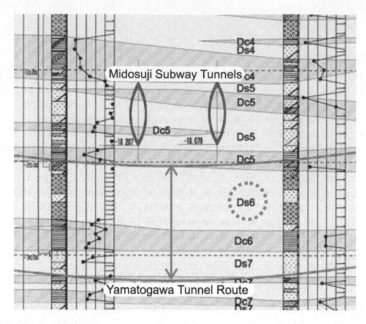

Figure 8. Longitudinal section of soil at the underpass point of the Yamatogawa and Midosuji tunnels clearly showing the multitude of Dc and Ds layers present at the site.

readings were then taken of the subsidence which occurred during the passing of the Shield machine, and continually until 1 day after the machine had passed. An initial measurement of 1.5 mm of subsidence was observed, increasing to 2.6 mm in 1 day. A secondary effect of the subsidence was an overall increase of thrust force of the shield machine during the drive to 144,000 kN, 70–80% of the maximum thrust of the machine (Shima et al., 2015). This field data influenced the general contractor to inject Clay Shock simultaneously via 8 injection ports on the machine during the drive under the Midosuji tunnel. (Figure 9).

The construction plan for the Yamatogawa Tunnel accounted for a maximum subsidence of -10.2 mm while passing under the Midosuji tunnels, but by injecting Clay Shock around the machine and managing the earth pressure correctly, the Yamatogawa Tunnel was driven under the Midosuji line with a final measurement, once the EPB Shield machine had driven

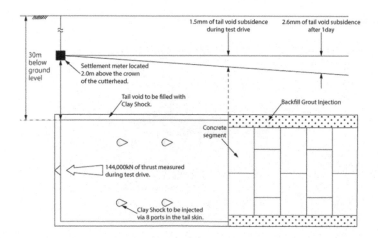

Figure 9. Measuring the actual settlement and thrust force without use of Clay Shock.

1684

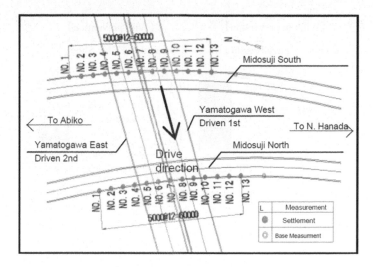

Figure 10. Settlement meter placement, for Yamatogawa Tunnel.

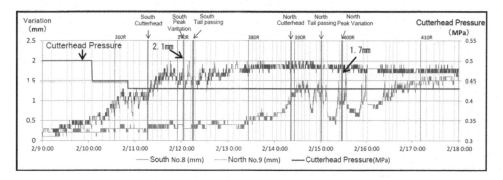

Figure 11. Settlement meter reading during drive under Midosuji Tunnel.

through and the segments were in place, that showed a maximum level increase of +2.1 mm for the North bound tunnel, and +1.7 mm for the South bound tunnel. (Shima et al., 2015) The increase had been due to the cutterhead pressure, and once the cutterhead had passed, the Clay Shock had held the ground in place until the void was subsequently filled with backfill grout as soon as the segment had left the rear body of the machine, (Figures 10–11). Undoubtedly, the application of Clay Shock had been a resounding success resulting in zero ground subsidence in the most sensitive part of this project.

3.3 *Use of Clay Shock for slurry pressure management with Slurry TBMs.*

Applications of Clay Shock didn't stop at EPB machines. Slurry Shields, inherently thought to be stable due to the flow of slurry around the machine, have been shown as they increase in diameter to have an inherent imbalances of slurry pressure the further away from the cutterface, and the further towards the crown of the machine pressure is modelled. The vertical gradient of pressure generated by slurry density as shown by (Van Eeklen, Van der Berg, Bakker, 1997), as well as the pressure of slurry in such a shield system (Ngoc-Anh, Dias, Oresta and Djeran-Maigre, 2013) indicates the need for a filler material, which can maintain pressure in larger diameter machines. The pressure differentials in Slurry shield machines make Clay Sand based, waterproofed fillers a viable solution to this reduce this differential, (Figure 12).

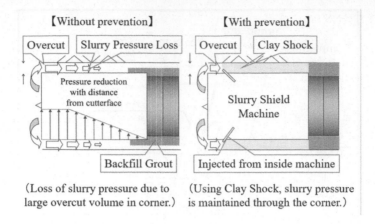

【Without prevention】 **【With prevention】**

| Overcut | Slurry Pressure Loss | | Overcut | Clay Shock |

Pressure reduction with distance from cutterface

Slurry Shield Machine

Backfill Grout   Injected from inside machine

(Loss of slurry pressure due to large overcut volume in corner.)   (Using Clay Shock, slurry pressure is maintained through the corner.)

Figure 12.   Use of Clay Shock to reduce pressure differential of Slurry Shield Machines.

Clay Shock use in Japan, in Slurry Shield for this purpose is well accepted, with Maeda Construction Corporation using such a system for a Kawasaki Heavy Industries made 10.30m Slurry Shield Machine used to excavate a 990m flood prevention tunnel for Tokyo Water Board in Tachiaigawa in 2008. This tunnel passed through three R=30m curves, two of which were S-curves. Slurry pressure loss through these sensitive turns was controlled by the injection of Clay Shock, which in turn limited subsidence in the middle of heavily populated area in downtown Tokyo. (Masuda, 2011)

4   FUTURE APPLICATIONS

4.1   *Hardening Clay Shock for pipe jacking.*

With a quarter of the 1400 or so construction sites in Japan and overseas where TAC has been active in the last 40 years having adopted the Clay Shock Method, there is a strong amount of data to back up the efficacy of the method and material itself. The real asset the material has been its ability to adapt into numerous situations and be effective in providing real solutions to real problems in tunneling. As with any product, it is the evolution of the product, and the search for the "next big thing" which drives progress. The next evolution for Clay Shock has been in producing hardening variants of what had become known as a non-hardening material.

Clay Shock has generally been associated with Shield machine applications, however pipe jacking, where frictional resistance should be low, but once the mining is complete the filler should solidify, is a very interesting future field to explore. Generally, this kind of construction has been completed using a two-step approach, using a low friction gel component, which is then replaced by a grout type component in a two-part process. With a hardening type of Clay Shock, it was thought that this process could be reduced to only one step. The following mix design was tested in the laboratory, see Table 3.

Table 3.   Clay Shock Hard Mix Design and UCS and Viscosity test results.

Mix Design	A Liquid for 1m³ batch				B Liquid	UCS Test N/mm²		
	Slag	Clay Sand	Retarder	Water	Sodium Silicate	$\sigma 28d$	$\sigma 56d$	$\sigma 91d$
Mix 1	120kg	350kg	10kg	814L	25L	0.30	0.60	1.20
Mix 2	120kg	350kg	10kg	814L	50L	0.14	0.25	0.30

5 CONCLUSIONS

This paper has attempted to show the progression over many years of one single product with in the tunneling industry in Japan, from inception to final acceptance and on into its future research and development by using real world examples to paint a picture of research and testing in Japan. The hope is this has deepened an interest in Japanese solution-based engineering, which can possibly lead to collaborative research in the future.

The Clay Shock Method was developed at a time when the economic bubble was just about to burst. When it did burst, it sent the burgeoning Japanese construction and civil engineering sector into free fall. On the back of this, many large general contractors were looking for more cost-effective ways to construct tunnels for less money, but without cutting back on safety and reliability, Clay Shock became a very viable alternative to other ground conditioning methods for reducing subsidence in tunneling, especially at sites where it would prove impractical to shut down roads, such as in the center of major cities. In particular, Clay Shock has been successfully applied to complex geology, hard rock and loose soil mixed layers, as well as diluvial clay and diluvial sand split-layer tunneling projects with great success in Japan and overseas. In Singapore, Clay Shock has been used for Land Transport Authority projects not only as a subsidence limiting material, but also as a face stabilization material for both EPB and Slurry shield machines, where stabilization of the face is essential in preventing subsidence during interventions. (Shirlaw, 2008)

The evolution of applications for the use of Clay Shock is based on the feedback of General Contractors, Project Directors, Engineers, Designers and City Officials who have supported this method and challenged TAC to develop more complex and innovative solutions for the problems which occur in tunneling. It is this kind of collaborative atmosphere which helps innovation thrive, not only in Japan but worldwide. Facing future on-site challenges head on, with a spirit of inquisitiveness and a passion for doing the seemingly impossible, is what will enable the further evolution of new materials within the field of tunnel engineering fueling the next generation of growth and success for us all.

REFERENCES

Ngoc-Anh, D. Dias, D. Oreste, P. Dejeran-Maigre, I. 2013. *3D Modelling for Mechanized Tunneling in Soft Ground-Influence of the Constitutive Model.* American Journal of Applied Sciences. 10 (8): pp.863–875

Masuda, M. 2011. *Sewage Tunnel Construction with a Large Diameter (φ 10.3 m) Shield Machine and Sharp Curve Excavation.* Maeda Construction Company presentation at 55th NPR, Shield & Tunnel Construction Technique Workshop, 2011. (in Japanese)

Shima, T. Minamikawa, S. Nishiki, O. Nishimori, F. Miyake, S. 2015. *Excavation Management of a Large Diameter (φ 12.5 m) Shield Tunnel, Constructed with a Minimum Separation of 2.2m Beneath a Subway Tunnel Structure.* Osaka City Transportation Bureau, Japan Civil Engineering Society, National Conference, 2015. (in Japanese)

Shirlaw, N. 2008. *Mixed Face Conditions and the Risk of Loss of Face in Singapore.* ICDE, Singapore, 2008.

Van Eekelen, S. Van de Berg, P. Bakker, K.J.C. 1997. *3D Analysis of Soft Tunneling.* Proceedings of the fourteenth international conference on solid mechanics and foundation engineering., Hamburg, 1997.

Tunnels and Underground Cities: Engineering and Innovation meet Archaeology,
Architecture and Art, Volume 5: Innovation in underground engineering,
materials and equipment - Part 1 – Peila, Viggiani & Celestino (Eds)
© 2020 Taylor & Francis Group, London, ISBN 978-0-367-46870-5

Durability of precast concrete tunnel segments

M. Bakhshi & V. Nasri
AECOM, New York, USA

ABSTRACT: In one-pass lining systems, the durability of tunnel structure is directly related to durability of concrete segments acting as both the initial support and the tunnel final lining. In this paper, most-frequent degradation mechanisms of concrete linings are discussed including chloride- and carbonation- induced corrosions, sulfate, acid and freeze-and-thaw attacks, and alkali-aggregate reactions. Mitigation method for each specific degradation mechanism is explained. A durability factor specific to railway and subway tunnel known as stray current corrosion is presented. Mitigation methods for this specific corrosion together with coupling effects with other conventional damage mechanisms are explained. Prescriptive approaches for durability design based on major codes and standards are explained and comparison is made between these methods. Exposure classes as the main inputs to the prescriptive approaches are elaborated and requirements specified by the codes and standards are presented and analyzed. The need to move on from prescriptive approach and embrace performance-based approaches for durability design of tunnel segmental lining is demonstrated and future studies are discussed.

1 INTRODUCTION

Tunnels as important underground structures are typically designed for a service life of more than 100 years. Mechanized tunneling method with Tunnel Boring Machines (TBMs), as the most common excavation method, is often associated with continuous installation of one-pass precast concrete segments in the form of rings behind TBM cutterhead. In these tunnels, durability of tunnel is directly related to durability of concrete segments acting as both the initial support and the tunnel final lining. In this paper, most-frequent degradation mechanisms of concrete linings are briefly discussed. This includes corrosion of reinforcement by chloride attack and carbonation, as well as sulfate, and acid attacks as major deterioration processes caused by external agents. Alkali-aggregate reactions caused by internal chemical reactions and frost attack and freeze-and-thaw damages are also explained. Stray current-induced corrosion as one major durability concern specific to railway and subway tunnel linings is discussed. Mitigation methods for different durability factor are discussed. Stray current corrosion mitigation method including use of FRC segments are presented and durability of segments under coupling effects of stray current with other conventional degradation factors are explained. Prescriptive approach for durability design based on European standard (EN 206-1:2013, Eurocode EN 1992-1-1:2004) and American Code ACI 318 (2014) is explained and comparison is made between two methods. Exposure classes related to environmental actions as the main inputs to both prescriptive approaches are explained separately. Conforming to these two major standards, recommendations made on concrete to ensure typical service life of tunnels are explained including concrete strength, maximum water-to-cement (w/c) ratio, minimum cement content and minimum air content.

2 DEGRADATION MECHANISMS IN TUNNEL LININGS

Possible degradation and damage mechanisms in bored tunnels include corrosion of reinforcement by chloride attack and carbonation, sulfate and acid attacks, alkali-aggregate reactions, freeze-and-thaw damages as well as stray current corrosion.

2.1 Chloride-induced reinforcement corrosion

Chloride-induced corrosion of reinforcement is the main cause of degradation in tunnels lined with reinforced concrete. Table 1 presents thirteen major tunnels that are significantly damaged and corroded due to chloride ingress before the year 1991 (ITA 1991; Abbas 2014). Chloride-induced corrosion is even a greater durability issue specifically in sub-sea, sea outfall, and road/rail tunnels. In sub-sea and outfall tunnels, and tunnels exposed to brackish groundwater, the intrusion of chloride ions present in sweater and salt water into reinforced concrete can cause steel corrosion. In cold region road/rail tunnels, major durability issue is the ingress of chloride ions present in deicing salts sprayed from vehicles during the snow fall. Chloride induced corrosion due to water infiltration initiates from the lining extrados, while corrosion due to de-icing salts sprayed from vehicle tires starts from lining intrados. Rust as the reaction product has a greater volume than the steel and cause expansion resulting in excessive tensile stresses, cracking, delamination, and spalling in the concrete (Figure 1).

2.2 Carbonation -induced reinforcement corrosion

Carbonation-induced corrosion in general is considered as a minor durability issue in reinforced concrete structures compared to chloride-induced corrosion. This is mainly due to limited impact area of carbonation and reduced strength zone limited to the extreme outer layer. In bored tunnels, carbonation is unlikely to occur due to the fact that generally extrados of tunnel lining is limited impact area of carbonation and reduced strength zone limited to the extreme outer layer. It is well-known that high rates of carbonation occur when the relative humidity is maintained between 50% and 75% (PCA 2002). In a lower relative humidity, the degree of carbonation is insignificant and above this range, moisture in concrete pores restricts penetration of CO_2 (ACI 201.2R 2016). In tunnels, only portal areas and entrance zones can maintain a relative humidity in the aforementioned range as lining in such areas is exposed to cyclic wet

Table 1. Damaged/Corroded tunnels due to chloride ingress (ITA 1991; Abbas 2014).

Tunnels	Location	Tunnel type	Diameter	Completion year
Basel/Olten Hauenstein	Switzerland	Railway	-	1916
Northern Line Old Street to Moorgate	U.K.	Metro	3.5 m	1924
Shimonoseki/Moji Kanmon	Japan	Railway	-	1944
Mikuni National Route 17	Japan	Highway	7.6 m	1959
Uebonmachi-Nipponbashi	Japan	Railway	10 m	1970
Dubai	U.A.E	Roadway	3.6 m	1975
Tokyo Underground	Japan	Roadway	-	1976
Berlin Tunnel Airport	Germany	Roadway	-	1978
Second Dartford	U.K.	Roadway	9.6 m	1980
Mass Transit Railway	Hong Kong	Metro	5.6 m	1980
Ahmed Hamdi	Egypt	Roadway	10.4 m	1980
Stockholm Underground	Sweden	Metro	-	1988

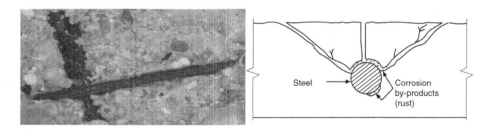

Figure 1. Loss of reinforcement section and cracks caused by chloride-induced steel corrosion (PCA, 2002; Romer, 2013).

and dry conditions. Also high rate of carbonation requires elevated atmospheric carbon dioxide (CO_2) levels which is only a case in heavily trafficked road runnels because of CO_2 emission from car exhaust. Therefore, carbonation is a major durability factor in portal areas and entrance zones of heavily trafficked road runnels. Carbonation can also occur in tunnel linings exposed to bicarbonate (HCO_3) ground water which often formed by the reaction of carbon dioxide with water and carbonate bedrocks such as limestone and dolomite.

Chloride- and carbonation-induced corrosion can be mitigated using low w/c ratio, high compressive strength and high cement content. This in conjunction with considering a sufficient concrete cover over reinforcement provide with a high quality and dense concrete that can delay the initiation time of corrosion also known as propagation time beyond the service life of structure. Other effective mitigation methods that are not in the codes include using cements with high amount of C_3A, and addition of corrosion inhibitors to the concrete mix.

2.3 *Sulfate attack*

Sulfate attack is a major durability issue for concrete structures in contact with soil or water containing deleterious amounts of water-soluble sulfate ions. Tunnels as underground structures, regardless of their specific use, can be exposed to external sulfate attack from common sources such as sulfates of sodium, potassium, calcium, or magnesium found in in the surrounding ground or dissolved in natural ground water. Ancient sedimentary clays and the weathered zone (< 10m) of other geological strata, as well as contaminated grounds and groundwater generally contain significant sulfate concentrations (BTS 2004). In tunnel linings exposed to such conditions, sulfate attack is a major concrete degradation mechanism. In tunnels, usually ettringite and gypsum can be produced as a result of a sulfate attack which in turn results in expansion of cement (e.g. ettringite volume is ~2.2 times higher in volume than the reactants). As a result concrete cracks and loses strength. It is expected that damages in tunnel linings due to sulfate attack start on segment extrados and at the interface between lining and the ground where sulfate from ground or groundwater can penetrate the concrete.

Sulfate attack can be mitigated by using cements with low amount of C_3A (<8%), use of high content of active mineral components, low w/c ratio and use of blended cements with pozzolans. Codes and standards recommendations to mitigate sulfate attack are based on using a concrete with low w/c ratio, high compressive strength and high cement content. In addition codes require use of sulfate-resisting cements such as type II portland cement (ASTM C150 2017) or in severe cases type V (ASTM C150 2017) plus pozzolan or slag cement.

2.4 *Acid attack*

Acid attack is a chemical attack that can be a major durability issue when concrete structure is exposed to high concentrations of aggressive acids with high degrees of dissociation. The deterioration of concrete by acids is primarily the result of decomposition of the hydration products of the cementitious paste (ACI 201.2R 2016). Sulfuric and hydrochloric nitric acids are main inorganic (mineral) acids, and acetic, formic and lactic acid are main organic acids with rapid rate of attack on concrete at ambient temperature. Acids reduce the pH or alkalinity of the concrete, and once the pH reduces to less than 5.5 to 4.5, severe damages are imminent as cement hydration products such as Portlandite (CH, $Ca(OH)_2$) and C-S-H starts to decompose when pH drops to around 12 and 10, respectively (ACI 201.2R 2016). This is the main reason that no concrete materials have a good resistance to acids.

In tunnels, the rapid deterioration of concrete only normally occurs when concrete is subject to the action of highly mobile acidic water (BTS 2004). With external acidic groundwater this is rarely the case, since ground waters are not usually highly mobile. Regarding internal sources for acid attack, flow of acid-containing runoff from outside the tunnel is not a major concern. However, sulfuric acid solutions result from decay of organic matter by bacterial action in sewage and wastewater tunnels is the primary mechanism of degradation in these tunnels. This is due high attack rate of sulfuric acid and continuous movement of the acidic materials inside the tunnel as gravitational flow of sewage in these tunnels is always guaranteed. Sewage is not aggressive to concrete buy itself but hydrogen sulfide produced by anaerobic bacteria reaction with the sludge is subsequently oxidized by aerobic bacteria to form sulfuric acid. In addition to decomposition of the cement hydration products, sulfuric acid is particularly aggressive to concrete because the calcium sulfate formed from the acid reaction may drive sulfate attack of adjacent concrete that was unaffected by the initial acid attack (ACI 201.2R 2016; PCA 2002).

Acid attacks can be mitigated with providing a dense and high quality concrete by lowering w/c ratio and increasing compressive strength and cement content. Codes and standards provide specific limits to achieve very high density and relatively impermeable concrete to reduce the damage due to acid attack. Type of cement has an insignificant role on mitigation of acid attacks. When concrete is exposed to very server acid attacks, a surface protection method such as coatings, waterproofing membranes or a sacrificial layer should be considered.

2.5 *Alkali-Aggregate Reaction (AAR)*

AAR as a chemical attack can be a major durability concern when concrete aggregates contain materials that can be reactive with alkali hydroxides in cement phase. The AAR generates expansive products and may result in damaging deformation and cracking of concrete over a period of years. AAR has two main forms of alkali-silica reaction (ASR) and alkali-carbonate reaction (ACR). ASR is often major concern compared to ACR as aggregates containing reactive silica are more common (PCA 2002) whereas aggregates susceptible to ACR are less common and usually unsuitable for use in concrete. Reactive forms of silica can be found in aggregates such as chert, volcanic glass, quartzite, opal, chalcedony, and strained quartz crystals. Damage to concrete only normally occurs when concrete alkali content is high, aggregate contains an alkali-reactive constituent, and concrete is under wet conditions (BTS 2004). ASR reactions can be summarized as:

Alkalis + Reactive Silica → Gel Reaction Product
Gel Reaction Product + Moisture → Expansion

PCA (2002) reports the internal relative humidity of 80% as a threshold, below which the alkali-silica reactivity can be virtually stopped. AAR does not depend on the specific use of each tunnel. Sub-sea tunnels may be more susceptible due to exposure to warm seawater containing dissolved alkalis which may aggravate alkali-silica reactivity. AAR can be mitigated by using inert aggregate, controlling the amount of soluble alkalis in concrete, and using blended cements with pozzolans.

2.6 *Frost attack and freeze-and-thaw damages*

Frost attack and freeze-and-thaw damages are durability concerns in concrete structures built in cold regions. Water expands by about 9% when it freezes and as a result, the moisture in concrete capillary pores exerts pressure on the concrete solid skeleton. This leads to development of excessive tensile stresses in the concrete and rupture of cavities. Successive cycles of freeze-thaw can disrupt paste and aggregate and eventually cause significant expansion and cracking, scaling, and crumbling of the concrete (PCA 2002). Frost damage is considerably accelerated by deicing salts (ACI 201.2R 2016).

Surface scaling is the only frost damage that can possibly occur in precast tunnel segments. Since the increase in volume when water turns to ice is about 9%, more than 90% of capillary pores volume must be filled with water in order for internal stresses to be induced by ice formation (BTS 2004). Moisture content near saturation level is usually the case for tunnel linings as often times tunnels are built under the water table and concrete lining can be near saturation level. However, along most of tunnel alignment, the temperature rarely falls under the freezing point because tunnel is embedded in the ground. Tunnel entrances, portals and shafts are parts of tunnel system that should be designed for exposure to cycles of freezing and thawing because of saturation level and exposure to freezing temperature.

Freeze–thaw attacks are mitigated by controlling w/c ratio, compressive strength and cement content. Controlling air content in the mix to a minimum 4% using air-entraining admixtures is the most effective mitigation method. Codes and standards often provide limits for maximum w/c ratio and minimum compressive strength, or require frost-resistant aggregates (EN 206-1 2013).

2.7 Stray current corrosion

Stray current corrosion is a type of corrosion specific to rail tunnels where corrosion is caused by traction current resulting in accelerated oxidation of metals and rapid migration of the chloride ions (ITA 1991). Inspection of removed segments from tunnels with high conductivity between running rails and lining reinforcement such as Bucharest Metro has shown an extensive corrosion of the outer reinforcement layer (Buhr et al. 1999).

Government agencies around the world are promoting electric trains and all modern railway systems take advantages of railway electrification. Power transmission is provided by overhead catenary wire or a conductor rail also known as third rail. The running rail connected to nearby substations is often used as traction loop through which the return circuit is made. The running rail has a limited conductivity, and insulation between the rail and the ground is sometimes reduced or constructed poorly form the beginning. This causes a fraction of the traction current to leave the rail, leak into the ground and flow back along the running rail on the return path to the traction substation by the earth diversion, which is referred to as stray current. Figure 2(a) shows a simplified electronic circuit of the electric railway system for modeling the stray current. In this figure, I_T, I_R and I_s represent the train (overhead catenary system) current, stray current and the current flowing through the running rail, respectively. R_R is the running rail resistance, R_s is the ground resistance at the traction substation, and R_T is the ground resistance as seen at the train. It's evident that reduced R_T or R_s results in increased stray current. When train runs in a lined tunnel, stray current leaks to the tunnel lining and through the concrete reinforcement. This is shown schematically in Figure 2(b) with a cathode formed at reinforcement where stray current enters the rebar and an anode is formed where stray current leaves the rebar and flows back to substation. Figure 2(c) shows that in cathode, rebar is disengaged from the concrete due to trapped hydrogen isostatic pressure, and in the anode, the rebar is oxidized in contact with electrolytic material, i.e. concrete, and accumulation of corrosion products exerts excessive pressure leading to cracking (Wang et al. 2018). This type of corrosion is not limited to rebar in concrete lining but also metal utilities and steel pipelines embedded in the ground in the proximity of tracks.

General mitigation methods for this type of corrosion are based on reducing the amount of stray current by decreasing rail resistance, improving rail to ground insulation using isolated rail

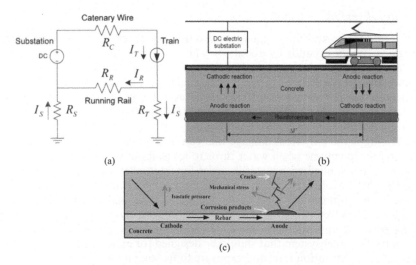

Figure 2.   a) Modeling stray current leakage with simplified electronic circuit (Niasati & Gholami 2008); b) Schematics of stray current from a train overhead catenary system picked up by reinforcement in concrete (Bertolini et al. 2007), c) corrosive effect of stray current (Wang et al. 2018).

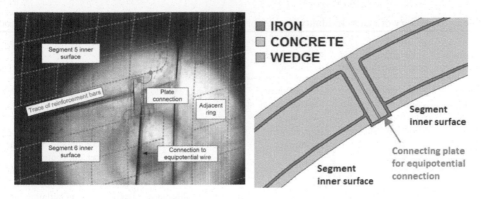

IRON
CONCRETE
WEDGE

Segment 5 inner surface

Trace of reinforcement bars

Plate connection

Adjacent ring

Segment 6 inner surface

Connection to equipotential wire

Segment inner surface

Segment inner surface

Connecting plate for equipotential connection

Figure 3. Stray current mitigation using equipotential connection provided by copper plates/straps connecting reinforcement cages of segments by Dolara et al. (2012).

fastening systems or pads, keeping the substation as close to the point of maximum current as possible, developing monitoring systems, devices and measurement apparatus (Brenna et al. 2010).As shown in Figure 3, stray current corrosion can be also mitigated using equipotential connection provided by copper plates/straps connecting reinforcement cages of segments. The equipotential connections between the reinforcing bars of all segments in a ring constitute a path with extremely low electrical resistance that allows the current to flow from a segment to the adjacent one without passing into the ground. The equipotential connection reduces bar-to-ground voltage to values well below the standard limits (EN 50122-2 2010) for corrosion initiation, and provides an effective method to prevent stray current corrosion in segments.

Another mitigation method for stray current induced corrosion is use of fiber-reinforced concrete (FRC) segments (Tang 2017). Results of studies on stray current corrosion of FRC show that steel bars are more likely to pick up current than short steel fibres under same conditions (Edvardsen et al. 2017). This can be due to the fact that the chloride threshold for the corrosion of steel bars in concrete is between 0.15–0.6% by mass of cement (ACI 318 2014). However, steel fiber-reinforced concrete demonstrates a higher corrosion resistance compared to steel bar reinforced with a chloride threshold level for corrosion at 4% by mass of cement (Tang 2017). The discontinuous and discrete nature of steel fibres or the length-effect is the main factor to be accounted for this higher corrosion resistance as fibers rarely touch each other and there is no continuous conductive path for stray currents through the concrete (ACI 544.1R-96 2009).

3 DURABILITY UNDER COUPLED DURABILITY FACTORS

Precast concrete tunnel segments may be subjected to the coupling effects of degradation factors such as carbonation, sulfate and chloride-induce corrosion of steel bars by groundwater and surrounding ground. For subway tunnel, stray current is another major factor that accelerates the steel corrosion. A summary of these different major degradation mechanisms together with their mitigation methods are shown in Table 2. A literature review on experiments conducted on coupled effect of stray current and other degradation factors reveals that the majority of previous works (Xiong 2008) has been focused on the material scale level which cannot truly reflect the durability aspects of full-size concrete members (Zhu & Zou, 2012; Geng and Ding, 2010). Coupling effects of multi factors on durability of segments in large-scale were studied by Li et al. (2014). These factors include carbonation, sulfate, chloride ion penetration and stray current corrosion. Study was conducted on segments made of concrete with water–cement ratio of 0.28–0.33, compressive strength of 60–70 MPa, and steel bars of 6.5mm diameter with yield and ultimate strengths of 472 and 586 MPa. Extrados side of segments was immersed in solutions of 3.5% NaCl for simulation of exposure to chloride ions, and 3.5% NaCl + 5% Na_2SO_4 for simulation of exposure to both chloride and sulfate ions. Segment intrados was exposed to a carbonation setup simulating CO_2 environment with concentration of 20%, temperature of 20°±5, and relative

Table 2. Summary of major durability factors for tunnel linings, their sources and mitigation methods.

Degradation mechanism	Type of tunnels susceptible to this factor	Main sources of degradation	Specific location of tunnel prone to this factor	Mitigation method

Chloride-induced corrosion	- Sub-sea tunnels
	- Sea outfall tunnels
	- Transportation tunnels in cold region
		- Sea/salt water
		- Sea/salt water
		- Deicing salts sprayed from vehicle tires
			- Lining extrados	Delay corrosion initiation by:
				- Sufficient cover over rebar
				- Dense/high quality concrete:
				  . Low w/c ratio
				  . High compressive strength
				  . High cement content
				- Cement w/high C_3A content
				- Use of corrosion inhibitors

Carbonate-induced corrosion	- Heavily-trafficked roadway tunnels	CO_2 emission from car exhaust	- Lining intrados near portals, entrance zones, shafts	Delay corrosion initiation by:
				- Sufficient cover over rebar
				- Dense/high quality concrete:
				  . Low w/c ratio
				  . High compressive strength
				  . High cement content

	- All types of tunnels embedded in carbonate bedrock such as limestone or dolomite	Bicarbonate (HCO3) groundwater formed by reaction of water & carbonate bedrock	- Lining extrados	- Sufficient cover over rebar - Cement w/high C_3A content - Use of corrosion inhibitors

External sulfate attack	- All types of tunnels embedded in ancient sedimentary clays or
	- All types of shallow tunnels exposed to weathered zone (<10 m) of other geological strata
	- All types of tunnels exposed to sulfate contamination
		Formation of ettringite due to sulfate reacting with calcium aluminates or $Ca(OH)_2$	- Lining extrados	- Dense/high quality concrete:
				  . Low w/c ratio
				  . High compressive strength
				  . High cement content
				- Cement w/low C_3A (<8%)
				- Pozzolans/Blended cement

Internal acid attack	- Sewage/Wastewater tunnels	Formation of H2S and oxidization to sulfuric acid	- Lining intrados	- Dense/high quality concrete - Coatings - Sacrificial layers - Use calcareous aggregates
Alkali aggregate reaction (AAR)	- All types of tunnels built with reactive silica aggregate	Volcanic glass Opal/chalcedony Deformed quartz	No specific location	- Use inert aggregate - Control amount of soluble alkalis in concrete - Pozzolans/Blended cement
	- Sub-sea tunnels	warm seawater containing dissolved alkalis	- Lining extrados	

Frost attack/ Freeze-thawing	- All types of tunnels in cold region	Surface scaling due to increase in volume when water turn to ice near saturation	- Lining intrados near portals, entrance zones, shafts	- Dense/high quality concrete:
				  . Low w/c ratio
				  . High compressive strength
				  . High cement content
				- Air-entraining admixtures

Stray current corrosion	- Subway tunnels - Railway tunnels	OCS current leaking into lining when returning from running rail	- Near rebar	Reduce amount of current: - Decrease rail resistance - Improve rail/ground insulation - Substation close to max current Use of straps connecting bars Use fiber reinforcement

humidity of 70%±5%. Figure 4a shows the dimension of segments and arrangement of reinforcement. Figure 4b shows schematics of the test setup demonstrating series connections between 6 segment samples in a group to ensure equal current flow of 1A among all segments. Segment reinforcement as anode and stainless steel tube placed in corrosion pools as cathode were connected to DC power supply to simulate the stray current. After casting and standard curing, samples were immersed in corrosion solution for 18 days and were exposed to carbonation setup for 28 days. Free chloride ion was determined from powders collected using drilling at 5mm intervals along the segment thickness. Results show that stray current accelerates the migration of chloride ions. Stray current corrosion also changes the penetration distribution of chloride ion in the section resulting in the largest concentration of chloride ions to be at the reinforcement level instead of the exposure surface which is the case for general chloride ion exposure without stray current. Also chloride ion concentration is higher for segments immersed in chloride solutions (Cl^-) than the ones immersed in solution with both chloride and sulfate ($Cl^- + SO_4^{2-}$). This can be due to the concrete pores that may be filled by ettringite produced by reaction of SO_4^{2-} ions in the solution with hydration products, resulting in obstruction of channels through which chloride ions migrate. Another argument is that SO_4^{2-} ions firstly react with C_3A producing ettringite while decreasing the opportunity of integration of Cl^- ion and C_3A. As a result it is more likely to be absorbed by C-S-H, altering the combination form of Cl^- ions (Li et al. 2014). Another major outcome is that carbonation depth is only 1–4 mm, leading to a conclusion that carbonation is not a controlling durability factor for concrete segments compared with the chloride and sulfate ions and stray current corrosion. A typical corroded steel cage under coupled durability factors using this setup is shown in Figure 4d. Another major conclusion specific to reinforcement is the more significant corrosion of the reinforcement layer near the extrados compared to intrados, and more corrosion damage on stirrups than main transverse reinforcing bars. Considering results of this study, one can conclude that coupling factors of chloride ion penetration and stray current has the most detrimental effect on the durability of concrete tunnel segments.

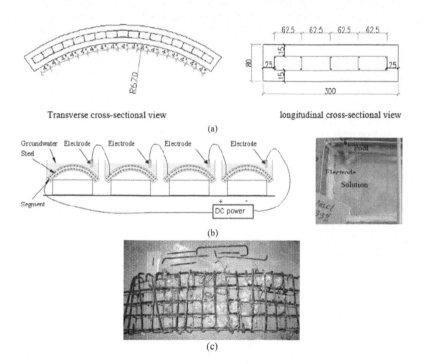

Figure 4. Li et al. (2014) study on stray current coupled with multi degradation factors: a) size and layout of experimental segment samples and reinforcement (in mm), b) schematics of test setup for combined effect of corrosion solution and stray current, c) Segment reinforcement corrosion.

4 PRESCRIPTIVE-BASED APPROACHES

Durability design according to prescriptive approaches is the most-common method in tunnel and concrete industries that are performed in accordance with major national and international structural codes (EN 206-1 2013; EN 1992-1-1 2004, ACI 318 2014). Inputs to these methods are environmental exposure classes and outputs to these methods are required concrete characteristics such as concrete strength and maximum w/c ratio. Following Euro standards (EN 1992-1-1 2004), for specific case of tunnel linings, suggested exposure classes according to Helsing and Mueller (2013) for CO_2 carbonation are XC3 to XC4, for seawater chloride-induced corrosion is XS2 to XS3, for deicing salt chloride-induced corrosion is XD2 to XD3, for frost exposure is XF3 to XF4 and for harmful ions other than chloride (Mg^{2+}, SO_4^{2-}) is XA1 to XA3. This is mainly due to the fact that this standard includes assumption of design service life of 50 years, and exposure classes should be increased by 2 in order to consider 100 years as minimum tunnel service life. Exposure category C2 in ACI 318 (2014) can be compared to above-mentioned XC, XD/XS exposure classes in EN 206-1 (2013) and EN 1992-1-1 (2004), and categories F2 and F3 with XF3 and XF4. Similarly, considering sulfate attack, ACI 318 (2014) exposure categories S1 to S3 can be compared with EN categories XA1 to XA3. A case example for an extreme exposure to chloride-induced corrosion (ACI category C2 vs. EN class XS3/XD3) shows that ACI would require a maximum w/c ratio of 0.4 and a minimum compressive strength of 35 MPa. On the other hand, EN would require a maximum w/c ratio of 0.45, a minimum compressive strength of 35 MPa, and a minimum cement content of 340 kg/m^3. Concrete cover specified by ACI 318 (2014) as minimum 38 mm (1.5 in) for reinforcement size of No. 19 or smaller (<imperial #6) can be compared with 45 mm required by EN 1992-1-1 (2004). This indicates that concrete requirements set forth for concrete by either ACI or EN would be very similar and most likely would result in a concrete specification with similar if not identical quality.

The major flaw of prescriptive apaches is lack of connection between the limiting requirements and main source of degradation mechanisms for each specific type of concrete damage. In contrast, performance-based design approaches despite all challenges related to these methods provides significant benefits to designers by focusing on the specific sources of concrete damages in a project-specific fashion (Swiss Standard SIA 262 2003). In order to achieve a performance design, rapid, easy, and reliable test methods are needed to assess properties of structural concrete. Future studies are needed to develop a performance-based design approach with reference to different major tunnel segment projects. While this design approach is not discussed in this paper, studies such as Rashidi & Nasri (2012), Sigl et al. (2000) and Li et al. (2015) provide important new insight into such design method for wastewater, subway and sub-sea road tunnels, respectively. In addition, durability recommendations of national and international tunnel segment guidelines should be analyzed and compared in future studies.

5 CONCLUSION

Degradation mechanisms of concrete linings include chloride- and carbonation- induced corrosion, sulfate and acid attacks, alkali-aggregate reactions, and frost attack. All these mechanisms are introduced and mitigation methods are explained. Stray current-induced corrosion as one major durability concern specific to railway and subway tunnel linings is discussed. Mitigation methods for stray current corrosion including use of FRC segments are discussed and durability of segments under coupling multi-factors is explained. Coupled factors of chloride ion penetration and stray current are presented as the most detrimental for service life of tunnel precast segments. Prescriptive approaches for durability design based on major codes and standards are presented and compared. The need for developing a performance-based design approach for tunnel linings is explained. Future studies should include also comparison of current durability recommendations by national and international tunnel guidelines.

REFERENCES

Abbas, S. 2014. Structural and durability performance of precast segmental tunnel linings. PhD Dissertation, The University of Western Ontario, London, Ontario, Canada.

ACI 201.2R. 2016. Guide to durable concrete. American Concrete Institute.

ACI 318. 2014. Building code requirements for structural concrete and commentary. American Concrete Institute.

ACI 544.1R. 2009. Report on fiber reinforced concrete (Reapproved 2009). American Concrete Institute.

ASTM C150/C150M-17. 2017. Standard Specification for Portland Cement, ASTM International, West Conshohocken, PA, 2017, www.astm.org.

Bertolini, L. & Carsana, M. & Pedeferri, P. 2007. Corrosion behaviour of steel in concrete in the presence of stray current. Corrosion Science, 49(3):1056–1068.

Brenna, M. & Dolara, A. & Leva, S. & Zaninelli, D. 2010. Effects of the DC stray currents on subway tunnel structures evaluated by FEM analysis. Power and Energy Society General Meeting: IEEE:1–7.

BTS. 2004. Tunnel lining. Design guide. British Tunnelling Society (BTS).

Buhr, B. & Nielsen, P. V. & Bajernaru, F. & McLeish, A. 1999. Bucharest metro: dealing with stray current corrosion. Proceedings of the Tunnel Construction and Piling Conference, 99.

Dolara, A. & Foiadelli, F. & Leva, S. 2012. Stray current effects mitigation in subway tunnels. IEEE Transactions on Power Delivery, 27(4):2304–2311.

Edvardson, C. & Müller, S. & Nell, W. & Eberli, M. 2017. Steel fibre reinforced concrete for tunnel lining segments – design, durability aspects and case studies on contemporary projects. Proceedings of STUVA Conference 2017, Stuttgart, Germany: 184–189.

EN 206-1. 2013. Concrete – Specification, performance, production and conformity. European Norm (EN).

EN 1992- 1-1 Part 1. 2014. Eurocode 2: Design of concrete structures - Part 1-1: General rules and rules for buildings. European Norm (EN).

EN 50122-2. 2010. Railway applications - Fixed installations - Electrical safety, earthing and the return circuit Part 2: Provisions against the effects of stray currents caused by d.c. traction systems. European Norm (EN).

Helsing E, & Mueller U. 2013. Beständighet av cement och betong i tunnelmiljö (Resistance to cement and concrete in tunnel environment). Seminarium Vatten i anläggningsbyggande (Seminar on Water in Construction). Göteborg, Sweden. Nov 27, 2013.

Geng, J. & Ding, Q.J. 2010. Transport characteristics of chloride ion in concrete with stray current. Journal of Building Materials, 1:121–124.

ITA. 1991. Report on the damaging effects of water on tunnels during their working life. Tunnelling and Underground space technology, 6(1):11–76.

Li, K. & Li, Q. & Wang, P. Fan, Z. 2015, September. Durability assessment of concrete immersed tube tunnel in Hong Kong-Zhuhai-Macau sea link project. Concrete Institute of Australia Conference, 27th, 2015, Melbourne, Victoria, Australia.

Li, Q. & Yu, H. & Ma, H. & Chen, S. & Liu, S. 2014. Test on durability of shield tunnel concrete segment under coupling multi-factors. Open Civil Engineering Journal, 8:451–457.

Niasati, M. & Gholami, A. 2008. Overview of stray current control in DC railway systems. International Conference on Railway Engineering - Challenges for Railway Transportation in Information Age, ICRE, Hong Kong, 2008:1–6.

PCA. 2002. Types and causes of concrete deterioration. PCA IS536.

Rashidi, S. & Nasri, V. 2012. Mitigation of the corrosion risk for large concrete sewer tunnels. ITA-AITES World Tunnel Congress (WTC) 2012, Bangkok, Thailand.

Romer, M. 2013. Durability of underground concrete. Workshop on Underground Infrastructures: Challenges and Solutions, June 27, 2013, LTA, Singapore.

Sigl, O. &, Raupach, M. & Rieker, L. 2000. Durability design of concrete tunnel lining segments. 25th Conference on Our World in Concrete and Structures: 23–24 August 2000, Singapore.

Swiss Standard SIA 262. 2003. Concrete structures. Swiss Society of Engineers and Architects.

Tang, K. 2017. Stray current induced corrosion of steel fibre reinforced concrete. Cement and Concrete Research, 100:445–456.

Wang, C. & Li, W. & Wang, Y. & Xu, S. & Fan, M. 2018. Stray current distributing model in the subway system: A review and outlook. International Journal of Electrochemical Science, 13:1700–1727.

Xiong, W. 2008. Study on deterioration characteristics of reinforced concrete in the presence of stray current and chloride ion. PhD dissertation, Wuhan University of Technology, Wuhan, China.

Zhu, Y.H. & Zou, Y.S. 2012. Influence of stray current on chloride ion migrates in concrete. Journal of Wuhan University of Technology, 7:32–36.

Tunnels and Underground Cities: Engineering and Innovation meet Archaeology, Architecture and Art, Volume 5: Innovation in underground engineering, materials and equipment - Part 1 – Peila, Viggiani & Celestino (Eds)

Relieved specific energy estimation using FLCM and PLCM linear rock cutting machines and comparison with rock properties

C. Balci, H. Copur & D. Tumac
Mining Engineering Department, Istanbul Technical University, Maslak, Istanbul, Turkey

R. Comakli
Mining Engineering Department, Nigde Omer Halisdemir University, Nigde, Turkey

ABSTRACT: Specific energy is defined as the amount of work required to break a unit volume of rock and used to predict the performance of mechanical miners. This value can be obtained from full-scaled laboratory linear (FLCM) or portable linear rock cutting (PLCM) experiments at different cut spacings and depths, respectively. For this purpose, full-scaled linear and portable linear rock cutting experiments are performed on 5 different blocks of rock samples including Beige marble, Kufeki limestone, Travertine, Sandstone and Limestone. Cutter forces acting on a cutter in three orthogonal directions (cutting force, normal force, and sideway force) and, specific energy values are measured during testing. In addition, some physical and mechanical property testing are carried out and the relationships between optimum specific energy values and rock mechanical properties are analyzed using regression analysis. Statistical analyses suggest that the relieved specific energy values can be predicted reliably from rock mechanical properties to select the most efficient mechanical miners for a given rock or mineral.

1 INTRODUCTION

Prediction of the excavation performance of any mechanical excavator such as roadheaders, continuous miners and shearers for any geological formation is one of the main concerns in determining the economics of a mechanized mining and/or tunneling operation. There are several methods of performance prediction and the best approach may be the use of more than one of these methods. These methods may be generally classified as full-scale linear cutting test, small-scale cutting test (core cutting), empirical approach, semi-theoretical approach and field trial of a real machine.

Empirical performance prediction models are mainly based on the past experience and the statistical interpretation of the previously recorded field data. Collection of field data is very important for developing empirical performance prediction models. The accuracy and reliability of these models depend on the quality and amount of the data. It is usually difficult to collect large amount of and high quality data in the field (Balci, 2009; Balci and Bilgin 2007; Balci and Tumac, 2012).

It is widely accepted that efficiency of mechanical excavators such as roadheaders, continuous miners, and shearers are measured based on specific energy value for a given rock and cutter type. Specific energy concept provides a realistic measure of rock cuttability and a simple method for a quick and informative performance (production) capacity of all types of mechanical excavators. The specific energy is best obtained from full-scale rock cutting tests, which fulfills the gap and weaknesses of the theoretical and empirical models. The basic disadvantage of full-scale rock cutting tests is that it requires large blocks of rock samples (around 1x1x0.6 m), which are usually difficult, too expensive or impossible to obtain. Therefore, the core sample based cuttability tests are preferred in many cases, even though their predictive abilities are lower than full-scale rock cutting tests. Specific energy values were correlated, in the past, with rock properties by different researchers. However, these researches were usually limited to one rock type, one machine type and/or one index rock property. Therefore, this study focuses on the prediction of specific energy based on core based index tests including a wide range of rock, mineral and ore types. Portable and Full-scale rock cutting tests and physical and mechanical property tests are performed on the rock samples collected from different mine and tunnel sites. The relationships between the specific energy and rock properties such as compressive strength, tensile strength values are investigated by using regression analysis.

2 FLCM AND PLCM LINEAR ROCK CUTTING TESTS AND PROCEDURES

Laboratory testing program includes portable and full-scale rock cutting tests and physical and mechanical property tests. The tests are performed in the laboratories of Istanbul Technical University, Mining Engineering Department. The testing equipment, procedures and parameters are introduced and discussed in this section.

Full scale linear rock cutting machine (FLCM) is presented in Figure 1. Cutting (drag or rolling) force, acting parallel to the surface being cut and tool travel (cutting) direction, is directly related to the torque requirement of a mechanical miner, and used to estimate specific energy, which is defined as energy (work) required to excavate a unit volume or mass of rock (Pomeroy 1963, Roxborough 1973). Normal (thrust) force, acting perpendicular to the surface being cut and tool travel direction, is used to estimate required effective mass and thrust of the excavator to keep the tool in a desired depth of cut (penetration). Sideways force, acting perpendicular to the tool travel direction and the direction of normal and cutting forces, may be used along with normal and cutting forces to balance tool lacing for minimizing machine vibrations.

One of the portable rock-cutting devices was developed in the Mining Engineering Department of Istanbul Technical University (Bilgin et al. 2010). In the new generation of PLCM all mini scale cutters (mini discs, mini conical and chisel cutters) can be used in the rock cutting experiments (Balci et al.2016, Comakli et al. 2015). The cutter is attached to the dynamometer with a tool holder. Experiments are performed at 3 cm/s cutting speed and forces are measured with a data logger system. Before performing the experiments, surface of the rock is trimmed or conditioned with the cutters. All experiments are replicated 3 or 4 times. In Figure 2, a photographic view of the testing machine is given.

Figure 1.   Full-scale linear cutting machine (FLCM) in ITU Laboratories.

Figure 2.   A new generation portable linear rock cutting machine in ITU Laboratories.

A groove is cut on the surface of a rock sample with a small disc cutter with a certain depth of cut. The table of the portable linear rock-cutting device is moved by a hydraulic cylinder. Block rock samples in 20 × 20 × 10 cm in size or core samples split into two pieces (two halves) are attached to the table with a special mechanism or cast in a sample box with a flat surface on top to cut the rock with a minidisc. The forces acting on the cutter and specific energy values are measured using triaxial force transducer (dynamometer). Dynamometers equipped with strain gauges have been designed and developed for this special application, reaching a precision in the order of 1 kN and covering a range from 0 to 100 kN. The force dynamometer is calibrated with a hydraulic cylinder by applying known forces. The tests should be replicated at least 3 times for more reliable results in a given rock type. Using the new generation PLCM different cutter tools can be used for rock cutting tests. CCS type mini discs used with PLCM are shown in Figure 2.

2.1   Estimation of optimum specific energy from laboratory cutting experiments

Specific energy concept provides a realistic measure of rock cuttability and a simple method for a quick and informative performance (production) capacity of all types of mechanical excavators. The specific energy is best obtained from small-scale or full-scale rock cutting tests, which fulfills the gap and weaknesses of the theoretical and empirical models.

The production rate of a given mechanical miner is calculated by the following equation:

Net cutting rate can be calculated using Equation 1 below:

$$ICR = k.\frac{P}{SE_{opt}} \qquad (1)$$

where ICR = net cutting rate in m^3/h; P = power consumed in optimum conditions in kW; k = energy transfer ratio from cutterhead to tunnel face usually taken as 0.85–0.90, (Rostami et al., 1994); SEopt = optimum specific energy, kWh/m^3.

The predicted net cutting rate by using Equation 1 is valid for competent rock conditions and does not include the effect of rock mass properties (Bilgin et al., 2014). Energy transfer ratio (k factor) is also selected for competent rock to be 0.85 -0.9 (Rostami et al., 1994). It is obvious that the geological discontinuities will increase or decrease the net excavation rate to a certain level. Low RQD or water income in rock formation with high amount of clay will decrease the daily advance rate due to regional collapses, face instability and chocking the cutters etc.

The effect of the spacing between the cuts and depth of cut on the cutting efficiency/specific energy is also explained in Figure 3. If the line spacing is too small (a), the cutting is not efficient due to over-crushing the rock. If the line spacing is too wide (c), the cutting is not

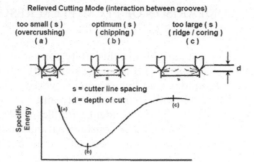

Figure 3.   Effect of line spacing and depth of cut on specific energy.

efficient since the cuts cannot generate relieved cuts (tensile fractures from adjacent cuts can not reach each other to form a chip), creating a groove deepening situation or forming a bridge/rib between the cuts. The minimum specific energy is obtained with an optimum spacing to depth of cut ratio (b). The optimum ratio of cutter spacing to depth of cut varies generally between 1 and 5 for pick (drag) cutters. The specific energy (SE) is calculated as follows using Equation 2 below:

$$SE = \frac{FC}{Q} \tag{2}$$

where, SE = specific energy in MJ/m³; FC = mean cutting (drag) force in kN; Q – yield per unit length of cut in m³/km. Cutting force and yield are functions of rock properties, bit type and cutting geometry.

3   PHYSICAL AND MECHANICAL PROPERTIES OF ROCKS

The physical and mechanical properties of the rock sample is determined according to ASTM (2005) for acoustic velocity, for the rest of the physical and mechanical property tests standards of ISRM (Ulusay and Hudson, 2007). Physical and mechanical property tests include uniaxial compressive strength, Brazilian tensile strength, static elasticity modulus, acoustic velocity test (dynamic elasticity modulus, Poisson's ratio). Uniaxial compressive strength tests are carried out on grinded core samples with a length to diameter ratio of 2. The stress rate applied to core samples is 0.5 kN/s. Brazilian tensile tests are performed on core samples with 0.25 kN/s stress rate and the ratio of a length to diameter of 1. The results of physical and mechanical properties of the beige marble is given in Table 1.

Table 1.   Results of the rock mechanics tests.

	Travertine	Limestone	Beige marble	Sandstone	Kufeki limestone
UCS (MPa)	9.10 ± 1	70.2 ± 12.5	160.7 ± 10	120.6 ± 21.5	15.6 ± 3.3
BTS (MPa)	3.17 ± 0.26	7.73 ± 1.33	7.82 ± 1.82	11.94 ± 1.32	1.14 ± 0.16
γ (gr/cm³)	1.92 ± 0.06	2.7 ± 0.01	2.7 ± 0.01	2.8 ± 0.01	2.16 ± 0.02

γ = Natural unit weight (density), UCS = Uniaxial compressive strength, BTS = Brazilian (indirect) tensile strength.

4 SPECIFIC ENERGY COMPARISON OF FLCM, PLCM RESULTS AND ROCK PROPERTIES

The independent linear cutting test variables in this study consist of five major constant variables: rock type (five blocks of samples), cutting mode (relieved), cutter penetration (1, 1.5, 2, 3, 4, 5 mm for relieved cutting mode), and line spacing (relieved mode) and cutter type (432 mm in diameter CCS type disc cutter with 18 mm tip width and 144 mm in diameter CCS type mini disc cutter with 2 mm tip width). The dependent variables are average cutter forces (rolling and normal force), and specific energy. The constant parameters through the testing program are cutting sequence (single-start), cutting speed (12.7 cm/s in FLCM 3 cm/s in PLCM), and data sampling rate (2,000 Hz). Each cut is replicated at least 3 or 4 times. Mean forces (rolling and normal forces) and yield (the volume of rock obtained per unit length of cut cutting) are recorded in each cut. Specific energy is obtained by dividing mean cutting force to yield. Results of the rock cutting tests with FLCM with 432 mm in diameter CCS type disc cutter and 144 mm in diameter CCS type mini disc cutter in unrelieved cutting modes is presented in Table 2.

The results of the specific energy obtained from FLCM and rock properties were correlated with the result of the specific energy obtained from PLCM and rock properties for144 mm in diameter mini disc cutter in unrelieved cutting conditions (Balci et. al, 2017). Relationships between UCS and specific energy for PLCM and FLCM in relieved cutting conditions at 5 mm depth of cut are given in Figures 4 and 5.

Table 2. Results of the rock cutting tests obtained from FLCM and PLCM in relieved cutting conditions.

Rock formation	PLCM (CCS)			FLCM (CCS)		
	SE (kWh/m^3)	FN (kN)	FR (kN)	SE (kWh/m^3)	FN (kN)	FR (kN)
Beige marble	12.07	11.91	2.03	5.58	142.94	4.50
Kufeki limestone	4.13	5.29	0.86	1.18	26.34	2.57
Travertine	6.41	13.77	3.07	1.81	66.26	8.47
Sandstone	10.62	12.00	3.11	5.09	108.82	9.77
Limestone	8.94	13.76	3.27	3.37	61.28	5.60

FN = Experimental mean normal forces; FR = Experimental mean rolling forces; SE = Specific energy.

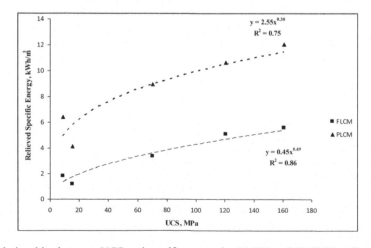

Figure 4. Relationships between UCS and specific energy for PLCM and FLCM in relieved cutting conditions at 5 mm depth of cut.

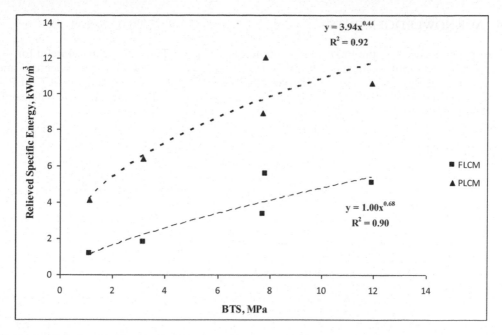

Figure 5. Relationships between BTS and specific energy for PLCM and FLCM in relieved cutting conditions at 5 mm depth of cut.

It is seen from the figures that the best-fitted relationships are found to be best represented by power functions. The correlation coefficients between unrelieved specific energy from FLCM and PLCM and uniaxial compressive at 5 mm depth of cut values is found to be 0.75 and 0.86, respectively. The correlation coefficients between unrelieved specific energy from FLCM and PLCM and Brazilian tensile strength at 5 mm depth of cut values is found to be 0.92 and 0.90, respectively.

5 CONCLUSION

This paper summarizes the results obtained with a new developed portable linear cutting machine (PLCM). A recently developed rock cutting testing device named as Portable Linear Cutting Machine (PLCM) is used in this study to measure normal and rolling forces acting on mini-scale cutting tools for mini disc, and specific energy. This machine minimizes the disadvantages of FLCM and can be used for predicting performances of the mechanical excavators used for hard rocks such as tunnel boring machines (TBMs), roadheaders, surface miners and, shearers. PLCM tests can be performed faster compared to the other full scale linear rock cutting machines and do not require too much manpower and large blocks of rock samples; thus it is comparatively cheaper.

The results indicate that specific energy values obtained from PLCM and FLCM by using disc cutters are very well-correlated with UCS and BTS for five different rock types. Relieved specific energy values can be predicted by BTS and UCS parameters reliably. The performance parameters of mechanical excavators such as thrust force, cutterhead power, and net cutting rate of known machines are theoretically estimated with using the PLCM test results.

The studies are still continuing with some other rock types, other cutter types (conical and chisel), and different experimental conditions in both laboratory and field for developing more reliable predictive models and verification purposes. Scale effect due to the cutters geometry (size, cutter ring tip width, the cutter ring diameter, cutting velocity etc.) is under analysis for further investigations.

6 ACKNOWLEDGEMENTS

This paper is based on the PhD thesis of Ramazan Comakli. The authors are grateful for the support of Istanbul Technical University (ITU) Research Foundation, the Scientific and Technological Research Council of Turkey (TUBITAK 112M859), the company of E-BERK Tunneling and Foundation Technologies for supplying mini discs and all research people involved in this project.

REFERENCES

Balci, C., 2009. Correlation of rock cutting tests with field performance of a TBM in a highly fractured rock formation: a case study in Kozyatagi-Kadikoy Metro Tunnel, Turkey. Tunnelling and Underground Space Technology 24, 423–435.

Balci, Comakli, R., Bilgin, N., Copur, H., Tumac, D., Avunduk, E. 2017. Unrelieved Specific Energy Estimation from PLCM and FLCM Linear Rock Cutting Tests and Comparison with Rock Properties, ITA-AITES World Tunnel Congress 9.-15, Bergen, Norway.

Balci, C., Bilgin, N., Copur, H., Tumac, D., 2016 (a). Development of a portable linear rock cutting machine for selection of mechanical miners by investigating the cuttability of some rocks. TUBITAK Project No: MAG 112M859, Report submitted to Tubitak. (in Turkish).

Balci, C., Tumac, D., 2012. Investigation into the effect of different rocks on rock cuttability by a V type disc cutter. Tunnelling and Underground Space Technology 30, 183–193.

Bilgin, N., Balci, C., Tumac, D., Feridunoglu, C., Copur, H., 2010. Development of a portable rock cutting rig for rock cuttability determination. Proc. European Rock Mechanics Symposium (EUROCK 2010), Editors: J. Zhao, V. Labiouse, J.P. Duth and J.F. Mathier, ISBN: 978-0-415-58654-2, June 15-18, Lausanne-Switzerland, CRC Pres/ Balkema, Taylor and Francis Group, pp. 405–408.

Bilgin, N., Copur, H., Balci, C., 2014. Mechanical Excavation in Mining and Civil Industries. CRC Press, Taylor and Francis Group, ISBN-13: 978-1466584747, 366 p.

Comakli, R., Balci, C., Polat, C., Tumac, D., Avunduk, E., Copur, H., Bilgin, N., 2015. "Cutter Forces Measurement with (PLCM©) Using Mini Disc Cutters: Comparison with the Theoretical Models", ITA World Tunnel Congress (WTC 2015) and 41st General Assembly, p. 10, Dubrovnik, Croatia.

Pomeroy, C.D., 1963. Breakage of coal by wedge action – Factors affecting breakage by any given shape of tool. Colliery Guardian, Nov. 21, pp. 642–648; Nov. 28, pp. 672–677.

Rostami, J., Ozdemir, L., Neil, D.M. 1994. Performance prediction: a key issue in mechanical hard rock mining, Mining Engineer, November, 1263–1267.

Roxborough, F.F., 1973. Cutting rocks with picks. The Mining Engineer, June, pp. 445–455.

Ulusay, R., Hudson, J.A., 2007. The Complete ISRM Suggested Methods for Rock Characterization, Testing and Monitoring: 1974-2006. Suggested Methods Prepared by the Commission on Testing Methods, International Society for Rock Mechanics, Compilation Arranged by the ISRM Turkish National Group, Ankara, Turkey, p. 628.

Tunnels and Underground Cities: Engineering and Innovation meet Archaeology, Architecture and Art, Volume 5: Innovation in underground engineering, materials and equipment - Part 1 – Peila, Viggiani & Celestino (Eds)
© 2020 Taylor & Francis Group, London, ISBN 978-0-367-46870-5

Performance based design for tunnels

K. Bergmeister
Brenner Base Tunnel BBT-SE, Innsbruck, Austria

ABSTRACT: Performance-based design includes all the features needed for a high-quality tunnel. A holistic approach is used for the performance-based design in order to provide a useful framework for analyzing and sizing tunnels and underground structures. In the following paper, some general remarks are presented and the holistic approach for performance-based design of tunnels will be presented. For this reason, not only the ultimate limit state, serviceability verifications and durability aspects are included, but also robustness, fire resistance und sustainability. Some examples of the performance-based design approach for the Brenner Base Tunnel, the world's-longest tunnel, will be described.

1 INTRODUCTION

In the past design codes included three different key elements:

– the check of the ultimate limit state
– the verifications of the serviceability limit states, taking into account deformations, crack width, water tightness,
– the provisions in terms of durability. Traditionally, durability aspects are included in the structural details by using certain minimum geometrical requirements and material properties. For the future it is certainly relevant to also verify the limit states associated with durability.

As regards robustness, Nassim Nicholas Taleb (2012), the New York times bestselling author of the „Black Swan"stated that „antifragility is beyond resilience or robustness". A resilient system resists shocks and unforeseen events. The aim must be to strengthen the system in order to move from a fragile toward a robust structure. Therefore, at least for bridges and tunnels, unforeseen collapses must be avoided by verifying the robustness of the structural system.

As specifically concerns tunnel linings, fire resistance plays an important role and therefore specific concrete admixtures und concrete covers have to be adapted. Also sustainability issues during the construction and operational phases are getting more important.

For the performance based design of tunnels at least the following relevant 6 key performance elements must be verified.

– Ultimate limit state
– Serviceability limit state
– Durability
– Robustness
– Fire resistance
– Sustainability

2 PERFORMANCE BASED DESIGN APPROACH

Performance based design takes into account the uncertainties and the time-dependent behaviour of the loads and the material resistance; coupled with the theory of structural reliability, provides a useful framework for analyzing structures with uncertainties [Bergmeister, 2016]. Many of the performance-based design criteria have not been explicitly standardized on the basis of a numerical risk assessment. Using a holistic approach, the criteria implicitly include acceptance as well as cultural threshold values in terms of the public's risk tolerance. Some of the performance parameters used for the design of the Brenner Base Tunnel are presented in the following chapters.

2.1 Safety format

The performance of structural systems is often evaluated on the basis of probability of failure criteria. A simplified approach is the use of system reliability index values.

In the Eurocode EN 1990 (revised in 2018) the reference period of 50 years and the relevant target reliability values have been used in derivation of the partial safety factors. The reliability (complementary to failure probabilities) $\Phi(\text{ß}_n)$ related to the reference period of n years is determined from the annual target reliability $\Phi(\text{ß}_1)$, where the safety index ß_1 is related to one year, as the product of n annual reliabilities, thus in average as $\Phi(\text{ß}_1)^n$. Therefore the safety index ß_n can be calculated from the index ß_1 using the following expression (Holicky et al., 2018).

$$[\Phi(\text{ß}_1)^n] = \Phi(\text{ß}_n). \tag{1}$$

For tunnels a reference period of a minimum of 100 years should be used. For the Brenner Base Tunnel a life time of 200 years have been assumed. In following Table 1 the values for the safety index ß related to the service life and the limit state is presented.

The performance of structural systems is often evaluated on the basis of probability of failure criteria. A simplified approach is the use of the reliability index ß. The evaluation of the probability of failure is related to the analysis of the consequences of failure. Therefore, performance-based design distinguishes between different impacts of failures, building occupancy categories, risk acceptance and short or long term evaluation.

In tunneling, specifically, the deformations induced by the stresses on the rock mass are time-dependent. The deformations are continuously or periodically measured and they are strongly dependent on rock mass type and behavior. The deformations of the ground load (action) S have to be smaller than (or in an equilibrium with) the possible deformation of the concrete structure R of the outer shell. Assuming normal distributions for R and S, the reliability index ß can be calculated taking into account the variability (coefficient of variation: v) of R and S:

$$\text{ß} = (\mu_R - \mu_S v) / \left\{ \sigma_R \left[1 + (\mu_S/\mu_R)^2 \right]^{1/2} \right\} \tag{2}$$

Table 1. Safety index ß for different reference periods.

	ß for the reference period of 1 year for a life time of 100 years	ß for the reference period of 1 year for a life time of 200 years	After the life time of 100/ 200 years
Ultimate limit state (200 years)	5.2	5.8	3.7
State of deformation equilibrium for temporary tunnel shells (10 years)	3.0	3.0	2.5
Serviceability limit state for the Inner Lining (200 years)	3.0	3.7	2.0
Durability	1.8	2.0	1.5

The partial safety factor for lognormal distributed basis parameters e.g. concrete can be derived as follows:

$$\gamma_c = \left[m_c \exp\left(-k_R v_c - 0,5\, v_c^2\right)\right] / \left[m_c \exp\left(-\alpha_c \beta v_c - 0,5\, v_c^2\right)\right]$$
$$= \exp[v_c(\alpha_c \beta v_c - k_R)] \tag{3}$$

Considering a coefficient of variation v=18%, an unlimited quantity of samples for the 5%-fractile, $k_R = 1.645$ and assuming a sensitivity factor $\alpha_c = 0.8$, the partial safety factor for a life time period of 200 years $(\beta = 5, 8\, or\, Pf = 10^{-8}/year)$ can be calculated as follows:.

$$\gamma_c = \exp[0,18\ (0,8 \times 5,8 \times 0,18 + 1,645)] = 1,56 \tag{4}$$

Therefore for the Brenner Base Tunnel, a partial safety factor for concrete of $\gamma_R = 1.6$ is applied (see Table 2).

2.2 *Ultimate limit state design*

High-quality building materials, excellent construction management and constant monitoring are of fundamental importance to ensure a long life span of the structures.

These are the main life-span-reducing factors in tunnel construction:

- Changes in rock mass pressure
- The presence of substances that are highly destructive to concrete, such as sulphates, pyrite (FeS2) etc.
- Frost aggression on the external elements

For the design of the inner lining the following loads S should be considered:

- ground pressure resulting from the 2d or 3d geotechnical simulations,
- ground pressure redistributions due to adjacent tunnel excavations,
- water pressure as radial pressure to the lining (for drained tunnels the pressure is considered only radially at the invert arch below the drainage),
- dead loads,
- inner loads caused by facilities and track structure,
- creeping of concrete,
- shrinkage of concrete,
- swelling pressure based on experimental investigations,
- temperature-related constraints; temperature depends on the distance from the portal,
- traffic loads.

Additionally extraordinary loads should be taken into account:

- impact load,
- fire load and concrete spalling,
- extraordinary pressure/suction loads of $\Delta_{ppre} = +11kN/m^2$ and $\Delta_{psuc} = -9kN/m^2$ due to uncommon high-speed train collectives in the tunnel and
- seismic loads in loose ground conditions (in rock this extraordinary load does not have to be taken into account).

For complex geometries, difficult geological conditions and temperature development only 3d simulations including the excavation process and a non-linear analysis of the lining considering fracture-plastic material models will give an appropriate representation of the structural behavior in these cases.

For the Brenner Base Tunnel, in order to ensure the resistance R of the concrete structures the following main measures were taken:

- the thickness of the concrete was increased,
- the composition of the concrete and the exposure classes were carefully determined,

- the partial safety factors were increased,
- the min concrete cover was determined as c_{nom} = 50 mm and
- some constructive details were defined.

Based on an ageing model, the following partial safety factors were elaborated for the design of the Brenner Base Tunnel related to a life time of 200 years:

Table 2. Partial safety factors for the design of the Brenner Base Tunnel.

Construction material	Life time	Partial safety factor
Normal concrete	50 years	$\gamma_c = 1,5 a = 0,85 - 1,0$
Normal concrete	>200 years	$\gamma_c = 1,6 a = 0,85$
Concrete with tunnel spoil as aggregate (variation coefficient f_c < 30%)	50 years	$\gamma_c = 1,5 a = 0,75$
Concrete with tunnel spoil as aggregate (variation coefficient f_c < 30%)	>200 years	$\gamma_c = 1,6 a = 0,75$
Global coefficient for the non-linear calculation of concrete	50 years	$\gamma_R = 1,3$
Global coefficient for the non-linear calculation of concrete	>200 years	$\gamma_R = 1,4$
Reinforcement steel	50 years	$\gamma_s = 1,15$
Reinforcement steel	>200 years	$\gamma_s = 1,2$

For the calculation of tunnel linings, finite element codes or a simplified shell approach with radial embedment can be used.

In finite element modelling, triangular or rectangular elements with multiple shape functions are often applied. Geometric non-homogeneities or concentrated load distributions should be calculated with finer finite element net divisions and more specified shape functions.

When modelling the springs between the rock mass and the tunnel linings, the embedding reaction is only permitted in the direction of the ground/rock mass. This radial embedding in the rock mass or in the soil can be determined, as a rule, by considering the elasticity modulus E and the coefficient of expansion on the transverse axial v of the rock mass or the soil and the system line of the corresponding inner lining radius R, as follows (see also Hofmann 2019):

$$K_R = E \times \frac{(1 - v)}{(1 + v)(1 - 2 \times v) \times R} = \frac{E_s}{R} \tag{4}$$

Here,
K_R = radial embedding of the inner lining – rock mass [MN/m³]
E = Elasticity modulus of the surrounding soil or rock mass
E_S = Rigidity modulus of the surrounding soil or rock mass
v = coefficient of expansion on the transverse axial of the surrounding soil or rock mass
R = tunnel radius system line

2.3 *Serviceability limit state*

The most significant serviceability limit state parameters of tunnel linings are deformations and as a consequence crack widths.

The characteristic line method is an aid in estimating the radial displacements of the tunnel wall on the longitudinal profile for the rock mass and the system behaviour. It uses numerous simplifying assumptions and is based on the theory of an isotropic elastic disk with a circular hole under plane deformation in homogeneous initial strain conditions. For this method, two characteristic lines are used, one for rock mass (ground reaction curve) and one for the preliminary lining (support reaction curve).

A simplified calculation of the maximum deformation of the shotcrete lining can be carried out, as follows (Wittke, 2005; Kainraht-Reumayer, 2009; Vlachopoulos et al., 2009):

$$u_{c,\max} = \frac{p \cdot r_0^2}{E_c \cdot t_c} = \frac{f_c \cdot r_o}{E_c} \tag{5}$$

and the deformations before the support are assumed by (Hoek, E. et al. 2008)

$$u_{c,0} = u_{\max} \frac{1 \cdot}{3} e^{-0,15\frac{R_{pl}}{R}}. \tag{6}$$

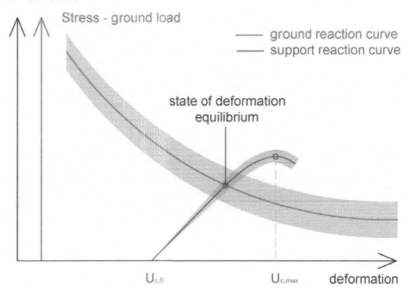

Figure 1.   Ground reaction curve and the support reaction curve.

The intersection of the ground reaction curve and the support reaction curve fulfills the kinematic compatibility and the equilibrium between rock mass and preliminary lining. It gives the final convergence and pressure at the preliminary lining.

The calculated crack width of the inner lining should be smaller than 0,3 mm (0,4–0,6 mm is still acceptable for certain exhibition classes). For watertight concrete linings without sealing, a maximum crack width of 0,1 mm is considered.

2.4   *Durability*

Traditionally, durability aspects are included in the structural details by using certain minimum geometrical requirements and material properties. Specifically for tunnels, durability is governed by the choice of material composition. Therefore the following requirements should be fulfilled.

Table 3. Partial safety factors for the design of the Brenner Base Tunnel.

Design parameter	Tunnels (100 years life cycle)	BBT (200 years life-cycle)
Concrete cover (c_{nom})	45 mm	50 mm
permeability	$(3 - 10) \times 10–16 \, m^2$	$10 \times 10–17 \, m^2$
water penetration	< 40 mm	< 40 mm

2.5 Robustness

Robustness can be considered as the capability of performing without failure under unexpected conditions. The practical measure of redundancy and robustness which was proposed by Ghosn et al. (1998) can be adapted also for tunnels. Redundancy is defined as the capability of the system to redistribute and continue to carry loads after the failure of one main member. As a provision of this capacity, a robust structure has additional structural capacity and reserve strength, allowing it to bear higher unforeseen actions than anticipated. The measures of robustness for tunnel linings can be described as follows:

$$R_u = LF_u/LF_1 \leq 1,5 \tag{7}$$

$$R_f = LF_f/LF_1 \leq 1,2 \tag{8}$$

where: LF_1 is the load that causes the failure of a first element (lining segment or a portion of the shotcrete/concrete lining); LF_u is the load that causes the collapse of the system (whole supporting structure, etc.); LF_f is the load that causes the functionality limit state of the initially intact structure to be exceeded.

2.6 Fire resistance

A fire inside a tunnel is one of the most catastrophic events that can occur in land transportation of goods and people. Tunnel fires represent a great risk to human life and cause extended infrastructural damage. The very high temperatures, the limited chance of escape and the difficulties in the dispersion of smoke and heat make this type of event very dangerous.

In the assessment of the structure response to the fire, it is crucial to properly evaluate the thermal load. Usually, a prescriptive approach is adopted to this end (see Bergmeister, Cordes et al. 2018). It is based on the definition of certain heating profiles which should represent, more or less correctly, the temperature evolution in the tunnel during a fire, in particular in the layer of gases close to the structure surfaces.

However, temperature-time curves, though easily applicable in professional practice, are less significant from a physical point of view: a fire may have very different characteristics depending on a considerable number of factors. Obviously, by changing the thermal input, that is the thermal load, the structure response will also vary, sometimes drastically.

The heating profiles used in the design of the Brenner Base Tunnel are the RWS curve and the RABT-ZTVcurve (EUREKA curve), respectively for the Italian and the Austrian sides of the tunnel. Both these two fire curves are considered as among the most representative for the description of the thermal load in a tunnel during a fire (Bergmeister, Schrefler et al. 2018).

The profile RWS (Figure 2), adopted in Italy with the code UNI 11076, is widely used in Europe. Almost all countries using the RWS curve, including Italy, have decided to limit the thermal program to two hours, as it is assumed that after that time the firemen will be able to approach the source of fire and begin to extinguish it.

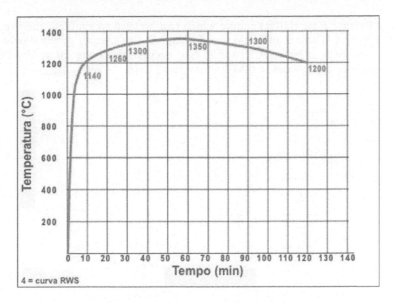

Figure 2. profile RWS - UNI 11076 (taken from Bergmeister, Schrefler, 2018).

The **RABT-ZTV/EUREKA** curve (Figure 3) has similar characteristics to RWS only for the first five minutes, showing the same concept of rapid temperature rise at the start of combustion. Then, it assumes a flat and constant shape at 1200 ° C up to 30 minutes, subsequently showing a linear decrease to room temperature by 140 minutes.

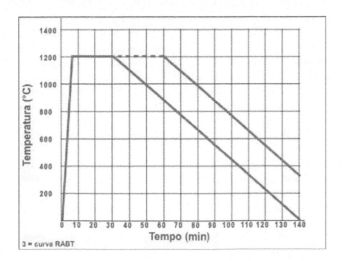

Figure 3. profile RABT-ZTV/EUREKA (taken from Bergmeister, Schrefler, 2018).

This purely descriptive approach of temperature profiles has to be applied to the safety assessment of significant partial structures of a tunnel and a direct simulation of the fire must be addressed with application to whole tunnel stretches.

For the simulation and the modelling of a fire, several approaches are available. The most accurate numerical model is based on computational fluid dynamics (CFD) (Pachera, 2017).

The "Heat Release Rate (HRR)" curve which can be used to simulate the fire scenario is presented in Figure 4. There is an initial phase where the fire's power grows slowly and then it increases up to 60 MW in 600 s. Afterwards, heat energy stays constant at 60 MW for 600 s until 1200 s and later it decreases to 0 MW in the last 2400 s.

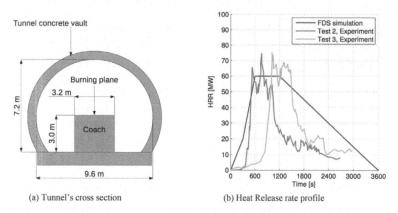

(a) Tunnel's cross section   (b) Heat Release rate profile

Figure 4.   Fire scenario investigated with the coupled approach (taken from Bergmeister, Schrefler, 2018).

The use of polypropylene fibers, in the amount of 2 kg/m³, is strongly recommended considering the effects they have on the pore pressure field which is reduced for a considerable extent (Murr 2019a). The pressure peaks decreases by over 50% to values around 0.4 MPa, that is, to values compatible for a standard concrete structure. This range of values is not usually dangerous for concrete structures, at least for explosive or otherwise violent forms of spalling.

2.7 *Sustainability*

As a parameter for the sustainability the CO_2 – balance during the construction process including the transportation of the construction materials can be used. Because of the high demand of energy, especially in its production process, the cement dominates the CO_2-balance of concrete significantly. In addition also the aggregate and the transportation of the components and of the concrete itself cause CO_2-emissions. The aggregate has a high percentage of the mass of the produced concrete; its part of the CO_2-emissions is only about 2 %, mostly caused by the transport.

The investigation of the CO_2-balances connected to the production of concrete for tunnel constructions shows the dominating influence of cement. Caused by the high demand of electrical energy for the cement production approx. 75 % of the CO_2-emissions are caused by cement. The potentials for optimization can be found in the reduction of the volume of concrete but also in the use of composite cements, by the use of fly ash and blast furnace slag, by reducing the cement portion and also by reducing the minimum cement mass per m³ of concrete.

It can be stated, that the material transport by use of electrical powered trains, vehicles and conveyor belts have significant advantage compared to the transport by vehicles with diesel engines. The CO_2-emissions can be influenced also by the choice of the energy producer, especially they can be reduced by minimizing the use of fossil material for the production of electrical energy.

Finally the re-utilization of excavation materials for aggregates is another important issue to improve the sustainability of tunnels.

For the Brenner Basis Tunnel the CO2-balances were investigated [Eurac, 2011]. Most significant for the CO2-emissions are the transport of the excavated material and the influence of the minimum cement portion per m3 as well as the different consideration of fly ash in the concrete mixtures [Keuser et al. 2014]. The autarkic aggregate supply with recycled tunnel spoil at the lot

E52 of the Brenner Base Tunnel (Murr 2019b) improved the life cycle assessment (e.g. CO2, SO2, non-renewable energy and mineral resources according to (BBT-intern, 2015)

3 CONCLUSIONS

The performance-based design approach for tunnels on the basis of the theory of structural reliability includes the verification of the ultimate limit state, the serviceability limit state, durability, robustness and fire resistance. The design must be based on a certain life-cycle, which is normally 100 years. For the Brenner Base Tunnel, a life period of 200 years was chosen.

REFERENCES

Bergmeister, K. 2015. Brenner Base Tunnel - Life Cycle Design and innovative Construction Technology. *Swisstunnel Congress.*

Bergmeister, K. 2016. Performance based design of Bridges and Tunnels. *Proceedings of the fib Congress* Capetown.

Bergmeister, K.; Cordes, T.; Lun, H.; Murr, R.; Reichel, E. 2018. Beton unter hoher Temperaturbeanspruchung - Brandschutz und Rettungssysteme in Tunneln. In *Betonkalender 2018 Bautenschutz, Brandschutz*, Ernst&Sohn Verlag, Berlin.

Bergmeister, K.; Schrefler, B.A.; Pesavento, F.; Pachera, M; Brunello, P. 2018. Simulation of fire and structural response in railway tunnels: the case of the Brenner Base Tunnel. ACI-Publication under review.

Brenner Base Tunnel internal report 2015. E52 Life Cycle Assessment (LCA), Ökobilanzstudie Variantenuntersuchung Spritzbeton.

Carranza-Torres, C. 2004. Elasto-plastic solution of tunnel problems using the generalized form of the hoek-brown failure criterion. *International Journal of Rock Mechanics and Mining Sciences* 41, Supplement 1, pp. 629–639.

EURAC Research 2011. Ausbau Eisenbahnachse München-Verona-Brenner Basistunnel. Untersuchung der Nachhaltigkeit des BBT im Hinblick auf seine CO2-Emissionen. Innsbruck.

Feder, G. and Arwanitakis, G. 1976. Zur Gebirgsmechanik ausbruchsnaher Bereiche tiefliegender Hohlraumbauten (unter zentralsymmetrischer Belastung), *Berg- und Hüttenmännische Monatshefte* 121, No. 4.

Ghosn, M.; Moses, F. 1998. Nchrp Report 406, Redundancy in Highway Bridge Superstructures. Washington. DC: *Transportation Research Board*, National Academy Press.

Hoek, E., Carranza-Torres, C., Diederichs, M.S. and Corkum, B. 2008. Integration of Geotechnical and Structural Design in Tunnelling. Proceedings Univ. of *Minesota 56th Annual Geotechnical Engineering Conference*. Minneapolis, 29th February 2008, pp. 1–53.

Hofmann, M.; Cordes, T.; Bergmeister, K. 2019. The subgrade reaction modulus method in tunneling, *ITA-AITES World Tunnel Congress 2019*, Naples, Italy.

Holicky, M.; Diamantidis, D.; Sykora, M. 2018. Reliabiliy levels related to different reference periods and consequence classes. In: *16th International Probabilistic Workshop*, Vienna, Specia Issue Ernst&Sohn – Wiley Online Library, pp. 22–27.

Kainrath-Reumayer, S. 2009. The convergence confinement method as an aid in the design of deep tunnels, *Geomechanics and Tunneling*, 2(5), pp. 553–560.

Keuser, M.; Bergmeister, K.; Otto, J.; Lückmann, H. 2014. Sustainability in Tunneling – CO2- Balances derived from investigations concerning the Brenner Base Tunnel. In: *Proceedings of the 10th CCC Congress*, Liberec.

Murr, R.; Bergmeister, K. 2019a. Fire resistant high strength concrete concept for inner tunnel linings at the Brenner Base Tunnel. *ITA-AITES World Tunnel Congress 2019*, Naples, Italy.

Murr, R.; Cordes, T.; Hofmann, M.; Bergmeister, K. 2019b. Autarkic aggregate supply with recycled tunnel spoil at the Brenner Base Tunnel. *ITA-AITES World Tunnel Congress 2019*, Naples, Italy.

Pachera, P.: Numerical simulations of fires in road and rail tunnels with structural and fluid dynamic analysis 2017. Ph.D. thesis, Univeristy of Padova.

Taleb, N.N. 2012. Antifragile. Things that gain from the disorder. Random House New York.

Vlachopoulos, N. and Diederichs, M.S. 2009. Improved longitudinal displacement profiles for convergence-confinement analysis of deep tunnels. *Rock Mech. and Rock Eng.*, pp. 131–146.

Wittke, W. and Wittke-Gattermann, P. 2005. Tunnelstatik. *Betonkalender 2005*, pp. 495–498. Hrsg. Bergmeister, K.; Wörner, J., Verlag Ernst & Sohn, Berlin.

Tunnels and Underground Cities: Engineering and Innovation meet Archaeology,
Architecture and Art, Volume 5: Innovation in underground engineering,
materials and equipment - Part 1 – Peila, Viggiani & Celestino (Eds)
© 2020 Taylor & Francis Group, London, ISBN 978-0-367-46870-5

Novel semiprobabilistic tunnel lining design approach with improved concrete mixture

K. Bergmeister, T. Cordes & R. Murr
Brenner Base Tunnel BBT-SE, Innsbruck, Austria

ABSTRACT: The 64 km long Brenner Base Tunnel (BBT) between Tulfes/Innsbruck, Austria, and Fortezza, Italy will become the world's longest railway tunnel. The service life was set at 200 years for the tunnel structures with a regular annual inspection and maintenance of the inner linings. The design for a life-cycle time of more than 200 years considers the ultimate limit state with a probability of failure of $P_f=10^{-7}$. The partial safety coefficients γ_R have been defined by FORM-analysis according to $\gamma_C=1.6$ (concrete) and $\gamma_S=1.2$ (reinforcing steel). A life-cycle time of more than 200 years and a high availability of the tunnel require concrete mixed with an eye to durability. Besides the concrete compressive strength, water penetration, permeability and gas porosity have also been tested on site. With an optimized concrete mixture, crack development due to shrinkage constraints as well as due to temperature development has been achieved.

1 INTRODUCTION

The total tunnel system of the BBT includes 230 km of tunnels built for a life-cycle time of 200 years. Half of these are used as single track railway tunnels and the other half is used for exploration, emergency, ventilation and access purposes. Depending on the excavation method and the requirements the design of the lining is different. Drill and blast tunnels are built with a double-shell lining method according to the schematic cross section in Figure 1a). The first constructed primary lining, in shotcrete, secures the cavity. The inner secondary lining archives the load carrying capacity, the structural safety and the serviceability requirements. In this case the load bearing capacity of the primary linings is not considered for the permanent design.

Tunnels with fewer requirements in terms of serviceability, such as for example water-tightness, durability, surface quality or availability are supported if possible with single permanent shotcrete linings. In this case the permanent shotcrete lining must secure the cavity and archives the load carrying capacity, the structural safety and the serviceability requirements (Bergmeister, 2018). In unfavorable geology and hydrogeology conditions the double-shell lining method is applied.

The machine driven railway tunnels are built mainly with the single shell lining method. Here the requirements in terms of load carrying capacity, the structural safety and the serviceability must be met. For structural safety, the construction progress and block loading in partly embedded rings must be additionally verified. In sections with high rock loads and high water pressure the double-shell lining method is applied for machine-driven railway tunnels.

2 LIFE-CYCLE-BASED DESIGN

In the case of the Brenner Base Tunnel, the life spans were defined based on the technical life spans of the various elements. High-quality building materials, excellent construction

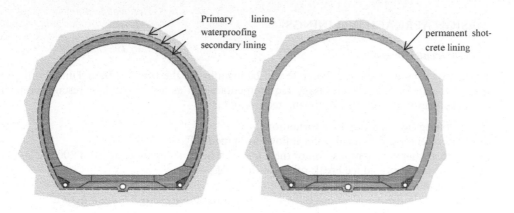

Figure 1. Schematic cross-section on the left for, a) double-shell lining and on the right for b) single permanent shotcrete linings.

management and constant monitoring are of fundamental importance to ensure a long life span of the structures. These are the main life-span-reducing factors in tunnel construction:

– Changes in rock mass pressure
– Presence of substances that are destructive to concrete, eg. sulphates, pyrite (FeS2)
– Frost aggression on the external elements

To ensure the resistance of the concrete structures to these factors for the prescribed technical life span, these main measures were taken:

– the thickness of the concrete was increased,
– the composition of the concrete and the exposure classes were carefully determined,
– the partial safety factors were increased,
– the minimum concrete cover was determined as c_{nom} = 50 mm and
– construction details were defined.

Based on an ageing model, the following partial safety factors were elaborated for a technical service life of 200 years (Bergmeister, 2015):

Table 1. Partial safety factors for the design of the Brenner Base Tunnel.

Construction material	Technical service life Years	Technical life span -
Normal concrete	50	γ_c = 1.5, α = 0.85 – 1.0
Normal concrete	< 200	γ_c = 1.5, α = 0.85
Concrete witch tunnel spoil as aggregate (variation coefficient f_c < 30%)	50	γ_c = 1.5, α = 0.75
Concrete witch tunnel spoil as aggregate (variation coefficient f_c < 30%)	> 200	γ_c = 1.6, α = 0.75
Global coefficient for the non-linear calculation of concrete	50	γ_R = 1.3
Global coefficient for the non-linear calculation of concrete	> 200	γ_R = 1.4
Reinforcement steel	50	γ_S = 1.15
Reinforcement steel	> 200	γ_S = 1.2

The level of possible degradation of the construction materials or systems depends decisively on regular and appropriate inspections and maintenance.

3 DESIGN APROACH OF LININGS

3.1 *Permanent primary shell*

A permanent primary lining in shotcrete will be installed in the Brenner Base Tunnel only for auxiliary structures with certain usage requirements (such as access tunnels, rescue tunnels). These structures must meet the following requirements:

– stable geological conditions - deformations,
– low water penetration - local water inflows are piped off,
– non-critical chemical composition of the water (eg. Max. sulphate content of 200 mg/l),
– no ice formation (avoid ice break-off from the roof),
– defined monitoring intervals,
– no impact on railway operations in case of additional monitoring or repair activities.

The following requirements and certifications for the construction of the permanent shotcrete lining:

– minimum thickness 25 cm, double-layer reinforcement
– Strength class C25/30
– Shotcrete and/or monitoring class ÜK3
– increased concrete coverage of 5.5 cm
– Load-bearing ability and serviceability once deformation has stabilized

To evaluate the durability of permanent shotcrete linings, porosity and gas permeability were assessed, besides compression resistance and water penetration depth. Comprehensive 3D laser scans provide complete coverage of the lining thickness and its homogeneity. The mechanical properties of the fresh shotcrete, such as the E-module, shrinkage and creep were assessed using samples from the excavation (Neuner et al. 2017). The stresses on shotcrete tunnel linings were determined with new material models via back-analyses (Marcher er al. 2019, Cordes et al. 2019).

3.1.1 *Experimental test programme*

In order to verify the achievement of the material parameters for the permanent shotcrete lining, a special test programme was set up (for example for the rescue tunnel near Innsbruck), which also evaluated the results of the conformity and identity tests. In the excavation sections, 3 measurement sections were established with 2 drill cores of 100 mm in diameter and 2 drill cores of 150 mm in diameter taken from the upper arch of the dome. The following tests were done on the cores, as can be seen in Figure 2:

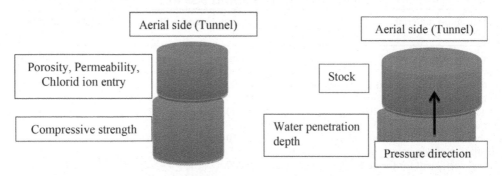

Figure 2.   Positioning of the samples from the drill cores.

– Test of compression resistance and bulk density on drill cores from the area toward the rock mass,
– Test of structure density based on water penetration depth on drill cores from the area toward the rock mass,
– Test of open porosity on drill cores from the area toward the tunnel,

- Test of gas permeability on drill cores from the area toward the tunnel,
- Test of chloride penetration on drill cores from the area toward the tunnel.

The required durability and shotcrete quality was verified based on the calculated parameters and with reference to state-of-the-art technology (RILEM (1999), ÖNORM EN 1936, 2007). Compression resistance, water penetration depth/structure density and porosity/permeability were determined based on the drill cores. If it becomes clear that, in on-site conditions (as assumed from the drill cores), the required values are not being reached in a certain stretch of tunnel, then either a specific shotcrete reinforcement or an inner lining must be applied.

3.1.2 *Shotcrete recipe*
To produce the shotcrete lining, the following types of concrete SpC 25/30(56)/II/J2/XC4/XF3 and/or SpC 25/30(56/III/J2/XC4/XF3 with aggregate size 8 and/or 11 mm were used:

Table 2. Shotcrete recipe.

Component	Quantity of material kg/m³	%
Cement: CEM I 52.5 N sb Eiberg	380	
Hydraulically prepared effective additives: Fly ash	40	
Total water content	208	
Actual water content	199	
Dry aggregates	1706	
Sand 0/4: Ahrenberg (60%)	1024	
Gravel 4/8: Ahrenberg (40%)	682	
Air content		4
Bulk density	2334	

3.1.3 *Verification of the values of the installed shotcrete lining*
The technical and mechanical results from the identity tests were used (Tables 3 and 4) and were checked for compliance with the required compression resistance values. Careful attention was paid to the development over time of the compression resistance of the shotcrete (John et al. 2003).

Table 3. Fresh concrete values.

	Total water Content kg/m³	Density kg/m³	Temp. °C	Spread cm	Air content %
No. of tests	25	25	25	25	25
Average	204	2326	22	58	3.6
± standard deviation	± 8	± 29	± 3	± 3	± 0.9
Maximum	222	2400	27	65	5.2
Minimum	188	2283	16	54	1.5

Table 4. Values for hardened concrete.

	Compressive Strength 7d MPa	Compressive Strength 28d MPa	Compressive Strength 56d MPa	Loss of firmness %	Penetration-depth mm
No. of tests	16	16	10	5	7
Average	33	41.3	43.4	28	19.7
± standard deviation	± 3.9	± 3.8	± 3.6	± 3	± 6
Maximum	40.9	49.5	50.2	31	29
Minimum	24.7	35.8	36.2	25	13

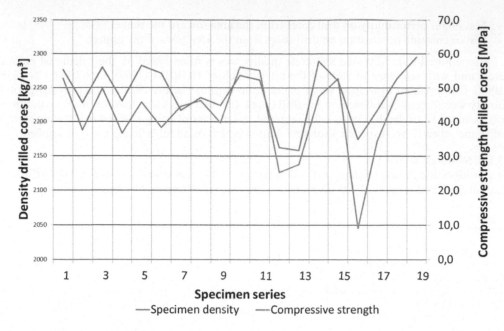

Figure 3. Water penetration depth determined on the drill cores.

3.1.3.1 Compression resistance

The average compression resistance of the drill cores was about 50 N/mm². The values taken directly from the tunnel structures were therefore significantly higher than the required shotcrete class of SpC25/30.

3.1.3.2 Water penetration depth/Structure density

The water penetration depths were between 15 mm and 31 mm, as can be seen in Figure 3, a total average of 23 mm. The 15 cm drill cores were tested on the opposite side as well and it was thereby possible to confirm that there were no cracks running through the cores and that for this reason, the water penetration depths were far less.

3.1.3.3 Porosity/Permeability

Open porosity was around 20% (ÖNORM EN 1936 (2007). The permeability of the shotcrete samples was determined using oven-dried drill core slices. The tests give values from $6x10^{-17}$ m² to $6.5x10^{-16}$ m². Only one test in the series showed a higher porosity of $1.3x10^{-15}$ m². Based on the permeability values, the following classifications in terms of potential durability for the shotcrete were assigned according to Baroghel-Bouny (2007) (Table 5):

Table 5. Durability potential for shotcrete.

Durability	Permeability m²
high	1 to 10 $x10^{-17}$
medium	1 to 10 $x10^{-16}$
low – very low	1 to 10 $x10^{-15}$

Tests according to RILEM (1999) show a link between carbonation depth and permeability. After one year of storage of the concrete samples, the carbonation depth in laboratory conditions was determined and compared to the measured permeability value. A logarithmic regression gave the following connection, based on the measured permeability values and carbonation depths:

$$\log(y) = k1 \cdot x \cdot \log(x) + c \tag{1}$$

where y = carbonation depth in mm, x = permeability in m², k1=gradient = 0.57041 and c= constant = 0.31971. The achieved coefficient of determination (correlation factor), as can be seen in Figure 4, is 98%. With the permeability values measured in an oven-dried state, the carbonation depth after 1 year in storage can be calculated using equation (1). According to the square-root-t law (2), the coefficient k2=6.68 was defined for a one-year storage period. With the coefficient k, via (2) carbonation progress y can be foreseen at any point in time.

$$y = k2 \cdot t^{0,5} \tag{2}$$

where y= carbonation depth in mm, k2= factor in mm/t 0.5 and t= time in years.

Figure 5 shows the extrapolation of the carbonation front progression with the calculated minimum and maximum k-factors of the external drill core slice for the shotcrete as lower and upper limits.

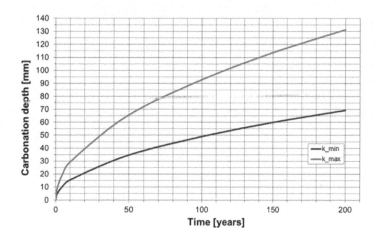

Figure 4. Relationship of depth of carbonation and permeability coefficient according to Cordes et al. (2018).

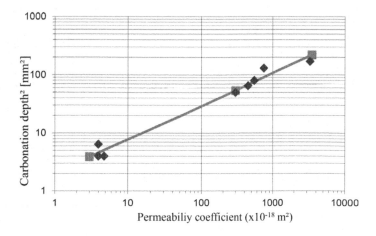

Figure 5. Development of carbonation depth up to 200 years.

3.1.3.4 Summary of the test results for the shotcrete: Hardness, water penetration depths, permeability and porosity

The hardness levels found were significantly higher than the levels for the required shotcrete class 25/30(56). The water penetration depths found, taken as a basis for structure density, showed higher structure density in the tested drill cores with values close to those in class XC4. Open porosity was about 20% and was therefore as expected for shotcrete. Permeability of samples in an oven-dried state reached values between $6 \times 10\text{-}17 \text{ m}^2$ and $6.5 \times 10\text{-}16 \text{ m}^2$. These values indicate high or medium durability of shotcrete (Baroghel-Bouny, 2007). This method can be used to determine the quality of a permanent primary lining and thus the load-bearing capacity, the serviceability and durability of the lining.

3.2 *Secondary lining*

Unreinforced inner shells in tunnel construction are still an advantageous and durable solution for permanent tunnel lining. Under dynamic, fatigue loads, such as those caused by the pressure and suction loads of high-speed railway traffic, the fatigue strength of the concrete on tension shall be taken into account.

The fatigue design is a decisive criterion for tunnels with high-speed traffic, both in terms of load-bearing capacity and serviceability. By taking into account the fatigue, the formation of bending cracks and the development of crack branches and intersections at the block joints should be prevented. They can lead to the detachment of concrete fragments and corresponding disturbances in train operation. Unreinforced concrete linings are being installed in the Brenner Base Tunnel only for sections with lower requirements with no train traffic and no dynamic loading by train pressure waves.

For the design of unreinforced secondary linings, for example in Figure 6, only the simulation of the structure with non-linear material behavior, including crack initiation and propagation can verify the compatibility of deformations and rotations in an appropriate way.

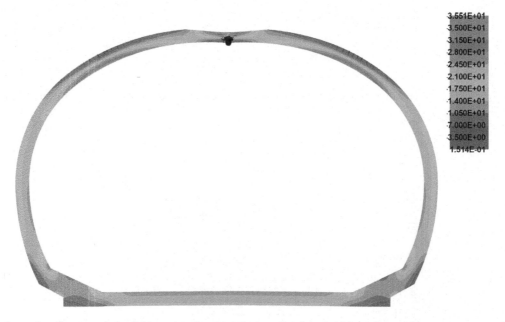

Figure 6.   Cross-section of the connection tunnels near Innsbruck with an incomplete concrete tunnel roof, minimal principal stresses in N/mm², crack development in the crown to 0.9m.

3.3 Concrete with minimum reinforcement

Structural concrete may also be less reinforced than the code-based minimum reinforcement requires. In order to reduce the minimum reinforcement, special care has to be taken with concrete technology and the curing process of the concrete. A research project has been carried out for prevention structures (on the basis of the Austrian code ÖNORM 24802), since the height of the structural element increases the minimum reinforcement (Strieder et al., 2017).

The amount of the low reinforcement A'$_{s,min}$ can be calculated as follows

$$A'_{s,min} = \sqrt{\frac{A_{s,GZT}}{A_{s,min}}} \cdot A_{s,min} \qquad (3)$$

A$_{S,min}$ = minimum reinforcement based on code provisions (e.g 0.002 x b$_c$ x h$_c$)
A$_{S,GZT}$ = reinforcement resulting from the cross section forces for the ultimate limit state

In order to use lower reinforcement as foreseen in the EN 1992-1-1 the following prescriptions should be considered:

– The stepwise reduction of the reinforcement should not exceed 50%, in order to avoid a severe increase in stress within the concrete
– The stepwise reduction of the reinforcement should be spaced with a length of more than 100 cm or 2 d (d= height of the St. Venant-region)
– The concrete mixture must be optimized in order to reduce the energy exposure (De Schutter, G., 2003)

3.4 Reinforced concrete, where the section forces (normal force and moment) govern the amount of reinforcement

Higher rock loads and difficult cross sections directly govern the amount of reinforcement. Especially changes of curvature in the upper sidewalls of low-roofed cross section will lead to high bending moments and high degrees of reinforcement. In tunnel sections with higher requirements, for example main tunnels with rail traffic, the recommended EC2 minimum reinforcement must be applied.

4 CONCLUSION

The structural design of the tunnel lining for an accepted service life of 200 years for the Brenner Base Tunnel requested a novel design approach. Linings are designed using unreinforced concrete, low or minimum reinforcement as well as ultimate limit state designed reinforced concrete. Numerical nonlinear analysis have been carried out in order to understand rock mass behaviour during the construction phase and to optimize the performance of the structural elements.

For the permanent shotcrete lining a specific experimental program has been worked out in order to verify the in-situ quality of the concrete. Besides the concrete compressive strength, water penetration, permeability and gas porosity have also been tested. With an optimized concrete mixture, crack development due to shrinkage constraints as well as due to temperature development has been achieved.

REFERENCES

Bergmeister, K. 2018. Anwendung von maßgeschneidertem Spritzbeton beim Brenner Basistunnel. *Grazer Betonkolloquium*, University Graz
Bergmeister, K. 2015. Brenner Base Tunnel - Life Cycle Design and innovative Construction Technology. Swisstunnel Congress

Bergmeister, K. 2017. Optimized design of the Brenner Base Tunnel through numerical modeling. *ECCO-MAS Computational Methods in Tunnelling EURO:TUN*, Innsbruck, Austria

Baroghel-Bouny V. 2007. Durability Indicators: relevant tools for an improved assessment for RC durability. Toutlemonde, F. (Ed.), *Proc. the 5th International Conference on Concrete under Severe Conditions, Environment and Loading*. Paris, (LCPC), pp. 67–84.

Cordes, T; Dummer, A., Neuner, M., Hofstetter, G., Bergmeister, K. 2019. The load-bearing capacity of primary linings with consideration of time-dependence at the Brenner Base Tunnel, *ITA – AITES WTC*, Neaples

Cordes, T.; Hofmann, M.; Murr, R.; Bergmeister, K. 2018. Current developments in shotcrete technology at the Brenner Base Tunnel: In: Kusterle, W. (Hrsg.): *Spritzbeton – Tagung*, Alpbach, Eigenverlag.

De Schutter, G.; Kovler, K. 2003. Short term mechanical properties. Report of RILEM Technical Committee TC 181-EAS: Early age shrinkage induced stresses and cracking in cementitious systems, A. Bentur, Ed.

John, M.; Mattle, B.; Zoidl, Th. 2002. Berücksichtigung des Materialverhaltens des jungen Spritzbetons bei Standsicherheitsuntersuchungen für Verkehrstunnel. Taschenbuch für den Tunnelbau, S. 149–188. Essen: Verlag *Glückauf GmbH*

Neuner, M.; Cordes T.; Drexel M.; Hofstetter, G. 2017a: Time-Dependent Material Properties of Shotcrete: Experimental and Numerical Study. *Materials*, 10 (9),1067, doi:10.3390/ma10091067, Basel, Switzerland.

Neuner, M.; Schreter, M.; Unteregger, D.; Hofstetter, G. 2017b: Influence of the Constitutive Model for Shotcrete on the Predicted Structural Behavior of the Shotcrete Shell of a Deep Tunnel. *Materials*, 10 (6),577; doi:10.3390/ma10060577, Basel, Switzerland, 2017

ÖNORM EN 1936: Prüfverfahren für Naturstein – Bestimmung der Reindichte, der Rohdichte, der offenen Porosität und der Gesamtporosität, *ON Österreichisches Normungsinstitut*.

ÖVBB: Guidline Shotcrete 2009, Austrian Society for construction and technology. Austria.

Pöttler, R. 1990. Time-dependent rock-shotcrete interaction - A numerical shortcut. *Comput. Geotech.* 9, 149–169.

RILEM TC 116-PCD 1999. Concrete durability - An approach towards performance testing. *Materials and Structures*, Vol. 32, April 1999, pp 163–173.

Schädlich, B.; Schweiger, H.F.; Marcher, T.; Sauer, E. 2014: Application of a novel constitutive shotcrete model for tunneling. Eurock 2014, *Rock Engineering and Rock Mechanics: Structures in and on Rock Masses*, Taylor & Francis Group, p. 799–804.

Strieder, E., Hilber, R., Murr, R., Bergmeister, K. 2017. Entwicklung der Materialeigenschaften im jungen Massenbeton - Bestimmung des zeitlichen Verlaufs der Materialparameter in der Hydratationsphase als Basis für die Modellierung der Risswahrscheinlichkeit im Massenbeton, *Beton- und Stahlbetonbau* 112, Heft 1, Page 41 to 49

Tunnels and Underground Cities: Engineering and Innovation meet Archaeology,
Architecture and Art, Volume 5: Innovation in underground engineering,
materials and equipment - Part 1 – Peila, Viggiani & Celestino (Eds)
© 2020 Taylor & Francis Group, London, ISBN 978-0-367-46870-5

Formwork for road tunnel in Indian Kashmir

E. Bertino
Cifa SpA Business Development Manager, Senago, Italy

ABSTRACT: Z-Mohr Tunnel will be one of the main strategic tunnels allowing continuous passage from Laak district to Kashmir, in Northern India. This 6.5 km tunnel under ZoJi-la pass (at an altitude of 3.528 m) will make Srinagar and Kargil passable all year. Currently the road is closed 7 months a year, due to the snow.

The tunnel is currently in advanced digging phase by JV APCO- TITAN and the final linings will be realized with hydraulic self-propelled steel formworks. To keep the executive programs of this major work, we planned casting sequences on several work fronts, considering the operative recommendations of the designers.

All the steel formworks are designed and realized with the most modern and tested technologies integrating tunnelling equipment, to keep and compress the final lining cycles of the work.

1 THE PROJECT

Among the big projects started in India in the last 10 years, there is the improvement of the current road network which stretches for 5,4 million km. It is in fact the second for development among the world network, but it does not have an easy and continuous access to the villages scattered on a varied territory. It is not only due to a complex territory with large mountain ranges, but also due to instable politic situations with the territories on the border that projects have been started on a priority basis. Among these, one of the most important projects is the connection of the northeastern province of Ladakh, in Jammu & Kashmir with its neighbouring regions and with the capital Delhi.

The roads connecting this province and its main town Leh with Srinagar are not accessible for about 7 months a year, due to heavy snow. In this winding Himalayan road various construction sites have been opened to insert medium and long-length tunnels.

The Z-Mohr Tunnel, whose project began in 2015 and excavation in 2017, will be connecting Gagarigir with Sonamarg.

Figure 1.   West Portal.

Figure 2.   East Portal.

The Main Tunnel, 6,5 km long with radius of 5,4m, has two bidirectional lanes and a total track width of 8m, an escape tunnel radius of 3,7m at the side, with useful track of 5m of width.

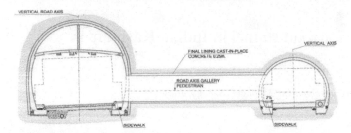

Figure 3. Cross section.

The two tunnels are connected every 250m by pedestrian cross passages and every 750m by vehicular cross passages in the middle of the Lay-By 40m long, which is positioned in line with the Main Tunnel with total track of 13,60m.

The two tunnels, besides the two main access portals, are served by an Adit Tunnel positioned at 3.5 km distance from the West portal. This access tunnel of about 3 km of length will be used to supply the ventilation chamber with fresh air, which reaches all areas open to traffic through the ducts obtained from the horizontal slab in the Main Tunnel.

The state Company National Highway and Infrastructure Development Corporation (NHIDCL) has assigned this track to Srinagar Sonamarg Tunnelways LTD., which has awarded all the construction works of the Z-Morh Tunnel to the international JV between APCO Infratech Pvt. Ltd. e Global Indian Company For Energy Highways, Industrial and Urban Infrastructure and to the Turkish Company TITAN Limited, with a long experience in the tunnel construction and construction equipment manufacturing.

This important work represents the 4th road tunnel for length in India, considering those completed and the ones under construction and it crosses the formation of Pirpanjal and Zozilla, composed mainly of Metabasites/Quartztic/Shale and Slates.

2   CONSTRUCTION WORKS OF THE TUNNELS

JV Apco-Titan started the excavation works for the construction of the Adit and the Main Tunnel from the western portal in the year 2017. At the same time, negotiation began with CIFA to finalise the formwork system for the different linings, with particular care to the planning and to the work fronts opened in the different tracks.

In the first phase, due to access difficulties and problems of soil, it was advisable for both the Main Tunnel and the Escape Tunnel to open a front from the Western portal and two fronts from the Middle Adit. Afterwards, a second phase will be opened also from the Eastern portal front, where the advancement phases will be slower.

During the finalization of the complete formwork set and the installation programme on the job-site, the two teams of engineers of CIFA and APCO have analysed different technical solutions and, at the end of 2017, succeeded in preparing a final list with the technical specifications of the different equipment, membrane gantries, reinforcement carriers and formwork with relevant accessories.

Based on this list, the negotiations started and an agreement between CIFA and APCO was reached in February 2018, following to which the design phase started.

With this contract, CIFA assumed all the executive design of the formwork for the construction in Indian workshops chosen by the JV and the supply of all the mechanical, hydraulic, electrical and compressed air components. The first set of formworks under construction, including ITALY for components and INDIA for the steel structural work, is composed of two steel forms 12 m long with relevant transport traveller for the ESCAPE TUNNEL.

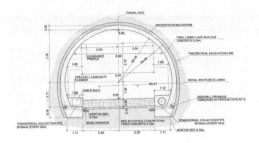

Figure 4.   Cross section Escape Tunnel.

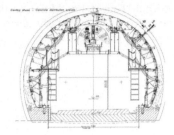

Figure 5.   Escape Formwork.

These will be the first equipment units to start the lining in October/November 2018.

These formworks, which are anchored to the concrete kicker with service sidewalk, are equipped with a wall vibration system and a semi-automatic concrete distributor with a push-button panel with cable.

The reduced dimensions of the tunnel and the need to safeguard the free transit of the job-site's vehicles with a maximum height of 3.20 m, required to optimize at the best space and dimensions of the structure.

For the whole track of the Escape Tunnel the installation of a waterproof and reinforced membrane with differentiated meshes is foreseen, which will be kept installed thanks to a special multi-level gantry 6 m long, equipped with service platform. This equipment will be installed in front of the formwork and shall slide on the same rails located on the kicker.

The same principle has been used to construct the formwork for the Main Tunnel, which included the passage of the ventilation pipe of diameter 2.30 m inside the equipment, as well as a template for the transit of all the job-site's vehicles up to a height of 4,30 m. For this type of work, the customer preferred to set 4 work fronts and, therefore, 4 steel forms 12m long with relevant transport travellers. In this case, the formwork is equipped with special shapes, to create the side support for the ventilation slab and the positioning for the suspension tie-rods in the cap center.

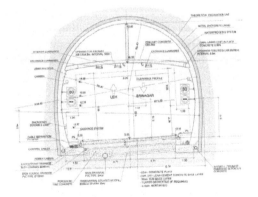

Figure 6.   Main Tunnel Cross section.

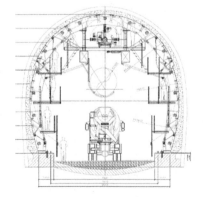

Figure 7.   Main Tunnel Formwork.

On dimensioning the formwork, high concrete filling capacities have been considered, by increasing the concrete injection spots thanks to the tight presence of reinforcement meshes. To comply with the cycle times and the restricted work areas accessible by all means also for the Main Tunnel, the company opted for the purchase of n. 4 gantry for membrane laying and reinforcing rods.

In the Main Tunnel, in addition to the two pipelines for longitudinal movements for fresh air and polluted air, every 500 m of track side niches have been foreseen for the opposing installation of a pair of Jet Fans. These niches require a long connecting section between the axial fan installation section and the standard tunnel sections. The total length of the Jet Fan niche is 31.64 m.

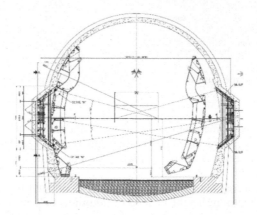

Figure 8.   Jet Fan Niche Formwork.

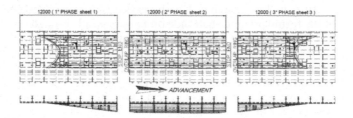

Figure 9.   Longitudinal view of Jet Fan Niche.

For the realization of a complete Jet Fan Niche, 3 concrete blocks of 12 m will be necessary by installing every time a double set of steel forms with different sections. In order to facilitate the installation and removal of these side niche forms, the main formwork has been foreseen with proper drillings on the mantle and access doors in the deepest areas of the niche.

All the operations of installation, use and disassembly are widely described in the manuals provided to the site and on specific tables illustrating all the step by step operating sequences.

Along the 6.4 km of tunnel no. 7 Lay-By of 40 m were planned by the Spanish designers at whose center starts the cross passage for vehicles for the emergency connection with the Escape Tunnel.

For the lining of this section of more than 15 m of width, specific formworks 8.328 mm long made in two elements of 4.164 mm, will be used, moved by a special traveller provided with 4 heavy-duty construction tires. The advantage of this self-propelled traveller with Diesel engine is in the easy transfer from one enlargement to the next one, by rotating the closed formwork of 4.164 mm by 90° and passing through the standard section of the already lined Main Tunnel.

The Lay-By formwork has features similar to the ones of the standard section with a much wider cap. The casting sequence within a 40 m Lay-By foresees 6 different blocks. This to respect the full cover of the middle intersection cross passage and of the "Transformer Room", which has the same section of the Cross Passage.

Other important phases to complete the lining of the Main Tunnel are, in sequence, the casting of the ventilation slab, as well as of the upper partition wall, which will create the two ducts for fresh and exhausted.

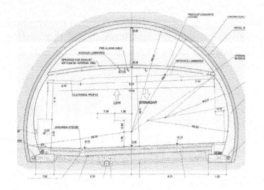

Figure 10. Lay-By Cross section.

Figure 11. Lay-By Formwork.

All surfaces of these parts of the work will have a great finishing accuracy to avoid pressure losses, which would greatly affect the running costs of the work and its safety.

For the casting of the horizontal slab with a slightly curved profile, 3 sets of equipment have been foreseen, each one consisting of two formworks 12m long and one traveller for the telescopic movement of the formworks. With this solution it will be possible to leave one formwork with the casting under curing and, in the front part, to leave a free formwork for the assembly of the reinforcing bar, so as to proceed afterwards with the concreting.

Leaving the formwork in this position, proceed with the dismantling of the rear formwork using the traveller. Afterwards, proceed with the dismantling of the rear formwork by means of the traveller and, then, passing under the formwork in setting-up position, you will position it in the front part to re-start the complete cycle.

After the concrete casting and curing of the slab, you will proceed with the construction of the vertical wall. This is the final lining phase of the main tunnel.

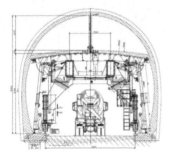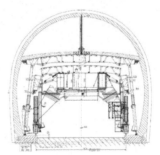

Figure 12. Telescopic Deck slab formwork.

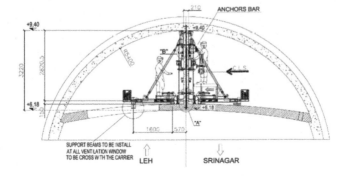

Figure 13. Vertical Wall formwork.

For these concrete casts of 12 m, steel formworks 6+6 m long with mechanical movements on traveller will be used. Particular care is given to the upper filling by means of the casting pipes and the exhaust pipes, which discharge the concrete in excess. For the horizontal slab and the vertical casting, the first casting operations are scheduled in March 2019.

3 CONCLUSIONS

This important Indian project for the construction of a road pass tunnel of considerable length is an example of great planning of all the design and construction activities of the whole work.

The contracting entity that will manage the work in concession and the construction company have taken advantage of their own experience in the design and construction gained in recent similar projects

The final target is to create a work that could guarantee functionality and safety well above the current Indian infrastructure standards.

The all-round collaboration between the engineers of APCO and CIFA would permit to accelerate the construction of all the lining works of the main and service tunnels, in order to achieve the final delivery of such an important work for the whole Indian Kashmir.

Tunnels and Underground Cities: Engineering and Innovation meet Archaeology,
Architecture and Art, Volume 5: Innovation in underground engineering,
materials and equipment - Part 1 – Peila, Viggiani & Celestino (Eds)
© 2020 Taylor & Francis Group, London, ISBN 978-0-367-46870-5

Foam soil interaction and the influence on the stability of the tunnel face in saturated sand

A. Bezuijen
Ghent University, Ghent, Belgium/Deltares, Delft, The Netherlands

T. Xu
Ghent University, Ghent, Belgium

ABSTRACT: The face stability of an EPB-shield in saturated sandy soil is investigated. Recent research on the behavior of foam at the interface is combined with the knowledge on the influence of excess pore water pressures in the soil on the face stability. A simple model is described to illustrate this influence. It appears that the processes at the tunnel face for an EPB are quite comparable with the processes for a slurry shield. 'Clean foam' leads to more reduction of the permeability at the face than a foamy sand that can be expected during drilling. Micro instability is not very likely, except for large diameter tunnels and when drilling in a confined aquifer. Excess pore water pressures in front of the EPB can require a roughly 2 times higher pressure difference between the mixing chamber and the pore pressure far away from the tunnel than calculated with traditional methods.

1 INTRODUCTION

Excess pore pressures measured in front of a slurry TBM have consequences for the stability of the tunnel face (Bezuijen et al. 2001; Broere, 2003; Dias & Bezuijen, 2015). It is shown that during drilling there is no immediate plastering by the slurry and consequently the pressure in the mixing chamber, which is higher than the pore water pressure in the surrounding soil, causes a groundwater flow from the TBM to the soil. This groundwater flow stabilizes the soil, but this stabilization is less effective than will be achieved by a plastered tunnel face. During standstill, there will be an external or internal cake building (Xu & Bezuijen, 2018, Xu, 2018), resulting in a lower permeability at the face and a better stability.

Excess pore water pressure in front of a tunnel face is also measured in front of an EBP shield (Hoefsloot, 2001). However, until recently the permeability of the foam-sand muck in the mixing chamber and is influence of the tunnel face is less well described.

Recent research has shown that the overall permeability of a tunnel face is only a bit lower than the permeability of the original sand (Bezuijen & Dias, 2017). At what location at the tunnel face this reduction was realised could not be measured. Infiltration experiments with foam and a foam-sand muck (Xu et al., 2018, Xu, 2018) have improved our understanding what happens at the tunnel face for an EPB shield.

This paper will present an overview of literature describing what determines the permeability at the tunnel face for an EPB. This will appear of importance to describe the stability of the tunnel face. The stability will be described by wedge and silo formula, as proposed by Janscecz & Steiner (1994). This is to give an idea of the influence of the permeability distribution on the stability. The concept can also be used in a numerical calculation on the stability.

2 STABILITY OF THE TUNNEL FACE

Well known is the calculation of the macro stability of a tunnel face, see Figure 1. It is assumed that there is a silo loading on the surface DEFC and the stability of the wedge ABCDEF is calculated, taking into account the friction and cohesion of the soil material. In the standardized calculations as the German DIN, the influence of the excess pore water pressure is not taken into account. Broere (2003) and Dias & Bezuijen (2015) illustrated how this can be done. Aime et al. (2004) and Kaalberg et al. (2014) showed for field conditions that knowledge of the excess pore water pressures in front of the tunnel can be essential for the success of a project. Since the method is not standardized, different methods deviate in details. These methods are developed for a slurry shield. It will be shown that the principle can also be used for an EPB.

Apart from the macro stability, also the micro stability has to be taken into account. Van Rhee and Bezuijen (1992) have shown that a vertical slope of a sandy material in saturated conditions can only be stable when there is a hydraulic gradient perpendicular and inward directed to that slope of around 3. For gradients between 1 and 3 the sand grains will continuously rain from the slope, for gradients less than 1, the slope will be unstable.

To create a stable tunnel face, it is therefore necessary to have a pressure in the mixing chamber that is high enough to stabilize the wedge shown in Figure 1, but also to have a sufficient hydraulic gradient at the tunnel face to avoid micro instability. This last criterion is not automatically fulfilled when the first is fulfilled. If no plastering at all is assumed, or for an EPB there is no extra flow resistance at all at the tunnel face, then the hydraulic gradient into the face is approximately (Bezuijen 2002):

$$i = \phi/R \tag{1}$$

Where ϕ is the piezometric head at the tunnel face assuming that the piezometric head in the soil far from the TBM is zero and R is the radius of the tunnel. For a tunnel with 10 m diameter and an excess pressure in the mixing chamber of 50 kpa (= 5 m difference in piezometric head), this means that the gradient close to the tunnel face will be just 1 and is therefore

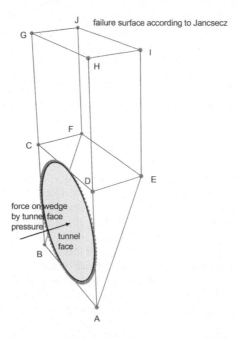

Figure 1.   Possible failure surface in front of a tunnel (Bezuijen et al. 2001).

not enough to avoid micro instability. This corresponds with well-known knowledge from practice: you cannot drill in saturated sand with just water as a drilling fluid.

If, due to plastering, or foam bubbles, there is a small zone at the tunnel face where the permeability is 3 times lower than in the sand itself, the gradient in that zone will be around 3 and micro stability is assured. For a slurry shield, this criterion is more or less automatically fulfilled since the viscosity of the bentonite slurry is much higher of than the viscosity of water. The hydraulic gradient over the bentonite slurry penetrated into the sand for small penetration depths (as is normally the case during drilling) will be approximately μ_s/μ_w higher than the value in the soil as calculated with Eq. (1), where μ_s is the viscosity of the slurry and μ_w the viscosity of water. Furthermore, the yield strength of the slurry will contribute to the stability. For an EPB shield this mechanism is less obvious and will be investigated in the next section.

3   THE PERMEABILITY OF A FOAM-SAND MIXTURE

In theory, two permeabilities are of importance. The permeability of the muck and the permeability of the tunnel face itself, thus at the boundary between the muck and the original sand. Infiltration experiments, using saturated sand and a foam-sand mixture to simulate the conditions in the mixing chamber, have shown (Xu, 2018) that for a high-density muck there can be a considerable pressure drop over the muck, indicating that there is a flow resistance in the muck itself).

The permeability of the tunnel face depends on the density of the muck. Two situations are shown in Figure 2.

One with 'clean foam' without sand in the mixing chamber (left sketch) and one with sandy foam during excavation (right sketch). For the clean foam, the situation at the boundary foam/sand can be described as follows: Without rotation of the cutter head and an excess pore water pressure in the mixing chamber, the water and foam will flow into the sand. However, at the sand surface the foam bubbles are blocked by the sand grains, creating an area with a lower permeability and thus a higher-pressure gradient. This higher-pressure gradient deforms

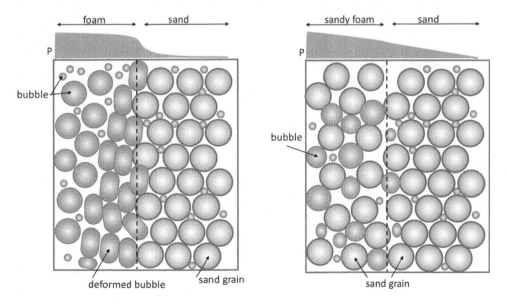

Figure 2.   Sketch bubble and pressure distribution for 'clean foam' (left) and sandy foam at a tunnel face. The sand on the right of each plot is the original sand before excavation. The left side represents the mixing chamber.

the bubbles as indicated in Figure 2 on the left side, which leads to a further reduction of the permeability. In case of sandy foam in the mixing chamber as can be expected during excavation, there is a different situation: The flow from the mixing chamber will also transport the sand grains to the tunnel face. These sand grains form a grain skeleton in front of the tunnel faces causing there a decrease in pressure. The pressure gradient is thus not only taken by the bubbles, but also by the un-deformable sand grains reducing the loading on the bubbles that will therefore not deform. Consequently, the permeability remains higher although still significantly lower than for clean sand and there is a pressure gradient but no 'steep' pressure drop at the tunnel face.

It may seem a somewhat theoretical assumption to assume clean foam at the tunnel face, where sand is removed continuously during drilling. However, due to the groundwater flow from the mixing chamber, the excess pressure decreases and regularly foam has to be injected during stand still (Bezuijen & Dias, 2017). Measurements have shown that due to this foam injection the vertical pressure gradient decreases which implies that the sand concentration decreases, see Figure 3).

Hoefsloot (2001) proved the existence of a low permeability area in front of the tunnel face during standstill. He showed measurements of the pore water pressure in the soil in front of the tunnel face, see Figure 4.

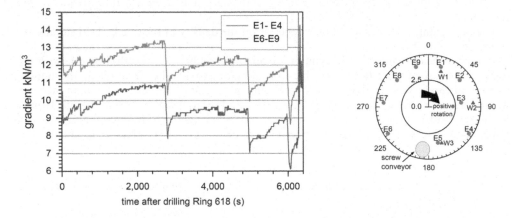

Figure 3. Vertical gradient determined with pressure gauges left and right of the TBM. The sharp decreases in the gradient indicate foam injections (Bezuijen & Dias, 2017). The figure at the right shows the positions of the various instrument looking with face to the tunnel front.

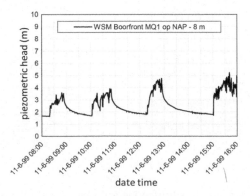

Figure 4. Pore pressures measured in sand 6 to 3 m in front of a TBM (PPT is fixed the distance becomes shorter as the TBM approaches). (Hoefsloot, 2001).

The plot shows that the pore pressure in front of a TBM increases during drilling and decreases during standstill, as for a slurry shield (Bezuijen et al., 2001). The pore pressure decrease is measured, because for an EPB shield as well as for a slurry shield there is a low permeable layer in front of the TBM. This layer is removed during drilling (when the drilling speed is higher than the pore velocity of the water) leading to an increase in pore water pressure.

4 MICRO STABILITY

It was mentioned that the gradient at the tunnel face is around 1 for a 10 m diameter tunnel drilled with an excess pressure of 50 kPa in the mixing chamber compared to the pore pressure in the surrounding soil.

Bezuijen & Dias (2017) have shown that during drilling of the Botlek Rail Tunnel (an EPB tunnel) the apparent permeability of the tunnel face during the standstill periods is on average only a factor 2 to 3 lower than the permeability of the original sand as measured during the soil investigation. However, the foam penetrated less than 0.5 m into the sand for the conditions of the Botlek Rail Tunnel, which means that at the tunnel face, over a thin layer the permeability has to be considerably less. This implies that during stand-still the micro stability is not an issue. The permeability at the tunnel face is sufficiently reduced to have a gradient high enough to prevent micro instability at the tunnel face.

However, during drilling, the low permeable layer of foam bubbles and sand is removed and then the micro stability can become important. Hoefsloot (2001) mentioned that for a situation where the drilling velocity is higher than the pore velocity, the piezometric head measured in front of the tunnel face in the sand is more or less equal to the face pressure, which means that the drilling removes a low permeable layer before it is formed. Quite comparable as the way it is described for a slurry (Bezuijen, 2001). However, also comparable to the way it is described for a slurry, the foam creates more or less immediately a thin layer in the sand with a lower permeability. During drilling this layer is thin when the drilling velocity is larger than the pore velocity of the water and it does not influence the overall pressure distribution. However, it does influence the local gradient at the tunnel face. This can be seen from infiltration experiments, where the discharge is plotted as a function of time. An example is shown in Figure 5. Where saturated sand with a d_{50} of 160 μm was infiltrated with foam with a FER (Foam Expansion Ratio) of 15 with an excess pressure of 50 kPa.

The plot shows that immediately after the start of the experiment, the discharge decreases one order of magnitude, which shows that the foam creates a low permeable layer from the very beginning of the experiment.

From the results described above it can be concluded that under normal conditions micro-instability is not an issue, the less permeable foam in the mixing chamber will create a layer with high flow resistance in which the hydraulic gradient is well above 1. Situation can be

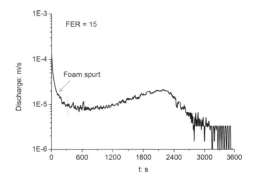

Figure 5.   Discharge as a function of time in an infiltration experiment, using foam (Xu et al. 2019).

1733

more complicated when the permeability of the sand is low, the tunnel is drilled in a confined or semi-confined aquifer or the diameter is large. For the first two situations, the groundwater flow from the tunnel into the sand will be very limited and may be insufficient to create a layer with reduced permeability at the tunnel face. In the last situation, the gradient can be low (see Eq. (1)) and may be insufficient to create a stable tunnel face.

In conclusion, it should be realized that the permeability of a foam-sand mixture at the tunnel face is higher than the permeability of a pure mixture and this will lead to higher excess pore pressures than expected based on filtration tests with 'clean foam'.

5 CONSEQUENCES FOR STABILITY CALCULATIONS

The consequences for stability calculations for an EPB in saturated sand conditions are:

- The groundwater flow situation has to be incorporated in a stability calculation.
- The risk that stability problems occur due to micro instability is limited, as it is for a slurry shield. This can only be a problem for large diameter shields, a low permeability sand, when drilling in a semi-confined aquifer with a long leakage length and a thickness that is not much larger than that of the tunnel (Bezuijen & Xu, 2018).
- The overall stability has to be calculated taking into account that there will be excess pore water pressure in front of the tunnel for an EPB as well as for a slurry shield. Several of these calculation methods are mentioned in Section 2.

A way to qualitatively the influence of the excess pore water pressure on the stability is shown in Figure 6. This figure shows a 2D representation of the ground wedge that is shown in Figure 1. If there is no excess pore water pressure in the ground in front of the tunnel, as shown in the left part if this figure, the excess pressure in the mixing chamber will stabilize the wedge. However, if there is an excess pore water pressure in the ground in front of the tunnel, as shown in the right hand side, then the effective force that is acting on the wedge is only the average initial pressure times the area of the tunnel fase minus the average excess pore pressure on the other side of the wedge times that area. This means that the stabilizing force on the wedge is less when the same face pressure is applied and a higher face pressure is necessary to create a stable face.

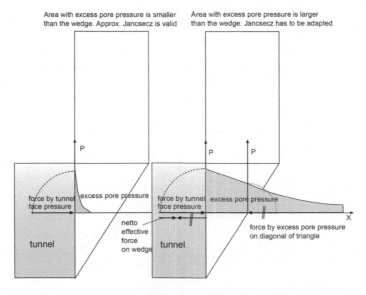

Figure 6.   Influence of excess pore water pressures on stability of tunnel face.

A first estimation on the influence of the pore water pressure on the stability can be made following the sketch in Figure 6. In case of a tunnel made in an unconfined aquifer, the excess pore pressure in front of the tunnel can be approximated by:

$$\varphi = \varphi_0(\sqrt{1 + (x/R)^2} - x/R) \tag{2}$$

In the equation R is the diameter of the tunnel, x is the distance in front of the tunnel, φ the piezometric head at the location x and φ_0 the piezometric head at the tunnel face. This formula calculates the worst-case situation that there is no plastering at all by the foam. The critical failure surface has to be determined by assuming different angles however it will be in the order of the active failure surface so the angle with the horizontal will be around $45/2+\phi$ degrees where ϕ is the friction angle of the soil in front of the tunnel face. With Eq. (3) this means that the piezometric head in the middle of the tunnel can be written as:

$$\varphi = \varphi_0(\sqrt{1 + \tan^2(45/2 - \phi)} - \tan(45/2 - \phi)) \tag{3}$$

Assuming a friction angle of 30 degrees this means that φ is 0.58 times φ_0 and thus that only a fraction of 0.42 times the original pressure contributes to the stability of the tunnel face. As mentioned before, this is a worst-case calculation. The influence of the sidewalls is not included nor the reduction of the vertical loading by the excess pore water pressures. However, the simple method described above shows clearly, what the influence can be. Numerical calculations (Bezuijen et al. 2001) show an influence in the same order of magnitude.

It should be realized that the model uses the flow model for an unconfined aquifer. The results can be on the unsafe side when drilling in a semi-confined or confined aquifer, in which higher excess pore pressures can be expected.

6 CONCLUSIONS

The foam-soil interaction at the face of an EPB shield is investigated and the consequences for the stability of a tunnel face. It appears that the foam-soil interaction is comparable with that of a slurry. This led to the following conclusions:

1. Infiltration experiments with 'clean foam' in saturated sand lead to a low-permeable boundary at the transition between the foam and the sand. This is comparable as the external cake formation in a slurry shield. This mechanism does not occur (or has less influence) when a foam sand mixture is used. This implies that also at the face of a TBM the sand concentration in the foam influences the permeability at the face.
2. The permeability at the foam-sand transition is normally low enough to avoid micro instability at the tunnel face. However, problems with micro instability cannot be excluded for low permeable soils, large diameter tunnels and/or tunnelling in a confined or semi-confined aquifer.
3. The permeability at the tunnel face leads to excess pore water pressures in the soil in front of the tunnel. These has to be taken into account when calculating the overall stability of the soil in front of the tunnel. A simple calculation is presented that explain the mechanism how the stability is influenced. This model shows that a more than two times higher excess pressure, above the pore pressure, in the mixing chamber is necessary when taking into account the excess pore water pressures than according to a traditional calculation without the influence of the excess pore water pressures in the ground due to tunnelling. Comparable numbers were found in numerical calculations.

ACKNOWLEDGEMENT

The second author would like to acknowledge the scholarship funded by China Scholarship Council and the CWO Mobility Fund of Ghent University.

REFERENCES

Aime, R., Aristaghes, P., Autuori, P. & Minec, S. 2004. 15m Diameter Tunneling under Netherlands Polders. *Proc. Underground Space for Sustainable Urban Development (WTC Singapore)*, Elsevier.

Bezuijen, A. & Xu, T., 2018, Excess pore water pressures in front of a tunnel face when drilling in a semi-confined aquifer, *Proceedings WTC 2018*, Dubai.

Bezuijen, A. & Dias, T.G.S., 2017, EPB, chamber pressure dissipation during standstill, *EURO:TUN 2017*, Innsbruck University, Austria.

Bezuijen, A., 2007. Bentonite and grout flow around a TBM, in: Barták, Hrdina, Romancov, Zlámal (Eds.), *ITA World Tunnel Congress 2007* - Underground Space – The 4th Dimension of Metropolises. Prague, Czech Republic, pp. 383–388.

Bezuijen, A., 2002. The influence of soil permeability on the properties of a foam mixture in a TBM, in: *Geotechnical Aspects of Underground Construction in Soft* Ground - *4th International Symposium (IS-Toulouse)*. Toulouse, France, pp. 221–226.

Bezuijen, A., Pruiksma, J.P. & van Meerten, H.H., 2001. Pore pressures in front of tunnel, measurements, calculations and consequences for stability of tunnel face, in: Adachi, T., Kimura, M., Tateyama, K. (Eds.), *International Symposium on Modern Tunneling Science and Technology*. Kyoto, Japan.

Broere, W., 2003. Influence of Excess Pore Pressures on the Stability of the Tunnel Face. *Reclaiming the Underground Space ITA*, Amsterdam, pp. 759–765.

Dias, T.G.S. & Bezuijen, A., 2015. TBM Pressure models: Calculation tools, in: Kolíc, D. (Ed.), *ITA World Tunnel Congress 2015 - SEE Tunnel - Promoting Tunnelling in SEE Region*. Dubrovnik, Croatia. doi: 10.13140/RG.2.1.1780.4962

Hoefsloot, F.J.M. 2001. Pore pressures in front of the bore front: A simple hydrological model (in Dutch), *Geotechniek* October. pp. 26–33.

Jancsecz, S. & Steiner, W., 1994. Face support for a large Mix-shield in heterogeneous ground condtiontions, *Tunneling' 94*, 531–550.

Kaalberg, F.J., Ruigrok, J.A.T. & De Nijs, R., 2014. TBM face stability & excess pore presssures in close proximity of piled bridge foundations controlled with 3D FEM, in: Yoo, C., Park, S., Kim, B., Ban, H. (Eds.), *Geotechnical Aspects of Underground Construction in Soft Ground - 8th International Symposium (IS-Seoul)*. CRC Press/ Balkema,Leiden, Seoul, South Korea, pp. 555–560.

Van Rhee, C. & Bezuijen A., Influence of seepage on stability of sandy slope. *Journal of Geotechnical Engineering-Asce*, 1992. **118**(8): p. 1236–1246.

Xu T. Bezuijen A. & Zhao L., 2019, Pressure infiltration of sandy foam during EPB shield tunnelling in saturated sand, to be published *Proceedings*, WTC *2019*, Naples.

Xu, T, 2018, *Infiltration and Excess Pore Water Pressures in front of a TBM, Experiments, Mechanisms and Computational models*, PhD thesis, Ghent, 2018.

Xu, T. & Bezuijen, A., 2018. Pressure infiltration characteristics of bentonite slurry. *Géotechnique*. https://doi.org/10.1680/jgeot.17.T.026.

Xu, T., Bezuijen, A. & Thewes, M., 2018. Pressure infiltration characteristics of foam for EPB shield tunnelling in saturated sand – Part 1: clean foam. *Géotechnique*. In preparation.

Precast tunnel segments with glass fiber reinforced polymer composite cage and sulphoaluminate cement

M. Bianchi & F. Canonico
Buzzi Unicem SpA, Casale Monferrato, Italy

N. Giamundo
ATP srl, Angri, Italy

A. Meda, Z. Rinaldi & S. Spagnuolo
Tunnelling Engineering Research Centre, University of Rome "Tor Vergata", Rome, Italy

ABSTRACT: During tunnel excavation by means of a tunnel boring machine (TBM), the lining consists of precast structural elements, placed by the TBM during the excavation process and used as support elements during the advancing phase. The use of concrete based on sulphoaluminate binders allows the production of the elements to be speeded up, a reduction in the number of segments stacked waiting for the required strengths to be achieved and provides a more eco-sustainable process avoiding the use of steam curing. In order to verify the feasibility of the proposed solution, two full-scale tests (bending test and TBM jacks thrust test) on metro tunnel precast segments were carried out. The internal reinforcement consisted of a next-gen Glass Fibre Reinforced Polymer (GFRP) cage. The use of GFRP reinforcement, to replace traditional steel, in tunnel segments provides several advantages mainly related to durability aspects or when the use of a provisional lining is foreseen.

1 INTRODUCTION

Tunnel construction using a mechanized excavation technique by means of a Tunnel Boring Machine (TBM) has been widely used, especially over the last thirty years, allowing for the construction of underground structures. Furthermore, this technique appears to be more effective in urban environments where the excavation has a greater impact, mainly on the surrounding structures.

The production of the precast tunnel segments, which form the final lining, is generally carried out in prefabrication plants, often located some distance from the location where they are installed.

Based on this consideration, it is necessary to take into account the amount of formwork, curing time and storage areas, ensuring a sufficient amount of segments are available to meet demand due to the possible unforeseen acceleration of the excavation speed.

The formwork opening to remove superficial flaws, the subsequent steam curing of the concrete segments (generally 5–6 hours until the concrete strength, f_{ck}, equal to $12 \div 15$ MPa useful for demoulding is reached), are critical operating steps with high environmental and economic impacts.

Once the segments are demoulded, they are stored until the required strength is reached. From that moment, the segments can be installed and will be able to counteract the TBM thrust during the excavation advancement, without suffering structural damage.

To speed up the production process, to reduce the number of segments stored waiting for the required strength to be reached and to obtain sustainable processes without steam curing of the concrete, a new sulphoaluminate cement-based concrete has been developed, able to reach high mechanical performance in a short time.

Sulphoaluminate cements (CSA) are innovative hydraulic binders obtained by grinding clinker, cooked to 1300 °C, which is based on raw materials such as limestone, bauxite and gypsum (in approximately equal proportions). These clinker can be used with the addition of appropriate quantities of gypsum or anhydrite, or can be mixed with Portland cement. Based on these combinations, it is possible, to quickly achieve high mechanical strengths by maintaining very low hygrometric retreats of the concrete.

Buzzi Unicem SpA received the CE marking for its sulphoaluminate binders line, named Buzzi Unicem Next, also paving the way for the use of these binders in the structural field.

A different hydration process compared to the common Portland cements characterizes Sulfoaluminate binders. The hydration of Portland cement is dominated by the reaction of tricalcium silicate (C_3S - Alite) which, in its reaction with water, develops calcium silicate hydrates and calcium hydroxide. The ettringite is the main hydration product of the Sulphoaluminated cements (Bullard et al. 2011).

The pH generated by means of sulphoaluminate cement hydration is higher than 12. Although lower than that of Portland cements (pH > 13), it is sufficient to passivate the reinforcement. However, this characteristic can be exploited to effectively combine these binders with fiberglass reinforcements that are able to resist low-pH values (Bertolini et al. 2015).

The use of fiberglass reinforcing bars, combined with this new concrete, has been investigated.

The use of fiberglass re-bars (Glass Fiber Reinforced Polymer – GFRP), as reinforcement in concrete structures, is an innovative solution that can be used as an alternative to common steel reinforcing bars, especially in conditions where high resistance to environmental attack is required (Almusallam et al. 2006, Chen et al. 2007, Benmokrane et al. al. 2002, Gooranorimi and Nanni 2017). In fact, compared to traditional steel reinforcement bars, fiberglass reinforcements are not affected by corrosion problems, and have high tensile strength and lower weight (Micelli and Nanni 2004; Chen et al. 2007; Nanni 1993; Alsayed et al. 2000). The material is also non-conductive and non-magnetic (Almussalam et al. 2006).

The use of fiberglass reinforcements leads to different advantages such as an increase in durability, the absence of corrosion, the possibility to create dielectric joints and the reduction of concrete cover, limiting cracking problems during transitional phases until the final installation of the lining (Spagnuolo et al. 2017).

In light of the advantages of using this new technology in underground structural fields, over the last few years, experimental campaigns have been carried out both for the characterization of the physical-mechanical properties, and the durability of fiberglass reinforcement with curvilinear axis (Spagnuolo et al. 2018).

Furthermore, several full-scale tests on typical precast segments for different tunnel linings (hydraulic, metro and highway tunnel geometries) were carried out to demonstrate the validity of this new reinforcement to replace traditional steel re-bars (Spagnuolo et al. 2014; Caratelli et al. 2016; Caratelli et al. 2017).

The research was focused on the characterization of an innovative concrete based on sulphoaluminate cement, testing its chemical and mechanical properties on an industrial scale for the production of precast tunnel segments. Based on the results obtained on fresh concrete, two full-scale metro tunnel precast segments were cast and tested. The segments were reinforced with GFRP re-bars. Two different full-scale tests (bending test and point load test) were carried out to verify and validate the structural capacity and materials effectiveness, respectively.

2 EXPERIMENTAL CAMPAIGN

2.1 Materials and test methods

For concrete packaging, a sulphoaluminate cement-based binder, CE marked (Next binder - SL05) was used, dosed at 380 kg/m^3.

The aggregates, used in the mix design, were of calcareous origin originating from the UNICAL concrete mixing plant in Santa Lucia (Rome). The aggregates consisted of the following particle size fractions: 0–4 mm sand, 8–15 mm gravel, 15–25 mm crushed stone.

The fluidifying additive used was Sika CC39/P22.

The concrete mix was carried out by means of a concrete mixer at a room temperature of 20 °C. Tests were also carried out at an external temperature of 30 °C to evaluate the workability time and the strength development at higher temperatures, respectively.

Concrete properties such as consistency class (in accordance with Standard UNI EN 12350-2), maintenance, density (in accordance with Standard UNI EN 12350-6), compressive cubic strength (by means of 150x150x150 mm samples aged at 20 °C and 95% RH, in accordance with Standard UNI EN 12390-3) were defined. The compressive cubic strength evaluation was carried out at 3-4-5-6-24 hours, 7 and 28 days, respectively.

The precast segments were cast by means of a truck mixer, by using the same aggregates and mix design (Figure 1). For the cast, a concrete volume equal to 4 m³ was mixed.

Table 1 shows the results of the tests carried out on the concrete, depending on the temperature.

Both precast segments were reinforced with GFRP re-bars. Each rebar consisted of boron-free E-CR fiberglass 4800tex and matrix thermoset vinyl ester resin-based.

The fiber volume fraction was equal to 60%, for which unidirectional tensile strength and tensile Young's modulus were 1200 MPa and 48 GPa, respectively.

Concrete shrinkage was measured, over time, by means of deformeter EDU by Huggenberger AG. The measurement was carried out by means of two steel plates, drowned or glued to the concrete, fixed at a distance equal to 250 mm (Figure 2a). The instrument resolution was equal to 0.001 mm with ±0.005 mm accuracy. An Invar reference bar was used as auto-zero comparator. The main advantage, of using a deformeter, consisted s of concrete shrinkage measurements that can be carried out starting a few hours after the casting, directly on the structure.

The experimental campaign on the structural elements was performed by carrying out full-scale tests to simulate the most severe conditions to which the tunnel lining, made by mechanized excavation, would be subjected to during the transitional and serviceability phases.

The two tests were:

- Three points bending test, in order to understand the segment behaviour during the transitional phases (which range from demoulding to installation by means of TBM erector) and to evaluate the maximum bending capacity at the ultimate limit state;
- Point load test to simulate jack thrust exercises by the TBM on the segment during the excavation phase and its progress.

Figure 1.   GFRP reinforcement detail and concrete cast.

Table 1. Properties of the mix design, prepared with superplasticizer CC39T22 at different temperatures.

Mix design	Unit	20°C	30°C	Ref.
Binder (NextSL05)	kg/m^3		380	380
CEM IV/A(PV) 42.5N	kg/m^3		380	380
Sand	kg/m^3		892	892
Rubble	kg/m^3		445	445
Gravel	kg/m^3		503	503
Additive CC39/P22	%		0.7	0.7
Water	l/m^3		160	160
Volumetric data				
Yield	%	99.8	100.6	101
Actual water	l/m^3	162	159	161
Air content	%	2	1.9	2.0
Density	kg/m^3	2384	2382	2395
Workability loss *				
t_0	mm	195	200	200
t_{30}	mm	190	160	200
t_{60}	mm	180	-	190
Compressive strength**				
3 h	MPa	3.7	10.0	-
4 h	MPa	10.0	16.1	-
5 h	MPa	15.4	18.0	-
6 h	MPa	18.2	20.3	-
24 h	MPa	39.9	44.2	-
7 days	MPa	54.6	51.9	41.9
28 days	MPa	66.7	64.4	52.6

Note
* t_0 refers to time zero; t_{30} refers to 30 minutes after casting; and t_{60} refers to 60 minutes after casting.
** Compressive strength refers to cubic strength R_{cm}.

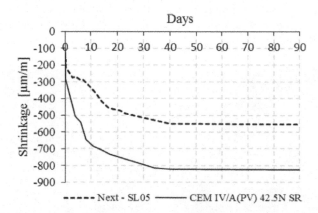

Figure 2. Shrinkage evolution.

Figure 2 shows the results of the concrete shrinkage. The shrinkage test was carried out taking into account the same mix design, water content and binder dosage. CEM IV/A(PV) 42.5N SR pozzolanic cement, normally adopted to cast tunnel segments, was used as the reference binder.

Concrete shrinkage measurement was carried out starting from 2.5 hours for the Next-SL05 mix and 7 hours for the reference mix after the concrete casting, respectively.

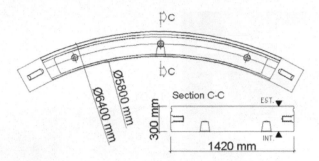

Figure 3.   Segment geometry.

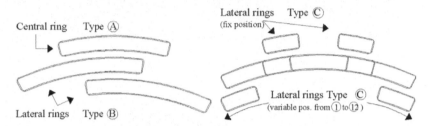

Figure 4.   GFRP reinforcement details.

For the full-scale application two metro tunnel segments were cast. Each precast segment was 300 mm in thickness, with 5800 mm internal diameter and 1420 mm in depth (Figure 3).

The reinforcement consisted of a GFRP cage for which composite re-bars were designed in order to optimize the length of each closed-loop ring (Figure 4). The optimization allowed the achievement of both performance aspects that segments would need to ensure during the serviceability phase, and technical aspects due to mechanical components located on the structural elements such as cups, connectors, etc..

Each segment consisted of 12 equivalent Ø13 mm base closed-loop rings (longitudinal direction), each of which in turn, contained superimposed rings (A, B and C types) of different length, according to layout as shown in Figure 4.

The equivalent area of each GFRP ring was equal to 43.9 mm^2.

The cage, in transverse direction, was confined by means of 16 equivalent Ø8 mm GFRP closed-loop stirrups.

2.2   Three points bending test

A full-scale bending test was carried out in displacement control, according to the testing set-up shown in Figure 5. The segment was placed on two cylindrical supports with a wheelbase of 2 m. Load was applied by means of a electromechanical jack with a load cell of 1000 kN and 0.2% accuracy, by imposing 10 μm/sec running speed. Load, in turn, was distributed, by means of a frame system, along the segment mid-span on the extrados surface (Figure 5a).

During the test, in addition to applied load, displacements and crack widths, at the mid-span of intrados surface, were recorded by means of three wire transducers and two Linear Variable Differential Transformers (LVDTs), respectively (Figure 5b). Increasing the load, new cracks developed, whose width were recorded by means of a crack width ruler.

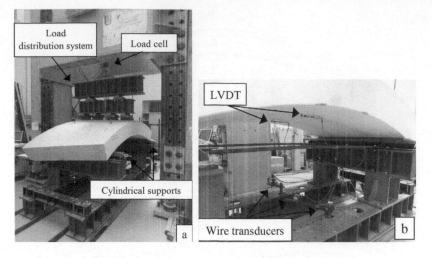

Figure 5.   Testing set-up. Bending test: a) Load system; b) Instrumentation.

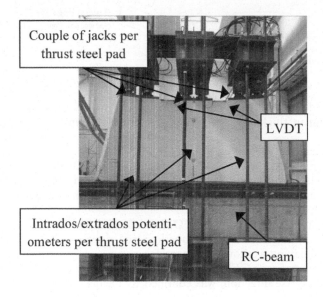

Figure 6.   Testing set-up: point load test (TBM).

2.3   *Point load test (TBM)*

A point load test, simulating TBM thrust exerted on segments (previously installed on the back of the TBM shield) during its progress, was carried out applying the load according to the testing set-up shown in Figure 6.

The load, during testing, was applied on each steel pad according to TBM geometry and config-uration by means of two 2000 kN hydraulic jacks (max 4,000 kN per steel pad). The testing set-up, for the configuration shown in Figure 6, was able to provide a maximum load of 12,000 kN (1200 Tons). The segment was placed on a correctly designed reinforced concrete (RC) beam.

As shown in Figure 6, besides the load, vertical displacements below the thrust steel pads and crack widths between the same were recorded by means of potentiometers (two for each steel pad – intrados/extrados) and two Linear Variable Differential Transformers (LVDTs), respectively. In addition to data acquisition, increasing the load step by step, crack patterns were mapped out. For each significant crack, its width was measured by means of a crack width ruler.

3 RESULTS

3.1 *Concrete*

Based on the results obtained, CSA binders can be used for concrete packaging as can be done with the common Portland cements with the advantage of being able to reduce the binder content (from 400 kg/m^3 to 380 kg/m^3) and to avoid steam curing of the segment to speed up the demoulding process.

CSA cements have a reduced working time compared to pozzolanic cements used for this application, but it is sufficient to be used for the casting of precast elements. The workability time can be modulated depending on the casting temperatures and needs with the addition of retardant additives.

A concrete mix based on Next-SL05 sulphoaluminate binder allows for a compressive strength higher than 15 MPa to be achieved as early as 5 hours after casting with a workability time of 90 minutes at 20 °C.

The influence of temperature considerably modifies the workability time (40 minutes at 30 ° C compared to 90 minutes at 20 °C), however leaving time for the concrete to be cast in pre-fabrication. Useful demoulding strengths were already achieved three hours after casting.

The development of a high strength, in the early hours, avoids the thermal cycle of steam curing, making the precast segment production process easier both from an operational point of view and from a logistics point of view of the plant realization.

The results of shrinkage show how mix design based on sulphoaluminate cement is characterized by a shrinkage lower than that obtained with pozzolanic cement, -550 μm/m and -820 μm/m, respectively.

3.2 *Three points bending test*

A full-scale bending test was carried out 26 hours and 40 minutes after segment casting, for which the average compressive strength (R_{cm}) achieved was equal to 41 MPa.

The first crack (red line in Figure 7) occurred at a load level of 130 kN with a maximum crack width equal to 0.30 mm. Increasing the load, several cracks developed both on the intrados and lateral surfaces of the segment. Figure 7 shows the crack pattern at the end of the test.

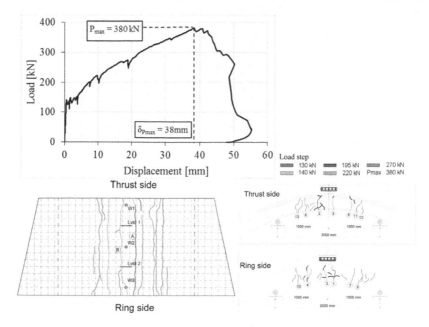

Figure 7.   Bending test results.

Table 2. Bending Tests: Crack widths.

Load [kN]	Load level				
	130 kN	140 kN	195 kN	220 kN	270 kN
Crack color					
Crack n.					
1	0.25÷0.30	0.40÷0.45	1.00÷1.20	1.50	2.00÷2.50
2	0.25	0.35	0.90	1.50	2.00
3	-	0.35	1.00	1.25	2.50
4	-	0.30	0.45	0.35	0.40
5	-	0.45	1.25	1.50	2.50
6	-	0.15	0.45	0.70	1.00
7	-	-	-	0.30	0.30
8	-	-	-	0.15	0.35
9	-	-	-	-	0.70
10	-	-	-	-	0.70
11	-	-	-	-	0.60
12	-	-	-	-	0.20
13	-	-	-	-	0.40

The maximum load achieved was 380 kN with a displacement equal to 38 mm at the mid-span of the segment intrados surface, as shown in Figure 7, where the load versus average displacement curve is plotted.

Table 2 shows, for each load step, the width of the different crack formed step-by-step.

3.3 Point load test (TBM)

A full-scale point load test was carried out 30 hours and 30 minutes after segment casting.

The test was carried out according to two loading/unloading cycles with 250 kN increasing steps. The cycles were:

- Cycle I: 0–1580 kN (TBM service load);
- Cycle II: 0–2670 kN (Unblocking thrust/TBM pushing capacity).

The first crack occurred at a load level of 750 kN (for each steel pad) with maximum crack width less than 0.05 mm between green/red steel pads at the thrust side surface. As the load was increased, several cracks occurred, as shown in Figure 8. With reference to two maximum imposed load levels (1580 kN for service load and 2670 kN for unblocking thrust load), the maximum crack width was equal to 0.50 mm and 0.80 mm, respectively.

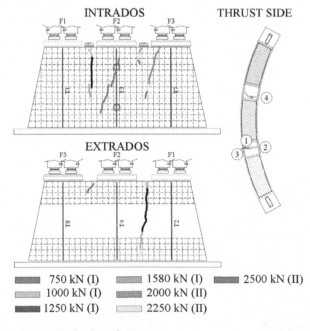

Figure 8. Point load test (TBM): final crack pattern.

Table 3. Crack's width (Cycle I: service load).

Phase	Cycle I (Service)					
	Loading				Unloading	
Load [kN]	750	1000	1250	1580	1000	100
Crack color						

Crack n.	Crack width [mm]					
1	<0.05	0.10	0.30	0.50	0.40	0.15
2	-	0.05	0.20	0.40	0.30	0.15
3	-	0.05	0.30	0.40	0.40	0.15
4	-	-	-	0.05	0.05	Closed
5	-	-	-	-	-	-
6	-	-	-	-	-	-

Table 4. Crack widths (Cycle II: TBM pushing capacity).

Phase	Cycle II (TBM pushing capacity)								
	Loading						Unloading		
Load [kN]	1580	1750	2000	2250	2500	2670	1580	0	0 after 5'
Crack color									

Crack n.	Crack width								
1	0.40	0.45	0.50	0.60	0.70	0.80	0.80	0.25	0.15
2	0.35÷0.40	0.35	0.45	0.45÷0.50	0.50	0.60	0.50	0.20	0.15
3	0.40	0.40	0.50	0.60	0.70	0.80	0.80	0.20	0.15
4	0.05	0.10	0.10÷0.15	0.15	0.20	0.25	0.15	0.05	0.05
5	-	-	0.10	0.15	0.20	0.20	0.10	0.05	0.05
6	-	-	-	-	0.10	0.15	0.10	<0.05	<0.05

At the end of the test, when the unload occurred, the maximum residual crack width was equal to 0.15 mm. Tables 3 and 4 show, for the two loading/unloading cycles, the crack widths measured.

3.4 GFRP reinforcement

The performance of GFRP reinforcement was in line with the values already observed in other experimental situations, both in terms of maximum load achieved (bending test) and in relation to the crack width. With reference to the maximum crack width (cracking load and unloading for bending test and point load test, respectively), these values are within the ranges provided by Regulations and Codes.

4 CONCLUSIONS

In the paper, the results of the experimental full-scale tests on metro tunnel segments consisting of sulphoaluminate concrete reinforced with GFRP reinforcement are presented and discussed. Both bending test and three points load test, simulating the TBM thrust, were carried out. Based on the results obtained with experimental full-scale tests, developed by University of Rome Tor Vergata – Tunnelling Engineering Research Centre (TERC), the following aspects can be remarked upon:

- Based on several full-scale tests carried out over the last few years, the obtained results show, once again, the effectiveness of the proposed reinforcement with GFRP bars for tunnel lining. The new GFRP reinforcement optimization has allowed the resolution of the problem of crack width containment due to the low bond strength of the GFRP rebars in both cases: bending actions under service load and during the excavation under TBM thrust.
- The combination of concrete based on sulphoaluminate and GFRP reinforcement represents a synergistic solution which may lead to a new way forward to problems and challenges, such as: the speeding up of production, the reduction of the number of segments stacked waiting for the required strengths to be achieved, and the achievement of a more eco-sustainable process, avoiding steam curing. Finally, the reduction of stacking space could allow for on-site prefabrication.

REFERENCES

Bullard, J.W., Jennings, H. M., Livingston, R. A., Nonat, A., Scherer, G. W., Schweitzer, G. S., Scrivener, K. L. & Thomas, J. J. 2011. Mechanism of cement hydration. *Cement and concrete research* 41 (12): 1208–1223.

Bertolini, L., Canonico, F., Buzzi, L., Carsana, M. & Bertola, F. 2015. Steel corrosion behaviour in real-size concrete elements prepared with sulpho-aluminate cements. *19° IBAUSIL 16–18 septembre 2015 Weimar.*

Almusallam, T. H. & Al-Salloum, Y. A. 2006. Durability of GFRP Rebars in Concrete Beams under Sustained Loads at Severe Environments. *Journal of Composite Materials* 40(7): 623–637.

Chen, Y., Davalos, J. F. & Kim, H-Y. 2007. Accelerated aging tests for evaluations of durability performance of FRP reinforcing bars for concrete structures. *Composite Structures* 78: 101–111.

Benmokrane, B., Wang, P., Ton-That, T.M., Rahman, H. & Robert, J.F. (2002. Durability of glass fiber reinforced polymer reinforcing bars in concrete environment. *Journal Compos. Constr.* 6 (3): 143–153.

Gooranorimi, O. & Nanni, A. 2017. GFRP reinforcement in concrete after 15 years of service." J. of Composites for Construction, Vol. 21 (5) October 2017.

Micelli, F. & Nanni, A. 2004. Durability of FRP rods for concrete structures. *Construction and Building Materials* 18(7): 491–503.

Nanni, A. 1993. Fiber-Reinforced-Plastic (GFRP) Reinforcement for Concrete Structures: Properties and applications. *Elsevier Science. Developments in Civil Engineering.*

Alsayed, S.H., Al-Salloum, Y.A. & Almusallam, T.H. (2000). "Performance of glass fiber reinforced plastic bars as a reinforcing material for concrete structures." *Composite Part B: Engineering* 31: 555–567.

Spagnuolo, S., Meda, A., Rinaldi, Z. & Nanni, A. 2017. Precast Concrete Tunnel Segments with GFRP Reinforcement. *ASCE J. Compos. Constr.* 21(5): 04017020.

Spagnuolo, S., Meda, A., Rinaldi, Z. & Nanni, A. 2018. Curvilinear GFRP bars for tunnel segments applications. *Composite Part B: Engineering* 141: 137–147.

Spagnuolo, S., Meda, A. & Rinaldi, Z. 2014. Fiber glass reinforcement in tunneling applications. *Proc. The 10th fib international PhD symposium in civil engineering. 21–23 July, Québec, Canada.*

Caratelli, A., Meda, A., Rinaldi, Z., & Spagnuolo, S. 2016. Precast tunnel segments with GFRP reinforcement. *Tunnelling and Underground Space Technology* 60: 10–20.

Caratelli, A., Meda, A., Rinaldi, Z., Spagnuolo, S. & Maddaluno, G. 2017. Optimization of GFRP reinforcement in precast segments for metro tunnel lining. *Composite Structures* 181: 336–346.

Meda, A., Rinaldi, Z., Caratelli, A. & Cignitti, F. (2016). Experimental investigation on precast tunnel segments under TBM thrust action. *Engineering Structures* 119: 174–185.

Gastaldi, D. Canonico, F.Capelli, L.Boccaleri, E. Milanesio, M. Palin, L. Croce, G. Marone, F. Mader, K. & Stampanoni, M. In situ tomographic investigation on the early hydration behaviors of cementing systems, *Constr. Build. Mater.* 29: 284–290.

Irico, S., Gastaldi, D., Canonico, F. & Magnacca G. Investigation of the microstructural evolution of calcium sulfoaluminate cements by thermoporometry, *Cem. Concr. Res.* 53:239–247.

Paul, G., Boccaleri, E., Buzzi, L., Canonico, F. & Gastaldi, D. Friedel's salt formation in sulfoaluminate cements: A combined XRD and 27Al MAS NMR study, *Cem. Concr. Res.* 67: 93–102.

Gastaldi, D., Paul, G., Marchese, L., Irico, S., Boccaleri, E., Mutke, S., Buzzi, L. & Canonico F. Hydration products in sulfoaluminate cements: evaluation of amorphous phases by XRD/solid-state NMR. *Cem. Concr. Res (in press)*

Winnefeld, F. & Lothenbach, B. Hydration of calcium sulfoaluminate cements – Experimental findings and thermodynamic modeling, *Cem. Concr. Res.* 40: 1239–1247.

Aranda, M.A.G. & de La Torre, A.G. 2013 Sulfoaluminate cement, in: F. Pachego-Torgal, S. Jalali, J. Labrincha, V.M. John (Eds.) *Eco-Efficient Concrete, Woodhead Publishing,* Cambridge.

Bernardo, G., Telesca, A. & Valenti G. L. A porosimetric study of calcium sulfoaluminate cement pastes cured at early ages, *Cem. Concr. Res.* 36: 1042–1047.

Canonico, F., Buzzi, L. & Schäffel P. 2012. Durability properties of concrete based on industrial calcium sulfoaluminate cement. *International Congress on the Durability of Concrete, Trondheim, Norway.*

Bertolini, L., Bianchi, M., Buzzi, L., Canonico, F. & Capelli L. 2015. Corrosion behaviour of steel embedded in calcium sulfoaluminate-cement concrete. *International Congress on the Chemistry of cement, Beijing, China.*

Canonico, F., Bertola, F., Bertolini, L., Buzzi, L. & Carsana M. 2015. Steel corrosion behaviour in real-size concrete elements prepared with sulpho-aluminate cements. *Ibausil, Weimar Germany.*

Tunnels and Underground Cities: Engineering and Innovation meet Archaeology,
Architecture and Art, Volume 5: Innovation in underground engineering,
materials and equipment - Part 1 – Peila, Viggiani & Celestino (Eds)
© 2020 Taylor & Francis Group, London, ISBN 978-0-367-46870-5

Fiber reinforced concrete segmental lining for Catania Metro Tunnel

N. Bona
CMC, Ravenna, Italy

A. Meda
Tunnelling Engineering Research Centre, University of Rome "Tor Vergata", Rome, Italy

ABSTRACT: The use of Fiber Reinforced Concrete segments was proposed by CMC for the Stesicoro – Airport part of Catania Metro Tunnel. The tunnel has an internal diameter of 9.80 m with a lining thickness of 320 mm. The proposed solution forecast a light traditional steel cage, mainly along the segment perimeter, and the use of 40 kg/m^3 of 4D steel fiber. A process for obtaining the authorization of the Italian Authorities of the use of Fiber Reinforced Concrete was followed, according to the Italian Regulations. CMC obtained the first authorization issued in Italy for the use of Fiber Reinforce Concrete in new constructions. In order to evidence the effectiveness of the proposed solution, full scale tests were performed at the TERC (Tunnelling Engineering Research Centre) of the University of Rome "Tor Vergata". Furthermore, a quality control process was defined in order to ensure the segment performances during the production.

1 INTRODUCTION

In the last few years the adoption of fiber reinforced concrete (FRC) in precast tunnel segments, has encountered a great interest, as witnessed by theoretical and experimental studies (Plizzari and Tiberti, 2008; Caratelli et al., 2011; Liao et al., 2015), and actual applications (Kasper et al. 2008, De La Fuente et al., 2012, Caratelli et al., 2012).

The use of Fiber Reinforced Concrete (FRC) in precast segments for tunnel lining is growing in the last years and this is becoming one of the principal applications of this material (Figure 1). Typically, two solutions are proposed: a hybrid solution, where fiber can partially substitute the steel cage, and a FRC only solution, where the ordinary reinforcement is totally substituted by the fiber reinforcement.

The main advantages linked to the use of FRC in tunnel segments are related not only to the cost, but also to the possibility of enhancing structural and durability performances.

After the publication of a series of National Codes, the publication of key documents as fib Model Code 2010 (2013), certainly boosted the use of FRC application in tunnels. Already several tunnels have been designed adopting Model Code 2010 as references document. Furthermore, a specific document for precast tunnel segments in fibre-reinforced concrete was recently issued (fib Bulletin 83, 2017).

Since the use of FRC in segmental lining tunnels is an application that can be nowadays considered common, CMC proposed this technology for Metro Catania tunnel. According to the Italian Codes it was necessary to authorization of the Italian Authorities of the use of Fiber Reinforced Concrete.

With this project, CMC obtained the first authorization issued in Italy for the use of Fiber Reinforce Concrete in new constructions.

FRC & RC/FRC precast tunnel segments: case studies over the years

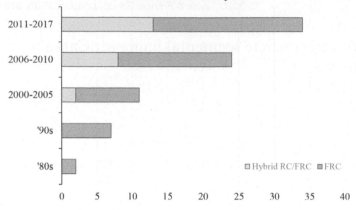

Figure 1.   FRC tunnel application in the recent years (from fib Bulletin 83).

2   CATANIA METRO: STESICORO-AEROPORTO LINE

Stesicoro- Aeroporto line is part of Metro Catania project. The line is under construction and it will connect Catania Airport with the part of the line already in service. CMC was in charge for the construction of 2177 m of the line between the station of Stesicoro and Palermo.

Figure 2.   Catania Metro.

A single tunnel is forecast for the line, with an internal diameter of 9.80 m with a lining thickness of 320 mm (Figure 3). The tunnel will be excavated with a TBM-EPB for a total length of 2315 m (since some provisional parts are present).

The precast segmental lining is characterized by 6 segments plus 1 key segment (6+1) configuration.

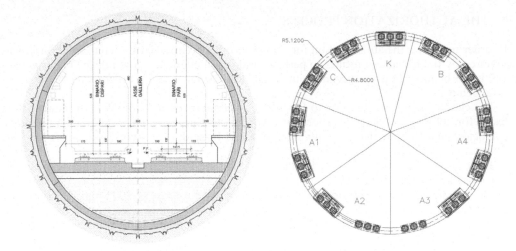

Figure 3. Catania Metro: tunnel lining.

3 FIBER REINFORCED CONCRETE SOLUTION

The design of the FRC segments was made adopting fib Model Code 2010 are reference document. It was decided to adopt an hybrid solution, with fiber reinforcement coupled with a light perimeter reinforcement (Figure 4).

The FRC performance was defined by classified the material as 4c according to fib Model Code 2010. As a consequence, the residual characteristic tensile strength were $f_{R1k} > 4.0$ MPa and $f_{R3k} > 3.6$ MPa measured with EN 14651.

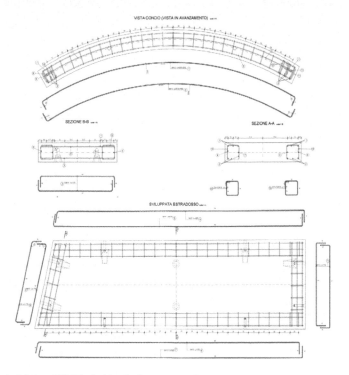

Figure 4. Catania Metro: FRC hybrid solution segment.

4 THE AUTHORIZATION PROCESS

In order to obtain the authorization for using the Fiber Reinforced Concrete solution, it was required by the Italian authorities to perform and initial characterization of the material with EN 14651 bending tests on specimens and a full-scale bending test in order to verify the structural behavior.

The characterization according to EN 14651 is summarized in Table 1. 12 specimens were tested for defining the characteristic value according to Eurocode 0, considering an unknown coefficient of variation. These results were obtained by adding 40 kg/m^3 of Bekaert 4D steel fiber to the

The bending tests were performed with the loading set-up illustrated in Figure 5, by adopting a 4000kN hydraulic jacket.

The segments were placed on cylindrical support with a span of 3000 mm and the load, applied at the midspan, was transversally distributed be adopting a steel beam as shown in Figure 5. The measure devices, consisting in four potentiometer wires transducers and two LVDTs, are shown in Figure 6.

Figure 7 shows the results of the bending test in terms of load versus displacement diagram.

Figure 8 shows the crack pattern at different loading steps. Figure 9 shows the opening of the principal crack.

In order to compare the experimental behavior with a traditional reinforced concrete segment, it was request to CMC to test a segment produced for another part of the line where the FRC was not forecast (Nesima-Misterbianco line, already excavated). The segment has the same geometry of the FRC segment but it is reinforced with only a steel cage (Figure 10) was designed for higher load level. For this reason in the comparison of Figure 11, the ultimate design load for the two cases is reported.

The comparison between the two segments evidenced an interesting result: referring to a crack opening value of 0.2 mm, this limit was reached at a load level of 188 kN for the FRC segment and at a load level of 150 kN for the traditional reinforced concrete segment. The crack opening was significantly smaller in the FRC segment respect to the traditional RC one.

After the test campaign CMC was authorized by the Italian Authorities to adopt FRC in the project.

Table 1. Material characterization.

	fl	fR1	fR3
	4.6	5.2	6.2
	5.5	7.6	7.8
	6.3	6.5	6.4
	6.5	7.9	7.4
	5.9	7.8	9.3
	5.0	5.8	8.2
	6.1	6.7	8.3
	5.3	5.2	6.5
	6.0	6.5	8.6
	5.1	5.6	6.7
	5.7	7.5	8.2
	5.7	6.9	8.0
Average	5.6	6.6	7.6
Standard dev	0.56	0.98	0.99
Coef. var.	0.10	0.15	0.13
Kn	1.87	1.87	1.87
	flk	fR1k	fR3k
	4.53	4.67	5.70

Figure 5.   Bending test set-up.

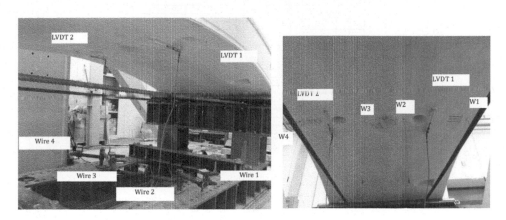

Figure 6.   Bending test instrumentation.

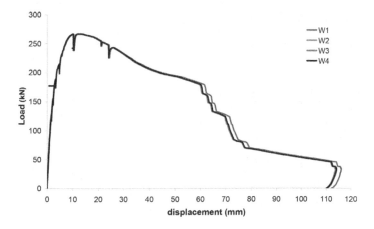

Figure 7.   Bending test results.

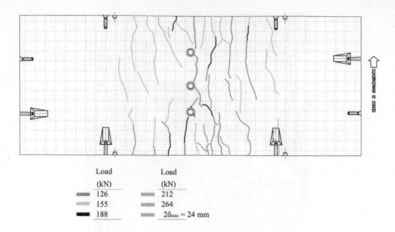

Load (kN)		Load (kN)	
	126		212
	155		264
	188		$2\delta_{max} = 24$ mm

Figure 8.   Bending test crack pattern.

Crack	Load level [kN]				
	126	155	188	212	264
1	0.1	0.05	0.20	0.3	0.90-1.00
2	0.15	0.05	0.05	0.10	0.40
3	<0.05	0.05	0.10	0.20	0.20
4	-	0.05	0.05	0.05	0.15-0.20
5	-	<0.05	<0.05	0.05	0.05
6	-	<0.05	0.05	0.10	0.10
7	-	<0.05	<0.05	<0.05	0.05
8	-	<0.05	0.05-0.10	0.10-0.15	0.90-1.00
9	-	<0.05	<0.05	<0.05	<0.05
10	-	<0.05	<0.05	<0.05	0.05
11	-	-	0.05	0.10	0.20
12	-	-	<0.05	0.05	0.10
13	-	-	-	<0.05	0.05
14	-	-	-	0.05	0.15
15	-	-	-	0.10	0.10
16	-	-	-	<0.05	<0.05

Figure 9.   crack opening.

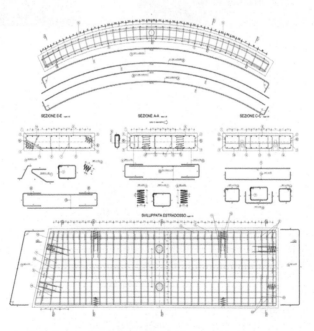

Figure 10.   Traditional reinforcement in Nesima-Misterbianco segment.

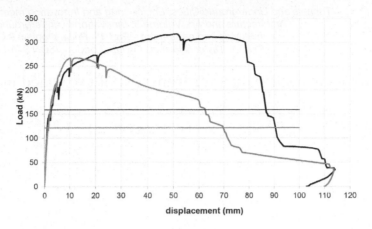

Figure 11. Comparison between traditional and FRC solution.

This authorization was the first one issued in Italy for the use of FRC

5 CONCLUSIONS

CMC proposed the use of FRC segment for the Catania Metro Stesicoro-Aeroporto line. The tunnel lining was designed according to fib Model Code 2010.

In order to obtain the authorization, according to Italian Code a process was followed, by performing full scale tests.

The tests highlighted the good performances of the FRC solution and the authorization was issued.

REFERENCES

Caratelli, A., Meda, A., Rinaldi, Z., Romualdi, P. 2011. Structural behaviour of precast tunnel segments in fiber reinforced concrete. *Tunnelling and Underground Space Technology* 26: 284–291.

Caratelli, A., Meda, A., Rinaldi, Z. 2012. Design according to MC2010 of a fibre-reinforced concrete tunnel in Monte Lirio, Panama. *Structural Concrete* 13(3): 166–173.

Cosenza, E., Manfredi, G., and Realfonzo, R. 1997. Behavior and Modeling of Bond of FRP Rebars to Concrete. *Journal of Composites for Construction* 1(2): 40–51.

De la Fuente, A, Pujadas, P, Blanco, A, Aguado, A. 2012 Experiences in Barcelona with the use of fibres in segmental linings. *Tunnelling and Underground Space Technology* 27(1): 60–71.

Di Carlo, F., Meda, A., Rinaldi, Z. 2016. Design procedure of precast fiber reinforced segments for tunnel lining construction. *Structural Concrete* 17(5): 747–759.

EN 14651. 2005. Test method for metallic fibre concrete – Measuring the flexural tensile strength.

fib Bulletin No. 83. 2017. Precast tunnel segment in fiber-reinforced concrete.

fib Model Code for Concrete Structures 2010. 2013. Ernst & Sohn.

Liao, L., De La Fuente A, Cavalaro S, Aguado A. 2015. Design of FRC tunnel segments considering the ductility requirements of the Model Code 2010. *Tunnelling and Underground Space Technology* 47: 200–210.

Meda, A., Rinaldi, Z., Caratelli, A., Cignitti, F. 2016. Experimental investigation on precast tunnel segments under TBM thrust action. *Engineering Structures* 119(15): 174–185.

Plizzari, GA, Tiberti, G. 2008. Steel fibers as reinforcement for precast tunnel segments. *Tunnelling and Underground Space Technology* 21: 3–4.

*Tunnels and Underground Cities: Engineering and Innovation meet Archaeology,
Architecture and Art, Volume 5: Innovation in underground engineering,
materials and equipment - Part 1 – Peila, Viggiani & Celestino (Eds)*
© 2020 Taylor & Francis Group, London, ISBN 978-0-367-46870-5

Gronda di Genova slurryduct for transport of excavated material

S. Bozano Gandolfi
Scamac S.r.l, Genoa, Italy

M. Bringiotti
GeoTunnel S.r.l., Genoa, Italy

ABSTRACT: SPEA Engineering, has prepared the Definitive Project for the adaptation of the Genoa motorway junction for the A7/A10/A12 motorways, also named "La Gronda di Genova". The construction of the Gronda involves, among the various activities planned, the treatment of approximately 6.8 million cubic meters of potentially asbestos-containing debris deriving from the excavation work carried out mostly by two Tunnel Boring Machines (TBM). The excavated material will be temporarily stored, analyzed and, if having low asbestos content, reutilized for the enlargement of the airport landing strip.

This material will be transferred in the form of a slurry through a special pipeline having a length of 9 km, For reducing wear and pump power consumption the transporting media will be a bentonite based fluid.

1 INTRODUCTION

In September 2017, the Italian Ministry for Infrastructures and Transport approved the project for the construction of "Gronda di Genova"(Figure 1). The final project was approved and the works recognized as of public interest. Works are expected to start in October 2018 and last for 10 years with an investment of nearly euro 4.3 billion. The project is the result of a long-term design and discussion process that had involved also the Regional and the Genoa City Authorities.

The new infrastructure will connect to the existing junctions just outside the urban area (Genova Est, Genova Ovest and Bolzaneto) and link up with the Motorway A 26 in Voltri and the Motorway A 10 in Vesima. This infrastructure will not only relieve traffic congestion on the Motorway A10,but also improve Genoa's potential and logistic competitiveness and it represents an opportunity for growth for the entire North-West region.

Here below are the figures that provide an idea of the complexity of the project:

– 54 km in tunnel, nearly 90 % of the entire infrastructure;
– 23 tunnels;

Figure 1.   General view of "Gronda" project.

- 13 new viaducts;
- 11 existing viaducts to be extended;
- 10,000,000 m^3 of excavation material.

Particular attention was paid to environmental sustainability, due to the orographic complexity, the high urbanization of the area and the presence of asbestos in the rock formations that will be subject to excavation.

The excavation material will be treated in accordance with the asbestos contained and the geotechnical characteristics. Most of the excavation material will be used for:

- The construction of an infrastructure on the seaside that will allow developing a 50 m wide backfilling to ensure the adjustment of the shoulder of the runway strip of the Genoa airport.
- filling of the motorway tunnel invert.

The material which is not suitable for the above listed applications will be destined to a waste disposal area.

2 MATERIAL HANDLING

The main part of the material, nearly 8,000,000 m^3, will be excavated by means of TBMs, the remaining by means of conventional method such as drill and blast and hydraulic breaker. Two TBMs, EPB type, will be used and their relevant excavation will start from Bolzaneto site (named C1) towards East up to Vesima. The excavation work to be carried out by conventional method will be for the tunnels on the west side of Bolzaneto towards Genoa East and West and some minor sites which are located close to Voltri area. What above means that almost all the material excavated has to pass through Bolzaneto site which becomes the natural heart of the transport system. All this material has to be transported to its final destination and many difficulties have to be overcome for what concerns:

- Asbestos. The risk related to the occurrence of natural asbestos is a major issue affecting the highway plan, which is expected to cross potentially asbestos-bearing rock units. The asbestos is supposed to cross the path of the Tunnels located in Voltri and Vesima areas which will be excavated mostly by means of the two TBMs and partially by conventional method. The asbestos presence is a big problem for what concern environmental aspects and health aspects of workers who might be in contact with this material. Furthermore, considering that the main working sites are located in the metropolitan area, it has been necessary to scout the optimal solution to carry out basic and simple operations such as crushing, loading, unloading and, of course, define the proper and suitable transport mean which has to be environmentally safe and does not interfere with an already congested local road network. Furthermore, the excavated material has to be characterized in order to evaluate the asbestos content and based on this value to dispose it to the final destination.
- Lack of space. The area available to stock, analyze and prepare the excavated material before transport to the final destination site, is very limited. This means that all the activities to be carried out on the material excavated, including transport to the final destination, must rely on suitable systems having a very high availability since there is no possibility to have a bigger buffering area.
- Transport. As soon as the material excavated has been characterized, it has to be transported to the final destination site. The biggest portion of the material will be used to create a safe area aside the runway strip of the Genoa airport which is located on the seaside. Between the working sites and the airport area there is Genoa city, which, in case a road transport solution would be adopted, has to be crossed through. Genoa has a really congested traffic system which would be incapable to absorb the extra traffic as the road transport for the material excavated would generate. Furthermore it has to be considered that the airport strip has to be enlarged on the sea side and this means that whatever

transport system would be adopted this cannot interfere with the limitation of the airport activity.

– Between Bolzaneto and Genoa Airport, the only free area to be used for transporting the material, without using the existing road network, is the river bed of Polcevera. Furthermore the transport system, in accordance to the agreement taken with the local authorities, have to be environmentally safe in order to avoid any type of pollution due to asbestos fibers and material spillage. Polcevera river bed is crossing all the urban area and furthermore has many bridges and height limitations.

Considering what above, among the possible transport solutions, the transport system chosen is the following:

a) Convey/Transport the biggest part of the excavated material to Bolzaneto;
b) Place in Bolzaneto site a plant properly designed in order to fit into the small space available and suitable to fulfill to the following targets:
   – capability to characterize the excavated material;
   – capability to host all the process systems necessary to treat the material excavated in accordance to its final use;
   – Having a stocking and reloading capacity suitable to allow the characterization of the material and a minimum buffering to absorb the material production in case some problems in the transport system chosen might occur.
c) Use a slurry system utilizing sea water as transport medium from Bolzaneto site to Airport site. By means of this solution the following targets will be achieved:
   – No contact of the material with the environment;
   – No impact on the road network of the city;
   – Easy adaptable to the area available.

3 DESCRIPTION OF THE SYSTEM

The system, as a base, will operate as per simplified Flowsheet indicated in figure 2. The quantities indicated have been based on the theoretical maximum advance rate of the two TBMs, 15 m/day each. Considering a tunnel section of 168 m^2 the total volume produced per day per TBM will be roughly 2550 m^3/day, to this has to be added also 1300 m^3 produced from the tunnels excavated by conventional methods (drill and blast or hydraulic breaker). Based on what above the total capacity required will be 6400 m^3/per day which corresponds to an average hourly production of 270 m^3/h.

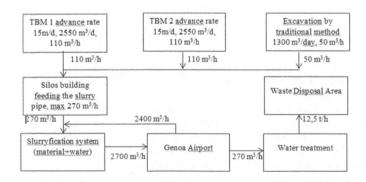

Figure 2. Slurry duct system simplified mass flow.

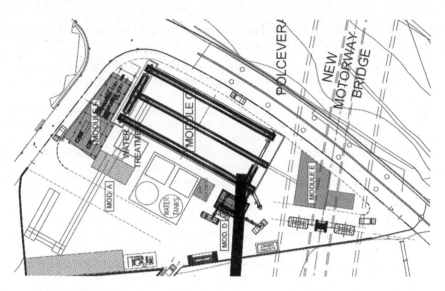

Figure 3. Bolzaneto plant area.

The total area available for the plant of Bolzaneto (Figure 3) is only 15,000 m^2 and furthermore there is also a height limitation since the area will remain under the new Gronda bridge which will cross Polcevera bridge

As a base the Bolzaneto working site will be composed of the following main 5 modules:

– Module A: Crushing system for the reduction to desired size of the material excavated in traditional method and transported to the site by trucks; sampling section for material characterization;
– Module C: Made of 24 silos. Each silo will have a capacity of 900 m^3. The stored material, in accordance to the analysis carried out, will be conveyed to the inertisation for filling the tunnel invert, or bagging to be sent to a waste disposal area, or sent to slurry section to be pumped down to the Airport site;
– Module D: Inertisation of the material having a content of asbestos higher than 1gr/kg will be threated but which can be reused for the Tunnel consolidation work;
– Module E: Slurry preparation. The material is mixed with sea water (asbestos content lower than 1gr/kg) in order to be pumped down to the airport area;
– Module F: The material is bagged, sealed and sent to dedicated waste disposal areas.

4 SLURRY TRANSPORT SYSTEM

4.1 *Slurry transport system by sea water*

A deep analysis has been dedicated on the slurry transport system, which, as a base, is the critical part of the system. The solid transport via pipeline is rather common in the mining industry, many processes in mineral treatment use water (milling, flotation, screening, magnetic separation etc) and water allows to move the material through the various steps of the process in an easy and environmentally safe way. Furthermore pumping of solid is commonly used for handling the tailings of mineral processing as for transporting the mineral ore concentrated to the final destination. By this transport system, many million tons of material have been conveyed for distances which are by far higher than the one required to the Gronda Project. On the other side it has to be considered that the Gronda Project has some peculiarities that standard applications do not have. The main differences are the following:

- The material to be transported has a high geological variability, as it will be explained in the following section, while in a mineral processing projects the system is sized for almost a constant type of feed;
- The top size dimension in a mineral processing slurry transport is typically not more than 5-6 mm while in our case the aim is to transport a slurry having a dimension equal to the top size produced by the two TBMs which is 75 mm;
- The material to be transported is really poor of fines since the fraction below 75 microns is not separated from the bentonite separation system placed on the TBM's slurry system; fine particles act like a medium for coarser material increasing the ability to keep the material in suspension.

Solid transport via slurry pipe are used in other industrial processes, the main ones are oil sand and sea dredging. In the first case the material to be transported has a maximum dimension up to 100-120 mm, but the application, more than a transport system, is a process since the pumping is used for dissolving the crust of oil sand in order to minimize its dimension and to facilitate the separation process between oil and sand and the slurry characteristics got modified during the pumping. The solid pumping in the dredging field, is used also for long conveyance, the biggest difference with our system is that the water to be added to the slurry is practically endless and this has, as a consequence, that in case of increasing of the head loss it is sufficient to lift the pump and dilute the slurry to flush the pipe without the constrain of pumping a certain amount of solid continuously.

The study has been commissioned to one of the most known engineering company active in this sector: a first study has been carried out considering a solid fraction having a top size of 8mm, but the solution has been immediately discarded since there was no space available in Bolzaneto for placing the necessary crushing and screening plant. A second study has been analyzed considering a top size of 25 mm. and the results achieved are as per table 1.

As an explanation of the results indicated in table 1 what follows has to be considered

Critical velocity: in the preliminary steady state slurry flow modelling report, P&C calculated a critical stationary speed within a 450 NB pipe of 3.7 m/s. Due to the very coarseness of the slurry to be pumped, P&C therefore recommend operating at 1.5 times this stationary speed. The increase in minimum design stream flow to prevent particle settling has a big impact on all parts of the pipeline design, i.e.:

- Increases/decreases stream flow rate vs solids concentration vs solids mass flow rate;
- Significantly increases total dynamic head over the length of the pipeline;

Table 1. Solid pumping data sheet for max size of 25 mm.

Parameter	Value
Particle Top Size (D_{100})	25 mm
Solids Flow Rate	270 m^3/h
Differential height	+55 m
Pipe length	11000 m
Total Dynamic Head	1119 m
Pipe Specification	500NB MSRL
Maximum Pipeline Pressure	21.9 bar(g)
Pump type	20/18
Number of Pumps	22
Gland seal fresh water required	43 m^3/h
Total Installed Power	20,9 MW

- Limits increasing pipe size as larger pipe diameters increases required stream speed to prevent settling;
- Increases risk of pipeline failure or pumps from excessive abrasive wear.

Pump Impeller tip speed: as an industrial standard, it is recommended that pump impeller tip speeds for metal impellers to be limited to approximately 28 m/s. The reason for this limit is to prevent excessive wear on the impeller by the abrasiveness of slurries. The larger particle sizes in this application further motivates to lower speeds to prevent excessive wear and therefore maintenance costs. By limiting impeller tip speed to approximately 28 m/s, the limit of hydraulic head that each pump stage is capable of adding is limited given a particular flow rate. For a 20/18 pumping 3450 m3/h this limited each stage to 50 m of hydraulic head per stage.

Gland seal water on slurry pumps is preferred over mechanical seals or expellers, this is pump manufacturer recommendation. The reasons are the following:

- An expeller cannot be used given the series pumping arrangement required at each booster station and the large discharge head at each final stage.
- Mechanical seals may be specified if the seal manufacturer is willing to guarantee their life. They may be prohibitively expensive given the size of the 20/18 pump and the use of sea water as a transport medium would also necessitate special materials to limit corrosion from high chloride concentration. Mechanical seals require continuous maintenance to ensure the fine particles do not ingress and damage the sealing arrangement. Maintaining the 2x20 mechanical seals on these seals may be a complex task requiring potentially long downtime.
- Provided clean, potable water, used as gland seal water, can be supplied via gland seal pumps on site at the correct pressure and flow rate. Gland seal water also provides a suitable barrier between the slurry and shaft seal.

The results obtained are by far worse than it was foreseen in the preliminary study carried out. The reasons why this solution has been discarded are listed below:

- too high power demand for the local energy network system;
- too many boosters pumps to be installed on the river bed and impossibility to find areas easily accessible to allow for maintenance of the equipment;
- high consumption of fresh water for the pump gland seal;
- very high wear on pumps due to the slurry speed, 5,5 m/s, necessary to avoid sedimentation in the pipeline;
- necessity to install a crushing and screening system suitable to reduce the top dimension of the material fed to the plant down to 25 mm. Problem connected to the space available, to the geological variety of the material to be fed, to the impossibility to have a second crushing line in standby position;
- value of investment and running cost higher than what foreseen;
- last but not least the results obtained have to be considered theoretical since there is not any simulation program suitable to evaluate the data of a slurry duct pumping material having dimension higher than 10 mm and a pilot test is necessary to validate the data.

4.2 *Slurry transport system by bentonite*

Alternative solutions to sea water slurry system have been scouted and it has been decided to adopt the same system used by the TBMs for transporting the excavated material from the working face to the tunnel adit. In this project the two TBMs are pumping 110 m^3/h of solid at a distance of roughly 16 km using as transporting media a mix of water and bentonite. To apply this technology to our case is not that easy, first of all for the capacity required which is by far higher than what done up to this moment using this type of transport system, secondly for a series of different aspects which are completely new such as the slurryfication procedure and the separation system to be used. Another point considered is that there is not a real well proven calculation to validate the sizing of the system. The slurry transport by using bentonite as media has been used for decades in well drilling and later in other fields such as diaphragm walls, piling and finally in TBMs. The bentonite, mixed with water, creates a viscous mud which allows to reach the following targets: maintains hydrostatic pressure against the borehole/excavation wall keeping it from collapsing; seals the excavation walls; cools and lubricate the drilling bits and, finally, the muds helps remove cuttings from the bore-hole/excavation. This last point is the characteristic which has the bigger impact on our application. The following example will better explain how the process works: if we consider to have a tube filled with water and drop inside a stone in the tube, the stone will have a settling velocity of a certain value. If the tube is filled with oil the settling speed of the same stone will be lower even though the oil density is lower than water. What above is connected to another parameter of the fluid which is viscosity.

This characteristic, if applied to our case, increases the dragging force of the medium which means that, using a bentonite mixture, it is possible to apply a lower speed than what is necessary by using water only and the consequences are relevant for what concern power consumption and, of course wear.

Obviously there are some modifications to be applied to the previous project, the material transported by means of sea water as a medium can be discharged directly in the basin to be filled, if bentonite is used a separation plant has to be installed at the discharge end of the slurryduct to separate the bentonite from the excavated material. To this have to be added a series of equipment such as fresh bentonite preparation plant, tanks for the used and fresh bentonite and finally tank for empting the slurryduct pipeline. The area neccssary for all this has to be created close to the airport strip area taking into account all the limitations in terms of heights and activities imposed from Airport Authorities. Many simulation have been carried out and the results obtained have been in general positive. The system has been sized assuming a total solid transportation of 350 m3/h which is 30% higher than the value indicated for the system using sea water as medium. The pipeline length has been reduced since it was not possible any more to pump the material up to farthest area to be filled, but to the separation plant which has been foreseen almost in the middle of the filling area.

The data obtained are per table 2.

The advantages connected to this solution are various:

– it is possible to transport material having a top size of 75 mm. By this system it is not necessary anymore to install a crushing and screening plant in Bolzaneto saving space and, even more important, removing a possible cause of low dependability since a single crushing line was foreseen;
– significant increase of solid percentage transported in the slurry;

Table 2. Slurry pumping data sheet, solid max size of 75 mm, medium bentonite.

Parameter	Value
Particle Top Size (D100)	75 mm
Solids Flow Rate	350 m^3/h
Slurry flow	2170 m^3/h
Solid concentration in volume	16,2%
Bentonite medium density	1,1 t/h
Differential height	+55 m
Pipe length	9100 m
Total Head	358 m
Pipe Diameter	450 mm
Pump type	20/18
Number of Pumps	5
Total Installed Power	5,5 MW

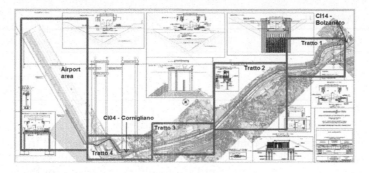

Figure 4.   Slurry duct lay out.

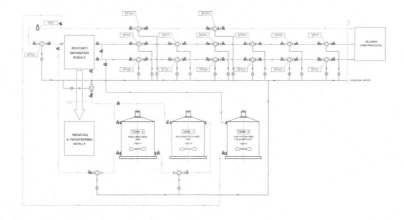

Figure 5.   Piping and Instrumentation Diagram of slurry duct.

– production suitable to handle the maximum production of the two TBMs;
– well proven system utilized for material transports in hundreds of application and able to handle solids having different geological characteristics;
– by the separation system all the material having dimension lower than 75 μm is removed. In general the asbestos fibers are mostly included in this material fraction;
– less fines amount increase the settling speed of the material in the landfill operation;
– possibility to separate the transport phase from the land fill operations creating a buffer in the between.

In order to grant a high dependability to the transport system two complete slurryducts, will be built. In the figure 4 is represented the slurryduct layout and in figure 5 a simplified P&ID.

5   SLURRY SEPARATION SYSTEM AND LAND FILL

One of the main topics to be tackled with has been the separation plant necessary to separate the bentonite from the excavated material and the system to be used for distributing the excavated material for the land fill.

As previously mentioned the land fill has to be done aside of Genoa Airport and it is indicated in green in figure 4. By means of concrete elements to be sunk and filled with material, will be created three artificial basins having dimensions as per table 3.

Table 3. Artificial basin dimension.

Basin	Length (m)	Width (m)	Depth (m)	Asbestos content
A1	805	150	10	> 1gr/m^3
A2	1040	150	10	< 1gr/m^3
A3	1475	150	10	< 1gr/m^3

The basins will be completely compartmented one from the other and from the sea as well. Between basins 1 and 2, will be installed the "Receiving and Separation plant". To the plant will be transported all the material to be utilized for filling the basins for the land reclamation activity. The material will be transported mainly by the slurry duct, but also by means of conveyor and rubber tire vehicles. At this reference it has to be considered that the material, excavated by conventional method (drill and blast) in small working sites, will be transported mainly by truck to the Airport area (Figure 6).

The plant (Figure 7) is composed of two main modules: a) separation module; b) receiving and transferring module. The Separation Module treats the slurry to separate the excavated material, biggest than 75 microns, from the bentonite medium which is recovered and pumped back to Bolzaneto site. The plant is composed of screens, pumps and cyclones working in three parallel production lines. The second module is located aside of the first one and it is an

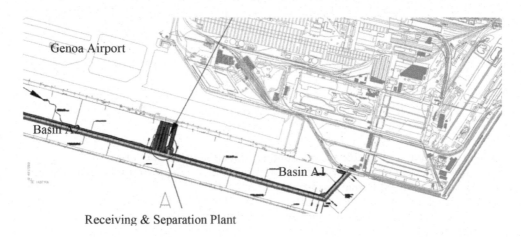

Figure 6. Basin Plant closed to Genoa Airport.

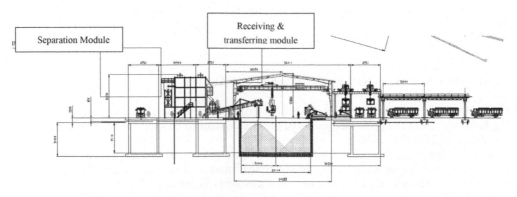

Figure 7. Detailed view of the treatment plant.

1762

industrial building containing two basins in which the material, separated by means of the separation module, is discharged. In the basins is discharged also the material transported by means of the rubber tire vehicles. The two basins have a volume of approximately 7500 m³ and each one is serviced from a dedicated double girder overhead crane complete of submergible pump.

The material transported by rubber tired vehicles is loaded inside closed containers. The vehicles, each one transporting two containers having a capacity of 8 m³, are unloaded by means of automatic cranes and, via horizontal movers, the containers are discharged inside the basins. A double door system grant the total insulation of the building from the environment.

The first basin (A) is dedicated to the material containing more than 1gr/m³of asbestos, the second one (B) for the material lower than 1 gr/m³. In order to avoid any material contamination, the building is divided by a separation wall. The complete building, as well the separation module, are kept slightly depressurized to avoid that asbestos fibers might be liberated in the environment.

The pumps, moved by the overhead crane, convey the material from the basins inside the building to the land fill basins. The water overflowing from the landfill basins, either keeps constant the water level inside the basins inside the "Receiving and separation plant", either feed a water treatment plant. The material discharged is then distributed in the basin by means of dredgers and delivery self-propelled units.

By this system it has been created a good buffering between the transport and the land fill activities so that it is possible to grant a high availability of the whole system.

In the following figure (Figure 8) is represented a simplified P&ID of the system.

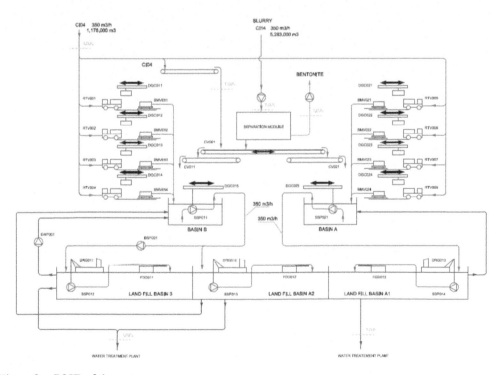

Figure 8.   P&ID of the system.

6 CONCLUSIONS

The system studied from Spea is a miscellaneous of different know-how which are used in really different working fields such as mineral processing, dredging and maritime works and solid transport via pipeline. To this it has to be added that the geological variability of the material to be transported generates really big difficulties in sizing the proper transport and stocking system.

It has to be considered that many of the processes, which will be applied, are by far bigger in size and performances than what it has been done up to now. Nevertheless the innovation of the project lies in the use of these technologies all together to face all the limits connected to the application.

*Tunnels and Underground Cities: Engineering and Innovation meet Archaeology,
Architecture and Art, Volume 5: Innovation in underground engineering,
materials and equipment - Part 1 – Peila, Viggiani & Celestino (Eds)
© 2020 Taylor & Francis Group, London, ISBN 978-0-367-46870-5*

The relevance of gaskets for the segmental design

C. Braun, G. Mauchamp & O. Pasemann
Datwyler Sealing Technologies, Waltershausen, Germany

ABSTRACT: The paper will examine the experiences in the TBM tunneling and give some recommendations for better interaction of compression, precision and tightness of a tunnel. In order to build safer tunnels, the right product has to be selected, which has a huge influence on tightness and durability. The current trend in tunneling moves towards thinner segments to reduce the total costs and to increase the TBM speed. To follow this trend, we need more knowledge to be able to take right decision regarding the specified gaps, offsets and position of the gaskets on the corners of the segment. The next generation of gaskets is already waiting and the influence of the mathematical analysis as the Finite Elements Analysis (FEA) will improve the selection of the right product.

1 INTRODUCTION

The year 2019 will be marked with a special celebration, the 50th anniversary of the first TBM tunnel with EPDM gaskets, the Elbtunnel in Hamburg, Germany (The company Phoenix delivered the first EPDM gaskets in 1969. Phoenix was merged in 2012 with Datwyler group from Switzerland).

Figure 1. Elbtunnel in Hamburg, Germany [source: NDR].

In the meantime, more than 2,500 tunnels have been built with the mechanical technology of the TBMs and sealed by EPDM gaskets worldwide. From Europe to America, from Asia and Africa to Australia, you will find this technology for safely and speedy driven tunnels. In order to ensure a tight connection between the segments, to generate safety contact joints, EPDM gaskets are a suitable product to be used. There are many solutions in the market where traditional and new companies are successfully competing in development of the best solution with the highest sealing properties. Many companies want to improve the interaction of the concrete segments and the rubber seals and to accelerate the TBM tunneling industry.

This paper will examine the latest experiences and the influences of the components involved in sealing technology of the tunneling industry.

2 EXPERIENCES IN DEEPER TUNNELS

For the next generation of tunnels, there has been many inquiries for water tightness tests (WT tests) with more than 30 bar and test procedures longer than 7 days. This poses new challenges to gasket and testing equipment manufacturers. As an example for higher water pressure, the Oslo Follo Line project in Norway may be mentioned with 33 bar as test water pressure.

Figure 2.   Water tightness test rig [source: Datwyler].

2.1   Water tightness of the tunnels

The following conditions shall be fulfilled when speaking about the tightness of a tunnel: an interaction of compression of the gaskets according to the placing, a correct position surrounding the concrete segments, the forces of the connectors and the resistance of the rubber seals.

2.2   Compression of the gaskets

The ability of compression of the gaskets depends on the inner structure of the gaskets. The first generation was developed for gluing at the surface of the concrete segments by brush or by spraying gun (Figure 3). With the second generation of anchored gaskets (Figure 4), the new structure is more filigree with the sealing lips at the sides and the rubber feet, which can have an influence of the risk of concrete cracks. There are different solutions in the market, interesting one could be the one with the groove with an additional relieve area (Figure 5) to avoid overfilling the groove area.

2.3   Connection of the segments

In the beginning, the screw system was popular for connection of the concrete segments. In most of the projects, the screws were reused for the ring building. During the last years, the kind of connectors has changed: the dowel systems are being used more and more frequently.

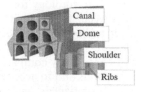

Figure 3.   Glued-on gasket [source: Datwyler].

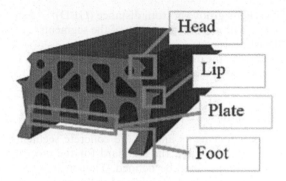

Figure 4. Anchored gasket [source: Datwyler].

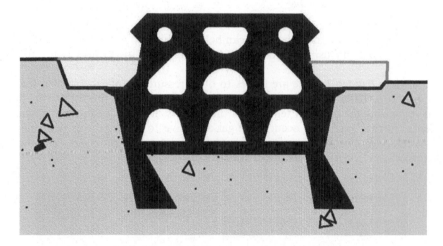

Figure 5. Relieve area (marked in yellow) of anchored gaskets [source: Datwyler].

The dowels help to install the rings in a more precise way. In most of the tunnel projects, the specification is limiting the tolerances of the gaps and the offsets for the segmental lining. The tolerances can be much smaller by using the dowel systems. The tunneling process can be executed easier and speedier too.

To achieve high water pressure, the right groove bottom distance is required. To reach that, we need well-fitted segments, which can be done by dowels. In projects with screws, it is easier to close the gap by compression of the gaskets, but in general, some higher offsets can be noticed.

2.4 Placing of the gaskets

According to the experiences in critical tunnel projects, it is recommended to pay more attention to the position of the gasket at the corners of the segments. For projects with high loads, the position of the gaskets shall be checked very carefully. Proper geometrical, physical and interactive parameter shall be applied to evaluate the right gasket for the project.

The following parameters are recommended:

• Distance from the axis of the gasket to the extrados/intrados surface:

At least 150% of the gasket's groove bottom distance (GBD)
- Additional spalling tests with the real geometry of the segments
- Relieve areas at the gasket grooves
- Packers in combination with the dowel system
- Ball joints in combination with the crew system

2.5 Precision as standard

The precision of the tender documents and of the concrete segments is the key element of water tightness for (deep) tunnels. First, the standard for the precision of concrete elements and for the products on the joints shall be defined. That includes the connector system, the packers, the moulds, the shrinking and creeping of the segments and the tolerances for the gasket system.

2.6 Forces during the ring building

Different forces are working in a segment. The erectors of the TBM press the segments into the right position. In some cases, the rams can help for this step and bring an additional force on the contact point. By using the screw systems, some additional forces for fixation will appear.

In case of dowel systems, there are forces to overpressure the plastic teeth of the dowel in the right position. Later on, the grouting process and the water pressure from the underground is increasing the different tensions in the gaskets. In case of settlements, earthquakes or accidents, further loads can occur. The gaskets have to resist all this deformations and forces. It is relevant to obtain the information and understand the processes and interaction of all these forces.

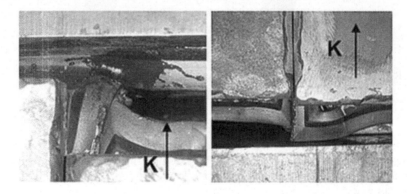

Figure 6.   Situation at the corner [source: Datwyler].

2.7 Loss of flexibility

Most of the gaskets are made of EPDM rubber quality. The material offers similar properties as the wheels of the cars: durability over long product lifetime, flexibility for any weather conditions and on each road surface, resistances against chemical exposures, flexibility and a "memory behavior" of the original structure in case of displacements.

Therefore, the safety factor for the long-term relaxation in all international standards is very high: 2.22 in Germany and 2.85 in UK. All products in the market should fulfill that high requirement. This should be an important part of calculation for the safety of the tunnel. The French tunneling association, AFTES, recommends the calculation for water tightness by this individual values according to the selected products.

To construct deeper tunnel in the future, more knowledge of all this interactions and influences will be needed. The gasket manufacturers have to start the processes of investigation of all this properties, influences and the common work in the joints of the segments.

3 GASKET SYSTEMS FOR SAFER TUNNELS

To build safer tunnels, the right gasket system in combination with the concrete segments and all the installation parts shall be selected. All details of a segment such as

- dimensions and groove design
- specification for gap and offset
- fixation of the segments
- properties of the concrete

have a huge influence on the tightness and the durability.

3.1 *Groove bottom distance (GBD)*

The width of a gasket profile depends on the size of the tunnel, as segment thickness is a function of tunnel diameter. The following gasket profile widths are recommend with regard to the tunnel diameter as the current industry practice:

- 1 m < Tunnel diameter < 4 m gasket width of 20 mm
- 4 m < Tunnel diameter < 7 m gasket width of 26 mm
- 7 m < Tunnel diameter < 11 m gasket width of 33 mm
- >11 m Tunnel diameter gasket width of 44 mm

3.2 *Gap and offset*

The gasket size is also related to the erection tolerances, which in turn depends on the diameter of the tunnel (segment size) and the connection system. The connection system with bolts usually allows offsets up to 15 mm, which can be reduced to 5 mm by high precision.

Dowels, however, are the connection systems with reduced tolerances. Therefore, reducing the tolerances has a considerable effect on the gasket system. First, gasket needs to cover a smaller offset range and therefore a narrower gasket profile can be selected. Second, due to reduction of gasket offset, required gasket resistance pressure is reduced. In addition to smaller and therefore lower cost of a gasket. Other advantages of a sealing system with reduced tolerances include:

- reduction of TBM erector forces
- reduction of induced forces in connectors and in turn reduction in size of the system
- reduction of designed space for connection and gasket systems

Most of tunnel project specifications allow for 5 mm gap and 10 mm offset for segment gaskets. Dowels can easily provide this requirement. However, bolts are the predominant connection system in longitudinal joints.

The wider the gasket, the higher the profile.

The higher the profile, the more spring distance for more gap.

- 26 mm gaskets with ~10...16 mm height
- 33 mm gaskets with ~14...20 mm height
- 44 mm gaskets with ~20...24 mm height

The wider the profile, the more overlapping area at the surface.

The more overlapping surface, the more capacity for more offset.

- 26 mm gaskets with ~20...25 mm surface

- 33 mm gaskets with ~28...32 mm surface
- 44 mm gaskets with ~35...40 mm surface

The wider the gasket, the more rubber at the cross section.
The more rubber, the more capacity for restoring forces.

- 26 mm gaskets with ~30...60 kN as max load deflection
- 33 mm gaskets with ~40...70 kN as max load deflection
- 44 mm gaskets with ~50...80 kN as max load deflection

The wider the gasket, the more rubber structure.
The more rubber structure, the more capacity for higher water pressure.
For example: gap/offset is 5 mm/10 mm

- 26 mm gaskets for ~25 bar water pressure
- 33 mm gaskets for ~45 bar water pressure
- 44 mm gaskets for ~80 bar water pressure

3.3 *Gasket recommendation*

To summarize, the use of the following gasket sizes (GBD) is recommended depending on the following tunnel diameters:

Gasket size	Diameter recommendation	
•20 mm	OD < 4 m	< 5 bar
•26 mm	3 m < 9 m	< 25 bar
•33 mm	4 m < 12 m	< 45 bar
•44 mm	6 m < 15 m	< 80 bar

Each type of gasket has its own triangle of gap, offset and water pressure. The reaction forces of the gasket type depend on the GBD and how much the gasket will be compressed by gap/offset. The gasket design has to be determined together with the groove design and segment design.

4 TUNNEL SEGMENTS BECOMES THINNER

As mentioned above, the trend in tunneling is towards thinner segments to reduce the total costs and to speed up the TBM drive. Regarding the gaskets, they have to move further to the outside of the segments. What has changed in the tunnel projects over the last years and what experiences have been gathered to reach thinner segments?

4.1 *Segment details*

The ring length is increasing up to 2.20 m to reduce the length of the joints in the segmental lining and to minimize the risk of leakages. This is possible in rail or road tunnel projects with long radii.

In the same time, the designer wants to reduce the length of the ramps in between the working area for the installation of the segments and the TBM head by twisting the longitudinal joints of the K stone segment. It has a strong influence for the design of the segment moulds and of the precision of the gasket corners.

The first segments of TBM tunneling had 90° shapes for the regular segments. More and more projects have been using the dowel connectors, which changed the shape into 80°/100°. It reduced the friction at the surface of the gaskets during installation of the segments.

K stone width is longer and its dimensions are quite similar to the regular segments to get a more homogenous structure along the ring.

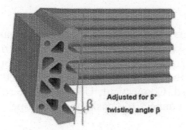

Figure 7.   Twisted corner [source: Datwyler].

The connection of segments has changed from bolts into dowel systems with guiding rods. That can accelerate the TBM drive and make the ring building process more safe. Smaller offsets can be noticed and the specified gaps reached better. But for the interaction of gaskets and dowels, more investigations are needed as well as further tests to reach the aim of the specified gap/offset scenarios.

The requirement of thinner segments is driven by economic reasons. The first segments for TBM tunnels had thicknesses of ~500 mm and more. In current projects, we can find ~300 mm of thickness.

4.2   *Gasket position*

Therefore, the position of gaskets has to move closer to the corner of the segments. This is only possible by using all the experiences of the interaction of the inserts and the concrete segment details. Further tests, investigations and better test equipment should help to guarantee the quality of the tunnel.

5   NEXT GASKET'S GENERATION

Over the last 20 years, the market has been growing up very speedy. The "industrialization" of the mechanized tunneling business has thrived with much more tunnels, longer tunnel distances and demands for optimization and efficiency.

5.1   *Performance corner*

The optimization of the segment thickness, the concrete quality and the gasket corners lead to new problems, which had never occurred before. All producers of gaskets try to handle this issue very sensitive and carefully.

The answer could be a PERFORMANCE CORNER (developed by Datwyler Germany, current reference projects in Europe and America), which offers the same tightness by using less rubber volume in the vulcanization process of the corners. The goal is to get a balance for the same tightness of the gasket systems and to protect the concrete segments from cracking.

Figure 8.   Corner shapes [source: Datwyler].

5.2 Fiber anchored gaskets

The third generation of gaskets – starting from the glued-on gaskets in the 1960s (used for the Elbtunnel in Germany, gaskets made by PHOENIX Germany), through the (rubber) anchored gaskets ~15 years ago (used for the Lee Tunnel in London, UK) and ending up with the latest innovation of the fiber anchored gaskets (developed by Datwyler Germany, reference project in South Hartford, Connecticut, US). Here the rubber "legs" were changed into thousands of small fibers to connect the gaskets in a proper and safe way on the concrete surface.

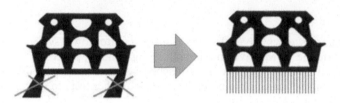

Figure 9.   Fiber anchored gaskets [source: Datwyler].

5.3 Next generation of spalling tests

Every manufacturer willing to learn more about the "dancing" of the interaction of the gaskets and the concrete segments, shall increase the knowledge about it. In case of companies like Datwyler, this has been directly translated into investments in moulds, vibration tools and equipment to produce concrete blocks and to conduct improved spalling tests.

First time in the market, this has given a chance to test anchored gaskets directly in concrete. The target was to estimate the distance for the gasket's position at the corner of the segments to avoid cracks. To develop new gasket shapes and for many important challenging projects, the exact project situation was simulated based on the precise evaluation of the drawings and specifications and the requirements of the project.

5.4 CAM by FEA

Finite element analysis (FEA) had a great impact on the optimization of the new products in the market. Thanks to mathematical simulations with the adaption of the measured results at the real gaskets, the simulation and optimization of the new gasket geometries has been done much speedier. (The company Datwyler with headquarters in Switzerland and about 6,000 employees worldwide has been developing solutions focused on different markets as Automotive, Medical health and Consumer goods and been using the FEA (Finite element analysis) over the last 10 years for all these applications).

The reason is to ensure the right volume of rubber in the structured geometry according to the project related requirements (For new projects, we can developed the geometry and test it virtually by the FEA. With the FEA results for the load deflection tests and the water tightness tests, the manufacturer can start the production of real gaskets.).

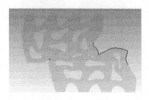

Figure 10.   WT test by FEA [source: Datwyler].

5.5 *Water pressure tests up to 150bar*

Tunnels will run longer. Tunnels will last longer. Tunnels will go deeper.

For the deeper tunnels as the Follo Line project in Norway (Datwyler gaskets M38928 OSLO) and the Rondout West Branch Bypass in New York (Datwyler gaskets M80157 RONDOUT), a special equipment and a new strategy had to be employed to test the water tightness of tunnels deeper than 100 m. At the moment, the test equipment for these tests was developed for 40 bar to 60 bar.

Higher requirements appear more and more often. According to this trend, we are preparing the possibility to conduct the tests until 150 bar.

6 CONCLUSION

What will happen in the future of the gasket market?

Tunnels are being built deeper and longer.

Modern gasket systems help to master the rising risks of water pressure up to 60 bar. Greater precision in the construction of the segments and during the installation of the segments help to further develop the industrial construction of tunnels. Dowels in combi-nation with sealing systems adapted to the respective projects are a driving force here.

To build safer tunnels, you have to select the right gasket system in combination with the concrete segments and all the installation parts. Each detail of a segment such as

dimensions and groove design,

specification for gap and offset,

fixation and properties of the concrete

have a huge influence on the tightness and the durability,

The trend in tunnelling is towards thinner segments for reduce the total costs and to accelerate of the TBM drive. Regarding the gaskets, they always have to move further to the extrados side of the segments. It has a huge influence for the parameter of the tunnel specifications as gap, offset and the connection details. The steps will be smaller and depending on the connection systems, the market has developed a huge range of well-fitting systems to waterproof the tunnels.

Tunnels and Underground Cities: Engineering and Innovation meet Archaeology,
Architecture and Art, Volume 5: Innovation in underground engineering,
materials and equipment - Part 1 – Peila, Viggiani & Celestino (Eds)
© 2020 Taylor & Francis Group, London, ISBN 978-0-367-46870-5

Moving material with belt conveyors in urban environment and for long tunnel construction in Italy: Metro Catania, Scilla's shaft, A1 highway and Brenner Base Tunnel

M. Bringiotti & G. Bringiotti
GeoTunnel S.r.l., Genoa, Italy

W. Aebersold & P. Rufer
Marti Technik AG., Moosedorf, Switzerland

ABSTRACT: Mucking out automatic systems have been nowadays widely exploited in practically all mechanized tunnelling projects. This mostly occurs due to the improvement of the related technologies that solve complex logistic-related problems, but also to the environment, safety, energetic aspects, CO_2 emission, manpower optimization. Also Owners have already been specifying such systems in their tender documents as Brenner alpine tunnel, Turin-Lyon, Gothard, Loetschbeg. The paper describes a particular case in a urban contest which is the Catania Metro. Therefore, starting from little capacity and short tunnels we will arrive to huge diameter TBMs, such as the 16m diameter in A1 highway, up to BTC-Brenner Mules, where about 80 km of very complicate systems are able to handle various excavation front faces.

1 WHY TO USE BELT CONVEYORS?

A. Answering to this question is not complicated, at all...
B. In nearly all mining procedures, conveyor belts (Fig. 1) are especially necessary to limits movement costs (in terms of incidence reduction per ton of produced prime material) of excavated material toward the plant which is able to process it. Cost reduction is a combination of various parameters but, although this is a primary reason to develop such a technology, it is not the only one. There is a list of positive aspects in the application of this

Figure 1.   Belt conveyor solution to move material out from a trench.

method, also projected in the tunneling world, in comparison with classical systems of transportation on wheeled vehicles. Here is a summarized list:

C. Environment and safety: limitation of the environmental impact of the entire production process (CO_2!), limitation of noise, elimination of dust production, more safety on the TBM concerning material management through an intrinsically extremely reliable system, general increase of safety parameters to prevent road accidents along the path to the land-fill and pure presence of electric engines and exclusion of endothermic engines.

D. Energy: considerable energy saving in comparison with gas oil consumption and dimensioning and consumption saving for the ventilation implant.

E. Work cycle and manpower: the construction mucking system is continuous and is therefore available at any moment, overall less workforce engagement (including management such as cafeteria, housing, etc.) and more flexibility for work cycles (less employees and use of less technicians as opposed to many drivers).

F. Benefits related to the use of TBM: material excavated from the TBM is continuously moved in the tunnel and/or from the excavation shaft, no necessity of a long back-up to load the whole quantity of a complete stroke on the mucking train, absence of a wagons rollover system at the portal tunnel, no interference on the road between vehicles going out (mucked material) and in (segment transportation, injection material and personnel), no auto-repair garage for assistance and maintenance of the locomotors, no tankers or re-fueling plants, optimization and minimization of the spare part area and fewer personnel to handle the muck as a whole (drivers, mechanics, maintainers, safety systems inspectors, etc.).

G. Infrastructures: limitation of measures on the streets (repaving, signalizing, traffic lights, etc.) and less disturbance to the collateral infrastructures (urban, provincial and highway streets).

H. Therefore, the prejudice according to which a conveyor belt implant's costs are too high is unreasonable. To consider the elements mentioned above in the right way, is enough to evaluate the benefits of choosing this solution.

I. There are several examples of standard, complex or peculiar jobsites, in which applying such technology resulted in a successful job.

2 WHAT CHARACTERIZES A CONVEYOR?

Various topics should be analyzed in the preliminary design phase and the main components should be chosen accordingly, as:

- Product that occurs transportation
- Route, slopes, misalignments, ...
- Capacity
- Cleaning (the cleaning of the belt is a major element in the design of the conveyors. Scrapers generally ensure this particular one with carburized blades of subjacent or tangential type. Turning systems usually serve on long conveyors).
- Motorization
- Belt tension
- Belt type
- Scraper
- Drive unit
- Feeding corridor
- Idlers/rollers (the idlers are the metal supports of the rollers and can be equipped with 2, 3, 4 or 5 rollers according, to needs; its lubrication is lifelong and its diameter is generally of 89, 133 or 159 mm. The load of products, idler spacing, and belt speed determine the rollers).
- Pulleys

The product which need to be handled has to be analyzed for it physical and chemical characteristics as: density, corrosive, powder (as cement), round (as aggregates), abrasive, liquid (muds), dimension (0 – 1000 mm), sticking, cutting, explosive, dangerous, . . .

The environment of the conveyor is also important (inside, outside, moisture, dusty, etc.).

The conveyor installation (functional lay out) is mostly important for the following:

– The calculation of power
– The installation of equipment of the conveyor and particularly the tensioning system

The conveyor is equipped with a tension system to:

– Allow the transmission of the power between the drive pulley and the belt
– Absorb the belt's length differences, caused by the elastic lengthening of the belt

There are three principal devices of belt tensioning:

– Tension (with screws or hydraulic) for conveyors of short distances between centres.
– Automatic tension with counterweight as a dancer, for higher distances between centres.
– Tension by capstan for small overall dimensions and high distances between centres.

The type of handled product will determine the width, speed and belt's coating.

The characteristics of conveying belts are mainly the type of carcass, of coating and structure elements.

There are three types of carcasses:

– Belts with metal carcass (Steel Cord), composed of steel wire ropes
– The multiple belt textile, equipped with various fabric folds (cotton, polyester, polyamide, aramid)
– The monopoly-belt textile, a carcass «mono-fold» of polyester or polyamide

The coating is the part of the belt in contact with the products. This can be: anti-abrasive, anti-heat, cold-resistant, anti-fat, anti-static, not easily flammable, impact-resistant, anti-clogging, . . .

Boring a tunnel is a specific application for belt conveyors, because their characteristics evolve as the boring goes on; therefore, a simultaneous extension of the belt conveyor is necessary.

There are two methods to bore a tunnel and two possibilities to realize the lengthening.

– If the tunnel is traditionally bored with explosive or road header, the belt conveyor generally extends per strokes, from 50 to 200 meters. The return pulley later usually assembles onto a structure equipped with skids, in order to translate it easily as the new structure has to erect on this side of the belt conveyor.
– If the tunnel is bored using a tunnel-boring machine (TBM), then the belt conveyor could extend continuously while the TBM progresses, without stop. In this case, the installation of a special frame elongation station onto the TBM's trailers is necessary, as well as the installation of a belt storage in the drive unit area.

In both cases, it is necessary to dimension the belt conveyor, taking into account the whole specificities encountered along the tunnel route, such as horizontal or vertical curves, slopes, etc.

Two types of belt storage are existing: horizontal or vertical; the vertical model is frequently useful for the city jobsites with little available space.

Before covering Italian experiences, we will describe an incredible job site in China, just to demonstrate the potential of the system on a large scale.

3 JINPING II – THE MOST DARING CONVEYOR BELT PROJECT EVER REALIZED

The Yalong River is one of the main tributaries of the Yangtze River and is located South-West from the Sichuan province. It is about 1600 km long with a 2800 m difference in level along several, very deep canyons. The Jinping II central will exploit along 120 km and generate 4800 MW.

Figure 2.   Complicate in-out lay out and interaction with the miner's buildings.

Figure 3.   Cavern intersection and overland installation.

The whole plant, with a huge underground power room, includes the realization of 4 pressurized tunnels, about 17 km long and with a 12,4 m diameter, also considered a drainage tunnel (about 6m diameter) running parallel to the other conducts, because these configuration is strongly karstic.

3 TBMs and 2 D&B equipped fronts produced about 20 million tons of waste material; conveyor belts moved and handled such quantities.

The power of the belt behind the TBM, in adit no. 3 for example, is in total 2800 kW, with an 1800 t/h output. This probably is one of the most high-performing installations ever made in the world, just overtaken from the Brenner-Mules one. This system must grant a TBM excavation productivity of 600 m each month.

However, the external belt system represents the true peculiarity of the whole system. It transports material excavated from the portal area to the terminal landfill and it is about 6200 m long. The total capacity established is 5600 t/h. Furthermore, the plant can work in reverse transportation for concrete aggregates with a capacity of 600 t/h.

The excessive muck serves to fill a lateral valley, which is at a height of about 300 m, and it is approximately 1 km long.

The challenge for this implant consisted of the rough terrain, the necessity of interaction not only with the "wild nature", but also with a settlement build for workers, which was sort of a small town (Fig. 2), together with a high capacity and the obligation to work without intermediate controls.

The whole structure inside the tunnel is suspended on the roof because of a clear project request which did not include installations that were fix on the ground, in order to leave the floor completely free to allow passage of vehicles and railways.

In order to realize a track compatible with the plant on a steep, mountainous land, the conveyor system was fundamentally elevated (Fig. 3). It is mainly composed of bridges with a 48 m span. In addition, it was necessary to create four bridges with spans between 62 and 200 m and a short tunnel to avoid a particularly tortuous track. Two of the bridges are de-signed in a suspended version, because of the width of the respective spans.

Various integrated systems composed of different elements such as transfer towers, intermediate belts, intermediate stocking volumes placed at a 60m height completed the whole installation . . .

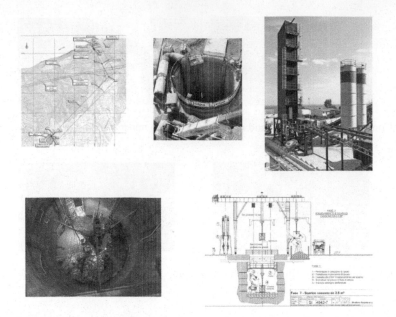

Figure 4. Some views and shaft installation sketch.

4 SCILLA'S TUNNEL AND MELIA'S SHAFT, CIPA S.P.A

The project consists of a sub-horizontal, 2842 m long tunnel, blind bore excavation and a vertical shaft approximately 300 m deep (Fig. 4). This work's function is to receive certain electrical submarine high tension cables from Sicily.

The TBM's diameter is 4100 mm, prepared to excavate in highly fractured, competent gneiss, in single and double shielded mode.

There were four main peculiarities of the disposal system:

1. A continuous plant, which is able to follow the TBM excavation even with elevated gradients and especially the limited altimeter-plan curve radius.
2. A compact plant in order to work in a 3,6 m internal diameter tunnel.
3. A regenerative plant which (13% slope), with no influence whatsoever on the electric consumption, is able to produce EE to be "pumped" in line.
4. A vertical belt storage, which limits space in plant as much as possible. The launching tunnel is located on a very nice beach!

We must mention the "vertical wise" excavation, record in Europe, when explaining this project. It is a 300m deep shaft with an excavation diameter of approximately 7 m.

The excavation of the first 80 m consisted of weak-consolidated sand, which foresaw "soil improvement" operations through jet grouting, reinforced with steel tubes.

After this stretch, the excavation was systematic "drill & blast", with different specifically designed equipment for this process.

The drilling machine is equipped with two booms, with a portal crane to support it. Appropriate stabilizers granted a stable and safe work cycle.

The lifting system is able to move longitudinally and transversally on appropriate rails. It bears a 4 rope winch to lift the working platform and a double rope to lift loads (muck and construction material).

The platform's central area can be subject to enlargement in order to allow the passage of necessary excavation and consolidation tools at the bottom of the shaft.

The designed system also includes a lift and an elevator for the transportation of personnel and various materials.

In order to handle the presence of copious water there are several intermediate niches, which are used to contain tanks and relaunching pumps.

In order to realize the final lining, an auto lifting platform started at the bottom of the shaft. During this operation, 3 pumps (5 l/s with a 300 m prevalence) have been installed with pipes inserted in the lining.

5 METRO CATANIA: HORIZONTAL AND VERTICAL SYSTEM ABLE TO MOVE DIF-FERENT KINDS OF GEOLOGICAL MATERIAL (VOLCANIC ROCK AND CLAY)

This is the extension of the railway on Catania's underground route from F.S. Main Station to the Airport – "Stesicoro-Airport route 1st Lot". Ravenna based CMC has acquired the job in current final execution phase.

It was necessary to realize an appropriate system to handle different rock types during excavation phase in this jobsite (Fig. 5). These types include stones and clays. Apposite additives and foaming handle these rocks in order to allow TBM-EPB mode excavations.

The major complexity is the material management integrated system. The geology has many different characteristics (volcanic rocks, clay, sand, fractured and weak material) and it has been necessary design a particular system able to efficient work especially during the vertical transportation phase.

The system composition is as follows:

1. Tunnel belt with belt storage (100): length 2500 m, width 1000 mm, type steel cord (EN 14793, Class A), capacity 800 t/h, speed 3 m/s. The system is compatible with the one in-

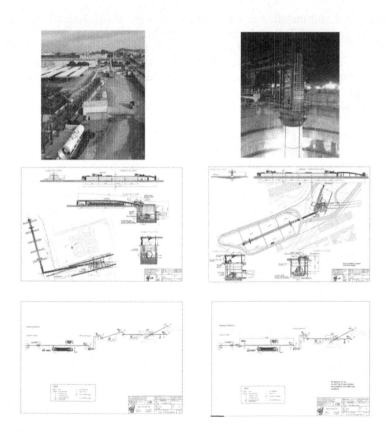

Figure 5.   Installation site and system lay out.

stalled on the Herrenknecht S454 machine, property of CMC, and equipped with emergency stop every 250 m (included design and construction of the structures meant for wagon no. 5 and 6).

2. General Characteristics of Tunnel Belt Storage: capacity 500 m, type horizontal with n. 14 return stations, length 62 m, width 2,4 m, height 4,5 m.
3. Vertical belt (Elevator 200: height 27,2 m, capacity 800 ton/h, bucket belt type XDE-SC 1000/6+2, belt width 1400 mm, presence of an automatic cleaning system (wash box) including double pulley and small service belt.
4. Surface belt (300): length 94 m, structure width 2,3 m, belt width 1.000 mm, capacity 800 ton/h, speed 3 m/s, belt type EP – Textile 400/4+2 (DIN 22102, Y).

6 HIGHWAY A1 FLORENCE-BOLOGNA: HIGH PERFORMANCE BELT CONVEYORS IN THE BASIS, SPARVO AND ST. LUCIA - Ø 16 M - TUNNEL

The Variante di Valico (Crossing Variant) is the alternative way of the Apennine section of the Autostrada del Sole Milano-Napoli; it is a 62,5 km long route between Bologna and Florence. Traffic peaks of up to 90,000 vehicles per day of which about 24,000 are heavy vehicles, characterize this part of the A1 highway. These numbers make the old infrastructure unsuitable for the current needs of transport. The new project will overcome the Apennines at a lower altitude than the previous track, with a path full of viaducts and tunnels that will make the new highway more modern and efficient.

The conventional method excavated Base Tunnel's section measures about 180 m². It is approximately 8500 m long and features two tubes. It has been realized by Company Todini Spa.

The belt, supplied and installed in the central portal by Marti Technik, was 800 mm wide with a minimum radius of 215 m. It could carry 500 tons of material per hour for a power demand of 450 kW. It moved a total of about 2 million tons of muck.

The muck feeds into a jaw crusher and later moves through an encapsulated belt conveyor with bridges for about 1.4 km and then reaches a temporary storage area with a wheeled movable stacker.

The plant allowed avoiding heavy traffic, noise, dust and pollution around the local areas.

The Sparvo tunnel (Fig. 6), realized by Company Toto S.p.A., is 2 x 2.6 km long and its excavation occurred with a record 15.6 m diameter TBM; at that time, it used to be the largest in the world. The excavated material is composed of sandstone and clay and presented different behaviors throughout the process.

A 1200 t/h Marti belt conveyor carried material for about 1,6 km and deposits it in the storage area through a special 315 m long tripper, which runs on rails. From here, trucks loaded with this muck transport it to a service area closed by the existing highway. The material was later discharge into a hopper of 30 m³ capacity and reaches the final stock area by a 500 m belt conveyor crossing a regional road. The muck finally spreads out through an extensible belt in combination with a telescopic crawler stacker (Fig. 7).

Figure 6.   The Sparvo Tunnel.

Figure 7.   Special belt conveyor to spread out the muck.

Figure 8.   Tripper car, overland conveyor and general lay out.

The last challenge, presently still on going, is the St. Lucia tunnel, where Pavimental S.p.A. is using an EPB-TBM having the world record diameter of 16 m; it's excavating a 7,9 km tunnel.

The just installed plant has a conveying capacity max 2000 t/h, average 1750 t/h, with rock density (loose) of 1700 kg/m^3, it can handle lump size max 0-300 mm (as very exceptional case, only single lump L= max. 800 mm), 3 m/s speed.

The belt width is 1400 mm, only the TBM belt has ca. 4000 kW power installed. The belt storage has been designed to host 600 m of belt, in order to let the TBM run for 300 m before the new belt recharge,

The excavated material of the TBM is transported by TBM conveyor onto the tunnel conveyor. The area where the material has to be fed, the return and extension stations for the continuous conveyor extending, are on the back up system on the TBM. The tunnel conveyor is directly connected with the TBM via the return station and can be continuously extended without interrupting the tunnelling.

The drive station is approximately 300 m before the portal. At the drive station of the tunnel conveyor, the excavation material is conveyed on the external belt conveyor which transports the muck to the dump conveyor.

A long overland conveyor (ca. 2 km), with curves, is bringing the material up to the dump conveyor where a tripper of 300 m length is filling the huge prefabricated dumping area. The tripper, having a belt width of 1.600 mm, is of reversible type and have a length of 12 m (Fig. 8).

There is no explosive atmosphere during regular operation expected - if such a case would happen, then only seldom and temporary. It has been therefore defined zone 2 according to ATEX 1000 m behind TBM, so the system is in a sort of evolving ATEX con-figuration while the TBM is excavating.

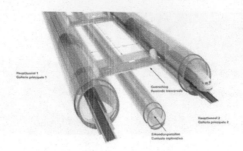

Figure 9. Lay out of the Brenner Base Tunnel: main tunnels and the exploration tunnel.

7 BTC-BRENNER BASIS TUNNEL

The Companies Astaldi, Ghella, PAC, Cogeis and Obersosler form BTC Consortium. Since early 2017, BTC has been involved in the construction of what will become the longest underground railway link in the world: the Brenner Base Tunnel, which is the central part of the Munich-Verona railway corridor.

The project as a whole consists of a straight, flat railway tunnel, which reaches a length of 55 km and connects Fortezza (Italy) to Innsbruck (Austria). The tunnel will interconnect with the existing railway bypass near Innsbruck and will therefore reach a total extension of 64 km.

The tunnel configuration includes two main single track tubes, which run parallel, 70 m across from each other throughout most of the track, and connecting side tunnels link them every 333 m (Fig. 9).

Between the two main tunnels, and running 12 meters below them, the plan is to construct a preliminary exploratory tunnel. Its main function during the construction phase is to provide detailed information about rock mass. Its position also allows performing important logistic support during the construction of the main tunnel for both the transport of the excavated material as shaft as of construction materials. During the operations, it will be essential for the drainage of the main tunnel.

The excavation process divides into 2 blocks. The first one is planned to bore with n. 3 TBMs (n. 2 for the main tunnels and n. 1 for the exploratory tunnel, direction North), and the second with traditional method, mainly including drill & blast in the competent material and special drilling techniques in the faulty zones, mostly going direction South and in some areas direction North. CIPA S.p.A. Company is currently performing most of the tunnel excavations in the lot named "Mules 2-3" with the French drilling partner Robodrill SA.

7.1 *The design principal*

The system on conveyor belts allows a straightforward, efficient mean of transportation for material going both in and out of concrete crushing, and in and out of mixing plants. It evolves within the carrying out of the project, and gradually implements congruently with the works in progress.

The conveyor system rationalized its complexity since the tender phase (Fig. 10), but always maintaining its high potential, flexibility and capacity. The differentiation of the two belts in Aica (belt 1 and belt 2) allows each single band to be allocated with a certain type of material; therefore belt 1 is for material A (which is good for concrete) and belt 2 is for material B+C (semi-good and not concrete appropriate material). The differentiated configurations avoid alternating different material types thanks to a temporary stocking buffer on the same conveyor. However, the system provides the possibility to switch the material types in Aica according to necessities. This type of realization allows an easier management of the whole system.

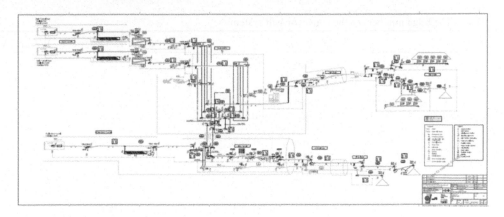

Figure 10. The complex conveyor system general lay out.

Figure 11. Inside and outside tunnel conveyor system.

We may identify such works into three major phases. This report's aim is to describe the decisions taken within the Construction Project and to list them congruently to time-related phases. Without taking into consideration the specific calculations regarding the Space-Time Diagram of the Executive Work Program, here follows a description of the three phases:

– Phase 1: excavation of the Exploratory Tunnel (CE) by traditional excavating system and TBM assembly chamber; excavation of the first part of the North Line Tunnels (GLN) by traditional system and excavation of the South Line Tunnels, always by traditional method (executed by Cipa S.p.A. with Drilling Partner Robodrill SA);
– Phase 2: continuation of the activities, following the previous phase; realization of "in cavern" assembly areas for the two TBMs, which will excavate the Line Tunnels (direction North), excavation prosecution of the South Line Tunnels by drill & blast system (GLS); and realization of the mechanized excavation of the CE Northwards.
– Phase 3: prosecution of previous activities, excavation of North Line by mechanized tunnelling (GLN) and subsequent finishing job site works.

The complex lay out reported shows the integrated phases.
All belts generally connected to the five excavation faces with the Logistic Knot (N). Subsequently, from Knot N:

– Material type A is carried towards the jobsites Mules and Genauen 2 (deposit area reached by a 180 m length bridge belt conveyor – 300 ton/h – which is crossing the A22 highway, the Isarco river, the National road S.S. 12 and the existing railway line). It is stocked and/ or crushed outside and/or gets back to the batching plat, so it can be used for spritz beton, and for the definitive lining in the gallery parts, excavated by traditional methods.
– Material type A/B+C is carried towards the Hinterrigger jobsite and is used to produce the precast elements and the pea gravel, which will fill up the segment over-space during excavation by TBMs. Concrete segments and pea gravel are carried back in through the CE on the dedicated rail system (Fig. 11).

Table 1. The huge numbers of the conveyor belt system.

Zone	Items	N.	Length, m	Power, kW
Mules logistic knot	Belts	6	211	206
	Chutes	2		
	Dosing units	2		
	Service platforms	3		
Tunnel North	Belt explorative tunnel	1	16,763	
	N-E belt	2	7195	
	Main tunnel N-E	1	13,182	
	Main tunnel N-W	1	13,223	
			50,363	5981
Tunnel South	Belt	1	50	22
Aica - Unterplattner - Hinterrigger	Aica belt	2	10,761	
	Unterplattner - Hiterrigger	1	1169	
	Hinterrigger	2	499	
	Movable stacker	1	48	
			12,477	2778
Adit Mules	Belts	3	1979	673
Site Mules	Belts	6	831	213
TOTAL - new installation			**65,911**	**9.873**

Figure 12. The belt conveyor installation in main the storage area.

7.2 *Key numbers*

In conclusion, the system has been developed in approx. 80 km (integrated with the existing one, ca. 14 km), with a total installed power of more than 10 MW, which is really a huge number, especially considering the complexity and different working sites (Fig. 12). Below is a synthetic table (Tab. 1):

Figure 13. Everything is under control!

7.3 The system's brain

The various MCCs (Motor Control Cabinets) are installed in a special container and equipped with a cooling system, which is able to safely handle peak temperatures (of the working environment and produced by the inverters), as far as potential dust.

The belt conveyor system, in its complexity, is fully equipped with a PLC system, which is able to control and command all functions. Touch screen panel designs allow an easy, friendly use.

All critical points have special devices installed, in order to detect potential problems that may happen in its quite long working lifetime, estimated to be ca. 4 years, ahead of time.

Ordinary maintenance planning settles with the various usage coefficients, daily time scheduled availability. In addition, a study of special algorithms allows planning revision caused by the high level of wear and tear, predicted to happen due to the extreme quartz content concentration in the moved rock (mainly granite).

The plan foresees no. 3 control rooms in the critical areas. These will all connect to one another.

The conveyor belt Supplier grants 24/7 Tele-Assistance, thanks to a Wi-Fi connection with the central office based in Switzerland, Moosseedorf-Bern, and an "in situ" job site assistance with electrical – mechanical – electronic & welding supervision, assembling, handling, maintenance and spare parts service (Fig. 13).

8 CONCLUSION

Handling such a variety of complex projects is purely a matter of Teamwork among highly professional Subjects. In no way can one think of this as a simple "purchase and install" procedure. A serious partnership between the Constructor and the Supplier is the key to its success.

REFERENCES

Bringiotti, M. 2002. *Frantoi & Vagli: trattato sulla tecnologia delle macchine per la riduzione e classificazione delle rocce*, Edizioni PEI, February

Bringiotti, M., Duchateau, J.B., Nicastro, D. & Scherwey, P.A. 2009. *Sistemi di smarino via nastro trasportatore - La Marti Technik in Italia e nel progetto del Brennero*, Convegno "Le gallerie stradali ed autostradali - Innovazione e tradizione", SIG, Società Italiana Gallerie, Bolzano, Viatec, 05/03

Bringiotti, M., Parodi, G.P. & Nicastro, D. 2010. *Sistemi di smarino via nastro trasportatore*, Strade & Autostrade, Edicem, Milano, Febbraio

Tunnels and Underground Cities: Engineering and Innovation meet Archaeology,
Architecture and Art, Volume 5: Innovation in underground engineering,
materials and equipment - Part 1 – Peila, Viggiani & Celestino (Eds)
© 2020 Taylor & Francis Group, London, ISBN 978-0-367-46870-5

Precision drilling for D&B and bolt installation: Analysis of relevant case histories

M. Bringiotti & G. Bringiotti
GeoTunnel S.r.l., *Genoa, Italy*

F. De Villneuve, A. Harmignies & F. Vernerey
Robodrill S.A., *Lyon, France*

ABSTRACT: A drilling jumbo is machine that has seen, along the last decades, some major evolutions as the introduction of accurate sensors in order to define the exact boom positioning, everything integrated with an automatic drilling PLC able to find and realize the exact holes, data recording and restitution, data analyser with even geological interpretation of the rock mass worked out, high frequency drifters and heavy duty undercarriage for a long life in the hard environment where frequently are used. The paper presents examples from relevant case histories: the tunnels to be constructed with partial sections in heavy congestioned urban areas as Metro Paris, drilling contemporary steel sub-horizontal umbrellas and radial bolts in faulty zone as in Brenner, installing long cable bolts or CT-Bolts automatically injected as in Sydney, drilling and splitting for launching shafts in granite as in Hong Kong.

1 INTRODUCTION

The new generation of fully computerized drilling jumbos have been designed and built with the intention to provide the Owner and the Tunneling Contractor with an efficient, Industry4.0 and safe tunnelling tool. The main improvements, tested and proved through various worldwide case histories, can be expected as follows:

- More accurate profile, which means less over blast, less damage to the surrounding rock, reduced support work, substantial concrete saving (in case of final lining).
- Excellent tool for drill pattern optimization and subsequent optimization of the explosives consumption.
- Facilitates longer rounds for water drainage and exploration drilling.
- Complete round documentation available through the logging facilities.

In the papers are presented case histories of examples managed with machine produced by Robodrill. The Company is designing and manufacturing "in series" and "highly customized" drilling equipment for the major worldwide tunnelling general Contractors, working in the 5 Continents. Main components are usually coming from CAT for what is related to the undercarriages, Montabert for the drifters, the robotized system is developed in house as the full system electric-meca-tronic engineering.

Figure 1. Example of different bolt types which can be installed automatically by the drilling system.

2 BOLTING, PIPING AND CABLING

Jumbos are especially known for Drill & Blast procedures but the "world of bolting" is so extended that it's difficult to have an idea and probably it so wider just because the modern machines are nowadays able to realize operation which previously have been impossible, leading to change the design philosophy linked to the support technology.

Normal jumbos are nowadays able to install, under an optic of industrialized tunnelling, in a very safe way, an incredible number of supports as (Fig. 1):

- Pipe umbrellas
- Combination Tube Bolts
- Cable bolts
- Self drilling spiles
- Grouted and resin bolts

3 CONNECTING SYDENY

The NSW Government is building a complete transport solution, including public transport and road infrastructure to ensure Sydney does not come to a grinding halt; one of them is called "Connex". WestConnex is Australia's largest transport infrastructure project, linking Sydney's west and southwest with the city, airport and port; it is critical to delivering an integrated transport solution to tackle congestion across the city and provides the catalyst for urban revitalisation throughout the corridor.

NorthConnex is a 9 km motorway tunnel under construction in northern Sydney, New South Wales, Australia. About 21 kilometres of tunnelling is being carried out for NorthConnex. The two main nine kilometre tunnels will carry motorists between the M1 Pacific Motorway and the Hills M2 Motorway and additional tunnels are being excavated for on and off ramps.

The deepest part of the tunnel will be around 90 metres, passing under the Sydney Metro Northwest tunnel at Beecroft; more than half the tunnel is more than 60 metres deep.

Robodrill delivered tens (ca. 40) new machines for Northconnex and Westconnex, designed for various applications including shaft drilling, face drilling, canopy tubes, roof bolting, long bolts and cables.

For NorthConnex (Fig. 2) the objectives were:

- 15 min travel time savings for Australians
- Better and more reliable trips
- 21 traffic lights bypassed
- Link M1 Pacific Motorway to the Hills M2 Motorway

Important figures are:

- 9 kilometres tunnel motorway with interchanges

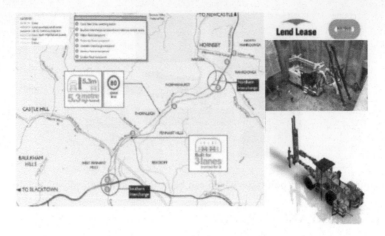

Figure 2.   North Connex lay out; shaft jumbo and 3-4-5 m CT-Bolt turret automatic (also injection) jumbo.

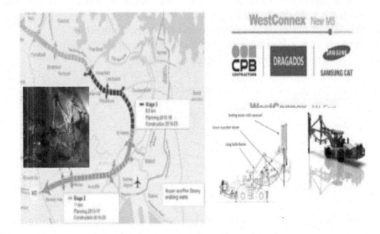

Figure 3.   West Connex layout; in the pictures the pendular cable jumbo.

– $3 billion project
– Includes $2.65 billion construction budget
– Tunnel height clearance of 5.3 m
– Max depth of 90 m

  For West Connex (Fig. 3) the objectives were:

– Provide quicker and more reliable trips between Western Sydney and the Sydney Airport
– Remove bottlenecks
– Provide a widened M4 from Church Street to Concord Road
– 52 traffic lights bypassed and 40 minutes saved per travel

  Important figures are:

– 33 kilometres connection project
– $3.3 billion project

4 NEW YORK: DELAWARE AQUEDUCT BYPASS

The contract will involve excavating a 4 km, 6.7 m diameter, TBM bored bypass tunnel under the Hudson River, between two shafts at Newburgh and Wappinger that are 274 m, 9,14 m finished diameter, and 213 m deep, Φ 9 m, respectively. The tunnel is designed to bypass a damaged section of the Rondout-West Branch Tunnel that is leaking approximately 56 to 130 million litres of water per day into the Hudson River. The project consists of two contracts, BT-1 and BT-2, for construction of the two new deep shafts and the bypass tunnel to replace an existing section of the Delaware Aqueduct, New York's leading source of drinking water (Fig. 4).

The order is placed by the Kiewit/Shea Joint Venture contractor. The new deep bypass will prevent leakage of up to 20 million gal/day through the fractured limestone geology of the 72-year-old existing tunnel as it runs under the Hudson River. The tunnel conveys more than half of New York City's drinking water. The contracting authority for the US$1 billion project is the New York Department of Environmental Protection (DEP). Design has been completed in-house.

Full scope of works includes TBM excavation of the segmentally-lined 4000 m long bypass of the Rondout Tunnel; installation of 2800 m of steel interliner along the limestone section of the bypass to prevent recurrence of the leaks; excavation by drill+blast of 46m-long tunnels to connect both the TBM launch and retrieval shafts of the new bypass to the existing tunnel; and grouting work during tunnel drain down in another lesser effected section of the main tunnel some 29 km north of the TBM drive, in Ulster County, New York.

Kiewit/Shea, being on target, started drill & blast excavation of the TBM bell-out chamber in summer 2016 (Fig. 5). This follows completion last month (in March 2016), under a separate US$ 101.6 million contract awarded to Schiavone, of the two deep shafts – 5B and 6B – at either end of the bypass. The chamber has been completed in 2017 ahead of arrival later in the year of the Robbins TBM.

The single shield machine has been launched from the shaft 5B which is situated on the Newburgh side of the Hudson River. From here it will pass 600 ft below sea level, some 100 ft below the river under a maximum anticipated hydrostatic pressure of up to 20 bar, through to the reception shaft on the Wappinger side. The shafts will remain in place following the completion of the TBM drives as permanent access structures.

Just like the original tunnel, the new section will include a steel inter-liner. Critically, however, at 2800 m long (70% of the total bypass length) it will be considerably longer than the

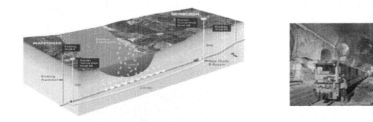

Figure 4. Delaware Aqueduct Bypass Tunnel cross section alignment (at right NY metro, Long Island East Side Access CM019, powered as well by RBDR).

Figure 5. Shaft jumbo able to perform vertical and horizontal drilling operations.

previous steel liner so as to prevent a recurrence of the leakage problem. "During original construction in the 1940s, tunnel workers dealt with huge inflows of water coming in at them while they were drilling," said Bosch. "Total inflows were recorded at approximately 2–4 million gal/day, with the largest single section of inflow coming in at roughly 1500 gal/minute. This was inflow from both groundwater and river influence, and managing water will also be a major challenge for this project."

To handle these expected inflows a grout curtain will be installed ahead of the TBM and the contractor will also be prepared to pump out water where necessary. Despite the highly pressurized conditions at the face, hyperbaric interventions are not expected. "We will be relying on drill probes of the ground ahead, installing a grout curtain where necessary ahead of the face, and there are expected to be adequate sections of competent rock in which the contractor can perform maintenance to the cutterhead and the cutting tools in free air," said Bosch.

5 ATLANTA: WATER SUPPLY PROGRAM PHASE 1

PC-Russell, a Joint Venture, has awarded Atkinson Construction an $81 million contract for the Atlanta Water Supply Program Phase 1 Extension. The Phase 1 Extension connects the old Bellwood Quarry (Fig. 6) to the Hemphill Water Treatment Facility. At the Bellwood Quarry site, Atkinson has constructed four shafts, approximately 1,000 linear-feet of adit connections joining the shafts to each other, and one main 400-foot deep tunnel. The main tunnel will run approximately 5,500 linear-feet northwest to the Hemphill site, where the project team will construct five additional small-diameter well shafts.

Atlanta's water supply program will transform a quarry into one of the largest reservoirs in North America. The average North American public utility has only a three-day back-up supply of clean drinking water. The overtaxed system, paired with the increasing risk of drought, prompted the city's Department of Watershed Management into action. In 2006, the Department took steps to purchase the Bellwood Quarry from Vulcan Materials Co., a 300 ft (91,4 m) deep, vertical-sided behemoth of a quarry where granitic gneiss was mined for a century to become structural blocks for Atlanta's buildings as well as crushed stone aggregate for roads. The USD 300M project would turn the inactive quarry into a 2.4bn gallon (9bn L) raw water storage facility, bolstering the city's emergency water supply to 30 days at full use and to 90 days with emergency conservation measures. To make the program a reality would require excavation of Georgia's deepest tunnel (more than 400 ft, 122 m), starting at the quarry and running under two treatment facilities for 5 miles (8 km) to an intake at the

Figure 6.   Job lay out and TBM assembly with Robo-drilling probe & bolt holes equipment.

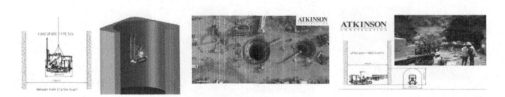

Figure 7.   Shafts excavation equipment for primary and lower level pump station.

Figure 8. Project idea and road header + bolter impression.

Chattahoochee River. It would also require construction of two pump stations at the Quarry and Hemphill Reservoir, five blind-bored pump station shafts at the Hemphill site up to 420 ft (128 m) deep, as well as two more pump station shafts, one riser shaft, and one drop shaft (Fig. 7). The quarry would ultimately store raw water before it is withdrawn for treatment at the Hemphill and or Chattahoochee water treatment plants, connecting the Quarry to the Hemphill Water Treatment Plant (HWTP), the Chattahoochee Water Treatment Plant (CWTP) and the Chattahoochee River. After construction, the area around the Quarry would then be turned into Atlanta's largest park totalling 300 acres (1.2km^2) complete with hiking and biking trails, baseball fields, and an amphitheatre.

The project schedule, primarily driven by the condition of the city's existing water infrastructure, compelled the city to consider Alternative Project Delivery (APD) instead of traditional design-bid-build. The project schedule required a start date for construction of January 2016 and a substantial completion date of September 2018. The method selected was construction manager at risk (CMAR), where the contractor acts as a consultant to the owner during the development and design phase and as a general contractor during the construction phase. The setup resulted in a unique process to start TBM manufacturing, in particular, before the tunnelling subcontractor was mobilised at the site. The decision to use a new TBM by the City of Atlanta was primarily risk-based. The PC Construction/HJ Russell (PCR) JV was selected as the CMAR for the project, who then purchased a 12.5 ft (3.8 m) diameter Robbins Main Beam TBM for the tunnel. The designer for the construction works including tunnel and shafts, JP2—consisting of Stantec, PRAD Group, Inc., and River 2 Tap—specified the hard rock TBM.

6 OTTAWA: LIGTH RAIL TRANSIT

The Confederation Line of the OLRT System is a 12.5 km east-west Light Rail Transit Project through the heart of Canada's national capital. The project will replace, enhance and extend the existing Bus Rapid Transit service corridor, and will serve over 9 million passengers annually. The project (Fig. 8) includes 13 new stations, 2.5 km of twin tunnels (with 3 of the 13 stations located underground through Ottawa's downtown business core), and possessed a capital construction of cost of nearly $ 2.1 Bn.

Its principal feature is a 3 stations and 2.5 km tunnel running underneath the downtown core.

The tunnels are excavated by mean of 3 road-headers backed by 5 bolting rigs supplied by Robodrill. The track-type bolters are used for roof bolting and installation of 12 m canopy tubes (tube diameters ranging from 139.7 mm to 168.3 mm).

The machines comply with the Canada CSA standards. They are fitted with a drilling boom with HC108 drifter and a 2 person basket-boom. They are operated and maintained by local operators and fitters who were initially trained by a Robodrill technician. In Fig. 8 project lay out and job site picture.

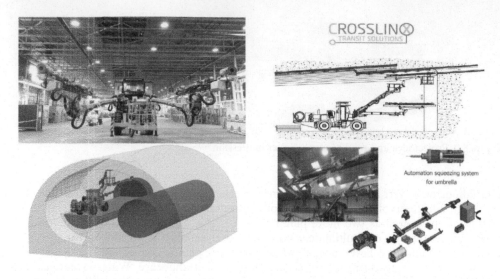

Figure 9. Jumbos provided with drifters and automatic squeezing system for DSI pipe umbrella.

7 TORONTO: EGLINTON CROSSTOWN LIGTH RAIL TRANSIT

ECLRT is commonly referred to as Crosstown, is a 19 km light rail transit (LRT) line being constructed from Kennedy Station to Mount Dennis (Weston Road) in Toronto, Canada. Approximately 10 km of the line will be located underground and up to 26 stations will be built along the stretch. Pre-construction works for the project commenced in mid-2011 and tunnel boring started in June 2013. The 10 km twin tunnels are located between Keele Street and Laird Drive, while the remaining sections of the LRT will include 1.5 km of cut-and-cover tunnels, 0.5 km of elevated guideway and 7 km of at-grade right-of-way surface transit guideway, with signalling provided at intersections. Up to 54 bus routes, three subway stations and several GO Transit lines will be connected to the Crosstown.

Of the overall 26 stations, 13 stations will be located underground. A maintenance and storage facility for the line's rolling stock will be located at Mount Dennis Station and emergency exit buildings will be located on certain sections of the tunnels.

The tunnels will have an internal diameter of 5.57 m and are being bored using four EPB tunnel boring machines (TBMs) each measuring 81 m in length. The TBMs, named Dennis, Lea, Don and Humber, were supplied by Caterpillar. The twin tunnels are being constructed under two separate sections, which were awarded to two different contractors. The first section covering 6,2 km will stretch from the west launch shaft area at Black Creek Drive to Yonge Street in the east section, and is being bored by the Dennis and Lea TBMs. The Eglinton Crosstown LRT will cut across central Toronto for 19 km, with about 10 km of tunnels and 25 stations or stops. The tunnel alignment runs close to more than 1200 existing buildings, as well as buried and surface utilities and equipment. A very narrow right-of-way requires us to pay close attention to issues of private property, treatment of utilities, and traffic management.

Special bolting equipment has been delivered especially for umbrella roof operation; DSI & Robodrill have developed in JV an automatic pipe squeezing device able to install with a continuous operation, fast and safe, long protection sub-horizontal tubes after the hole drilling procedure (Fig 9).

Figure 10. View of the complex down town job site location.

8 HONG KONG: MTR SHATIN TO CENTRAL LINK – CAUSEWAY BAY TYPHOON SHELTER TO ADMIRALTY TUNNELS – CONTRACT NO. 1128

Dragages was selected to construct one of Hong Kong's most technically challenging tunnels – part of the 6 km extension of the ongoing Shatin to Central Line – from Kowloon to the transit hub on Hong Kong Island. The main scope and challenge of the project is to construct the Eastern Tunnels (2 x 680 m) and Western Tunnels (2 x 480 m) by Tunnel Boring Machine (TBM), as both tunnels go through the city's busiest districts. Meticulous project planning is essential in aiding the TBMs to navigate through a large volume of pile obstructions and utilities with zero interruption of services to citizens. To cope with the complex geological conditions, two different types of TBMs are being employed for excavation – a slurry TBM and an Earth Pressure Balanced (EPB) TBM.

The construction of the South Ventilation Building requires careful effort as well, as the existing Police Officers' Club (POC) needs to be demolished and re-provisioned on the top of the South Ventilation Building. Delicate drilling operations through shafts and in very narrow sections have been successfully performed (Fig. 10).

9 HONG KONG: HATS SEWAGE CONVEYANCE SYSTEM

Mainly Gammon Company has performed this huge job (Fig. 11), using 10 drilling jumbos. In few words:

– Approximately 12 km of tunnel by Drill & Blast
– 4 Production shafts North Point, Wanchai, Sai Ying Pun & Stonecutters' Island
– Tunnels are at ~ 160 m underneath the Victoria Harbour
– Tunnel area varies 12 – 26 m², 1600 – 2600 m long drives (size varies depending on support class)
– Mainly solid granite

Figure 11. Lay out of the "in the sea" tunnels.

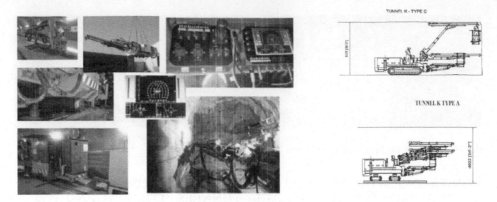

Figure 12.   Jumbos switchable from wheels to rail undercarriage.

- Typical rock support is steel fibre reinforced shotcrete with epoxy resin rock bolts
- Some fault zones may require pipe roof and arches
- Stringent water ingress requirement – significant pre-excavation grouting is foreseen
- The water that leaks into the tunnel is very corrosive

Excavation has been be done by rail mounted gear, typical equipment per face is:

- 2-Boom rail jumbo (Fig. 12)
- Haggloader with excavator arm (single) and hydraulic scaling hammer
- Hagglund shuttle-cars 1×3 nos
- Rail-mounted Meyco shotcrete robot
- Rail-mounted grouting platform
- Schoema locomotives, flatcars and man-rider car

A fully equipped workshop at each shaft bottom has been provided, included inspection recess and wash basin.

The construction works has taken place 24 hours a day and 7 days a week, typically on 6 days on/3 days off scheme.

Explosives can only be delivered on normal working days (Monday to Saturday 9 a.m. to 6 p.m.), Sundays and statutory public holidays excluded. Explosives charging can only be done by local shotfirers; bulk emulsion explosives and Nonel detonators have been be used.

10   ROCK SPLITTING IN HONG KONG

In urban sites can be frequent the impossibility to use hydraulic hammer or road header for vibration or no production motivations (too hard rock in function of the excavation section), for similar reason drill & blast as well. From these consideration a new technology has been developed, called Drill & Split. In few words a dedicated jumbo is used to perform short (1-1.5 m) and closed axis holes (20 - 50 cm) in order to be enlarged by a suitable hydraulic rock splitter. The working principle of the rock splitter is based on two fixed matching steel probes being inserted into a pre drilled hole. A hydraulic cylinder mounted on the attachment pushes out a hardened steel wedge between the probes thereby causing the assembly to expand in the hole forcing the rock to crack and break away.

Of course high drilling precision is required, enough boom cinematic is necessary, fast drilling positioning procedure as well correct rock splitter handling are the key of the production success.

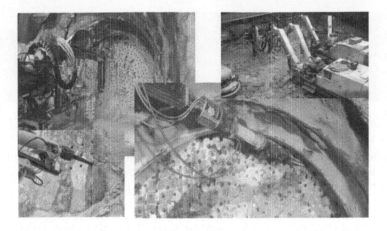

Figure 13.   Rock splitting in HK.

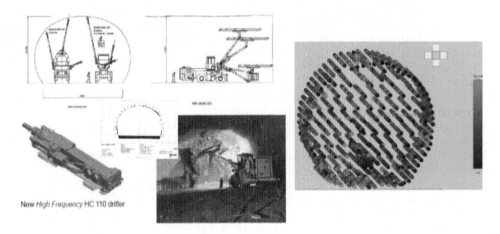

New *High Frequency* HC 110 drifter

Figure 14.   Drilling in granite in Hong Kong.

Robodrill has also designed a special jumbo with a view to providing an increased and varied range of articulation to aid the rock splitter (usually called Super Wedge) to be positioned for entry into pre drilled holes at the rock face. In this kind of equipment an horizontal slideway designed and manufactured by Robodrill has been incorporated into the design so as to give parallel advance and retract of the Super Wedge, thereby reducing jamming in the hole as would be expected if that movement was facilitated by the telescopic boom only.

In Hong Kong similar systems have been used in different projects as, for example, MTR contract 703, to drill shafts and tunnel lines in the hard granite (Fig. 13).

11   HIGH FREQUENCY DRIFTERS IN HONG KONG

Of course talking in general about hard granite, very frequently have been as well used high frequency drifters, the new Montabert HC110.

Example in HK can be referred to MTR - XRL Contract 820/821 (Fig. 14).

Figure 15. Long holes in faulty zones with 4 booms robotized jumbo and self-drilling bolt carousel.

These drifters have performed as expected with a very fast drilling cycle, optimum excavation profile, and are as well presently used in Brenner, Italy in combination with a sophisticated and performing PLC system.

12 BRENNER BASIS TUNNEL: USAGE OF SPECIAL DRILLING JUMBOS FOR LONG PROTECTION SELF-DRILLING REINFORCEMENT AND UMBRELLA

The Brenner Base Tunnel is a central element of the Corridor AV/AC Berlin-Monaco-Verona-Bologna-Palermo, crossing 10 regions of Italy, passing by the Brenner Pass, continuing along the Tyrrhenian regions, to arrive to Sicily; it represents a fundamental connecting route for transporting goods at long distance N-S Europe.

The Brenner Base Tunnel, about 55 km long, will consist of two single-track main tubes, connected by a tunnel cross every 333 m, with a distance between 40 and 70 m, with a circular cross section of ca. 4.05 m radius. The speed of the project, in line with European standards for the high speed lines, will be 250 km/h.

When fully operational, it will be crossed by at least 400 trains per day, of which 320 loaded with goods.

The north portal of the Brenner base tunnel is located just before the entrance station of Innsbruck, while the south portal is located at the entrance of the station Fortezza, in Italy. The portal of Mules is one of the lateral adits. It's a natural tunnel with a length of 1.8 km that reaches a maximum coverage of more than 1200 m, with a slope of about 8%. Along the tunnel line the structure of the rock is mainly massive granite which does not constitute a problem about the advancement as the excavation has been realized by Drill&Blast, using high performance Montabert drifters (when not excavated by TBM).

A heavy fault of ca. 250 m has obliged to use a particular consolidation technique constituted by protection umbrellas plus radial and front face R38 and R51, 12-18 m long, self-drilling anchors. Robodrill SA provided several conventional and robotized jumbos with 2/3 booms for the granite drilling operations to various Companies as Salini & Strabag, PAC – Oberosler – Cogeis, Europea 92 and CIPA S.p.A. In the fault some special jumbos, 4 booms (2 for face drilling, 1 for radial drilling + basket) have been supplied; the 2 front face feeds have been equipped with special rod handling systems in order to automatize the entire drilling procedure for the 12-18 m length operations. The system engineered have optimized drilling operations and speed up the fault passage in a considerable way (Fig. 15).

13 GRAND PARIS EXPRESS METRO

It consists in a fundamental rethink, redesign and focus on the public transport network on the scale of the metropolitan area. The purpose of this exercise is to avail Grand Paris with multimodal transport solutions, more integrated transport services, hence supporting a model of polycentric development. Grand Paris Express in main figures can be so defined:

- 4 additional lines
- 200 km of new railway lines
- 68 brand new interconnected stations

Figure 16.   Customized jumbos for special application in Gran Paris.

- 2 million passengers every day
- a train every 2 to 3 minutes
- a 100% automatic metro system
- 90% of lines will be built underground

Grand Paris Express, as an automated transit network, is the new metro of the Capital Region. With its 68 new stations and 200 kilometres of additional tracks, Grand Paris Express consists of a ring route around Paris (line 15) and lines connecting developing neighbourhoods (lines 16, 17 and 18). Additionally, Grand Paris Express also involves the extension of existing metro lines. Its 4 new lines circle the capital and provide connections with Paris' 3 airports, business districts and research clusters. It will service 165,000 companies and daily transport 2 million commuters.

At the moment of the paper submission Robodrill is working in the following sites:

- Prolongement du Métro Ligne 11 à l'Est – GROUPEMENT ALLIANCE
- EOLE – Prolongement du RER E vers l'ouest - Tronçon Saint-Lazare – Nanterre la Folie – Gare la Défense – Groupement EDEF
- Prolongement du Métro Ligne 4 – Lot T01 Montrouge Bagneux
- Prolongement du Métro Ligne 14 - Lot T02
- Ligne 15 - Puits de reconnaissance – ZAC Seguin Boulogne
- Eole Entonnement GC Haussmann Saint Lazare (HSL)
- Grand Paris Express/Ligne 15/Gare FIVC/Clamart

Works to be done have been various, as can be imagined; especially the drilling experience has been in the field of supports being them in fiber glass GFRP (mainly produced by the Italian MAPLAD) and in steel (DSI).

Special jumbos (Fig. 16) have been designed also for long injection holes, umbrella canopies or tunnel excavated per partial sections. Usually all the equipment have been calibrated in order to work in very narrow spaces; down town working procedures, metro access shafts between old historical buildings, tunnelling in weak materials (from the geological point of view but also for the potential existing foundation interferences) is part of the daily job.

14   CONCLUSIONS

The use of a modern drilling technology is a key for modern tunneling. The presence of a specialized drilling company is the key of the success not only for a complex job site; customizations capacity, flexibility, hardware & software top quality and perfect service are the necessary qualities for reaching a successful project.

*Tunnels and Underground Cities: Engineering and Innovation meet Archaeology,
Architecture and Art, Volume 5: Innovation in underground engineering,
materials and equipment - Part 1 – Peila, Viggiani & Celestino (Eds)
© 2020 Taylor & Francis Group, London, ISBN 978-0-367-46870-5*

The longest Robodrilled tunnel in the world

M. Bringiotti, V. Fleres & M. Manfredi
Cipa S.p.A., Rome, Italy

F. De Villeneuve
Robodrill S.A., Lyon, France

F. Serra
Geotunnel S.r.l., Genoa, Italy

ABSTRACT: The excavation in the lot named "Mules 2-3", being the biggest construction lot of the BBT Italian side, has been done by the "traditional method", which is including drill & blast and parts of consolidation & excavation under faulty zones. The works are related to the excavation and lining of the Access Gallery at the Trens Emergency Stop, Exploratory Tunnel, Main Tunnel south direction east and west tube and 19 Connecting Side Tunnels linking the two main tubes. In total ca.12 km. The use of Robodrill has improved the technologies adopted in order to fulfil the challenging needs of the job site. The paper will report about various data related to the production and costs in the different working sections, taking in account, for example, synthetic information about explosive consumption, drilling material, spritz beton and consequent lining concrete overconsumption.

1 INTRODUCTION

The Brenner Base Tunnel forms the central part of the Munich-Verona railway corridor. The project as a whole consists of a straight, flat railway tunnel, which reaches a length of 55 km and connects Fortezza (Italy) to Innsbruck (Austria); next to Innsbruck, the tunnel will interconnect with the existing railway bypass and will therefore reach a total extension of 64 km.

The tunnel configuration includes two main single-track tubes, which run parallel and 70 m across each other in most of the track, and connected to each other every 333 m by side tunnels.

Between the two main tunnels, and running 12 meters below them, an exploratory tunnel preliminary construction. The exploratory tunnel's main function during the construction phase is to provide detailed information about the rock mass; its position also allows it to perform important logistic support during the construction of the main tunnel, for the transport of the excavated material as well as for the transport of construction materials. During the operations, it will be essential for the drainage of the main tunnel.

In the next years, CIPA will perform most of the tunnel excavations in the lot named "Mules 2-3" by "traditional method". In addition to being the biggest construction lot of the Brenner Base Tunnel, it is the main part of the route on the Italian side, between the border of the State on the North (km 32.0+88 tube East) and the adjacent lot "Isarco river Underpass", on the South (km 54.0+15 tube west). Main tunnel configuration are shown in Fig. 1.

The works Cipa is responsible for are:

– Excavation and lining of the Access Gallery at the Trens emergency stop and the Central Tunnel, with a total length of approx. 4,500 m;

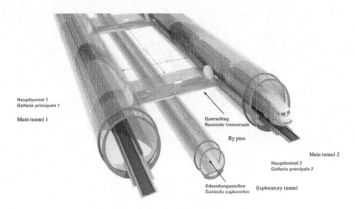

Figure 1. Main Brenner Basis tunnels configuration.

- Excavation and lining of the Exploratory Tunnel by traditional method, with a total length of approx. 830 m;
- Excavation and lining of the Main Tunnel, South direction East tube and West tube, in single track section, with a total length of approx. 7,320 m;
- Excavation and lining of the Main Tunnel, South direction East tube and West tube, in double track section, with a total length of approx. 2,590 m;
- Excavation and lining of 19 Connecting Side Tunnels linking the two main tubes, with a total length of approx. 900 m.

2 GEOLOGICAL DESCRIPTION

As anticipated, the Brenner Base Tunnel is the high-speed rail link between Italy and Austria, and therefore to Northern Europe. It consists of a system of tunnels, which include two one-track tubes, a service/exploratory tunnel that runs 12 m below and mostly parallel to the two main tunnels and bypasses between the two main tubes placed every 333 m, and three emergency stops located roughly 20 km apart from each other. The bypasses and the emergency stops are the heart of the safety system for the operational phase of this line.

Overburden is, on average, between 900 and 1000 m, with the highest points about 1800 m at the border between Italy and Austria.

The excavation will drive through all the geological formations that make up the eastern Alpine Area. Most of these are metamorphic rocks, consisting of Phyllites (22%), Schist (Carbonate Schist and Phyllite Schist, 41%) and Gneiss of various origin (14%). In addition, there are important amounts of plutonic rock (Brixen Granite and Tonalite, 14%) and rocks with various degrees of metamorphism, such as marble (9%).

Among the tectonic structures in Italy, we find the above-mentioned Periadriatic Fault. As stated, the unknown characteristics of the rock masses along this stretch lead to excavate the exploratory tunnel long before the main tunnels, allowing identification of the lithological sequences of the various types of rock mass within this heavily tectonized zone, as well as a study of their responses to excavation. These studies were subsequently useful to adjust the consolidation and support measures both for the exploratory tunnel and for the main tubes. The excavation of the exploratory tunnel inside the Periadriatic Faults allowed determining the actual sequence of lithologies within this area.

Fig. 2 shows a summary of the encountered rock sequences.

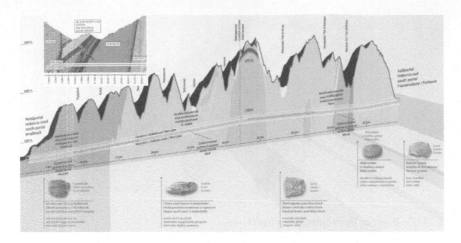

Figure 2. The geological sequence.

3 ACCESS TUNNEL AND TRENS CENTRAL CROSS PASSAGE

Access tunnel (GA), ending in the Trens Emergency Stop, whose length is approx. 3,805 m, starts in the Mules Adit (in a diversion tunnel at 1,4+79 m). After a long parallel and straight stretch to the West Main Tunnel, GA ends at the starting point of the Trens Central Cross Passage (CcT), ca. 680m long and linked with the Transversal Trens Cavern.

GA is excavated in full section, variable from 85 up to 105 sqm (Fig. 3), while CcT section has from 88 to 170 sqm but it's excavated partialzed.

Fresh air pumps through the Access Tunnel along the technical areas and in the Emergency Stop. In case of need, exhaust fumes are available for use from the Emergency Stop, too. The realization of the design of the separation of the two airflows used an intermediary slab (35 m thickness, 830 m length).

The tunnel then lines with a 41 cm thick concrete for definitive train use.

Access Tunnel (GA) and Trens Central Cross (CcT), from South to North, cross the following homogenous geological rock formation sectors: Brixen Granite (km 3+800 - 3+150 circa GA), Pusteria's Fault (km 3+150 - 2+950 circa GA), Mules' Tonaliti (km 2+950 - 2+350 GA), Mules Valley Fault, South (km 2+350 - 1+780 GA), Austrian-Alps Parascists and Anfibolites (km 1+780 - 0+200 GA), Superior Schieferhülle Calcescists and Anfibolites (km 0+200 GA - km 0 CcT). First drilling (blasting) and supporting are shown in Fig. 4.

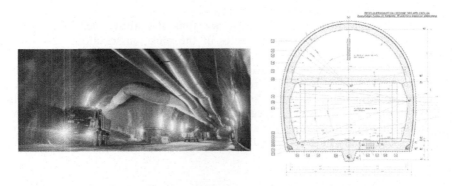

Figure 3. GA typical section and mucking out in the same tunnel.

Figure 4. First drilling (blasting) and supporting operations with Robodrill in GA Trens.

3.1 *GA Trens – consumptions and overconsumptions*

Synthetic data reported:

- Explosive consumption index: 1,30 kg/m^3
- Explosive type: Nonel detonators, emulsion and gelatin
- Explosive cost index: approx. 8,20 euro/m^3
- Average over excavation compared to the theoretical design profile: approx. +76 %
- Drill & blast consumable costs index: approx. 0,45 euro per drilled meter
- Spritz beton overconsumption compared to the theoretical designed thickness: approx. +25 %

There is to note that the average excavation radius increased is in the range of 25 cm, the cause relies on the geological conditions. Spritz beton overconsumption is very low due to the frequent steel arches absence.

4 EXPLORATORY TUNNEL IN ROBODRILLED METHOD

The final design states that the excavation and lining of the Exploratory Tunnel in the section between pk10+419 (connection with the Mules emergency access) and pk13+060, is carried out with the "traditional method". In this section, in fact, the excavation crosses the fault of "Val di Mules" first, and then a rock mass with better geo mechanical characteristics characterized by the presence of Parascists (the covers in this section vary from a minimum of 600 m up to a maximum of 1,135 m).

The excavation of the tunnel section between pk10+419 and pk12+459 already occurred within the Lot Mules 1. CIPA has been entrusted with the excavation and lining of the section from pk12+459 to pk13+060 (for a total length of approx. 600 m) with an excavation front section between 30 to 39sqm, two logistic chambers between pk12+580 and pk12+605 and between pk12+930 and pk12+955 (with an excavation section of approx. 80 sqm), as well as a cavern between pk13+000 and pk13+060 (with an excavation section of 137 sqm) inside which the first shielded TBM has been assembled for the mechanized excavation of the next section included in the lot "Mules 2-3".

The excavation works began in February 2017; during the first 8 months, CIPA almost entirely completed the planned section of approximately 600m, with an average production of 2.5 m/day and with the following typical sections:

- In the section between pk12+459 and pk12+524 ("Val di Mules" fault) section type CT-5 (Fig. 5), excavation with excavator and demolition hammer, maximum steps of 1.50 m, laying of steel ribs with profile type 2 x IPN160, fiber-reinforced shotcrete 5cm thick +

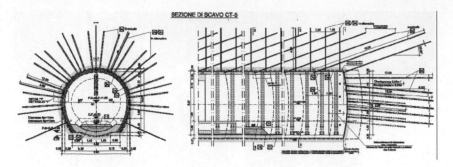

Figure 5. Typical CT-5 section.

25cm at the edge and 5cm at the front, pre-consolidation at the front with 26 self-drilling bars R51, pre-consolidation at the edge with 16 self-drilling bars R51 and radial consolidation with 11 self-drilling bars R38.

- In the section between pk12+459 and pk12+524 ("Val di Mules" fault) section type C-T5, excavation with excavator and demolition hammer, maximum steps of 1.50 m, laying steel ribs with profile type 2 x IPN160, fiber-reinforced shotcrete 5cm thick + 25cm at the edge and 5cm at the front, pre-consolidation at the front with 26 self-drilling bars R51, pre-consolidation at the edge with 16 self-drilling bars R51 and radial consolidation with 11 self-drilling bars R38.
- In the section between pk12+524 and pk12+569 section type C-T4, blasting excavation with maximum steps 1.5 m, laying steel ribs profile type 2 x IPN160, fiber-reinforced shotcrete 5cm thick + 25cm in the crown and 5cm at the front, radial consolidation with 11 rock bolts type Swellex PM24 L = 4.5 m. In Fig. 6 are shown some operative pictures in confined space.

For the excavation works and primary lining works, the following are the used equipment in the tunnel front:

- 1 Jumbo ROBODRILL with two booms for the drilling, for the blasting and for the rock bolting;
- 1 crawler excavator Hitachi ZX135 for digging, loading and unloading operations in the current section, equipped with a demolition hammer, bucket and hydraulic shears for bolt cutting at the front of the excavation;
- 1 crawler excavator Hitachi ZX240 for digging, loading and unloading operations in the bigger sections as logistic chambers and cavern (Fig. 7).
- 1 wheel loader Hitachi ZW310 and 4 trucks with 4-axle (which transport the excavated material to the logistic chambers for primary crushing);
- 1 shotcrete pump Cifa CSS3 for spraying the shotcrete;
- 1 telescopic handler Pegasus with man basket Fops.

Figure 6. Excavation in a narrow and confined space and self drilling anchors with a 2 boom Robo Jumbo.

Figure 7. Mucking out in the Exploratory tunnel after the drilling for blast execution.

The realization of the final lining of the entire sections (except for the invert in the fault zone that built along with the excavation) will occur by means of a formwork panels and shoring towers only after the excavation with the TBM.

4.1 *Exploratory tunnel – consumptions and overconsumptions*

Summarized data reported:

- Explosive consumption index Sez. CT3 (ca. 29 m^2): 2,06 kg/m^3
- Explosive type: Nonel detonators, emulsion and gelatine
- Explosive cost index: approx. 13,50 euro/m^3
- Explosive consumption index Sez. PL-C-T3 (ca. 72 m^2): 1,06 kg/m^3
- Explosive cost index: approx. 8,10 euro/m^3
- Explosive consumption index Sez. CMC-T3 (ca. 138 m^2): 1,06 kg/m^3
- Explosive cost index: approx. 7,10 euro/m^3
- Average over excavation compared to the theoretical design profile: approx. +80 %
- Drill consumable costs index (in the self-drilling zone): approx. 1,10 euro per drilled meter
- Drill & blast consumable costs index: approx. 0,27 euro per drilled meter
- Spritz beton overconsumption compared to the theoretical designed thickness: approx. +22 %

There is to note that the average excavation radius increased is in the range of 25 cm, the cause appointed to the geological conditions. Spritz beton overconsumption is very low due to the frequent steel arches absence.

5 MAIN TUNNELS, SOUTH DIRECTION

The sections of the Main Tunnel, which are to be excavated within the Lot Mules 2-3 with traditional method, are placed between km 49+083.7 and 54+015 (East Tube) and between km 49+058 and 54+042 (West Tube).

The project includes a single track for a first section of 3,500 m of tunnel for the East tube and approx. 3,800 m for the West tube (excavation area of approx. 65 sqm). Follows an enlarged section with double track (excavation area of approx. 114 sqm) with a length of approx. 1,400 m for the East tube and 1,200 m for the West tube. Within this section, 19 side tunnels connecting the two tubes (transversal tunnels with an average length of approx. 45 m); therefore requiring a further 900 m of excavation and lining connect the two tubes (Fig. 8).

The Connecting Side Tunnels have the following basic functions: connection of the two tubes of the main tunnel, escape and rescue routes in case of emergency, space for technical installations and drainage of the infiltration waters into the Exploratory Tunnel.

A single lithological unit, the "Bressanone Granite", divided into 11 homogeneous geomechanical sectors, characterizes this area.

In order to meet the requirements of the execution phase, the designs of the two main tubes and the connecting side tunnels include a final concrete lining (generally not reinforced) with a nominal thickness of 40 cm.

The excavation and lining of these 9,900 m of tunnel is going to be entirely with traditional method by CIPA, starting from the entrance of the logistic chambers and going south towards the end of the lot.

The excavation works began in April 2017, and will be brought to completion by the end of October 2020: the challenge is therefore to excavate an average of almost 9 m/day nonstop.

To date (31.07.18), 2,655 m of excavation have already been completed, consisting of the excavation section GL-TRB-TER, which considers an excavation with explosives with maximum steps of 3 m, fibre-reinforced shotcrete (and also reinforced with electro-welded mesh) 5cm thick + 10cm at the edge and 5cm at the front, radial consolidation with rock bolts type Swellex PM24 L = 4.5 m.

In addition, the first five connecting side tunnels (total length of approx. 220 m) were executed according to section CT1-TRB (Fig. 8). This considers an excavation with explosives with maximum steps of 1.5 m, fiber-reinforced shotcrete (and reinforced with electro-welded mesh) 5cm thick + 10cm at the edge and 5cm at the front, radial consolidation with rock bolts type Swellex PM16 L = 3.0 m.

For the execution of the excavation and of the primary lining in the two work fronts of both Main Tubes heading south, the following equipment has been used:

- 1 Jumbo ROBODRILL with 3 booms for drill and blast operations and for rock bolting (main tunnel);
- 1 spare quite aged jumbo Atlas as spare machine
- 1 Jumbo ROBODRILL with 2 booms for the drill and blast operations and for rock bolting (connecting side tunnels);
- 1 crawler excavator Case 370 + 1 crawler excavator CAT 330 for the scaling operations (refer to Fig. 9 for this and the next point);
- 1 wheel loader Hitachi ZW310 equipped with side unloading bucket and 10 4-axle trucks (which transport the excavated material to the logistic chambers for primary crushing);
- 2 shotcrete pumps Cifa CSS3 for spraying the shotcrete;
- 2 telescopic handlers Pegasus with man basket Fops.

The final lining of both tubes will be done simultaneously with the excavation with of 2 self-reacting formworks mounted on 12.5 m long tracks and 2 formwork systems wall/slab movable on wheels (Fig. 10).

5.1 *Main tunnel GL South – consumptions and overconsumptions*

Summarized data reported:

- Explosive consumption index Sez. GLTRb-Ter (ca. 65 m^2): 1,66 kg/m^3
- Explosive type: Nonel detonators, emulsion and gelatin
- Explosive cost index: approx. 9,00 euro/m^3

Figure 8.   Excavation section CT1-TRb, by-pass and Main Tunnel direction south single track.

Figure 9. Scaling and muck tipping loading operations.

Figure 10. Systems and formworks in use and ... St. Barbara!

- Explosive consumption index Sez. CT1-TRb (ca. 29 m^2): 2,2 kg/m^3
- Explosive cost index: approx. 14,00 euro/m^3
- Average over excavation compared to the theoretical design profile: approx. +60 %
- Drill & blast consumable costs index: approx. 0,35 euro per drilled meter
- Spritz beton overconsumption compared to the theoretical designed thickness: approx. +26 %

There is to note that the average excavation radius increased is in the range of 25 cm, the cause appointed to the geological conditions. Spritz beton overconsumption is very low due to the frequent steel arches absence.

Just a note: GL excavation along the South direction in D&B is the result of a design variation along the initial project phase; the first idea was to use the two TBMs, which were planned afterwards to excavate in northwards.

6 ROBODRILL: THE DRILLING PARTNER

Robodrill has been part of the Brenner's project almost ever since the beginning; 3 years with PAC involved in the excavation of all the logistic down the Mules adit, currently also working with Europea92 and the Consortium Salini-Impregilo & Strabag in Isarco projects, too.

Robodrill has been able to serve in a quite difficult environment various jobs sites and Companies for years, providing jumbos, manpower, elettro-mechanical services.

It has also contributed to develop new drilling strategies designing efficient systems, for example machines (with 4 drilling booms), which are able to work with self-drilling anchors in difficult sections, as well 2 booms, computerized, able to D&B in narrow niches in some of the main sections.

Through the collaboration with CAT and Montabert new heavy-duty jumbos have been designed and used on hard granite as well as on faulty areas. An excellent and updated computerized system has been able to limit drilling positioning time and over breaks at the maximum possibilities. Some visual images can be seen in Fig. 11.

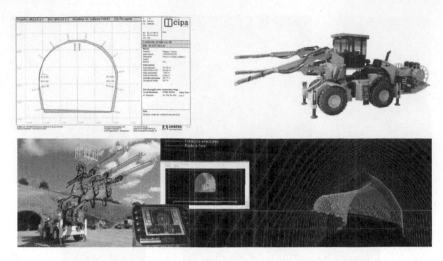

Figure 11.   Accuracy in Brenner tunnel, new CAT jumbos and I4.0 PLC Robo-Geo-Controlled system.

7   DRILLING PERFORMANCE IN THE SITE

Cipa's Company has been able, in such a complex situation, to perform at the limit of human possibilities. As follows, a synthesis of the production data up to end July 2018 (Fig. 12,13).

EXCAVATION MONTHLY PRODUCTION GL EAST SOUTH				
REF [month]	MONTHLY PRODUCTION [m]	DAILY RATE on MONTHLY PRODUCTION [d.a.c.]	MONTHLY CUMULATIVE PRODUCTION [m]	DAILY RATE on MONTHLY CUMULATIVE PRODUCTION [d.a.c.]
ottobre-17	162,20	5,23	162,20	5,23
novembre-17	128,10	4,27	290,30	1,00
dicembre-17	91,70	2,96	382,00	4,15
gennaio-18	130,70	4,22	512,70	4,17
febbraio-18	116,80	4,17	629,50	4,17
marzo-18	110,50	3,66	740,00	4,06
aprile-18	103,50	3,45	846,50	3,39
maggio-18	155,90	5,03	1.002,40	4,13
giugno-18	146,60	4,89	1.149,00	4,21
luglio-18	163,70	5,28	1.312,70	4,32

EXCAVATION MONTHLY PRODUCTION GL WEST SOUTH				
REF [month]	MONTHLY PRODUCTION [m]	DAILY RATE on MONTHLY PRODUCTION [d.a.c.]	MONTHLY CUMULATIVE PRODUCTION [m]	DAILY RATE on MONTHLY CUMULATIVE PRODUCTION [d.a.c.]
maggio-17	52,20	1,68	52,20	1,68
giugno-17	109,20	2,64	161,40	2,85
luglio-17	135,70	4,38	297,10	2,23
agosto-17	90,90	2,93	388,00	2,15
settembre-17	154,30	5,14	542,30	2,74
ottobre-17	165,10	5,33	707,40	2,94
novembre-17	142,90	4,76	850,20	2,97
dicembre-17	83,40	2,69	933,60	3,01
gennaio-18	134,60	4,35	968,00	3,37
febbraio-18	122,90	4,39	1.180,90	3,54
marzo-18	129,60	4,35	1.257,60	3,69
aprile-18	123,00	4,10	1.380,60	3,73
maggio-18	153,30	4,94	1.533,90	3,82
giugno-18	144,00	4,80	1.651,50	3,89
luglio-18	159,30	5,14	1.896,80	3,98

Figure 12.   Productivity in GL East and West South tubes.

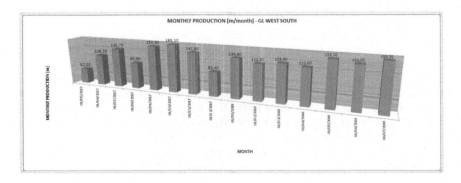

Figure 13.   Productivity in GL East and West South tubes.

				Year 2017			Year 2018						
				October	November	December	January	February	March	April	May	June	July
GL South - Tube EAST	Explosive	Rate (kg/cm)	Partial month	1,68	1,78	1,73	1,87	1,7	1,56	1,89	1,69	1,78	1,67
			Progressive	1,68	1,72	1,72	1,76	1,75	1,72	1,74	1,73	1,74	1,73
	Spritz beton	Extra quantity (cm)	Partial month	60,89	109,92	101,39	118,91	120,05	152,3	151,78	98,12	131,88	91,58
			Progressive	60,89	170,81	272,2	391,11	511,16	663,46	815,24	913,36	1045,24	1136,82
		Percentage (%)	Partial month	20,36%	20,93%	27,43%	29,12%	31,03%	39,56%	44,33%	15,48%	21,87%	13,36%
			Progressive	20,36%	20,73%	22,60%	24,75%	26,17%	28,53%	30,67%	24,10%	23,79%	22,39%
	Final lining concrete over consumption calculated	Extra quantity (cm)	Partial month	929,6	845,13	709,51	661,18	725,17	822,31	261,46	1510,66	822,7	412,2
			Progressive	929,6	1774,73	2484,24	3166,42	3890,59	4712,9	4974,38	6485,04	7307,74	7719,94
		Percentage (%)	Partial month	55,00%	64,00%	76,00%	51,00%	60,00%	78,00%	78,00%	73,00%	52,00%	50,00%
			Progressive	55,00%	59,03%	62,99%	60,04%	60,13%	62,67%	63,32%	65,41%	63,58%	62,66%
GL South - Tube WEST	Explosive	Incidenza (kg/mc)	Partial month	1,68	1,78	1,73	1,87	1,7	1,56	1,89	1,69	1,78	1,67
			Progressive	1,68	1,72	1,72	1,76	1,75	1,72	1,74	1,73	1,74	1,73
	Spritz beton	Extra quantity (cm)	Partial month	63,5	100,16	163,6	114,31	136,52	149,64	161,56	97,89	96,25	144,02
			Progressive	63,5	163,66	327,25	441,56	578,08	727,72	889,28	987,17	1083,42	1227,44
		Percentage (%)	Partial month	21,49%	17,27%	47,76%	25,02%	30,50%	32,64%	36,61%	16,10%	15,84%	21,99%
			Progressive	21,49%	18,70%	26,87%	26,30%	27,26%	28,27%	29,54%	24,94%	23,73%	23,51%
	Final lining concrete over consumption calculated	Extra quantity (cm)	Partial month	953,2	838,79	425,73	656,64	1081,01	1111,65	859,53	1388,52	637,46	538,81
			Progressive	953,2	1791,99	2217,72	2874,36	3955,37	5067,02	5926,55	7315,07	7952,53	8491,34
		Percentage (%)	Partial month	58,00%	57,00%	50,00%	48,00%	88,00%	96,00%	104,00%	75,00%	53,00%	61,00%
			Progressive	58,00%	57,47%	55,80%	53,83%	60,00%	61,69%	64,14%	65,37%	64,51%	64,37%
Transversal tunnel	Explosive	Incidenza (kg/mc)	Partial month	Explosive incidence calculated in average rate GL SOUTH - specific rate approx 2,22 kg/cm									
			Progressive										
	Spritz beton	Extra quantity (cm)	Partial month	10,51	17,24	-2,34	36,52	0	39,58	0	1,67	13,65	0
			Progressive	10,51	27,75	25,41	61,93	61,93	101,51	101,51	103,18	116,83	116,83
		Percentage (%)	Partial month	11,00%	21,00%	-16,00%	45,00%	0,00%	54,00%	0,00%	12,00%	21,00%	0,00%
			Progressive	11,00%	14,00%	11,95%	21,13%	21,13%	27,70%	27,70%	27,17%	26,30%	26,30%

Figure 14.   Consumption and spritz & final lining concrete overconsumption rate.

8 CONCLUSIONS

Let's talk about overconsumption! The tunnel is not lined with steel arches which means that the spritz rebound should be quite limited, in the range of 10%. In reality, due to geological conditions, even if in presence of competent granite (but frequently bursting, spalling and not only . . .), the job site suffered 10% higher spritz beton consumption (mainly for profile regularization) plus an average of ca. 50-60% concrete over consumption for the final lining (data reported in Fig. 14), costs which should need to be professionally discussed (and absorbed by the Client). An average of excavated radius increase, in the range of 20-25 cm, has been systematically detected, whatever kind of action taken in order to limit it.

REFERENCES

Bringiotti, M. 2003. *Guida al Tunnelling – l'evoluzione e la sfida, 2001*, Edizioni PEI, Parma
Fuoco, S., Zurlo, R. & Lanconelli, M. 2017. *Tunnel deformation limits and interaction with cavity support: The experience inside the exploratory tunnel of the Brenner Base Tunnel*, Proceedings of the World Tunnel Congress – Surface challenges – Underground solutions, Bergen, Norway
Rehbock, M., Radončić, N., Crapp, R. & Insam R. 2017. *The Brenner Base Tunnel, Overview and TBM Specifications at the Austrian Side*, Proceedings of the World Tunnel Congress – Surface challenges – Underground solutions, Bergen, Norway

Tunnels and Underground Cities: Engineering and Innovation meet Archaeology,
Architecture and Art, Volume 5: Innovation in underground engineering,
materials and equipment - Part 1 – Peila, Viggiani & Celestino (Eds)
© 2020 Taylor & Francis Group, London, ISBN 978-0-367-46870-5

Motivations to swap from trench cutter to hydraulic grab: Case histories from Florence, Rome and Milan

M. Bringiotti & D. Nicastro
GeoTunnel S.r.l., Genoa, Italy

S. Bechter
Liebherr-Werk Nenzing GmbH, Nenzing, Austria

G. Franco
SAOS S.r.l., Frosinone, Italy

G. Manuele
Maplad S.r.l., Catania, Italy

ABSTRACT: Frequently new Metro lines have to cross their path with existing ones, combined with the frequent need to encounter better geological conditions and preserve archeological heritages; due to this tunnels, stations and technological shafts are often designed to be realized at deeper depth. Another topic is related to the very frequent limited construction space available with space constraint, interferences with existing buildings and infrastructures. In the last 2 decades trench cutter technology evolution has been quite and as a matter of fact the world is looking for something new since a long time. What has been really, technologically speaking, evolving in the last decade is the "marriage" between heavy duty crawler crane and both mechanical and hydraulic grab. The well-known traditional diaphragm wall excavation has experienced new applications also in urban metro projects; deeper depth, higher accuracy, top joint quality, easy and fast mob & demob phases in limited site spaces.

1 INTRODUCTION

For the construction of diaphragm walls, grabs, both mechanical or hydraulic in function of the jaws closing system, are frequently used. It can be named that it is more and more an international practice for the construction of very deep walls (and not only) to use combined systems: crawler excavator and hydraulic grab in its different technical declinations; this combination represents a real modern evolution for the excavation of diaphragms walls.

A lifting and excavating equipment, such as a crawler crane class LWN HD-HS, powerful, silent, with generous, redundant and fast winches, with integrated controls coupled to a hydraulic grab (there are some types worth to mention) may really make a difference in a project that foresees high depths, cutting precision, work in urban environment, minimum pollution (also acoustic) and optimal management of the excavation materials

2 HYDRAULIC GRABS

In general, the diaphragm excavation grab may be of two different types: mechanical and hydraulic. The mechanical grab is the easiest to be manufactured and requires a crawler excavator equipped with 2 winches; the first able to lift the grab body and the second able to operate the mechanical kinematic of opening and closing the grab jaws. The mechanical grab, with freefall winch, although in certain situations can be slightly faster than the hydraulic one and may be efficiently used for repeated chiselling works (particularly resistant materials):

1. Of course requires a heavy duty type crane
2. The grab is particularly mechanically stressed
3. Definitely requires a very experienced operator
4. The operator must always be very concentrated during the entire working phase
5. It can rotate only by using the winding direction of the ropes by manually using the winches
6. It is not equipped with deviation correction systems
7. It cannot manage for high depths deviations
8. It cause excessive ropes consumption

The hydraulic grab, besides, has a whole series of undoubted advantages due to its management through suitable hydraulic circuits that allow it to tighten the excavations jaws, rotate, divert ... with safe, controlled, precise and "smooth" manoeuvres. For example, the grab model used for the excavation of some diaphragms walls in the Milan Metro has been equipped with a patented grab rotation device that allows the structure to be swing at 180° on the longitudinal axis. This system can also reduce the excavation deviation and, especially in the case of hard soils, considerably improves the production performance; the device allows to realize panels in narrow spaces or corners where the carrier cannot be perfectly positioned in front of the excavation (Fig. 1).

The asymmetry of the excavation teeth between the two jaws is compensated by the 180° rotation device. This equipment may be supplied with special thrust systems placed in various areas of the grab frame, used for the correction of the grab deviation during the excavation thus in order to allow a constant verticality (Fig. 2). Usually grabs are equipped with two sets of steering plates (flaps, pushing cylinders, . . .), which may be operated independently. This operation and the effect on the deviation is controlled and monitored by an inclinometer installed on the grab itself and displayed in the operator's cab. It may be directly linked to the crane through different connection systems, depending on the type of carrier, the type of hoses reeling and the size of the panel to be realized. The control system is usually designed to continuously monitor the verticality of the grab during the excavation. The position of the grab is displayed on the operators monitor in real time; an inclinometer

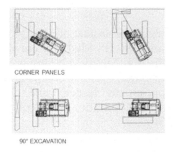

CORNER PANELS

90° EXCAVATION

Figure 1.   Some usual grab working position.

Figure 2.   Steering plates and rotating system to drive with the right verticality.

is installed in the frame in order to measure the inclination along two orthogonal axes (x and y).

The data transfer from the inclinometer to the operator's cabin is carried out via a particularly resistant power cable, which to follow the grab movements automatically winded or unwinded by a dedicated hydraulic cable reel. There are interesting variants that involve the information transfer through batteries properly installed on the digging body but which have the drawback of returning the information only at the end of each working cycle. Naturally data are displayed in a touch screen and can be stored and printed.

2.1 *Dragon white*

For some of the M4 Stations (Metro Milan), for example, it has been used an hydraulic grab called Dragon White (design and manufactured by Negrini S.r.l.) specifically configured according to the operational requirements, handled by a new Liebherr crane type HS 8100 HD in order to primarily excavate 2,8 x 1 m diaphragms wall. It is a modular hydraulic grab integrated with the operating machine without having to modify its support structure or having to re-certify the crane boom and neither the boom head. It may be manufactured with several optional accessories to increase its performance and it is mainly composed by a central body and an excavation set divided into: jaws, ejector, jaw supports (by simply replacing jaw supports it may be possible to achieve a wide range of excavation dimensions: from a minimum of about 500 x 2500 mm up to a maximum one of about 1800 x 3600 mm) and hulls. By fitting a special hull handling device, it may be possible to adjust the excavation verticality using the guided hulls. The hulls may be remotely managed (therefore not using electrical cables) directly from the operator's cabin; without getting out of the crane's cabin, by means of a transmitter integrated in the "machine system" and the supplied joysticks, it may be possible to handle all the excavation operations. The receiver is integrated into a suitable enclosure fitted inside the top frame of the grab. All accessories are interchangeable between hydraulic and mechanical grab; a significant economic advantage because different excavation set may be used on both type. A specific extension can increases the grab height, allowing greater verticality of the excavation at high depths. In hard soils or particularly high depths, finally, it may be possible to apply additional counterweights, adding weight and lowering the center of gravity of the grab, resulting – as for the Milan Metro – in an effective and versatile device.

3 THE CRAWLED EXCAVATOR LWN HS 8100 HD

The crane subject of this brief description is part of a range of machines specifically designed for special foundations application. The HS 8100 HD crawler excavator, which for example has been used for the works of the M4 (Metro Milan), has been designed to perform various foundations works such as: trench cutter, dynamic compaction, drag line, several kinds of seaport works, with the help of appropriate "leader": pile, CFA, vibropiling, vibrofloatation, bored piles grab and casing oscillator, application frequently used for the construction of deep piles of large diameter, . . . (Fig. 3)

Figure 3. The HS8100 with oscillator.

3.1 Operating weight: 115 ton

This parameter has a significant importance since the stability of the entire working group has a close connection with the mass of the undercarriage and all the components = > greater weight, greater stiffness, less vibrations and consequently less deviations (in addition to minor possible breakages for fatigue)

3.2 Motor power and consumption: 390 kW

The engine power is a feature that clearly links all the machine hydraulic functions; having more power means not only be faster in all functions but also to generate spare power in potential critical situations i.e., for example, at high depths. Moreover, the greater power can be useful to manage different kind of works (trench cuter, vibrators, ...). The Liebherr engine is of the latest generation Tier 4i; it is provided with DPF regenerative particle filters, which greatly reduce the environmental impact of machines and allow them to be used in sites particularly environmentally sensitive.

The abundant power available means having an engine that works at a lower rpm, despite having spare power available. LWN has developed low-consumption engine, integrated through the electronics with the hydraulic system, having an innovative concept called ECO-Mode able to limit the fuel consumption; summarizing, it is a new generation 8 cylinders engine with an electronic system that optimizes or disconnects all the hydraulic services when not used. Moreover, thanks to the automatic thermoregulation of the fan system (which allows to work in any climate) helps to reduce fuel consumption. It is reasonable to say that the hourly consumption of the Liebherr combination (Engine-Electronic-Hydraulic) is lower at least of about 20% lower compared to the consumption of comparable machines employed for similar kind of works, per lineal meter of excavated panel. For instance, for the Port Said Project, the average consumptions for 100 h of work has been equal to 16,2 l/h!

3.3 Noise emissions

With regards to noise emission, it has been developed a specific system called ECO.SILENT-Mode; it is a function that when activated by the operator limits the number of rpm of the engine as well as of the cooling fan, ensuring a significant reduction of noise emission, with a negligent loss of power. This is a very important feature especially in urban areas operation. (i.e. construction works of diaphragms wall for metro stations, such lastly Milan, Rome, Paris, London or Warsaw). In a construction site, it is clearly perceived the difference between HS 8100 HD and all any other machine available on the same site.

3.4 Winches and depth

The winches are outstanding in their compact design and easy assembly. Clutch and braking functions on the free-fall system are provided by a compact designed, low wear and maintenance-free multi–disc brake (Fig. 4).

Figure 4.   Well known LWN high quality double winches.

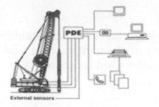

Figure 5. The communication net system Li-tronic + PDE.

Figure 6. Some of the main brand of foundations rig with grab.

The drag and hoist winches use pressure controlled and variable flow hydraulic motors. This system features sensors that automatically adjust oil flow in order to provide max. winch speed depending on load. The rated speed of the 2 synchronized winches (stnd. 25 ton capacity) is equal to 88 m/min with approx. 40 m of rope in the first layer. It is reasonable to be said that the pure speed of the LWN HS 8100 HD is at least 20-30% higher when compared to similar machines and the winch speed is one of the key points for excavation productivity! The depth achievable in standard configuration is equal to 120 m.

3.5 *Telematics System for assistance*

The integrated Li-Tronic + PDE system allows not only the transmission of operational data but also the remote assistance from the manufacturer headquarter; this "architecture" apart from being a modern feature is also an important operational advantage in the construction sites and a great savings in case of problem occurrence, easing a fast assistance service knowing in advance the cause of any problem, frequently been able to remotely fix it (Fig. 5).

3.6 *Crane's boom and compatibility with any kind of grab*

The duty cycle crane's boom, which is mainly used with grab and other multifunction tool for deep foundation works (Fig. 6), shall be able to withstand significantly higher loads of a standard crane boom. The loads are generated essentially by the horizontal forces due to the lateral grab acceleration, by the potential lateral flexion forces, by the dynamic forces connected to the operation of the grab itself and by the auxiliary tools applied to the boom; the lateral stresses components are also added to the axial forces generally stressing the boom. The structure is therefore dimensioned for heavy loads thus in order to absorb all these actions. For the record, it is worth to mention that the LWN system is compatible with any kind of grab: mechanical, hydraulic, kelly and semi-kelly.

4 INTEGRATION OF DUTY CYCLE CRANE AND GRAB

It should be noted that there is an important integration between the excavator and the excavator tool (the grab, in this case); in short, there is a hydraulic and electronic dialogue platform that transfers information, commands, data returns (in visual, graphic and system

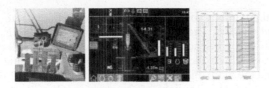

Figure 7. Li-Tronic System on board.

format) and can transmit them to the User and to the Mother House with the other potential related analysis and active interventions, such as maintenance, if needed.

4.1 *Integration, control and data management*

The crane/grab system can be integrated with the LWN electronic system called Li-Tronic. The crane management system is the centerpiece for precise and reliable crane operation (Fig. 7). Li-Tronic is a unique one-system-solution which unites the best technologies available for reliable crane control; it is the one-level interface between crane and driver. Various analysis tools provide relevant information on the operation. The system empowers the driver to efficiently control the crane and to optimize turnover.

4.2 *Grab chiseling*

Chiseling with an hydraulic grab is not really a usual practice, nevertheless LWN system provides a specific integrated function allowing to perform chiseling with grab operating in free-fall at a certain height. This option can be used in various scenario, for instance: breaking through supporting slabs, excavating lenses of compact material at variable depths, headings in the bed rock ...

4.3 *Grab safety and double winch availability*

The fact that these machines work with 2 synchronized winches and consequently with 2 ropes simultaneously guarantees a redundant system not only for the loads lifting but also providing a high degree of safety for the grab itself. It has occurred more than once, having a single grab lifting block (i.e. crane equipped with a sole winch only) that rope got broken subject to excessive fatigue or wearing, particularly in the area of the return pulley on top of the grab lifting block (the lower pulley, very delicate zone!). Moreover, these types of machines are first of all cranes; their system, manufactured considering the strict referred design standards, takes care of monitoring the lifting loads as well as the position of the boom, putting them in closed relation. There are excavation machines suitably designed for such operations (lifting), which may not be provided with any load safe indicator when used just as excavators (it's also an accuracy and integrated stress calculation problem) and not as lifting equipment, a clear limit and a safety issue handling extraordinary unbalanced loads at high depths.

4.4 *Ropes tensioning and wearing*

The load indicator device is also used to calibrate the pressures of the winches in order to keep the rope in tension (without affecting the grab) during the excavation phase; this ensures no loosening, unloading of the pulleys, irregular wear, offload operation of the winches, tears, undesired turbulence along the panel walls, ... This specific function aims to reduce the deviation tendency during the operational phase, a further factor minimizing the danger of deviation from the design verticality. A further consideration on the ropes wearing shall be made on the geometry of the reeling drums of the winches; the greater is the drum length or the diameter, the greater will be the reeling length in the first layer of the rope, the less will be the wearing.

Therefore, it is essential, in order to optimize the useful ropes lifetime, to provide larger winches (as well as powerful and reliable). The useful life of a lifting rope is in direct function of the type of material to be excavated (presence of stones hanging on the pulleys, anomalous wearing cause by friction, ...) and of the excavation depth (numbers of reeling cycles and multi-layer reeling, ...). Considering the mechanical grab, the technology foresees 2 ropes having 2 different tasks: the first winch bears the full weight of the grab (while with the hydraulic version the weight is equally distributed on both ropes), being this element particularly stressed. The second rope is engaged for the excavation jaws opening and closing procedure; it passes through a series of suitably dimensioned pulleys and through a numbers of referral mechanism (with short curvature radii) as well as the presence (quite frequent) of soil between the groove of the pulley and the rope itself, resulting in a reduced useful life. Moreover, the ropes consumption is not symmetrical, being subjected to anomalous strains. It is reasonable to say that the ratio of the ropes lifetime for the mechanical grab vs the hydraulic grab is 1:5 (indicatively 500 h: 2500 h); steel ropes not only have a cost and are related to operational safety, but also provide long stop for their replacement ... at the expense of production.

5 THE FLORENCE HIGH SPEED RAILWAY LINK PROJECT

The works related to the Florence High Speed Railway Link (Fig. 8) and the realization of the new high speed train station represents a kind of urban revolution of the famous city; it provides for a vertically distributed development (a "negative" skyscraper), more like a large metro stop than a railway station. The new structure has been designed by Norman Foster & Partners; the platform level in the new station is located 25 meters below ground. The station chamber consists of a single volume, 454 meters long and 52 meters wide, built using cut-and-cover techniques similar to those deployed at Canary Wharf Station in London. The station will be linked to the existing Santa Maria Novella station, 2 km to the south, with the aim of realizing a multimodal hub sensitive to its historic location, but looking forward in its use of energy and other resources, offering a model for contemporary rail travel.

The diaphragms walling has been realized providing complex reinforcement having the slab rebars directly link to the bulkheads, the latter working in the entire structure and, by virtue of that choice, no longer having the need to build a foundation supporting wall, but more likely a make-up lining of the existing one.

For the realization of the diaphragm walls n. 5 duty cycle cranes, LWN HS-HD class, have been deployed, owned and operated by SAOS S.r.l., a well-known company specialized in deep foundation works; three working crews has been arranged for the excavation of 40 m deep diaphragm walls handled by two HS 855 HD (450 kW power) and one HS 845 HD both equipped with hydraulic grab, while a HS 833 and a HS835 cranes have been employed as service cranes, handling the 36 ton rebar cages. The site manager, Geom. Fabio Rossi, reported: "Negrini has provided us with special bucket jaws and grab shoulders for the existing grabs and with Liebherr we have designed and implemented a new software solutions with the aim to increase the effectiveness of the diaphragm slurry walls realization. We may say

Figure 8.   Florence High Speed Railway.

that it has been possible to pursue a wide range of experimentation benefitting of the unlimited horizons provide by Liebherr's research and development on the heavy cycle machines. Without this assumption, we could not have been able to give such a decisive impetus to the planned intervention times of this works."

6 METRO LINE 4 IN MILANO AND SAOS COMPANY

For some years, the city of Milan is increasingly projected into a European scenario, particularly from the point of view of services designed to ensure the best sustainable mobility. One of the latest improvements involving the city infrastructure is Metro 4 (M4 or "Blue Line"), which will allow the Lombard capital to compete with other European capitals in terms of transport as well (Fig. 9). Among the stated goals of M4 S.p.A. - the company that manages the works and which was set up in 2014 by the Milan City Council as a majority shareholder - is the decline in wheeled traffic for the benefit of public transport, with a forecast of 30 million moves in less annually and reducing pollutant emissions by 2%, about 16 million ton less fuel (figures which, however, also include the contribution provided by the new M5 service).

Line 4 crosses the city transversely by connecting the east and west boundaries and passing through the historic center along the Cerchia dei Navigli; by importance of the connected and crossed areas can be compared to Line 1. Given its track and intersections with the other lines and the Passive Railroad will be of fundamental development at the metropolitan and regional TPL. By 2022 the metro network in Milan will be the sixth in Europe; with M4, the metro of Milan will have a total of 118 km of line, 136 stations and an increase in the network in 10 years by 34 %, a pace of development among the highest in Europe.

The main features of the M4 are: 15 km of extension, 21 stations, 86 million passengers a year, 6 interchange points (Sant'Ambrogio M2, San Babila M1, Crocetta/Policlinico Sforza M3, Bypass Dateo, St. Christopher FS, Forlanini FS), 1 workshop in San Cristoforo; frequency in peak/hours 90 seconds, maximum speed 80 km/h, automated without driver.

The features of the "center" stations are (Fig. 10):

– Central Stations are defined as "deep typological stations" (about 30 m in depth)
– Located in the city center, in the oldest and most urbanized part, these stations have greater constraints than those of the outer trails

Figure 9.   The Line 4 in Milan.

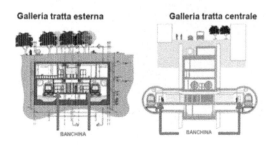

Figure 10.   Typical view section of the Line 4.

- The solution adopted combines the small spaces on the surface with the necessary distance of the galleries where the trains run, through the creation of a deep central shaft between the two tunnels
- Inside the tunnels they find the train travel routes and the related platforms.

To carry out this kind of intervention, the Contractor has entrusted an important number of foundational works to S.A.O.S. (Società Appalti Opere Specializzate) S.r.l. - a company specialized in foundations and which plays a role as interlocutor of large general construction companies for the creation of special foundations.

SAOS used its existing LWN diaphragms crane fleet and also purchased a new HS 8100 HD crawler excavator and a special hydraulic grab from Negrini.

7 DIAPHRAGM WALLS AND JOINTS FOR THE BLU LINE

The project of the Metro Milan Line foresees the construction of reinforced concrete diaphragms walls, side by side eventually bonded by struts and tie back rods, as supports structures for the excavations of deep shaft stations and line tunnels, or in any case, in general, for underground works. The drilling technique is normally based on the use of bentonite slurry (Fig. 11). "Dry" excavation, without the use of stabilization slurry, is not allowed. Excavation shall not, in any case, compromise or worsen the mechanical characteristics of the ground surrounding the diaphragm; therefore, it shall be avoided or at least minimized: any possible variations of the groundwater levels, the reduction of the relative density of the incoherent layers, the decrease of the effective horizontal tensions of the natural soil state and the adhesion reduction between the diaphragm and surrounded ground due to an improper use of the bentonite slurry. Realizing diaphragms slurry walls in urban area, means operating: in presence of road traffic, close to existing buildings and their foundations, balconies, canopies and any other protruding structures, street lighting, electric overhead lines, sewer pipeline and any other subservice, it is therefore required to conduct such a works by using proper equipment taking care of ensuring the full functionality of these services, constantly monitoring them during the whole the duration of the works. The employed equipment shall also comply with the stringent regulation and standards related to noise emission and vibration transmission to the surrounding environment. Diaphragm walls shall be realized bearing into consideration their precise geo-localization and in compliance to the parameters and tolerances given by the designer (with exception of stricter limitations indicated in particular situations) such as:

- Floor plan: ± 3cm
- Diaphragm head: ± 5cm

Figure 11.   Excavation with slurry grab in down town Milano.

- Depth: ± 25cm
- Verticality: ± 0,5% and, in any case, not to involve, by the work performed, an off-axis of the internal line of the plane in relation to the vertical one through the path of the track exceeding 10 cm (for total diaphragm length ≤ 20 m)
- Joints complanarity: ± 1% which implies a restriction in the permissible tolerance for the absolute verticality of the panels; if a panel vertical offset is equal to 1%, the adjacent panel shall bear the same vertical offset following the same direction, or it shall be perfectly vertical.

During the excavation of the diaphragm panels, the machine systems and operators shall be able to monitor and record verticality, stratigraphy, the actual depth reached by the excavation tool, bentonite slurry consumption (thus in order to record any slurry loss due to anomalous ground condition) and as well as any other relevant information. The excavation technology could be chosen by the Client and the Contractor between the various types of grabs or trench cutter; of course, with high depths (> 50 m), potential presence of resistant hard materials, proximity to existing (historical) buildings, space and environmental constraints as well as the need to comply and guarantee high construction tolerances, it could have driven Everybody to the trench cutter technology; this has not happened!

The typical phases for the realization of diaphragm walls are as follows (Fig. 12):

- Preparation of the ground surface
- Topographic alignment
- Realization of guide walls
- Consolidation injections where required
- Preparation of the stabilizing fluid (bentonite slurry)
  Excavation with grab in bentonite slurry
- Installation of the rebars cages;
- Concrete placing and retrieval of the bentonite slurry
- Diaphragm wall top level cropping
- Head concrete beam construction

The diaphragms walls shall be realized assuring a perfect hydraulic seal (in Milan the spring line is everywhere at very low level); all its elements shall be perfectly in contact and seamlessly compenetrating. Therefore, realizing a waterproof diaphragm to ensure a perfect hydraulic joint, before the casting at both ends of the primary panel, a stop-end shall be placed, which usually consists in a:

- Form pipe (stop hand) with diameter equal to the diaphragm thickness
- Suitable shaped elements (semicircular, sheet pile, etc.)

Project's specification requires that these elements are to be removed after the pouring (after concrete setting), leaving in the casted element their semi-circular footprint, interfacing groove. During the excavation of the secondary panels, it is necessary to clean the groove of both panel ends, thus is order to produce a rectangular shaped groove in order to have a perfectly matching joint during the casting of the secondary panel. Alternatively, different types of joints may be

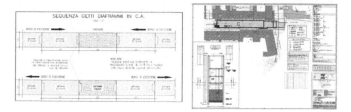

Figure 12.   Usual phases to realize panels by diaphragm method.

realized, i.e. by fitting tubes or plastic formwork panels. In this case, during the excavation of the secondary panels, the plastic material of the formwork needs to be taken out by the grab; this will result in a concave impression similar to that described above.

Diaphragm wall cannot be realized continually for a very long section due to various limitations; the wall is usually constructed in alternative section. Two stop end tubes will be placed at the ends of the excavated trench before concreting; the designed technology foreseen that the tubes are withdrawn at the same time of concreting so that a semi-circular end section is formed. Wall sections are formed alternatively leaving an intermediate section in between. The in-between sections are built similarly afterward but without the end tube. At the end a continual diaphragm wall is realized with the panel sections tightly joined by the semi-circular groove.

7.1 *MM Line 4: Usage of GFRP Rebar Cages for Tunnel Boring Machine "Soft-eye" openings*

Metro Line 4 will be serving the densely populated areas in city center of Milan. In order to minimize disruption caused by construction activities, it has been designed to be compatible with other modes of transport and maintain sufficient groundwater level. Metro Line 4 will have twin tunnels with single tracks in each direction. Extensive use of tunnel boring machines (TBM) will be required. Metro Line 4 will have a total of 21 stations, including interchange stations on Lines 1, 2 and 3. The 21 stations, including the terminal, are San Cristoforo FS, Segneri, Gelsomini, Frattini, Tolstoi, Washington-Bolivar, Foppa, Parco Solari, S. Ambrogio, De Amicis, Vetra, S. Sofia, Sforza-Policlinico, San Babila, Tricolore, Dateo, Susa, Argonne, Forlanini FS, Q.re Forlanini and Linate Airport. The future stations will be built in open construction pits: An open central shaft and blind-hole side tunnel technique will be implemented to facilitate passage of the TBM and minimise excavation.

TBMs cannot cut through steel-reinforced concrete drilled shaft walls as the steel bars get caught in the shovels of their shield. In addition, the steel bars cannot be cut into pieces small enough to allow their transport by the TBM's conveyor belt system. As a result, the conventional construction method with steel-reinforced drilled shaft walls needs the manual removal of the steel reinforcement in the path of the TBM. Not only is this time-consuming and expensive in itself, it also required the stoppage and retraction of the TBM in front of each shaft wall. Finally, to ensure that neither the soil nor potential groundwater outside the shaft wall would collapse into the opened hole, complex and expensive soil stabilization measures are required outside the wall.

All these time-consuming and costly measures are not required when the areas of the launch shaft head walls to be penetrated by the TBM are reinforced with glass fibre-reinforced polymer (GFRP). Even though these bars have a much higher tensile strength than steel rebars, they are easily machined and can be broken down into small bar segments by the cutter head of the TBM (Fig. 13). These segments can then be transported by the machine's conveyor system together with the excavated soil. The TBM does not have to be stopped, and soil

Figure 13.   Using the Maplad i-BAR glass fibre reinforced polymer bars.

stabilization measures are not required, as the soil is always stabilized by the TBM. The resulting savings in the overall construction time and cost are substantial. Construction of the first two shafts for the project, at Argonne and Frattini Stations, was opened for bids in January 2015. In both cases, GFRP reinforcement was specified in the bid documents. In early July 2015, MAPLAD was awarded the contract to deliver the soft-eyes GFRP rebar cages.

8 THE ROME METRO PROJECT EXPERIENCE AND THE JOINTS MODIFICATION FOR THE MILAN METRO PROJECT: GRAB VS HYDROMILL

After a careful analysis of the excavation methodology and process specifications for the construction of the "Lot T4 - Line 4 Milan Metro" and been familiar with the Milan's ground morphology, SAOS has proposed an alternative construction method stated, keeping the same quality standards, beneficial time and economically wise. SAOS with a consolidated experience particularly in the construction of diaphragms walls at considerable depths, proposed a solution, which has been accepted, able to ensure a successful result; this technology was already used by SAOS for the "WHSD" in St. Petersburg. The equipment indicated for the excavation of the diaphragms wall is an hydraulic operated grab manufactured with a special long frame; this device allows to reach high depths (over 60 m), guaranteeing speed and verticality comparable to a trench cutter.

The willing of using a grab in place of a cutter was referred to the previous experience in Rome where, for the construction of Metro Line C San Giovanni Station (diaphragms 2,8x1,2 m, depth 56 m), in order to avoid leakages of bentonite slurry and ground collapses (such as the one occurred during the construction with the trench cutter of the neighbor Lodi Station), it has been successfully chosen the grab option.

Here following the report draft prepared in collaboration with Metro C Roma Metropolitano: "... the choice of using an *intelligent grab* have been taken after the experience gained during the construction of the slurry wall central panels of Lodi station. As reported, during the excavation with cutter, a rapid loss of bentonite slurry occurred within a layer of sands and gravel. This phenomenon is attributable to the presence of confined groundwater with reduced hydrostatic pressure within the gravelly layer not sufficient to counter balance the vertical pressure of the bentonite. To make the situation worse the use of the cutter it has been required a continuous recirculation of slurry not properly mixed with correct amount of bentonite affecting the chiseling capability of the cutter. The diaphragms, of the new San Giovanni Station, has a height of about 56 m and crossing entirely a sand and gravel layer with a thickness of about 13 m, heading for about 10 m in a layer of Pliocene's Clays. Given the geometry of the diaphragm walls and the geological nature of the ground, the phenomenon, observed at Lodi's jobsite, would be amplified in San Giovanni jobsite. The design choice pursued involves the use of an "intelligent hydraulic grab" to replace the trench cutter, in order to limit the bentonite slurry recirculation therefore reducing the risk of leakages of the excavation stabilizing fluids, having the benefit of using thicker slurry mix without affecting the excavation capability ...".

Excavation verticality is guaranteed by the considerable length of the grab, approx. 10 m, and the ability to rotate by 180 degrees, while the speed control is given by the crane's lifting configuration. A key feature of this type of equipment is the heavy weight (in configuration 2,8x1,2 m it is about 20 tons), easing the excavation at high depth. The verticality of the diaphragm, in this case, has been constantly monitored in the operator's cab also by means of a dedicated Jean Lutz monitoring device. At the end of each excavation cycle, it has been possible to print reports, having most of the time excavation verticality in a good range minor of 0,5%.

Experience in the long story of Metro of Roma Linea C (Fig. 14) has therefore foreseen 4 different technologies; panels excavated with trench cutter, primary panels with grab and secondary with cutter and then diaphragms solely excavated with grab, mechanical first, hydraulic afterwards. In terms of precision, productivity, cost and time the experience has slowly indicated the hydraulic grab excavation methodology as the most reliable one.

Figure 14.   The "Eternal City" frequently under heavy grabbing ....

Figure 15.   Positioning of joints.

8.1   *Operational methodology for the joints realization*

Given the considerable height of the diaphragms walls to be constructed in Milan (50 m depth) and the need to realize a water tight joint, it has proposed the "virgin-interlocking" methodology with sheet pile stop-end (Fig. 15). This methodology consists in realizing "virgin" or "primary" panels by installing at both ends trapezoidal 36 m long sheet pile in one go. Two cranes of suitable capacity have been used to lift them.

The first one, supporting vertically the sheet pile placed on end and the second, which acting as a support, placed at about 2/3 of the sheet pile length. The sheet pile head is equipped with a dedicated lifting anchor, while the second crane is using omega shaped clamp. Once the two cranes are in position, the lifting operations are carried out; the tail of the sheet pile acts as a fixed point resting on the ground and the second crane holds the sheet pile in position, in order to avoid the bending. The first crane moves forward lifting the sheet pile by keeping its position perpendicular to the lifting point until it is fully lifted in a vertical position. When the lifting operations are finished, the sheet-pile is then lifted into the bore-hole. Once two adjacent primary panels have been completed, the "interlocking" or "second-ary" panel are excavated. After the excavation phase, the two sheet piles previously installed are removed with the help of a vibrator. The demolding operation is very fast (ca. 4 minutes per sheet pile) without causing any damage to the realized structure. The advantage of this operation is to guarantee the absolute cleanliness and integrity of the hydraulic coupling, result not easily obtained by using extraction type sheet or pipe stop ends.

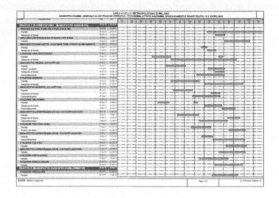

Figure 16.   Job time schedule; target achieved!

8.2   *Compliance with the project schedule*

The works part of SAOS subcontract for the M4 Metro Line foreseen the realization of various size of diaphragm walls having thicknesses ranging from 600 up to 1.500 mm, at a depths up to ca. 50 m.

Given the variety of works to be realized, versatile and modular/adaptive equipment has been required; the collaboration between Liebherr LWN & Negrini has shown its strength. The time schedule of the contractualized foundation works has been successfully achieved: technological Building San Vittore: 56 days, Station San Ambrogio: 60 days, Technological building de Amicis: 73 days, Station de Amicis; 120 days, technological building Ticinese: 76 days, station Vetra: 116 days, technological Building Vettabia: 68 days, Stazione Santa Sofia: 60 days, technological building San Calimero: 70 days, station Sforza Policlinico: 125 days, technological building Augusto: 57 days, station San Babila: 78 days and technological building San Damiano: 18 days.

The synthetic schedule of the previous lot is reported (Fig. 16), just to give a picture of the working phases complexity, especially from the organizational point of view and of simultaneous workings activities to be conducted with suitable equipment in a cost effective and punctual manner.

9   CONCLUSIONS

The first SAOS experience in the Metro of Rome line C has taught us that in many of the Italian soil, at even high depths (> 50 m), diaphragms walls can be realized with hydraulic grab with excellent results in terms of cutting precision, productivity and costs; the Florence Railway link and the Metro of Milan are the clear successful evidence. Nevertheless, it is pretty clear that in order to achieve such a result Contractors shall be equipped with state of the art machineries and tools as well as supported by experienced technical operators.

REFERENCES

Bringiotti, M. 2003. *Consolidamenti e Fondazioni*, Edizioni PEI, Parma, Italy
Bringiotti, M. 2010. *Geotecnica & Macchine da perforazione*, Edizioni PEI, Parma, Italy
Comune di Milano, 2010. *Capitolato speciale opere strutturali*, N. M4-0730
Finotto, A. 2011. *Dal cantiere Liebherr – squadra ad Alta Velocità*, PF, SCI Editrice, RA, I, September - October
M4 La linea Blu, 2016. *Cerchia dei Navigli*, Pubblicazione interna
SAOS, 2016. *Tratta T4 Linea4 - Proposta migliorativa diaframmi*, Documento contrattuale
Vitali, S. 2017, *On Site Construction*, Capoverso Editrice, RA, Italy, September

Tunnels and Underground Cities: Engineering and Innovation meet Archaeology,
Architecture and Art, Volume 5: Innovation in underground engineering,
materials and equipment - Part 1 – Peila, Viggiani & Celestino (Eds)
© 2020 Taylor & Francis Group, London, ISBN 978-0-367-46870-5

EPDM gaskets; new materials and procedures for segments and other application in tunneling

M. Bringiotti & D. Nicastro
GeoTunnel S.r.l., Genoa, Italy

F. Garate & S. Villoslada
Algaher S.A., Alfaro, La Rioja, Spain

K. Pini
CP Technology S.r.l., Milano, Italy

ABSTRACT: Sealing systems for various applications such as TBM's segments, shield driven tunnels and various types of pipe jacking have been developed; they are used for roads, rail and metro infrastructures, as well for cable, service, water and sewer projects. One of the latest product which has been designed is a seal that's anchored into the concrete, performing segment production and gasket installation easier and faster without manpower added and adhesive cost, shortened the process due to time savings for the grooved area cleaning, with relevant corner keystone protection advantages, with a quality insurance granted in line not only with the modern times but also with the new Industry 4.0 construction process integration. The article will describe also a different tunnel application in Metro Milano (starting seal) and with the biggest EPB-TBM presently running in the world St. Lucia (culvert seal), referring as well about a special device related to the 30 ton culvert handling procedure.

1 INTRODUCTION

Since many years, Algaher has been delivering a wide range of sealing profiles for various tunnel excavation techniques, following not only the STUVA recommendations but also new Standards under development in EU and USA which are regulating the usage of new composite materials.

Sealing systems for various applications such as segments for tunnel boring machines (TBM), shield driven tunnels and various types of pipe jacking have been developed, as well gaskets for TBM starting, emergency, door and caisson seals. They are used for roads, rail and metro infrastructures, as well for cable, service, water and sewer projects.

One of the latest product which has been designed, in several models and with different optional added devices, is a seal that's anchored into the concrete during the tunnel segment production process. This system is performing segment production and gasket installation easier and faster without manpower added and adhesive cost, shortened the process due to time savings for the grooved area cleaning, integrated link between concrete and seal with relevant advantages on the corner keystone protection, with a quality insurance granted in line not only with the modern times but also with the new Industry 4.0 construction process integration.

Algaher Group is a family company founded in 1984 with the purpose of manufacturing rubber profiles; it is specialized in the manufacture of segmental lining sealing gaskets suitably designed for all kind of tunnels. It develops solutions depending on the needs of each project by four main types of product (Figure 1):

1. Glued gaskets
2. Integrated Gaskets

Figure 1.   Some rubber profiles.

3. Gaskets with hydrophilic profile
4. Co-extruded gasket

2   WHAT IS EPDM?

EPDM rubber (ethylene propylene diene monomer rubber), a type of synthetic rubber, is an elastomer characterized by a wide range of applications. This is an M-Class rubber where the 'M' in M-Class refers to its classification in ASTM standard D-1418; the *M* class includes rubbers having a saturated chain of the polyethylene type. Dienes used in the manufacture of EPDM rubbers are dicyclopentadiene (DCPD), ethylidene norbornene (ENB), and vinyl norbornene (VNB). EPDM rubber is closely related to ethylene propylene rubber: ethylene propylene rubber is a copolymer of ethylene and propylene, whereas EPDM rubber is a terpolymer of ethylene, propylene and a diene-component (Figure 2).

The ethylene content is around 45% to 85%. The higher the ethylene content, the higher the loading possibilities of the polymer, better mixing and extrusion. Peroxide curing these polymers gives a higher crosslink density compared with their amorphous counterpart. The amorphous polymer is also excellent in processing. Processability is very much influenced by their molecular structure. The dienes, typically comprising from 2.5% to 12% by weight of the composition, serve as sites of cross-links when curing with sulphur and resin; with peroxide cures, the diene (or third monomer) functions as a coagent, which provides resistance to unwanted tackiness, creep, or flow during end use.

EPDM is compatible with polar substances, e.g. fireproof hydraulic fluids, ketones, hot and cold water and alkalis. It is incompatible with most hydrocarbons, such as oils, kerosene, aromatic, gasoline as well halogenated solvents. EPDM exhibits outstanding resistance to heat, ozone, steam and weather. It is an electrical insulator.

Typical properties of EPDM vulcanizates are given below. EPDM can be compounded to meet specific properties to a limit, depending first on the EPDM polymers available, then the processing and curing method(s) employed. EPDMs are available in a range of molecular weights [indicated in terms of Mooney viscosity ML(1+4) at 125 °C], varying levels of ethylene, third monomer, and oil content.

The main mechanical and thermal EPDM's properties are:

– Hardness: 40–90 Shore A
– Tensile failure stress, ultimate: 25 MPa
– Elongation after fracture in %: ≥ 300%
– Density: can be compounded from 0.90 to >2.00 g/cm^3

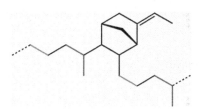

Figure 2.   Idealized EPDM polymer, red = ethylene-derived, blue = propylene-derived, black = ethylidene norbornene-derived.

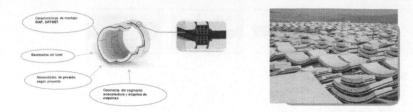

Figure 3. EPDM gaskets and seals in tunneling and tubing sector.

- Coefficient of thermal expansion, linear: 160 μm/(m·K)
- Maximum service temperature: 150 °C
- Minimum service temperature: −50 °C
- Glass transition temperature: −54 °C

A common use is in vehicles: door seals, window seals, trunk seals and sometimes hood seals. Frequently, these seals are the source of noise due to movement of the door against the car body and the resulting friction between the EPDM rubber and the mating surface (painted sheet metal or glass). Other uses in vehicles include cooling system circuit hoses where water pumps, thermostats, EGR valves, EGR coolers, heaters, oil coolers, radiators and degas bottles are connected with EPDM hoses, as well as charge air tubing on turbocharged engines to connect the cold side of the charge air cooler (intercooler) to the intake manifold.

EPDM rubber is used in seals, for example in cold-room doors since it is an insulator, as well as in the face seals of industrial respirators in automotive paint spray environments. EPDM is also used in glass-run channels, radiators, garden, and appliance hose, tubing, pond liners, washers, belts, electrical insulation, vibrators, O-rings, solar panel heat collectors and speaker cone surrounds. It is also used as a medium for water resistance in electrical cable-jointing, roofing membranes (since it does not pollute the run-off rainwater, which is of vital importance for rainwater harvesting), geomembranes, rubber mechanical goods, plastic impact modification, thermoplastic, vulcanizates and many other applications. Colored EPDM granules are mixed with polyurethane binders and troweled or sprayed onto concrete, asphalt, screenings, interlocking brick, wood, etc. to create a non-slip, soft, porous safety surface for wet-deck areas such as pool decks and as safety surfacing under playground play equipment (designed to help lessen fall injury).

EPDM application in tunneling is related to whatever seal (gasket) in the tubbing and segment sector (Figure 3). In case of hydrocarbons presence (for example heading in contaminated areas) it can be protected with a sort of shield, entering in the field of a composite product.

3 PRODUCTION PROCESS

Algaher has a 5000 m² plant of manufacture on a 15,000 m² extension, equipped with the most advanced technology for the production of this kind of profiles. The area of production is composed by 3 lines of extrusion and by 10 injection equipment, which provides the company of a really relevant production capacity.

The lines of extrusion have a length up to 70 meters where the raw material pass different phases of production (Figure 4):

1. Extrusion and molded
2. Oven for radiation (microwave)
3. Oven for convection (gas)
4. Machine of automated cut

The whole process of extrusion is governed by automatic systems to assure the control of thickness in all the profiles manufactured.

Figure 4.   Production line.

The injection equipment has a capacity up to 350 cc per cycle and are provided with all the necessary controls for the manufacture of this kind of profiles:

1. Control of temperature of the mould
2. Control of temperature of mixture
3. Control of vulcanization time.

4   PRODUCTION CONTROL

Internal quality control is a key topic in all kind of industries; Algaher has an own laboratory where the tests and quality controls are realized in all manufactures produced under usual norm of application as the EN 681-1:1996. The internal LAB, besides other equipment, is provided with certified Micrometer, Heater, Durometer and Dynamometer.

The technical department is the one in charge of realizing Analysis of Stress FEM (Method Finite Elements) to estimate the behavior of the internal designs and to come up to the configuration that offers the values target.

It is also clear that the whole traceability of all the products has to be guaranteed during the production processes (Figure 5).

ALGAHER has spaces in order to realize tests of water tightness with different GAP and OFFSET, according to STUVA's recommendations (Figure 6).

All the tunneling water tightness gaskets have (and should have) these internal tests:

– Test: Load – Displacement # Method: STUVA
– Test: Water tightness # Method: STUVA

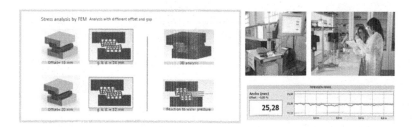

Figure 5.   FEM analysis, traceability and Lab.

Figure 6.   STUVA test devices and results.

- Test: Stress relaxation at 23°C and 70°C # Method: Iso 3384
- Test: Hardness IRHD # Method: ISO 48
- Test: Traction resistence # Method: ISO 37
- Test: Elongation at break # Method: ISO 37
- Test: Remaining deformation by Compression # Method: ISO 815
- Test: Variation after aging in air durig 7 days at 70 °C (Hardness – Resistence – Elongation) # Method: ISO 188
- Test: Stress relaxation in compression (7 days at 23 °C - 100 days at 23 °C) # Method: ISO 3384
- Test: Volume variation after water immersion (7 days at 70 °C) # Method: ISO 1817
- Test: Ozone resistance to the at 40 °C, 50 ppcm 48 hours # Method: ISO 1431-1

5 WATER TIGTHNESS IN CORNERS

Each Manufacturer, usually designs their products in function of customers' needs and in relation to its own experience and they should be calculated to reach high sealing characteristics. One of the gasket key technical point is related to the water tightness in corners; they are designed and manufactured in Algaher and have some technical and competitive advantages:

1. Specific sharped to create a perfect interface between corners.
2. Central nerve specially designed to confer to the "corner" a high resistance.
   As result, the gasket corners reach values very raised in parameters of water tightness.
3. Corners with "Twisted Angle";
   They must adapt perfectly to any obtaining a perfect union between mould and gasket.
4. Corners injected with profile cut at 45 °.

In this way it is obtained a minor quantity of material in the injection without changing either the force of retention or the performance of water tightness (Figure 7).

Another leading factor is related to the special treatment anti-friction "NO GRIP". All water tightness gaskets lodged at the key piece (K) and in the adjacent segments, are provided with a special lubricant thus in order to ease the slide between the gaskets with the purpose to avoid damages between them (Figure 8).

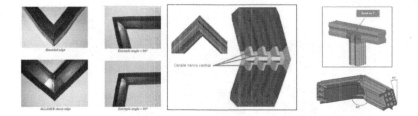

Figure 7. Water tightness in corners.

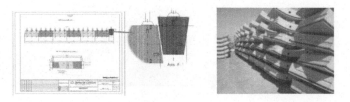

Figure 8. Anti friction "NO GRIP" in Metro Caracas.

6 STARTING TBM SEAL IN METRO MILANO

Metro Blu, consortium between Salini-Impregilo and Astaldi S.p.A., incharged for the excavation of various new Milan Metro line stretches, had and is having the problem to launch various TBMs in a down town environment with the big disadvantage of water table presence at around ground zero! Start boring with a TBM in such a condition is not an easy job.

A special water proof system needed to be designed, using a steel frame and starting seals which consist usually of three parts: seal profile, filler profile and U-channel (Figure 9).

Starting seals are mounted on a special support frame on the tunnel starting ring, as a watertight seal at the gap between the tunnel excavation line and the TBM. For greater water-tightness, several starting seals can be installed in series. When this is done, profile butt joints should be staggered from one seal to the next. Starting seals are fastened into a U-channel without screws or bolts.

They are designed in function of the pressure which have to resist; normal values are starting from 0,5 up o 6 bars and the choice depends on the technical requirements and tunnel diameter.

It is most practical to mount the starting seal before installing the support frame. The seal is laid out to mate with the support frame and pressed into the U-channel, one portion (ca. 50m) at a time. If used, filler profile is pressed in simultaneously, locking the seal profile into position.

Usually when all but the last 75 cm of the starting seal has been installed, the remainder is cut to an extra length of 50 mm and installed. This additional length creates pressure on the profile butt joint, providing an adequate seal.

For eventual added water-tightness, adhesive is applied to the butting surfaces, following the adhesive manufacturer's application instructions. It is recommended to position the butt joint at the top of the support frame to minimise hydraulic pressure on the joint during boring.

Various type of Sealing Ring Lip Seals, in accordance to the Client's specification and under Algaher's experience suggestions, have been successfully installed (Figure 10).

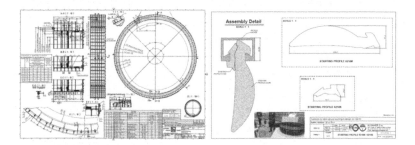

Figure 9.   Steel frame design and starting seal.

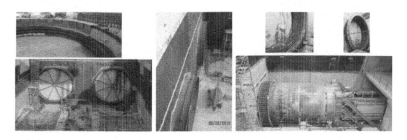

Figure 10.   Installation and start boring in different stations.

7 CULVERT SEAL IN ST. LUCIA

On the highway A1 (Milan-Napoli), huge tunnelling jobs are under execution since years, in order to modernize and power the old route. Santa Lucia tunnel represent the principal job which is creating a new highway line in the South direction. Few main data are:

- Owner: Autostrade per l'Italia S.p.A
- Main Designer: SPEA S.p.A.
- Contractor: Pavimental S.p.A.
- Tunnel length: 7548 m
- TBM excavation diameter: 15,965 mm
- Excavation surface: 199 m^2
- Excavated volume: 1,480,000 m^3 fully handled automatically trough a complex belt conveyor system designed and manufactured by the well known Marti Technik.AG
- Lining type: universal precast ring in 9 + 0 elements, length: 2.2 m, thickness: 55 cm and 16 ton of weight for each element
- Geology: Monte Morello limestone and Sillano clay, with potential presence of Methane gas.
- Job duration: 1852 days
- Job start: 15th March 2016
- Job finish: 9th April 2021
- TBM excavation average speed: 9 m a day (28 months is the finishing time foreseen)

The TBM, in EPB mode, is the biggest in the world presently in operation and the n. 2 ever built. The weight is in the range of 4800 ton.

The 200 m^2 section will host 3 highway lines (3.75 m width), plus platform and plants and a huge precast safety tunnel has been designed above the vehicles running deck.

No. 2830 bi-cell precast concrete culverts measuring 2.83 m long x 7458m wide x 2.90m high having a 32 tons weight are currently, manufactured at the same casting yard of the tunnel lining, to be fitted under the road deck.

The seal designed is the Delta 24, having in section 24 x 42 mm dimensions, 14,67 m in length, with round corners, 45°±5° sH, under EN 681-1/CE norms. The essential characteristics, Performance and Harmonised technical specification are:

- Dimensional: L3 – E1, ISO 3302
- Hardness: 45 ± 5, UNE EN 681-1:1996
- Tensile strength: Min. 9 MPa, UNE EN 681-1:1996
- Elongation at break: Min. 375 %, UNE EN 681-1:1996
- Compression set 72 h. at 23 °C: ≤ 12, UNE EN 681-1:1996
- Compression set 24 h. at 70 °C ≤ 20, UNE EN 681-1:1996
- Compression set 72 h. at -10 °C ≤ 40, UNE EN 681-1:1996
- Ageing 7 d. a 70 °C, Hardness change max: +8/-5, UNE EN 681-1:1996
- Ageing 7 d. a 70 °C, Tensile strength change max. -20 % UNE EN 681-1:1996
- Ageing 7 d. a 70 °C, Elongation at break change max: +10/-30, UNE EN 68-1:1996

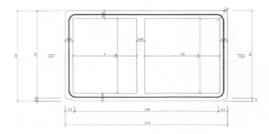

Figure 11.   Culvert dimension in section.

– Stress relaxation, max. 7 days at 23 °C: 16 %, UNE EN 681-1:1996
– Stress relaxation, max. 100 days at 23 °C: 23 %, UNE EN 681-1:1996
– Volume change in water, 7 days al 70 °C max:. +8/-1 %, UNE EN 681-1:1996
– Ozone resistance: No cracking, UNE EN 681-1:1996

7.1 Culvert gasket & handling procedure

The Culvert prefabrication, seals application, culvert handling and installation is quite complex and very well designed by the technical team lead by Maccafaerri Tunnelling S.r.l., company which is realizing such elements for Pavimental S.p.A, Algaher SA and CP Technologies S.r.l. who has been the brain engine to design and manufacture all the mechanical devices to let the full system work perfectly (besides the fact that CP Technologies has as well delivered the full carousel, segment moulds, external structure, lifting devices, aerial concrete distributor lines, electronic control . . . in a full integrated Industry 4.0 environment).

First of all, the elements are casted with a stationary hydraulically operated mould, allowing the casting and demoulding operation aimed by a very special demoulding and tilting device. A safe and controlled Algaher's procedure is used to fix the gasket in the right position (Figure 12)..

A very interesting matter is the culvert's handling procedure from the prefabrication area, which is ca. 50 km far away from the Santa Lucia job site. The culvert once cast, demoulded, turned at 90 degrees (Figure 13), finished with installation of the sealing gasket is than loaded on a truck to be transported at site.

Here it is offloaded by a portal crane and transported in the tunnel by MSVs. CP-Technology has developed a self-propelled culvert laying gantry with integrated vehicles ramps (Figure 14) easing at the same time the transit of heavy pay load vehicles, suitably designed to be driven inside lined tunnel, thus to lay behind the operating TBM the prefabricated concrete culverts.

Figure 12.   Culvert cast in situ, rotation device and gasket installation.

Figure 13.   Culvert demoulding and turning device.

Figure 14.   Culvert handling procedure from the casting yard to the job site and along the tunnel.

Separating the culvert-laying gantry from the TBM backup is a new idea. The pros and cons of adopting such a technique are:
   Pros:

– Individual design approach avoiding the need to raise the road to achieve adequate cover
– Typical pre-stressed RC units - Not innovative material = reasonable cheap
– Good fire performance of RC elements
– Quick construction Prefabricated units "ready to use"
– Simple and fast installation
– Culvert-laying gantry directly behind the TBM
– Separate TBM production from culvert-placing production meaning Installation contemporary to the excavation
– Continuous Access/Traffic to the excavation front
– Limited areas of concrete mass fill on sides only
– No shutter (Formwork) needed

   Cons:

– Thick Elements
– Multiple Short Spans
– Slow construction progress due to cast in-situ topping

8.1   *Culvert installation procedure*

In details the installation procedures involve the following activities:

Bedding, intended to level out any remaining irregularities on the tunnel lining invert portion and ensuring uniform support under the full width and length of the box culvert including:
   a. Lay a thin flat apron of unreinforced lean mix concrete on the invert portion of the lining which has been well prepared to a uniform firmness
   b. Lay of thin bed of sand (Figure 15).

Transport of the box culvert (Figure 16) in longitudinal position along the axle of the tunnel, by means of MSV, and stop under the gantry crane.
Lifting of the box culvert by means of the gantry crane and opening of the folding ramp in order to allow the laying of the culvert (Figure 17).
Transport of the box culvert to the installation area by displacing the gantry crane trolley and rotation of the culvert in its final position (Figure 18).
Laying of the box culvert on the already prepared invert surface and closing of the joints by pushing the culvert against the one previously laid. The complex operation is conducted by keeping the weight of the culvert jointed on the crane thus to reduce the frictional resistance at the base of the culvert, easing sufficient compression load to compress the culvert seal, therefore closing the joint to the specified nominal gap (Figure 19).

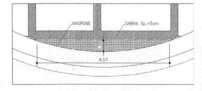

Figure 15.   Installed culvert.

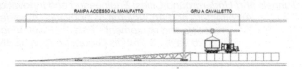

Figure 16.   Transport of the culvert by MSV.

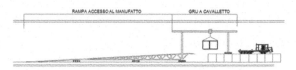

Figure 17.   Lifting of the culvert.

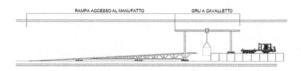

Figure 18.   Transport of the box culvert to the installation area.

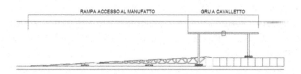

Figure 19.   Positioning the culvert into the invert surface.

6. Lifting of the folding ramp and displacement of self-propelled culvert laying gantry to the next bay.
7. Finally Culvert sides shall be Backfilled to the level of the top of the box culvert working evenly on each side using selected backfill material well compacted in layers.

9   CONCLUSIONS

EPDM is a world for Specialists which are called to work in Team; from the "gasket" producer, where it is requested a high level of professionality and a big effort for research and development in order to find always new solutions and advanced material applications in a full quality optic, to the Contractor who is requested to deep analyze problematics, times and methods.

Other "Subjects" are called to play this type of "game"; for example, as we have seen, the Self-Propelled Culvert Gantry operates in accordance to methods and procedures given by the Contractor (studied in Team work session) for the installation of precast concrete box culverts taking careful attention to details, leading to safer working activities, a smoother flow of operations and a higher standard of finished culvert installation.

The complex circle is what is called, at the end, "Industry 4.0"; pull of Specialists, fully integrated for a perfect final product.

Tunnels and Underground Cities: Engineering and Innovation meet Archaeology,
Architecture and Art, Volume 5: Innovation in underground engineering,
materials and equipment - Part 1 – Peila, Viggiani & Celestino (Eds)
© 2020 Taylor & Francis Group, London, ISBN 978-0-367-46870-5

New results from laboratory research to reduce the clogging potential for EPB-Shields

C. Budach
ELE Beratende Ingenieure GmbH, Essen, Germany

E. Kleen
MC Bauchemie Müller GmbH, Bottrop, Germany

ABSTRACT: Tunnel drives with EPB-Shields in fine grained soils could lead to clogging e.g. in the excavation chamber. In these soils water is usually being used to change the consistency index of the support media in the excavation chamber. In addition, additives are often used in the suspension to reduce the risk of further clogging. Systematically investigations to reduce the clogging potential of fine grained soils using additives for EPB-Shields are unusual. Laboratory tests were carried out to investigate the adhesion tension in order to describe the clogging potential for EPB-Shields using water. In the laboratory tests an additive was also used to change the consistency index and to simulate the soil conditioning of EPB-Shields. Comparable investigations carried out to determine the influence of the additive. It can be shown that a reduction of the adhesion tension is possible and the correct use of additives could have a significant influence.

1 INTRODUCTION

Earth Pressure Balance (EPB-) Shields were originally used in fine-grained soils like clay and silt. According to (Maidl 1996) the support medium of an EPB-Shield in fine grained soils should have a consistency index between $I_C = 0.4$ and $I_C = 0.75$ to allow an adequate workability. For tunnel drives in these soils clogging on parts of the tunnel boring machine can occur. For example clogging can occur at the cutting wheel, in the excavation chamber or in the screw conveyors and can have a significant influence on the performance and thus on the economic efficiency of a project. In general, qualitative assessments are made in advance of a tunnel drive with an EPB-Shield to determine the clogging potential based on typical soil properties.

Additives are often used – in fine grained soils predominantly as part of the suspension – to change the properties of the excavated soil and reduce the clogging potential. For the determination of the clogging potential as well as the determination of the effectiveness of additives in tunneling no general determination methods currently exist. The clogging potential is estimated in common using different investigation methods and the effectiveness of additives is usually determined during tunnel drives on the job sites.

2 STATE OF THE ART FOR THE DETERMINATION OF CLOGGING IN FINE GRAINED SOILS FOR SHIELD MACHINES

Investigations were carried out at the RWTH Aachen University, Germany, with a conical tensile test to assume the clogging potential, see inter alia (Ziegler et al. 2015). The test procedure provides that a cone is pushed into a soil sample and, after the application of a compression, the cone is pulled out of the material. Subsequently, the mass of the adherence of the material on the cone can be determined and a clogging potential could be estimated, see Figure 1. For these investigations, which were designed for EPB-Shields, different soils with different plasticity and consistency indices were tested. The clogging potential according to (Ziegler et al. 2015) depends on the mass of adherence on the cone and high clogging potential could be between $I_C = 0.1$ and $I_C = 0.8$ according to the type of soil, see red area in Figure 1.

The plasticity index as well as the water content and the consistency index of soils play an important role in the investigations of the clogging potential for EPB-Shields after (Hollmann & Thewes 2011). Investigations have led to a general evaluation diagram for the assessment of possible critical changes of the soil properties (see Hollmann & Thewes 2013). This evaluation diagram could be used for EPB-Shields in closed mode and for open mode. It shows among other things various consistency ranges, including the expected effects on the clogging potential (see Figure 2). The clogging potential could be caused by the change in consistency index due to the processing technology, for example by the access of water in open mode using an EPB-Shield. The evaluation diagram was validated by various tunnel drives with EPB-Shields, see (Hollmann & Thewes 2013).

Further results of other investigations e.g. with clay-sand-mixture simulating mixed-face conditions at the tunnel face are presented inter alia in (Oliveira et al. 2017 and Oliveira et al. 2018) and results on clogging of fine grained soils in (Zumsteg & Puzrin 2012). In these investigations a mixer with a paddle was used to mix soil or different soils. Due to the

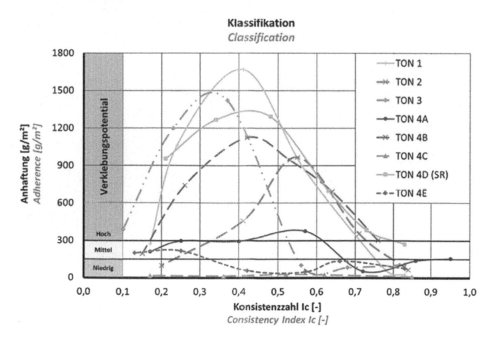

Figure 1.   Results of conical tensile test (Ziegler et al. 2015).

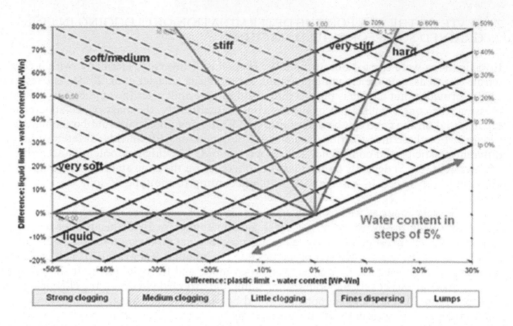

Figure 2. Evaluation diagram for the assessment of possible critical changes of the properties of soils (Hollmann & Thewes 2013).

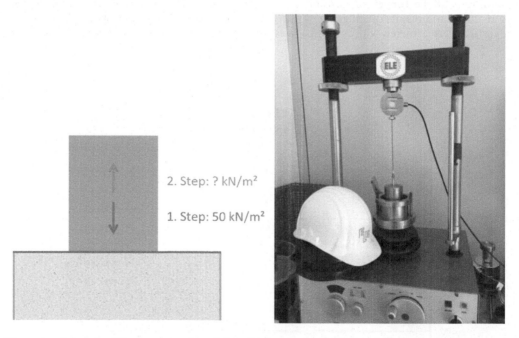

Figure 3. Principle of the determination of the adhesion tension (left) and test equipment (right).

determination of the mass of adherence on the paddle it should be possible to assume the clogging potential.

In the investigations according to (Thewes 1999) the adhesion tensions of various fine grained soils were determined. For this purpose, selected soils were placed in a container, the surface of the soils were smoothly planed and moistened with water and a steel cylinder of defined weight was placed on the soil for a specified time. Subsequently, the steel cylinder was pulled vertically upwards and the force to lift this cylinder respective the adhesion tensions was determined (see Figure 3). The wetting with water thereby represents the influence of the supporting medium of shield machines with slurry face support.

Additional tests and test results with this test equipment are presented in (Burbaum 2009).

The summary of the laboratory research on clogging shows that the clogging potential for shield machines is often assumed by the adhesion tension or by the mass of soil that stays on a paddle or cone. For our own laboratory research the test to determine the adhesion tension with different test boundaries was used due to the positive results simulating the process technology of hydroshields with wetting the surface with water and additives, see (Budach & Kleen 2017).

3 METHOD OF DETERMINATION OF ADHESION TENSIONS FOR EPB-SHIELDS

Using an EPB-Shield in fine grained soils the clogging potential could be reduced in common by decreasing the consistency index. Based on experiences on clogging of EPB-Shields the consistency index should be lower than $I_C = 0.5$ to reduce the risk of clogging, see Figure 2. Based on experiences of the properties of the support medium of an EPB-Shield, the support medium should have a consistency index between $I_C = 0.4$ and $I_C = 0.75$ see (Maidl 1996). Based on these two boundaries, the support medium should have a consistency index of between $I_C = 0.4$ and $I_C = 0.5$ to reduce the risk of clogging and guarantee an adequate workability.

Transferring the process technology of an EPB-Shield on the laboratory test to determine the adhesion tension the surface of the soil sample should not be wetted with water because the support medium in the excavation chamber of an EPB-Shield has not got the properties of a hydroshield-suspension. So during the procedure of our own laboratory tests no water was placed on the surface of the soil sample

During the process of an EPB-Shield drive, the liquid to change the consistency index could be water and additives. These two different liquids were tested in the laboratory research, too.

According to the investigations on clogging potential of hydroshields it was possible to categorize the clogging potential subject to the measured adhesion tension with similar boundary conditions (Budach & Kleen 2017). The quantitative categorization of clogging potential using the adhesion tension is listed in table 1. Assuming that the adhesion tension strongly influence the clogging potential for EPB-Shields the description of the clogging potential of hydroshields could also be transferred on EPB-Shields.

Table 1. Clogging potential for fine grained soils using hydroshields.

Description of the clogging potential	Adhesion tension [kN/m²]
Low potential of clogging	0 – 5
Medium potential of clogging	5 – 10
High potential of clogging	> 10

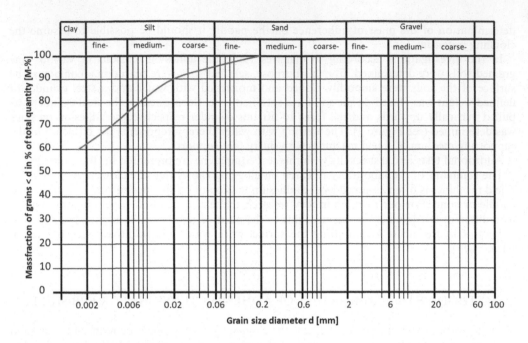

Figure 4.   Grain-size-distribution curve of tested soil.

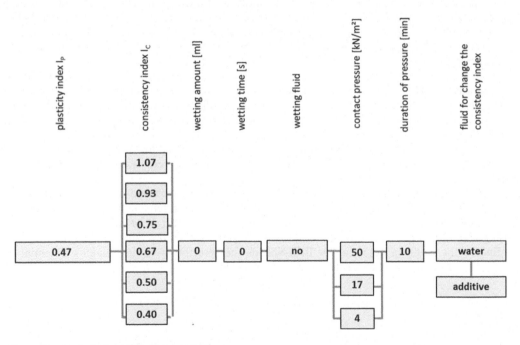

Figure 5.   Boundaries of the test procedure.

For the own laboratory research the soil was used that has also been tested during the clogging potential tests for hydroshields. The plasticity index of this soil was 0.47 and the grain-size distribution curve of this soil is shown in Figure 4. Further properties of this soil are described in (Budach & Kleen 2017, see soil D).

The boundaries of the own test procedure and its variation are shown in Figure 5. The tested soil had a plasticity index $I_P = 0.47$. Different consistency index starting from $I_C = 0.4$ (lower value of application range for support medium after (Maidl 1996)), $I_C = 0.5$ (lower value of application range for clogging after (Hollmann & Thewes, 2013)), $I_C = 0.67$ (approximately upper value of application range after (Maidl 1996) and for clogging after (Hollmann & Thewes 2013)), $I_C = 0.75$ (upper value of application range after (Maidl 1996) and for clogging after (Hollmann & Thewes 2013)) as well as $I_C = 0.93$ and $I_C = 1.07$ as a comparison to the results for hydroshields, see (Budach & Kleen 2017). Representing an EPB-Shield no water was placed on the surface of the sample. As contact pressure in common 50 kN/m² were used. If this was not possible and the stamp couldn't be placed on the sample due to low consistency indices, lower loads like 17 kN/m² or 4 kN/m² were used for the tests. Water as well as an additive was used to change the consistency/consistency index of the sample to factor in their influences.

4 RESULTS OF THE LABORARTORY RESEARCH

The test results carried out with a contact pressure of 50 kN/m² (hydroshield) and without wetting the sample surface (EPB-Shields) are shown in Figure 6 as function of the consistency index and coloured areas of the consistency.

The blue coloured line (test results with moisturized surface) shows increasing values of the adhesion tension with increasing consistency index. These results fit well with the clogging potential of hydroshields, see (Thewes 1999 or Budach & Kleen 2017). The results of the tests without wetting the surface of the sample are shown in red. The maximum value of the adhesion tension is lower than the maximum value of the adhesion tension of a moisturized sample surface.

The development of the values of the adhesion tension without wetting the sample shows a maximum with the consistency index of $I_C = 0.75$. The value of the consistency index of $I_C = 0.67$ is only minor lower. Tests with consistency indices lower 0.67 were not possible using 50 kN/m² contact pressure, because the surface was deformed by the stamp.

The values without wetting the surface go along with the description of the clogging potential after (Hollmann & Thewes 2013), because the highest values of the adhesion tension were determined between $I_C = 0.67$ and $I_C = 0.75$, which results in strong clogging (see Figure 4). With increasing consistency index the adhesion tension decreases, so that this development could fit to the medium clogging (between $I_C = 0.75$ and $I_C = 1.00$) and lumps (between $I_C = 1.00$ and $I_C = 1.25$).

The contact pressure of 50 kN/m² led to a deformation of the sample and non-reproducible results, so that the contact pressure was reduced to 17 kN/m² as well as 4 kN/m² to enable the determination of the adhesion tension of samples with low consistency indices. The test results of these investigations are shown in Figure 7.

In general it can be summarized that the lower the contact pressure at a defined consistency index the lower the adhesion tension. Using 17 kN/m² contact pressure it was possible to test a soil with a consistency index of $I_C = 0.5$ and with 4 kN/m² contact pressure it was possible to test a soil with a consistency index of $I_C = 0.4$. The development of the adhesion tensions over the consistency index show as maximum between $I_C = 0.4$ (contact pressure = 4 kN/m²) and $I_C = 0.75$ (contact pressure = 50 kN/m²). These results fit with the range of strong clogging after (Hollmann & Thewes 2013) and it could be possible to describe strong

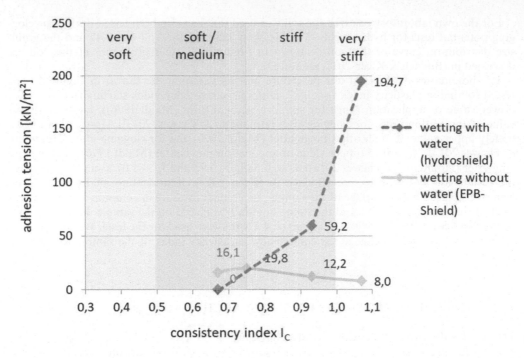

Figure 6. Test results with and without wetting the surface of the sample and contact pressure of 50 kN/m².

clogging after (Hollmann & Thewes 2013) for this soil, if the value of the adhesion tension (using a contact pressure of 17 kN/m² in the laboratory research) is higher than e.g. 10 kN/m². This value fits with the qualitative description of clogging potential for fine grained soils using hydroshields, see table 1.

The adhesion tensions of different consistency indices using a contact pressure of 4 kN/m² only show minor differences, therefore the influence of the test procedure on the results could be high. Due to this and to enable a large band of possible consistency indices a contact pressure of 17 kN/m² should be chosen in further tests to determine the influence of the consistency index on the adhesion tension.

The results of (Ziegler et al. 2015) have shown that different soils could lead to different masses of the adherence and that the peak of the adherence will not be ever between $I_C = 0.4$ and $I_C = 0.75$. These results could go along with the results of the own laboratory research as presented above.

Although the test results of the own laboratory research fit very well to the results of the investigations of (Hollmann & Thewes, 2014) further tests should be done in the future with different soils to verify these results.

Because of the application of additives (foam or slurry) in order to change the consistency index and the soil properties, additional tests were carried out in the laboratory. The consistency index was changed by an additive instead of water to simulate the influence of soil conditioning during a tunnel drive with an EPB-Shield. The soil samples contain a defined amount of water (corresponding to the natural water content) and an amount of additive according to the change of the consistency index. The results with different loads of the contact pressure as well as soil samples with a consistency index of $I_C = 0.4$ and $I_C = 0.67$ are

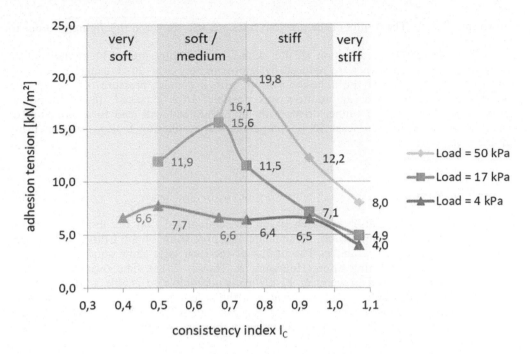

Figure 7. Test results without wetting the surface of the sample and different contact pressures.

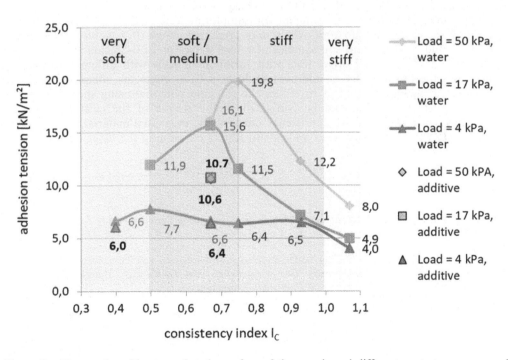

Figure 8. Test results without wetting the surface of the sample and different contact pressures and results using an additive to change the consistency index.

shown in Figure 8. The results of the previous tests are also shown in this figure in order to compare.

The high values of the adhesion tension with a contact pressure of 17 kN/m² and 50 kN/m² without an additive were reduced by approximately over 30 % using an additive. Due to the low values of the adhesion tension with a contact pressure of 4 kN/m² no significant reduction using an additive was possible. It is expected, that an optimal choice of additives during tunnel drives with an EPB-Shield will lead to a reduction of the adhesion tension of at least 30 % compared to the adhesion tension without using an adequate additive.

5 CONCLUSION

Based on the own laboratory research it can be summarized that the test to determine the adhesion tension is applicable to simulate the procedure of an EPB-Shield by not wetting the surface of the sample. The results of the own tests show good consensus to the band of strong clogging after (Hollmann & Thewes 2013). The own research has also shown that a contact pressure of 17 kN/m² is acceptable to enable a wide range of consistency indices as well as high differences in the adhesion tension subject to the consistency index. A first benchmark of strong clogging can be derived from the test results with adhesion tension over 10 kN/m², but this benchmark should be verified in further investigations with other soils.

The test results using an additive have shown that the adhesion tension can be reduced. This reduction should help to decrease the clogging potential during tunnel drives and to make tunnel drives with EPB-Shields in fine grained soils more efficient.

Further laboratory tests with different soil samples as well as investigations on the job site should be carried out to verify the results presented above. If possible, these investigations should be done with soil samples of tunnel drives of EPB-Shields in fine grained soils. In front of a tunnel drive, tests with variation of soils with different consistency indices and the use of different additives could help to estimate the adhesion tension as well as the clogging potential or to reduce the clogging potential using special additives. The results of the tests should be correlated to the torque of the cutting wheel and screw conveyor and the in-situ consistency index of the excavated material, that a correlation between the laboratory tests and in-situ results concerning the clogging potential is possible.

REFERENCES

Budach, C. & Kleen, E. 2017. Reduction of Clogging in Mechanized Tunneling using Additives based on Laboratory Research, *Proceedings of the World Tunnel Congress (WTC 2018)*, Dubai, VAE, April 2018

Burbaum, U. 2009. Adhäsion bindiger Böden an Werkstoffoberflächen von Tunnelvortriebsmaschinen, Dissertation, *Institut für Angewandte Geowissenschaften vom Fachbereich Material- und Geowissenschaften der Technischen Universität Darmstadt*, Darmstadt

Hollmann, S. & Thewes, M. 2011. Bewertung der Neigung zur Ausbildung von Verklebungen und zum Anfall von gelöstem Feinkorn bei Schildvortrieben im Lockergestein, *18. Tagung für Ingenieurgeologie und Forum für junge Ingenieurgeologen*, S. 237–244

Hollmann, F. & Thewes, M. 2013. Assessment method for clay clogging and disintegration of fines in mechanised tunneling; *Tunnelling Underground Space Technology*, 37, 96–106, DOI: 10.1016/j.tust.2013.03.010

Maidl, U. 1996. Erweiterung der Einsatzbereiche der Erddruckschilde durch Bodenkonditionierung mit Schaum, Dissertation, Ruhr-Universität Bochum

Oliveira, D.; Diederichs, M.; Thewes, M; Freimann, S.; Aguiar, G. 2017. EPB Conditioning of Mixed Transitional Ground: Investigating Preliminary Aspects. *Proceedings ITA World Tunnel Congress*, Bergen, Norway

Oliveira, D; Thewes, M.; Diederichs, M.; Langmaack, L. 2018. Proposed Methodology for Clogging Evaluation in EPB Machines. *Proceedings of the World Tunnel Congress (WTC 2018)*, Dubai, VAE

Thewes, M. 1999. Adhäsion von Tonböden beim Tunnelvortrieb mit Flüssigkeitsschilden, Dissertation, *Bericht aus Bodenmechanik und Grundbau*, Bergische Universität Wuppertal, Fachbereich Bauingenieurwesen, Band 21, Shaker Verlag

Ziegler, M., Feinendegen, M., Englert, K. 2015. Verklebungen beim maschinellen Tunnelvortrieb: Bewertungsverfahren und deren bautechnische Aussagekraft sowie juristische Aspekte zu vertraglichen Regelungen, *Forschung + Praxis 46: STUVA-Tagung 2015*, Ernst & Sohn, S. 163 – 171

Tunnels and Underground Cities: Engineering and Innovation meet Archaeology,
Architecture and Art, Volume 5: Innovation in underground engineering,
materials and equipment - Part 1 – Peila, Viggiani & Celestino (Eds)
© 2020 Taylor & Francis Group, London, ISBN 978-0-367-46870-5

Energy absorption tests on fibre-reinforced-shotcrete round and square panels

N. Buratti, A. Incerti, A.R. Tilocca & C. Mazzotti
CIRI – Buildings and Construction, University of Bologna, Bologna, Italy

M. Paparella & M. Draconte
Istrice Fibres – Fili & Forme s.r.l., San Cesario sul Panaro, Italy

ABSTRACT: The present paper presents the results of an experimental campaign aimed at investigating the mechanical performances of different FRC shotcrete mixes by means of tests on square (EN 14488-5) and circular panels (ASTM C1550-12). In particular, two different concrete mixes were tested, they featured the same aggregates, but one of the two mixes contained silica fumes a replacement of a fraction of cement. Different dosages of macro synthetic fibres (3 to 8.5 kg/m^3) were used in each of the two mixes as well as a single dosage of steel fibres (30 kg/m^3). Both sprayed and cast panels were considered. In addition to these panels cast notched prisms were tested in three-point bending. The paper discusses the mechanical behaviour of the various specimens, analysing the influence of concrete mix and fibre dosage, and presents correlations among different tests.

1 INTRODUCTION

Fibre Reinforced Shotcrete (FRS) is a special sprayed concrete with the addition of either steel or synthetic fibres. Fibres allow concrete to support loads in cracked conditions. The most common application of FRS is ground stabilization in tunnelling and mining projects. FRS applications increased in the last 40 years because of its benefits, mainly in terms of efficiency and reduction of construction time.

The standard design procedure in these applications is based on the use of Barton's chart (Barrett and McCreath, 1995; Bernard, 2004; Grimstad et al., 2002; Grimstad and Barton, 1993; Papworth, 2002). An empirical approach that links the toughness of the FRS with the spacing between anchoring bots, rock quality and other geotechnical parameters. The toughness of FRS gives an estimate of the capacity of maintaining ground support by absorbing the energy produced in case of large deformations due to crack opening. The main experimental procedures used to estimate this energy absorption capacity are the EFNARC square panel test (Comite Europeen de Normalisation (CEN), 2006) and the Round Determinate Panel (RDP) tests (American Society for Testing and Materials (ASTM) International, 2012). Given the importance of the energy absorption capacity in design of FRS for tunnel stabilization, engineers need information on the relationship between fibre dosage and this parameter.

The present paper analyses the energy absorption capacity of FRS square and round panels containing different dosages of a macro polymeric fibre. It discusses the correlation between the results of the two different panel tests, the correlation with bending tests on notched prims, investigates the effect of fibre dosage and proposes empirical relationships for the FRS considered.

2 MATERIALS AND METHODS

2.1 *Material used and specimen preparation*

The experimental campaign was devoted to analyse the flexural behaviour and the energy absorption capacity of various FRS specimens, produced using two concrete mixes (indicated as I and E in the following) and different dosages of fibres. The nominal properties of the fibres used are listed in Table 1. Four different dosages of the macro Polymeric (P) fibres were used (3 to 8.5 kg/m^3) as well as a single dosage of Steel (S) fibres (30 kg/m^3). The ingredients of the concrete admixture are reported in Table 2, while the grading curve for the aggregates used in the concrete mixes is reported in Figure 1. Three kinds of specimens were used: 600 mm x 600 mm x 100 mm square panels, Ø800 mm × 75 mm round panels (RDP) and 150 mm x 150 mm x 600 mm prisms. Table 3, lists the number of specimens considered for each geometry for the different concrete mixes. Specimens were produced using different techniques; all the beams as well as some of the RDPs were cast, while another group of RDPs and all the square panels where sprayed.

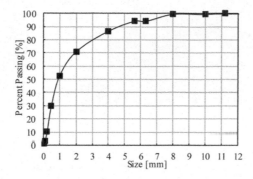

Figure 1.   Aggregate grading curve.

Table 1.   Main properties of the fibres used as declared by manufacturers. T_s: tensile strength.

Fibre name	Material	Code	Length	Diameter	E_m	T_s
			mm	mm	GPa	Mpa
MapefibreIT39NV	Polyolefin compound	P	55	0.91	3.9	560
DE 35/0.55 N	Steel	S	35	0.55	200	1100

Table 2.   Concrete mixes considered. C: Cement CEM II 42.5 R, A: Aggregates, SF: Silica Fume, W: Water, P: Powder, SP: Superplasticizer (Dynamon SX-N), SR: Set Retardant (Mapetard SD-2000), ACC: Accelerator (Mapequick AF 2000).

Mix	C	A	SF	W	W/P(<63μ)	Fiber type	Fiber dosage	SP	SR	ACC
	kg/m^3	kg/m^3	kg/m^3	l/m^3	-	-	kg/m^3	l/m^3	l/m^3	kg/m^3
I-P3	450	1595	-	204	0.44	P	3.0	5.8	1.3	20.3
I-P5	450	1589	-	204	0.44	P	5.0	6.1	1.3	20.3
I-P7	450	1583	-	204	0.44	P	7.0	6.1	1.3	20.3
I-P8.5	454	1571	-	205	0.44	P	8.5	6.1	1.8	20.4
I-S30	450	1594	-	204	0.44	S	30.0	6.1	1.3	20.3
E-P3	420	1610	10	204	0.46	P	3.0	5.8	1.3	18.9
E-P5	420	1605	10	204	0.46	P	5.0	5.8	1.3	18.9
E-S30	420	1609	10	204	0.46	S	30.0	5.8	1.3	18.9

Table 3.   Number of specimens for the different concrete mixes.

Concrete mix	Prisms (P)	RDP Cast (RC)	RDP Shot (R)	Square (S)
I-P3	4	3	5	5
I-P5	4	3	5	5
I-P7	3	3	5	5
I-P8.5	-	-	-	5
I-S30	4	3	5	5
E-P3	4	-	3	3
E-P5	4	-	3	3
E-S30	4	-	3	3

Figure 2.   Preparation of shotcrete specimens.

A Meyco Suprema concrete pump was used to this aim (see Figure 2). In the following the specimens will be indicated as follows [Concrete mix]-[Batch]-[Specimen type][Specimen Number], i.e. I-P5-1-R2 indicates the sprayed RPD #2 produced form the batch #1 of the concrete mix I-P5. All specimens were cured in controlled environmental conditions for 28 days. The test procedure adopted for the various types of specimens will be described in the following sections.

2.2   Three point bending tests on prisms

The flexural behaviour of the prismatic specimens was characterized by means of three-point bending tests, carried out according to EN-14651 (Comite Europeen de Normalisation (CEN), 2005). A servo-hydraulic machine with a maximum capacity of 500 kN (Figure 3), was used. During the tests, a COD transducer measured the CMOD while a 50 kN load cell measured the force applied to the specimens. As required by EN 14651 the testing procedure was performed under closed-loop CMOD control with a CMOD opening rate of 0.05 mm/min up to CMOD = 0.1 mm and then increased to 0.2 mm/min. Before testing, the samples were prepared with a 25 mm deep notch at the bottom of the mid-span cross-section. Prisms were notched at 25 days using a 4 mm wide wet saw. They were rotated over 90° around their longitudinal axis before sawing (i.e. the notch is on one side of the specimen considering the casting direction). In the following, the results of these tests will be reported as CMOD versus nominal residual flexural tensile strength (σ_N) curves, computed as the moment at mid-span divided by the section modulus of the notched section.

2.3   Round Determinate Panel tests

Energy absorption tests on RDPs were carried out according to ASTM C1550-12 (American Society for Testing and Materials (ASTM) International, 2012). A servo-hydraulic actuator with a 500 kN load cell was used to load the specimens (Figure 4), which were supported at three points by means of spherical hinges. During the tests, a LVDT transducer with 50 mm

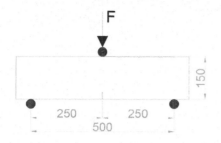

Figure 3.   Three-point bending test setup according to EN 14651. Lengths are in mm.

Figure 4.   Round Determinate Panel test setup according to ASTM C1550-12. Lengths are in mm.

of stroke measured the deflection at the centre of the panels while three LVDTs with 10 mm stroke monitored the support displacements. These latter were then used to compute the net mid-span deflection. Finally, three potentiometers measured the crack opening perpendicularly to the principal directions of cracking. As required by ASTM C1550-12 the testing procedure was performed in displacement control with a displacement rate equal to 4 mm/min.

2.4   *Square panel tests*

Square panels were tested according EN 14488-5 (Comite Europeen de Normalisation (CEN), 2006). The panels were supported by a 20 mm thick rigid square support all around the perimeter and loaded trough a square load block with contact surface of 100 mm x 100 mm. A servo-hydraulic actuator with a 500 kN load cell was used to load the specimens (Figure 5). During the tests, a LVDT transducer with 50 mm stroke measured the deflection at the centre of the panels. As required by 14488-5 the test was performed under displacement control with a displacement rate equal to 1 mm/min.

Figure 5.   Square panel test setup according to EN14488-5. Lengths are in mm.

3 EXPERIMENTAL RESULTS

3.1 *Mechanical behaviour of the specimens*

As an example of the behaviour of the specimens, Figures 6–8 shows experimental results for all the specimens of the mix I-P5. Their mean behaviour is also reported and compared to that of the specimens of the mixes I-P7 and I-S30. In particular for both square and round panels force vs deflection curves are reported, while for prismatic specimens σ_N versus CMOD curves are plotted. Furthermore, the typical cracking pattern at the end of the tests is shown in Figure 9.

Observing the force versus deflection curves for RDPs with P fibres (Figure 6), it is possible to notice that they feature a first peak corresponding to the onset of cracking, followed, by a second smoother peak. The three cracks (Figure 9) that characterize the normal behaviour of RDPs normally form in rapid sequence, and are indicated by steep reductions in the bearing capacity of the panels, in the neighbourhood of the main peak. The second smoother peak corresponds to the maximum crack-bridging capacity of the fibres. After this peak, the bearing capacity of the panels decreases, quasi-linearly, until the end of test. The fist peak mainly depends on the features of the concrete matrix and is not strongly affected by fibre dosage, while the second peak strongly depends on this parameter. The behaviour of square panels with P fibres (Figure 7) is in general similar but with some important differences. Typical curves feature two steep peaks in their very first part, corresponding to the opening of two orthogonal cracks, at the center of the panels, and parallel to their edges (see Figure 9). Curves then increase again up to a peak corresponding to the maximum capacity of fibres. Being square panels loaded in a redundant structural scheme, additional cracks can open during the tests. These are indicated by sudden drops in force capacity. The behaviour of the notched-prisms in three point bending typically features a clear first peak, mainly dependant on the tensile capacity of the concrete matrix, followed by a drop of bearing capacity which then increases again as the CMOD augments.

It is interesting to compare the behaviour of P and S fibres. As all the tests indicate the S fibres used herein are particularly effective for small crack openings (e.g. the reduction in bearing capacity after the first peak is smaller than for P fibres) but then, because of their limited length, their capacity rapidly decreases as the width of cracks increases. On the other hand, force-deflection curves for specimens with P fibres - being these less stiff and less effectively anchored to concrete than S fibres - feature a more significant strength drop after the firs peak. The maximum effectiveness of P fibres is achieved at about 10 mm deflection for RDPs, 6 mm deflection for square panels, and 4.1 mm CMOD for prisms. One might compare the behaviour of prisms and round panels considering the relationship between crack opening and deflection δ proposed by Ciancio et al. (Ciancio et al., 2016, 2014):

$$w = \frac{\delta \cos(60°)}{r} \cdot 2t \tag{1}$$

where r indicates the radius of the circle passing through the three supports (375 mm) and t the thickness of the panels (75 mm). According to Eq. (1) a deflection of 10 mm corresponds to a crack opening of 3.46 mm. On the other hand the crack opening at the top of the notch of the prims can be easily computed as $w = CMOD \cdot (125/150)$, therefore a CMOD of 4.1 mm corresponds to a crack opening of 3.41 mm. These simple calculations show the high consistency between the different tests. It should be noticed that the crack opening at the end of both RDP and square panel tests is larger than at the end of three-point bending tests, therefore the decrease in bearing capacity observed in the last part of panel curves is not visible in prism tests.

3.2 *Energy absorption capacity*

Table 4 reports peak forces, F_{max}, and energy absorption capacities, computed according to ASTM C1550-12 rules, for both cast and sprayed RDPs. Considering the F_{max} parameter, it is possible to observe that cast panels feature higher values than shot specimens, probably

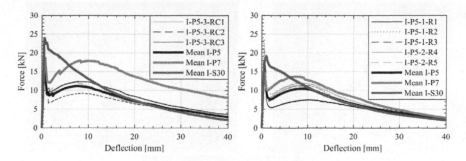

Figure 6. Force deflection curves for cast (left) and sprayed (right) RDPs.

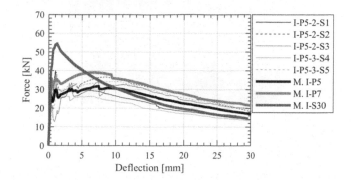

Figure 7. Force deflection curves for square panels.

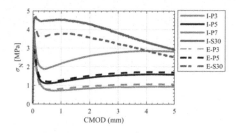

Figure 8. Nominal residual flexural tensile strength – CMOD curves for notched prisms.

because of a higher density of their concrete. Figures 10 a-b, show energy absorption curves. These confirm that cast panels have better performances than sprayed panels, probably because fibres in these latter panels have worse bond, due either to a different distribution/ orientation or to a lower bond to the concrete matrix. In particular, the difference is higher for higher fibre dosages, e.g. for 7 kg/m^3 of P fibres, sprayed panels featured on average 310 J wile cast panels 450 J. This difference might be explained considering the reduction of concrete workability and pumpability with the increase of fibre dosage.

The main results for square panels are reported in Table 5. For these specimens both the maximum bearing capacity and energy absorption are higher than for RDP specimens, because of the different test setup. The representative values for nominal flexural residual tensile strength, as defined by EN-14651, for prismatic specimens are listed in Table 6.

As clearly suggested by force-deflection curves discussed in the previous Section specimens with S fibres tend to dissipate relatively more energy at smaller deflections than P fibres which are more effective at larger crack openings.

Figure 9.   Typical crack pattern at the end of tests for RDPs (a), square panels (b) and notched prisms (c).

Table 4.   Peak force and energy absorption capacity for cast (RC) and sprayed (R) RDPs.

Sample	$F_{max\,(RC)}$	$CoV(F_{max(RC)})$	$F_{max\,(R)}$	$CoV(F_{max(R)})$	$E_{(RC)}$	$CoV(E_{(RC)})$	$E_{(R)}$	$CoV(E_{(R)})$
	kN	%	kN	%	J	%	J	%
I_P3	20.0	7.98	16.3	23.03	148	10.79	144	8.75
I_P5	19.0	27.05	18.6	24.83	252	13.18	231	6.93
I_P7	20.7	7.24	16.3	13.22	450	18.12	310	9.71
I_S30	26.3	1.92	22.2	7.89	312	4.58	269	25.92
E_P3	-	-	14.0	11.91	-	-	122	28.98
E_P5	-	-	16.0	22.75	-	-	240	22.79
E_S30	-	-	22.7	15.11	-	-	297	23.26

Table 5.   Peak force and energy absorption capacity for square panels.

Sample	$F_{max\,(S)}$	$CoV(F_{max})$	$E_{(S)}$	$CoV(E)$
	kN	%	J	%
I_P3	36.5	14.95	503	10.45
I_P5	37.0	7.96	662	14.47
I_P7	41.3	17.39	808	7.32
I_P8.5	50.3	7.97	1000	4.83
I_S30	58.6	6.23	738	10.11
E_P3	31.7	7.34	533	18.21
E_P5	38.7	10.07	687	27.53
E_S30	66.0	0.53	893	8.25

3.3   Analysis of results

Figure 11 shows the relationship between energy absorption capacity and P fibre dosage for sprayed RDPs (a) and square panels (b). The correlation between these parameters is very high as clearly visible in the plot and, as the trend-line suggest, might be assumed linear in the

Table 6. Nominal residual flexural tensile strengths for prismatic specimens.

Sample	$f_{L\ (P)}$	f_{R1}	f_{R2}	f_{R3}	f_{R4}
	MPa	MPa	MPa	MPa	MPa
_I_P3	3.47	0.77	0.79	0.89	0.92
I_P5	3.05	1.11	1.32	1.50	1.58
I_P7	3.88	1.90	2.41	2.70	2.83
I_S30	4.57	4.47	4.45	4.06	3.56
E_P3	3.08	0.80	0.90	1.01	1.06
E_P5	2.96	1.21	1.42	1.60	1.70
E_S30	3.67	3.65	3.72	3.41	3.06

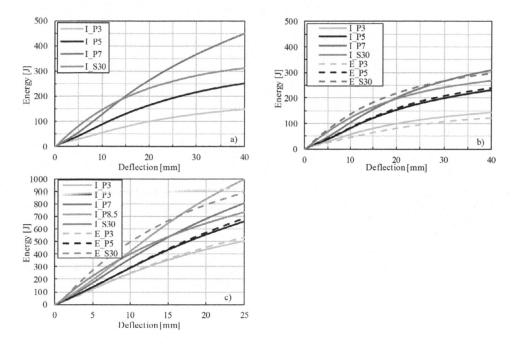

Figure 10. Mean energy absorption curves for cast RDPs (a), sprayed RDPs (b) and square panels (c).

range of dosages considered herein. In particular, both the trend-lines were fit considering the intercept as unknown, clearly, in order to obtain a better estimate of this parameter, one would need to have data on the energy absorption capacity of plain concrete specimens.

It is of interest also to investigate the correlation among the results of different tests. To this aim, Figures 12a-b report the correlation among the residual nominal flexural tensile strengths f_{R1} and f_{R3}, obtained from bending tests on prims, and RDP energy absorption capacity. In this figure only the average result for each mix is reported. If we consider the specimens containing P fibres only, a very strong correlation can be noticed between the aforementioned parameters. Specimens with S fibres seem to follow a different relationship but, since only one fibre dosage was considered here, it is not possible to define a trend-line. The different relationship between absorbed energy and either f_{R1} or f_{R3} for P and S fibres can be easily justified considering: i) the very different shape of the curves in Figure 8 for P and S fibres; ii) the different crack opening at the and of panel and prims tests, i.e. about 14 mm in the first case, 0.5 mm for f_{R1} and 2.5 mm for f_{R3}. Therefore, a much better correlation among these different tests can be obtained through an extrapolation of the σ_N – CMOD curves to larger crack opening values. In particular, defining the toughness T of the prisms as the area under the

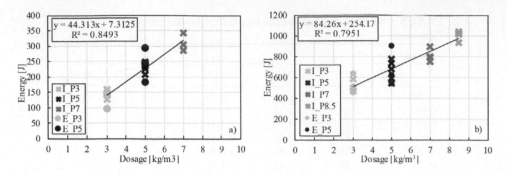

Figure 11. Correlation between absorbed energy and dosage of P fibres for sprayed RDPs (a) and square panels (b).

aforementioned curve, neglecting the contribution due to the first peak and assuming a linear σ_N decrease to zero for a crack opening equal to one half of the fibre length l_f (i.e. average value of the bonded length), T can be estimated as:

$$T = f_{R1} + f_{R2} + f_{R3} + 0.5 f_{R4} \left(1.2 \cdot 0.5 \cdot l_f - 2.5\right). \tag{2}$$

Figure 12c shows the correlation between the values given by Eqn. (2) and the energy absorbed in RDP tests. It can be clearly noticed that, since this parameter takes into account the shape of the σ_N – CMOD curves now data for P and S fibres follow the same trend-line.

Finally 12d shows the correlation between absorbed energies in RDP and square panel tests. The data reported in this figure suggests that P and S fibres follow the same relationship, even though data for S fibres is limited. Two trend-lines are reported in Figure 12d, the first

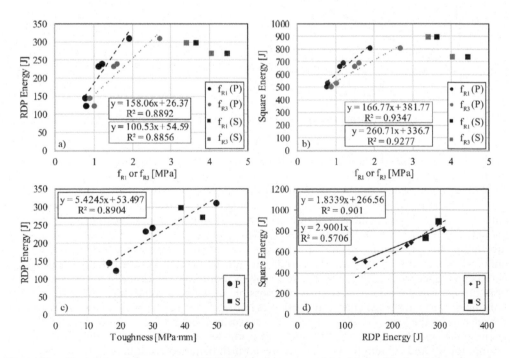

Figure 12. Correlation between mean RDP absorbed energy and f_{R1} (a), f_{R3} (b), toughness of prisms as defined in Eqn. (2) (c), and square panels absorbed energy (d).

line has zero intercept while the second was obtained considering the intercept as a regression parameter. Obviously, there are theoretical reasons that justify a zero intercept, even though given the range of energy absorption capacities investigated here it is not possible to draw any conclusion on the shape of the trend line, for this reason the trend line with non-zero intercept was considered. Bernard (Bernard, 2002) suggested a slope of the trend line equal to 2.50.

4 CONCLUSIONS

The present paper presented the results of an experimental campaign aimed at understanding the energy absorption capacity of FRS containing different dosages of a polymeric fibre (3 kg/m^3 – 8.5 kg/m^3). In particular, both round determinate panels (RDP) and square panels were tested as well as notched beams. After a discussion on the behaviour of panels in comparison to that of specimens with steel fibres, the paper illustrated the relationship between fibre dosage and energy absorption capacity and suggested that in the range of dosages considered herein this relationship can be assumed as linear. Furthermore, the paper presented an analysis of the correlation among the results of the three types of tests carried out. This analysis shows that the results obtained from panels are obviously strongly related. On the other hand, because of the smaller crack opening achieved during the tests, there is a lower correlation with three point bending tests on notched prisms. Therefore, the paper proposes to extrapolate these latter to larger crack openings, thus leading to a noticeably higher correlation.

ACKNOWLEDGEMENTS

The authors would like to acknowledge the financial and technical support of Fili & Forme Sil and MAPEI S.p.A; the technical support of Mr. D. Michelis (UTT - Underground Technology Team, MAPEI S.p.A.) and the technical support of Mr. D. Erra (CIRI – B&C).

REFERENCES

American Society for Testing and Materials (ASTM) International, 2012. Standard Test Method for Flexural Toughness of Fiber Reinforced Concrete (Using Centrally Loaded Round Panel).

Barrett, S.V.L., McCreath, D.R. 1995. Shotcrete Support Design in Blocky Ground: Towards A Deterministic Approach. *Tunneling Undergr. Sp. Technol.* 10: 79–89.

Bernard, E.S. 2004. Design performance requirements for fibre reinforced shotcrete using ASTM C-1550. *2nd Int. Conf. Eng. Devlopments Shotcrete.*

Bernard, E.S., 2002. Correlations in the behaviour of fibre reinforced shotcrete beam and panel specimens. *Materials and Structures* 35(3): 156–164.

Ciancio, D., Manca, M., Buratti, N., Mazzotti, C. 2016. Structural and material properties of Mini notched Round Determinate Panels. *Construction and Building Materials* 113: 395–403.

Ciancio, D., Mazzotti, C., Buratti, N. 2014. Evaluation of fibre-reinforced concrete fracture energy through tests on notched round determinate panels with different diameters. *Construction and Building Materials* 52: 86–95.

Comite Europeen de Normalisation (CEN) 2006. *EN 14488–5 testing sprayed concrete – Part 5: Determination of energy absorption capacity of fibre reinforced slab specimens.*

Comite Europeen de Normalisation (CEN) 2005. *Test method for metallic fibered concrete - Measuring the flexural tensile strength (limit of proportionality (LOP), residual).*

Grimstad, E., Barton, N. 1993. Updating of the Q-system for NMT. *Proc. of Int. Symp. on Sprayed Concrete - Modern Use of Wet Mix Sprayed Concrete for Underground Support, Fagernes 1993.*

Grimstad, E., Kankes, K., Bhasin, R., Magnussen, A.W., Kaynia, A., 2002. Rock mass quality Q used in designing reinforced ribs of sprayed concrete and energy absorption, *4th Int. Symp. on Sprayed Concrete, Davos,* 2002.

Papworth, F., 2002. Design Guidelines for the Use of Fiber-Reinforced Shotcrete in Ground Support. *Shotcrete Magazine* 1, 16–21.

Tunnels and Underground Cities: Engineering and Innovation meet Archaeology,
Architecture and Art, Volume 5: Innovation in underground engineering,
materials and equipment - Part 1 – Peila, Viggiani & Celestino (Eds)
© 2020 Taylor & Francis Group, London, ISBN 978-0-367-46870-5

Rapid automatic inspection system for tunnel lining based on a mobile vehicle

J.L. Cai, X.Y. Xie & B. Zhou
Tongji University, Shanghai, China

Y.X. Zhou & W.C. Zeng
Yunnan Communications Investment & Construction Group Co., Ltd, Kunming, Yunnan Province, China

ABSTRACT: Periodic inspection is needed for the maintenance of tunnel and increase of service life. Traditional inspection is operated manually and low-efficient. Besides, human error tends to produce inaccurate results. This paper presents a fast tunnel inspection system based on a mobile vehicle, composed of machine vision detection and ground penetrating radar (GPR) for the detection of visible and invisible defects of tunnel structure respectively. The inspection system enables continuous and highly-detailed measurements at a speed up to 40km/h without interrupting the traffic. A method is also developed for the rapid automatic extraction of defect features. The fast record and assessment of defects greatly improve the efficiency of inspection and minimize manual intervention. Finally, the inspection results and assessment of defects in a road tunnel are provided in this paper.

1 INTRODUCTION

Tunnels are playing more and more important role in the expressway in western areas in China. However, with the increase of traffic volume and service time, the service performance of tunnels degraded inevitably due to human and natural factors and great efforts have been made on the inspection and maintenance of tunnel structure. Periodic inspection, assessment and maintenance are required to ensure that these tunnels remain in safe condition and continue to provide reliable levels of service (Menendez et al. 2018). Common diseases of tunnel can be divided into visible and invisible defects. Visual inspection is an essential method for the visible defects. While nondestructive evaluation methods are widely used for the detection of invisible defects, including impact-echo test, ultrasonic method, electrical method, magnetic method, etc. Traditional inspections mentioned before are mostly operated manually, low efficient and cost a lot. Unfriendly environment in the tunnel not only increases the operative difficulty and estimation error, but also harm the inspectors. Besides, traffic has to be interrupt due to the inspection.

Recently, disadvantages of traditional inspection have made more attention payed to the research of rapid automatic detection technology. Sasama et al. (1998) developed a system named ConSIS to take continuously scanned images of tracks, rolling stock and tunnel walls in railways with a line-sensor camera. Yao et al. (2003) described an automatic concrete tunnel inspection system by an autonomous mobile robot. 24 ultrasonic range sensors and six video cameras mounted on a mobile robot can detect deformed inner walls at divisions of 14 mm when the robot moves at 20 mm/s. Yu et al. (2007) proposed a mobile robot system with a CCD camera for inspecting and measuring cracks in concrete structures to provide objective crack data. Menendez et al. (2018) presented ROBO-SPECT system composed of robotic system, ground control station and control room. The system can complete assessment of cracks and other defects of tunnels autonomously with one-lane closure. Huang et al. (2018)

applied a rapid detection and analysis system to survey the damage of tunnel lining of Metro Line 2 in Changsha, China. Multiple area array Charged Couple Device (CCD) cameras and intelligent analysis method were used to identify and quantify the damage. Attard et al. (2018) presented a computer vision system, TInspect, that used a robust hybrid change detection algorithm to monitor changes on the large hadron collider tunnel linings. The system achieves a high sensitivity of 83.5% and 82.8% precision, and an average accuracy of 81.4%. To some extent, the inspection system mentioned before realize automatic and continuous detection of tunnel diseases. However, the systems are limited to the detection of surface diseases of the tunnel lining, but could do nothing about the invisible defects, such as the void behind the lining. Automatic assessment of inspection results still needs to be developed for high efficiency and high precision.

This paper presents a rapid automatic inspection system for tunnel lining composed of machine vision detection and GPR for the detection of lining crack and void behind the lining respectively. The inspection system is mounted on a mobile vehicle and enables continuous measurements at high speed without interrupting the traffic. A method is also developed for the rapid automatic extraction of crack features, which greatly improves the efficiency of analysis and minimize manual intervention. Finally, the inspection results and assessment of defects in a road tunnel are provided in this paper.

2   MODULAR DESIGN OF INSPECTION SYSTEM

According to the investigation and analysis, the tunnel diseases include cracks, spalling, falling of lining, dislocated construction joint, lining deformation, leakage of water, deterioration of reinforced concrete, voids behind the lining and so on (Lu & Chen 2010). As for the existing systems, most are aimed at the surface diseases of tunnels. Some typical surface diseases are shown in Figure 1. These not only weaken the service performance of the tunnel, but also threaten the safety and property of the users. Besides, defects hidden from view, such as voids and deterioration of reinforced concrete, should also not be ignored. However, it is difficult to develop a system which can detect both visible and invisible defects due to various devices and sensors will be involved. In order to reduce the complexity and enhance the applicability, the modular design is introduced in this section. The inspection system (Figure 2) for tunnel lining is composed of three units, visual detection unit, GPR unit, monitoring and positioning unit, mounted on a motor vehicle.

2.1   *Visual detection unit*

The visual detection unit is located in the front of the carriage and acquires images from the surface of tunnel lining with high resolution. And then characteristic parameters of visible defects will be rapidly and automatically extracted based on machine vision.

20 area array CCD cameras are used for the acquisition of images and distribute uniformly around the carriage. Each camera is equipped with a motorized zoom lens and a rotation motorized stage, which enables the unit to adapt automatically to different tunnel section

Cracks   Spalling   Falling of lining   leakage of water

Figure 1.   Photographs of typical tunnel diseases.

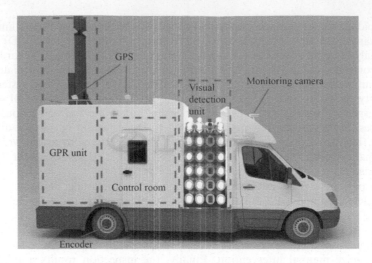

Figure 2. Inspection system for tunnel lining mounted on a motor vehicle.

geometry. Due to the high speed of the vehicle in the inspection, the number of frames per second should be high in order to acquire images of the whole tunnel, which has high requirement of lighting. However, the tunnel itself is insufficiently illuminated. Given this, 60 LED lamps divided into 3 rows are adopted as auxiliary lighting to ensure the image quality.

In addition, a distributed storage system is developed to store a large amounts of detection data and satisfy the high-speed storage requirement. There are 5 storage cells in the system and each cell is connected to 4 cameras via a switch.

2.2 *GPR unit*

The GPR unit uses radar pulses to image the subsurface structures and detect defects hidden from view. GPR as an effective nondestructive evaluation method is widely used in the inspection of civil engineering structures (Kilic & Eren 2018). However, electromagnetic radiation mostly reflects off the interface between the air and structure and little can transmit into the structure or the deeper area. It is needed to hold the antenna up manually to make it close to the structure as far as possible and it is hard to inspect the high place of structure. Thus, traditional inspection based on GPR cannot meet the demands for efficiency and convenience.

In view of this, a robot arm and lift platform are designed to replace manual personnel (Figure 3). The actuator and turnplate can make the arm rotate 90° in the vertical plane and 360° in the horizontal plane. The maximum strokes of the lift tower and pallet are 2600mm

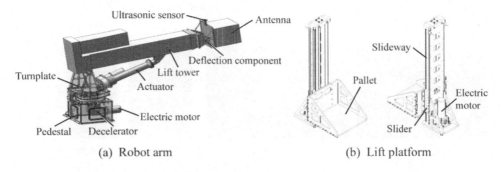

(a) Robot arm   (b) Lift platform

Figure 3. Robot arm and lift platform.

and 1100 mm respectively. The GPR inspection in the case can be used for the large section tunnel with maximum height of 7800 mm. The deflection component allows the antenna to have a slight rotation to adapt to the surface of structure. The ultrasonic sensor is equipped to capture installations in the tunnel so that the robot arm can response automatically at the speed of 10 km/h.

High frequency radio waves transmit farther than low frequency ones in the subsurface material but have lower resolution. To solve the contradiction a dual channel GPR system with 400 MHz and 900 MHz antennas is developed. The 900 MHz antenna is used to measure the space of rebars and thickness of reinforcement cover in the shallow part and the 400 MHz antenna is used to detect the defects like voids in the deep.

2.3 *Monitoring and positioning unit*

The monitoring and positioning unit marks the inspection data with position information to realize the location of defects. The monitoring video is assistance for the identification of defects. The unit is composed of Real Time Kinematic (RTK) GPS, encoder and monitoring cameras. RTK GPS can provide centimeter-level accuracy outside of the tunnel with GPS and GPRS signals. In the tunnel, the encoder produces a pulse at a specified interval with the vehicle moving and the resolution is up to millimeter level.

2.4 *Other devices*

A 10kw silent generator is selected to supply power for the all devices in the inspection system. Uninterrupted Power Supply (UPS) is used to improve the quality of electric current and prepare for the emergency.

3 WORKFLOW OF THE INSPECTION

The workflow of the inspection is illustrated in Figure 4. Image fusion technology is adopted to combine the images collected by 20 cameras into the image of the whole tunnel. In the longitudinal direction of tunnel, cameras take pictures at a specified interval according to the pulses produced by the encoder. In the transversal direction, the angles and zooms of lens can

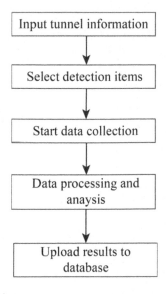

Figure 4. Workflow of the inspection.

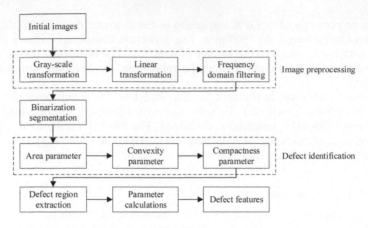

Figure 5. Flow chart of the feature extraction algorithm.

be changed automatically according to the geometry of tunnel section benefit from the motorized zoom lens and rotation motorized stages. Visual and GPR inspections can be both or alone selected. The angle of robot arm should be specified if the GPR inspection is selected. When the inspection starts, the data collected will be stored in the distributed storage system. The inspection will stop when the vehicle movement reaches the length of tunnel. A method based on machine vision is developed to process and analyze the image data. Both visual defect identification and feature extraction are carried out intelligently by machine. The processing and analysis of GPR data still depend on experienced engineers. Finally, the results are uploaded to the database. It is convenient to compare inspection results from different periods.

4 FEATURE EXTRACTION

Traditional analysis of inspection data usually depends on qualified personnel, which is low-efficient and cannot satisfy the requirements of rapid automatic inspection system. In addition, human intervention tends to produce inaccurate results. In order to minimize human operations in the analysis, an algorithm of feature extraction for surface defects is presented in this paper.

Figure 5 is the flow chart of the feature extraction algorithm. For the image preprocessing, the algorithm combines grayscale transformation algorithm and frequency domain filtering algorithm, which effectively enhance the target region and smooth the noise region. Based on the image segmentation by means of gray threshold, and according to the shape and direction feature of defects, the algorithm extracts the defect region by using area, convexity and compactness parameters. On this basis, we can calculate the geometric parameters of defects and quantify the defect characteristic, which provides evidence for evaluating structure security.

5 CASE STUDY

5.1 Background

Wuding-Yimen Expressway located in Yunnan Province extends from Wuding to Yimen by passing through four tunnels, which is an important section of expressway network of Yunnan. Fengshou Tunnel is one of the four tunnels and designed as a three-lane tunnel with width of 15.367m and height of 7.75m. The total length of the tunnel is 603–620m. Figure 6 and Figure 7 illustrate the plane layout and cross section of Fengshou Tunnel.

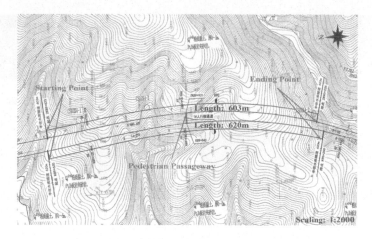

Figure 6. Contour map of Fengshou Tunnel in Yunnan Province, China.

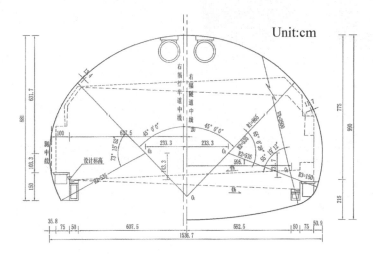

Figure 7. Diagram of Fengshou Tunnel section.

5.2 *Processing and analysis of data*

The rapid automatic inspection system presented in this paper was tested in Fengshou Tunnel and visual detection was selected. Figure 8 illustrates the processing of data with the feature extraction algorithm. Finally, the region of crack was successfully extracted by machine. The geometry of crack tends to be irregular, but slender. Considering this, we define the length of crack bounding rectangle as the length of crack, area of crack divided by the length as the width of crack. Table 1 lists the values of parameters extracted by machine.

5.3 *Assessment of defects*

Technical Specifications of Maintenance for Highway Tunnel (JTG H12-2015) gives the evaluation criterion of lining cracks, which were divided into cracks in propagation and uncertain cracks, as shown in Tables 2 and 3. D is the worst grade. In this case, the crack is not in propagation and the grade is B.

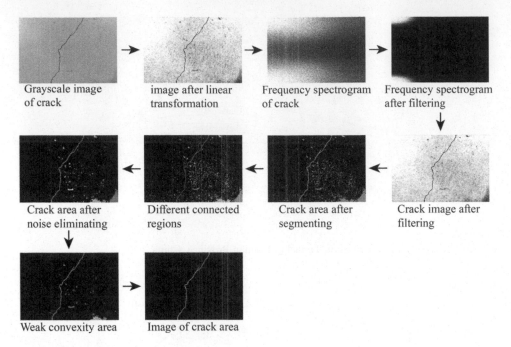

Figure 8. Processing of data with the feature extraction algorithm.

Table 1. Parameters of crack extracted by machine.

Parameters	Area/mm^2	Width/mm	Length/mm
Values	1244.1	3.1	401.3

Table 2. Criterion of lining cracks in propagation.

L/m b/mm	L>5	L≤5
b>3	C/D	B/C
b≤3	B	B

*L is the length of crack and b is the width of crack

Table 3. Criterion of uncertain lining cracks.

L/m b/mm	L>10	10≥L≥5	5≥L
b>5	C/D	B/C	B/C
5≥b>3	C	B/C	B
3≥b	A/B	A/B	A/B

*L is the length of crack and b is the width of crack

6 CONCLUSION

The rapid automatic inspection system presented in this paper is demonstrated to be effective and have good applicability. The crack feature extraction algorithm reduces human intervention and improves the processing of data. There is no doubt that continuous rapid inspection and automatic processing by machine are in favor of the maintenance of tunnel in the future. However, the processing and analysis of GPR data still depend on qualified personnel. The improvement of data processing algorithm for higher resolution and accuracy deserves more efforts. It becomes easy to collect data via the rapid automatic inspection system. Traditional assessment methods cannot make full use of such big data. Application of artificial intelligence and deep learning in the assessment of tunnel defect is the research in the future.

ACKNOWLEDGMENTS

This work was supported by the National Key Research and Development Plan of China (2018YFC0808702) and Science and Technology Project of Yunnan Transportation Department in 2016. Gratitude also goes to Construction Headquarters of Yunnan Wuyi Expressway, Yunnan Communications Investment & Construction Group Co., Ltd.

REFERENCES

Attard, L., Debono, C.J., Valentino, G. & Castro, M.D. 2018. Vision-based change detection for inspection of tunnel liners. Automation in Construction, 91, 142–154.

Chongqing Traffic Committee, 2003. Technical Specifications of Maintenance for Highway Tunnel. Beijing: China Communications Press.

Huang, Z., Fu, H.L., Chen, W., Zhang, J.B. & Huang, H.W. 2018. Damage detection and quantitative analysis of shield tunnel structure. Automation in Construction, 94, 303–316.

Kilic, G. & Eren, L. 2018. Neural network based inspection of voids and karst conduits in hydro–electric power station tunnels using GPR. Journal of Applied Geophysics, 151, 194–204.

Lu, Y.M. & Chen, L.W. 2010. Disease classification and cause analysis of existing tunnel. Railway Engineering, 11, 46–49.

Menendez, E., Victores, J.G., Montero, R., Martínez, S. & Balaguer, C. 2018. Tunnel structural inspection and assessment using an autonomous robotic system. Automation in Construction, 87, 117–126.

Sasama, H., Ukai, M., Ohta, M. & Miyamoto, T. 1998. Inspection system for railway facilities using a continuously scanned image. Electrical Engineering in Japan, 125(2),52–64.

Yao, F.H., Shao, G.F., Takaue, R. & Tamaki, A. 2003. Automatic concrete tunnel inspection robot system. Advanced Robotics, 17(4),319–337.

Yu, S.N., Jang, J.H. & Han, C.S. 2007. Auto inspection system using a mobile robot for detecting concrete cracks in a tunnel. Automation in Construction, 16(3),255–261.

Tunnels and Underground Cities: Engineering and Innovation meet Archaeology,
Architecture and Art, Volume 5: Innovation in underground engineering,
materials and equipment - Part 1 – Peila, Viggiani & Celestino (Eds)
© 2020 Taylor & Francis Group, London, ISBN 978-0-367-46870-5

Construction TBM tunnel – Operating data real time – Red Line Metro Tel Aviv

A. Capobianco
ELPA, Rome, Italy

M. Kalimyan
NTA, Tel Aviv, Israel

P. Merlanti
RINA Consulting S.p.A./ELPA, Milan, Italy

L. Romagnoli
RINA Consulting S.p.A., Genova, Italy

E. Santucci
ELPA, Rome, Italy

ABSTRACT: Elpa Project is a joint effort of Elpa and Rina Consulting S.p.A (appointed by NTA Ltd that is the responsible for the design and construction of a mass transit system for Tel Aviv) to establish a clear classification of the findings of Red Line Metro in Tel Aviv and clearly communicate them. The tunnel extends from Bat Yam to Petach Tikva in Tel Aviv and is one of the most heavily used traffic corridors in that Metropolitan area. 8 TBMs of internal diameter of 6,50 m will complete the underground Metro excavations adopting EPB methodologies in calcareous sandstone under sea water table in a very urbanized area. Elpa effort produced a novel approach to the analysis and the communication of findings and corrective actions to Contractor and Engineer during excavation monitoring. This method merges a strong expertise in TBM analysis with state of art tools for Big Data management.

1   INTRODUCTION

The Red Line Metro in Tel Aviv extends from Bat Yam to Petach Tikva via Tel Aviv, Ramat Gan and Bnei Brak and is one of the most heavily used traffic corridors in the Tel Aviv Metropolitan area. The Red Line has a total length of approximately 24 km and has a central underground section of approximately 12 km. The Red Line will be double track through all its length.

N.8 TBMs of internal diameter of 6,50 m will complete the underground Metro excavations adopting EPB methodologies in calcareous sandstone under sea water table (Porat, Wintle & Ritte 2014) in a very urbanized area (Figure 1).

N.6 TBMs operate on the Western side: the activity of contractor has been assigned to the China Railway Tunnel Group; N.2 TBMs operate on the Eastern side: the activity of contractor has been assigned to the China Civil Engineering Construction Corporation.

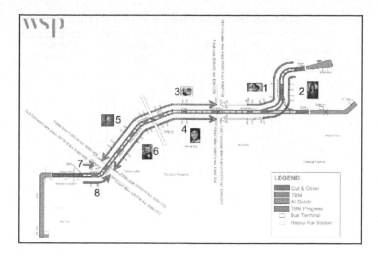

Figure 1. Red Line – TBMs route.

2 BRIEF DESCRIPTION OF RINA AUDIT ACTIVITY

Rina Consulting is participating, as technical advisor, in the Red Line Project in Tel Aviv since 2016 and is currently involved in several review and control activities.

The assurance management and technical auditing services provided by RINA are related to the all the activities of the Metropolitan Tel Aviv Transit System. The overall project value amounts to approximately EUR 1.5 billion.

3 TEL AVIV RED LINE METRO TUNNELLING TBM OPERATING DATA MONITORING

In tunneling works there are many important daily tasks to perform: the contractor needs to check the quality of work; the project manager needs quality assurance; monitoring data produced by TBM machines need to be collected by the project manager, analyzed and delivered to the contractor in the form of reports.

TBM monitoring data can be difficult to read just as it can be difficult to measure the deviation between design parameters and machine specific construction parameters.

NTA's senior management (the owner) need to understand the practical meaning of the main operating parameters of the TBM and need to capture the gap between design and construction parameters in order to make recommendations and carry out corrective actions.

The objective of the real-time monitoring of the TBM has been defined by the owner NTA in light of the need to communicate quickly, easily and comprehensively the excavation activities (standard progress as well as critical situations) to its senior management and to give indications to both Contractors and site engineers. This effective communication helps to meet contractual requirements.

Data from the TBMs sensors are transferred daily from the Contractors to the TBM's Auditors and are neither modified nor elaborated: only the original outputs produced by the TBM are transferred.

Together with TBM data, surface settlement (Bilotta, Russo & Viggiani 2002) above the TBM shield are also collected and analyzed in order to correlate findings related to the excavation activities with the geological and geomechanical characteristics of the excavated soil (Gvirtzman, Shachnai, Bakler & Ilani 1984).

All operational data collected during TBM excavation monitoring are archived to have the possibility to use them in case of future legal disputes between NTA and other parties.

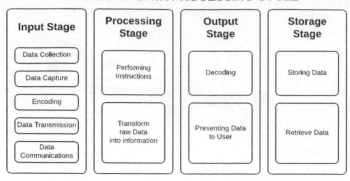

STAGES OF DATA PROCESSING CYCLE

Input Stage	Processing Stage	Output Stage	Storage Stage
Data Collection	Performing Instructions	Decoding	Storing Data
Data Capture			
Encoding			
Data Transmission	Transform raw Data into information	Presenting Data to User	Retrieve Data
Data Communications			

Figure 2.   Elpa process.

The weekly and monthly reports sent to NTA by Rina Consulting/Elpa auditors allow to the owner to conduct discussions on a monthly basis to clarify the reasons of the findings and to indicate how to correct them.

The Red Line project is only the first of others LRT and Metro lines projects that are to be built in the near future: Green Line and Purple Line, for example, are currently at the design stage.

4   THE PROBLEM WITH TBM'S DATA ANALYSIS

One of the problems that is encountered when working with TBMs is that generally there is some proprietary software that is able to report *simple* metrics only coming from the huge set of sensors related to the instrument.

On the other side there can be engineers or specialists with a background in construction theory but with poor experience in the management of TBMs sensors readings.

In this context there is often space for some home-made solution based on Excel to aggregate data: this kind of approach is not scalable, it is, in general, difficult to manage and, without a good analyst on the reading side, can be pointless.

Even good data visualization tools, however, require the experience of a good analyst: only a specialist that knows the mechanics, structure and problems that can result from a TBM's management can make sense of its data flow (Figure 2).

Moreover, there are no synthetic indicators that are able to convey important findings, or to communicate the needs for adjustment to a non-technical audience on the contractor side.

Finally, in case of legal disputes it is not simple to show what went wrong or to collect evidences of operator's mistakes or of problems with the TBM manufacturer.

5   ELPA METHOD FOR TBM'S DATA ANALYSIS

The main characteristics of Elpa TBMs' Data Analysis are: acquisition, cleaning, validation, storage, real time and batch analysis of Big Data coming from TBMs sensors, the analysis of status and synthetic reporting of critical observations to the Contractors (Figure 3).

Sensor data extracted from the TBMs are used to report progress of excavation, status of rings' construction; reports are collected on a monthly and weekly basis (Figure 4).

Elpa analysis made it possible to detect non-conformities related to work-site activities and give indications of the recommended corrective actions.

MAJOR PARAMETERS

RING	DATE	ADVANCE SPEED	CUTTER HEAD TORQUE	FORCE MAIN THRUST	EARTH PRESSURE (CROWN)	MUCK WEIGHT
n.	[DD/MM/YY hh.mm.ss]	[mm/min]	[KNm]	[Kn]	[bar]	[ton]
1262	01/02/2018 13:30:06	33,64	4130,38	21315,48	0,65	84,66
1263	01/02/2018 13:30:06	33,64	4130,38	21315,48	0,65	84,66
1264	02/02/2018 08:40:09	38,98	5315,88	18965,67	0,74	82,56
1265	02/02/2018 10:25:17	48,17	4497,18	14639,71	0,62	92,66
1266	02/02/2018 11:39:48	49,47	4152,86	14258,59	0,51	94,52
1267	02/02/2018 13:45:17	46,27	3103,71	15769,62	0,57	98,95
1268	02/02/2018 15:32:57	36,80	3995,71	20031,33	0,58	95,77
1269	02/02/2018 17:25:15	41,45	4570,46	18709,36	0,60	101,67
1270	03/02/2018 00:13:06	32,19	4634,94	19312,77	0,60	90,74
1271	03/02/2018 02:11:14	33,06	3807,80	18304,99	0,65	94,88
1272	03/02/2018 03:38:12	33,90	3407,66	17889,52	0,63	102,00
1273	03/02/2018 06:40:20	37,72	5031,35	19581,51	0,58	99,30
1274	03/02/2018 13:04:20	49,22	3841,21	17783,54	0,50	105,20
1275	03/02/2018 14:42:16	36,39	4125,52	17713,22	0,54	121,45
1276	03/02/2018 15:47:55	45,88	3855,66	16813,94	0,53	112,63
1277	03/02/2018 17:14:52	45,01	3842,32	17403,89	0,56	110,43
1278	03/02/2018 18:33:52	48,81	4325,26	16195,68	0,59	110,11

Figure 3.   Table with values of major parameters.

3: EARTH PRESSURE ANALYSYS: FACE EARTH PRESSURE - SINGLE LINE/THEORICAL PRESSURE MIN AND MAX

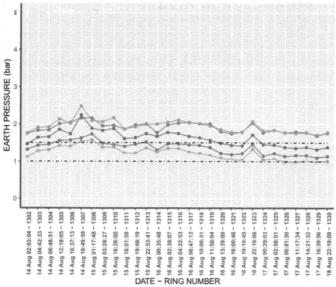

Figure 4.   Example of weekly report graph

5.1   *"Traffic Light" reports*

Monthly and weekly reports give a lot of synthetic information on the behavior of the TBM but are not enough to take prompt decisions. Elpa created a new way to alert the Contractor based on the semantic of traffic lights (Figure 5)

Like a real traffic light Elpa delivers a particular report aimed at highlighting how much the behavior of excavation metrics deviates from the expected. This report helps to show a particular

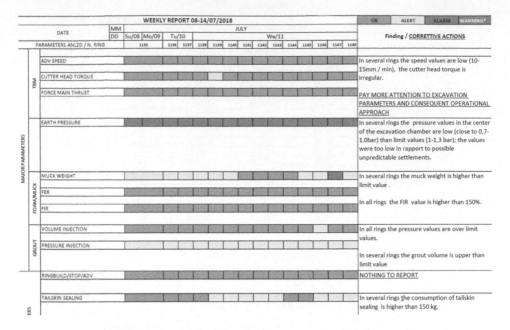

Figure 5. "Traffic Light" reports.

dimension of the ring advance (related to torque or foam or progress for example) with a simple three color pattern: a ring is red when something needs attention in order to take decisions on work in progress (ALARM); a red starred code is used in case of immediate necessity of intervention (ALARM*); yellow provides a warning but the work can keep going (ALERT), and green is used if everything is going fine (OK). Alerts are categorized in minor and major parameters so that a site manager can act after a quick glance to a single dashboard. This kind of construction made *very easy to communicate* dangers, unusual behaviors and suggest proper adjustment (by mean of a short text description in the rightmost column of the report).

5.2 *Back analysis*

Another contribution of Elpa is the long term storage of data after validation, analysis and report production. This allows the *back analysis* of all the TBM's excavation (Standing 2002). Since Elpa operates as an *autonomous and independent third party* during the progress of work, stored data can be retrieved in order to perform legal analysis or to assess if a problem, occurred during excavation, if it was due to problems in manufacturing of the TBM or to a misuse of the TBM.

5.3 *Technical details*

Data produced by more than 300 sensors was collected on a daily basis with a traffic of 170 Mb per day per TBM (i.e. 8 TBM produced about 40 Gigabytes in a 30 day period). The monitoring was up and running for a long period of time (up to 250 days for TBM4). Elpa organized and cleaned data in order to deal with: different columns ordering and headers (some in Chinese, some in English), deletions, file of reduced size and retransmissions. This first effort is mandatory and not trivial to have a very good and clean data-set to analyze. Software used to process Big Data is the state-of-art in Computer Science for Statistics and for data manipulation. Even if each TBM has its own software to produce reports (in fact it is possible to see values related to different rings and metrics), there is no possibility to see relations between metrics or to carry out statistical operations (average, minimum, maximum...).

Table 1. Major and minor parameters analyzed.

TBM Components*	Major parameters	Minor parameters
Cutter Head	Rotation/Torque Advance	Penetration
Cylinder	Force main thrust	Thrust cylinder press
Screw conveyor		Rot. Speed/Torque Earth press.
Exc chamber	Earth pressure	
Muck	Weight	Volume
Foam	FER	Volume
	FIR	Flow press. Flow liq. Flow air Flow/Vol. polymer Vol. bentonite
Tailskin	Grout Volume Grout Pressure	Grease vol./weight Grease Pressure
TBM guidance	Ringbuild Mining	Roll Pitch

* In addition the analysis of the topographic data near the TBM axis is performed

Elpa developed a proprietary software in order to fine-tune reports and metric. The software gave the possibility to rapidly inspect data in order to assess their validity or to identify peculiar behaviors. Finally, the software is used to identify relations between metrics and to predict anomalies to be communicated to the client.

5.4 *Further Improvements*

Elpa is developing a Machine Learning based software in order to predict anomalies. In the common approach an analyst explores and analyzes a given set of data in order to assess the anomalies and produce reports and suggestions. Generally speaking, the Elpa reporting solution shortens the access to critical data but remains not scalable. In fact, even with a good data visualization technique, the task becomes more difficult as the number of TBM or the size of the dataset increase.

As further improvements Elpa is developing a system to process data in an automatic way in order to classify anomalies with an Artificial Intelligence approach. Anomalies found by this classifier will help the analyst to start looking to more meaningful data instead of spreading his attention over the complete domain. This approach does not eliminate the need for human intervention, which remains a key point of analysis but will be a useful tool to reduce the complexity of data space to analyze and to spotlight some events faster.

5.5 *Site Inspections*

As requested by NTA, every month Rina – Elpa Technicians carry out site visits to the Tel Aviv construction sites in order to check directly for each TBM, the actual implementation of the corrective actions indicated in the weekly reports. In a climate of active collaboration with Contractors, many construction items are discussed in order to improve the excavation activities.

In case of launching or relaunching operations of TBMs from shafts or portals, the Rina – Elpa technicians were also in charge of auditing the Site Acceptance Tests made by the Contractors and the site engineers on the following but not exhaustive list of items:

Figure 6.   Picture taken during a site inspection before a relaunching.

Table 2.   Launching of TBM - List of items.

TBM items	Parameters/elements	Functioning
Cutter Head	Rotation	General condition
	Overcutter	System testing
Excavation chamber	Face pressure sensors	System testing
	Air valve	Connected and clean
Screw conveyor	Rotation	General condition
	Gate	opening/closing
Segment erector	Vacuum	General condition/System testing
Segment feeder		General condition/System testing
Cylinders		Movements/alignments
Tailskin	Grease Lines	Grease pump
	Grout Lines	Grout pump
	Brush ring	General condition
Foam system	Lines	System testing
Bentonite system	Lines	System testing
Conveyor belt	scale(s) belt	Cleaned and calibrated
Fire curtain		System testing

5.6   *Monthly meetings*

In order to finalize the Audit activity and maximize the sharing of knowledge, monthly meetings of the Rina – Elpa team of auditors with contractors and site engineers are held at the NTA offices.

In these meetings, the monthly data graphs, correlated with critical factors like settlements or TBM excavation problems, are discussed in order to improve the TBM performance. The Contractors are requested to provide a written response to the recommendations presented by the auditors.

The monthly reports sent to the client allow him to open discussions with the project manager or with the contractor to find the most appropriate solutions to understand the reasons for and reduce the gap between design and operational values.

6   RISK ANALYSIS ON SPECIFIC ITEMS

As an example of the possible applications of Elpa's method, in this section we present the back analysis that Elpa has conducted with the aim of assessing the potential for long-term sinkholes (depressions extending to the ground surface generated by the collapse of cavities created during the TBM excavation).

To understand the possibility of having a sinkhole (Shirlaw, Ong, Rosser, Tan, Osborne & Heslop 2003) in the completed sections of the tunnel, Elpa carried out a detailed analysis of the excavation parameters for each single installed ring. All the most significant parameters strictly correlated to the potential occurrence of sinkholes have been collected: Major parameters (Stop of TBM, Muck weight, Earth pressure), (Lunardi, Mancinelli & Zimbaldi 2015) (Chiriotti, Jackson & Taylor 2010) and Minor Parameters (Cutter head torque, advance speed, grout volume), (Bezuijen & Talmon 2006).

For each parameter a *risk factor* has been calculated based on the divergence between the value measured by the sensor and a set of reference values provided by the Contractor in the TBM Shift Instruction.

Finally, the set of *risk factors* calculated for each parameter was combined and weighted to create a synthetic indicator called *risk index percentage*. With this analysis Elpa was able to evaluate the probability of a sinkhole event in a given area. This analysis led to the identification of potentially dangerous areas along the tunnel route and to take actions to ensure the safety for the public.

During excavation Elpa was also able to produce another type of back analysis after stopping the work of the TBM due to force majeure: the cutter head was blocked by excavated material.

The analysis of the available data (Cardu & Oreste 2011) substantially confirmed what was observed during the excavation phases: the front pressure was fluctuating, which increased the difficulty of maintaining the front pressure according to the operating parameters; the values of thrust force and torque increased until they reached the safety shutdown value; it was difficult to maintain a constant speed; the two belt scales that measure the weight of the excavated muck showed very different values from the beginning of each push.

Based on this particular analysis several corrective actions were recommended before the restart of the TBM: calibration of scales, study of the pressure parameters at the front, study of the conditioning of the materials for possible improvements, intervention in the excavation chamber. Only after the corrective actions were carried out the TBM was allowed to restart.

7 CONCLUSIONS

The Elpa method aims to speed up, simplify and to render more intelligible the communication of findings and the performing of corrective actions during an excavation with a TBM. It configures itself as the missing link between the analysis of raw data from TBM sensors and the high level of operability of construction management. It enhances the control (on behalf of the client) of the work of those responsible for construction and works management. The power of this solution is in its full-stack, comprehensive approach: on one side, the development of software to manage BigData and a strong knowledge of the mechanical structure of TBM, on the other side the production of a set of reports and diagrams of differing granularity to allow effective communication. An example of practical implications of Elpa methodology is the assessment of potential long-term sinkholes in days (not at the end of the excavation) reducing the cost of corrective actions, saving money and reducing damage to TBM or Tunnels. This kind of approach allows being quickly aware of anomalous situations, find the root causes and act to take counter measures with a clear understanding at any level of the organization of the actor involved. The vision behind the Elpa method is to be able in the future to shorten the lifecycle of excavation monitoring to analyze/act in real time in this complex task by improving the proactivity of this solution (Machine Learning based approach) and working on the formalization of reports that will be clear to non-experts of an excavation.

REFERENCES

Bezuijen, A. & Talmon, A.M. 2006. Grout properties and their influence on back fill grouting. *Proceedings of the Symposium on Geotechnical Aspects of Underground Construction in Soft Ground, Amsterdam, pp 187–195.*

Bilotta, E., Russo G. & Viggiani C. 2002. Cedimenti indotti da gallerie superficiali in ambiente urbano. *XXI Convegno Nazionale di Geotecnica: Opere geotecniche in ambiente urbano, L'Aquila.*

Cardu M. & Oreste P. 2011. Tunnelling in urban areas by EPB machines: technical evaluation of the system. *Earth Sci. Res. J.vol.15 no.1 Bogotá Jan./June 2011*

Chiriotti, E., Jackson, P. & Taylor, S. 2010. Earth Pressure Balance tunnel boring machines, experience in mixed face conditions. *Proceedings of the ITA-AITES World Tunnel Congress entitled "Tunnel Vision Towards 2020", Vancouver.*

Gvirtzman, G., Shachnai, E., Bakler, N. & Ilani, S. 1984. Stratigraphy of the Kurkar Group (Quaternary) of the coastal plain of Israel. *Geological Survey of Israel Current Research 1983–1984, pp. 70–82*

Lunardi, G., Mancinelli, L. & Zimbaldi, A. 2015. Copenhagen Cityringen Metro: EPB-TBM head pressure definition. *ITA WTC 2015 Congress, Dubrovnik, Croatia.*

Porat N., Wintle G. A. & Ritte M. 2014. Mode and timing of kurkar and hamra formation, central coastal plain, Israel. *Isr. J. Earth Sci.; 53: 13–25.*

Shirlaw, J.N., Ong, J.C.W., Rosser, H.B., Tan, C.G., Osborne, N.H. & Heslop, P.J.E. 2003. Local settlements and sinkholes due to EPB tunnelling. *Proc. ICE, Geotechnical Engineering I 56, October Issue GE4, pp 193–211.*

Standing, J.R. 2002. Ground movements caused by tunnelling: measurements and back-analysis. *Geotechnical Aspects of Underground Construction in Soft Ground, Kastner, Emeriault, Dias, Guilloux (eds), Lyon. pp 37–48.*

Tunnels and Underground Cities: Engineering and Innovation meet Archaeology,
Architecture and Art, Volume 5: Innovation in underground engineering,
materials and equipment - Part 1 – Peila, Viggiani & Celestino (Eds)
© 2020 Taylor & Francis Group, London, ISBN 978-0-367-46870-5

Tilt sliding test to assess the behaviour of conditioned soil with large amount of cobbles

A. Carigi, A. Luciani, D. Martinelli, C. Todaro & D. Peila
Politecnico di Torino, Turin, Italy

ABSTRACT: The excavation in alluvial soils is an important field of application of EPB shields since it corresponds to the conditions that can be found in many cities. If conditioning is not optimized, cobbles that are frequently present may move inside the conditioned mass and settle down thus reducing the homogeneity of the muck with an irregular pressure at the face and along the screw and increasing the risk of excavation tool breakage due to the collisions with cobbles. For these reasons it is important that the conditioned soil moves homogeneously. This problem has not been deeply investigated and additional researches could be important for design and machine users. A laboratory test procedure and device that allows to check the interaction between the matrix made of the fine part of the conditioned soil and the cobbles embedded in it is presented and the preliminary test results are discussed.

1 INTRODUCTION

EPB technology is largely used all over the world in a wide variety of soils and it is expanding in number of applications and in new and more difficult soils. This kind of TBM uses a mix of excavated soil and conditioning agents in order to apply a stabilizing pressure on the excavation face and to waterproof the bulkhead. The ability to manage different types of soils is obtained with a proper soil conditioning (Hollmann & Thewes 2013; Peila et al., 2016; Martinelli et al., 2017). As the coarse fraction of the excavated soil grows inside the bulk chamber it may cause problems to the excavation since it is difficult to properly apply the pressure to the excavation face (Figure 2), to waterproof the bulkhead (Figure 3) and to correctly extract soil through the screw conveyor (Figure 3) if the cobbles sediment in the chamber. This phenomenon occurs when conditioned mass is not able to conglobate the cobbles winning the gravitational pull on the cobbles themselves.

Many studies among them it is worth to highlight the researches on the conditioning of soils and clay through foams and polymers to investigate the properties of the mixtures through slump tests, evaluating the behavior under pressure, and extraction tests carried out by Maidl (1995), (Quebaud et al. (1998), Jancsecz et al. (1999), Houlsby & Psomas (2001), Peña (2003), Borio et al. (2007), Peila, et al. (2009), Merritt & Mair (2008), Vinai et al. (2008), Borio & Peila (2011), Thewes et al. (2012), Peila (2014), Talebi et al. (2015), Mooney et al. (2016), Galli & Thewes (2016), Todaro (2016) and Martinelli et al. (2017). These researches are usually developed using soils with particle with small diameter, usually with a size of 2 to 4 cm, due to geometrical limits of the used test devices.

To investigate the effect of the cobbles in the mass and the ability of the conditioning to create an homogeneous muck it is necessary to develop a new test procedure. For this reason a specific test has been designed to evaluate if a cobbles-rich conditioned material can flow homogeneously or not inside the bulk chamber.

The test should be able to quantify in a clear way the flow properties of the soil with the cobbles inside since the most used test for conditioning assessment is slump test is not able to verify this property. Using this device a set of tests on different soils has been carried out and the obtained results have allowed to verify the feasibility of the procedure and to propose some test indexes.

Figure 1.   Examples of different soils A) Soil suitable for slump test, B) and C) soils used in the tests carried out, D) original to be tested.

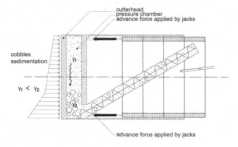

Figure 2.   Example of force applied on the cutterhead due to sedimentation of cobbles inside the chamber. A different density of material can also be observed with a not homogenous pressure in the chamber.

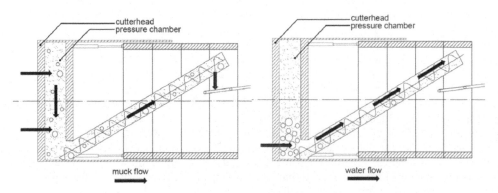

Figure 3.   Left: Flow path of the excavated material in the EPB machine with an homogeneous distribution of the cobbles in the muck. Right: Possible water inflow through the screw conveyor due to sedimentation of cobbles and incomplete filling of the screw conveyor.

2 PROPOSED METHOD

In the following a description of the proposed procedure and test device is presented.

2.1 Description of the test device and the test procedure

The device is made of a trapezoidal chute with a total length of 1500 mm. At 500 mm from the minor base there is the housing for a guillotine that can be removed and that creates the sample slot where the material is stored before the test. All the lateral barriers have an height of 100 mm as shown in Figure 4. The device is covered with a layer of hot rolled steel. During the test the conditioned soil is set in the sample slot and the chute is tilted to 20° with reference to the horizontal. After the guillotine is removed to allow the flow of the material.

At the end of this first dynamic phase the two components, fines and cobbles, of the material in the collection tank located at the bottom of the chute are weighted. The test then foresee an increment of the slope of the chute by steps of 5° with a constant speed of 0.5°/s. After each step the material stops to flow and the two components in the collection tank are weighted. This process is repeated till the chute has a slope of 35° when also the material remained on the chute is weighted.

The test is carried out with the following operational scheme:

- the soil sample is conditioned following the procedure described in the papers by Vinai et al. (2008) and Salazar et al. (2018);
- the conditioned soil is put in the sample slot;
- the chute is tilted to 20° with reference to the horizontal, the guillotine is removed and at the end of the dynamic phase the soil and the cobbles in the collection tank are weighted;
- the chute is tilted to 25° with a speed of 0.5°/s and at the end of the dynamic phase the soil and the cobbles in the collection tank are weighted;
- the chute is tilted to 30° with a speed of 0.5°/s and at the end of the dynamic phase the soil and the cobbles in the collection tank are weighted;
- the chute is tilted to 35° with a speed of 0.5°/s and at the end of the dynamic phase the soil and the cobbles in the collection tank are weighted;
- at the end of the test the soil and the cobbles on the chute are weighted.

2.2 Granulometric separation measurements

At every increment of slope part of the material flowed into the collection tank. For an homogeneous behavior the mass that slides is expected to have the same composition of the original sample and the percentage of cobbles in the collection tank valued on the total amount of cobbles (CCT expressed in percentage) and the percentage of fines in the collection tank valued

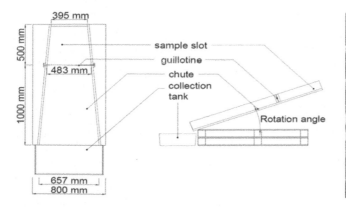

Figure 4. Top view and lateral view of the test device.

Figure 5. Example of the execution of the test with the flow of the material at different tilting angles.

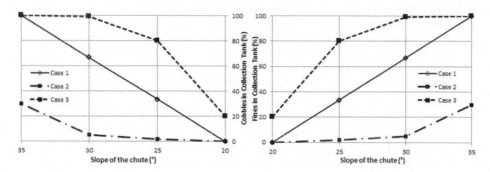

Figure 6. Example of the measured data referring to homogeneous behaviors. Case 1: the flow develops through the entire test. Case 2: the flow mainly develops for low slope value. Case 3: the flow mainly develops for high slope values.

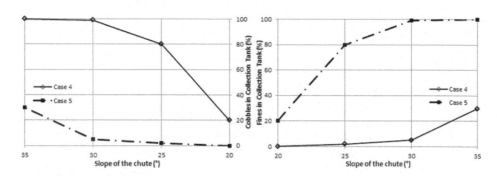

Figure 7. Example of the measured data referring to not homogeneous behaviors. Case 4: the cobbles slide mainly at low slope values while the fines mainly at high slope values. Case 5: the cobbles slide mainly at high slope values while the fines mainly at low slope values.

on the total amount of fines (FCT expressed in percentage) should be equal at every step (Figure 7). If the behavior is not homogeneous the CCT index and FCT index will be different at every step (Figure 6).

The flow, either if is homogeneous can be characterized by a development concentrated in a specific phase of the test or be distributed during the whole test (Figure 7).

Given the values CCT_i and FCT_i, respectively the percentage of cobbles in collection tank and the fines in collection tank at the i-th phase of the test (1:20°; 2:25°;3:30°;4:35°), to evaluate the behavior of conditioned soil three indexes have been proposed and designed:

– Asymmetry index (I_{skw})

$$I_{skw} = (|CCT_1 - FCT_1| + |CCT_2 - FCT_2|) + (|CCT_3 - FCT_3|) + (|CCT_4 - FCT_4|)/400 \quad (1)$$

– Cobble index ($I_{\sigma,c}$)

$$I_{\sigma,c} = [(c_1)^2 + \left(c_2 - \frac{1}{3}\right)^2 + \left(c_3 - \frac{2}{3}\right)^2 + (c_4)^2] \cdot \frac{9}{14} \quad (2)$$

– Fine index ($I_{\sigma,f}$)

$$I_{\sigma,f} = [(f_1)^2 + \left(f_2 - \frac{1}{3}\right)^2 + \left(f_3 - \frac{2}{3}\right)^2 + (f_4)^2] \cdot \frac{9}{14} \quad (3)$$

Where 400 and 9/14 are coefficients used to normalize the indexes between 0 and 1.

The asymmetry index, I_{skw}, gives information about how different are the flow of the two components and the indexes of deviation. ($I_{skw} = 0$ Homogeneous flow; $I_{skw} = 1$ Not homogeneous flow) while $I_{\sigma,c}$ and $I_{\sigma,f}$, give information about how far are the behavior of the single component from the condition of a regular flow during the whole test ($I_{\sigma,c} = 0$ Regular flow of cobbles; $I_{\sigma,c} = 1$ Not regular flow of cobbles; $I_{\sigma,f} = 0$ $I_{\sigma,f} = 0$Regular flow of fines; $I_{\sigma,f} = 1$ Not regular flow of fines).

3 CARRIED OUT TESTS

The results of the preliminary tests that have been carried out to verify the feasibility of the proposed test procedure are presented and discussed in the following. The obtained results show the feasibility of the procedure.

3.1 *Used soils and conditioning agents*

The tests have been carried out using an alluvial soil (Soil A) with grain size distribution given in Figure 8. From Soil A two different soils (Soil B and Soil C) have been artificially created adding cobbles to soil A (Table 1) and their granulometric distribution is shown in Figure 8.

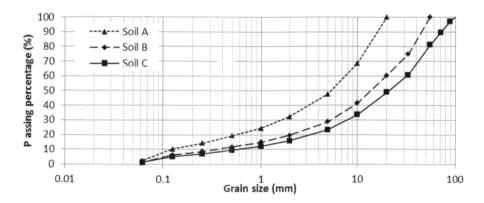

Figure 8. Grain size distribution of the soils used in the tests.

Table 1. Summary of geometrical properties of used soils.

Soil	Soil A	Cobbles	D_{min} cobbles	D_{max} cobbles	D10	D60	U
	%	%	mm	mm	mm	mm	-
A	100	0	-	-	0.125	7.5	60
B	60.3	39.7	20	55	0.25	20	80
C	49.0	51	20	100	0.5	23	60

Due to the different maximum diameter of the cobbles in soils, the weight of the used samples has been varied accordingly. The samples of soil B are of 27 kg while the samples of soil C are of 40 kg.

The assessment of the conditioning has been made through slump test on Soil A following the procedure described in the paper by Borio et al. (2007). The conditioning agents used for the test are the Polyfoamer ECO 100 and Mapedrill M1, commercial products of Mapei SpA, and the foam has been produced with the foam generator described by Peila et al. (2007).

The soil used in each test has been conditioned differently: in test 1 the soil contained 12% of water and no foam or polymers, in test 2 has been added 30% more of foam from the quantity used in test 1, in test 3 has been added 0.015% of polymer Mapedril M1 on the conditioning of test 2 and in test 4 an additional 0.015% of polymer Mapedril M1 has been added to the conditioning set of test 3.

In addition, in order to evaluate the effect of time on the behavior of the mass conditioned with the polymer some test have been carried out 1 hour after the addition of the polymer Mapedril M1 in the mix. The results are summarized by the indexes in Table 4 and Figure 10 and Figure 13. For concentration of polymer of 0.015% the delay of 1 hour reduces the indexes, i.e. a more homogeneous behavior is induced. For a concentration of polymer of 0.030% the conditioned soil slides more homogeneously.

The used conditioning sets are summarized in Table 2 while Table 3 reports the details of the carried out tests.

Figure 9. Left: Picture of the slump test of a well-conditioned mass with a maximum diameter 20 mm. Right: Picture of the slump test of a well-conditioned mass with maximum diameter 60 mm, 68% passing at 20 mm. The presence of cobbles in the test shown in the right influences the behavior of the mass by breaking the cone and do not allow the creation of a homogeneous mass.

Table 2. Description of conditioning sets.

Conditioning set	W_{tot}	c	FER	FIR	cM1
	%	%	-	%	%
CS1	12	-	-	-	-
CS2	12	1.5	10	30	-
CS3	12	1.5	10	30	0.015
CS4	12	1.5	10	30	0.030

Table 3. Description of used conditioning sets for the carried out tests.

Test	Soil type	Sample size kg	Conditioning set	Delay min
1	A	27	CS1	-
2	A	27	CS2	-
3	A	27	CS3	-
4	A	27	CS4	-
3d	A	27	CS3	60
4d	A	27	CS4	60
5	B	27	CS1	-
6	B	27	CS2	-
7	B	27	CS3	-
8	B	27	CS4	-
7d	B	27	CS3	60
8d	B	27	CS4	60
9	C	40	CS1	-
10	C	40	CS2	-
11	C	40	CS3	-
12	C	40	CS4	-

Table 4. Indexes obtained from the carried out tests.

Test	Soil	I_{skw}	$I_{\sigma,c}$	$I_{\sigma,f}$
1	A		-	0.233
2	A	-	-	0.059
3	A	-	-	0.121
4	A	-	-	0.153
3d	A	-	-	0.134
4d	A	-	-	0.395
5	B	0.362	0.113	0.182
6	B	0.323	0.076	0.170
7	B	0.346	0.044	0.251
8	B	0.242	0.026	0.339
7d	B	0.236	0.047	0.163
8d	B	0.158	0.197	0.515
9	C	0.126	0.043	0.075
10	C	0.274	0.039	0.247
11	C	0.264	0.004	0.235
12	C	0.165	0.060	0.038

3.2 *Experimental results*

In the following the main results of the carried out tests are presented and discussed.

As shown in Figure 10, for tests 1÷4 (i.e. the tests carried out on soil A, the soil without cobbles) the fine index shows that the homogeneity of the flux reach its maximum for the CS2 (material conditioned with water and foam). While adding the polymer (tests 3 and 4, CS3 and CS4 respectively) reduces the homogeneity of flux. The effect of the polymer increases after 60 minutes (tests 3d and 4d), this is more evident for higher concetrations (CS4).

The tests carried out on soil B (the soil with cobbles with diameter up to 55 mm, tests 5÷8) highlight that the addition of foam and of polymer to the mass allow to reduce the asymmetry index, i.e. a more homogeneous composition of the sliding mass is obtained. Is also possible to observe that the cobbles index decreases while the fines index increases. The interaction between cobbles and conditioned mass induces a regolarization of the cobble flux while for the fines the trend is the opposite. These results show that the use of foam lubricate the mass reducing the friction between grains as shown by Martinelli et al. (2017) using a modified shear box. On the other hand the use of polymer aggregates the soil mass therefore it moves as a lumped mass more or less all toghether.

The tests 3d and 4d made on soil A and the tests 7d and 8d made on soil B have been carried out to evaluate the effect of time on the mass conditioned with the use of polymer (CS3 and CS4 conditioning sets, results shown in Figure 10 and Figure 13). For low concentrations of the polymer the delay induces an increase of regularity of the flux of fines while for higher concentrations of the polymer it agglomerates much more the mass that slides more like a lumped mass than a flux (I_{skw} decreases, $I_{\sigma,c}$ and $I_{\sigma,f}$ increase) because the activation of the polymer requires some time.

For tests 9÷11 (i.e. the tests carried out on soil C, the soil with cobbles with diameter up to 100 mm) it was necessary to use a larger amount of soil to have a sample representative of the used granulometric distribution. Figure 12 shows that the addition of polymer reduces the asymmetry index, i.e. a more regular flow is achieved due to the higher interaction between fines and cobbles.

In conclusion it is possible to summarize that a proper conditioning allow to make the two components of the soil (cobbles and fines) to flow in a more homogeneous way since it reduces the intergranular friction. The values of the of the various indexes are summarized in Table 4. It is clear from the reduction of the cobbles index that the coarse component slide gets more regular during the phases of the test as the conditioning of the matrix gives a more plastic and viscous consistency to the mass. On the other hand is possible to see that for the fines the trend is opposite because of the interaction with the cobbles.

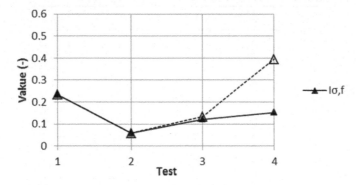

Figure 10.   Effect of conditioning on the fine index evalued on soil A (without cobbles) without delay (black) and with 60 minutes of delay (white). The tests 3d and 4d are given at the same position of tests 3 and 4 to enable an easy comparison.

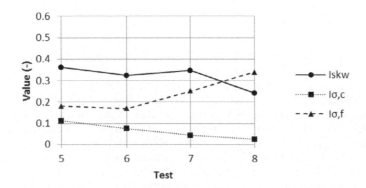

Figure 11.   Left: Indexes for the tests on samples of soil B (26 kg).

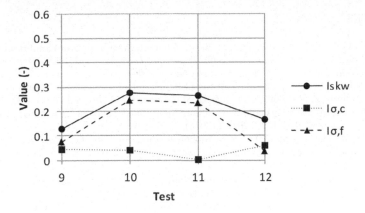

Figure 12. Indexes for the tests on samples of soil C (40 kg).

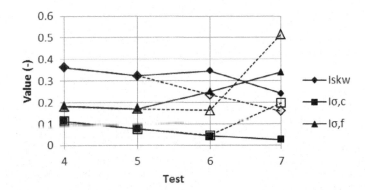

Figure 13. Effect of the delay on the indexes. Comparison between indexes without delay (black) and with 60 minutes of delay (white). The tests 6d and 7d are given at the same position of tests 6 and 7 to enable an easy comparison.

4 CONCLUSIONS

The proposed procedure (Tilt Sliding Test) was designed to provide a tool to assess the proper EPB conditioning for soils with a large amount of cobbles. New indexes have been proposed and a set of calibration tests has been carried out.

The tests made possible to assess if a cobbles-rich conditioned material can flow homogeneously when inclining a sliding surface. Through the proposed indexes is possible to define if the cobbles and the conditioned mass move independently and to make a comparison between different conditioning sets in the ability to form an homogeneous medium. This knowledge can used to understand how the conditioning affects the behavior of the mass in the bulk chamber and to set limits to the consistency of the mass in order to avoid the segregation of the coarser granulometric fractions in the chamber. In the paper different type of conditioning have been tested to preliminary assess the feasibility of the test.

In conclusion the TST has proven to be feasible to provide indication on the effect, positive or negative, of different conditioning sets when tunneling in a cobble-rich soil.

REFERENCES

Borio, L., Peila, D., Oggeri, C. & Pelizza, S. 2007. "Characterization of Soil Conditioning for Mechanized Tunnelling." In *International Society for Trenchless Technology - 25th No-Dig International Conference and Exhibition, Roma 07: Mediterranean No-Dig 2007.*

Borio, L. & D. Peila. 2011. "Laboratory Test for EPB Tunnelling Assessment: Results of Test Campaign on Two Different Granular Soils." *Gospodarka Surowcami Mineralnymi* 27 (2005): 85–100.

Galli, M. & Thewes, M. 2016. "Rheology of Foam-Conditioned Sands in EPB Tunneling." *ITA-AITES World Tunnel Congress 2016, WTC 2016* 3 (2016): 2386.

Hollmann, F. S. & Thewes, M. 2013. "Assessment Method for Clay Clogging and Disintegration of Fines in Mechanised Tunnelling." *Tunnelling and Underground Space Technology* 37: 96–106. https://doi.org/10.1016/j.tust.2013.03.010.

Houlsby, G. T. & Psomas, S. 2001. "Soil Conditioning for Pipejacking and Tunnelling: Properties of Sand/Foam Mixtures." *Proceedings Underground Construction*, 128–138.

Jancsecz, S., Krause, R. & Langmaack, L. 1999. "Advantages of Soil Conditioning in Shield Tunnelling. Experiences of LRTS Izmir." In *World Tunnel Congress.*

Maidl, U. 1995. "Erweiterung Der Einsatzbereiche Der Erddruckschilde Durch Bodenkonditionierung Mit Schaum." Ruhr-Universität Bochum, Germany.

Martinelli, D., Chieregato, A., Salazar, C.G.O., Peila, D. & Barbero, M. 2017. "Conditioning of Fractured Rock Masses for the Excavation with EPB Shields." *13th ISRM International Congress of Rock Mechanics* 2017–Janua (2017): 2018.

Martinelli, D., Winderholler, R., & Peila, D. 2017. "Undräniertes Verhalten von Grobkörnigen Böden Für Vortriebe Mit EPB-Schilden – Ein Neues Experimentelles Verfahren." *Geomechanik Und Tunnelbau* 10 (1): 81–89.

Merritt, A. S., & R. J. Mair. 2008. "Mechanics of Tunnelling Machine Screw Conveyors: A Theoretical Model." *Géotechnique.* https://doi.org/10.1680/geot.2008.58.2.79.

Mooney, M.A., Wu, Y., Mori, L., Bearce, R. & Cha, M. 2016. "Earth Pressure Balance TBM Soil Conditioning: It's About the Pressure." *Proceedings of the World Tunnel Congress*, no. 2016: 2433–2444.

Peila, D. 2014. "Soil Conditioning for EPB Shield Tunnelling." *KSCE Journal of Civil Engineering* 18 (3): 831–836. https://doi.org/10.1007/s12205-014-0023-3.

Peila, D., Picchio, A., Martinelli, D. & Dal Negro, E. 2016. "Laboratory Tests on Soil Conditioning of Clayey Soil." *Acta Geotechnica* 11 (5): 1061–1074. https://doi.org/10.1007/s11440-015-0406-8.

Peila, D., Oggeri, C. & Borio, L. 2009. "Using the Slump Test to Assess the Behavior of Conditioned Soil for EPB Tunneling." *Environmental and Engineering Geoscience* 15 (3): 167–174. https://doi.org/10.2113/gseegeosci.15.3.167.

Peña, M. 2003. "Soil Conditioning for Sands." *Tunnels and Tunnelling International* 35 (7): 40–42.

Quebaud, S., Sibai, M. & Henry, J.P. 1998. "Use of Chemical Foam for Improvements in Drilling by Earth-Pressure Balanced Shields in Granular Soils." *Tunnelling and Underground Space Technology.* https://doi.org/10.1016/S0886-7798(98)00045-5.

Talebi, K., Memarian, H., Rostami, J. & Gharahbagh, E.A. 2015. "Modeling of Soil Movement in the Screw Conveyor of the Earth Pressure Balance Machines (EPBM) Using Computational Fluid Dynamics." *Tunnelling and Underground Space Technology* 47 (2015): 136–142. https://doi.org/10.1016/j.tust.2014.12.008.

Thewes, M., Budach, C. & Bezuijen, A. 2012. "Foam Conditioning in EPB Tunnelling." *Geotechnical Aspects of Underground Construction in Soft Ground*, no. September: 127–135. https://doi.org/10.1201/b12748-19.

Todaro, C. 2016. "Analisi Della Penetrazione Nello Scavo Con EPB." *Geoingegneria Ambientale e Mineraria* 147: 49–52.

Vinai, R., Oggeri, C. & Peila, D. 2008. "Soil Conditioning of Sand for EPB Applications: A Laboratory Research." *Tunnelling and Underground Space Technology* 23 (3): 308–317. https://doi.org/10.1016/j.tust.2007.04.010.

Tunnels and Underground Cities: Engineering and Innovation meet Archaeology,
Architecture and Art, Volume 5: Innovation in underground engineering,
materials and equipment - Part 1 – Peila, Viggiani & Celestino (Eds)
© 2020 Taylor & Francis Group, London, ISBN 978-0-367-46870-5

Influence of set accelerating admixtures on the compressive strength of sprayed concrete

G. Cascino
Italferr S.p.A., Rome, Italy

G. Estrafallaces
Italferr S.p.A., Bologna, Italy

ABSTRACT: The purpose of this article is to illustrate the experience gained in the field of wet-sprayed concrete in the execution of tunnel construction work for high-speed railway lines in Italy, with particular regard to the influence exerted by the nature and dosage of accelerating admixtures on compressive strength over short and long curing periods. This paper therefore illustrates the results obtained from the comparison between mixtures of sprayed concrete made with silicate and with alkali-free accelerating admixtures, in terms of compressive strength, observed at time intervals of 2 and 28 days

1 FOREWORD

Sprayed concrete has gained importance over time as a replacement for traditional concrete lining. Originally used in tunnelling mostly for the temporary stabilisation of the cavity, in recent years sprayed concrete has become a main structural element (Figures 1 and 2).

This is because its resistance to compression and its durability have indeed proven to be essential requirements, crucial to guaranteeing the preservation in time of its mechanical features from aggressive agents that, transported by gravitational water present in the rock mass, permeate the coating surface.

In Italy, the sprayed concrete method is used mostly to pre-cover tunnels as a temporary construction tool, in order to control soil decompression and to ensure workforce safety.

Within this context, in order to ensure fast confinement of deformations of the cavity it is necessary to use types of concrete capable of quickly developing the required mechanical strength.

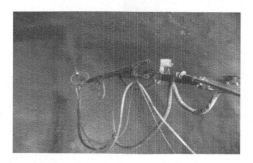

Figure 1. Application of sprayed concrete to the shell.

Figure 2. Spraying on piers.

2 ACCELERATING ADMIXTURE

Shortened setting time and faster rate of early strength development are technological and operational requirements for underground works met through the use of accelerating admixtures.

'Set accelerating admixtures for sprayed concrete', as defined in EN 934-5, are in fact admixtures designed to develop very early setting and very early hardening of the sprayed concrete over short curing periods.

There are two most commonly used types of set accelerating admixtures: alkali-free and sodium silicate (Figure 3).

2.1 *Alkali free sprayed concrete set accelerating admixtures (herein after 'AF')*

These are sprayed concrete set accelerating admixtures with an alkali content not exceeding 1% by mass of the admixture.

They have a pH between 4÷6 and they are not toxic. They confer a thixotrophic effect to the mixture and they drastically reduce the release of alkali in underground water. Their price, significantly higher than that of sodium silicate admixtures, is compensated by the need for lower doses required in the spraying process (about 8% in weight of concrete for AF, versus approx. 15% for SS).

With wet mixes, the accelerating action is instantaneous because the product is already dissolved. With dry mixes, the performance of the product is conditioned by the kinetics of the dissolving process.

As regards setting time activation, attention must be paid to its compatibility with the concrete. For better performance, they should be used at temperatures above 5°C.

2.2 *Sodium silicate sprayed concrete set accelerating admixtures (herein after 'SS')*

Also known as 'water glass', they are the accelerating admixtures most commonly used due to their lower cost, faster setting time and strength development over a short curing period.

High percentage w.r.t. concrete mass ratios are commonly used also in order to exploit the colloidal effect of the silicate that improves adhesion to the support.

Doses depend on various factors: application point (e.g.: walls, crown), application conditions (wet or dry surface), construction requirements.

In presence of calcium hydroxide, the mixture jellifies, creating a bonding effect. The pH of sodium silicates is lower than 12. They are compatible with almost every type of concrete (performance is the best with Portland concrete, and the worst with pozzolan concrete).

The accelerating effect decreases as the temperature lowers. This means that in winter the doses must be increased.

Figure 3.   1000 litre tanks containing accelerating admixtures.

Table 1. Loss of strength of sprayed concrete as a function of sodium silicate dosage (as percentage to cement mass ratio) [S. Tavano, G. Zambetti: "Il calcestruzzo proiettato strutturale e durevole"].

Sodium silicate dosage (%)	Strength reduction (%)
0	0
10	35
15	50
20	60
25	75
50	95

Table 2. Acceptance limits of sodium silicate based accelerating admixtures used in tests.

Requirements	Acceptance limits
SiO_2/Na_2O mass ratio	> 3.3
SiO_2 % (as is)	28
Na_2O (as is)	8
Baumé at 20°C	38 ÷ 42
Density	1.35 ÷ 1.40
Dry weight %	35 ÷ 37
Chlorides %	absent
pH	> 11
Soluble in H_2O	yes

The shortcomings connected with the use of this admixture consist in increased withdrawal, increased risk of reaction with aggregates, and especially a significant reduction in mechanical strength over longer curing periods, which reduces further as the dosage increases. (Table 1)

The table below shows the requirements for sodium silicate based accelerating admixtures used in the qualification tests of the various mixtures of sprayed concrete (Table 2).

3 EXPERIMENTATION

In order to better understand the influence exerted by the nature and the dosage of accelerating admixtures on mechanical compressive strength over short and long curing periods of sprayed concrete mixtures, widespread experimentation at work sites involved in the construction of tunnels for high-speed railway lines has been conducted.

The conformity of the sprayed concrete mixtures and of the mixture composition procedure has been verified through the execution of appropriate field tests to define the qualification of the process (Table 3).

The results of the tests were analyzed and compared.

This paper shows the data resulting from the experimental tests carried out on mixtures of wet-mix sprayed concrete with sodium silicate based and alkali-free admixtures.

The comparison between the mixtures with both types of accelerating admixture added and the basic mix was carried out in terms of compressive strength at 2 days and at 28 days.

Table 3. Tests on hardened sprayed concrete conducted during experimentation.

Tests	Specimen types
Compression on basic mix	cubes (failure at 2 days and at 28 days)
Compression on plate-derived specimens	cubes (failure at 2 days and at 28 days)
Energy absorption capacity'	plate

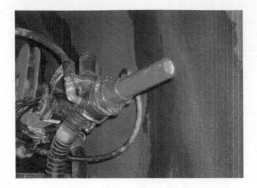

Figure 5. Percentage of accelerating admixture supplied.

Figure 4. Concrete spraying nozzle.

For this purpose, more than 120 variously formulated mixtures were taken into consideration, made with cements such as IIA-LL 42.5R and IVA 42.5R and at doses ranging from 450 to 500 kg/m^3.

In particular, 55 mixtures were made with alkali-free admixtures and slightly more than double that with sodium silicate.

All mixtures were made with a max. 0.50 w/c ratio and S5 consistency class measured by slump test in compliance with EN 12350-2.

In the formulation of each mixture, at least 3 aggregate classes (Dmax 10 mm) were used, in order to improve aggregate rating assortment.

The accelerating admixtures, all in liquid form, were added to the lance during the spraying of the mixture at the dosage of 15% for sodium silicates (65 ÷ 75 kg/m3) and about 7% for alkali-free (30 ÷ 35 kg/m3) in order to facilitate setting and strength development over a short curing period. (Figure 4).

The accelerating admixtures were fed by a manually adjustable dosing device by means of a special handwheel with a graduated scale so as to keep constant the percentage used as the pumped concrete flow rate changed. (Figure 5).

During spraying, the potentiometer of the admixture dispenser was set to the value corresponding to the dosage chosen for the tests.

The potentiometer used during spraying is subject to continuous adjustments by the operator. The consequent changes in dosage of accelerating admixture with respect to the theoretical value, mainly related to the application to the crown and to the areas affected by dripping, are the main cause of drops in strength over long curing periods.

In the case of SS admixtures, the dosage of 15% was made according to the theoretical stoichiometric amount of silicate necessary to neutralize the calcium sulphate present in the cement and to ensure the acceleration of the set of the sprayed concrete considering a 36% concentration of dry substance.

Compressive strength properties were derived from:

• cylindrical specimens (h/d = 1, l = 100 mm, h = 100 mm) taken by coring from specifically cast plates;
• cubic specimens cast from the mixes following the composition of the basic mix (Table 4).

At the batching plants equipped with a feed divider, additional mixtures were made using low carbon fibre.

Table 4. Summary of results of compression tests performed on basic mix and on mixtures with AF and SS.

Days	basic mix	SS	basic mix	AF
2 d	17.5	17.6	23.5	19.4
28 d	46.8	29.3	45.3	38.0

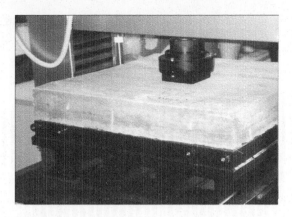

Figure 6.   Test plate 'Energy absorption capacity'.

In these cases, tests 'energy absorption capacity' were also carried out on plates measuring 600 x 600 x 100 mm for the calculation of the strain energy (area under the load-strain curve up to the 25 mm deflection). (Figure 6).

The concrete was sprayed onto the plates orthogonally to the laying surface at a distance between 0.50 and 1.50 m from the lance.

For the laying of the sprayed concrete, piston pumps were used, activated by electric and/or diesel engines, equipped with a movable arm with a lance that can be swung at the end, controlled remotely by a control panel to orient the spraying.

On the chassis, instead, the accelerator tank was positioned together with the dose dispenser to deliver the admixture in the quantity required for spraying.

Spraying operations were preceded by checking:

- consistency and temperature of the basic mix
- room temperature
- compressed air pressure and flow
- setting of the theoretical dosage of accelerating admixture

4   GETTING STARTED

The development of the compressive strength of the sprayed concrete mixtures was investigated based on the basic mix and on the mixture containing the set accelerating admixture added at the lance (AF and SS).

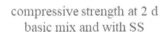

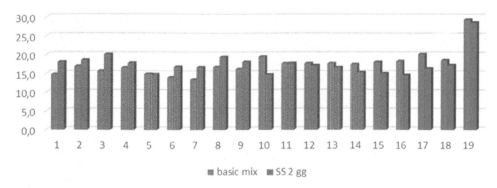

Figure 7.   Compressive strength at 2 d – basic mix and with SS.

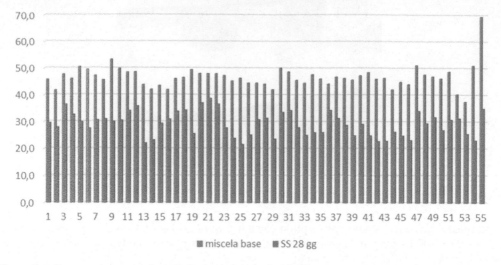

Figure 8. Compressive strength at 28 d – basic mix and with SS.

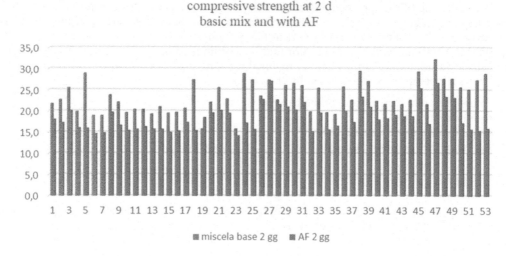

Figure 9. Compressive strength at 2 d – basic mix and with AF.

The compression tests carried out during the experimentation showed a decrease in strength over long curing periods (28 days) by about 40% in the case of mixtures containing SS, and by about 16% for mixtures containing AF, compared to the failure values of the specimens made with the basic mix (BM).

The decrease in compressive strength at 28 days of the mixtures with SS was almost double that of the AF mixtures.

The mixtures with SS, even if dosed with percentages (15%) lower than those actually used at the construction site, showed compressive strength properties at 28 days that were much lower than those of AF mixtures containing about half of the AF admixture (approx. 7%).

After 2 days of curing, the mixture sprayed with SS showed the same value of compressive strength as that of the basic mix, while the value of the mixture with AF was significantly lower.

Overall, the concrete sprayed with AF admixture showed a lower initial strength at 2 days followed by an increase in strength at the 28 days mark.

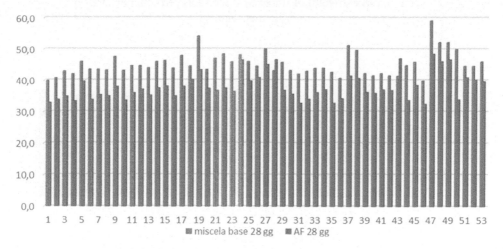

Figure 10.   Compressive strength at 28 d – basic mix and with AF.

On the other hand, the behaviour of the mixtures with SS was completely different and opposite, with compressive strength substantially equal to that of the basic mix at 2 days and significantly lower at 28 days.

The difference in strength between the value obtained from the compression failure of cubic concrete specimens cast from the basic mix and that obtained from the failure of the cylinders extracted from specifically cast panels, is essentially attributable to two aspects:

• change in the composition of the mixture due to spraying;
• influence of the accelerating admixture, especially over long curing periods.

5   COMPRESSIVE STRENGTH (FIBRE-REINFORCED SPRAYED CONCRETE

Investigations on the influence of the type and dosage of set accelerating admixture have been extended to sprayed concrete mixtures reinforced with metal fibres.

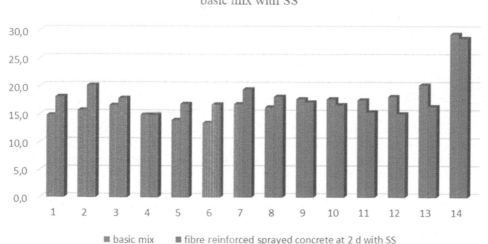

Figure 11.   Compressive strength of fibre reinforced sprayed concrete at 2 d – basic mix and with SS

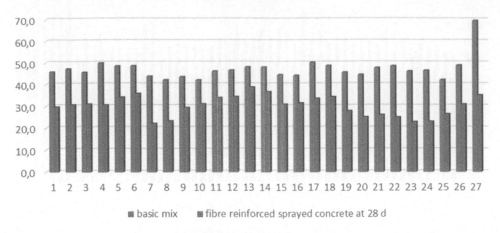

Figure 12. Compressive strength of fibre reinforced sprayed concrete at 28 d – basic mix and with SS.

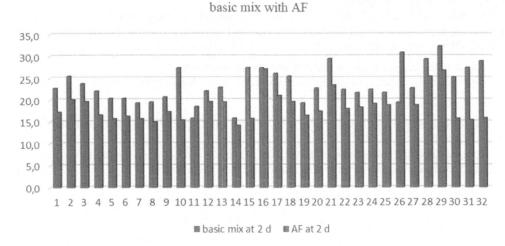

Figure 13. Compressive strength of fibre reinforced sprayed concrete at 2 d – basic mix and with AF.

The observations made regarding the strength at 2 days and 28 days of the sprayed concrete mixtures with metallic fibres are the same as those made previously for mixtures with SS and with AF.

The accelerating admixture (regardless of what type it is) causes a decrease in compressive strength of fibre-reinforced sprayed concrete.

In particular, the compressive strength value at 2 days of curing was found to be of the same order of magnitude as that of the basic mix and of the mixture with SS, while the mixture with AF showed a 15–20% decrease compared to the BM.

The tests 'energy absorption capacity' carried out at the end of 28 days on plates made with the same quantity of fibres showed that the mixtures with AF gave results higher by 15% compared to those with SS.

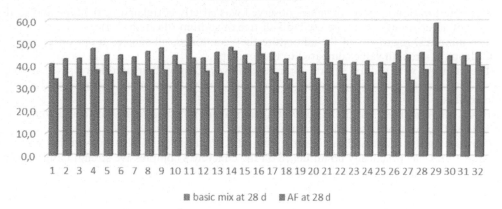

Figure 14.   Compressive strength of fibre reinforced sprayed concrete at 28 d – basic mix and with AF.

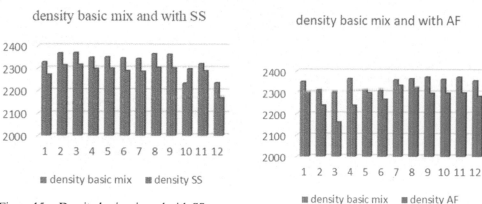

Figure 15.   Density basic mix and with SS.

Figure 16.   Density basic mix and with AF.

6 DENSITY

The differences in density may be attributed to the different methods of composition of the specimens. The cylinders extracted from the panels have undergone compacting action due to the impact of the mixture on the application surface, while the basic mix specimens have been compacted to refusal on a vibrating table.

Compared to the density value resulting from the basic mix tested in compliance with the UNI EN 12350-6 standard at 28 days, the value obtained from the cylinders was less than 4% for mixtures with SS and approx. less than 2% for mixtures with AF.

7 CONCLUSIONS

In this article we examined sprayed concrete mixtures with different doses and components, containing alkali-free and silicate accelerating admixtures used in railway works carried out underground in our country in order to illustrate the influence exerted by the type and by the dosage of accelerating admixture on the compressive strength over short and long curing periods.

The deterioration in strength values observed was strictly correlated to the type and to the dosage of accelerating admixture used (alkali-free or sodium silicate based).

The mixtures, regardless of the admixture they contained, all showed more or less significant drops in strength value compared to the basic mix.

The mixtures with sodium silicate admixture, however, showed greater decreases in strength over long curing periods compared to those of sprayed concretes containing alkali-free admixtures.

It was also observed that both accelerating admixture types (sodium silicate and alkali-free) were less effective when used with mixture cements.

Confirming the results indicated in the literature, the test results also showed that the addition of fibres has practically no influence on the concrete's compressive strength and anyhow not such as to induce significant increases.

REFERENCES

EN 14487-1 – '*Sprayed concrete. Part 1: Definitions, specifications and conformity*'
EN 14488-5 – '*Testing sprayed concrete. Part 5: Determination of energy absorption capacity of fibre reinforced slab specimens*'
Austen S. 1995. '*Sprayed Concrete*', Sprayed Concrete Association,
Estrafallaces G. 2006. '*Sul comportamento del conglomerato cementizio proiettato con fibre*', Ingegneria Ferroviaria, CIFI, n°6/2006
Ruffert G., Brux G., Badzong H. J. 1995. '*Herstellung Prufung und Anwendung von Spritzbeton – Abwicklung von Spritzbetonarbeiten*', Expert Verlag,

Tunnels and Underground Cities: Engineering and Innovation meet Archaeology,
Architecture and Art, Volume 5: Innovation in underground engineering,
materials and equipment - Part 1 – Peila, Viggiani & Celestino (Eds)
© 2020 Taylor & Francis Group, London, ISBN 978-0-367-46870-5

Steel fibre reinforced concrete for segmental lining – crack mitigation measures at design phase

G. Castrogiovanni
AECOM, New York, USA

G. Busacchi
COWI A/S, Doha, Qatar

T. Léber
COWI A/S, Lyngby, Denmark

G. Mariani
L.S.A.B JV, Stockholm, Sweden

ABSTRACT: Regardless of type of reinforcement, cracks in the precast segmental tunnel lining often create concerns during the design and construction phases. Due to its limited tensile strength, the Steel Fibre Reinforced Concrete (SFRC) requires early implementation of the project requisites and design criteria in order to avoid subsequent and expensive repairs. Based on the Authors' experiences in some of the world's landmark mechanized tunnelling projects, especially in the Middle East, the application of SFRC in sewer, storm water, and metro tunnels is reviewed and the mitigation measures to control potential crack propagations are summarized. The importance of the decisions in the early design phase, e.g. ring and segment geometry, general arrangement, number and position of push rams, are discussed along with the radial and circumferential joint design. Finally, the recommended calculations, which are required to minimize the risk of cracking, are presented.

1 INTRODUCTION

Precast segmental lining has become a common solution in sewer, storm water, and metro tunnels for its inherent advantages. The associated risks, however, must be understood and it is, therefore, the Designer's responsibility to identify the shortcomings of the TBM technology in the early design phases to achieve the appropriate performances and project-specific requirements.

The recommended mitigation measures to be carried out at the design phase are described below to reduce the crack phenomena in the TBM tunnel lining. The benefits and disadvantages of the use of SFRC in the tunnel lining are further discussed in Castrogiovanni et al. (2018).

2 PROJECT REFERENCES

The principles presented below are based on the Authors' experience in approximately 50 km of SFRC tunnels, particularly in the following Middle East projects, designed and constructed without the use of rebar reinforcement:

– Strategic Tunnel Enhancement Program (STEP), Abu Dhabi, UAE;
– Abu Hamour Water Drainage Tunnel, Doha, Qatar;
– Doha Metro, Red Line North Underground, Doha, Qatar.

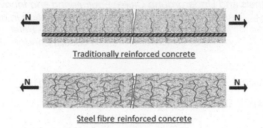

Traditionally reinforced concrete

Steel fibre reinforced concrete

Figure 1. Crack and reinforcement distribution exemplification.

Table 1. Crack control – comparison of the two technologies.

	Rebar Reinforced Concrete (RC)	Steel Fibre Reinforced Concrete (SFRC)
micro-crack control	possible only if cracks cross the steel bar, otherwise widening cannot be prevented	micro cracks are within the limit of steel fibre bridging capacity, hence due to the distributed position of steel fibres further widening can be prevented
macro-crack control	wider cracks can be controlled, although only in the location of steel bar	only smaller cracks can be controlled
tensile stress distribution	concentrated and high in stress	area of plastic zone can be extended but smaller in stress

3 CRACK DISTRIBUTION AND MECHANISM

3.1 Crack control

Post-cracking tensile strength of the concrete can be provided by traditional steel rebar or steel fibre reinforcement (Figure 1.). The latter provides better spalling resistance because some typical tunnel segment damages (e.g. concrete edge spalling) cannot be mitigated by steel bar installation due to the durability requirements (i.e. concrete cover).

The random distribution of steel fibres enables more favourable crack propagation and tensile stress distribution in the concrete. Additionally, the steel fibre application prevents crack propagation from micro-cracks to macro-cracks, helps distribute localized tensile stresses, improves the fatigue resistance and reduces cracks caused by concrete shrinkage.

On the other hand, the steel fibre reinforcement provides lower post-cracking tensile capacity than steel bars, and due to its limited crack bridging capability, the cracks need to be controlled. SFRC fails only if the steel fibres break or are pulled out of the cement matrix.

3.2 Major causes of segment cracks

Most of the cracks in the tunnel segments occurs during the lining installation. To identify the potential crack mitigation measures, the major causes of crack occurrences are identified and listed in Table 2. The recommended mitigation measures in the design, casting and construction phases depend on the requirements and crack width limitation.

4 DESIGN CONSIDERATIONS TO MITIGATE RISK OF CRACKING

Thanks to the experience gained by the projects mentioned above, the ring and segment geometry, the number and position of the push rams as well as the joint design have been identified as key elements to control concrete cracks and to provide a successful application of the technology. The mitigation of cracks involves all the design phases and in the following sections, the main aspects are presented.

Table 2. Crack sources during project life and their impacts.

Phase	Event	Consequences
Prior to installation	Damage in production plant due to casting (temperature, shrinkage, etc.)	To be repaired, segment with high severity (unacceptable/unrepairable) damage is not allowed to install
	Damage on construction site (surface damage, broken edges, damage around erector socket)	To be repaired, segment with high severity (unacceptable/unrepairable) damage is not allowed to install
During installation	High thrust ram forces	Concrete splitting, local surface cracks between push ram plates. Concrete bursting, localized inside the concrete; no trough going crack allowed
	Poor ring assembly	Occurs in the early stages of TBM operation, potential risk to water ingress and impact on the structural integrity of the segments
After installation	Permanent loads	Localized cracks occur, no trough going cracks allowed

4.1 Typical tunnel sections and mitigation by reinforcement alternatives

Due to variable loads acting along the TBM alignment, characteristic sections shall be selected to study the stress level in the tunnel ring.

The verification of the lining may prove the capacity of steel fibre reinforcement sufficient, however, the design might also identify sections where rebar reinforcement is deemed necessary, for example in cases where local conditions provide high tensile stresses that cannot be managed solely by steel fibre reinforced concrete.

Recent project experiences show that the steel fibre reinforced concrete rings can be installed in excess of 80% of overall alignment length, with the exclusion of sections at cross-passages, or when asymmetrical geotechnical or loading condition are present and in close proximity to other underground structures.

4.2 Early design phase - Recommended aspect ratio limits to mitigate cracks

Most of the crack issues can be prevented by careful selection of the basic ring parameters in the early design phase. The first step is to define the main geometrical data of the tunnel ring, involving case histories of existing tunnels.

The influences of the steel bar and steel fibre reinforcement on the basic lining parameters (thickness, diameter, segment length/segmentation, segment thickness) are presented in the figures below (Figure 2. to Figure. 6). The diagrams depict the general applicability of the two technologies. The SFRC (in dashed line) is in several instances equivalent to the RC (in dotted line) and, for typical applications, an effective substitute.

Where;

Di	internal tunnel diameter;
T	segment thickness;
B	segment width;
L	segment length;
Di/t	ring slenderness;
L/t	segment area aspect ratio;
B/t	segment aspect ratio;

It is evident from the charts above that the lining thickness and the number of segments shall increase with the tunnel diameter and, in case of big tunnel diameters, the application of steel

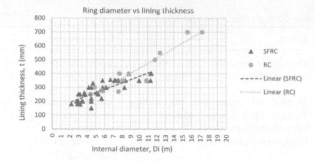

Figure 2. Influence of ring diameter and thickness ratio on the type of reinforcement.

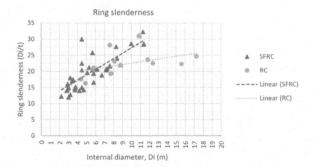

Figure 3. Influence of ring slenderness on the type of reinforcement.

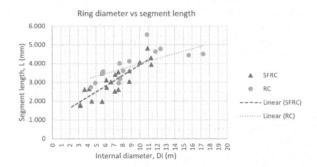

Figure 4. Influence of ring diameter and length ratio on the type of reinforcement.

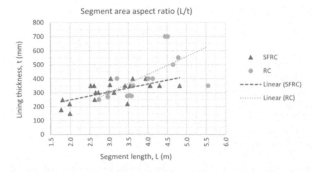

Figure 5. Influence of the segment aspect ratio on the type of reinforcement.

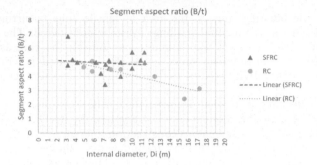

Figure 6.   Influence of the segment aspect ratio on the type of reinforcement.

fibre is not advantageous for the tunnel lining, instead the use of traditional reinforcement is more beneficial due to the significantly higher tensile stresses occurring in the tunnel lining.

4.3   *Considerations on the radial joints design*

For the radial joints (Figure 7.), both the compressive and tensile capacities of the section shall be verified. To minimize the risk of cracks so far as is reasonably practicable, a detailed study of the radial joint is required in terms of the joint geometry and analysis of potential stresses. The contact surface must be maximized to allow the most favourable tensile stress distribution near the joints.

Due to the limited contact surface between the joints, the axial forces being transmitted through the lining, converge towards the area of bearing. Compressive stress zone also develops directly under the loaded area. At a certain distance from the contact surface, the compression forces evenly distribute across the entire cross section. In the radial direction, tensile forces are induced by the compressive stress distribution (Figure 8.).

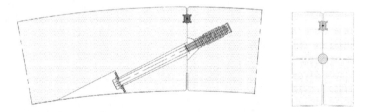

Figure 7.   Typical radial joint configuration. In the common practice the guide rod or the bolt are used depending if the ring is pinned in the circumferential joint or fully bolted.

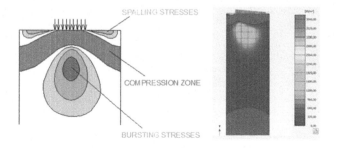

Figure 8.   Typical distribution of stresses.

In addition to the bursting stresses, the area of tensile zone occurring at the concrete surface, and that can cause potential spalling, must be considered. The design must also account for reasonable tolerances that might occur during the ring assembly (e.g. segment misalignment, improper use of bolts, etc.). The associated potential risks must be evaluated in terms of sensitivity studies evaluating the effect determined by different tolerance ranges.

In general, regardless of type of reinforcement, the tensile stress regime in the concrete must be carefully investigated and the residual tensile strength of the steel fibres should only be considered for exceptional loads or anomalous construction tolerances.

The impact of long terms loads, such as ground and groundwater actions, can be critical on the structural response of the radial joints. The design has to verify every relevant load condition in each geometrical and geotechnical configuration considering the actual joint stiffness that influences the stress distribution in the lining. The presence of significant eccentricities in the radial joint can reach unacceptable stress levels in the SFRC concrete.

The key in crack limitation is the determination of the effective contact area in the radial joint considering the actual load eccentricity and mitigate any potential ring assembly error that would result in a reduction of contact area and unexpected stress in the joints.

In the specific, the large application of universal rings with dowels (pins) in the circumferential joints and guide rods in the radial joints (Figure 9.) increases the risk of not perfect assembly of the ring (compared to the fully bolted/dowelled solution) and adds extra risk of segments cracks.

The deformation of the tunnel ring caused by eccentric thrust rams loads during the segment installation, shall not be neglected and must be compared to the results provided by the calculation model. The better overall cracking behaviour can be also related to the stiffer connection due to the bolts connection of the radial joints. The opportunity of additional (potential) bolts in the radial joints should be considered in design phase to mitigate the segment crack.

Based on the experience of the Authors, the benefits of these additional bolts in the radial joint can actively be used if cracks are present and must be considered as a foreseeable option to mitigate and control the potential construction errors, also when the only use of the guide rod is anticipated (Figure 10.).

4.4 *Considerations on the circumferential joints design*

The circumferential joints are loaded by the thrust jacks and in most projects this load provides the highest risk for cracking. Hence the design must involve a detailed study taking all the potential influence on the circumferential joints into account, including the number of thrust ram shoes, their planned and accidental positions along the segment axis.

Figure 9. Typical segment provided with dowels for circumferential joints and guide rods for radial joint.

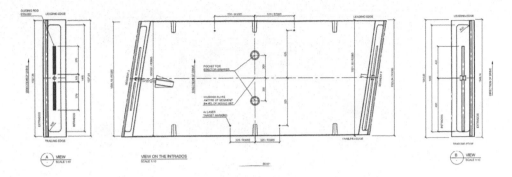

Figure 10. The guide rod is generally split into two sections to accommodate the bolt socket.

The pairs of thrust jack cylinders can be positioned in different ways on the segments, resulting in different spread of compression (and induced tensile) stresses (Figure 11.). The jack force applied on the segment introduces additional spalling stresses between the ram plates, causing potentially serious cracks with large depths, that should be avoided.

The segment layout has an impact on the allowable total thrust ram force (assuming the same thrust ram configuration). The verification of the circumferential joint shall show if the most relevant tensile stresses occur at the segment surface between the pads or as bursting stresses in the interior of the segments. If the bursting stresses exceed the tensile strength, cracks take place and activate the steel fibres. The bursting stresses induced by the ram loads shall be analyzed based on different load cases in two directions, as shown in Figure 12. and Figure 13.

The allowable construction eccentricity derived from the misposition of the push rams is limited to an acceptable value, generally around 2 cm. The most demanding situation occurs when the thrust cylinders are out of position along the lining thickness resulting in a reduced loaded area. Consequently, high tensile stresses occur at the surface of the segment between the ram shoes. The stresses shall not exceed the characteristic tensile strength of the concrete, otherwise cracks may develop.

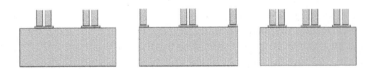

Figure 11. Typical ram configurations.

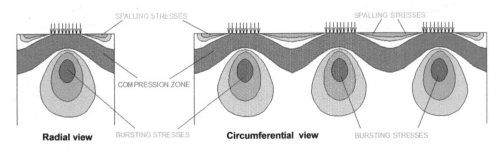

Figure 12. Compression and tensile zone induced by push ram loads applied on a single segment.

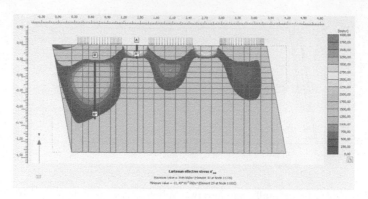

Figure 13.   Tensile zone induced by push ram loads applied on a single segment.

To reduce the risk of cracks, the tensile limit of plain concrete should be consistent with the operational thrust ram loads. A sensitivity analysis, evaluating larger predictable design tolerances (i.e. greater than 2 cm), is strongly recommended and should be carried out to check the limit of steel fibre application.

5   CONCLUSIONS

The use of SFRC lining was successfully applied by the Authors in some of the world's landmark mechanized tunnelling schemes, including the three aforementioned projects in the Middle East.

The experience gained have confirmed that the concrete segments are very sensitive to tolerances during ring assembly. The use of the Steel Fibre Reinforced Concrete (SFRC) has proven to be a valid alternative to traditional rebar cages, although its sensibility to intensive material testing and high standard quality control may appear less attractive at first sight.

The technology has an inherent limited tensile strength in its composite material and it requires the implementation of possible mitigation measures to avoid expensive repairs.

Today the application of SFRC lining in sewer, storm water, and metro tunnels is successfully applied, nevertheless design mitigation measures should be taken into account at the design phase and an expert Contractor should be involved.

In present paper, the major causes of lining cracks and recommendations for implementation of the mitigation measures in the early design phase are described. In particular:

– Need for alternative reinforcement solutions. The design should provide a traditional reinforcement solution for singularity along the alignment.
– The designer should consider existing experiences for defining the aspect ratio and geometry of the tunnel ring, including diameter, thickness, length, number of segments, etc.
– The importance of the radial joint shape and stress evaluation are highlighted as critical to mitigate the cracking risk.
– The circumferential joint issues related to the thrust jack cylinders are identified as very critical risks for cracking. The elements that should be carefully evaluated are identified.
– The radial joints should have the necessary arrangements for bolting also in case universal pinned rings are used, to minimize the assembly errors and mitigate the segment cracks.
– Suggestions on the tensile stress limits in the SFRC design are addressed.

REFERENCES

Castrogiovanni, G. & Busacchi, G. & Mariani, G. 2018. Steel Fibre Reinforced Concrete (SFRC) for TBM Tunnel Segmental Lining – Case Histories in the Middle East. *North American Tunneling Conference 2018*.

CEB-FIB Model Code 2010 - Volume 1. Section 5.6. Fibres/fibre reinforced concrete.

CEN EN 1992-1-1 (2004): Eurocode 2: Design of concrete structures - Part 1-1: General rules and rules for buildings.

ITAtech, 2016. Guidance for Precast Fibre Reinforced Concrete Segments - Vol 1 Design Aspects.

Leonhardt, F. 1986. Vorlesungen über Massivbau – Teil 2 – 3. Aufgabe, Abschnitt

Muir Wood, A. M., Curtis, D. J. 1975. The circular tunnel in elastic ground. *Géotechnique*, Volume 25 Issue 1, March 1975, pp. 115–127.

Tunnels and Underground Cities: Engineering and Innovation meet Archaeology,
Architecture and Art, Volume 5: Innovation in underground engineering,
materials and equipment - Part 1 – Peila, Viggiani & Celestino (Eds)
© 2020 Taylor & Francis Group, London, ISBN 978-0-367-46870-5

Variable support method: An innovative approach for deep excavations in the Middle East

G. Castrogiovanni
AECOM, New York, USA

G. Busacchi
COWI A/S, Doha, Qatar

C. Maier
COWI A/S, Lyngby, Denmark

M. Mancini
COWI NA, Springfield, USA

ABSTRACT: Underground construction in the Middle East presents some unique conditions that Design & Build contracts have to overcome in order to avoid the cumbersome and
time-consuming iterations of short-term changes implied by the contractor during execution
of the works. Extremely variable ground conditions of the sedimentary rock in the Middle
East, the shallow groundwater table and tight construction schedules are a few of the challenges that have to be coped with. An efficient solution, particularly for large open cuts and
shaft excavations, is the Variable Support Method. It defines a series of excavation classes and
associated support systems based on variable rock mass conditions in the design phase. The
excavation classes are presented in a matrix form, based on the expected ground conditions.
The contractor is then allowed, with the assistance of the Designer, to choose and apply the
most suitable support class within the defined design framework.

1 INTRODUCTION

In common practice, deep excavations in variable grounds are designed considering a design
geotechnical section (DGS), i.e. a design stratigraphy with associated design ground parameter.

The high variability of ground conditions and stratigraphy in combination with the usually
very limited design time results either in:

- design approach a); a conservative design for worst case conditions with a low risk of
  changes during execution (low risk of delay, but at higher cost), or
- design approach b); a more optimistic design but with a higher risk of changes during execution (lower cost, but higher risk of delay).

Shoring systems for deep excavations with concurrently applied support, e.g. shotcrete support with rock bolting, allow for visual inspection and therefore deviations from the design
assumption (DGS) are relatively easy to identify.

In case the above mentioned design approach b) with a more optimistic design is chosen,
adverse deviations in ground conditions could result in a re-design of the excavation support
with relevant impact in terms of cost and time.

This situation becomes even more evident and critical in a Design-Built delivery system in
which the design duration is particularly limited and in the critical path of project completion.

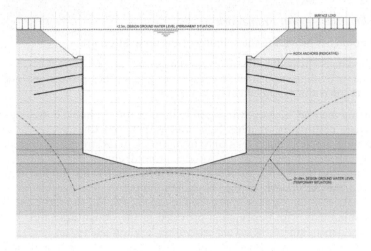

Figure 1. Typical Design Geotechnical Section of a deep open cut excavation.

As the Designer is responsible for a safe and robust design, above listed approach a) needs to be selected.

The traditional approach to the design of retaining walls/shoring systems is therefore a conservative DGS in order to have reserves in support capacity in case the site conditions differ from the design assumptions. However, substantial time and resources can be saved with a flexible design.

The solution adopted in deep open cuts excavated in the Middle East dominated by carbonatic rocks was to use a Variable Support Method. This method is an innovative approach which considers a series of support configurations for different site conditions to be readily available during excavation operations depending on the actual ground conditions determined by documented face-mapping.

The method is regularly used in the conventional tunnelling by defining different support classes for different ground conditions and ground behaviour. However, a crucial difference is that in the conventional tunnelling each advance is stabilized individually by the support measures and is practically independent of the previous or subsequent advance. In contrast, for the Variable Support Method in vertical cuts, the support applied in previous excavation steps has an influence on the excavation stability until reaching the final excavation level.

In the following paragraphs, the basis and principles of this method are illustrated.

2   DESIGN PROCESS

The conservative design process, i.e. scenario a), overcomes the generally limited time constraints imposed during the excavation of shafts and big open cuts, effectively preventing any room for potential optimization of the shoring system in real time. This can have a significant impact in terms of success of the construction within tight schedules but requires considerable higher construction cost.

The Variable Support Method allows to bypass the normally unavoidable compromise between higher costs for a conservative design or higher risk of delay for a more optimistic design, and instead provides a tailor-made design for the actual conditions within the expected frame-work.

The new implemented design approach foresees a support system consisting of a combination of shotcrete and rock-bolts, based on GSI rock mass characterization, that considers the supporting class installed in the previous excavation stages (i.e. pre-existing conditions to the new excavation stage).

The quality of encountered geomaterial is expected to be defined at each excavation stages by systematic face logging, which becomes the key element in the choice of the support system.

2.1 Establishing the Support Framework

The main components providing the stability of the excavation are:

– the rock mass itself (providing the main resistance),
– the shotcrete (providing supplementary resistance and surface protection), and
– the rockbolts (providing supplementary resistance).

The supporting method, also known as soil-nailing, is fully suitable in weak carbonatic rock conditions, where continuum or mass behaviour is expected to activate the shotcrete while rock-bolts provide supplementary resistance in case rock mass strength is exceeded.

Due to the expected high variability of the ground conditions in terms of weathering degrees, the intention of the design concept is to provide a valid support scheme for those varying conditions within clearly defined boundary limits.

The determination of the support system defines an excavation support class framework (see Figure 3) in conjunction with information provided in the design drawings. This outlines the required support depending on the actual excavation level on-site (stress level) and the outcome of the face logging procedure that provides the ground type (intact UCS) and GSI value.

It shall be noted that in all the instances where the method was applied, no clear bedding patterns or set of discontinuities were identified. Therefore, the chosen design approach and constitutive model assuming continuum behaviour of the rock mass was assumed to be appropriate. No pronounced failure of blocks separated by inclined jointing/discontinuities were expected from the existing ground investigation and local experience. However, as a general mitigation measure, if such unfavourable discontinuity configurations were encountered during excavation on site, local rock bolting would be applied in order to avoid such discontinuity-induced failure modes. The decision of such local spot bolting shall be taken on site by the responsible site geologist.

Ground conditions not covered by the design and not covered in the support framework shall be 'hatched out' in the matrix without providing support information. Therefore, the Contractor shall stop the excavation in the relevant area and contact the Designer to define any further actions.

The support scheme is determined from an extensive series of numerical analysis considering the limitations above in order to keep the calculation effort in balance with the likely boundary conditions on-site as per the available ground information (i.e. the geotechnical framework).

The validity of the developed support class framework is subsequently checked and fine-tuned by applying it on the actual design geotechnical sections, before direct application on site.

Under the uncertainties of the rock mass condition at shallow depths, mandatory support configurations may also be instructed to manage local high degree of weathering.

2.2 Numerical Modelling

The FEM simulations for the excavations models must aim primarily at verifying that the overall stability is fulfilled.

The model(s) must prove that the selected excavation support, meaning the rock mass enforcing rock bolts, are fit for purpose based on the assumptions and simplifications.

Secondly, in the light of this, yield estimations on deformations and other serviceability limit states must be checked.

2.3 System Application

The Variable Support Method system application is relatively simple. Supposed the excavation class frameworks as in Figure 4 is being defined, assumed the excavation stage/level has reached the indicated row "A", with the face logging identifying the geomaterial type (i.e. intact UCS and GSI value) to be as in the column "B", the required support type would be the one in cell "C" (e.g. Class 2 in our case).

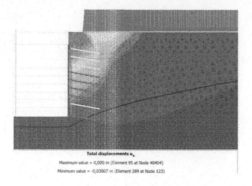

Total displacements u_x

Maximum value = 0,000 m (Element 95 at Node 40404)

Minimum value = -0,03007 m (Element 289 at Node 123)

Figure 2. Typical horizontal deformations in final excavation stage of the SLS.

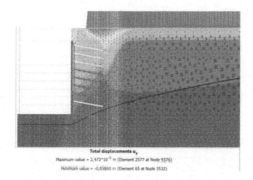

Total displacements u_y

Maximum value = 2,472*10⁻³ m (Element 2577 at Node 9376)

Minimum value = -0,03860 m (Element 65 at Node 3532)

Figure 3. Typical vertical deformations in final excavation stage of the SLS.

EXCAVATION CLASSES FROM FACE LOGGING:

EXCAVATION STAGE	FROM LEVEL	TO LEVEL	SIMSIMA LIMESTONE				MIDRA SHALE AND MIDRA LIMESTONE
			"WHEATHERED" 5MPa≤UCS<20MPa 30≤GSI<45	"LESS WHEATHERED" UCS≥20MPa 30≤GSI<45	"LESS COMPETENT" UCS≥20MPa 45≤GSI<60	"COMPETENT" UCS≥20MPa GSI≥60	UCS≤5MPa GSI≤50
EXCAVATION STAGE 00	GSL	-2.3					
EXCAVATION STAGE 01	-2.3	-4.3	SUPPORT CLASS 1A/B	SUPPORT CLASS 1A/B	SUPPORT CLASS 1A/B	SUPPORT CLASS 1A/B	/////
EXCAVATION STAGE 02	-4.3	-6.3	SUPPORT CLASS 1A/B	SUPPORT CLASS 1A/B	SUPPORT CLASS 1A/B	SUPPORT CLASS 1A/B	/////
EXCAVATION STAGE 03	-6.3	-8.3	SUPPORT CLASS 1A/B	SUPPORT CLASS 1A/B	SUPPORT CLASS 1A/B	SUPPORT CLASS 1A/B	/////
EXCAVATION STAGE 04	-8.3	-10.3	SUPPORT CLASS 3A	NO STRUCTURAL SUPPORT	NO STRUCTURAL SUPPORT	NO STRUCTURAL SUPPORT	/////
EXCAVATION STAGE 05	-10.3	-12.3	SUPPORT CLASS 3A	SUPPORT CLASS 2	NO STRUCTURAL SUPPORT	NO STRUCTURAL SUPPORT	/////
EXCAVATION STAGE 06	-12.3	-14.3	SUPPORT CLASS 3A	SUPPORT CLASS 2	SUPPORT CLASS 2	NO STRUCTURAL SUPPORT	/////
EXCAVATION STAGE 07	-14.3	-16.3	SUPPORT CLASS 3A	SUPPORT CLASS 2	SUPPORT CLASS 2	NO STRUCTURAL SUPPORT	/////
EXCAVATION STAGE 08	-16.3	-18.3	/////	/////	SUPPORT CLASS 2	NO STRUCTURAL SUPPORT	/////
EXCAVATION STAGE 09-I	-18.3	-19.3	/////	/////	SUPPORT CLASS 3B	NO STRUCTURAL SUPPORT	SUPPORT CLASS 3B
EXCAVATION STAGE 09-II	-19.3	-20.3	/////	/////	SUPPORT CLASS 3B	NO STRUCTURAL SUPPORT	SUPPORT CLASS 3B
EXCAVATION STAGE 10-I	-20.3	-20.7	/////	/////	SUPPORT CLASS 3B	NO STRUCTURAL SUPPORT	SUPPORT CLASS 3B

Figure 4. Excavation support needed for different rock formations and conditions.

The support class details could be described in a separate design drawing for site application. The application of this support class already considers the potential installation in the previous excavation stages of any of the possible indicated support classes, excluded those hatched out.

The flexibility of the Variable Support Method is clear.

3 SITE APPLICATION

The two case studies shown below in Figure 5 to Figure 7 (Case 1), and Figure 8 to Figure 10 (Case 2) are a realistic application of the method in two large open cuts, each of several hundreds of square meters of surface exposed and face logged.

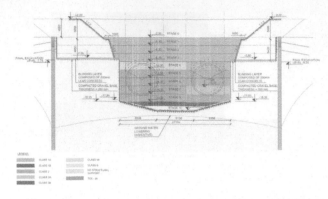

Figure 5.   Site application of excavation support classes – Case 1. Front excavation support.

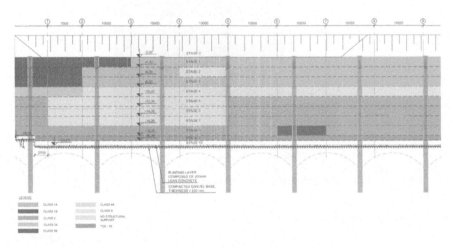

Figure 6.   Site application of excavation support classes – Case 1. Lateral excavation support.

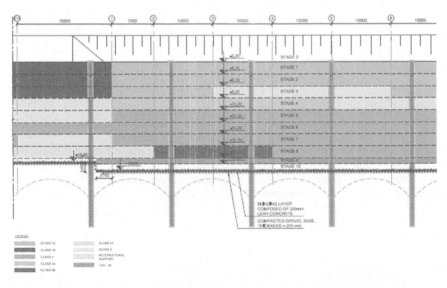

Figure 7.   Site application of excavation support classes – Case 1. Lateral excavation support.

The different shades in the elevation views of the walls indicate different support classes identified during the face logging and documented in Required Excavation and Support Sheets (RESS), for each excavation phase and stage.

The two cases are benchmarked against the scenario when a conservative design for the worst-case conditions would have been considered.

3.1 Case Study 1

In this example, the method allowed to save approximatively 30% in terms of rockbolts quantities installation (kg/m²), and 10% in terms of required shotcrete quantities (theoretical thickness) by implementing the appropriate excavation support class at each stage, as opposed to using the design approach a) with the worst excavation support class.

As a comparison, if the design approach b) was used with the most optimistic support class at each stage, the changes during execution would have affected more than 46% of the excavated stages. Due to unsafe excavation support in large portions of the open cut, re-design of the shoring system would have been required in several subsequent stages, causing considerable delays in the execution of the works.

3.2 Case Study 2

In this example and again in comparison to design approach a), the method has allowed to save approximatively 45% in terms of rockbolts average quantities installation (kg/m²), and 30% in terms of required shotcrete quantities (theoretical average thickness).

If the design approach b) were used with the most optimistic support class at each stage (in this specific case 'stable conditions' without no rockbolts and structural shotcrete), the changes during execution would have affected some 50–55% of the excavated surface. Large sectors of the excavation would have been unsafe, requiring re-design of the shoring system in several subsequent stages and causing substantial delays in the execution of the works.

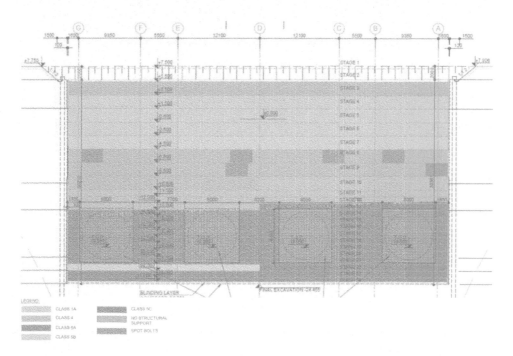

Figure 8. Site application of excavation support classes – Case 2. Front excavation support.

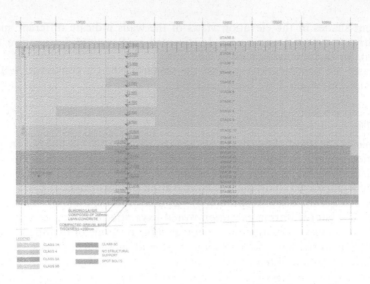

Figure 9. Site application of excavation support classes – Case 2. Lateral excavation support.

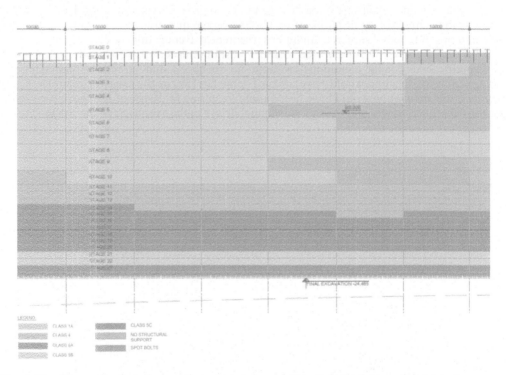

Figure 10. Site application of excavation support classes – Case 2. Lateral excavation support.

4 CONCLUSIONS

The Variable Support Method has been proven to be an efficient design approach for excavations with concurrent installation of the support system as e.g. shotcrete with rock bolts in variable weak rock conditions which allow for detailed face mapping.

The process of face mapping and agreement of the appropriate support class from the support matrix within the wider geotechnical framework was easy to conduct and did not slow down/adversely affect the construction process.

Variations in ground conditions could be addressed in real time with the support matrix.

The Variable Support Method revealed great benefit in terms of time and cost saving during construction.

In this paper, the major benefits of implementing a similar methodology have been described. In particular:

- The constraint schedule for the excavation of shafts and stations may not permit to optimize the shoring system. This can have a significant impact in terms of success of the design.
- The Variable Support Method allows to avoid the higher costs for a conservative design without imposing the higher risk of delay for a more optimistic design and, instead, provide a tailor-made design for the actual conditions within the expected frame-work.
- Shafts and Metro stations shoring systems may be designed considering a limited number of different geotechnical models, identifying reasonable different scenarios of rock mass conditions, to allow change in the support system (i.e. class) during the excavation stages.
- The design should include the generic shoring and a matrix that identify the support class to be applied on the basis of the rock mass characteristic (e.g. intact UCS and GSI), for simple application on site.
- The experience was gained by the application of the approach to challenging ground conditions on a fast-tracked projects in the Middle East. But it can be general applicable for other, similar projects with similar boundary conditions and variable ground conditions.

REFERENCES

Marinos, P. & Hoek, E. 2000. GSI: A Geological Friendly Tool for Rock Mass Strength Estimation. *Proceedings of the GeoEng 2000 at the International Conference on Geotechnical and Geological Engineering, Melbourne, 19–24 November 2000*, 1422–1446.

Marinos, P., Marinos, V. & Hoek, E. 2005. The Geological Strength Index: Applications and limitations. *Bulletin of Engineering Geology and the Environment, January 2005.*

Marinos, P., Marinos, V. & Hoek, E. 2007. The Geological Strength Index (GSI): A characterization tool for assessing engineering properties of rock masses. *Underground works under special conditions, Taylor and Francis.*

Marinos, V., & Carter, T. 2018. Maintaining geological reality in application of GSI for design of engineering structures in rock. *Engineering Geology, Volume 239, 18 May 2018*, Pages 282–297.

Tunnels and Underground Cities: Engineering and Innovation meet Archaeology,
Architecture and Art, Volume 5: Innovation in underground engineering,
materials and equipment - Part 1 – Peila, Viggiani & Celestino (Eds)
© 2020 Taylor & Francis Group, London, ISBN 978-0-367-46870-5

Advanced 4-channel scan system for tunnel inspection

A. Cereyon, W. Weber, R. Wissler & A. Wittwer
Spacetec Datengewinnung GmbH, Freiburg, Germany

ABSTRACT: Since many years Spacetec provides tailor-made tunnel scanner solutions for maintenance purposes. The latest development is a four-channel scanner (TS4) which records high resolution colour images during the laser scan. Advantages are significantly faster driving speed during the recording and an easier interpretation of data with colour information.

Today, laser scanning is commonly used to support manual tunnel inspections. The current state of the art is marked by the TS3 system, a three-channel tunnel scanner, which combines noncontact visual, 3D and thermographic data in a single measurement. The surface covering laser scanner data establish a basis to implement the colour images as an additional channel for further usage.

This paper covers the scan method of the TS4 scanner and the operation of the system in the field. The benefits of high-resolution colour imaging for tunnel-inspection are discussed based on examples.

1 INTRODUCTION

The development of TS3 to TS4 is the answer to requests that are frequently expressed by the users of laser scanning for tunnel-inspection:

- Colour imaging for better recognition of damages like corrosion or mineral deposit
- Higher resolution for better recognition of fine details, particularly cracks
- Faster recording speed

Currently available scanners based on LIDAR technique have arrived already at a very high technical level. The technology is constantly being developed and improved, but revolutionary development leaps are not to be expected.

Reasons for the limited development potential are:

- Point size smaller than 3 mm at 5 m distance seems to be unrealistic due to physical limits in focusing the laser beam.
- Today, point rates of about 3 MHz are feasible in operational systems. The development goes on, but improvements are moderate.
- LIDAR tunnel scanners use moving mirrors. For scanning tunnels, rotating mirrors are commonly used. The mechanical components are exposed to high stresses due to centrifugal forces. This limits the achievable speed.
- Due to the monochromatic nature of laser light, colour imaging is not feasible with only one laser source. Until today, no feasible solution of a colour laser scanner for traffic tunnels has been presented.

For achieving the objectives mentioned above, therefore, it requires additional technology. Using a completely different technique, however, may also lose benefits of the proven technique. The idea behind the TS4 solution is maintaining the advantages of the proven technique and supplement it with other techniques. When integrating the additional recording technology, care was taken to ensure that both parts complement each other optimally, so that the respective advantages come into play and individual limitations are overcome.

2 TS3 – THE STARTING POINT

The TS3 tunnel scanner has been specially developed and optimized for the inspection of traffic tunnels. It was first introduced in 2004, and since steadily improved. It has proven its power and quality in measuring thousands of tunnel kilometres.

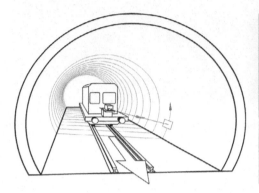

Figure 1.   Scanning principle of TS3 (left) and scanner survey in Gotthard Base Tunnel (right).

A fast rotating laser beam scans a tunnel circumference for each rotation. The complete visible surface of the structures is recorded by a single pass at a typical speed of 4 km per hour.

2.1   *The most important features in brief*

-360° recording angle: The all-round angle of view delivers a full and complete recording of the entire surface of the object in a single recording scan.

-Resolution 10,000 pixels at 360°: This pixel density is the same for all three recording channels. This also allows recordings for the purpose of crack identification.

-Mirror rotating speed 300 Hz: The mirror rotating speed is one of the crucial features for the measuring speed. The TS3 is capable of high resolution 10,000 pixels recording at this mirror speed. That results in a pixel rate of 3 MHz.

The electrical and mechanical design allows this high operating speed in continuous operation. Data storage capacity is large enough for uninterrupted operation for a full working shift.

Figure 2.   TS3 visual image (left) and thermal image (right).

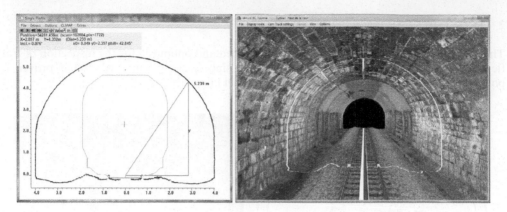

Figure 3.   Cross-sectional profile (left) and 3D display of the point cloud textured with the visual image (right).

-Ultimate flexibility: The TS3 tunnel scanner can be installed in virtually every vehicle which offers enough space and load-carrying capacity for the scanner head and the operator console. These can be road vehicles such as mini vans or off-road cars, but also track vehicles of any description. This makes the TS3 suitable for any application, whether on roads or on tracks. Special applications such as power plant tunnels or water mains are also feasible.

-Three simultaneous recording channels: In a single recording sweep, the TS3 generates an image recording of the tunnel surface, a thermographic image and a three-dimensional measurement survey. Computer-aided evaluation of the three data channels in context opens a wide variety of different analysis options.

Figure 2 illustrates the added value of thermal data. The phenomena visible on the right side do not show up in the visual nor the profile data. The orange coloured signatures indicate drainage pipes under the shotcrete layer. The blue areas show moisture and wet surfaces.

3   TS4 – CONCEPT

The TS4 solution leaves the TS3 setup unchanged. It just adds additional sensor technique to the setup. Cameras are added to the setup as additional sensors.

The system is open to a variety of camera types and specifications. The system can be configured with area cameras or line scan cameras, both technologies having their own advantages.

In any case, bright and even lighting is the key to good recording results. Creating a suitable lighting therefore was one of the core tasks of the development.

The development covered a wide field of tasks:

- Selection of suitable cameras in respect to speed of data acquisition and image quality
- Development of bright and even lighting. A key feature for fast data acquisition and good image quality
- Data acquisition and storage system. Covering the surface of tunnels with high resolution image date creates huge amount of data. Continuous streaming and lossless storage are a real challenge.
- Data processing techniques. Seamless integration of the additional colour image data with the already existing data channels was realised.
- Data evaluation. The existing software tools have been amended for handling the new colour channel in the same intuitive way as the previously existing channels.

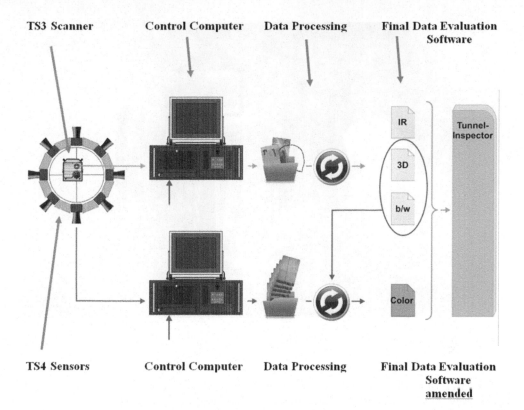

Figure 4.   TS4 system configuration.

4   TS4 – SETUP AND OPERATION IIN THE FIELD

Two setups have been realized.

1.1   *Combination of the TS3 scanner with area scan cameras*

For this setup type, the TS3 scanner is combined with area scan cameras. Each camera is integrated in a camera module together with illumination, and control electronics. The modules connect to a central unit which controls the camera modules and provides data storage for the captured images.

All system functions are controlled via this central unit. An operation software provides feedback on correct operation of all components and an easy-to use interface for the operation functions.

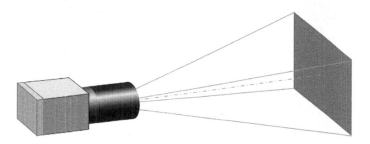

Figure 5.   Area scan camera.

Figure 6.   TS4 area camera solution in the field.

The TS4 system with area scan cameras can be set up for road tunnels and railway tunnels equally well. Using 6 camera modules, the field of view covers the complete tunnel circumference. The advantage of this variant is its compactness, light weight and low power consumption.

While the scanner records one continuous picture, covering the complete tunnel surface, the cameras take series of frames which later are assembled into a continuous image. As shown, the raw images from the camera are distorted considerably because of the tubular shape of the tunnel surface. Advanced software tools have been developed for combining the series of images seamlessly to a complete image, which is in perfect match with the scanned b/w image.

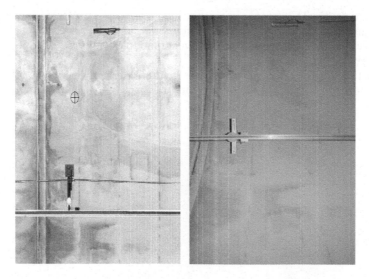

Figure 7.   scanned image (left) and area camera image before processing (right).

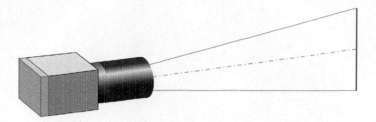

Figure 8.   Line scan camera.

For this setup type, the TS3 scanner is combined with line scan cameras. The cameras are positioned in a circular setup around the TS3 scanner such that the whole 360° field of view is covered.

Line scan cameras require bright and even line light illumination. For this special application, a white light ring light has been designed and realized, that illuminates the whole tunnel circumference along the line which is defined by the combined field of view of the circular camera setup.

This ring light is a key component for the line scan camera version of the TS4 system. It ensures good quality of results and high speed of operation.

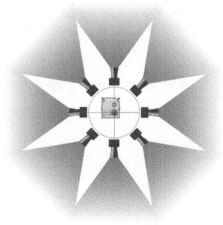

Figure 9.   TS4 line scan camera setup.

The setup of cameras and ring light is dimensioned for usage in traffic tunnels of any kind and dimension. It works equally well in an old single-track tunnel made of masonry or a new highspeed tunnel with comparatively large cross-section. The image below shows a test run in a three tracks railway tunnel. The average wall distance to all sides is about 8 metres. The examples given in chapter 5 below show the quality of results obtained with the TS4 solution.

Figure 10.   TS4 line scan camera solution in the field.

5   SAMPLES OF RESULTS

Both TS4 setups have been tested in a variety of tunnels. In the following, we show a few samples taken under typical conditions. The images are taken as screenshot from the evaluation software packages TuView, TuDrive and Tunnel-Inspector. Their enhancement was part of the TS4 development.

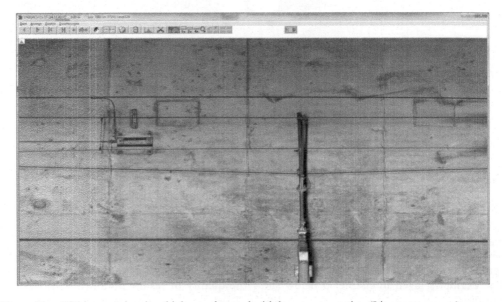

Figure 11.   TS4 image taken in a high-speed tunnel with large cross-section (Line scan camera).

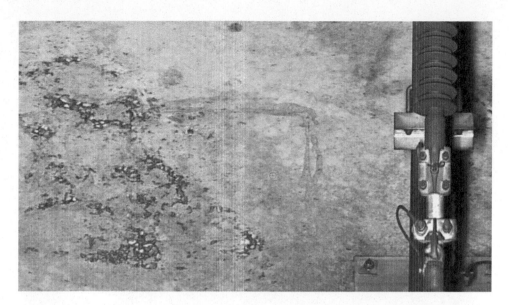

Figure 12.   Detail of TS4 image. High speed tunnel with large cross-section. Wall distance ca. 8 m.

Figure 13.   Old masonry tunnel. Virtual drive through the textured 3D model in TuDrive.

Figure 14. TS4 result: old masonry tunnel showing mineral deposit caused by water ingress (Area camera).

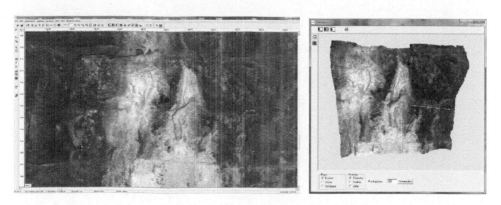

Figure 15. TS4 result: 2D and 3D detail display of mineral deposit.

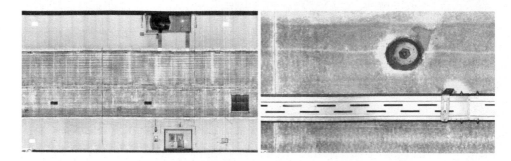

Figure 16. TS4 result: Motorway tunnel. Detail view shows corrosion on bolt and cable conduit.

Figure 17. TS4 result: The high resolution together with color helps to better recognize damages.

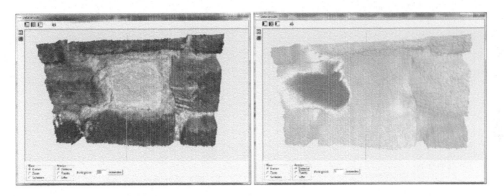

Figure 18. TS4 result: 3D detail. Colour and thermal channels.

6 CONCLUSION

TS4 development takes the digital capture of tunnels to a new level. It brings about revolutionary improvements in resolution and measurement speed. Colour image is an invaluable help for recognition of damages like corrosion, mineral deposits and spalling.

REFERENCES

Sandrone, F., Wissler, R., 2011. Laser scanning images analysis for tunnel inspection. Proceedings of the 12th International Congress on Rock Mechanics, Beijing, China.
Malva, R., Wissler, R., 2014. Laser scanning images analysis for tunnel inspection. Proceedings of the World Tunnel Congress 2014 – Tunnels for a better Life. Foz do Iguaçu, Brazil.

Wittwer, A., Wissler, R., 2015. Use of thermography for tunnel-inspection. Proceedings of the World Tunnel Congress 2015. Dubrovnik, Croatia.

Wissler, R., Weber, W. 2017. High resolution laser scanning of the Gotthard Base Tunnel (GBT). Proceedings of the World Tunnel Congress 2017. Bergen, Norway.

Tabrizi, K., Celaya, M., Miller, B.S., Wittwer, A. and Ruzzi, L., 2017. Damage Assessment of Tunnel Lining by Mobile Laser Scanning, Pittsburgh, Pennsylvania, Implementation Phase of FHWA SHRP 2 R06G Project

Tunnels and Underground Cities: Engineering and Innovation meet Archaeology, Architecture and Art, Volume 5: Innovation in underground engineering, materials and equipment - Part 1 – Peila, Viggiani & Celestino (Eds)
© 2020 Taylor & Francis Group, London, ISBN 978-0-367-46870-5

Fiber-reinforced concrete segments: Comparison of simplified analyses and FE simulations for the 3RPORT CSO Project, Fort Wayne, Indiana, United States

D. Chapman & J. Parkes
Schnabel Engineering, Morristown, New Jersey, USA

G. Venturini & R. Comini
SWS Engineering Spa, Trento, Italy

L. Waddell
Lane Construction Corporation, Fort Wayne, Indiana

ABSTRACT: The design of steel fiber reinforced concrete (FRC) precast segmental liners was recently developed for the 3RPORT Project following American Concrete Institute (ACI) requirements and European standards. Nine load cases were analyzed including final ground and water loads with up to 8 bar of external hydrostatic pressure, as well as fabrication, handling, and TBM thrust loads. The TBM thrust produces concentrated forces that cause bursting and spalling stresses. Simplified analysis methods cannot describe the development of spalling stresses between TBM jack pads and do not account for the ductility of the steel fibers. Consequently, reinforcement bars may be unnecessarily prescribed based on analyses using such methods. Advanced 3D numerical models were developed including the use of a constitutive law for FRC. This allowed evaluation of post-cracking behavior during TBM thrust, crack width progress and consequent stress redistribution. These results justified exclusive use of steel fiber reinforcement.

1 INTRODUCTION

The Three Rivers Protection & Overflow Reduction Tunnel (3RPORT) is a combined sewer overflow (CSO) tunnel project located in Fort Wayne, Indiana. The tunnel and the associated pipe network are part of the City's Long-Term Control Plan for reducing the amount of combined sewage that is discharged into Fort Wayne's rivers every year.

The tunnel will be constructed in rock at depths up to about 70m below grade using a pressurized-face (slurry) tunnel boring machine (TBM). The projected life for the 7,600m, 4.88m internal diameter (ID) tunnel is 100 years.

The approximately $188 million project, when completed, will reduce combined sewage overflows to the St. Mary's and Maumee Rivers by 90 percent from over 70 per year to just four. The project was awarded to the Salini-Impregilo/S.A. Healy Joint Venture, now Lane construction (Lane), for construction in April, 2017.

2 TUNNEL LINING REQUIREMENTS

2.1 Specification Requirements

The tunnel lining will be constructed using bolted and gasketed precast concrete tunnel lining segments. Schnabel-SWS was engaged by Lane and by their precast segment supplier, CSI-Forterra Fort Wayne JV to perform the design of the precast segmental lining.

The design was based on requirements in the project specifications, drawings, and external loads provided in the GBR. The primary requirements pointed to a design using steel fiber reinforcement in lieu of traditional steel reinforcing bars. Key design requirements included:

– Design to be based on ACI 350 (2006). Do not increase allowable stresses for temporary loads, or reduce load factors to account for temporary structure design.
– Avoid cruciform joints and ensure that TBM jacking shoes do not cross radial joints.
– Minimum ID of 4.88m and minimum thickness of 305mm.
– Minimum 28-day concrete compressive strength of 55 MPa
– Tunnel lining to be designed using steel fiber, minimum dosage 21 kg/m^3, with additional requirements:
  – Minimum flexural tensile strength at 28 days of 670 psi per ASTM C1609. This was interpreted to be equivalent to the first peak strength, f_1, (ASTM 1609).
  – Minimum post-crack equivalent residual flexural tensile strength at any deflection at or beyond the deflection of L/600 of 460 psi per ASTM C1609. This is interpreted to be equivalent to the residual strength at or beyond a deflection of L/600 per ASTM C1609, f^D_{600}, per paragraph 3.2.14 of ASTM C1609.

– Gaskets to be designed for a minimum working pressure of 8 bars above atmospheric with additional performance requirements for any gap between adjacent segments, and for 150 percent of the maximum tail void grouting pressure specified.
– Bolted connections at radial joints; bolts or dowels for circumferential joints.
– Minimum concrete cover to steel reinforcement and minimum steel ratio of 0.5% of gross concrete area radially and circumferentially specified for possible rebar alternative.

2.2 Segment Manufacturer Requirements

The segment ring consists of 6 segments as shown in Figure 1.

CSI-Forterra provided the following additional parameters of the planned segment design:

– Nominal segment ring width of 60 inches with a maximum ring taper of 1.25 inches, in order to meet the curve radii of the geometric design of the tunnel route.
– Basic ring geometry: 6 segments (four parallelogram segments, each with 67.5 degrees of arc length and two trapezoidal key segments, each with 45 degrees of arc length).
– Radial and circumferential joint surfaces will utilize a compressional packing sheet to enhance distribution of thrust between segments and minimize concentrated point loads.
– The 4-hour concrete strength is 2,000 psi, suitable for segment demolding and stacking.

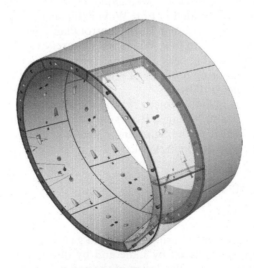

Figure 1.   Design Rendering of Precast Concrete Segment Rings for 3RPORT Tunnel.

Table 1. Summary of load cases for TBM segment analysis.

Load Case Numbers	Load Case Description
Production/transient stages	
Load Cases 1-4	Segment stripping, storage, transportation, & handling
Construction stages	
Load Case 5	TBM thrust jack forces
Load Case 6	Tail skin back-grouting pressure
Load Case 7	Localized back-grouting (secondary grouting) pressures
Services Stages	
Load Case 8	Earth pressure, groundwater pressure, and surcharge loads
Load Case 9	Longitudinal joint bursting pressure

2.3 *ACI Code Requirements*

ACI 350 (2006) was designated as the concrete code for design of the precast concrete segments. The code governs the design of liquid-retaining structures that "...are subject to different loadings, more severe exposure conditions, and more restrictive serviceability requirements than non-environmental building structures." Its objective is production of dense and durable concrete, suitable for the exposure conditions, limiting deflections and controlling crack width through requirements for concrete composition and spacing, size and distribution of reinforcement. However, it does not contain requirements for elements using steel fiber reinforcement.

2.4 *ACI 544.7R-16 Guidelines*

ACI 544.7R-16 (2016) – Report on Design and Construction of Fiber-Reinforced Concrete Tunnel Segments, was used to address requirements for design of fiber-reinforced segments, as was information from various European codes for concrete design using fibers. This ACI report notes the increased use of fiber reinforcing due to its significantly improved post-cracking behavior and crack control characteristics. These characteristics are ideal for producing a robust product that meets the performance requirements of ACI 350 (2006), which is more resistant to handling stresses and has better long-term durability than conventionally reinforced concrete. The ACI 544 report provides a procedure for structural analysis and design based on governing load cases; and a description of material parameters, test, and analyses required for design. The load cases used in the subject design are listed in Table 1.

ACI 544.7R-16 also sets the framework for determination of concrete properties to be used in the analysis, as derived from flexural strength testing performed on cast beam samples in accordance with the requirements of ASTM C1609. This test determines a load-deflection curve showing peak and post-peak behavior for third-point loading on a simply-supported beam over a defined range of deflections. Typical beams used in the test are 6 inches square and 18 inches long, and test loads at peak strength and at deflections of L/600 (0.75 mm) and L/150 (3.0 mm) are reported. These test loads are used to calculate the first peak flexural strength, f_1 and the residual flexural strengths at deflections of L/600 and L/150, e.g., f^D_{150}.

ACI 544.7R-16 (2016) notes that in elastic analysis, the key parameter is residual tensile strength, σ_p, which is calculated from flexural strength using an adjustment factor of 0.33 to 0.37. The nominal resistance bending moment, Mn, is calculated using either $Mn = \sigma_p bh^2/2$ or $MN = f^D_{150} bh^2/6$. ACI 544 also gives recommended load factors for the various load cases and strength reduction factors (ϕ) for various behavior modes (flexure, compression, shear, and bearing).

3 DESIGN APPROACH

3.1 *Overall design approach*

The design was organized to address the ACI 544.7R-16 (2016) load cases listed above. The load cases are grouped in Table 1 by stages from production through construction, to service.

The segments were designed to meet the required load factors of ACI 350 (2006), some of which are higher than those of ACI 544.7R-16. The analysis also proceeded in a different sequence than indicated by the case number in order to define the segment geometry based on the most significant loadings.

3.2 *Load Case 8 – Earth, groundwater, and surcharge loadings*

This load case includes loadings on the tunnel lining due to the pressure of overlying rock and soil, groundwater, and surcharge loads on the ground surface. The external loads were developed using the finite element (FE) code PLAXIS. The tunnel profile was reviewed in context with the GBR and selected borings and test data to identify critical sections for analysis and to develop overburden and rock mass properties for use in the PLAXIS modeling.

ACI 544.7R-16 (2016) load factors are lower than the 1.6 required by ACI-350 (2006) for earth and groundwater loadings. Accordingly, the PLAXIS output loadings were factored by 1.6 and the results were evaluated without attempting to separate the effects of the various components of the loads. This was performed using a bending moment-axial force (M-N) interaction diagram for fiber-reinforced concrete, constructed in accordance with procedures described in Appendix A of ACI 544.7R-16.

3.3 *Load Cases 6 and 7 – Grouting Loads*

Case 6 determines the concentric loads from the annular tail void grouting process.

Because the baselined groundwater pressure is 6.5 bars, and the anticipated grouting pressures are up to 2 bars above that, a grouting pressure of 8.5 bars was used for analysis. Because the segmental ring is surrounded by semi-liquid and fresh grout during application of grouting pressure, no interaction is considered between the segmental ring and the ground. Radial pressure is applied linearly from minimum grout pressure at the crown to maximum at the invert, with only the lining self-weight of the lining and the grout pressure considered as shown in Figure 2a.

Load Case 7 evaluates the effects of localized secondary contact grouting. This grouting is performed to verify closure of the annular void and is modeled by applying the differential grout pressure of 2 bars over 10 percent of the segment ring at the tunnel crown, as shown in Figure 2b. The tunnel lining is considered to be in full contact with the surrounding ground except in the area where the secondary grouting is performed. The interaction between the lining and the surrounding ground is modeled using radial springs. The results were checked using the higher load factor of 1.6 per ACI 350 (2006), using an M-N interaction diagram for FRC.

3.4 *Load Cases 1-4 - Production, Handling, and Transportation Loads*

Load cases 1-4 are evaluated using a spreadsheet for calculating bending moments from simple segment and handling system geometry. Case 1 involves segment demolding, gripping using cast-in pockets in the segment faces, and rotating the segment from the casting mold to stacking and storage. Case 2 involves stacking the six segments of one ring using timber blocking between the segments. These cases are evaluated based on concrete strength at four hours.

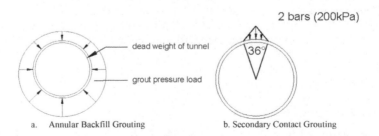

Figure 2.   Model for Cases 6 and 7 – annular backfill grouting and secondary contact grouting loads.

Cases 3 and 4 are for transportation of the segments from the storage yard into the tunnel and handling of the segments for placement during tunneling. These analyses utilize the flexural strength parameters from concrete tested at 28 days, and they incorporate an impact factor of 2.0 in addition to the normal load factor to account for the nature of the construction process.

Code checks were also performed for shear around the cast-in segment rotator pockets, and tensile pullout and shear around a center-pin lifting insert. Cases 1-4 and the related checks all showed adequate capacity in the segments as designed.

3.5   *Load Case 5 – Circumferential Joint Bursting Stresses*

Load Case 5 evaluates the effects of the TBM thrust jacks against the most recently completed ring of segments. High compressive stresses develop under the jacking pads and give rise to significant bursting tensile stresses within the segment. Spalling tensile forces also act between adjacent jacking pads along the circumferential joints.

ACI 544.7R-16 (2016) presents simplified equations for initial analysis of these forces based on the ACI 318 (2014) concrete code and the DAUB segment guideline published by the German Tunneling committee (2013). The ACI code is based on consideration of bearing on concrete by the anchorage of a post-tensioning strand using a strut and tie model.

TBM thrust forces were estimated considering the weight of the TBM and back-up, the baseline groundwater pressure of 6.5 bars at the tunnel face, the maximum excavation force for each cutter, and the number of cutters. Frictional loads on the tunnel shield were estimated using a coefficient of friction of 0.15 for normal conditions and 0.40 for upset conditions. A load factor of 1.2 was used for normal friction per ACI 544.7R-16 (2016) and 1.0 was used for upset friction. The normal condition with 1.2 load factor resulted in a maximum thrust of 34,493 kN, which is similar to the maximum thrust capacity of the TBM. Although ACI 350 (2006) requires a live load factor of 1.6 because these loads may have significant uncertainty, this was not considered applicable because the maximum TBM thrust load was considered to be more predictable, and a load exceeding that is considered unlikely. A capacity reduction factor of 0.65 was used per ACI 350 because the action of the load for this case is compressive in nature.

The simplified equations of ACI 544.7R-16 (2016) were evaluated using spreadsheets. The results of these simplified analyses showed a high demand for strength required to resist bursting stresses, which was not satisfied by the residual tensile strength of the FRC segments. Therefore, based on the simplified analyses, conventional steel rebar reinforcement would be required in addition to fiber reinforcements.

In order to further explore whether rebar was actually needed, a FE analysis was performed using the commercial software Straus7 (2015), distributed by Strand 7 Pty Ltd, Sydney, Australia. The intent of the FE analysis was to examine more closely the concrete stress-strain behavior and extent of possible cracking that may occur. This included development of a constitutive model based on flexural beam testing information to represent the entire range of compression and tension behavior of the FRC segments. The constitutive law incorporates the capacity reduction factor $\phi = 0.65$ because of the compressional nature of the loading.

3.6   *Load Case 9 – Longitudinal Joint Bursting Forces*

Load Case 9 addresses bursting forces in the longitudinal joints due to service earth and groundwater loadings. The methodology for the assessment is essentially the same as for Load Case 5.

4   DEVELOPMENT OF CONSTITUTIVE MODEL

The stress-strain relationship for the FE analysis was chosen to simulate the expected behavior of the FRC before and after flexural yield. The constitutive law aims to reproduce actual stress-strain response of uniaxial tests; however, experimental data are usually obtained from flexural tests due to the difficulty of gaining significant results from uniaxial tensile tests. Consequently, the "flexural tensile" parameters need to be converted into "equivalent tensile" parameters.

A literature review indicated that one of the most widely-used approaches is that of RILEM Technical Committee 162, Test and Design Methods for Steel Fiber-Reinforced Concrete (2003). In this approach, tensile stress parameters are derived from flexural test results, after scaling by proper stress coefficients. Relationships connecting tensile stress values with flexural tensile values vary depending on the type of test beam used in different test practices.

ACI 544.7R (2016) suggests scaling the post-crack (residual) flexural strength parameters by an adjustment factor to convert the residual flexural parameters from the standard flexural test to residual tensile strength for design. This scale factor is used to account for the nonlinear stress distribution in the critical section and is only applied to the residual strengths. The first peak flexural strength is not scaled because it occurs before the onset of cracking and a linear stress distribution is expected at this stage.

RILEM TC 162 (2003), proposes a stress coefficient of 0.45 for the first residual strength at CMOD1 (crack mouth opening displacement =0.5mm), which approximately corresponds to the residual strength at L/600 according to ASTM C1609 (2012). RILEM proposes a coefficient equal to 0.37 for the residual strength at CMOD4 (crack mouth opening displacement =3.5mm), which approximately corresponds to the residual strength at L/150 according to ASTM C1609.

The tensile branch of the design stress-strain is determined as follows:

− The first peak design strength is obtained by multiplying the specified value by the concrete strength factor φ;
− The residual design strength at L/600 is obtained by multiplying the specified value by the concrete strength factor φ and the scaling factor 0.45;
− The residual design strength at L/150 is obtained by multiplying the specified value by the concrete strength factor φ and the scaling factor 0.37.

A parabolic pattern is assumed for the compressive stress-strain relationship until a deformation of 0.2%, after which the stress remains constant. The maximum compressive strain is taken 0.3%, instead of the 0.35% value suggested by RILEM (2003), in order to respect the convention according to ACI 318 (2014). The derived constitutive law is shown in Figure 3.

This constitutive law corresponds to an ultimate limit state (ULS) because it has been reduced by the partial material factor for concrete.

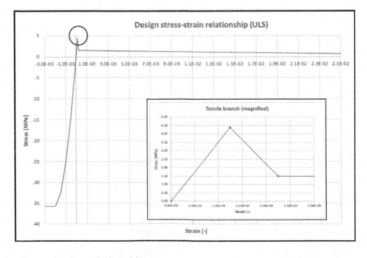

Figure 3.   Derived constitutive relationship.

5 NUMERICAL (FINITE ELEMENT) ANALYSIS

5.1 *Considerations for finite element analysis*

FE analysis offers several advantages compared to the simplified equations. In addition to the bursting tensile stresses, it can determine the spalling stresses expected in the areas between the jacking pads. It can follow the post-crack behavior of FRC if the first peak strength is exceeded, which allows evaluation of stress redistribution capacity due to stable crack propagation. Simplified equations, by contrast, only compare the maximum elastic stress with the FRC tensile strength to evaluate if conventional reinforcement is required. Finally, FE analysis enables obtaining more precise stress propagation because the actual segment and jack shoe geometry, and the eventual load eccentricity, are all included in the model.

5.2 *Model Details*

Modeling was performed for the larger rhomboidal tunnel segment, which is contacted by three, equally-spaced jack shoes. The segment geometry was simplified for modeling by straightening the angled sides out; this simplification does not affect the results. The model mesh size is approximately 30 mm, resulting in a total of about 50,000 brick elements. Stress relief recesses near the segment intrados and extrados were modeled by removing corresponding brick elements to facilitate stress distribution from TBM jack pressure deep into the segment to avoid minor concrete edge spalling that was not the focus of this analysis.

For the restraints, the longitudinal degree of freedom of all segment joints on the rear segment face in contact with the previously installed ring was fixed. The segment is conservatively left free to move in the circumferential direction, whereas in reality it is partially restrained by adjacent segments. This will result in more conservatively modeling bursting and spalling stress effects and actual behavior should be better than estimated from the numerical analysis.

The TBM force was applied as equivalent normal pressure on the jack shoe. The factored force per shoe is 2155.8kN. After division by the shoe area, a contact pressure approximately equal to 20 MPa is found, as shown in Figure 4, which also shows the simplified segment geometry. The total design force on the segment is about 6467kN. This design force is not applied all at once, in order to ease convergence, but rather over 20 simulation steps with a 5% load increment.

5.3 *Model results – Bursting stresses*

Initial results were encouraging, with the pattern of stresses associated with application of 60% of the design pressure shown in Figure 5, the pattern strongly resembling that of the example analysis described in ACI 544.7R-16 (2016) for this load case.

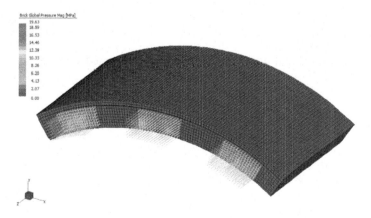

Figure 4.   FE model of segment with TBM shoe forces applies.

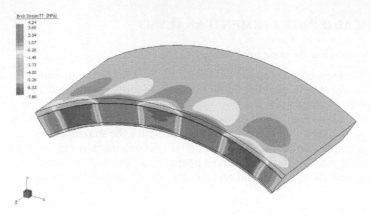

Figure 5.   Model of segment showing stresses 60% load application.

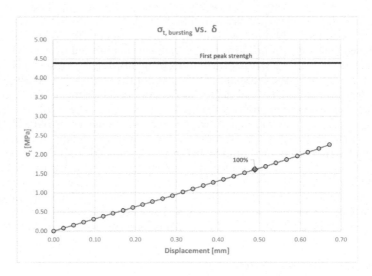

Figure 6.   Graph of bursting forces from the FE FRC segment model.

Figure 6 shows that estimated bursting stresses are well below the first peak strength for the portion of analysis past the first peak load which occurs between load Steps 13 and 14, and for the portion past 100% of the TBM thrust load. Thus, they do not require evaluation against residual flexural or tensile strength values. This also shows that the results from the more precise FE model are much less than those estimated using the simplified analyses, the results of which indicated that the bursting strengths were exceeded and rebar reinforcing was necessary.

5.4 *Spalling tensile stresses*

Besides bursting stresses, spalling tensile stresses are also considered in the analysis. Contours of tangential stress and strain at the next load simulation step, the 65% load step, show the beginning of crack formation in the joint, at the midpoint of the spaces between the jacks, exactly where spalling tensile stresses are predicted, as shown in Figure 7.

This is further illustrated in Figure 8 showing how the tangential stress develops in the spalling area during the TBM thrust load application. The spalling stress increases until Step 13, the 60% load step, where the peak stress defined in the constitutive law is reached. In the

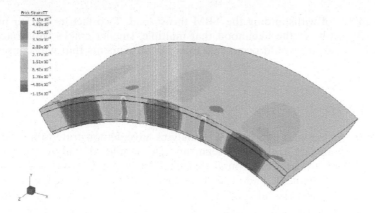

Figure 7. FE Segment model showing the development of crack formation between TBM shoes.

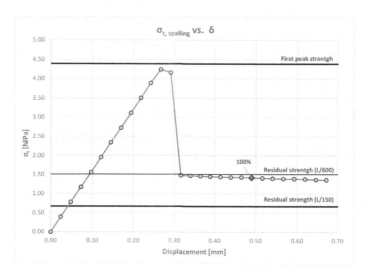

Figure 8. Tangential stresses in spalling area during TBM thrust load application.

following step, which marks the onset of cracking, a sharp decrease in stress occurs and there is a corresponding strain increase. Then, until the total TBM thrust is reached the element stress falls off slightly at a magnitude near the residual tensile strength.

Using this information, the maximum expected crack width at the full TBM thrust load was estimated and compared with an acceptability limit. The analysis considers a single crack, which in reality could be several smaller cracks in the area rather than one at the midpoint. The crack affected a single element width, so the crack width is estimated by multiplying the maximum strain by the element size:

$$W_{max} = 30 \ mm \ x \ 2.22E - 03 = 0.07 \ mm \qquad (1)$$

It should be noted that this analysis, as an ULS analysis, would not normally be used for studying crack width, which is normally considered at the service limit state (SLS). Both U.S. and European codes limit crack width to about 0.3 mm for ordinary exposure conditions, with values of about 0.10 mm for environmental water-retaining structures, for an SLS analysis. As the designed segments met this criterion for a ULS analysis, the segment was considered capable of withstanding the TBM thrust load. Two factors further mitigating the estimated crack width are the likelihood that multiple, smaller cracks would actually occur, and the confinement that would occur due to adjacent segments that was conservatively not modeled.

6 CONCLUSIONS

Both simplified analyses and numerical FE analyses have demonstrated that the TBM thrust force can be safely applied to the FRC segments. The simplified analyses were used to check the majority of the load cases and found that the FRC was adequate. For load cases where the simplified analyses indicated additional reinforcement may be necessary, the FE analyses was used to further assess the adequacy of the FRC and demonstrate that rebar reinforcing was not needed. The numerical model has been validated through comparison with analytical solutions and analogue numerical results found in the literature.

The bursting stresses estimated in the FE analyses remain below the first peak strength, indicating that under 100% of the TBM thrust these stresses are still in the elastic range. The most critical aspect during TBM thrust consists of the spalling stresses between adjacent jack pads. Until approximately 60% of the design thrust, the stresses are within the first peak limit. Above this load value, the peak strength is reached and very small cracks (less than 0.1mm) are expected to develop in the spalling area; however, the crack width is far below the acceptability limit given in both European and American Standards. The crack width is also a maximum value, estimated for one crack, whereas the actual condition may be multiple smaller cracks.

Because the predicted maximum crack width under ULS conditions is well below the allowable crack width, the use of conventional steel rebar reinforcement is not considered necessary. Therefore the use of FE modeling has demonstrated that the FRC segment of 304.8mm thickness is sufficient to withstand the expected loads without need of adding conventional reinforcement.

REFERENCES

American Concrete Institute (ACI). 2016. *Report on Design and Construction of Fiber-Reinforced Precast Concrete Tunnel Segments*. ACI 544R-16. Farmington Hills, MI

American Concrete Institute (ACI). 2014. *Building Code Requirements for Structural Concrete*. ACI 318-14. Farmington Hills, MI

American Concrete Institute (ACI). 2006. *Code Requirements for Environmental Engineering Concrete Structures*. ACI 350-06. Farmington Hills, MI

ASTM International (ASTM). 2012. *C1609 Standard Test Method for Flexural Performance of Fiber-Reinforced Concrete (Using Beam with Third-Point Loading)*. West Conshohocken, PA.

Deutscher Ausschuss fur unterirdisches Bauen e. V. (DAUB) German Tunneling Committee (ITA-AITES). 2013. *Recommendations for the Design, Production, and Installation of Segmental Rings*. Cologne, Germany.

Reunion Internationale des Laboratoires et Experts des Materiaux (RILEM), Technical Committee 162. 2003. Test and Design Methods for Steel Fiber-Reinforced Concrete. In *Materials and Structures* Vol. 36: 560–567. Paris, France.

Tunnels and Underground Cities: Engineering and Innovation meet Archaeology, Architecture and Art, Volume 5: Innovation in underground engineering, materials and equipment - Part 1 – Peila, Viggiani & Celestino (Eds)
© *2020 Taylor & Francis Group, London, ISBN 978-0-367-46870-5*

A simple approach to characterise tunnel bore conditions using pipejacking data

W.-C. Cheng
Xi'an University of Architecture and Technology, Shaanxi, China

J.-S. Shen
Shanghai Jiao Tong University, Shanghai, China

A. Arulrajah
Swinburne University of Technology, Victoria, Australia

ABSTRACT: Despite several well-established jacking force models being available for determining the jacking loads, their ability to characterise the tunnel bore conditions is very limited. In this study, a simple approach to characterise the tunnel bore conditions is proposed and applied to a case history where four wastewater pipelines were constructed in soft alluvial deposits to verify its validity. Four jacking force models are reviewed. By the given soil properties and pipe dimensions as well as buried depth of pipeline, the normal contact pressure in each model and the measured frictional stress in each section of baseline which is constituted by the minima of total jacking load are used for back-analysis of the frictional coefficient μ_{avg}. The μ_{avg} values outside the 0.1–0.3 range suggested for lubricated drives can be linked to the increasing pipe friction resulting from excessive pipe deviation or ground closure or the gravel formation not being long enough to develop lower face resistance or total jacking load.

1 INTRODUCTION

Jacking force prediction highly correlates with the design and selection of a pipejacking system. Despite many well-established jacking force models available for the provision of jacking capacity, their ability to characterise the tunnel bore conditions is very limited (Pellet-Beaucour & Kastner 2002, Barla et al. 2006, Chapman & Ichioka 1999, Shen et al. 2016). It is first proposed that the tunnel bore conditions can be characterised through assessing the soil (lubricant)-pipe interface behaviour by comparing the back-analysed frictional coefficient μ_{avg} to the suggested value. The soil pressure, hereafter referred to as normal contact pressure σ', acting upon the pipe crown can be calculated using the jacking force models. A line constituted by the minima of jacking forces, as also referred to the baseline of jacking forces (Cheng et al. 2017b, 2018b, c), is used to determine the average jacking force f_{avg} and subsequently the frictional stress τ. The frictional stress τ divided by the calculated normal contact pressure σ' leads to the back-analysed frictional coefficient μ_{avg}. The higher μ_{avg} value indicates that the surrounding geology may be in direct contact with the pipe string and can be caused either by ground closure likely to be taken place while tunnelling in permeable ground or by increasing pipe friction resulting from excessive pipe deviation greater than a threshold value. This proposed simple approach enables the tunnelling parameters to be finetuned during pipejacking works reducing the potential of geo-hazards (Cheng et al. 2017a, c, 2018a, d, Wang et al. 2018).

In this study, pipejacking data from a project involved with four wastewater pipelines excavation in soft alluvial deposits in Taipei County, Taiwan were analysed using the baseline technique. The objectives of this study are (i) to review four well-established jacking force models for producing a most appropriate calculation of the normal contact pressure, (ii) to propose a simple approach that is capable of characterising the tunnel bore conditions, and (iii) to apply this proposed simple approach to a case history to verify its validity and applicability.

2 REVIEW AND RESEARCH METHODOLOGY

2.1 Review of jacking force models

Four jacking force models; that are, (i) Japan Microtunnelling Association (JMTA 2000), hereafter referred to as JMTA, (ii) Ma Baosong (Ma 2008), (iii) Shimada and Matsui (Shimada & Matsui 1998), and (iv) Pellet-Beaucour and Kastner (Pellet-Beaucour & Kastner 2002), were reviewed, as summarised in Table 1. Since the friction resistance F_s in the Pellet-Beaucour and Kastner model is calculated based upon the vertical soil stress σ_{EV} founded on the active trap-door experiment (The Ministry of Construction of China 2002, German ATV rules and standards 1990, ASTM International 2011, British Standards 2009), the empirical parameters in various standards are listed in Table 2.

2.2 Research methodology

A simple approach to characterise the tunnel bore conditions is detailed as follows: (1) establish the baseline of jacking forces which is constituted by the minima of jacking forces, (2) calculate the normal contact pressure based upon the given soil properties, pipe dimensions, and buried

Table 1. Review of jacking force models.

Model	JMTA	Ma Baosong	Shimada and Matsui	Pellet-Beaucour and Kastner
F_0	$F_0=10\times 1.32\pi\times D_e\times SPT\text{-}N$	$F_0=1/4\times\pi\times D_e^2\times[K_0\Sigma(\gamma_i'h_i)+\gamma_w h_w]$	$F_0=P_w\times A$ where P_w is slurry pressure and A is the area of tunnel face	$F_0=$initial jacking or first load
F_s	$F_s=\pi\times D_e\times\tau\times L +\omega\times\mu\times L$	$F_s=1.2\times[\mu\times(2P_V+2P_H +P_B)]$ where $P_V=K_p\times\gamma'\times h\times D_e\times L$, $P_H=\gamma'\times(h+D_e/2)\times D_e\times L\times\tan^2(45\text{-}\phi'/2)$, and $P_B=\omega\times L$	$F_s=[p\times b\times\mu_1+P_w\times\mu_2\times(\pi D_2\text{-}b)]\times L$ where b (contact width)$=1.6\times(P_u\times k_d\times C_e)$ for which $P_u=(D_1^2\text{-}D_2^2)\times\gamma_c\times0.25\times\pi$, $k_d=(D_1\times D_2)/(D_1\text{-}D_2)$, and $C_e=(1\text{-}n_1^2)/E_1+(1\text{-}n_2^2)/E_2$	$F_s=\mu\times L\times D_e\times(\pi/2)\times[(\sigma_{EV}+\gamma D_e/2)+K_2(\sigma_{EV}+\gamma D_e/2)]$ where $K_2=$ thrust coefficient of soil arching
σ'	$\sigma'=\gamma'\times(D_e/2)/\tan\phi'+\omega/(\pi\times D_e)$	$\sigma'= K_p\times\gamma'\times h+\gamma'\times(h+D_e/2)\times\tan^2(45\text{-}\phi'/2)+\omega/(2\times D_e)$	$\sigma'=2P_u\times[1\text{-}(x^2/a^2)]^{0.5}/(\pi\times a)$ for which $a=b/2$ and $x=$distance to either side of centerline of the area of contact	$\sigma_{EV}=b\times(\gamma\text{-}2\times C/b)/(2\times K\times\tan\delta)\times(1\text{-}e^{-2\times K\times\tan\delta\times(h/b)})$ where $C=$cohesion

Table 2. Empirical parameters used in each standard.

Empirical parameters	Initial Terzaghi's definition	GB 50332	ATV A 161	ASTM F 1962	BS EN 1594
b (Silo width): m	$D_e\times[1+2\tan(45°\text{-}\phi/2)]$	$D_e\times[1+\tan(45°\text{-}\phi/2)]$	$D_e\times3^{0.5}$	$1.5\times D_e$	$D_e\times[1+2\tan(45°\text{-}\phi/2)]$
δ (Angle of wall friction): deg.	ϕ	30	$\phi/2$	$\phi/2$	ϕ
K (soil pressure coefficient)	1	$\tan^2(45°\text{-}\phi/2)$	0.5	$\tan^2(45°\text{-}\phi/2)$	$1\text{-}\sin\phi$

1928

depth of pipeline, (3) extract the measured frictional stress of each baseline section, (4) back-analyse the frictional coefficient μ_{avg} of each baseline section, and (5) determine the tunnel bore conditions by comparing the μ_{avg} value to the suggested value for lubricated drives.

3 PIPEJACKING PROJECT

3.1 *Background*

The slurry shield was adopted to perform the tunnel excavation of four pipejacking drives in soft alluvial deposits located in Shulin district in Taipei County, Taiwan. During pipejacking works, the excavation face was kept stable by bentonite slurry with unit weight of 10.8 kN/m^3. This bentonite slurry also transported tunnelling spoils to decantation chambers which separate coarse particles from the tunnelling spoils through a slurry circulation system. Since the tunnels were excavated using the 1500-mm diameter cutting wheel, an overcut annulus of 30 mm was formed by introducing the smaller concrete pipe of 1440 mm in diameter. The four drives characteristics are detailed in Table 3.

3.2 *Engineering geology*

The stratigraphy is established by five 15-m deep geological boreholes arranged along the tunnel alignment. Above the poorly-graded to well graded sand and gravel layer at approximately 6 m from the ground surface, backfill and silty sand layers are deposited.

Table 3. Characteristics of four pipejacking drives.

Pipejacking drive	A	B	C	D
Length: m	73	126	75	102
Depth: m	10.3	10.3	10.8	10.8
Alignment	Straight	Straight	Straight	Straight
Phreatic surface: m	4.5	4.5	4.5	4.5
Cutter wheel diameter: m	1.5	1.5	1.5	1.5
Pipe diameter D_e: m	1.44	1.44	1.44	1.44
Soil cover h: m	9.6	9.6	10.1	10.1
h/D_e	6.7	6.7	7	7
Measured initial face resistance: kPa	388	499	555	555
Theoretical overcut annulus: litre/m	138	138	138	138
Avg. volume of injected lubricant: litre/m	378	381	552	534
Lubricant nature	Bentonite slurry (2% polymer)	Bentonite slurry (2% polymer)	Bentonite slurry (2% polymer)	Bentonite slurry (2% polymer)
Soil nature based upon baseline section	Section 1: gravel (1–5 m); Section 2: clayey gravel or clay (5–67 m); Section 3: gravel (67–70 m)	Section 1: gravel (4–29 m); Section 2: gravel (29–45 m); Section 3: clayey sand (45–114 m); Section 4: gravel (114–123 m)	Section 1: gravel (2–8 m); Section 2: gravel (8–21 m); Section 3: gravel (21–40 m); Section 4: clayey gravel (40–75 m)	Section 1: gravel (6–11 m); Section 2: clayey gravel (11–24 m); Section 3: clayey gravel (24–31 m); Section 4: clayey gravel (31–102 m)

4 ANALYSIS AND DISCUSSIONS

4.1 Development of baseline of jacking forces

In Drive A, the slurry shield traversed 73 m including an initial 5 m section of gravel, a 62 m section of fine soil governed gravel or sand deposit and a further 3 m section of gravel, and a final 3 m section of clay at a depth of 10.3 m below the surface, as shown in Figure 1. The minima of jacking forces lead to three baseline sections; that are, the initial 1–5 m section, 5–67 m section, and final 67–70 m section, with the avg. jacking forces f_{avg} of 36.8, 1.6, and 32.7 kN/m, respectively. The avg. jacking forces f_{avg} divided by the circumference of pipe yield the frictional stresses τ of 8.1, 0.4, and 7.2 kPa, respectively. Drive B spanning over a 126-m long jacking distance was undertaken in an alternating layer of gravel formation and clayey gravel or sand deposit at a depth of 10.3 m, as shown in Figure 2. The baseline due to the minima of jacking forces is constituted by four sections; that are, the first 4–29 m section, 29–45 m and 45–114 m sections, and final 114–123 m section. Their avg. jacking forces f_{avg} measured at 0.4, 3.1, 2.1, and 5.4 kN/m, respectively, resulting in the frictional stresses τ of 0.09, 0.7, 0.5, and 1.2 kPa, respectively.

During the pipejacking of Drive C, the shield spanned over a 75-m distance within an alternating layer of gravel formation (1–34 m distance) and clayey gravel or sand deposit (34–75 m distance) at a depth of 10.8 m, as shown in Figure 3. The minima of jacking forces were used

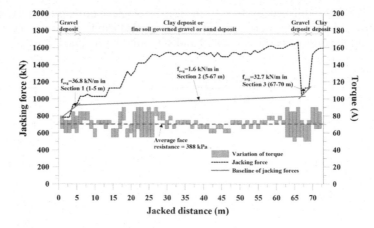

Figure 1.   Pipejacking activities at Drive A.

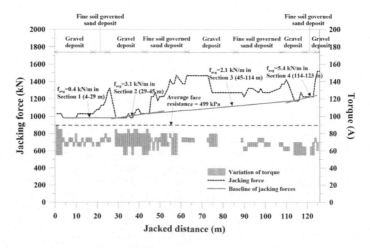

Figure 2.   Pipejacking activities at Drive B.

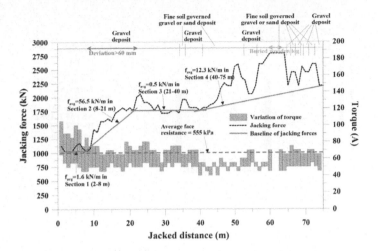

Figure 3.   Pipejacking activities at Drive C.

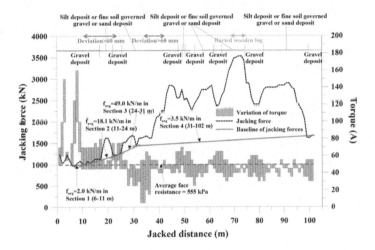

Figure 4.   Pipejacking activities at Drive D.

to establish the baseline of jacking forces consisted of the first 2–8 m section, 8–21 m and 21–40 m sections, and final 40–75 m section. Their avg. jacking forces f_{avg} measured at 1.6, 56.5, 0.5, and 12.3 kN/m, respectively. While their frictional stresses τ were calculated to be at 0.4, 12.5, 0.1, and 2.7 kPa, respectively. The pipejacking of Drive D traversed a 102-m distance from an alternating layer of gravel and sand (1–17 m distance) through a successive alternating layer of clayey gravel and silty sand (17–99 m distance) into a final layer of gravel (99–102 m distance) at a depth of 10.8 m, as shown in Figure 4. The pipejacking results determined the constitutes of the baseline of jacking forces, which corresponded to the initial 5 m section (from 6 to 11 m), 11–24 m and 24–31 m sections, and final 31–102 m section. Their avg. jacking forces f_{avg} measured at 2.0, 18.1, 49.0, and 3.5 kN/m, respectively. The associated frictional stresses τ were calculated to be at 0.4, 4.0, 10.8, and 0.8 kPa, respectively.

4.2   Calculation of normal contact pressure

One single stratum has been assumed to calculate the normal contact pressure σ' for simplicity. The σ' value at the four drives can be calculated, respectively, through the associated equation in the models by substituting the parameters listed in Table 4. The maximum and minimum of

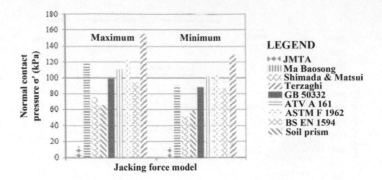

Figure 5.   Comparison of the calculated normal contact pressure σ' between the jacking force models.

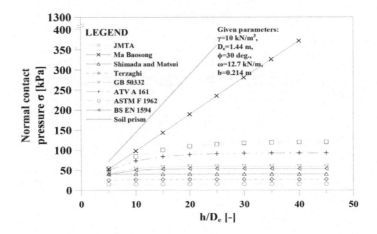

Figure 6.   Variation of the normal contact pressure σ' against h/D$_e$ ratio.

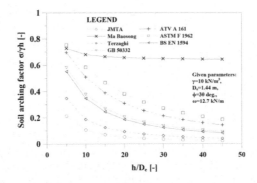

Figure 7.   Variation of the soil arching factor σ'/γh against h/D$_e$ ratio.

the σ' values in each model are depicted in Figure 5, together with the upper bound value resulting from the soil prism pressure acting upon the pipe crown. As can be seen, the σ' values varying from 13.8 to 14.8 kPa in JMTA were the minima amongst all the models, while in ASTM F 1962, the σ' values varying from 101.7 to 122.2 kPa were the maxima. Figure 6 gives a calculation sample for further clarification. It can be seen from Figure 6 that the σ' value in ASTM F 1962, ATV A 161, GB 50332, and BS EN 1594 increased a bit at smaller h/D$_e$ ratios and maintained at a constant value at larger h/D$_e$ ratios. The σ' value in Ma Baosong was linearly increased with the increasing h/D$_e$ ratio. The σ' value in Terzaghi and JMTA as well as Shimada

Table 4. Parameters input in the jacking force models.

JMTA	ϕ' (deg.)	c' (kPa)	γ (kN/m³)	D_e (m)
	34 (Sandy soil); 26.3 (Clayey soil)	0	10.09 (Sandy soil); 8.03 (Clayey soil)	1.44
Ma Baosong	K_p (kN/m³)	h (m)		ω (kN/m)
	0.3 (Sandy soil); 0.5 (Clayey soil)	9.6 (Drives A and B); 10.1 (Drives C and D)		12.7
Shimada and Matsui	$k_d=(D_1 \times D_2)/(D_1-D_2)$	D_1 (m)	D_2 (m)	E_2 (kPa)
	36	1.5	1.44	4e7
	E_1 (kPa)	n_1	n_2	b (m)
	22,373 (Sandy soil); 8,829 (Clayey soil)	0.35 (Sandy soil); 0.49 (Clayey soil)	0.2	0.21 (Sandy soil); 0.32 (Clayey soil)
	$C_e=(1-n_1^2)/E_1+(1-n_2^2)/E_2$ (kPa⁻¹)	x (m)		$p_{max}=2P_u \times [1-(x^2/a^2)]^{0.5}/(\pi \times a)$
	4e-5 (Sandy soil); 9e-5 (Clayey soil)	0		75.4 (Sandy soil); 50.9 (Clayey soil)
Pellet-Beaucour and Kastner	c' (kPa)	ϕ' (deg.)	γ (kN/m³)	h (m)
	0	34 (Sandy soil); 26.3 (Clayey soil)	10.09 (Sandy soil); 8.03 (Clayey soil)	9.6 (Drives A and B); 10.1 (Drives C and D)

and Matsui maintained at constant values of 26.9 and 15.3 as well as 40.4, respectively, in all the h/D_e ratios. The two constant values from Terzaghi and JMTA, respectively, are due to the fact that the soil arching factor in Terzaghi and JMTA converged sooner than the other models (see Figure 7) and the effect of soil arching thus quickly had a steady influence on the σ' value. It is evident that Shimada and Matsui assumed a stable tunnel bore and the calculation of σ' value was not linked to the h/D_e ratio, but linked to the values of P_u and b. It can also be seen from Figure 6 that as $h/D_e<7$, ASTM F 1962 was the maximum, while JMTA was the minimum. These results matched with those shown in Figure 5. As ASTM F 1962 made a more conservative assumption that reduces the mobilised shearing resistance along shearing bands by taking only half the friction angle into account, it thus gave the highest estimation. On the other hand, JMTA calculated the σ' value by introducing the reduced weight of soil prism that takes the overburden height equal to half the outer pipe diameter into account and considered fully developed shearing bands. The aforesaid two assumptions are considered as the two key factors producing the lowest estimation.

4.3 Back-analysis of frictional coefficient

The friction from the difference in the jacking force at both ends of a baseline section divided by the section length leads to the average jacking force f_{avg}. The f_{avg} value divided by the circumference of pipe results in the frictional stress τ. Thus, the frictional coefficient μ_{avg} can be back-analysed by introducing the Mohr-Coulomb equation. The f_{avg}, τ and σ' values are listed in Table 5. The back-analysis results are listed in Table 6. In the first and final sections at Drive A, the μ_{avg} values in JMTA were back-analysed to be 0.59 and 0.52, respectively, largely in excess of the upper limit of 0.3 suggested by Stein et al. (1989). In contrast with JMTA, the others reported the μ_{avg} values smaller than the lower limit of 0.1. In fact, the higher μ_{avg} values of 0.59 and 0.52 were caused by the increasing pipe friction resulting from ground closure.

Table 5. Summary of the calculated f_{avg}, τ and σ' values for the four pipejacking drives.

Drive	A	B	C	D
f_{avg} (kN/m)	1: 36.8; 2: 1.6; 3: 32.7	1: 0.4; 2: 3.1; 3: 2.1; 4: 5.4	1: 1.6; 2: 56.5; 3: 0.5; 4: 12.3	1: 2.0; 2: 18.1; 3: 49.0; 4: 3.5
τ (kPa)	1: 8.1; 2: 0.4; 3: 7.2	1: 0.09; 2: 0.7; 3: 0.5; 4: 1.2	1: 0.4; 2: 12.5; 3: 0.1; 4: 2.7	1: 0.4; 2: 4.0; 3: 10.8; 4: 0.8
σ' from JMTA (kPa)	1: 13.8; 2: 14.8; 3: 13.8	1: 13.8; 2: 13.8; 3: 14.8; 4: 13.8	1: 13.8; 2: 13.8; 3: 13.8; 4: 14.8	1: 13.8; 2: 14.8; 3: 14.8; 4: 14.8
σ' from Ma Baosong (kPa)	1: 89.8; 2: 115.8; 3: 89.8	1: 89.8; 2: 89.8; 3: 115.8; 4: 89.8	1: 92.8; 2: 92.8; 3: 92.8; 4: 119.4	1: 92.8; 2: 119.4; 3: 119.4; 4: 119.4
σ' from Shimata and Matsui (kPa)	1: 75.4; 2: 50.9; 3: 75.4	1: 75.4; 2: 75.4; 3: 50.9; 4: 75.4	1: 75.4; 2: 75.4; 3: 75.4; 4: 50.9	1: 75.4; 2: 50.9; 3: 50.9; 4: 50.9
σ' from Terzaghi (kPa)	1: 58.9; 2: 65.3; 3: 58.9	1: 58.9; 2: 58.9; 3: 65.3; 4: 58.9	1: 59.4; 2: 59.4; 3: 59.4; 4: 66.1	1: 59.4; 2: 66.1; 3: 66.1; 4: 66.1
σ' from GB 50332 (kPa)	1: 97.7; 2: 86.0; 3: 97.7	1: 97.7; 2: 97.7; 3: 86.0; 4: 97.7	1: 99.8; 2: 99.8; 3: 99.8; 4: 87.7	1: 99.8; 2: 87.7; 3: 87.7; 4: 87.7
σ' from ATV A 161 (kPa)	1: 108.4; 2: 99.5; 3: 108.4	1: 108.4; 2: 108.4; 3: 99.5; 4: 108.4	1: 111.1; 2: 111.1; 3: 111.1; 4: 102.0	1: 111.1; 2: 102.0; 3: 102.0; 4: 102.0
σ' from ASTM F 1962 (kPa)	1: 118.9; 2: 101.7; 3: 118.9	1: 118.9; 2: 118.9; 3: 101.7; 4: 118.9	1: 122.2; 2: 122.2; 3: 122.2; 4: 104.3	1: 122.2; 2: 104.3; 3: 104.3; 4: 104.3
σ' from BS EN 1594 (kPa)	1: 92.4; 2: 84.6; 3: 92.4	1: 92.4; 2: 92.4; 3: 84.6; 4: 92.4	1: 94.2; 2: 94.2; 3: 94.2; 4: 86.3	1: 94.2; 2: 86.3; 3: 86.3; 4: 86.3

Note: 1, 2, 3, and 4 represent the numbering of baseline section.

At Drive B, the μ_{avg} values derived from all the models were far below the lower limit of 0.1, indicating that Drive B was very well-lubricated. In the second section at Drive C, the μ_{avg} value in JMTA was 0.9, largely in excess of the upper limit of 0.3, suggesting that it was not adequately lubricated. The μ_{avg} values from the other models were within the limits of 0.1 and 0.3 recommended by Stein et al. (1989) for lubricated drives. This highest μ_{avg} value (0.9) was most likely linked to the increasing pipe friction resulting from the excessive pipe deviation greater than a threshold value of 60 mm. While the other models resulted in a misleading interference for estimation of the μ_{avg} value.

In the second section of clayey gravel at Drive D, the μ_{avg} value of 0.3 in JMTA matched with the upper limit of 0.3, indicating that it was moderately lubricated. Whereas the other models reported the μ_{avg} values smaller than 0.1. The μ_{avg} value of 0.3 was attributed to the increasing pipe friction resulting from the excessive pipe deviation greater than 60 mm. Similar

Table 6. Summary of the back-analysed μ_{avg} values for the four pipejacking drives.

Drive	A	B	C	D
μ recommended by Stein et al. (1989):	1: 0.3-0.4; 2: 0.2-0.3; 3: 0.3-0.4	1: 0.3-0.4; 2: 0.3-0.4; 3: 0.2-0.3; 4: 0.3-0.4	1: 0.3-0.4; 2: 0.3-0.4; 3: 0.3-0.4; 4: 0.2-0.3	1: 0.3-0.4; 2: 0.2-0.3; 3: 0.2-0.3; 4: 0.2-0.3
Back-analysed μ_{avg} by JMTA	1: 0.59; 2: 0.03; 3: 0.52	1: 0.007; 2: 0.05; 3: 0.03; 4: 0.09	1: 0.03; 2: 0.9; 3: 0.007; 4: 0.18	1: 0.03; 2: 0.30; 3: 0.73; 4: 0.05
Back-analysed μ_{avg} by Ma Baosong	1: 0.09; 2: 0.003; 3: 0.08	1: 0.001; 2: 0.008; 3: 0.004; 4: 0.01	1: 0.004; 2: 0.14; 3: 0.001; 4: 0.02	1: 0.004; 2: 0.03; 3: 0.09; 4: 0.007
Back-analysed μ_{avg} by Shimada and Matsui	1: 0.11; 2: 0.008; 3: 0.10	1: 0.001; 2: 0.009; 3: 0.01; 4: 0.02	1: 0.005; 2: 0.17; 3: 0.001; 4: 0.05	1: 0.005; 2: 0.08; 3: 0.21; 4: 0.02
Back-analysed μ_{avg} by Terzaghi	1: 0.14; 2: 0.006; 3: 0.12	1: 0.002; 2: 0.01; 3: 0.008; 4: 0.02	1: 0.007; 2: 0.21; 3: 0.002; 4: 0.04	1: 0.007; 2: 0.06; 3: 0.16; 4: 0.01
Back-analysed μ_{avg} by GB 50332	1: 0.08; 2: 0.005; 3: 0.07;	1: 0.001; 2: 0.007; 3: 0.006; 4: 0.01	1: 0.004; 2: 0.13; 3: 0.001; 4: 0.03	1: 0.004; 2: 0.05; 3: 0.12; 4: 0.009
Back-analysed μ_{avg} by ATV A 161	1: 0.08; 2: 0.004; 3: 0.07	1: 0.001; 2: 0.006; 3: 0.005; 4: 0.01	1: 0.004; 2: 0.11; 3: 0.001; 4: 0.03	1: 0.004; 2: 0.04; 3: 0.11; 4: 0.008
Back-analysed μ_{avg} by ASTM F 1962	1: 0.07; 2: 0.004; 3: 0.06	1: 0.001; 2: 0.006; 3: 0.005; 4: 0.01	1: 0.003; 2: 0.10; 3: 0.001; 4: 0.03	1: 0.003; 2: 0.04; 3: 0.10; 4: 0.008
Back-analysed μ_{avg} by BS EN 1594	1: 0.09; 2: 0.005; 3: 0.08	1: 0.001; 2: 0.008; 3: 0.006; 4: 0.01	1: 0.004; 2: 0.13; 3: 0.001; 4: 0.03	1: 0.004; 2: 0.05; 3: 0.13; 4: 0.009
Tunnel bore condition deduced by JMTA	1: Ground closure; 2: Stable tunnel bore; 3: Ground closure	1: Stable tunnel bore; 2: Stable tunnel bore; 3: Stable tunnel bore; 4: Stable tunnel bore	1: Stable tunnel bore; 2: Possible ground closure; 3: Stable tunnel bore; 4: Ground closure	1: Stable tunnel bore; 2: Possible ground closure; 3: Not applicable; 4: Stable tunnel bore

to the other drives, JMTA reported the μ_{avg} values more reliable than the other models. Additionally, the third section at Drive D indicated inadequate lubrication, with the μ_{avg} value of 0.73 (the second highest μ_{avg} value observed in this study). The possible reason to lead to this second highest μ_{avg} value was not due to unfavorable pipe friction, but due to the gravel formation at 32 m not being long enough to establish lower face resistance or total jacking load.

5 CONCLUSIONS

(1) JMTA provided the lowest estimation of the normal contact pressure σ by introducing the reduced overburden height and assuming fully developed shearing bands. As $h/D_e < 7$, ASTM F 1962 provided the highest estimation as it had a more cautious assumption that reduces the mobilised shearing resistance along the sliding places by considering only half the friction angle.

(2) In JMTA, the μ_{avg} value from the first and final sections of gravel at Drive A was 0.59 and 0.52, respectively, while the other models reported the μ_{avg} values smaller than 0.1. The higher μ_{avg} values were most likely because of the increasing pipe friction caused by ground closure.

(3) In JMTA, the μ_{avg} values of 0.9 and 0.3, respectively, incurred while pipejacking through the second section of gravel at Drive C and the second section of clayey gravel at Drive D were linked to the increasing pipe friction resulting from the excessive pipe deviation greater than the threshold value of 60 mm. While the other models reported the μ_{avg} values within the 0.1–0.3 range and smaller than 0.1, respectively.

(4) The pipejacking through the third section of clayey gravel at Drive D indicated inadequate lubrication, with the μ_{avg} value being equal to 0.73. The higher μ_{avg} value was not due to unfavourable pipe friction, but due to the gravel formation at 32 m distance not being long enough to develop lower face resistance or total jacking load.

REFERENCES

ASTM International. 2011. *Standard Guide for Use of Maxi-Horizontal Directional Drilling for Placement of Polyethylene Pipe or Conduit under Obstacles Including River Crossings* (ASTM F 1962-11). West Conshohocken. PA.

Barla, M., Camusso, M., Aiassa, S. 2006. Analysis of jacking forces during microtunnelling in limestone. *Tunnelling and Underground Space Technology* 21(6): 668–683.

British standards. 2009. *Gas supply system-pipelines for maximum operating pressure over 16 bar-functional requirements* (BS EN:1594-09): 76–78. Brussels.

Chapman, D.N., Ichioka, Y. 1999. Prediction of jacking forces for microtunnelling operations. *Tunnelling and Underground Space Technology* 14(1): 31–41.

Cheng, W.C., Ni, J.C., Arulrajah, A., Huang, H.W. 2018b. A simple approach for characterising tunnel bore conditions based upon pipe-jacking data. *Tunnelling and Underground Space Technology* 71: 494–504.

Cheng, W.C., Ni, J.C., Cheng, Y.H. 2017c. Alternative shoring for mitigation of pier foundation excavation disturbance to existing freeway. *Journal of Performance of Constructed Facilities* 31(5): 04017072.

Cheng, W.C., Ni, J.C., Huang, H.W., Arulrajah, A. 2018c. The use of tunnelling parameters and spoil characteristics to assess soil types: a case study from alluvial deposits at a pipejacking project site. *Bulletin of Engineering Geology and the Environment* doi: 10.1007/s10064-018-1288-4.

Cheng, W.C., Ni, J.C., Shen, S.L. 2017a. Experimental and analytical modeling of shield segment under cyclic loading. *International Journal of Geomechanics* 17(6): 04016146.

Cheng, W.C., Ni, J.C., Shen, S.L., Huang, H.W. 2017b. Investigation into factors affecting jacking force: a case study. *ICE – Geotechnical Engineering* 170(4): 322–334.

Cheng, W.C., Ni, J.C., Shen, S.L., Wang, Z.F. 2018a. Modeling of permeation and fracturing grouting in sand: laboratory investigations. *Journal of Testing and Evaluation* 46(5): 2067–2082.

Cheng, W.C., Song, Z.P., Tian, W., Wang, Z.F. 2018d. Shield tunnel uplift and deformation characterisation: a case study from Zhangzhou metro. *Tunnelling and Underground Space Technology* 79: 83–95.

German ATV rules and standards. 1990. *Structural calculation of driven pipes* (ATV-A 161 E-90): 18–20. Hennef.

Japan Microtunnelling Association. 2000. *Pipe-jacking Application*. Tokyo.

Ma, B. 2008. *The Science of Trenchless Engineering*. Beijing: China Communications Press.

Pellet-Beaucour, A.L., Kastner, R. 2002. Experimental and analytical study of friction forces during microtunneling operations. *Tunnelling and Underground Space Technology* 17(1): 83–97.

Shen, S.L., Cui, Q.L., Ho, C.E., Xu, Y.S. 2016. Ground response to multiple parallel microtunneling operations in cemented silty clay and sand. *Journal of Geotechnical and Geoenvironmental Engineering* 142(5): 04016001.

Shimada, H., Matsui, K. 1998. A new method for the construction of small diameter tunnels using pipe jacking. *Proc. of Regional Symposium on Sedimentary Rock Engineering*: 234–239.

Stein, D., Möllers, K., Bielecki, R. 1989. *Microtunneling: Installation and Renewal of Nonman-Size Supply and Sewage Lines by the Trenchless Construction Method*. Berlin: Ernst.

The Ministry of Construction of China. 2002. Structural design code for pipeline of water supply and waste water engineering (GB 50332-02): 11–12. Beijing.

Wang, Z.F., Cheng, W.C., Wang, Y.Q. 2018. Simple method to predict settlement of composite foundation under embankment. *International Journal of Geomechanics* in press.

Tunnels and Underground Cities: Engineering and Innovation meet Archaeology, Architecture and Art, Volume 5: Innovation in underground engineering, materials and equipment - Part 1 – Peila, Viggiani & Celestino (Eds)
© 2020 Taylor & Francis Group, London, ISBN 978-0-367-46870-5

Conditioning the clog: Advancing EPB technology through mixed transitional ground

E. Comis, W. Gyorgak & P. Raleigh
McMillen Jacobs Associates, Mayfield Heights, Ohio, USA

ABSTRACT: Earth Pressure Balance (EPB) Tunnel Boring Machines (TBMs) have a difficult time excavating through "mixed transitional ground". Soil conditioning plays a critical role in achieving the desired qualities of the excavated muck within the excavation chamber and the delicate balance between clogging and maintaining face support. Soil conditioning laboratory tests are often completed during the planning phase prior to TBM excavation in an effort to develop a baseline for soil conditioning, reduce risks, costs and improve EPB performance. Often these tests show considerable promise in the lab but don't deliver even where the excavated materials are relatively homogeneous and with even less success in so-called mixed transitional ground. Very slow advance rates, high torque and thrust and clogging of the cutterhead openings have been experienced on a number of recent projects as a result of varying amounts of clay in the face. This paper will present the observations made during the excavation of the Ohio Canal Interceptor Tunnel in Akron, Ohio, USA particularly with regards to very slow advanced rates, rapid tool wear and clogging of the cutterhead. It also discusses the limits of desktop versus empirical clogging evaluations with respect to actual clogging experienced during tunneling and mitigation measures that have been introduced during the excavation which helped decrease wear and increase advance rates.

1 INTRODUCTION

Because of alignment restrictions and increase in demand for tunnels with larger diameters, more and more tunnel boring machines are working in mixed face conditions. As mentioned by several authors and later summarized by Oliveira and Diederichs (2016), mixed face conditions still present a challenging scenario for TBMs which are summarized by the following:

1. Need to define the location of the transition zone prior to excavation,
2. Mitigate the impact loads on the cutterhead tool mounts and tools when passing from softer to harder materials;
3. Optimize muck characteristics to provide an effective support medium which can be readily conveyed by the muck handling systems
4. Adaptable to the addition of an increasing volume of rock chips together with excavated soil, below the water table which may lead to a highly permeable support medium, not suitable for an EPBM excavation;
5. Clogging due to clay present in the excavated material, or due to the fines in addition to high water content added in an effort to condition the soils or a combination of both.

Langmaak and Lee (2016) emphasize that a successful TBM drive in difficult ground conditions can only be obtained by combining:

• Adequate mechanical solutions (cutters, cutterhead opening ratio, mixing arms, screw conveyor)
• Suitable ground conditioners (chemical surfactants, polymers, water, etc.)

• Experienced Personnel (ground conditioning engineers, TBM operators, etc)

In EPB tunneling, the correct choice and use of soil conditioners can make a considerable difference for the success of a tunneling project – both in highly permeable grounds as well as in sticky clays.

To date, there is no standard procedure to evaluate clogging in either the field or laboratory, however stickiness and clogging potential has been well documented and investigated by Thewes et al. While this research is not new, contractors continue to struggle in the field despite having adapting several countermeasures which include: the use of water to reduce stickiness, reduction in advancement rate, use of compressed air to provide face support, design of the TBM cutterhead to avoid creating large lumps of soil, and of course through the use of conditioners and additives.

The Ohio Canal Interceptor Tunnel (OCIT) Project involved the construction of a 6,200 ft long conveyance and storage tunnel with a finished inside diameter of 27 ft (8.23m) to control combined sewer overflows for several regulators in the downtown Akron area. The OCIT Project was awarded to Kenny/Obayashi, a Joint Venture with a Notice to Proceed dated November 4, 2015.

The section of the GBR related to the OCIT identified three major reaches that were defined as distinctly different ground conditions: Reach 1 which primarily consisted of soft ground sandy soils, Reach 2 was a transitionary zone with soft ground overlying bedrock, characterized as "mixed ground conditions", and Reach 3 which was comprised of bedrock with two sections of low rock cover. The tunnel was excavated using a 30 ft (9.26m) bore dual mode type "Crossover" (XRE) Rock/EPB Tunnel Boring Machine (TBM), manufactured and supplied by The Robbins Company. The TBM design features adopted for the expected operating conditions are fully discussed in the paper, *Comis, E. et al, 2017*. As a combined rock/EPB TBM the machine could operate in Closed Mode (EPB) or Open Mode.

The working definition for Closed Mode is operation of the TBM with active face support above the ambient pressure and Open Mode is operation with some or no active face support below ambient pressure. The choice between the two modes of operation is dependent upon the stability of the excavation face, hence in granular materials below the ground water table Closed Mode was specified (Reach 1 and 2) and in the rock (Reach 3) open mode was permitted except in the Low Bedrock Cover Zones.

The TBM launched in October 2017 and completed the drive in August 2018. Reach 1 and 2, were a total of 810 ft (247 m) and presented a challenge for the contractor as these reaches were plagued by low penetration rate, high cutter consumption, difficulties in maintaining face pressure and low TBM utilization. Reach 3, was a total of 5,400 ft (1,647 m), and in contrast, distinguished by high TBM utilization and consistent advance rates.

This paper will discuss the soil conditioning tests that were carried out prior the beginning of the mining and the problems observed during Reach 1 and 2, mainly related to the "mixed transitional ground" or what is better described as a soil to rock transition zone.

2 SOIL CONDITIONING ASSESSMENT

As excavated materials enter the excavation chamber through the cutterhead they need to be rapidly conditioned from relatively stable in-bank material to an excavated material which is consistent with the EPB method including the following characteristics:

• Reasonable fluidity to pass through the muck handling system beginning at the cutterhead openings
• A level of homogeneity whereby water may not easily pipe or pass through open pores
• Sufficient density so that support pressure may be imparted to the excavation face

These characteristics of excavated soil are critical to maintaining active face support pressure and the metered discharge of excavated soil at a rate equal to the advancement of the TBM.

In the case of OCIT, a soil conditioning assessment was conducted by the foaming agent supplier prior to the commencement of tunneling to identify the baseline conditioners necessary to:

- Enhance "plastic flow" condition of the excavated materials
- Reduce permeability and promote homogeneity of the excavated materials
- Prevent adhesion of excavated materials to the cutterhead structure and openings.

In EPB tunneling, the correct approach to soil conditioning can make a considerable difference for the success of a tunneling project – both in highly permeable grounds as well as in sticky clays. Soil conditioning is determined by the parameters defined in EFNARC (2005), as follows:

- Foam Expansion Ratio (FER): is the ratio between the volume of foam at working pressure and the volume of the solution. The range of FER is typically between 5 and 30.
- Foam Injection Ratio (FIR): is the ratio between the injected volume of foam at working pressure and the in-bank volume of material to be excavated. The range of FIR is typically between 10% and 80%.
- Concentration of agent (C_f): is the volume of agent used as a percentage of water in the preparation of the conditioning mixture. The range of C_f is typically between 0.5% and 5%.

The anticipated ground conditions for the OCIT were the following:

1. Reach 1 consisted primarily of Unit 3 Silty Sand (brown, silty fine sand with trace amounts of clay) interbedded with Unit 4 Silt (brown silt or clayey silt) underlain by Glacial Till (brown to gray clay with varying amounts of gravel, cobbles, and boulders)
2. Reach 2 ranged from a full face of Units 3, 4, and weathered rock to a full face of bedrock
3. Reach 3 consisted of Shale and Siltstone, with minor amounts of Sandstone.

A total of six samples were tested for particle size distribution and moisture content. The soils were then grouped into either sands or silts. All but one contained larger gravels and cobbles. Silty soils revealed a higher moisture content (18.2% to 21.6%) than the sandy soils (5.3% to 8.6%).

Soil conditioning testing varied according to the soil type. For the sandy soils, testing for mobility was made using a tradition slump test method. For the silty soils, testing included adhesion (using a variation of the Langmaack adhesion test (NAT 2000, Langmaack)),

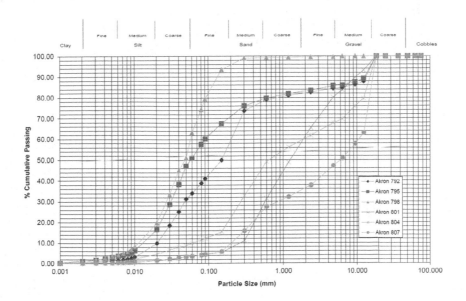

Figure 1.   Samples Particle Size Distribution.

slump, static flow, dynamic flow and rheometry. There were no tests carried out for the mixed ground conditions or for the shale.

Based on these tests the recommendation from the supplier was the following:

1. Soft Ground Closed Mode (Reach 1): A polymer-based foaming agent. If more than an expected portion of clay was found in the excavated soil, an anti-clay polymer could be added.
2. Mixed Face Closed Mode (Reach 2): The same polymer-based foaming agent was considered suitable for the mixed face ground with the change of ground conditions closely monitored and conditioners to be adjusted as follows;
   a. If the ratio of fine particles in the ground increased, water could be injected into the chamber to increase fluidity and to adjust viscosity of the excavated materials.
   b. If ratio of fine particles in the ground decreased, foam, polymer or bentonite could be injected into the chamber to keep necessary fluidity and viscosity of the excavated materials. Liquid type polymers were also considered suitable for this purpose.

3. Rock (Reach 3) Open Mode: The final setup for Reach 2 ground conditions were considered sufficient for the bedrock zone. Additionally, the excavation chamber could be filled by a mix of bentonite and sand before entering the second Low Bedrock Cover Zone to obtain necessary earth pressure balance for Closed Mode operation. The mix could be determined with the actual soil conditions when the TBM entered the low cover zone.

The suggested soil conditioning parameters were FER 15, FIR varying from 30 to 80% depending on moisture content, and C_f of 3.0.

3 IMPACT OF MIXED TRANSITIONAL GROUND IMPACT ON TBM PERFORMACES

The OCIT TBM weekly production for the whole drive is shown in Figure 2. An average weekly advance rate of 34.6 ft/week has been recorded for reach 1 and 2, and 242 ft/week for reach 3. Reach 1 and 2 were plagued by low penetration rate, high cutter consumption and difficulties in maintaining face pressure.

Lack of production (high thrust with low penetration rate) prompted several cutterhead inspections, which showed material packed in the rear of the housings as well as the buckets. Buckets were filled with clay and sand, easily removable with a pressure washer. Most of the outer radial buckets were hard packed and needed the use of pneumatic tools to remove hard packed material. Material was also found compacted in the disk cutter housings preventing rotation of the disk cutters and causing their failure due to flat-spotting. The cutterhead continued to clog developing high temperature in the plenum, baking the clay and sand material and steaming muck which was observed being conveyed outside the tunnel (Figure3).

The initial foam produced was of low-quality when checked at the foam generators on the TBM with a generally liquid consistency and no real "stand-up" time. The conditioner supplier ran multiple tests to detect if there was some defect with the product or the with mixing ratios being recommended and the results of these tests did not show a problem in the laboratory. Next, the foam generation medium in the generators was evaluated and modified from the initial setup of loosely packed larger metal cylinders to a tightly packed plastic top hat and then to a tightly packed steel wool. The result was a foam of good consistency and quality (Figure 4). Despite getting both the desired quality and quantity of foam, the cutterhead still experienced clogging.

Focus then shifted to the foam parameters and the adjustment of the various ratios. The initial soil conditioning evaluation suggested an FER of 15 with C_f between 3% and 5%. However even with a FIR that has been as high as 200-300% the clogging was not mitigated. The FER was then reduced to 2 and the FIR set between 60% and 80%. This finally allowed the completion of Reach 2 however at an unsatisfactorily slow pace, the penetration rate increased from 5mm/min to 15min/min suggesting that this approach to treatment was improving the

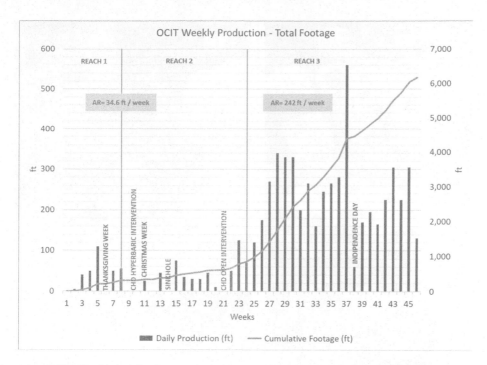

Figure 2.   OCIT Weekly Production.

Figure 3.   Clogging inside the TBM and steam in the tunnel.

characteristics of the excavated materials. However, the increase in penetration rate coincided with an increase in rock ratio at the excavation face (Figure 5) making that finding somewhat dubious.

Average Instantaneous Penetration Rate and Cutter Consumption are shown in Figure 6 for each single reach. Reach 1 and 2 required the TBM to be operated in Closed mode. During these reaches the penetration rate was very low to active maintain face pressure within the calculated limits. The extended push times increased the temperature of the plenum and affected the ground conditioning by baking the clogged material into the cutterhead buckets, creating an negative feedback effect.

Figure 4.   Foam Quality vs Generator Mediums.

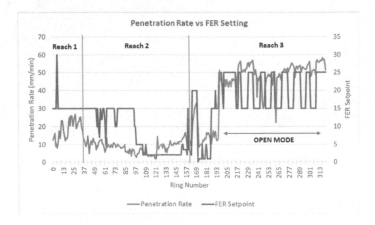

Figure 5.   Penetration Rate vs FER Setting.

Clogging resulted in high thrust and torque level as well, as shown in Figure 7. The observation was that the excavated material should have been better conditioned at the face, since water and bentonite were added in the plenum, they did not help in cleaning the cutterhead buckets or face openings. For a future large diameter project in similar

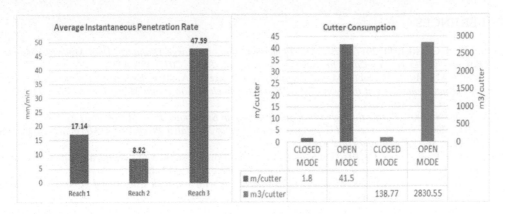

Figure 6.   Average Instantaneous Penetration Rate and Cutter Consumption.

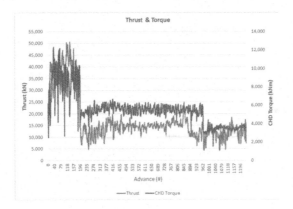

Figure 7.   Torque and Thrust for the whole drive.

conditions it is recommended to increase the number of foam outlets on the face (in this case 5) to minimum of 9.

4   CONCLUSIONS

The OCIT project involved excavation in dry sandy soil and transitional mixed ground. In these conditions the TBM experienced low penetration rates caused by the clogging of the cutterhead. The foaming agents selection and parameters suggested by the preliminary soil conditioning assessment didn't prove to be effective and the contractor was forced to experimental trials during the boring operation to keep the TBM advancing. To date, there is no standard procedure to evaluate clogging in the real operation conditions, however stickiness and clogging potential has been well documented and investigated by Thewes et al. While this research is not new, contractors continue to struggle in the field. The authors suggest including a more extensive stickiness and clogging potential analysis to the preliminary soil conditioning assessments, including conditioning of clayey rock conditions as shale.

REFERENCES

Comis, E. & Chastka, D. 2017. *Design and Implementation of a Large-Diameter, Dual-Mode "Crossover" TBM for the Akron Ohio Canal Interceptor Tunnel*; Proc. RETC 2017: 488–497.

Comulada, M., Maidl, U., Silva, M.A.P., Aguiar, G. & Ferreira, A. 2016. *Experiences gained in heterogeneous ground conditions at the twin tube EPB shield tunnels in São Paulo Metro Line 5*; Proc. ITA 2016 WTC, San Francisco: 1–11.

Della Valle, N. 2001. *Boring through a rock-soil interface in Singapore*; Proc. RETC 2001: 633–645.

EFNARC, 2005. *Specification and Guidelines for the Use of Specialist Products for Mechanized tunneling (TBM) in Soft Ground and Hard Rock*.

Langmaack, L. & Lee, K.F. 2016. Difficult ground conditions? Use the right chemicals! Chances–limits–requirements. *Tunnelling and Underground Space Technology 57*: 112–121.

Oliveira, D.G.G. & Diederichs, M. 2016. *TBM interaction with soil-rock transitional ground*; Proc. TAC 2016, Annual Conference, Ottawa: 1–8.

Thewes, M. & Burger, W. 2004. Clogging risks for TBM drives in clay. *Tunnels & Tunnelling International, June 2004*: 28–31.

Zhao, J., Gong, Q.M., & Eisensten, Z. 2007. Tunnelling through a frequently changing and mixed ground: a case history in Singapore. *Tunnelling and Underground Space Technology 22*: 388–400.

Tunnels and Underground Cities: Engineering and Innovation meet Archaeology,
Architecture and Art, Volume 5: Innovation in underground engineering,
materials and equipment - Part 1 – Peila, Viggiani & Celestino (Eds)
© 2020 Taylor & Francis Group, London, ISBN 978-0-367-46870-5

Maintenance costs for cutting tools in soft ground gained by process simulation

A. Conrads, A. Jodehl & M. Thewes
Institute for Tunnelling and Construction Management, Ruhr-University Bochum, Germany

M. Scheffer & M. König
Chair of Computing in Engineering, Ruhr-University Bochum, Germany

ABSTRACT: In shield tunnelling, the maintenance of cutting tools is an expensive and time-consuming process, where the tools are inaccessible during the advance process. A robust planning of the maintenance processes is necessary to get a predictable amount of downtime and maintenance costs. Even though there are first quantitative wear prediction models, the in-situ wear behaviour is influenced by a great number of uncertain input factors. These uncertainties must be considered for the evaluation and comparison of different maintenance strategies. This contribution shows the implementation of an empirical wear prediction model for hydro-shield projects into a process simulation framework. By using process simulation methods, the number of replaced tools and accesses into the excavation chamber and the time needed for each maintenance stop becomes predictable. Thus, the material as well as time-dependent costs are estimated considering the uncertainties of the input parameters.

1 INTRODUCTION

In mechanized tunnelling, the maintenance of the cutting tools has a great influence on the success of a tunnelling project. Insufficient maintenance may lead to sever wear and damages of the tunnelling machine, resulting in a long downtime and a significant increase of the project costs. Especially in soft ground, where a support of the tunnel face is always necessary, an efficient scheduling of the maintenance stops is mandatory. However, holistic monitoring concepts for tool condition during the advance process still do not exist. While several prediction models for the wear of cutting tools in hard rock are published in the literature and already applied in the industry (e.g. (Gehring 1995) or (Bruland 2000)), only few practicable approaches for soft ground wear prediction can be found.

First approaches are empirical wear prediction models, that are based on data of finished projects (Jakobsen 2014; Amoun et al. 2017; Köppl ct al. 2015a; Li et al. 2017). These models attempt to evaluate wear and thus maintenance intervals for either EPB or hydro-shield machines within certain boundary conditions. Using these models, the influence of the uncertain input parameters, especially for the soil properties, are taken into account as well in order to gain a robust maintenance strategy. Furthermore, evaluation criteria like the anticipated costs can be used to evaluate and compare different projects setups or maintenance strategies.

Earlier research of the authors shows, that by implementing an empirical wear prediction model into a process simulation model, uncertainties of the input parameters can be considered and evaluated using Monte Carlo Simulation and the robustness for the maintenance cost were analysed (Conrads et al. 2018). The following study is based on this approach.

2 BACKGROUND

Cutting tools are in direct contact with the tunnel face, thus are subjected to a continuous wear process. The wear process is highly related to prevailing ground conditions, e.g. abrasiveness of the soil, and steering parameters, e.g. face support pressure or penetration. Since it is hardly possible to monitor tool conditions, wear prediction models are used for maintenance planning. The prediction of wear is widely discussed, and several different approaches exist. For this research the wear prediction model proposed by (Köppl et al. 2015a) is applied, since it is enables a quantitative estimation of the wear of each cutting tool and is applicable for hydro-shield machines, which are in the focus of the research presented here. To consider uncertainties of the parameters and evaluate the chosen maintenance strategy, this method has been integrated into a process simulation model.

Unlike the maintenance of other machine parts, the production processes must be interrupted during the maintenance of cutting tools. This causes a noticeable amount of downtime. Nevertheless, the maintenance of cutting tools is indispensable for the operability of a TBM.

2.1 *Wear prediction model*

The quantitative approach of (Köppl et al. 2015a) is an empirical model to estimate the abrasive wear of cutting tools. It is based on soil properties and machine steering data from finished tunnelling projects (Köppl et al. 2015b). Using the soil abrasiveness index (SAI), which considers the equivalent quartz content eQu [%], the stresses using the shear strength of the soil τ_c [kN/mm²] and the grain size distribution of the ground using grain size D_{60} [mm], it is possible to define a maximum cutting path $s_{c,e(z)}$ [km] for different kind of tools.

$$SAI = \left(\frac{eQu}{100}\right)^2 * D_{60} * \tau_c \tag{1}$$

Cutting discs:

$$s_{c,e(z)} = 312.0 + \exp\left(-0.0048 * (SAI_{(z)} - 1398.2)\right) \tag{2}$$

Scrapers:

$$s_{c,e(z)} = 280.9 + \exp\left(-0.0050 * (SAI_{(z)} - 1300.7)\right) \tag{3}$$

During the advance process, each cutting tool follows a helix-shaped path. With the maximum cutting path, a corresponding maximum longitudinal length of the excavated tunnel section $L_{c(k)}$ [m] can be derived. The maximum longitudinal excavation length differs depending on the position of the tool on the cutting wheel. The overall maximum excavation length $L_{max(k)}$ of the whole machine is equal to the minimal excavation length of all tools. During a maintenance stops, tools, whose maximum cutting path do not exceed the next maintenance position $L_{d(k+1)}$, must be replaced preventively as shown in Figure 1.

To estimate the wear level of tools at a certain excavation position, (Köppl et al. 2015a) introduces the parameter $e_{cd,e(k)}$. It is defined by the ratio of the current driven cutting path to the maximum cutting path of the tool. The determination of the wear level simplifies the identification of tools, which will presumably reach their wear level before the next maintenance stop. Furthermore, it has to be used, if there is a change of the homogenous section of the soil before $L_{c(k)}$ is reached.

$$e_{cd,e(m)z} = \frac{s_{cd,e(k)}}{s_{c,e(z)}} \tag{4}$$

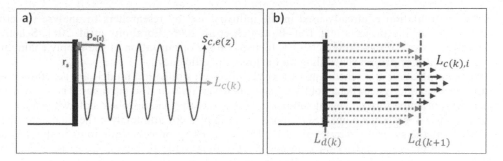

Figure 1. a) Helix shaped cutting path $s_{c,e(z)}$ and excavation length $L_{c(k)}$ of one cutting tool with the position given by the radius r_s and a penetration $p_{e(z)}$; b) Maximum longitudinal length of each cutting tool $L_{c(k),i}$. Cutting tools that have to be replaced preventively are marked in red (dotted line).

where $s_{cd,e(k)}$ = partial cutting path for a section k

2.2 Maintenance processes

Maintenance planning includes the determination of maintenance intervals, thus the maintenance positions. These intervals describe a certain length of tunnel excavation without a planned maintenance stoppage.

In this paper, maintenance planning for hydro-shield tunnelling projects are focused, where the excavation chamber is filled with a support medium to ensure the tunnel face stability. To perform the maintenance of the cutting tools, workers need to enter the excavation chamber, thus the support medium must be replaced by compressed air. Thus, the maintenance work is performed under pressurized conditions. This is riskier in terms of the stability of the tunnel face, endangerment of the workers and chance of a blow-out. Additionally, working time under pressurized condition is limited for reasons of work safety. Consequently, maintenance processes under pressurized conditions are more time- and cost-consuming (Thewes 2007). In unstable ground conditions, soil improvement measures like injections might even become necessary (Maidl et al. 2011). Therefore, maintenance stops in unstable areas are avoided as far as possible. To avoid maintenance stops under difficult conditions, predictive maintenance strategies are necessary.

The maintenance process duration can be roughly divided into three phases. During the mobilization phase, preparatory work for the tool change is carried out. The support medium is replaced, and the workers enter the excavation chamber through an air lock. Then the working platforms are installed, and the tools are cleaned. In the following replacement phase, the inspection of all tools as well as the replacement of the worn-out tools takes place. Loose bolts are tightened. The demobilization phase describes post-maintenance work processes, including demounting of working platform, decompression of workers and refilling the excavation chamber.

If tools exceed their critical wear level, there is a critical chance of sever damages of tool holders and even of the cutting wheel. This requires major reparation work (e.g. welding) and therefore leads to significant longer downtimes and respectively higher costs.

2.3 Process simulation

The evaluation of the system behaviour is facilitated using process simulation. Simulation models are abstracted simplification of a system to be analysed, in which the level of detail is significantly lower than in the systems they represent. However, the simplifications should not influence the fundamental system behaviour. Different system variants can be modelled, analysed and compared with each other. The simulation often supports the project planning and promotes a decision-making process. In the area of risk analysis, it is a comparably cost-effective means of assessing risks appropriately (AbouRizk 2010).

Process simulation is already used in the industry and by researchers to analyse the efficiency of tunnelling projects (e.g. (AL-Battaineh et al. 2006, Ebrahimy et al. 2011, Scheffer et al. 2016)). Most of these researches are focused on the optimization of the supply chain and on the influence of project scheduling on the logistical efficiency.

For the evaluation and scheduling of logistic processes, simulation models are generally used to analyse dynamic system behaviour with stochastic variables at discrete times. The models are often event-oriented (discrete-event simulation, DES), i.e. they change their state when an event occurs, or use a system-dynamic (SD) approach to implement and analyse the time dependent behaviour of the system. (März et al. 2011) However, using process simulation Mattern et al. (2016), Scheffer et al. (2016) and Conrads et al. (2017) made first approaches of evaluating maintenance strategies and the wear of cutting tools in soft ground.

3 SIMULATION MODEL FOR MAINTENANCE SCHEDULING

To evaluate and improve a maintenance schedule for hydro shield tunnelling projects, a simulation model was developed that includes the empirical wear prediction model of Köppl et al. (2015a) and enhanced by uncertain input parameters. Using a Monte Carlo Simulation approach, the variety of results gained from the deviation of the input parameters can be considered. Therefore, a more suitable estimation of the duration of the maintenance and the number of maintenance stops as well as the number of replaced tools can be gained. These values can be used to calculate time dependent and material costs for the maintenance processes of the whole project.

3.1 *Boundary conditions*

For the scheduling and evaluation of maintenance strategies, the boundary conditions of the regarded tunnelling project and the required input parameters are defined. By using the wear prediction model of Köppl et al. (2015a), the required parameter set is specified. The required parameters can be subdivided into soil parameters, machine design and steering parameters.

Especially soil and steering parameters are prone to uncertainties and should not be implemented into the model as deterministic values. Uncertainty is handled by distribution functions for these parameters. The distribution functions are gained by using back analysis methods and distribution fitting of documented data of finished projects. If the amount of data is not sufficient to gain a good correlation of the fitted distribution function, a distribution function can be assumed according to the expertise knowledge and with respect to the available data. Using a database of ground properties for similar soils, typical types of distribution functions can be found. For instance, Köppl et al. (2009) found that the uniaxial compressive strength of different types of rock has a skewed distribution to the left. Therefore, assuming a symmetrical distribution, e.g. normal distribution, would lead to wrong results.

The required machine design parameters are given by the design of the cutting wheel consisting of the amount, type and position of the cutting tools. This way, each cutting tool can be evaluated separately. Furthermore, time factors for the maintenance processes are required to calculate the duration of the maintenance work. Cost factors must be defined to determine the time-dependent and material costs. Together, they result in the total costs of the maintenance work. This value can later be used to evaluate the maintenance strategy.

The maintenance positions are defined beforehand and are implemented as fix values. However, it is not sufficient defining the maintenance positions only regarding the predicted wear of the cutting tools. Along the longitudinal section of the tunnel, there may be critical areas in which maintenance work is even more difficult to perform. A maintenance stop in an area with high ground water pressures, e.g. at positions below a river, or sensitive surface structure should be avoided if possible. In this case, the maintenance positions must be rescheduled. If the maintenance stop cannot be avoided, ground improvement measures become necessary, which lead to an increase in effort and costs.

3.2 Simulation model

The simulation model is implemented using the Java-based simulation framework AnyLogic, which has been proven useful for estimating the performance of tunnelling projects (Conrads et al. 2017; Duhme 2018; Rahm et al. 2016; Scheffer et al. 2016). The model bases on former research presented in Conrads et al. (2018). It is a simplification of the machine system focusing only on the production processes including wear and maintenance of the cutting tools. This way, the computing effort is reduced, while all necessary processes with an influence on the maintenance scheduling are still considered.

The model consists of four elements as shown in Figure 2 (left). The element *Project* contains all general information needed. The element *CuttingWheel* contains not only the different cutting tools, but also includes a state chart (Figure 2 right) representing processes of the machine. Thereby, *advance* and *ringbuild* are the main production processes. They are interrupted by *maintenance* or *repair*, which are performed after the ring building process is finished, and *technical failures,* that represents unplanned interruptions of the advancing processes. *repair* represents the corrective maintenance of the cutting tools, which becomes necessary, if the wear limit of the cutting tools is exceeded significantly. *starting* is a setup process that is necessary to build the model according to the input parameters.

The attributes of the cutting tools are the tool type and position on the cutting wheel. These attributes are needed for the calculation of the wear level of each tool at a certain point in time. Depending on the type of tool, several wear limits are defined. The first two limits are equal for each tool of the same type. The first limit is referred to as the normal wear limit, describing the amount of wear until either the wear protection is abraded, or the usability is restricted. The second limit is defined by the amount of wear until the tool holder or the cutting wheel will be subjected to wear, which necessitates more elaborate maintenance work. The third wear limit results from the need of preventive maintenance, in which cutting tools won't exceed the wear limit before the next maintenance stop. Hence, this limit depends on the interval length between the maintenance stops as well as the position of the tool on the cutting wheel.

Further, consequence of insufficient maintenance must be considered. Therefore, it is assumed that the excessive wear is detected, when the average wear level of the tools reaches 80% of the normal wear limit. If this limit is exceeded, an additional maintenance stop will be performed, causing additional downtime and costs.

The soil is defined according to the homogeneous sections, where all the necessary ground conditions are assumed to be approximately constant. Nonetheless, there is still a deviation of these values within one section, which is considered by implementing distribution functions for each parameter. These distribution functions are used to determine a new value of the soil and steering parameters for each excavated ring within the given boundaries. Similar, the penetration of the cutting wheel is determined by a predefined distribution function as well. The penetration is needed to calculate the cutting path for each tool, which directly influences the amount of wear. Furthermore, the air pressure that is needed to lower the support medium during the maintenance process for the homogeneous section must be given.

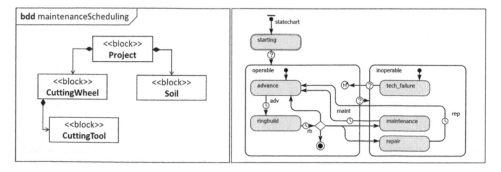

Figure 2.   Block definition diagram (bdd) of the model (left); state chart of the processes of the cutting wheel (right).

3.3 Evaluation criteria

The model is used to determine the duration of each maintenance stop considering the regulations for working under pressurized conditions. The total duration of one stop is divided into time for preparation, time for the replacement of tools and time for the post processes that are needed to restart the advancing processes. Hereby, especially the time to decompress the workers, who are leaving the excavation chamber, is significantly higher the higher the required pressure inside the excavation chamber.

In addition, the number and types of replaced tools can be derived from the model. These two factors are used to determine the maintenance cost by determining the time dependent and the material costs. Using the Mote Carlo Simulation, not only the amount of maintenance costs but also their deviation can be estimated and evaluated.

4 CASE STUDY

The presented model has been used to compare two designs of an exemplary metro tunnel project. The first scenario A considers two single-track tubes. Therefore, two smaller tunnels are excavated. The other scenario B considers the excavation of one double-track tunnel with a lager diameter as shown in Figure 3.

4.1 Input parameter

The total length of the tunnel is 3,000 m, divided into three homogeneous sections with different values for soil properties. The positions of the maintenance stops are determined using the wear prediction model for the caliber cutting tools, since they will be worn out first. Since the two scenarios differ in size of the cutting wheel, for scenario A there are six maintenance stops scheduled and for scenario B nine stops are scheduled as shown in Figure 4.

The input data of the three homogeneous sections are summarized in Table 1. For most of the parameter a uniform distribution is used. However, negative values are not feasible and

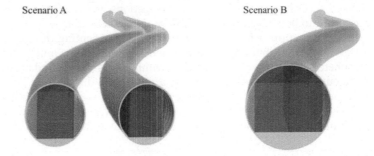

Figure 3. Layout of the two tunnel scenarios: A) two single-track tubes (D = 6.20 m); B) one double-track tube (D = 9.50 m).

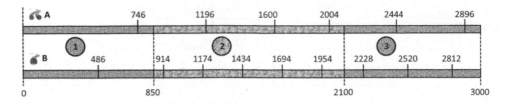

Figure 4. Homogeneous sections and positions of maintenance stops [m].

are excluded in the model. The value D_{60} as well as the penetration are given by a Weibull distribution. The function for D_{60} has been assumed with the purpose to gain a skewed distribution function that avoids negative values. The values of the penetration distribution function are gained by distribution fitting of project data. Furthermore, the air pressure inside the excavation chamber is set at 2.5 bar. Therefore, several accesses of the excavation chamber during one maintenance stop may become necessary.

The simulation model is used to calculate the number and duration of the maintenance stops and the amount of replaced tools. These values are combined using the cost factors presented in Conrads et al. (2018) and thus the maintenance costs are determined.

4.2 Results and discussion

For each scenario a Monte Carlo Simulation has been performed, conducting 10,000 simulation runs each. The maintenance costs of each simulation run are calculated. Since two tunnel tubes are excavated for scenario A, the maintenance costs had to be doubled to gain a comparable value.

The results are summarized using a histogram shown in Figure 5. It shows a clear difference between the total maintenance costs for both scenarios. Using two smaller tubes less maintenance stops are needed and less tools are replaced. Therefore, the total maintenance costs are significantly lower than for scenario B. The discontinuous distribution of values results from the low number of interventions per maintenance stop. For each additional intervention, the time, thus the cost, for the stop rises abruptly. For scenario A the number of interventions lies between 1–8 and for scenario B between 7–13.

In Scenario A six maintenance stops with a total duration of between 450-473 h and 272-284 replaced tools took place. Scenario B included nine maintenance stops with a duration of 578-593 h and 413-428 replaced tools. Due to the bigger diameter of the cutting wheel and the higher number of cutting tools mounted, more maintenance stops become necessary.

Table 1. Ground properties input data.

Homogeneous section	1	2	3
eQu [%]	normal* $\mu = 85; \sigma = 3$**	normal $\mu = 75; \sigma = 10$	normal $\mu = 78; \sigma = 6$
H_{TA} [m]	single value 3.0	single value 4.0	single value 5.0
W_{TA} [m]	single value 15.0	single value 16.0	single value 17.0
γ_i [kN/m³]	normal $\mu = 21; \sigma = 2$	normal $\mu = 22; \sigma = 2$	normal $\mu = 22; \sigma = 3$
γ'_i [kN/m³]	normal $\mu = 12; \sigma = 1$	normal $\mu = 15; \sigma = 1$	normal $\mu = 14; \sigma = 2$
c' [kN/m²]	single value 0.0	single value 0.	single value 0.0
φ' [°]	normal $\mu = 41; \sigma = 1.5$	normal $\mu = 35; \sigma = 5$	normal $\mu = 44; \sigma = 7$
D_{60} [mm]	weibull $\alpha = 1.5; \beta = 2.0; min = 0.01$	weibull $\alpha = 1.5; \beta = 10.0; min = 0.01$	weibull $\alpha = 1.5; \beta = 3.5; min = 0.01$
p [mm]	weibull $\alpha = 10.0; \beta = 5.0; min = 20.0$	weibull $\alpha = 10.0; \beta = 5.0; min = 20.0$	weibull $\alpha = 10.0; \beta = 5.0; min = 20.0$

* distribution function, **parameter of the distribution function
eQu: equivalent quartz content; H_{TA}: height of overburden above ground water table; W_{TA}: height of overburden below ground water table; γ_i: damp unit weight of the soil layer; γ_i': buoyant unit weight; c': cohesion; φ': friction angle; D_{60}: medium grain diameter of the sample at 60% mass friction; p: penetration

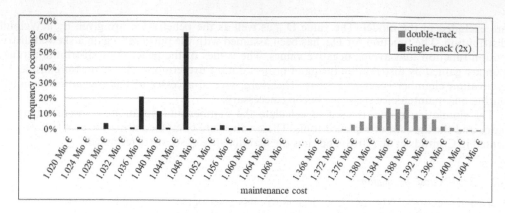

Figure 5. Histogram of the resulting maintenance cost for 10,000 simulation runs.

Furthermore, the number of cutting tools that must be replaced and the duration for each stop increases. Especially under pressurized conditions, the maintenance intervals for big diameter could be optimized with respect to the maintenance cost.

However, for both scenarios unplanned maintenance stops due to unexpected high wear of the cutting tools could be prevented using the proposed maintenance schedule. This shows, that with the help of the used wear prediction model, a reliable scheduling of the maintenance can be conducted within the given uncertainties. Nonetheless, if the assumptions that were made for this scheduling deviate from the actual conditions, the prediction must be updated and changed accordingly.

5 CONCLUSION AND OUTLOOK

The maintenance scheduling of cutting tools for mechanized tunnelling under unstable ground conditions is an important and complex task to ensure the project success and efficiency. For a good wear prediction, uncertain input parameters must be taken into account to evaluate the robustness of the proposed schedule. It has been shown by implementing an empirical wear prediction model for hydro-shield machines into a process simulation model, that the maintenance cost can be derived and uncertainties can be assessed. This way, maintenance strategies can be evaluated and different tunnel designs can be compared. These results can support the decision-making process.

Further optimizations can be conducted using the simulation model to improve the efficiency of the estimated maintenance strategy even though unknown deviations may occur during the realization of the project. Furthermore, the implemented wear prediction model only deals with abrasive wear. Sudden failure or other wear mechanism can still occur and must be taken into account during the planning phase as well. This shall be implemented into the model for further research, considering a probability of sudden failure of a tool. To use this probability, further research is needed, evaluating project data to gain particular input parameters.

ACKNOWLEDGMENTS

This paper presents the results from the subproject C3 "Simulation of production and logistic processes in mechanized tunnelling: Simulation-based maintenance and availability analysis", which is part of the Collaborative Research Center SFB 837 at Ruhr University Bochum. The authors would like to acknowledge the financial support for this research received from the German Science Foundation (DFG) within the framework of the SFB 837.

REFERENCES

AbouRizk, S. 2010. Role of Simulation in Construction Engineering and Management. *J. Constr. Eng. Manage.* 136 (10): 1140–1153.

AL-Battaineh, H.T.; AbouRizk, S.; Tan, J.; Fernando, S. 2006. Productivity Simulation during the Planning Phase of the Glencoe Tunnel in Calgary, Canada: A Case Study. In IEEE (Ed.): *Proceedings of the 2006 Winter Simulation Conference. Monterey, CA, USA, 3-6 Dec. 2006*: 2087–2092.

Amoun, S.; Sharifzadeh, M.; Shahriar, K.; Rostami, J.; Tarigh Azali, S. 2017. Evaluation of tool wear in EPB tunneling of Tehran Metro, Line 7 Expansion. *Tunnelling and Underground Space Technology 61*: 233–246.

Bruland, A. 2000. Hard Rock Tunnel Boring Vol. 3 - Advance Rate and Cutter Wear. *Doctoral theses*, Trondheim.

Conrads, A.; Scheffer, M.; König, M.; Thewes, M. 2018. Robustness evaluation of cutting tool maintenance planning for soft ground tunneling projects. *Underground Space 3 (1)*: 72–85.

Conrads, A.; Scheffer, M.; Mattern, H.; König, M.; Thewes, M. 2017. Assessing maintenance strategies for cutting tool replacements in mechanized tunneling using process simulation. *Journal of Simulation 11 (1)*: 51–61.

Duhme, R. 2018. Deterministic and simulation based planning approaches for advance and logistic processes in mechanized tunneling. *Dissertation*. Ruhr-Universität Bochum, Bochum. Department of Civil and Environmental Engineering.

Ebrahimy, Y.; AbouRizk, S.M.; Fernando, S.; Mohamed, Ya. 2011. Simulation modeling and sensitivity analysis of a tunneling construction project's supply chain. *Eng, Const and Arch Man 18 (5)*: 462–480.

Gehring, K. 1995. Leistungs- und Verschleißprognosen im maschinellen Tunnelbau. *Felsbau 13 (6)*: 439–448.

Jakobsen, P.D. 2014. Estimation of soft ground excavation tool life in TBM tunnelling. *Dissertation*. Norwegian University of Science and Technology, Trondheim. Faculty of Engineering Science and Technology, Department of Civil and Transport Engineering.

Köppl, F.; Frenzel, C.; Thuro, K. 2009. Statistische Modellierung von Gesteinsparametern für die Leistungs- und Verschleißprognose bei TBM Vortrieben. In DGGT Deutsche Gesellschaft für Geotechnik e.V., DGG Deutsche Gesellschaft für Geowissenschaften (Eds.): *17. Tagung für Ingenieurgeologie und Forum "Junge Ingenieurgeologen"*. Zittau, 06.-09. Mai.

Köppl, F.; Thuro, K.; Thewes, M. 2015a. Suggestion of an empirical prognosis model for cutting tool wear of Hydroshield TBM. *Tunnelling and Underground Space Technology 49*: 287–294.

Köppl, F.; Thuro, K.; Thewes, M. 2015b. Factors with an influence on the wear to excavation tools in hydroshield tunnelling in soft ground. *Geomechanik Tunnelbau 8 (3)*: 248–257.

Li, X.; Li, X.; Yuan, D. 2017. Application of an interval wear analysis method to cutting tools used in tunneling shields in soft ground. *Wear 392-393*: 21–28.

März, L., Krug, W.; Rose, O.; Weigert, G. 2011. *Simulation und Optimierung in Produktion und Logistik*. Springer-Verlag Berlin Heidelberg.

Maidl, B.; Herrenknecht, M.; Maidl, U.; Wehrmeyer, G. 2011. *Mechanised Shield Tunnelling*. 2nd ed.: Berlin.

Mattern, H.; Scheffer, M.; Conrads, A.; Thewes, M.; König, M. 2016. Simulation-Based Analysis of Maintenance Strategies for Mechanized Tunneling Projects. *Proceedings of ITA-AITES World Tunnel Congress 2016 (WTC 2016)*. San Francisco, California, USA, 22-28 April.

Rahm, T.; Scheffer, M.; Thewes, M.; König, M.; Duhme, R. 2016. Evaluation of Disturbances in Mechanized Tunneling Using Process Simulation. *Computer-Aided Civil and Infrastructure Engineering 31 (3)*: 176–192.

Scheffer, M.; Rahm, T.; König, M.; Thewes, M. 2016. Simulation-Based Analysis of Integrated Production and Jobsite Logistics in Mechanized Tunneling. *J. Comput. Civ. Eng. 30 (5)*.

Thewes, M. (2007): TBM Tunnelling Challenges - Redefining the State of the Art. In *Tunel, ITA-AITES WTC 2007*: 13–21.

Tunnels and Underground Cities: Engineering and Innovation meet Archaeology,
Architecture and Art, Volume 5: Innovation in underground engineering,
materials and equipment - Part 1 – Peila, Viggiani & Celestino (Eds)
© 2020 Taylor & Francis Group, London, ISBN 978-0-367-46870-5

Possibility of using vane shear testing device for optimizing soil conditioning

H. Copur, E. Avunduk, S. Tolouei, D. Tumac, C. Balci, N. Bilgin & A.S. Mamaghani
Istanbul Technical University, Mining Engineering Department, Istanbul, Turkey

ABSTRACT: The basic purpose of this study is to analyze the possibility of using vane shear testing device for optimizing soil conditioning. The soil sample was obtained during excavation of SK-15 shaft of the Ayvali 2 Wastewater Tunnel in Istanbul. The natural water content, specific gravity, size distribution and consistency limits of the soil sample were determined. An anticlay foaming agent was used in the tests. Mixing, cone penetrometer, flow table and vane shear (pocket type) tests were performed on the mixture of soil and foaming agent at certain water content. CF of 3% and FER of 16 were kept constant and FIR was varied. The experimental results were validated by field performance measurements. The laboratory and field studies indicated that optimum conditioning parameters and consistency of conditioned soils could be determined by vane shear testing. There were strong relationships between the results of classical testing methods and vane shear device.

1 INTRODUCTION

The excavation performance of EPB TBMs depends on basically soil conditioning through injecting special foams (air + water + foaming agents) to face, excavation chamber and screw conveyor. An inappropriate soil conditioning would result in very low performance, many stoppages and delays increasing the costs, as well as resulting in ground deformations/surface settlements. Therefore, selection, design and predicting the performance of these machines become very important in terms of feasibility, planning and project economics.

Conditioning of soil should provide for a homogeneous, pulpy and plastic muck (Thewes 2007a, 2007b, 2007c, Peila et al, 2007, Avunduk et al. 2017a, 2017b). Proper soil conditioning would enable higher penetration rate with lower torque and power requirement, lower clogging, transportation and stability problems, lower abrasive wear of TBMs (Jancsecz et al. 1999, Langmaack 2000, Thewes et al. 2010 and 2012).

The first systematic attempt to provide some suggestions on soil conditioning by laboratory testing belongs to Efnarc (European Federation of National Associations Representing Producers and Applicators of Specialist Building Products for Concrete) (Efnarc, 2005). Efnarc stated that these suggestions might not work for some soils. The soil conditioning design suggestions based on size distribution of soils given by some authors (Langmaack 2007, Budach & Thewes 2015) were also considered as insufficient (Thewes & Budach 2010, Thewes et al. 2012, Budach & Thewes 2015). The basic problem of the conditioning tests performed in laboratory is that the soil and the foam are mixed in the atmospheric pressure. This does not characterize the real behavior of foams under face pressure (Thewes et al. 2012, Mooney et al. 2016). However, the tests enable basic understanding of feasibility, processes and interactions during soil conditioning for EPB TBM operations (Galli & Thewes 2016, Avunduk et al. 2017a, 2017b). Galli & Thewes (2014) suggested using some testing devices, including vane shear strength device, other than commonly used testing methods.

The basic purpose of this study is to analyze the possibility of using vane shear testing device for optimizing soil conditioning. The soil sample was obtained during excavation of

SK-15 shaft of the Zeytinburnu Ayvali 2 Wastewater Tunnel in Istanbul. The natural water content, specific gravity, size distribution (sieve + hydrometer), consistency limits (liquid limit and plastic limit of only the fines) of the soil sample were determined in the laboratory. An anticlay foaming agent was used in the tests and the foam was generated by using a laboratory type foam generator. Mixing (with power and sticking measurement), cone penetrometer, flow table and vane shear (pocket type) tests were performed to find out optimum soil conditioning parameters for the mixture of the soil and foaming agent at a certain water content determined by clogging chart. Foam Concentration Ratio (CF) of 3% and Foam Expansion Ratio (FER) of 16 were kept constant and Foam Injection Ratio (FIR) was varied. The experimental results were validated by field performance measurements.

2 EXPERIMENTAL PROCEDURES

The soil sample was obtained during excavation of SK-15 shaft of the Zeytinburnu Ayvali 2 Wastewater Tunnel in Istanbul. Since the soil excavation face is stable in this part of the tunnel, no face pressure would be applied or EPB TBM would be used in open mode.

The natural water content, specific gravity, size distribution (sieve + hydrometer), consistency limits (liquid limit and plastic limit of only the fines) of the sample were determined in the laboratory. Mixing (with power and sticking measurement), cone penetrometer, flow table and vane shear (pocket type) tests were performed to find out optimum soil conditioning parameters for the mixture of the soil and anticlay foaming agent at a certain/constant water content determined by clogging chart (33% natural water content + 12% additional = 45%) suggested by Holmann and Thewes (2013). Foam Concentration Ratio (CF) of 3% and Foam Expansion Ratio (FER) of 16 were kept constant in the tests. Foam Injection Ratio (FIR) was varied between 50% and 300% with 50% steps. The anticlay used in the tests are the same as used in the field from where the sample was obtained.

Foaming agent used in the tests was the same as the one used in the Zeytinburnu Ayvali 2 Wastewater Tunnel. The foam was generated in the soil conditioning laboratory of the Istanbul Technical University by using a typical laboratory scale granular filled type foam generator. The foam was applied as soon as possible it was generated to keep its basic time dependent stability properties as suggested by Thewes et al. (2012).

The testing procedures and standards for characterizing the soil sample are: the natural water content (ASTM D 2216–10), specific gravity (ASTM D 854–14), particle size distribution by means of sieve analysis (ASTM D 422–63) for samples having grain size greater than 75 µm and hydrometer methods (ASTM D 1140–00) for samples having grain size smaller than 75 µm (No. 200 Sieve), and consistency (Atterberg) limits (liquid limit by BS 1377–2 and plastic limit by ASTM D4318-10E1). The soil class was determined based on the unified soil classification system (ASTM D2487-11).

Testing procedures and standards for characterizing the soil and anticlay mixtures are mixing test with power and sticking (to mixer blade) amount measurement (Efnarc, 2005), fall cone penetration (BS 1377-2), flow table (DIN EN 1015-3) and vane shear strength (ASTM D2573M-15).

The pocket type vane shear device is placed as perpendicular on a flat and homogenous soil surface (in laboratory or field) and zero correction is performed. The vane (blades) is slightly inserted to the soil at once by pushing and it is slowly rotated until the soil is sheared. Then, the shear strength reading is recorded (Figure 1).

3 EXPERIMENTAL RESULTS AND DISCUSSION

The natural water content of the sample is defined as 33%. The grain (particle) size distribution of the sample is seen in Figure 2. As seen, the sample consists of 80% clay+silt and remaining is sand and gravel.

Figure 1. Application of pocket type vane shear testing device in laboratory (left) and field (right).

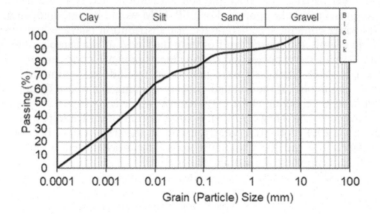

Figure 2. Grain size distribution of the sample, (Tolouei et al. 2016).

Table 1. Atterberg consistency limits, (Tolouei et al. 2016).

Liquid Limit (%)	56.2
Plastic Limit (%)	29.7
Plasticity Index	26.5
Consistency Index	0.87
Plasticity Level	Plastic
Specific Gravity	2.65
Natural Water Content (%)	33.0

Atterberg consistency limits of the clay+silt material of the sample are summarized in Table 1. The liquid and plastic limits of the sample are 56.2% and 29.7%, respectively. The plasticity and consistency index values of the sample are 26.5 and 0.87, respectively. Since the sample is plastic, there would be a clogging risk, which might reduce the performance of the EPB TBM. The place of the sample on the clogging risk chart and consistency chart given by Holmann and Thewes (2013) is seen in Figures 3 and 4 for EPB TBMs working in open mode, respectively. The clogging risk chart identifies that there is a medium-high clogging risk for the sample. The consistency chart suggest 12% water addition to the sample/soil to provide the required consistency. The water content of the sample is then set as 45% (33% natural water content + 12% additional water) during the soil conditioning experiments.

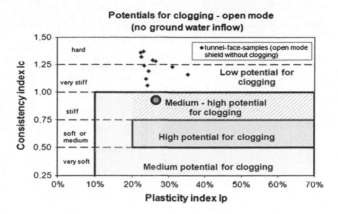

Figure 3. Place of sample on clogging risk chart for open mode EPB applications given by Holmann and Thewes (2013), (Tolouei et al., 2016).

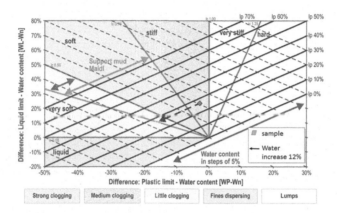

Figure 4. Place of sample on consistency chart for open mode EPB applications given by Holmann and Thewes (2013), (Tolouei et al., 2016).

The variations with FIR of net power, sticking amount, fall cone penetration, flow table and vane shear strength parameters performed at the constant values of water content of 45%, foam concentration of 3% and foam expansion ratio of 16 on the soil + water + anticlay mixture are summarized in Figures 5–9 for varied values of foam injection ratio between 50% and 300% with 50% incremental steps. The variations of net power, sticking amount, fall cone penetration, flow table and FIR parameters with vane shear strength are seen in Figures 10–14.

It is seen in Figures 5–9 that the power consumption and sticking amount become minimum at FIR values between 250% and 300%. Fall cone penetration test results indicate that the optimum consistency of the soil + water + anticlay mixture is reached at FIR value of around 270%. Falling flow table test results indicate that the optimum consistency is obtained at FIR values between 275% and 320%. These results give generally an optimum soil conditioning range of FIR value between 250% and 300% at 12% water addition to the natural water content of 33% (total 45%), 3% of foam concentration ratio and FER of 16 for the sample taken from Ayvali 2 tunnel.

It is seen in Figures 10–14 that the vane shear strength had good relationships with FER and the results of classical soil conditioning tests. The vane shear strength decreases with increasing FIR, flow table spreading diameter and fall cone penetration. Eventually, the power consumption in mixing test and sticking amount decrease with decreasing vane shear strength.

The variations of power consumption and sticking amount with vane shear strength indicate that vane shear strength should be less than 1.0 kgf/cm^2 for a suitable consistency. The variation of fall cone penetration with vane shear strength indicate that vane shear strength

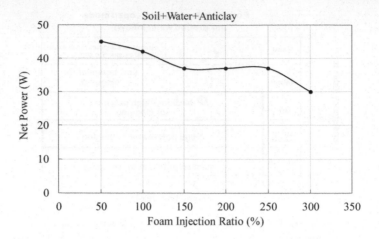

Figure 5.　Variation of net power consumption with FIR.

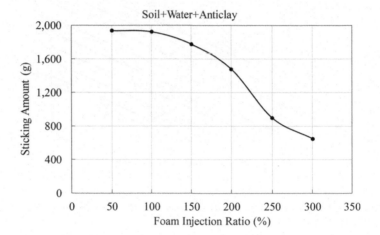

Figure 6.　Variation of sticking amount with FIR.

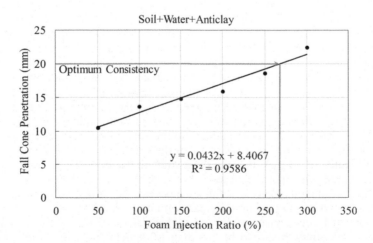

Figure 7.　Variation of fall cone penetration with FIR.

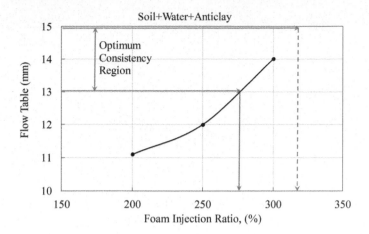

Figure 8.   Variation of flow table spreading diameter with FIR.

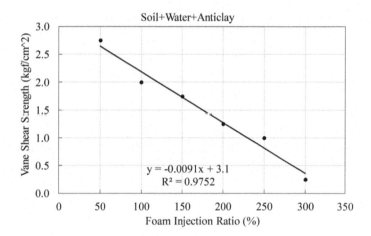

Figure 9.   Variation of vane shear strength with FIR.

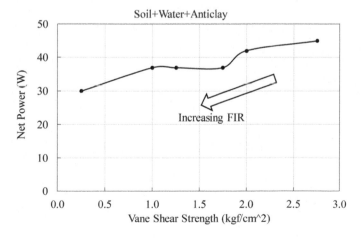

Figure 10.   Variation of net power consumption with vane shear strength.

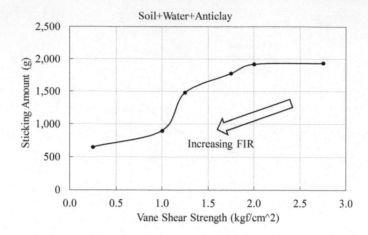

Figure 11.   Variation of sticking amount with vane shear strength.

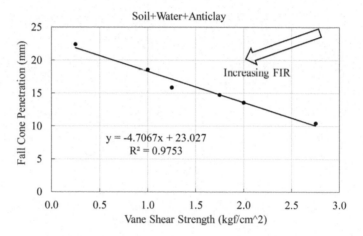

Figure 12.   Variation of fall cone penetration with vane shear strength.

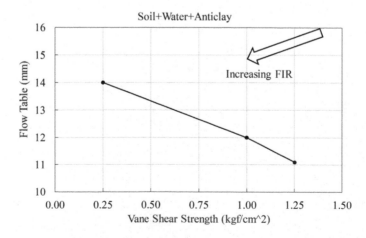

Figure 13.   Variation of flow table spreading diameter with vane shear strength.

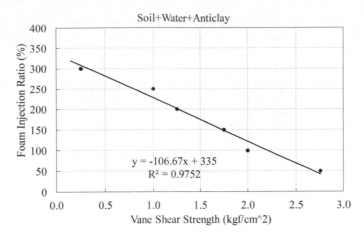

Figure 14. Variation of foam injection ratio with vane shear strength.

should be mostly at around 0.64 kgf/cm² for a suitable consistency. The variation of flow table spreading diameter with vane shear strength indicate that vane shear strength should be less than 0.62 kgf/cm² for a suitable consistency. If the FIR value of 270% giving the optimum consistency is considered, vane shear strength should be greater than 0.61 kgf/cm² for a suitable consistency. These results indicate that the best consistency can be obtained with a shear strength value of around 0.60–0.65 kgf/cm²; and it should be less than 1.0 kgf/cm².

Knowing an optimum consistency range based on vane shear test results, it is possible to decide the range of the amount of foam (FIR) to be applied in field by using the relationship given in Figure 14.

The coarse fraction (gravel) of the soil sample taken from Ayvali 2 was removed and then used in the soil conditioning tests, since the test procedures and equipment used in this study require such a removal.

The optimum conditioning range for flow table test given in Figure 5 was obtained for a soil sample obtained from another tunneling site, Akfirat Wastewater Tunnel in Istanbul, by using the relationship between flow table spreading diameter and slump height (Copur et al., 2017). Soil sample from Akfirat was clayey silty gravelly sand including 20% clay, 10% silt, and 40% sand. It is assumed in this study that the optimum flow diameter range found for the sample from Akfirat is also valid for the soil sample taken from Ayvali 2 wastewater tunnel.

It can be said that the vane shear testing device may be used for all types of soils except for gravels. However, this conclusion needs to be proved for soils having mostly sand fractions.

It can be inferred from the experimental results that the tests with pocket type vane shear device reflect consistency of soil + foam mixtures, as well as clogging potential, and this device may be used for optimizing soil conditioning parameters and performance of EPB TBMs. The device can also be used in field applications, such as on a tunnel face or on conditioned muck samples.

4 FIELD STUDIES AND VALIDATION

The cutterhead torque was considered high during excavation of Ayvali 2 wastewater tunnel and increased more after 150 m of excavation after starting the project due to clogging problems, although the cutterhead was half filled (not full) with muck since the face was stable. Thus, the operation stopped for remediation actions by the contractor representatives.

The grizzly bars on the cutterhead were first uninstalled to reduce sticking and cutterhead torque, and a new conditioning design was decided. The representatives of the contractor decided to apply a soil conditioning with foam (anticlay) concentration ratio of around 1.5%,

FER of 15–16, FIR of 150–200% and total water content of 50% (around 17% additional water to natural water content of 33%) and started applying it. The contractor used the half the experimental concentration ratio, but more than their previous application of 1%. They also added 17% water instead of experimental 12%.

It was observed that the clogging problem decreased significantly; the penetration rate increased from 19 to 30 mm/rev; the advance rate increased from ~50 to 90 mm/min. It was not possible to define the exact shares of the soil conditioning and removing grizzly bars on the performance improvement. The operator claimed that the share of soil conditioning on the performance improvement was around half.

Additional testing and field data are required for a more reliable inference on the use of vane shear testing and device for optimizing soil conditioning and performance of EPB TBMs.

5 CONCLUSIONS

The experimental results indicated that there were very strong and strong relationships between the results of the soil conditioning tests suggested by Efnarc and the vane shear strength test results. The optimum consistency of soil + foam mixtures were obtained with a shear strength value of around 0.60–0.65 kgf/cm^2; and it should be less than 1.0 kgf/cm^2. The tests with pocket type vane shear device reflect consistency of soil + foam mixtures, as well as clogging potential; and this device may be used for optimizing soil conditioning parameters and performance of EPB TBMs for all types of soils except for gravels. Field observations validated that the optimized soil conditioning parameters increased performance of the EPB TBM. Using more than one test method for optimizing soil conditioning parameters would provide higher confidence on results. Therefore, new testing methods for soil conditioning, such as pocket type vane shear device mentioned in this study, should be added to the classical testing devices.

6 ACKNOWLEDGEMENTS

This study forms a part of the unpublished PhD thesis by Emre Avunduk and MSc thesis by Sahand Tolouei. The authors are grateful to the Scientific and Technological Research Council of Turkey (TUBITAK, Project No: 213M487) and the Istanbul Technical University Research Foundation (ITU, BAPSO Project No: 37793) for their valuable supports. The authors would like to thank Mr. Muammer Cinar, and Mr. Sener Kahya for their assistance with the field studies. The authors also thank the job owner representatives of the Istanbul Water and Sewage Administration (ISKI) and the contractor Eferay-Silahtaroglu Construction JV for permitting us to conduct research in the Ayvali Wastewater Tunnel.

REFERENCES

ASTM, D 1140-00, 2007. Standard test methods for amount of material in soils finer than No. 200 (75-μm) sieve.
ASTM, D 2216-10, 2010. Standard test methods for laboratory determination of water (moisture) content of soil and rock by mass.
ASTM, D2487-11, 2011. Standard practice for classification of soils for engineering purposes (Unified Soil Classification System).
ASTM, D2573M-15E1, 2015. Standard test method for field vane shear test in saturated fine-grained soils.
ASTM, D422-63, 2007. Standard test method for particle-size analysis of soils.
ASTM, D4318-10E1, 2010. Standard test methods for liquid limit, plastic limit, and plasticity index of soils.
ASTM, D854-14, 2014. Standard test methods for specific gravity of soil solids by water pycnometer.

Avunduk, E. & Copur, H. 2018. Empirical Modeling for Predicting Excavation Performance of an EPB TBM based on Soil Properties. *Tunnelling and Underground Space Technology* 71: 340–353.

Avunduk, E., Copur, H., Tolouei, S., Gumus, U. & Altay, U. 2017a. Effect of Adhesion on EPB TBM Performance. In: Proceedings of the World Tunnel Congress, Bergen, Norway. 6p.

Avunduk, E., Tumac, D., Tolouei, S., Copur, H., Balci, C. & Bilgin, N. 2017b. Effect of Conditioning on Soil Workability determined by Mini-Slump and Flow Table Tests. In: Proceedings of the World Tunnel Congress, Bergen, Norway. 10 p.

BS, 1377–2, 1990. Methods of test for soils for civil engineering purposes. Classification tests.

Budach, C. & Thewes, M. 2015. Application ranges of EPB shields in coarse ground based on laboratory. *Tunnelling and Underground Space Technology* 50: 296–304.

Copur, H., Bilgin, N., Balci, C., Tumac, D., Avunduk, E. & Tolouei, S. 2017. Prediction and Optimization of Excavation Performance of Earth Pressure Balance Tunneling Machines (EPB TBM). Project Report submitted to the Scientific and Technological Research Council of Turkey (TUBITAK), Project No: 213M487, (in Turkish).

DIN EN 1015-3, 2007. Methods of test for mortar for masonry - Part 3: Determination of consistence of fresh mortar (by flow table).

EFNARC, 2005. Specification and guidelines for the use of specialist products for mechanized tunnelling (TBM) in soft ground and hard rock. European Federation Dedicated to Specialist Construction Chemicals and Concrete Systems.

Galli, M. & Thewes, M. 2014. Investigations for the application of EPB shields in difficult grounds. *Geomechanics and Tunnelling* 7 (1): 31–44.

Galli, M. & Thewes, M. 2016. Rheology of Foam-Conditioned Sands in EPB Tunneling. In: Proceedings of the World Tunnel Congress, San Francisco, 22–28 April.

Holmann, F.S. & Thewes, M. 2013. Assessment method for clay clogging and disintegration of fines in mechanised tunnelling. *Tunneling and Underground Space Technology* 37: 96–106.

Jancsecz, S., Krause, R. & Langmaack, L. 1999. Advantages of soil conditioning in shield tunneling, experiences of LRTS Izmir. In: Proceedings of the World Tunnel Congress, Oslo, pp. 865–875.

Langmaack, L. 2000. Advanced technology of soil conditioning in PB shield tunnelling. In: Proceedings of the North American Tunneling Congress, 16 p.

Langmaack, L. 2007. Soil Conditioning. In: TBM Conference Organized by BASF, Istanbul, Feb., pp. 7–9.

Mooney, M., Wu, Y., Mori, L., Bearce, R. & Cha, M. 2016. Earth pressure balance TBM soil conditioning: It's about the pressure. In: Proceedings of the World Tunnel Congress, San Francisco, 22–28 April.

Peila, D., Oggeri, C. & Vinai, C. 2007. Screw conveyor device for laboratory tests on conditioned soil for EPB tunnelling operations. *Journal of Geotechnical and Geoenvironmental Engineering* 133 (12): 1622–1625.

Thewes, M. 2007a. Mechanized urban tunnelling – Machine technology. The 3rd Training Course – Tunnelling in Urban Area. In: Proceedings of the World Tunnel Congress, Prague, 42 p.

Thewes, M. 2007b. TBM tunnelling challenges – redefining the state of the art. In: Proceedings of the World Tunnel Congress, Prague.

Thewes, M. 2007c. Shield tunnelling technology to mitigate geotechnical risks. In: Proceedings of the 2nd Symposium on Underground Excavations for Transportation, Istanbul Technical University, pp.49–56.

Thewes, M. & Budach, C. 2010. Soil conditioning with foam during EPB tunnelling. *Geomechanics and Tunnelling* 3 (3) 256–267.

Thewes, M., Budach, C. & Beziujen, A. 2012. Foam conditioning in EPB tunneling. In: Proceedings of the 7th International Symposium on Geotechnical Aspects of Underground Construction in Soft Ground, Roma, pp. 127–135.

Thewes, M., Budach, C. & Galli, M. 2010. Laboratory tests with various conditioned soils for tunnelling with earth pressure balance shield machines. In: Proceedings of the 4th BASF TBM Conference, London.

Tolouei, S., Avunduk, E., Copur, H., Tumac, D., Balci, C., Bilgin, N., Cinar, M. & Kahya, S. 2016. Foam optimization and relationship between foam use and EPB TBM performance. In: Proceedings of the 13th International Conference Underground Construction Prague 2016, 23–25 May, Prague, 9 p.

Tunnels and Underground Cities: Engineering and Innovation meet Archaeology,
Architecture and Art, Volume 5: Innovation in underground engineering,
materials and equipment - Part 1 – Peila, Viggiani & Celestino (Eds)
© 2020 Taylor & Francis Group, London, ISBN 978-0-367-46870-5

The load-bearing capacity of primary linings, considering time dependent parameters, at the Brenner Base Tunnel

T. Cordes & K. Bergmeister
Brenner Base Tunnel BBT-SE, Innsbruck, Austria

A. Dummer, M. Neuner & G. Hofstetter
University of Innsbruck, Innsbruck, Austria

ABSTRACT: The realistic assessment of the load-bearing capacity of shotcrete linings becomes necessary a) for early and large deformations in deep tunnels, b) for single lined permanent shotcrete supports and c) for the assessment of the load-bearing capacity of shotcrete structures subsequently stressed by nearby tunnel excavations. The nowadays used linear-elastic models of primary linings using a hypothetical shotcrete stiffness are not accurate enough.

This contribution presents a more detailed determination of the load-bearing capacity of the primary shotcrete lining. It captures the time-dependence of the loading history and of the shotcrete properties. The capability of this approach is shown by the back analysis of a deep tunnel section with an advanced shotcrete material model. The material parameters of the young shotcrete have been calibrated based on uniaxial tests on specimens gained during construction. The back analysis presented here combines the advanced material model with a new testing approach.

1 INTRODUCTION

The Brenner Base Tunnel (BBT) is a priority European infrastructure project on the Scandinavian-Mediterranean Corridor from Helsinki (Finland) to La Valletta (Malta). The corridor connects the economic centers and ports in Italy with those in Germany and Scandinavia and will be used especially for freight traffic. One main part of this corridor is the Brenner Base Tunnel. The two 64 km long, parallel, single-lane railway tunnels of the BBT enable a flat railway crossing of the Alps between Austria and Italy (Bergmeister 2011). In the future, freight and passenger trains will cross the Alps more ecologically on this route with higher loading capacities and shorter connection times without having to cross the 1,371 m high Brenner Pass. The total tunnel system of the BBT will comprise a total length of 230 km. Half of the tunnel length are railway tunnels, the other half are exploration, emergency, ventilation and access tunnels. The latter are tunnels with, in general, fewer requirements for the serviceability, such as water tightness, surface quality or availability. In some cases, a single permanent shotcrete lining is fulfilling these requirements and offers a very economic tunnel design. For the proof of the structural safety of permanent shotcrete linings of deep tunnels special attention should be paid to determine reliable rock loads and hence the load-bearing capacity of the lining.

2 DESIGN OF THE PRIMARY LINING AT THE BBT

For shotcrete linings two concepts are distinguished at the BBT (Cordes 2018):

a) Temporary shotcrete linings, which are built for a service life of at least 10 years for the construction phase and which are designed with partial safety factors for a service life of 50 years and

b) single shell permanent shotcrete linings, which consist of one monolithic shotcrete lining and which are designed with consideration of an increased reliability index accounting for a service life of 200 years. The requirements, the loadings and the resulting load-bearing capacities of the linings must be evaluated precisely. For this purpose, the real shotcrete lining thickness, the shotcrete properties and the deformations of the lining are monitored to evaluate the loading states and to ensure the structural safety and the serviceability.

Today the design of a shotcrete lining does only very roughly, take the development of convergences over time, the time-dependent shotcrete properties or the relaxation of the shotcrete into account. Usually, a softer, linear-elastic material behavior (30–50% of the normative stiffness, the hypothetical modulus of elasticity according to Pöttler (1990)) is considered in order to capture these effects. This hypothetical modulus of elasticity approximates the development of stiffness of the shotcrete and decisive creep effects, both in accordance with the occurring convergences. This widely used simple approach provides a design-friendly estimation of the load-bearing capacity of usual tunnel advances on the safe side. Slow and long-lasting convergences or fast-curing shotcretes are generally not covered by this approach. This approach is therefore insufficient for i) the simulation of shotcrete linings under early high loads in deep tunnels, ii) excavation-related rock relocations to existing shotcrete linings or iii) for determining the limit state of deep single shell permanent shotcrete linings by back analyses.

i) For deep tunnels, it can occur that temporary shotcrete linings are heavily loaded from the very beginning. This results in large strains and high stresses in the shell and can lead to early local crack development. Besides the evolution of the convergences the solidification of the shotcrete with the development of the shotcrete properties (e.g. Young's modulus, compressive and tensile strength, ductility, shrinkage and creep behavior) are of importance for these sections. The creep effects of very early loaded shotcrete are here of particular importance.

ii) Stress redistributions from nearby tunnel advances on existing shotcrete linings also require higher accuracy in simulations. The highly loaded shotcrete linings of deep tunnels often withstand only minor later deformations before damage occurs. Since the shotcrete in this state has already reached maturity, relaxation is only of little importance.

iii) The design of single shell permanent shotcrete linings requires a detailed analysis of the load-bearing capacity of the loaded tunnel structures. Considering the time-dependence of the material and of the loading leads to reliable results and to a realistic assessment of the load-bearing capacity and the factor of safety (Cordes 2017).

Considering time-dependent shotcrete models for back analysis of the load-bearing capacity shows clear benefits (Cordes 2017). The experience gained by the design of the permanent shotcrete linings at BBT (Skava 2018) and the differences between the two design approaches (Marcher 2019) display its application in detail and the differences. Numerical simulations with the shotcrete models after Schädlich et al. 2014 and Neuner et al. 2017a clearly depict the time-dependent behavior of the shotcrete and made the advantageous design for the aforementioned applications possible.

3   PERMANENT SHOTCRETE LINING OF THE EMERGENCY TUNNEL

Parallel to the bypass tunnel of Innsbruck, which was built in 1994, an emergency tunnel has been built as part of the BBT project to update the bypass tunnel to current safety standards. The emergency tunnel with a 47 m² excavation profile is 9.1 km long and is connected every 333 m by cross passages to the bypass tunnel. The tunnels are located at a distance of 30 m with a standard cross-section of the emergency tunnel according to Figure 1. In most sections the cross section was built without an invert.

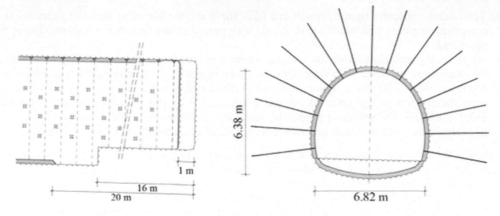

Figure 1.   Longitudinal view and cross-section of the excavation of the emergency tunnel.

In case of incidents, the rescue tunnel provides self-rescue routes every 333 m and allows fast access for emergency vehicles to the accident location. This also allows easy accessibility for, e.g., shortened inspection intervals or local remediation work. The requirements are less demanding compared to the requirements for a tunnel for railway operations. In the new emergency tunnel, the support system using a single shell permanent shotcrete lining met the serviceability requirements. The durability requirements are met using an increased concrete cover of 55 mm, the monitoring class 3 (ÜK3) and a maximum crack opening of 0.3 mm in the homogeneous geological sections. The planning design assumed a single shell permanent shotcrete lining with a thickness of 250 mm and two layers of reinforcement to meet the structural safety requirements. For fault zones and sections of higher convergences, the space for a secondary lining is kept within the stand-ard cross-section. For the planned determination of the load-bearing capacity by back analysis, the convergence monitoring was carried out with 5 monitoring points per cross-section with an initial measurement as soon as possible after construction of the first shotcrete layer. Additionally, 3D laser scanning at two construction stages was carried out to check the excavation pro-file and the shotcrete profile. Hence the built lining thickness was measured in this section with an average thickness of around 450 mm in the crown to 500 mm in the sidewalls (Skava 2018). The structural behavior of shells with irregular thicknesses is strongly influenced by the sections of lower thickness and stiffness. Hence the homogenous thickness for the simulation was assumed as 400 mm for the crown and sidewalls.

4   GEOLOGY

The emergency tunnel is embedded in Innsbruck quartz phyllite with intercalations of lime-stone, dolomitic marble and green schists. The eastern section (up to TM 4600) lies in the Silurian carbonate-sericite phyllite series and the western section in the Ordovician quartz phyllite-green schist series. The series are separated by the steeply incident Hasental fault. Brittle tectonic faults are common and the main fault system is the almost axis-parallel occurring Lavierental fault system. At the Tulfes portal there are Quaternary loose sedi-ments and further granular soil sections, for a total of 200 m, of ground moraine which were cut into during the tunnel excavation. In the predominant phyllite zones the schists breaks up easily into thin layers which consists of layered silicates and quartz. However, it is a dense rock due to its high proportion of mica, which results in low water conductivity and low water inflows to the tunnel.

During the construction of the bypass tunnel (1989–1993), 30 m close to the emergency tunnel, a total water flow of 10 l/s was documented. In the last 4 years an average rate of 5.1 l/s (min. 3.0 l/s to max. 10.2 l/s) was measured in the bypass tunnel. For the construction of the rescue tunnel running parallel to it over the complete length, it was assumed that the total amount of water due to the drainage of the existing bypass tunnel would be small and no major water ingress was to be expected. In case of dripping water ingress, the water is drawn off before shotcreting. In this area, the chemical analyses of the water samples show very low sulphate ion concentrations (e.g. sulphate: 41.0 mg/l). Due to the existing calcite saturation in the mountain water, deterioration of the concrete can be excluded.

5 CALIBRATION OF THE SHOTCRETE MODEL

The hydration process of shotcrete is influenced by the quantity and type of cement and its grinding fineness, the accelerator, the temperature and the aggregates. The matching of binders and additives enables the setting behavior to be adjusted to the necessary shotcrete requirements. There are two basic requirements for deep tunnels:

- a safe overhead application of shotcrete under all various project conditions obtained by a strongly accelerated start of solidification and
- a damage free lining under high convergences by a slower hydration process. The usage of less accelerator leads to slightly higher final strength.

Each shotcrete composition is slightly adapted according the project requirements considering the early strength classes J1, J2 and J3 from the Guideline for Shotcrete of the Austrian Concrete Society (öbv 2009). Due to limited and outdated data in literature (Neuner 2017a) the calibration of the shotcrete model was carried out employing a new testing method using tubular molds as form works. The evolution of Young's modulus, compressive strength, shrinkage and creep were determined for the calibration of the shotcrete model. The required shotcrete on site is a Spc25/30. The construction company applies the shotcrete recipe according to Table 1 and has clearly complied with the requirements.

For sampling the shotcrete specimens required for the lab tests, two different techniques were employed: For tests on shotcrete younger than 24 hours, a novel technique using sprayed specimens directly sampled from tubular molds was employed, whereas for tests on shotcrete older than 24 hours, both sprayed specimens and standard drill cores sampled from spray boxes were used. Sprayed specimens were unmolded safely 6 hours after casting, and lab tests were started subsequently on the evolution of the uniaxial compressive strength. For the sprayed specimens, special care must be taken with the spraying direction, which must be aligned coaxially with the axis of the tubular molds in order to minimize the number of faulty specimens due to shotcrete rebound and air pockets.

Subsequently, the evolution of Young's modulus was determined between a material age of 24 hours and 28 days, while uniaxial compressive strength was tested between 6 hours and 28 days. Shrinkage and creep tests were started simultaneously at three different ages of the material, i.e., 8 hours, 24 hours and 27 hours, and lasted for 56 days each. During the shrinkage and creep tests, the time-dependent strain at the center of the specimens was determined by measuring the time-dependent displacements over a distance of 200 mm, using three

Table 1.   Shotcrete composition employed in the emergency tunnel

Ingredient	Quantity	Unit
Water	203	kg/m^3
CEM I 52.5N	380	kg/m^3
Limestone sand (0/4)	1031	kg/m^3
Crushed limestone aggregates (4/8)	694	kg/m^3
Fluasit (hydraulically binding admixture)	40	kg/m^3
Accelerator *Mapequick 043 FFG*	7.5–8.5	%

Figure 2.   Formwork for the production of young shotcrete test specimens on site (left), stripped shotcrete specimen (center), and a sealed specimen during a creep test (right) (Neuner et al. 2017c).

Table 2.   Experimental calibrated material parameters employed for the SCDP model.

$E^{(1)}$(MPa)	$E^{(28)}$(MPa)	q_4(1/MPa)	v (−)	e^{flow}(−)	$f_{cu}^{(1)}$(MPa)	$f_{cu}^{(28)}$ (MPa)	f_{cv}/f_{cu}(−)
13 943	21 537	34×10^{-6}	0.21	1.22	18.56	40.85	0.1
f_{cb}/f_{cu}(-)	f_{tu}/f_{cu} (-)	ε_∞^{shr}(-)	t_{50}^{shr} (h)	$\varepsilon_{cmv}^{p(1)}$(-)	$\varepsilon_{cmv}^{p(28)}$(-)	$\varepsilon_{cmv}^{p(24)}$(-)	G_1^{28}(N/mm)
1.16	0.1	−0.0019	8645	−0.06	−0.0014	−0.0007	0.1

displacement transducers which were arranged along the perimeter of each specimen. Creep tests were conducted using a hydraulic creep test bench, and the magnitude of the applied compressive load was chosen to ensure exclusively linear viscoelastic material behavior. The specimens tested at 8 hours, 24 hours and 27 hours were loaded with compressive stresses of 1.9 MPa, 2.9 MPa and 2.7 MPa, and the applied load was held constant throughout the complete testing period. The employed tubular molds for the shotcrete specimens and the test setup for the creep tests are illustrated in Figure 2.

A new constitutive model for shotcrete, denoted as the SCDP model, was presented in (Neuner et al. 2017a) and extended in (Neuner et al. 2018). Based on the theories of plasticity, continuum damage mechanics and nonlinear viscoelasticity, it represents the time-dependent evolution of material properties like stiffness, strength, and ductility, hardening and softening material behavior, stiffness degradation due to damage, shrinkage and nonlinear creep behavior of shotcrete. For representing the shotcrete composition employed in the emergency tunnel, the model was calibrated based on the described experimental program. The performed calibration was done with the shotcrete applied in the rescue tunnel. The procedure is reported in (Neuner 2017c), and the resulting material properties are listed in Table 2. For the sake of brevity, the reader is referred to (Neuner et al. 2017a, Neuner et al. 2018) for a detailed description of the material model and the model parameters.

6   BACK ANALYSIS OF THE LOAD-BEARING CAPACITY OF A CROSS SECTION

The load-bearing capacity of the permanent shotcrete lining in the emergency tunnel is determined by means of a 2D finite element simulation in Abaqus (Abaqus 2017). In the simulation, a domain of 200m x 200m is considered. To represent the surrounding rock mass in the finite element model, a linear-elastic, perfectly-plastic material model based on the Mohr-Coulomb yield criterion is used. The geotechnical input parameters are adapted for a better agreement with the measured convergences. Accordingly, a Young's modulus of 500 MPa, a Poisson's ratio of 0.12, a cohesion stress of 0.4 MPa, a friction angle of 25° and a dilation angle of 0° are employed for the rock mass. A higher Young's modulus of 1500 MPa and a higher cohesion stress of 1 MPa are assumed in the region beneath the tunnel.

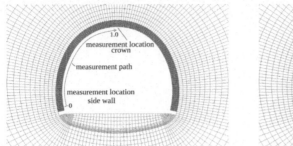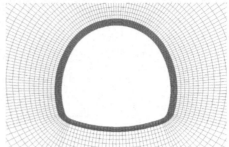

Figure 3. Finite element mesh during the excavation of the top heading and the bench (left) and after excavation of the invert and completion of the shotcrete lining (right).

For the shotcrete model, the experimental calibrated parameters listed in Table 2 are employed. In addition to the shotcrete lining, 15 rock bolts with a diameter of 654.5 mm² and a length of 6 m each (Figure 1) are modeled with truss elements. For the rock bolts, linear-elastic perfectly-plastic material behavior characterized by a Young's modulus of 210,000 MPa and a yield strength of 550 MPa is assumed. The simulation with an initial geostatic stress state characterized by a vertical compressive stress of 9.8 MPa and a lateral stress coefficient of $K_0 = 0.4$ align well with the measured convergences.

Due to the time-dependent material behavior of shotcrete, the time-dependent construction process consisting of a partial excavation sequence is represented in the simulation. On the construction site, the top heading and the bench were excavated simultaneously, whereas the invert was excavated 84 hours (3.5 days) later. The shotcrete lining with a thickness between 300 mm (invert) and 400 mm (top heading and bench) is discretized using fully integrated 8-node quadrilateral elements with 8 elements across the thickness. The employed finite element mesh is shown in Figure 3. From the geometry, the initial conditions and the boundary conditions follows symmetry with respect to the vertical axis.

For considering the time-dependent sequential partial excavation procedure, the simulation is performed according to the commonly employed convergence confinement method as reported in (Neuner et al. 2017b, Schreter et al. 2018) employing a sequence of 3 simulations steps: (i) Excavation of the top heading and the bench: an initial stress acting on the excavation boundary of the top heading and the bench and corresponding to the initial geostatic stress is decreased according to an initial stress release ratio of 80%; (ii) the upper part of the shotcrete lining and the rock bolts are installed in a stress free state, and the remaining stress is removed according to a step-by-step time-dependent stress release layout; (iii) excavation of the invert 84 hours (3.5 days) after excavation of the top heading and the bench, followed by the installation of the lower part of the shotcrete lining.

For simulation step (ii), the employed step-by-step stress release layout for excavation of the top heading and the bench follows an exponential decline, assuming 8 excavation steps in a longitudinal direction and assuming that the remaining stress will be completely removed within a time span of 48 hours. Accordingly, idle periods of 6 hours between the excavation steps are considered. The employed stress release layout is illustrated in Figure 4. In contrast, in simulation step (iii) the excavation of the invert 84 hours after excavation of the top heading and the bench is performed within a single excavation step within one hour, and no initial stress release prior to installation of the lower part of the lining is considered. To assess the influence of the time-dependent material behavior of shotcrete on the structural response, the time-dependent evolution of displacements and stresses is evaluated at two characteristic locations of the shotcrete lining, i.e., the top point at the crown and a point at the side wall. Both locations are indicated in Figure 3.

Figure 5 shows the predicted vertical displacement of the shotcrete lining determined at the crown and the horizontal displacement at the side wall. It can be seen that the largest displacement increments are accumulated during the first excavation steps within a time period of 24

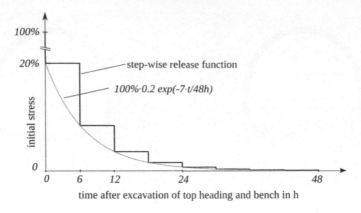

Figure 4. Assumed time-dependent stress release layout after excavation of the top heading and the bench.

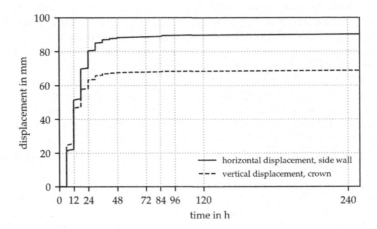

Figure 5. Predicted time-dependent evolution of the vertical displacement at the crown and horizontal displacement at the side wall.

hours. In addition, the displacements visibly increase during the idle periods between the excavation steps, due to creep and shrinkage.

Subsequent to the excavation of the invert, i.e., 84 hours after installation of the upper part of the shotcrete lining, the lining is completed by installing the lower part. However, it can be seen that the excavation of the invert has only a negligible influence on the structural response of the upper part of the lining, i.e., the displacements observed at the crown and the side wall remain virtually constant.

Geodetic measurements performed on site revealed average long term displacements of 71 mm and 54 mm at the crown and 79 mm at the left and the right side walls. The comparison with the respective results from the simulation obtained as 69 mm (crown) and 90 mm (side wall) indicates a reasonable agreement of the numerical model with the measurement data.

Figure 6 shows the predicted time-dependent evolution of the circumferential stresses acting on the lining determined at the rock-shotcrete interfaces and at the free surfaces of the crown and the side wall. It can be seen that comparatively high stresses in compression are predicted at the free surface of the crown, with the maximum value attained after the fourth excavation step, i.e., 24 hours after installation of the upper part of the lining. In contrast, comparatively low compressive stresses are predicted at the free surface of the side wall during the first excavation steps, which change to minor tensile stresses after 24 hours.

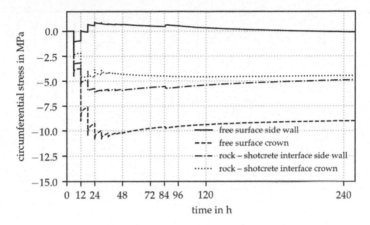

Figure 6. Predicted time-dependent evolution of the circumferential stresses at the rock-shotcrete interface and the free surface of the crown and the side wall.

Figure 7 shows the corresponding predicted longitudinal stresses for the crown and the side wall. For the crown, higher compression stresses occur at the free surface compared to the rock-shotcrete interface. In contrast, as compared to the rock-shotcrete interface at the side wall, lower compressive stresses are predicted at the free surface, which again attain the tensile regime after 24 hours. Beginning from 24 hours after installation of the shotcrete lining, the compressive longitudinal stresses are relaxing, whereas the tensile stresses at the free surface of the side wall are slightly increasing due to stress redistribution within the lining.

To assess the degree of utilization of the shotcrete lining, the circumferential stresses and the longitudinal stresses determined along the measurement path (Figure 3) are illustrated in Figure 8 for a material age of 24 hours. It can be seen that the circumferential compressive stresses at the free surface attain a magnitude of approximately 16 MPa with a local singularity of 17.5 MPa at the rock bolt, while the respective longitudinal compressive stresses do not exceed 4.5 MPa. The visible kinks in the stress distributions indicate the influence of the rock bolts on the stress state in the shotcrete lining. The comparison of the circumferential stress with the experimentally observed uniaxial compressive strength at 24 hours (18.56 MPa) reveals a considerable utilization of the shotcrete lining in compression. According to Figure 6 and 7, the compressive stresses decrease after 24 hours with progressing time due to stress relaxation, and the

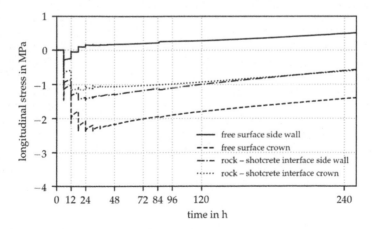

Figure 7. Predicted time-dependent evolution of the longitudinal stresses determined at the rock-shotcrete interface and the free surface of the crown and the side wall.

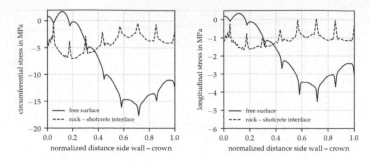

Figure 8. Predicted time-dependent evolution of the circumferential and the longitudinal stresses at the rock-shotcrete interface and the free surface of the crown and the side wall.

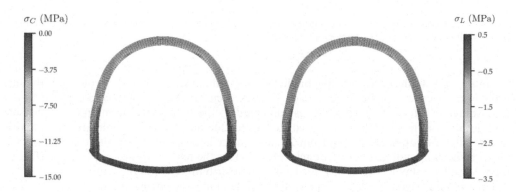

Figure 9. Contour plot of the circumferential stresses σ_C (left) and the longitudinal stresses σ_L (right) acting on the shotcrete lining 108 hours after excavation of the top heading and the bench (deformation scale factor 1).

uniaxial compressive strength increases. Hence the utilization of the shotcrete lining declines in time. In any case, the comparison is conservative because of the biaxial compressive stress state which is characterized by increased strength as compared to the uniaxial one.

Subsequent to the closing of the invert 84 hours after excavation of the top heading and the bench, the lower part of the lining is installed in a stress free state and the initial stress acting on the excavation boundary is decreased within one hour. Figure 9 shows the resulting distributions of the circumferential stresses and the longitudinal stresses acting on the completed lining 108 hours after installation of the upper part, i.e., 24 hours after installation of the lower part. Compared to the upper part of the lining, considerably lower circumferential stresses prevail in the lower part. Furthermore, minor tensile stresses in longitudinal direction act on the lower part. This phenomenon is a consequence of shrinkage in combination with low compressive stresses acting on the lower part of the lining, as explained in (Neuner et al. 2017b). The utilization of the shotcrete lining decreases at 108 hours. The circumferential compressive stresses at the free surface declines to a magnitude of approximately 14 MPa with a local peak of 15.2 MPa at the rock bolt and longitudinal compressive stresses of 3.4 MPa. The uniaxial compressive strength increases at 108 hours to 31.75 MPa.

7   CONCLUSION

Back analyses with time-dependent shotcrete models of built tunnel cross-sections allow detailed insights regarding the load-bearing capacities of the primary linings. Especially for early and highly loaded deep tunnel linings, permanent single shell shotcrete linings and later loadings of existing linings, these insights help to avoid overloading and damaging the linings. This article describes the successful execution of back analyses with a new developed shotcrete model. The new shotcrete damage plasticity (SCDP) model from the University of Innsbruck was calibrated with the test data obtained from a new experimental program. The specimens were sprayed directly on site and allowed testing at a very early age of the material. The results of the back analysis are in good agreement with the measured convergences and display the load-bearing capacity of the shotcrete lining in detail.

The need for further calibrations of the shotcrete material parameters to determine their characteristic values, the statistical scatter and finally the structural safety is therefore clear. In addition, non-linear creep tests, not yet performed, are important for calibration and are already planned. This further work is necessary for the application of the EC2 safety concept for design verifications.

REFERENCES

Abaqus 2017. Abaqus Standard, Simulia, Dassault Systèmes, France.

Bergmeister K. 2011. Brenner Basistunnel - Brenner Base Tunnel - Galleria di Base del Brennero. *Tappeiner Verlag*

Cordes, T. & Hofmann, M. & Murr, R. & Bergmeister, K. 2018. Aktuelle Entwicklungen der Spritzbetontechnologie und Spritzbetonbauweise am Brenner Basistunnel. 12. *Spritzbeton-Tagung*, Alpbach, Austria.

Cordes, T. & Schneider-Muntau, B. & Bergmeister, K. 2017. Inverse analysis of the loading state of a single permanent shotcrete lining at the BBT. *ECCOMAS EURO:TUN*, Innsbruck University, Austria.

Marcher, T. 2019. SEM – Single Shell Lining Application for the Brenner Base Tunnel *ITA-AITES World Tunnel Congress 2019*, Naples, Italy

Neuner, M. & Gamnitzer, P. & Hofstetter, G. 2017a. An Extended Damage Plasticity Model for Shotcrete: Formulation and Comparison with other Shotcrete Models. *Materials* 10(1) 82

Neuner, M. & Schreter, M. & Unteregger, D. & Hofstetter, G. 2017b. Influence of the Constitutive Model for Shotcrete on the Predicted Structural Behavior of the Shotcrete Shell of a Deep Tunnel. *Materials* 10(6) 577

Neuner, M. & Cordes, T. & Drexel, M. & Hofstetter, G, 2017c. Time-Dependent Material Properties of Shotcrete: Experimental and Numerical Study. *Materials* 10(9) 1067

Neuner, M. & Schreter, M. & Hofstetter, G. 2018. Comparative investigation of constitutive models for shotcrete based on numerical simulations of deep tunnel advance. *Numerical Methods in Geotechnical Engineering. 9th European Conference on Numerical Methods in Geotechnical Engineering*, CRC Press, pp. 103–1110

öbv – Austrian Society for Construction Technology 2009. *Guidline Shotcrete*, Vienna, Austria

Pöttler, R. 1990. Time-dependent rock-shotcrete interaction - A numerical shortcut. *Comput. Geotech.* 9, pp. 149–169

Schädlich, B. & Schweiger, H.F. & Marcher, T. & Sauer, E. 2014. Application of a novel constitutive shotcrete model for tunneling. *Eurock 2014, Rock Engineering and Rock Mechanics: Structures in and on Rock Masses*, Taylor & Francis Group, pp. 799–804

Skava 2018. Design of the permanent rescue tunnel, BBT H33, *BBT Internal*

Schreter, M. & Neuner, M. & Unteregger, D. & Hofstetter, G. 2018. On the importance of advanced constitutive models in finite element simulations of deep tunnel advance. *Tunnelling and Underground Space Technology* 80 pp. 103–113

Tunnels and Underground Cities: Engineering and Innovation meet Archaeology,
Architecture and Art, Volume 5: Innovation in underground engineering,
materials and equipment - Part 1 – Peila, Viggiani & Celestino (Eds)
© 2020 Taylor & Francis Group, London, ISBN 978-0-367-46870-5

Two-component backfilling grout for double shield TBM – The experience at Follo Line Project

E. Dal Negro, A. Boscaro, A. Picchio & E. Barbero
Underground Technology Team, MAPEI S.p.A., Milano, Italy

ABSTRACT: The Follo Line Project is one of the most important ongoing projects in Europe, connecting the Norwegian capital and the city of Ski with a high-speed railway. The main package consists in 2 tubes of 18.5 km each. The excavation has been carried out with 4 double shield TBMs. To backfill the gap between rock and concrete segments, the two-component system has been chosen and successfully used. Mapei worked with the contractor for the definition of the most cost-effective mix design, and the modification to the different conditions met during the tunnel excavation, have been some of the aspects of the successful application of the two component techniques. The paper describes the peculiarities of using two-component backfill grout in double shield TBMs, and the modifications to the mix design and to the injection procedure in conditions of relevant water inflow, one of the most challenging conditions for this technique.

1 INTRODUCTION

Two-component backfilling grout is a technique that is growing in the applications number all over the world, for the backfill of the annular gap between the segmental lining of a TBM driven tunnel, and the surrounding soil or rock. The many advantages of the two-component technology make it suitable in very different geological conditions and with different TBMs types. Usually applied in EPB Tunnelling, the two-component backfilling technology is now often applied also in hard rock tunnelling, with single-shield and double-shield TBMs.

The Follo Line Project in Oslo, Norway, is one of the most relevant examples of two-component backfilling grout in hard rock tunnelling. The Underground Technology Team of MAPEI, has followed the project from the very beginning, with specific products developed for the project, with support from the mix design study, through the site trials and the TBM launch, up to the study of specific solutions for difficult geological and hydrogeological conditions.

2 ADVANTAGES OF TWO COMPONENT BACKFILL GROUT

Two-component backfill grout has many advantages compared with the more conventional backfilling techniques. In the following paragraph the main advantages are described.

2.1 *Advantages of two component backfilling grout compared with pea gravel backfilling*

The main advantages of two component backfilling grout compared with pea gravel are:

- Immediate support: due to the quick gel time, the two-component backfilling grout provide immediate support to the tunnel lining;
- Impermeable backfilling: the continuous and homogeneous backfilling, provides a second waterproof shield behind the segments rubber gaskets;

- Complete filling: the pumping of a granular solid matter as pea gravel, makes difficult the complete filling of the annular gap up to the crown of the tunnel. The fluid consistency of two component backfill grout, makes easier the proper and complete filling of whole gap;
- Logistic advantages: two-component backfill grout may be pumped from the batching plant on the surface up to the TBM, through a pipeline without the use of a grout car;
- Safety advantages: lower risk of injection pipe damages and blasting.

2.2 *Advantages of two component backfilling grout compared with monocomponent backfilling mortar*

The main advantages of two component backfilling grout compared with single component mortar are:

- Immediate support: due to the quick gel time, the two-component backfilling grout provide immediate support to the tunnel lining;
- Raw materials consistency: the characteristics of the mortar depends also on the aggregates features, such as grain size distribution and moisture content, which may require constant adjustments of the mix design;
- Reduced wastage: when unforeseen TBM stops occurs, the A component of backfill grout, due to the workability that may be more than 72 hours, may be stored in the TBM tank. One component mortar has to be disposed within 6 – 8 hours from mixing.
- Logistic advantages: two-component backfill grout may be pumped from the batching plant on the surface up to the TBM, through a pipeline without the use of a mortar car;

2.3 *Advantages of two component backfilling grout compared with inert o semi-inert backfilling mortar*

The main advantages of two component backfilling grout compared with single component mortar are:

- Immediate support: due to the quick gel time, the two-component backfilling grout provide immediate support to the tunnel lining. Inert mortars do not provide support for the first 3 – 7 days;
- Raw materials consistency: the characteristics of the mortar depends also on the aggregates features, such as grain size distribution and moisture content, which may require constant adjustments of the mix design;
- Reduced wastage: as for the single component grout, in case of unforeseen TBM stops, the A component of backfill grout, due to the workability that may be more than 72 hours, may be stored in the TBM tank. One component mortar has to be disposed within 6 – 8 hours from mixing;
- Logistic advantages: two-components backfill grout may be pumped from the batching plant on the surface up to the TBM, through a pipeline without the use of a mortar car;

In general, the advantages of two component backfilling grout makes this technology the more flexible and reliable in variable conditions.

For the above-mentioned reasons two component backfill technology have been selected for one of the most important job sites in European continent: the Follo Line Project.

3   THE FOLLO LINE: PROJECT OVERVIEW

The Follo Line is one of the most important projects currently ongoing in Europe. the new railway will connect the Norwegian capital, Oslo, with the city of Ski with a high-speed railway, reducing the transit time of about 50%. The main package consists in 2 tubes of 18.5 km each, with an excavation diameter of 9.96 meters and a final internal diameter of 8.75m. The excavation has been carried out with 4 double shield TBMs.

Each tube of 18.5 km has been bored starting from the Åsland jobsite, where the 4 TBMs have been launched. Each TBM have been boring 9 km in about 2 years.

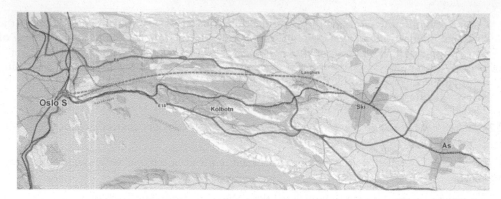

Figure 1. The Follo Line Project.

The main geology has been gneiss with a high variability of weathering conditions and hydrogeological characteristics.

Due to the main advantage of an immediate support to the tunnel lining, the tunnel design includes two-component technology to backfill the annular gap between the 40cm thick concrete lining and the rock mass. The average thickness of the void to be backfilled is about 20 cm with a total volume of about 250,000 m^3.

4 BACKFILL GROUT MIX DESIGN

4.1 *Laboratory study of the mix design*

The study of the mix design of the backfill grout have been carried out during several months before the TBM launch. The strictest technical requirements of the backfilling grout for the project were the compressive strength at 28 days of 3 MPa, and the elastic modulus development:

Based on the technical requirements of the project, and of the different raw materials evaluated by AGJV, the MAPEI research and development laboratory tested more than 200 mix designs. According to the choices of the contractor, the mix designs able to fulfill the requirements have been defined for each set of raw materials.

When the performances have been achieved, the economical optimization of the mix design have been carried out in order to fulfill the technical requirements, reducing the amount of raw materials (such as cement, bentonite, etc..).

Due to the elastic modulus development requirement, a specific alkaline accelerator (component B), have been specifically developed for the Follo Line Project: the MAPEQUICK CBS SYSTEM 3 HEM SS. The specific formulation of the accelerator, allows a faster strength development compared with the traditional backfill grout accelerators, reducing the quantity of cement needed.

In fact, to fulfill the requirement of elastic modulus development (UNI EN 13412-2006) with a CEM II/B-M 42,5 R, were necessary 365 kg/m^3 of cement, and more than 100 kg/m^3 of accelerator. With MAPEQUICK CBS SYSTEM 3 HEM SS, high performance setting accelerator for fast elastic modulus development, have been possible the reduction of the cement of about 7% and of the accelerator of 10%.

Table 1. Required Elastic Modulus development.

Curing time	Elastic Modulus Required [MPa]
2 h	50
24 h	100
7 days	1700
28 days	1900
56 days	2000

Table 2. Backfill grout quality control parameters.

Parameter	Reference values
Marsh funnel viscosity	30" – 45"
Grout Density	> 1.2 kg/l
Bleeding at 1h	< 0.5 %
Bleeding at 2h	< 1 %
Bleeding at 24h	< 5 %
Gel Time	6" – 12"
Strength at 1h	> 0.1 MPa
Strength at 2h	> 0.2 MPa
Strength at 24h	> 0.7 MPa
Strength at 7d	> 2 MPa
Strength at 28d	≥ 3 MPa

4.2 *Site trials for mix design and plant commissioning*

The mix design defined in laboratory, have been verified during the mixing plant commissioning phase. Every day during the commissioning phase and during the first TBM launch, a sample of component A have been collected from the mixer and tested to evaluate all the characteristics of the grout. The tests have been carried out in the site laboratory by the AG JV personnel trained by MAPEI technical service. The commissioning phase have been carried out in two phases:

- Plant calibration: verification of the right dosing and of the raw materials in the mixer during the batching process and check of the dosing consistency;
- Mix design verification: measurement of the characteristics of the grout produced on site in real scale mixer, at the fresh and cured stage.

The plant calibration has been carried out by the observation of the quantity of each raw material added in the mixer from the screw conveyor of the batching plant. The loading system at Follo Line Project is fully automatic. The automatic system is also able to adjust the quantity of each material at every batch, with a dichotomic logic.

The mix design verification has been carried out measuring the grout parameters during the commissioning phase. The same controls have been carried out during whole project for quality control purposes. The parameters measured are reported in the following table with the reference ranges of values.

The site trials confirmed the suitability of the mix design to fulfill the technical requirements. Site trials have been affected by site conditions (temperature, moisture, operator's ability, etc..). The arrangement of a proper site laboratory has been a necessary step for the right test execution, an especially for the proper curing of the hardened samples.

5 TWO COMPONENTS BACKFILL GROUT INJECTION

5.1 *Backfill grout injection sequence*

The injection of backfilling behind the segments was carried out in two phases during the mining operations. The injection of about 70% of the volume to be backfilled (called primary injection), have been filled during the regripping phase, so when the second shield and the TBM back-up move forward by pushing on the last ring built. The injection of this section has been carried out through the dedicated lines in the TBM tail. In this way the void is immediately filled after the TBM passage. The section backfilled in the first phase, is the lower one. For these reasons, the injected grout is immediately subject to the load of the lining ring. The second phase of the injection have been carried out during the excavation phase. The upper section of the annular gap in injected by pumping the grout through dedicated holes, present in the segments. The first injection phase allows the backfill of the void behind a single ring each time. The second phase injection instead, was carried out on more rings (3 – 5) at the same time (Figure 2).

Figure 2.   Secondary injection of adjacent rings

5.2   *Backfill injection parameters*

The primary injection has been carried out through 8 injection lines in the TBM tail skin. The lines are fed by 16 pumps. 8 pumps for A component connected to the grout tank, and 8 pumps connected to the accelerator tank. The injection system, as used at Follo Line, is able to pump the two components with the proportions set in the grouting system computer by the operator.

During the launch phase of the TBM the activity of the MAPEI technical service have been focused in the check of the system efficiency. The check of the 2 component proportions, both on the control panel and physical, is a must during the launch phase. The control of the injected volumes is instead necessary to verify the proper filling of the void. Together with the control of the injected volumes, the verification of the absence of grout leakages through the tail brushed and through the telescopic joint between the two shields, is fundamental to ensure that the injected volume have been filling the annular void.

In hard rock tunneling in fact, the main difference compared with the application in Earth Pressure Balance TBMs, is the lack the earth pressure in the front of the TBM, able to keep the grout in the designed position. The absence of counter pressure may result in leakages of backfill grout towards the tunnel face.

Table 3.   Backfill grout injection parameters.

Parameter	Normal values
Regripping speed	100 – 140 mm/min
Grout flow per line	100 – 140 l/min
Injection pressure higher lines	0 – 0.2 bar
Injection pressure lower lines	1.2 – 1.5 bar
Primary injection volume	7000 – 9000 l
Secondary injection volume	4500 – 2500 l

Figure 3.   Picture of the backfill in a cross-passage

For these reasons, the flow of grout has been adjusted during the injection, in order to guarantee the injection of the required volume of material to fill the void left by the TBM shield, and to keep the injection pressure as low as possible, to avoid the flow of grout towards the TBM shield and cutter head.

In normal conditions, the backfill grout injection parameters have been the following:

The injection pressure values may be affected by the maintenance status of the injection lines. If the injection lines are not clear or with debris reducing the pipe section, the injection pressure observed will be higher and may cause a lower injected volume with the risk of an incomplete filling of the void. Another parameter that affect the injection pressure is the regripping speed. A too high regripping speed, incompatible with the maximum flow of the grouting pumps, may cause a lower injected volume, with the risk of remaining voids. In this case the injection pressure observed will be low.

The control of the injected volumes, together with geophysical analysis and visual observations, especially in the cross-passages areas (Figure 3), have been used during the construction phase to assess the proper filling of the annular gap behind the segments.

6   TWO-COMPONENT BACKFILL GROUT INJECTION IN PRESENCE OF WATER

When the excavation has met fractured zones with presence of water, depending on the quantity of water, pre-grouting intervention from the TBM have been carried out to reduce the water inflow towards the tunnel.

In conditions of residual water, specific countermeasures have been studied from MAPEI to minimize the wash out of the grout.

In presence of relevant water flow, the two-component grout may be diluted and transported by the water towards the TBM shield, with the effect of reducing the effective grout injected and, therefore, causing the need of injecting more volume of grout to properly fill the void.

The specific solution studied for this condition involves both the mix design and the injection procedure.

6.1 Specific mix design for water inflow areas

A specific mix design of the A component and different injection proportion between components A and B have been studied. The results to be achieved by the backfill grout recipe modification are the following:

• a more viscous A component, able to be less diluted by water;
• a stiffer jellified grout immediately after the reaction;
• a faster strength development at the early stage.

To achieve the mentioned results, the mix design has been modified as follows:

• CEM II/B-M 42,5 R: quantity increase of 6%;
• Bentonite: quantity increase of 67%;
• Accelerator MAPEQUICK CBS SYSTEM 3 HEM SS: quantity increase of 10%.

The mix design has been previously studied in the MAPEI R&D laboratory in Norway, then tested on site with the available mixer. Unfortunately, the mixing plant has not been able to disperse such a high amount of bentonite. Even increasing the mixing time, the presence of a relevant amount of bentonite lumps was not avoidable (Figure 4).

Due to the limited quantity of bentonite that was possible to use, the actual mix design used for backfilling in conditions has been made with an increase of only 33% compared with the normal mix design.

The mix design for water inflow areas, have been regularly tested by Mapei personnel during the first days of tests with the modified grout. The grout was fulfilling all the technical requirements listed in Table 2.

6.2 Injection procedure for water inflow areas

The modifications of the injection parameters have been a key point of the procedure.

A relevant reduction of the regripping speed has been the main modification to the injection procedure. The lower regripping speed, allowed the injection with reduced flow and, therefore, reduced speed of the grout in the injection lines. The lower flow, was allowing to inject behind the

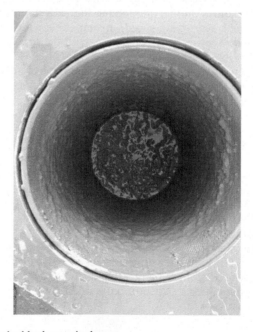

Figure 4.   Grout filter blocked by bentonite lumps

Table 4. Modified backfill grout injection parameters.

Parameter	Normal values
Regripping speed	75 – 95 mm/min
Grout flow per line	45 – 50 l/min
Injection pressure higher lines	0 – 0.2 bar
Injection pressure lower lines	1.2 – 1.5 bar
Primary injection volume	8000 – 9000 l
Secondary injection volume	3500 – 2500 l

segments, a grout at an advanced jellying reaction stage compared with the normal procedure. The advantage of having a grout closer to the jellification is the reduced time spent by the grout at the liquid stage when injected in the gap, with a reduced possibility to be diluted by the water and washed out. Reduced flow cause also reduced pressure, so lower risk for the grout to be pushed towards the TBM shield.

The secondary injection has been modified reducing the injection flow enough to allow the jellification of the grout immediately after the exit of the grout from the injection lance.

The injection process has been followed regularly by specialized Mapei personnel for the first week of test, training the operators and adjusting the parameters in order to achieve the best results.

6.3 Results of the specific injection procedure

After some weeks testing the modified backfilling procedure, visible improvements have been achieved:

- A reduction of the grout injection of about 20 % have been recorded. The reduced wash out of the backfilling grout, allowed to pump less material than before;
- Minimization of ring movements have been achieved without formation of evident steps between adjacent rings;
- Minimization of tertiary injection due to the presence of incomplete filling of the annular gap.

7 CONCLUSIONS

Two component backfilling grout is becoming a widely used technology for shielded hard rock machines, due to the technological, logistic and economic advantages. The Follo Line Project is a clear example of a successful application of this technology, in long tunnels with variable geological and hydrogeological conditions.

The definition of the most suitable mix design is a fundamental step to achieve the requirements of the project design, with a logistic and economic optimization. In hard rock tunneling the injection procedure have to be as much accurate as possible, to allow an effective injection that allows the complete filling of the annular gap, without wastage of materials.

As shown in the paper, the flexibility of the technology allows to face several different conditions, adapting the mix design and the injection procedure without the need of changing raw materials and plants, with a quick response to the site needs.

The follow up on site of a specialized technical expertise is an approach that Mapei apply to enhance the advantages of the two-component technology, with prompt and reliable solutions to the different situation that may occur during the excavation of a long tunnel in rock.

REFERENCES

Peila, D., et al. (2015) Long term behavior of two component back-fill grout mix used in full face mechanized tunnelling. *GEAM, Geoingegneria Ambientale e Mineraria – ISSN 1121-9041. 144:1, PP. 57–63.*
Peila, D., Borio, L., & Pelizza, S. (2011) The behavior of a two- component backfilling grout used in a Tunnel-Boring Machine. *Acta Geotechnica Slovenica – ISSN 1854-0171. PP. 5–15.*

Tunnels and Underground Cities: Engineering and Innovation meet Archaeology,
Architecture and Art, Volume 5: Innovation in underground engineering,
materials and equipment - Part 1 – Peila, Viggiani & Celestino (Eds)
© 2020 Taylor & Francis Group, London, ISBN 978-0-367-46870-5

Main advantages and operational aspects of two-components backfilling grout in mechanized tunnelling with hard-rock TBM: Example of the Saint-Martin-La-Porte TBM job-site at Lyon-Turin

E. Dal Negro, A. Boscaro, A. Picchio & E. Barbero
Underground Technology Team Mapei Spa, Italy

C. Acquista
Ghella Spa, Italy

G. Comin
Spie-Batignolles, France

ABSTRACT: The use of two-component grout to fill the annular void with a hard rock TBM is becoming more and more common. This is because of the two-component system, made up of a cement grout (Component A) and an accelerator (Component B), presents several technical and operative advantages. Compared with its use on EPBs or Slurry Shields, two-components grout application on hard rock TBMs presents several differences, the main one being the absence of a counter-pressure at the tunnel face. The grout mix-design must be specifically formulated and applied on the TBM taking into consideration these variations from the norm. The example of Saint-Martin-La-Porte job-site at Lyon-Turin project is presented: discussing the technical characteristics of the two-component system, this paper presents the solutions made at the job-site to prepare and test the grout and to optimize the injection procedures for proper back-filling injection through difficult geo-mechanic and hydrogeological conditions.

1 INTRODUCTION

During the advancing phase of a shielded TBM, an annulus void is created due to the small difference between cutterhead diameter and external shield one. That gap is created behind the tail shield between ground and segment lining and must be properly filled for a successful TBM drive.

A well filled annular void allows to minimize the settlements of the surrounding ground, to block the ring in its design position avoiding segments movements, to distribute in a homogeneous manner the load from the ground (soil, rock mass, water table, etc.) toward the lining and to improve the tunnel waterproofing reducing the water filtration that may occur.

Quicker the gap is filled and smoother the excavation process will be and so the filling material decision is a key parameter for the TBM project success. The two-components grout, which nowadays is the "standard" material selected for back-filling injection in EPB tunnelling, is a profitable choice also for hard-rock TBM tunnelling due to the several advantages of the system compared to cementitious grout, pea gravel, and all other techniques.

The application of the two-components grout on hard-rock TBM has some peculiarities due to the differences with its use on EPB TBMs (the main one is the absence of a counter-pressure at tunnel face), which should be considered since the grout mix-design study.

The two-components grout has been selected for the back-filling injection at SMP4 job-site, the first mechanized tunnelling lot started for the construction of the new railway line between

Turin and Lyon through the West Alps. The project is located at Saint Martin La Porte in the Savoy department of France close to the Italian border where a tunnel of about 9km is currently under construction as exploration tunnel and as part of the future base tunnel between the two countries.

The two-components grout has been well-designed by the Contractor SMP4 (J.V. between Eiffage TP, Spie Batignolles, Ghella Spa, CMC and Cogeis) in cooperation with UTT department of MAPEI group and is bringing several operative advantages for the tunnel construction.

2 TWO-COMPONENTS BACKFILL GROUT

In the tunnelling industry, the two-components grout is the most common solution adopted by tunnel designer, TBM manufacturer and contractor due to its well-known advantages and experienced implementation.

The two-components grout is a "system" made by two components (so-called Component A and Component B) which act together to fully satisfy the backfilling material scopes.

The system is made by:

• Component A: a cementitious grout made by water, bentonite, cement and a retarding-plasticizing agent (MAPEQUICK CBS SYSTEM 1 line). If well-designed, it is a super-fluid grout, with high volumetric stability, guaranteed workability over long periods (up to 72 hours) and pumpability for long distances (several kilometres)
• Component B: an accelerator admixture (MAPEQUICK CBS SYSTEM accelerators line), added to Component A immediately prior to its injection. When Component A and B are mixed together, a grout with "gel" consistency is reached in a few seconds and an immediate development of mechanical strength is achieved obtaining as fast as now technically possible a material with "hard" consistency.

For how the two-components grout is made, its mainly characteristics are divided in fresh state properties (the Component A features) and in hardened state ones (when Component A and B are mixed).

At fresh state the Component A has:

– very fluid consistency: slightly more viscous than water, it can be pumped without high efforts and even for long distances with the common pumps usually used in the TBM jobsites (piston pump as well as screw pumps or other types).
– very high volumetric stability: a proper mix-design study allows to obtain a uniform and stable grout which within the first 24hours after batching doesn't show any significant separation between water and solid content (cement, bentonite, etc.).

Figure 1–2.   Examples of annulus gap perfectly filled with two-components backfilling grout (on the left hard-rock TBM project for Belchen Tunnel in Swiss and on the right Extension of Line II of Warsaw Metro).

– very long workability maintenance: its super-fluid consistency can be kept up to 72 hours if the retarding-plasticizing agent (MAPEQUICK CBS SYSTEM 1 line) is correctly dosed.

When the Component B (MAPEQUICK CBS SYSTEM accelerator) is added to the grout at the proper dosage, the chemical reaction triggered allows to have a hardened material with almost "suddenly" changed properties:

– "gel" consistency in a few seconds: no more a fluid but a thixotropic material is reached.
– mechanical strengths and elasticity are developed since the first hours after Component A and B mixing.

3   ADVANTAGES OF TWO-COMPONENTS GROUT

The main advantages of the two-components grout compared to the other backfilling methods are recognized worldwide and are related to its characteristics both at fresh and hardened state:

– The super fluid consistency makes the grout very easy to be transported and/or pumped even for long distances. The possibility to pump the grout through a pipeline to transport it from the batching plant to the TBM storing tanks obviates the need of grout cars. The use of a super fluid grout is also an advantage from the operational and safety aspects due to the risk of injection pipe damages and blasting is minimized, especially if compared to some back-filling technique formerly very common in hard-rock TBM tunneling (i.e. pea gravel)
– The extremely long maintenance of workability (dosing correctly the retarding plasticizing agent MAPEQUICK CBS SYSTEM 1 line) allows to batch the grout (Component A) and store it independently from the TBM advance. The grout is still easy to be pumped and transported even several hours after its batching and therefore the logistic advantage compared to the other systems is huge.

Figure 3.   Marsh Cone and Bleeding test for fresh state properties measurement.

Table 1.   Laboratory tests carried out to investigate two-components grout characteristics.

Characteristics	Test
High fluidity	Marsh cone (4.76mm hole diameter)
High stability	Bleeding test
Long workability maintenance	Marsh cone (4.76mm hole diameter) up to 72hours
Very fast creation of a gel	\
Mechanical strength at early stages	Penetrometer and from 24hours compression test at press

Figure 4. Gel time test to measure the mixing time required to obtain a material with gel consistency.

- The great volumetric stability, the very low viscosity, the absence of aggregates and the highly extended workability minimise the risk of clogging of pipeline, storing tanks and injection pipes reducing the time loss due to extraordinary maintenance interventions. As it is commonly observed in the job-sites, the extraordinary activities are the main reasons of TBM production delays and moreover they are usually highly risky interventions.
- When Component B (MAPEQUICK CBS SYSTEM accelerator) is mixed with Component A, a thixotropic gel consistency material is obtained in a few seconds and so in time necessary to exit from the injection lines. Due to its very quick "gellification" time a material in an almost solid state is immediately placed behind the tail shield that it is paramount importance to reduce settlement, improve tunnel waterproofing and more in general to face difficult geotechnical and hydrogeological conditions that can be encountered during TBM tunnelling.
- Immediately after the complete mixing between Component A and B the chemical reaction for the mechanical strengths development starts and already in the first hours the material reached is typically hard enough to block the ring, when it exits from the shield, in its correct and design position and to distribute in homogeneously the ground and groundwater loads towards the segment lining which has the structural aim in a TBM tunnel.

These advantages are subjected to a proper mix-design study which has been made for each TBM project according to its peculiarities as geotechnical and hydrogeological conditions, TBM features, raw material availability in the job-site area, injection technique designed by the Contractor, the human factor, etc.

The mix-design study is made carrying out several laboratory tests (Table 1) that are later replicate at the job-site for the batching plant commissioning and for Quality Control of two-components grout characteristics.

4 TWO COMPONENTS MIX DESIGN STUDY FOR SMP4 JOB SITE

4.1 Laboratory test and preliminary mix-design proposal

The first step usually done for a correct two-components mix-design study is the evaluation of the raw material locally available and which are the main characteristics related to their application in the two-components grout system. Testing at laboratory scale the raw materials is possible to study the preliminary mix-design which should be tested and approved at the job-site batching plant.

For SMP4 job-site this stage has been carried at Research & Development headquarter laboratories of MAPEI group located in Milan (Italy) with the final aim to select for the Contractor the raw material (cement, bentonite, etc.) with best technical performances when applied for the two-components grout production and study the mix-design to prepare a grout with the required technical characteristics.

The parameters considered during the laboratory tests campaign are listed in the Table 2.

Table 2. Parameters for the preliminary mix-design study.

Characteristics	Test	Reference Values
High fluidity	Marsh cone (4.76mm hole diameter)	30"-45"
High stability	Bleeding test	<3% at 3hours < 8% at 24hours
Long workability maintenance	Marsh cone (4.76mm hole diameter) up to 72hours	30"-45"
Very fast creation of a gel	\	5"-15"
Mechanical strength at early stages	Penetrometer for early stages and from 24hours compression test at press	> 0.20 MPa at 1hour testing with penetrometer > 0.45 MPa at 3hour testing with penetrometer > 0.70 MPa at 24 hours testing at press > 1.20 MPa testing at 28days testing at press

After technical and commercial mix-design evaluation made by the Contractor, it has been verified and approved from the Polytechnic of Turin TUSC department, which is specifically dedicated to tunneling researching activities.

4.2 *Mix-design validation in the job-site*

When all tests and studies at laboratory scale finished, a real scale test has been carried out directly at SMP4 job-site using the batching plant dedicated to the Component A production.

The design solution implemented at SMP4 job-site is quite typical of a TBM project where two-components grout is used: Component A is produced at the plant located in the surface and pumped to the TBM storing tanks along dedicated transport pipelines while Component B (MAPEQUICK CBS SYSTEM accelerator) is stored in the surface within specific tanks and pumped along another dedicated pipeline up to the TBM storing containers.

As it is clear, all this system and process should be tested and commissioned before the TBM start to fully check its working stages and aspects. After the first test carried out at the batching plant of the job-site all laboratory tests parameters were complied, and so the industrial production of the two-components grout could start.

During the normal production the Contractor Quality Department periodically check the two-components grout parameters to keep high-standard quality of the material.

Figure 5–6. Two-components grout plant at SMP4 job-site and Component B storing tanks.

The Component A and B pumping distance, which is expected to reach about 12km (all tunnel length and access tunnel), is currently about 6km: such a long pumping distance is technically possible only with the two-components grout.

5 TWO COMPONENTS GROUT INJECTION

In the TBM, the two-components grout injection is usually carried out from the dedicated ports located in the rear part of the tail shield. Immediately before the ports, the Component A and B are mixed and consequently a gel consistency material is injected during the whole TBM advance to fill the annular void from the bottom up to the tunnel crown. The injection is usually carried out checking:

– Volume of grout
– Injection pressure of each single line

According to the common practice of EPB tunneling and what available on technical literature, about 10% more of the theoretical volume (the volume of the void according to the geometric difference between cutter-head and shield diameters) is usually injected and at a pressure a bit higher than the EPB pressure in the TBM chamber (usually 0.5 bars more than the EPB pressure in the upper sector of the chamber).

In a hard-rock TBM some specific aspects should be considered for a proper backfilling injection:

– The annular void volume could be much higher than the theoretical one, especially in case of very weathered rock mass or during the TBM passage through a fault zone.
– There is no counter pressure at the tunnel face and so none pressure value can be used as reference.

For the above reasons, two-components grout injection in hard-rock TBMs can be carried out with several methods and techniques.

At SMP4 job-sites the injection of Component A and B is carried out at two different stages:

– Injection through the tail shield ports during TBM advance to fill the bottom half of the annular void.
– Secondary injection from TBM back-up to fill the whole gap.

5.1 Injection through the tail shield

During the TBM advance, a grout operator is dedicated to the injection procedure and manages the pumping of the two-components grout through the dedicated ports located in the rear part of the tail shield. Managing the PLC (Figure 7) panel directly connected to the pumps, he can change the Component A and B flows.

As designed by SMP4 Contractor, the injection is usually carried out in semi-automatic mode: the injection operator changes, according to the TBM advancing speed and to the injection pressure, the flow of Component A at which is automatically related a Component B flow. The ratio Component B over Component A is set in the PLC in accordance to the Component B dosage used in the approved mix-design.

During the TBM advancing phase, the bottom part of the annular void is filled with this procedure injecting about 4.5m^3 of total grout (Component A and B).

5.2 Secondary injection from the TBM back-up

The filling of the upper part of the annulus void has been designed to inject the two-components grout from the TBM back-up through dedicated ports installed in the

Figure 7.   PLC panel at SMP4 job-site.

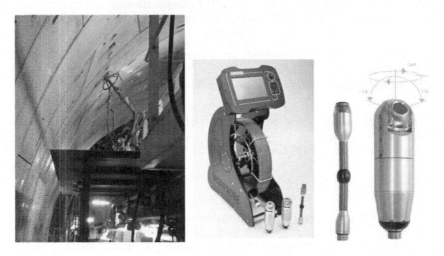

Figure 8–9.   Secondary two-components backfilling grout injection from TBM back-up and endoscope equipment to verify the void filling.

segments, which are specifically designed and manufactured for this purpose. This secondary injection is carried out from the TBM back-up some rings behind the last ring installed and it is made using dedicated pumps and specific injection nozzles designed and manufactured by the Contractor.

The same injection ports located in the segments are then used to visually check if a proper and complete filling of the void has been done. The verification of the complete void filling can also be made by the operators with an endoscope equipment which has been specifically selected for this purpose.

6 CONCLUSIONS

Tunneling with shielded TBM, the backfilling injection has extreme importance for the profitable and safe excavation of the tunnel. The proper filling of the annular void is mandatory to reduce the ground movements triggered by the TBM excavation, to guarantee the final lining design alignment and to enhance the tunnel waterproofing.

The selection of the backfilling material is a key design step and the Contractor should use what is nowadays the superior technical solution for all types of shielded TBMs: the two-components grout.

The two-components grout is already use as standard practice for EPB and Slurry TBMs while for its application in hard-rock TBM some specific solutions, related to the mix-design study and to the injection method in the TBM, are designed for the specific project.

The solution adopted at SMP4 job-site, part of the project for the construction of a new base tunnel between France and Italy, presents several aspects which highlight even more the advantages in the selection of the two-component grout.

REFERENCES

AFTES, (French National Tunnelling Association), *Choosing mechanized tunnelling techniques* (2005) Paris.
EFNARC (2005). *Specification and guidelines for the use of specialist products for Mechanized Tunnelling (TBM) in Soft Ground and Hard Rock* www.efnarc.org.
Dal Negro, E., Boscaro, A. & Plescia, E. (2014). *Two-component backfill grout system in TBM: The experience of the tunnel "Sparvo" in Italy*, Proceedings of TAC Congress 2014: "Tunnelling in a Resource Driven World", Vancouver, 26–28 October 2014.
Dal Negro, E et al. (2014). *Two-component backfill grout system in double shield hard rock TBM. The "Legacy Way" tunnel in Brisbane, Australia*, Proceedings of ITA-AITES World Tunnel Congress 2014: "Tunnels for a better life", Foz do Iguacu, Brazil, May 2014.
Guglielmetti, V., Mahtab A. & Xu, S. (2007) *Mechanized tunnelling in urban area*, Taylor & Francis, London.
Pelizza, S. et al. (2010). *Analysis of the Performance of Two Component Back-filling Grout in Tunnel Boring Machines Operating under Face Pressure*, Proceedings of ITA-AITES World Tunnel Congress 2010: "Tunnel vision towards 2020", Vancouver, 14 20 May 2010.
Thewes M., & Budach C. (2009). *Grouting of the annular gap in shield tunnelling – An important factor for minimization of settlements and production performance*, Proceedings of the ITA-AITES World Tunnel Congress 2009 "Safe Tunnelling for the City and Environment", Budapest, 23–28 May 2009.

*Tunnels and Underground Cities: Engineering and Innovation meet Archaeology,
Architecture and Art, Volume 5: Innovation in underground engineering,
materials and equipment - Part 1 – Peila, Viggiani & Celestino (Eds)*
© 2020 Taylor & Francis Group, London, ISBN 978-0-367-46870-5

Mucking with EPB machines in urban areas – considerations about the soil conditioning. Example of Line 14 project, lot T2 in Paris

E. Dal Negro, A. Boscaro, A. Picchio & E. Barbero
Underground Technology Team Mapei Spa, Italy

G. Broll
CSM Bessac, France

G. Calligaro
Bouygues TP, France

ABSTRACT: Tunnelling with EPB machines in urban environment, the mucking operation plays a key role for the project success. Often the mucking system made by a horizontal belt cannot be used since the beginning due to the lack of available areas for the whole TBM installation. To face these issues, the contractors can select different mucking techniques as piston pumps and/or vertical belts whose successful application is strictly related to the soil conditioning.

The paper presents the Metro Line 14 lot T2 project in Paris and the mucking solutions designed and applied by the J.V. (Bouygues TP, CSM Bessac, Soetanche-Bachy, Soletanche-Bachy-tunnel) with the cooperation of the MAPEI-UTT technical team.

The combination between a foaming agent and a dispersing agent (at a very low concentration) has been used to generate the foam, enhancing its lubricating properties with several advantages: soil consistency optimal both for the piston pump and the vertical belt.

1   INTRODUCTION

Line 14 North project in Paris has been one of the first job-sites started for the construction of the "Grand Paris Express", a huge metro network project made by about 180km of tunnels with the aim to improve the public transport system in the Paris area.

The project was divided in 2 main lots: the lot T1 from the existing "Saint-Lazare" station to "Clichy Saint-Ouen" station and the lot T2 from "Clichy Saint-Ouen" station to "Mairie de Saint Ouen" station including the civil works (shafts, station and tunnel) to connect the line 14 North to the existing line 13. Two EPB TBMS with 8.9m diameter have been selected for the tunnels excavation, one for the lot T1 and one for the T2.

Located in the crowded Paris district of Saint Ouen, the lot T2 had peculiar design and construction aspects: three tunnels had to be excavated from one access shaft within a very tiny environment and limited space for TBM, equipment and plant installation which had to be managed by the Contractor with some specific solutions for the excavation and soil extraction.

As all TBM projects, the soil conditioning had to be proper adjusted according to the geotechnical and hydrogeological conditions, to TBM characteristics and dimensions and to the specific job-site requirements. The selected mucking technique made the quality of the soil conditioning a key parameter for the tunnel excavation success even more important than other TBM projects.

The MAPEI conditioning products chosen as well as their parameters for the Line 14 North lot T2 had to be suitable for the site peculiarities and adjusted for the specific excavation modes and mucking procedures adopted from the Contractor. A deep laboratory test campaign had been carried out before the TBM start and further optimization in the job-site had to be made to fully satisfy the project requests.

2 PARIS GEOLOGY AND HYDROGEOLOGY

Mainly two geological formations have been encountered during the tunnels excavation:

– "Sables de Beauchamp": a very heterogeneous soil, from clay to silty sand with occasionally presence of some boulders. This type of soil was the most common along the tunnel alignment (Fig. 1) and due to its heterogeneous characteristics, in some sector coarse soil and in others cohesive one, had to be properly conditioned for an optimal TBM excavation.
– "Marno-Calcaire de Saint Ouen": weathered limestone characterized by cohesive properties and sticky behavior.

Both soil formations were mainly made by cohesive soils, quite plastic and sticky whose clogging rick had to be controlled and reduced for optimal TBM tunnelling.

From the hydrogeological point, the tunnel alignment was below the water table.

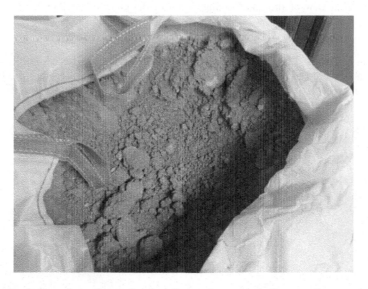

Figure 1. "Sables de Beauchamp" sample taken during the access shaft excavation.

3 LINE 14 NORTH LOT T2 MUCKING SOLUTIONS

The lot T2 TBM works consisted of the excavation of three tunnels, all starting from a narrow and small shaft which will be a future ventilation and emergency access. This type of design approach was due to the crowded and heavily urbanized environment where the use of the underground space was limited.

To face these critical aspects, the Contractor selected several smart solutions to guarantee a proper TBM excavation and extraction: mucking with a piston pump for the first TBM advances, installation of a horizontal belts system in the shaft and the installation of a vertical belt to exit the muck from the shaft.

These solutions were fitted for two different scenarios:

– <u>Scenario 1</u>: for the first advances (about 40rings) the TBM back-up could not be assembled completely and so the definitive mucking system was not available. During this excavation phase, the mucking system installed in the TBM was made by a horizontal belt located after the screw conveyor that ended in a wide and deep hopper installed over a piston pump which had been used to move the muck from the TBM back-up to the vertical belt which transports the spoil to the surface and in the final muck pit within the job-site.

Figure 2–3. Overall view of the initial configuration of the job-site installation with horizontal belt, hopper and piston pump.

– <u>Scenario 2</u>: when the TBM had completed the first advances, the whole back-up could be installed and entered in the tunnel. In this situation the mucking system had changed in a combination of horizontal and vertical belts installed along the tunnel and in the shaft.

Figure 4–5. Overview of the final configuration of the job-site.

The presence of such types of installations and solutions allowed a good excavation rate since the starting phase of the TBM tunnelling even though an optimal conditioned soil was mandatory to guarantee that both excavation and extraction process had the required synergy.

The most suitable soil conditioning products (POLYFOAMER FP/CC and MAPEDISP FLS) have been selected through preliminary conditioning tests at laboratory scale, really useful and importatant for the conditioning products selection and to have a first set of conditioning parameters to be further optimized in the TBM.

4 SOIL CONDITIONING TESTS

One of the most common and representative laboratory test to select the conditioning products and evaluate their parameters is the Slump test using the standardized Abrams Cone (Peila et al., 2009).

At MAPEI R&D laboratories, Slump test as well as other laboratory tests (Vane test, concussion table, etc.) are usually carried out on representative soil samples received from a project prior TBM start. The aims of the tests are to provide the proper conditioning solution (foaming agents or a combination of foaming agent and other products) and to have a first evaluation of the conditioning parameters (Foaming agent POLYFOAMER FP/CC concentration, FER, FIR, Dispersing agent MAPEDISP FLS concentration, WIR, PIR) which allow the correct soil characteristics in terms of mechanical behaviour ("pulpy" paste), reduced permeability and without excess of water and/or foam.

The selection of the correct soil conditioning parameters during the laboratory tests campaign is made adjusting the volume of water added to the soil (WIR), the foam characteristics (Foaming agent concentration and FER) and its amount added (FIR).

Laboratory tests for lot T2 of Line 14 North in Paris had been carried out considering also the peculiar aspects of the project: a more fluid soil was required for the correct work of the mucking pumps but not too much to affect the vertical belt efficiency.

During the laboratory tests carried out with representative "Sables de Beauchamp" samples, several solutions had been studied to properly treat the soil satisfying the project requirements and the Contractor requests. The solutions had been found out both for a mucking system with the piston pump (Scenario 1) and with only belts (Scenario 2).

Different combinations between foaming agents, dispersing agent and polymer have been tested to deeply evaluate the soil behaviour if treated with different chemical products.

- Foaming agent POLYFOAMER FP/CC in combination with pure water addition (WIR).
- Foaming agent POLYFOAMER FP/CC and dispersing agent MAPEDISP FLS.
- Foaming agent POLYFOAMER FP/CC and polymer MAPEDRILL M1 in "light" water solution (at 0.1%, 1l of polymer for 1000l of water).

Table 1. Laboratory tests results on "Sables de Beauchamp" soil samples.

Test 1: the soil is perfectly conditioned for an EPB TBM excavation and for mucking operations with only belts conveyor.

Cdispers agent	Cfoam. agent	FER	FIR	PIR	WIR	
0%	2%	10	50%	0%	10%	

Test 2: the soil is well conditioned for mucking with pumps and vertical belt as well as for the EPB excavation

Cdispers agent	Cfoam. agent	FER	FIR	PIR	WIR	
0.4%	2%	8	40%	0%	12%	

Test 3: the soil has the required behaviour for an EPB TBM and necessary fluidity: enough for a proper pumping but not too much for the vertical belt

Cdispers agent	Cfoam. agent	FER	FIR	PIR	WIR
0%	2%	8	40%	10%	0%

Test 4: the soil is ok for the pumping system while slightly too liquid for the vertical belt

Cdispers agent	Cfoam. agent	FER	FIR	PIR	WIR
0%	2%	8	40%	0%	20%

The results of the laboratory tests could be summed up as following:

– Scenario 1: mucking carried out using piston pumps and vertical belt.

Several combinations of soil conditioning products could be used for a proper soil conditioning for this excavation phase.

The final decision had been made considering the TBM characteristics and equipment installed: a solution for soil conditioning using foaming agent and dispersing agent had been adopted for the initial advances of the TBM.

– Scenario 2: mucking system with horizontal and vertical belt.

From the laboratory tests results, the required soil consistency could be reached using only a foaming agent and some pure water addition. Only in case of very sticky soil had been encountered, the use of dispersing agent could be necessary to reduce improve the lubricating effect of the foam.

5   SOIL CONDTITIONING IN TBM

5.1   *Scenario 1*

During the first advances of the TBM, the whole back-up could not be assembled and so the last gentries had to remain out of the tunnel. In this configuration, the mucking system made by the piston pump and the vertical belt had to be installed to allow as smooth as possible mucking operations.

As all TBM projects, the "learning" phase of TBM excavation is characterised by lower TBM production and a few delays which could increase because of the difficulties related to the mucking system. Optimizing the soil conditioning in accordance to the mucking system installed a lowed to reduce the time loss and have reasonable TBM production.

For the soil conditioning, in the TBM several foam exits were available as well as pure water injection ports and there was the possibility to integrate the foaming agent (POLYFOAMER FP/CC) solution with a dispersing agent (MAPEDISP FLS). All these options had been used to reach an optimal conditioned soil: injection of pure water, foam and the use of the dispersing agent.

The soil encountered during the starting of TBM was "Sables de Beauchamp" characterized by high presence of fine particles with sticky behaviour.

The following soil conditioning parameters had been used during the first advances (Table 2).

Table 2. Laboratory tests results on "Sables de Beauchamp" soil samples.

$C_{dispersing\ agent}$ (MAPEDISP FLS)	$C_{foaming\ agent}$ (POLYFOAMER FP/CC)	FER	FIR	PIR	WIR
0.2-0.4%	1.5-2.0%	8-10	40-50%	0%	10-12%

The soil conditioning parameters showed in Table 2 allowed to have a fluid material on the horizontal belt which could be pumped at the entrance of the vertical belt in a continuous way.

Figure 6–7. Soil on the horizontal TBM belt and at the exit of the piston pump.

The average TBM performance during the first advances using the piston pump were (Table 3).

Table 3. Average TBM data during the first advances.

Thrust force	(MN)	18-23
Contact force	(MN)	7-9
Cutterhead torque	(MNm)	1.5-1.8
Cutterhead rotational speed	(rpm)	1.5-1.9
Working pressure screw conveyor	(bar)	30-40
Screw conveyor rotational speed	(rpm)	0-2
Advancing speed	(mm/min)	15-25

Despite the narrow environment and the limited space, the TBM could start with good performance due to the solution designed by the Contractor and its synergy with the soil conditioning implemented.

Of course, the TBM production was below the expected one through such type of soil due to the limit imposed by the system horizontal belt-pump-vertical belt but in just a couple of weeks the Scenario 2 could be implemented and a further improvement in TBM performances could be reached.

5.2 Scenario 2

When the whole TBM had been installed in the tunnel, the mucking system made by only belts could be used improving a lot the efficiency of the excavation process.

Table 4. Laboratory tests results on "Sables de Beauchamp" soil samples.

$C_{\text{dispersing agent}}$ (MAPEDISP FLS)	$C_{\text{foaming agent}}$ (POLYFOAMER FP/CC)	FER	FIR	PIR	WIR
0-0.2%	1.5-2.0%	10-12	50-60%	0%	8-12%

The soils encountered in this phase were "Sables de Beauchamp" and "Marno-Calcaire de Saint Ouen". Both geological formations were characterized by cohesive soils which could exposed the TBM metallic parts to clogging rick that had to be avoided.

Figure 8–9. Optimal conditioned clayey soil on the belt.

Using the soil conditioning parameters showed in Table 4 both "Sables de Beauchamp" and "Marno-Calcaire de Saint Ouen" could be properly conditioned.

The average TBM performance recorded during the tunnel excavations were (Table 5).

Table 5. Average TBM data.

Thrust force	(MN)	22-25
Contact force	(MN)	8-9
Cutterhead torque	(MNm)	3.5-4.5
Cutterhead rotational speed	(rpm)	1.8-2.0
Working pressure screw conveyor	(bar)	30-50
Screw conveyor rotational speed	(rpm)	8-9
EPB pressure (top sensor)	(bar)	2.5-2.7
Advancing speed	(mm/min)	45-65

The TBM could obtain very good performance: high excavation rate with low energy consumption (low Cutterhead torque and low Screw conveyor torque), optimal soil consistency and low temperature of the muck in the chamber.

Continuous and smooth tunnelling could be reached optimizing the soil conditioning parameters according to the soil encountered.

6 CONCLUSIONS

Each TBM project has peculiar aspects that should be considered all together to successfully complete the tunnel excavation: geotechnical and hydrogeological soil characteristics, TBM features, project environment restriction, design solutions adopted by the Contractor, etc. To profitably face a TBM project the soil conditioning should work in synergy with all project phases where the soil is involved, from its excavation to the extraction and final disposal.

The construction of Lot 2 of Line 14 North in Paris has been an example of well-designed soil conditioning which helped the Contractor to optimally face the projects peculiarities related to the job-site environment and restrictions. The soil conditioning using POLYFOA-MER FP/CC and MAPEDISP FLS has been studied in cooperation with the Contractor's Technical Office since the beginning of the project throughout dedicated laboratory tests carried out at MAPEI R&D laboratory and MAPEI Technical Service presence in the job-site where a strong cooperation with Contractor's technician has been set up to assist the soil conditioning during TBM excavation.

REFERENCES

EFNARC (2005). *Specification and guidelines for the use of specialist products for Mechanized Tunnelling (TBM) in Soft Ground and Hard Rock* www.efnarc.org.

Guglielmetti, V., Mahtab A. & Xu, S. (2007) *Mechanized tunnelling in urban area*, Taylor & Francis, London.

Peila, D., Oggeri, C., Borio, L., (2009), *Using the slump test to assess the behavior of conditioned soil for EPB tunnelling*, Environmental & Engineering Geoscience, pp 167–174, 2009, Vol XV.

Pelizza, S., et al. (2011). *Lab test for EPB ground conditioning*, pp 48–50, Tunnels & Tunnelling, September 2011.

Tunnels and Underground Cities: Engineering and Innovation meet Archaeology,
Architecture and Art, Volume 5: Innovation in underground engineering,
materials and equipment - Part 1 – Peila, Viggiani & Celestino (Eds)
© 2020 Taylor & Francis Group, London, ISBN 978-0-367-46870-5

The 3D-BIM-FEM modeling of the Mairie des Lilas Paris metro station line 11 – from design to execution

F. De Matteis, C. Orci & S. Bilosi
ENSER FRANCE, Paris, France

G. Benedetti
Enser s.r.l., Faenza, Italy and DICAM, University of Bologna, Bologna, Italy

ABSTRACT: The extension of line 11 of the Paris metro required an adaptation of the current terminus of the line, the Mairie de Lilas station. Three new accesses are therefore to be built by means of conventional tunneling methods. Conventional excavation methods are adopted. All the challenges involved by underground excavation in urban environment must be dealt with, such as the interaction with the existing structures, the sensitivity of the adjacent buildings, the uncertainties related to underground infrastructures and to the soil characteristics. A robust traditional calculation approach is used, supported by finite element analyses and the BIM method. The observational method was implemented, coupled with alternative solutions designed beforehand. The continuous coordination between the construction site and the consultant allowed to face the risks and to quickly adjust the design solutions to the unexpected situations.

1   THE MAIRIE DE LILAS STATION PROJECT

The Grand Paris Express is one of the biggest infrastructure projects in Europe. It will expand the city's metro network by 200 km, with four new lines and the extension of two existing ones (11 and 14 lines).

The Mairie de Lilas station is currently the terminus station of the line 11 of the Paris metro. It is located below the Boulevard de la Liberté and it has been placed in service in 1937. Within the Grand Paris Express project, the line 11 will be extended and consequently the station needs to be adapted to the increased number of passengers. Three new accesses are therefore needed:

- The North access, composed by two lifts, serving the existing ticket office and the west-bound platform toward Châtelet;
- The South access composed by one lift serving the existing ticket office and the Eastbound platform toward Rosny-Bois-Perrier;
- The secondary East access composed by two stairs serving both platforms.

The North and the South accesses are being constructed, while the design of the East access is currently in progress.

2   THE GEOLOGICAL CONTEXT

The site is located in the Romainville Plateau overlooking the Marne River valley south, the Seine valley south-west and the Plaine Saint-Denis north.

The reference soil stratigraphy comes from the geological map of Paris 1/50000 of BRGM (Bureau de Recherches Géologiques et Minières), available in Figure 1, and from the archives of the RATP (Régie Autonome des Transports Parisiens).

The geological investigations have been carried out to define the soil profile and the geotechnical characterizations. The following soil layers were found:

- Anthropic fills: a heterogeneous layer of sands, calcareous blocks and bricks. This layer cannot be easily differentiated from the underlying sand layer (about 1.5m thick).
- Sables de Fontainebleau (SF): fine beige-brown sands (about 3–5 m thick).
- Marnes à Huitre (MH): greenish-blue-grey clay marls (about 5 m thick). This is the impermeable layer restricting the aquifer of the Sables de Fontainebleau.
- Calcaire de Brie (CB): heterogeneous marly limestones. The presence of rock blocks might lead to over-excavations. Grout injection is planned for the limestone to improve its mechanical characteristics.
- Argiles Vertes (AV): blue-green marly clay likely to swelling. This is the impermeable layer restricting the aquifer of the Calcaire de Brie.
- Marno-Calcaires de Pantin (MSGp): quite compact greenish beige marls, locally fractured (about 2–5 m thick).
- Marnes d'Argenteuil (MSGa): greenish-grey clays and marls. This is the impermeable layer restricting the aquifer of the Marno-Calcaires de Pantin (about 10 m thick).

As mentioned above, there are three aquifers:

- The water table of the Sables de Fontainebleu restricted by the MH layer,
- The water table of the Calcaire de Brie restricted by AV layer,
- The water table of the Marno-Calcaires de Pantin, restricted by the MSGa layer.

The geotechnical characterization has been possible through the on-site investigations and the laboratory tests.

Figure 1. On the left: extract of the geological map of Paris 1/50000 BRGM. On the right: the contact between the overlaying Calcaire de Brie and the Argiles Vertes at 10,5 m depth.

3 EXCAVATION METHODS

The project involves excavation of open pits and tunnels in an urbanized environment. The works are conducted by using conventional methods.

The open cut excavations are realized by means of the "puits blindé" method. The excavations support system is composed by:

- a retaining wall composed by wooden board, steel sheet piling or shotcrete,
- horizontal steel frames binding the shield against the excavation walls.

The shield and the steel frames are put in place at every step of the excavation of the pit. The steel frames are spaced along the shaft based on the geometric restraints, the soil characteristics and the uniformity of the load distribution. The connection between the frames is guaranteed by vertical suspension elements (typically UPN or L profiles).

The head frame is embedded with some support beams that convey the load to the ground.

The concrete final structure can be built inside the provisional retaining wall at the end of the excavation,

The tunnels are realized by the conventional method by using mechanical excavators. Conventional tunneling involves a cyclic process where every excavation step is followed by the primary support installation. The choice of the supports depends mainly on the soil/rock type faced during the excavation. In this case the primary supports are mostly composed by steel ribs and shotcrete.

Considering the short extension of the horizontal tunnels, conventional tunneling method was the natural choice. However, the real advantage of this method is its flexibility faced with contingencies which may lead to change in structural analysis or support measures.

4 THE EXISTING STRUCTURES

The Mairie de Lilas station has been built in the 1930s by conventional excavation method. Probably, partial excavation was adopted, by splitting the excavation face into sidewall tunnels, crown and invert.

The concrete was used for the side walls and the invert, while gritstone was used for the roof arch. Gritstone is a coarse-grained, hard, siliceous sandstone. Its mechanical characterization is subject by uncertainties due to its heterogeneity and to the state of preservation of the mortar.

The mechanical characterization of the existing structures is a challenging task also due to the uncertainties related to the excavation phases and the current state of the soil stress. In this project, to achieve reasonable results, the increment of stress is considered, instead of comparing the new state of stress with the supposed strength parameters of the structure.

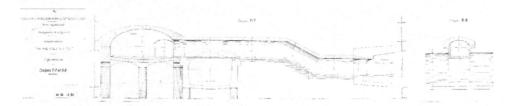

Figure 2. Section view of the Mairie de Lilas station (from RATP archives).

5 CALCULATION METHOD

Different finite elements software packages have been used to carry out the calculation of the structure.

The steel primary structures for the pits have been calculated by means of a frame element analysis carried out by the software Sap2000. The soil-structure interaction is taken into account through the hyperstatic reaction method. The structure is subject to active loads that are independent from the deformation of the structure, and to passive loads dependent on the deformation, namely the soil reactions.

The soil reactions are obtained by defining some line springs activated during the calculation only if compressed. The spring stiffness is computed based on the soil mechanical characteristics.

The tunnel primary supports have been designed by using the geotechnical FEM software Plaxis 2D. The additional purpose of the finite element model was to determine an initial estimate of settlements induced by the tunnel excavation and the deformation of the structure.

Plaxis 2D allows to study the non-linear elasto-plastic behavior of the soil, following the stress-strain evolution during the staged construction process. The hardening constitutive soil model has been used, and therefore the stress-dependency of the stiffness modulus is considered. The convergence-confinement method is applied to consider the 3D effects of the excavation.

The final concrete structures have been designed by Sap2000 too, performing a FEM analysis with shell elements.

A Plaxis 3D model has been set up for each access, in order to estimate the settlements induced by the works and to confirm the project feasibility. The modelling only concerns the temporary phase before construction of final structures.

The BIM technology has been implemented by using the Tekla structures software. The LOD (Level Of Development) system has been used to specify the reliability and the content of the model at various stages in the design and in the construction process. The level of development can be defined by using five levels. Being the object a detailed design, the fourth level (LOD400) is considered. All fabrication and construction schedules are therefore implemented.

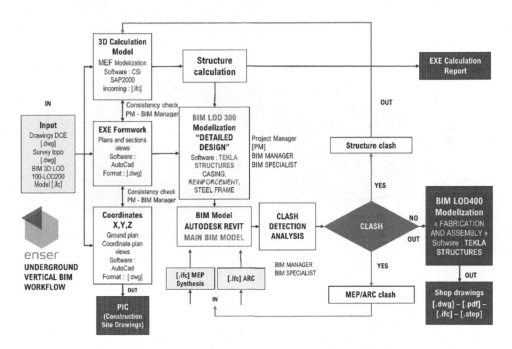

Figure 3.   Underground vertical BIM workflow.

A BIM model is realized for each access, including temporary supports, final structures and existing structures. A georeferenced coordinate system made the BIM to field exchanges faster. Moreover, construction stages have been implemented in the model achieving a 4D design approach.

6   DESCRIPTION OF THE PROJECT

6.1   *The North access*

The North access is located at the intersection between Rue de Paris and Boulevard de la Liberté, in front of the building n°114. This access is composed by two lifts. An opening in the existing wall allows to get into the existing ticket office hall. Similarly, an opening in the wall of the station will be realized to get into the platform toward Châtelet.

The construction of the lifts involves the excavation of a 21m depth shaft (Figure 5).

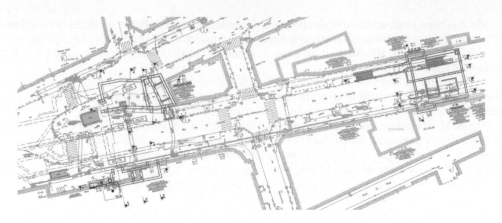

Figure 4.   Plan view of the project.

Figure 5.   On the left the BIM model of the North pit bracing system. On the right a picture from the
bottom of the shaft.

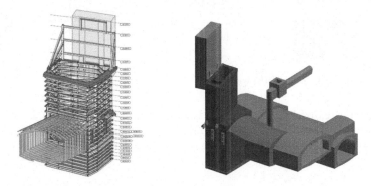

Figure 6.   BIM model of the North access: on the left the temporary support, on the right the final structures in blue with the existing ones in grey.

Being the pit very close to the building 114 Rue de Paris and being the front of the building
already weakened, a façade propping system has been realized to avoid further settlement-
induced damages. A·system of underpinning piles was constructed to respect the threshold
values of the building settlement.

A 13-meter-long tunnel was excavated from the pit at the level of the ticket office
(Figure 6). The section area is equal to almost 61m^2. The whole section of the tunnel was

excavated at once after the face stabilization by means of fiberglass bolts. The bottom part of the tunnel is in the layer of AV (Argiles Vertes) liable to swelling.

6.2 *The South access*

The South access will serve the existing ticket office by a lift and the platform toward Rosny-Bois-Perrier by opening the related wall.

A pit is excavated in front of the post office (n°6 Boulevard de la Liberté). At the bottom of the shaft the tunnel leading to the ticket office will be realized (in purple in Figure 7 on the left). This is the starting point of the excavation of the lift (in blue in Figure 7 on the left), which will lead to the platform towards Rosny-Bois-Perrier trough a deeper tunnel (in red in Figure 7 on the left).

The access is partially below the n°6 and n°8 Boulevard de la Liberté. The 6 Bd de la Liberté is the post office and it is composed by eight floors above ground and one underground floor.

A substructure excavation is therefore required under the post office through a shaft enlargement (Figure 7 on the right). This is a particularly delicate operation, given the building height and the uncertainties about its foundation system. Any deep foundation was indeed found.

The enlargement of the pit was realized as a tunnel suspended to a truss system able to drive the load to the pit top.

In this case, the purpose of the vertical elements is not only to sustain the pit during the excavation but also to reduce displacements related to the dissymmetric geometry and the dissymmetric application of the loads to the pit.

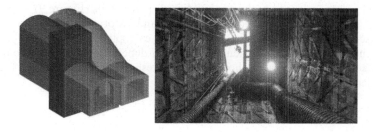

Figure 7. On the left the final structure of the South access: in grey the existing structures, in colors the new ones. On the right the bracing system of the South pit from the bottom of the excavation.

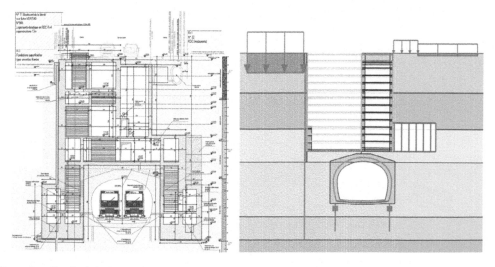

Figure 8. On the left the design section of the East access. On the right the Plaxis 2D model.

6.3 *The East access*

The secondary East access (Figure 8) will lead to both platforms by two stairs. The two stairs will join first the access to the service room on the line and then to the platform. The connection between the new structures and the existing station will require the wrecking of its wall.

The construction of the East access involves the excavation of three pits namely the so-called "small pit" (8m x 4m) and two "big pits" (8m x 7m each). Two tunnels depart from the pits. A subsequent deepening of the excavation is then necessary to get to the platforms.

7 MONITORING PROGRAM

Given the susceptibility of the construction work, due to the presence of the neighboring buildings and the existing underground structures, the application of the observational method was necessary.

The geological conditions must be recorded step by step during tunnels excavations, especially the excavation front, the soil profile, the water inflows, the proper installation of the supports.

Furthermore, a monitoring program is working during construction, including measurements of surface settlements, displacements of the basements of the buildings, deformation of the existing underground structures (i.e. the station and the line 11 tunnel) and deformation of the newly constructed structures.

An automatic topographic station with prisms has been installed to carry out the measurements.

In more detail, the measurements on surface are carried out by seven prisms, while twelve measurements points are used to monitor building displacements (Figure 9). Concerning existing structures, six prisms are installed on the station and six prisms on the line 11 tunnel. Two prisms are also put on a certain number of frames for each pit.

The frequency of these measurements is variable, depending on the outcome. Normally, measurements are taken every two hours. Some threshold values are set in the monitoring system in order to trigger an alert when exceeded.

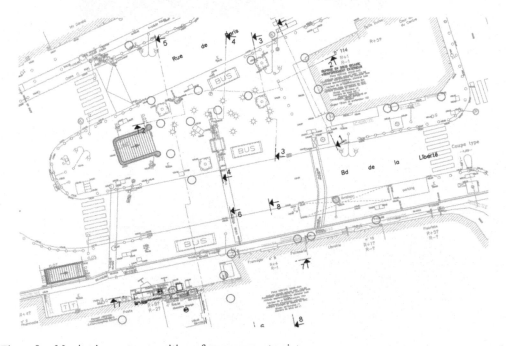

Figure 9. Monitoring system: position of measurement points.

8 CHALLENGES OF THE PROJECT

The main challenges of the project are the urban environment where works are inserted, the interaction with the existing structures, the tight work spaces available and the uncertainties as to the underground facilities.

Each access showed some critical points which made essential the application of the observational method.

During the construction of the North access great attention has been paid while starting the excavation of the tunnel from the pit. A reinforcement frame has been placed against the beginning of the tunnel, and some concrete blocks were built to prevent displacements of the pit due to the asymmetric loading condition (Figure 6 on the left). The tunnel excavation involved high risks, due to the extension of the excavation front (Figure 10 on the left).

The choice to excavate the whole section at once was supported by calculations; nevertheless, some precautionary measures were foreseen to face an unexpected behavior of the structure. Two supplementary columns were provided to be placed in case of unforeseen settlements or weak excavation front stability (Figure 10 on the right).

Furthermore, crackmeter measurements were carried out to check the conditions of the building 114 located in Rue de Paris.

The construction of the South access was particularly sensitive because of the substructure excavation of the post office. The design had been reviewed during the excavation because the effective basement level of this building was found some meters below the expected position. The BIM model was easily adapted, and the interferences were promptly identified (Figure 11).

After construction, the settlements below the post office resulted to be compatible with the thresholds.

The East access is the biggest and the most complex structure and is design is now in progress. Attention should be paid to the stress-strain conditions of the existing tunnel. During excavation, unloading will occur and this may induce some tensile stress in the structure.

The preliminary analyses were carried out to check the feasibility of the project. Plaxis 2D was used to evaluate the displacements and Sap2000 was used to compute the stresses in the tunnel.

The 3D Plaxis model is currently in progress and it will be used to confirm preliminary outcomes. The 2D analysis is typically conservative compared to a 3D analysis. Results are therefore expected to be encouraging. As explained before, a monitoring system is implemented on the existing structure. Some precautionary systems may be placed in case of unexpected behavior of the structure.

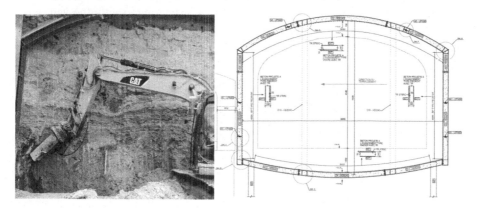

Figure 10. On the left the excavation front of the North tunnel. On the right, the section view of the steel ribs.

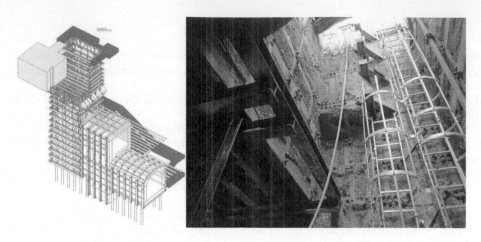

Figure 11.   On the left the BIM model of the temporary support of the South access. On the right the view of the excavation under the post office.

9   BACK ANALYSIS

As mentioned above, the 3D Plaxis model was useful to estimate settlements during construction. A comparison between the expected results vs the monitoring data has been produced.

In general, the observed system behavior was consistent with the expectations. The computed settlements and displacements sometimes showed a slight overestimation, always remaining below the threshold limits.

The steel ribs of the North tunnel at the point of maximum deflection were expected to have a 13mm displacement in the arch, while only 8mm of deflection has been observed (Figure 12). Besides, the ground surface predicted settlement along the position of the prisms was 11mm, but only 7mm was observed.

It is possible that the mechanical characterization of the soil parameters has been too conservative. Specifically, the cohesion of the Calcaire de Brie and the Argiles Vertes layers has been underestimated even if the soil tests may lead to higher values.

Therefore, a Plaxis 2D model has been used to analyze the effects of these parameters on the results. The results obtained with the conservative design parameters have been compared with results obtained with the measured values of soil strength.

The 2D analysis necessarily leads to overestimated displacements. A quick, rough assessment of the theoretical 3D results can be made by defining a conversion rate from the 2D to the 3D results. A 50% reduction of the displacements was observed in the 3D model.

If cohesion of the CB and the AV is modified by considering exactly the values obtained from the laboratory tests, the surface settlements returned by the 2D model show a slight but still appreciable decrease. The supposed 3D maximum surface settlement obtained using the

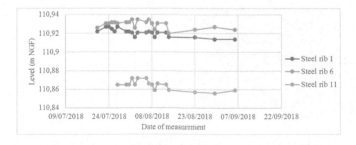

Figure 12.   Measured deflection of the crown of the steel ribs of the North tunnel.

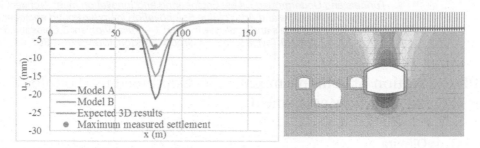

Figure 13. Plaxis models: expected vs measured surface settlements - Model A: conservative estimation of the soil parameters; Model B: measured mechanical soil parameters.

conversion rate, becomes 7,5mm (Figure 13). Discrepancy between predicted results and monitored data is therefore reduced.

Similar analyses can be conducted for the other measurements.

Furthermore, it is important to notice that discrepancy between expected results and FEM outputs is more significant in tunnel excavation, especially regarding displacements before the installation of the support. This might mean that the deconfinement of soil during excavation has not been exactly simulated.

A critical analysis of soil parameters used in calculations is currently in progress for the secondary access modelling.

10 CONCLUSIONS

The adaptation of the Mairie de Lilas station comprises the construction of three new accesses. The project involves underground excavations such as tunnels and shafts that are realized by conventional methods.

The observational method is used with success to take into account the uncertainties due to the urban environment, the natural soil variations, the unexpected presence of facilities.

Furthermore, a cool and responsive approach is implemented, by considering the unforeseen beforehand and designing in advance the alternative solutions easy to adapt to the real situation step by step.

Different technologies have been adopted to design structures, such as finite element analyses through the hyperstatic reactions method, 2D and 3D geotechnical finite elements analyses and BIM technology.

A critical comparison between the expected results and the monitoring data showed a global consistency of this design method. The design and the construction of the East access is currently in progress, this may induce some adjustments to the calibration of the finite elements models.

REFERENCES

Panet, M. 1995. *Le calcul des tunnels par la méthode convergence-confinement*. Paris: Presses ENPC.
Bouvard-Lecoanet, A. & Colombet, G. & Esteulle, F. 1992. *Ouvrages souterraines – Conception, réalisation, Entretien*. Paris: Presses ENPC.
Philipponnat, G. & Bertrand, H. 1997. *Fondations et ouvrages en terre*. Paris: Eyrolles.
Cornejo, L. 1989. Instability at the face: its repercussions for tunnelling technology. *Tunnels and Tunnelling*.
Robert, A. & Saitta, A. 1997. Modelisation numerique des effets du gonflement dans les ouvrages souterrains. *Tunnels et ouvrages souterrains*: 301–306.
Anagnostou, G., & Kovári, K. 1994. The face stability of slurry-shield-driven tunnels. *Tunnelling and underground space technology*, 9(2): 165–174.
Guilloux, A., Kazmierczak, J. B., Kurdts, A., Regal, G. & Wong, H. 2005. Stabilité et renforcement des fronts de taille des tunnels: une approche analytique en contraintes-déformations. *Tunnels et ouvrages souterrains*, (188): 98–108.
ITA Working Group 2009. Conventional Tunnelling. *General report on conventional tunnelling method*.

Tunnels and Underground Cities: Engineering and Innovation meet Archaeology,
Architecture and Art, Volume 5: Innovation in underground engineering,
materials and equipment - Part 1 – Peila, Viggiani & Celestino (Eds)
© 2020 Taylor & Francis Group, London, ISBN 978-0-367-46870-5

EPB excavation of cohesive mixed soils: Combined methodology for clogging and flow assessment

D.G.G. de Oliveira
Herrenknecht AG, Schwanau, Germany

M. Diederichs
Queens University, Kingston, Canada

M. Thewes
Ruhr University Bochum, Bochum, Germany

ABSTRACT: During an EPB drive it is essential to understand the behaviour of the material being excavated. There are numerous bodies of research concerning EPB conditioning that mostly focus on sandy or clayey soils, with relatively little published information regarding cohesive mixed soils. Terrains originating from the tropical weathering of rocks present these cohesive mixed soil characteristics. A laboratory routine was defined for investigating and characterising these cohesive mixed soils, their flow and clogging tendencies. This combined laboratory routine can be also applied to pure clayey soils. The laboratory routine is described here in detail, so it can be reproduced in any other facility, including tunnel project sites, or even using material from a borehole investigation, due to the low required amount of testing material. It is expected that this laboratory routine would make it feasible to more reliably assess the flow and the potential for clogging of cohesive soils.

1 INTRODUCTION

The fluidity of a soil and its potential for clogging are relevant characteristics to be detailed when excavating a cohesive soil with an EPB machine. The available methodologies to assess these above-mentioned characteristics of the cohesive ground are, at a certain extension, limited, especially when considering different percentages of the clay mineral portion. An example of a cohesive soils with different portions of clay minerals, with a varied grain size distribution, is the residual soil typical from the tropical weathering of rocks. The residual soil is the soft ground portion of what is called "mixed transitional ground" by Oliveira and Diederichs (2016). The most important aspect of the tropical residual soil is the grain size variability and distribution, occurring in a same specimen grain sizes varying from clay to gravels, frequently vertically and horizontally changing along a tunnel alignment, therefore, characterising a highly heterogenous ground.

Oliveira et al. (2018a) proposed the use of a flow table to obtain fluidity values from these soils, which can be later correlated with the machine parameters, such as the cutterhead and screw conveyor torque, the flow inside the chamber and screw conveyor, related indirectly to the tunnel face support, as well as the muck transportation, or muck pumping, and, lastly, to its final disposal.

The authors realised that cohesive soils with the same consistency did not always presented the same flow values, when tested with a flow table. The use of the flow table for assessing conditioned materials was first suggested by EFNARC (2005), but only applied to conditioned sands. Oliveira et al. (2018a) proposed the increase of the total amount of drops of the table from 15 to 40, obtaining the parameter Flow$_{40}$. Figure 1 shows the Flow$_{40}$ values for two different types of clay (kaolinite, in green; and bentonite, in orange and red), mixed with fine sand in different proportions.

As it can be observed and as expected, the samples with lower consistency, therefore, higher water content, flow more than the ones with higher consistency. Comparing both clay minerals, the samples mixed with kaolinite show higher flow values than the samples with bentonite, for all the clay fractions. For the kaolinite samples, a decrease in the clay fraction would mean an increase in the flow. The same was not observed with the bentonite samples, there was an initial decrease in the flow with the decrease of clay content up to around 60%, then the flow increased with the decrease of the clay fraction, consequently, the lowest flow observed for the samples mixed with bentonite and sand was around 65% of bentonite fraction.

The bulk density values are presented because this data can be used as a guide value during an EPB operation. The sensors inside an EPB machine do not measure consistency index, but pressures. The apparent density can be obtained by the pressure gradient between different levels of sensors. Consequently, once a target flow is defined, the requirements of the conditioning of the excavated material can be assessed by means of changes in density values.

Another important issue for any cohesive soil when excavated by an EPB machine, is the potential for the occurrence of clogging. Clogging happens when clay particles sticks to the front parts of the machine, obstructing the cutterhead openings or even the entire shield, in the worse cases (Thewes 1999, 2004, Sass and Burbaum 2009, amongst others). There are also limitations in the evaluation of the clogging potential, especially for cohesive mixed soils, with changing clay fractions, as mentioned by Oliveira et al. (2018b), especially when considering the sand contribution, with the addition of a frictional variable.

Oliveira et al. (2018b) conducted an extensive testing phase to evaluate the clogging potential of cohesive mixed soils. Due to its technique simplicity and easiness to acquire the device in the market, it was chosen to test the soils with the methodology proposed by Zumsteg and Puzrin (2012). This methodology provides a stickiness evaluation parameter (λ) by mixing the soil with a mixing machine (brand HOBART) and measuring the mass of the soil stuck in the mixing tool, a B-flat beater, in relation to the total mass of the soil. Depending on the obtained value, the clogging potential classifies as little, medium or strong.

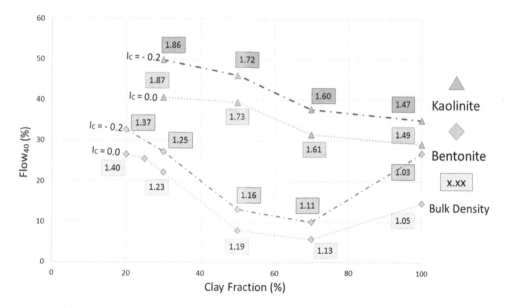

Figure 1. Comparison between the Flow$_{40}$ values with the clay fraction for samples mixed with kaolinite (green triangles) and with bentonite (orange diamond). The bulk density for samples are presented in the squares.

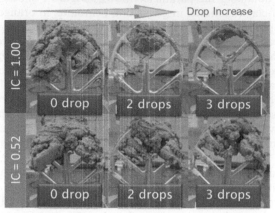

Kaolinite 30% + Fine Sand

Drop Increase

IC = 1.00

0 drop | 2 drops | 3 drops

IC = 0.52

0 drop | 2 drops | 3 drops

Figure 2. Initial tests when it was realized that the soil mass was not totally stuck at the beater. As the drop increases, there is a soil loss depending on the consistency index and properties of the soil.

Oliveira et al. (2018b) when testing mixed soils with different clay contents, realized that many times the material stuck in the beater could be easily removed by any light movement or drop of the beater. It also changed for the same soil, losing a certain quantity of soil mass after a certain number of drops depending on its consistency. Figure 2 shows an example of these comparisons between the mass of soil stuck with an absence of drop and, in different consistency indexes (I_C), loosing mass differently depending on the number of drops.

The lower the clay fraction, or the lower the soil plasticity, more evident these differences between the first measured empirical stickiness parameter (λ_0; no drop) and the following measurements taken after one or more drops (λ_1, λ_2, and so on). This led into an extra stage inclusion, the drop stage, with a new designed device, ATUR (from "Adhäsive Tone Untersuchung RUB-Queens", Clay Adhesion Evaluation RUB-Queens). Later, it was added a last variable which considers the removal and washing stages after the test with the HOBART mixer is conducted, the CB_{Factor} (Cleaning Beater and Bowl Adjusting Factor).

As a consequence, a new methodology to evaluate clogging potential, correlating the results with the well-known clogging evaluation chart from Hollman and Thewes (2012; 2013), was proposed by Oliveira (2018). Together with the flow behaviour methodology, it is feasible to improve the characterisation of a cohesive soil, mixed or not, as well as evaluating, at least qualitatively, the effectiveness of any soil conditioning.

This paper presents a resume of this combined methodology to evaluate flow behaviour and clogging potential of cohesive soils. The main advantage of this laboratory routine is the required amount of the testing specimen, which is less than one kilo per test, allowing preliminary evaluations even prior the tunnel excavation, at early design stages of a tunnel project. It also points out some insights regarding the CB_{Factor}, achieved while testing natural soils.

2 COMBINED METHODOLOGY

This combined laboratory routine was developed since December of 2016, starting with testing soils mixed in the laboratory, so all the variables could be controlled. In a later opportunity, natural soils have been tested, however, due to the confidential aspect of the location and data, only insights about this last phase can be here included. The assumptions here made should also be in future cross-checked with results from tunnel drives where this methodology would be applied.

Initially, the samples should be prepared, verifying if the target consistency of the soil is achieved. The test follows with the mixing of the soil with the HOBART mixer and the initial measurement of the stickiness parameter, as already proposed by Zumsteg and Puzrin (2012), obtaining λ_0. The sequential dropping of the beater with the ATUR device is the next step, measuring the λ_1, λ_2, λ_3 and λ_7 (1, 2, 3 and 7 drops, respectively). The beater and bowl should be cleaned, and observations should be made regarding the difficulties to remove the soil from it. Finally, whenever possible (soils with consistency below 0.6) the soil should be tested with the flow table, dropping it 40 times, obtaining the parameter $Flow_{40}$. The last stage is the squeezing of the samples, taking pictures and recording videos, which might be useful to later characterise the soil, and compare it with other soils or between different conditioning options. Figure 3 illustrates the stages of this methodology and it can be also assessed in detail in Oliveira (2018) and Oliveira et al. (2018a, 2018b).

The device ATUR has been assembled so far two times and still needs improvements, specially concerning the material of the base (Figure 4). Casagrande (1958) took several years to define the ideal material for his liquid limit device, the Casagrande cup. A similar process needs to be still conducted with ATUR, standardising all its parts, especially the thickness and material of the base. With the lack of this standardisation, it is required to preliminary conduct a calibration stage. This stage requires a pure clay material (or as close to pure as possible) with plasticity index between 20% and 40%. For both calibration events it was used a 95% pure kaolinite clay. As detailed in Oliveira (2018) and Oliveira et al. (2018b), the soil should be prepared in six different I_{Cs} (1.05, 0.95, 0.80, 0.70, 0.55 and 0.45). The test results should be correlated with the graph from Hollman and Thewes (2012), combining with the CB_{Factor} and using the equations 1, 2 and 3, as detailed ahead.

Once calibrated with the pure clay, other soils can be tested and directly compared with the pure clay sample. Summarily, the soil to be tested should be placed in the mixing bowl, weighting the total mass, mixing for three minutes in the speed 1 (around 100 rpm), as originally proposed by Zumsteg and Puzrin (2012). The soil mass stuck in the beater should be weighted, and divided by the total mass, providing the parameter λ_0 (Equation 1). Then the beater should be placed in the device ATUR and dropped once, twice, three times, every time

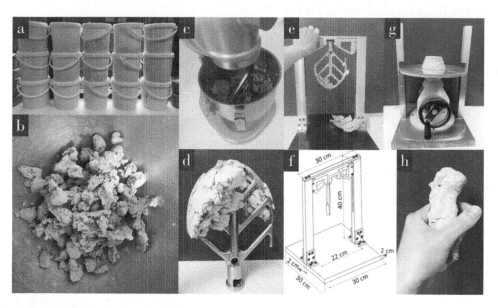

Figure 3. Illustration of parts of testing sequence: (a) soils were mixed with required mineral and water contents and stored in closed containers; (b) a reasonable quantity of sample was placed as lumps in the HOBART mixing bowl; (c) the HOBART mixer with soil specimen; (d) B-flat beater with stuck soil on it before any drop (λ_0); (e) B-flat beater placed in the ATUR device after one drop; (f) detail of the dimensions of the ATUR device; (g) flow table with soil specimen after 40 joltings of the table; (h) squeezing of soil specimen to assess muck consistency and flow (Oliveira 2018).

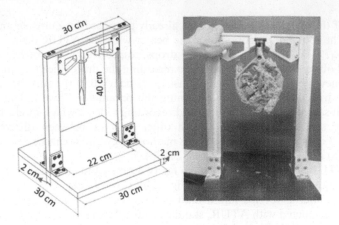

Figure 4. ATUR device: drawing with dimensions on the left and B-Flat beater with soil stuck to it on the right (Oliveira et al. 2018b and Oliveira 2018).

measuring the new mass of the stuck soil, providing the parameters λ_1, λ_2 and λ_3 (Equation 1), and at last, dropping four times in a roll, providing the parameter λ_7 (Equation 1). An average of all the stickiness parameters λ should be obtained (Equation 2).

The surface of ATUR should always be cleaned after each drop and pictures should be taken for each stage. The difficulties of removing the material stuck at the beater and in the bowl are going to provide the parameter CB_{Factor}, as resumed in the Figure 5. The added value of the CB_{Factor} would be used to define the final stickiness parameter λ_F, as defined by the Equation 3.

As mentioned, the calculation is done using the Equations 1, 2 and 3, where λ_x is the stickiness parameter (x replaced by 0, 1, 2, 3 or 7), G_{MTx} the mass of the soil stuck at the mixing tool, and G_{TOT} the total mass of the soil.

$$\lambda_x = \frac{G_{MTx}}{G_{TOT}} \tag{1}$$

$$\overline{\lambda_x} = \frac{\lambda_0 + \lambda_1 + \lambda_2 + \lambda_3 + \lambda_7}{5} \tag{2}$$

$$\lambda_F = \frac{\overline{\lambda_x}}{CB_{Factor}} \tag{3}$$

Initially, it was defined full numbers for the CB_{Factor}, but when testing different natural soils, it was realised that fractions would provide a better characterisation of the cleaning stage, especially helping to differentiate between soils in a similar area, or along a tunnel alignment. The differences where very clear when considering soils with only 30% of clay, for example, and a pure clay sample with a swelling mineral. The last would imply in a greater challenge to clean the tools, especially the bowl, where many times with the mixing procedure, the sticky clay would even "cook" in the bottom of the bowl, increasing the difficulties. This could also be correlated with expected cleaning issues during the interventions of a tunnel drive to clean the excavation tools and cutterhead.

Once defined the final stickiness parameter, the material can be characterised as little, medium or strong potential for clogging, with the following values, as shown in Figure 6. These values were defined after the calibration was done. Whenever a new ATUR is assembled, as mentioned before, a new calibration needs to be conducted, unless a final device is standardised. As well, whenever the λ_F is lower than 0.27 and the consistency is stiff ($I_C >$ 1.0), then the clogging potential should be classified as lumps, and not little.

Main category	Sub-category	Description	Added Value
St (stiff)	III	Practically, all stuck materials can be easily removed by hand; final cleaning requires only water (< 10 s).	3
	II	Considerable portion of stuck material can be removed by hand; remaining portion is slightly stiff and difficult to remove without water. Bowl needs spooning and washing (< 30 s) to remove stuck material at the bottom.	2
	I	Similar as above but requires longer than 30 s to clean beater and bowl, plus brushing and soap washing with lukewarm water.	1
So (soft)	III	Part of the soil can be removed by hand but soil gets easily stuck on them. Remaining material removed with water.	3
	II	Great portion of soil gets stuck on fingers; mainly, it can be cleaned with water but might require brushing for corners of beater. Bowl takes longer than 30 s to clean, maybe requiring sponging.	2
	I	Similar as above but more sticky and requires brush and soap to completely remove material	1
Vs (very soft)	III	Not possible to remove soil by hand; too soft, but requires only approximately 30 s to remove it only with water.	3
	II	Similar as above, but it needs sponge to remove all soil. It takes longer than 30 s.	2
	I	Similar as above, but needs brushing and soap washing to remove all soil. It takes longer than a minute	1

Ic – Consistency Index

IP – Plasticity Index

Figure 5. Cleaning beater and bowl factor (CB_{Factor}), with categories and sub-categories description on the top, including the respective values to be added to the equation to calculate λ_7 (Oliveira 2018).

This methodology to characterise clogging still needs to be improved and correlated with real tunnel drives, however, it can already provide useful insights about the real chances of facing issues in tunnel drives along cohesive soils. For example, between the natural tested soils, there was a well-known clay with high I_C and high plasticity. When the test was conducted, not a single mass of the clay was dropped, even after 7 drops. As a challenge, it was decided to continue dropping the beater in the device ATUR, and it took a total of 40 drops until the first piece of clay fell, clearly indicating that clogging issues should be expected when excavating this soil. Some of the other clayey soils would have loss of mass even in the first drop.

The next stage of the combined methodology would be to verify the flow, or fluidity of the soil, using the flow table. For the test is required a glass surface with the same shape as the flow base. This test is described in detail in Oliveira et al. (2018a) and Oliveira (2018). Summarily, the cone should be coated with oil, the soil placed inside the cone, in layers and slightly squeezing to remove any air bubbles trapped inside the soil layers, the cone lifted and the

λ_F	Clogging potential
≤ 0.27+St	Lumps
≤ 0.27+So;Vs	Little
0.27–0.48	Medium
≥ 0.48	Strong

Figure 6. Clogging potential categories, having the final stickiness parameter λF. The terms St, So and Vs means, respectively, stiff, soft and very soft (Oliveira 2018).

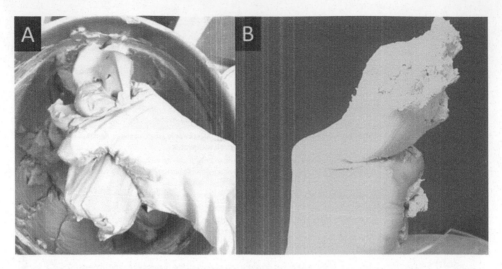

Figure 7. Comparison between two pure clay samples, bentonite (A) and kaolinite (B), with similar consistencies, I_C = 0.43 and 0.44, respectively, showing what would be the lowest boundary of the ideal EPB range (Oliveira 2018).

initial diameter of the base measured (m_0 in Equation 4). Then the table is jolted 40 times and the parameter m_{40} obtained by measuring the final diameter of the soil spread in the glass base, calculating the $Flow_{40}$ according to the Equation 4.

$$Flow_{40} = (\frac{m_{40} - m_0}{m_0}) * 100 \qquad (4)$$

At last, the material should be squeezed, and observations made, as well as pictures and, ideally, a video should be recorded. All this material can enrich the characterisation of the material and later will be useful for the tunnel design, operation, and, whenever necessary, for future stakeholder disputes. The squeezing stage, as shown in Figure 7, is already conducted spontaneously by the personnel checking the muck in the conveyer belt, to adjust the conditioning parameters. When squeezing the material is possible to qualitatively estimate the strength mand the consistency of the soil. The muck should never be too dry, not easy to be squeezed, therefore, not holding enough flow inside the chamber and screw conveyor, neither too liquid, falling by the side of the hands when squeezing it. If a visual correlation could be made with a toothpaste, that would be the ideal state of the muck, a new and perfect paste, not too dry and stuck inside the tube, neither too liquid.

3 EXAMPLES OF CONDUCTED TESTS

This section provides some examples of conducted tests, comparing pure clay materials with mixed clay-sand soils, as well as some examples of conditioned material. Figure 8 compares the results between samples with different clay fractions, 30 and 100%, respectively. It also compares the results between the original methodology of obtaining the stickiness parameter (λ_0), and the results after dropping the beater seven times (λ_7), which is still not the final stickiness parameter (λ_F). The sample with 30% of kaolinite presents less clogging potential, when considering λ_7, which is not the case with λ_0, really showing that only measuring the beater was not enough to produce reliable results. It also shows that is very unlikely to have any clogging occurrence with soils containing only 30% of kaolinite, as main clay. While removing the soil from the beater and cleaning the bowl was evident how easy and fast was to detach the

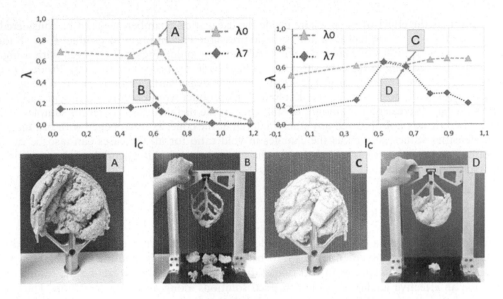

Figure 8. Comparison between two soils with different clay fraction, 30% (left) and 100% (right), showing how the extra drop stage provides different results for a certain consistency range, and, especially, for soils with not 100% of clay fraction contribution.

material, leading into $CB_{Factors}$ with added value of 3. The sample with 100% of kaolinite (left of the Figure 8) shows exactly what several authors have realized, that the highest clogging potential is between the consistencies 0.5 and 0.7, when considering the stickiness parameter λ_7. The main impression was that when a soil holds a potential for clogging to happen, the beater can be dropped, and the loss of soil is not significant, such as the case of the natural soil described in the previous item, that needed to be dropped 40 times until any loss would occur.

Tests with different conditioners were also done and can be assessed in Oliveira (2018). Figure 9 shows some examples of flow table tests with three different foam parameters, all conducted with a sample of 30% of kaolinite mixed with sand. The product described as A is only a foam and the product described as C is a foam with anti-clay polymer added. The first two images (Figure 9A and B) compares two different FER – Foam Expansion Ratios, and the first and last, compare two different products.

For the sample with 30% of kaolinite, a higher FER seemed to produce a "better quality" of soil mixture (see Figure 9 A and B), but it is difficult to say if it is better or worse for EPB

Figure 9. Flow of samples with 30% of kaolinite and two different target FERs (9.5 and 20), for two products A (only foam) and C (foam with anti-clay polymer). The cone on C shows the stuck material after lifting the cone, therefore, the product with anti-clay seemed to not have helped to decrease the stickiness properties of the soil.

operation, as the effect of the pressure inside the excavation chamber needs to be considered, as all the tests were conducted in atmospheric pressure. The addition of product C in all the samples leads to pieces of material being stuck in the mould (Figure 9 C), even with a high amount of oil coating in the cone surface, as described in the procedure (item 2). Product C is supposed to have anti-clay properties (polymer with dispersant functions of the clay aggregates). Thus, and most probably for kaolinite, this product would be functioning in quite the opposite way, increasing the amount of soil stuck in the metal surface. Even for samples with only water, it was not observed any stuck material in the cone, as it happened with product C. The densities values, as well as the changes in void ratio for all the cases can also be easily obtained and could serve as a guide for the tunnel drive and the effectiveness of the conditioning parameters, as detailed in Oliveira (2018).

4 FINAL CONSIDERATIONS

This paper briefed on a combined methodology which characterises cohesive soils, mixed or not, in terms of clogging potential and flow behaviour. As already mentioned, this laboratory routine should be validated with real tunnel drives, making feasible to refine the results and provide an effective guidance of the proposed solutions along an EPB excavation. Then the values of consistency and flow can be crossed with the excavation parameters. Also, a preliminary stage of muck characterisation can be conducted, and later, during the excavation, a back-analysis can be done, again refining the results.

An example of what could be a resulting product of this preliminary analysis done when conducting this laboratory routine, characterising and predicting the clogging potential and flow behaviour of cohesive soils, mixed or not, is illustrated in Figure 10. This figure did not use any real data from a project but is meant to provide an imaginary scenario for illustration purposes using the geological-geotechnical profile from the São Paulo metro line 4, phase 3, as presented in Oliveira et al. (2017), as a base. It is included also stretches of mixed transitional ground (MTG) with high amount of rock chips, as detailed in Oliveira (2018).

Something like this could always be implemented in the basic design stage of a tunnel project as soon as the borehole material is already available. A second stage can be later conducted including the material from the excavation of the tunnel, comparing both stages. By

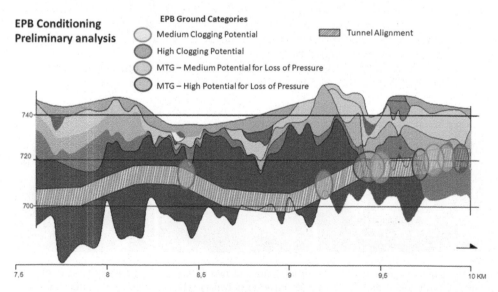

Figure 10. Example of an imaginary representation of a preliminary analysis regarding EPB flow behaviour and clogging potential in a longitudinal geological profile (modified from Oliveira et al. 2017).

doing that, we will be able to design, and construct tunnels excavated by EPB machines more efficiently, being feasible to overcome excavation challenges, especially related to conditioning of the material (e.g. face pressure, muck transportation and later disposal).

Most of the time the solutions of soil conditioning are done along the tunnel excavation, with a trial-and-error approach, which not always are successful, in the worst cases leading into major delays and accidents. By investigating and characterising the soil to be excavated prior the tunnel starts, can be provided a better knowledge of the material and its behaviour. Most importantly, it can be explored an optimisation of the conditioning solutions, many times avoiding unnecessary costs with high value products.

The primary value of this laboratory routine is the easiness of reproduction in any other laboratory facility, or job site, making future comparisons possible, and, perhaps, allowing a standardisation that can be applied in tunnel projects worldwide, establishing the choices of soil conditioning and tunnel operation in more scientific-based conclusions.

5 ACKNOWLEDGMENTS

Support for this research comes from the Natural Science and Engineering Research Council (NSERC), Canada. The authors would like to acknowledge the support of the Collaborative Research Centre SFB 837 "Interaction Modelling in Mechanised Tunnelling", funded by the German Research Foundation DFG. We are thankful to the laboratory team at TLB, at Ruhr University Bochum, for the support with the testing campaign.

REFERENCES

Casagrande, A. 1958, Notes on the design of the liquid limit device, Geotechnique 8(2): 84–91.

EFNARC. 2005. Specification and Guidelines for the use of specialist products for mechanised tunnelling (TBM) in soft ground and hard rock. EFNARC, UK.

Hollmann, F. and Thewes, M. 2012. Evaluation of the tendency of clogging and separation of fines on shield drives. Geomechanics and Tunnelling 5: 574–580.

Hollmann, F., and Thewes, M. 2013. Assessment method for clay clogging and disintegration of fines in mechanised tunnelling. Tunnelling and Underground Space Technology 37: 96–106. doi: 10.1016/j.tust.2013.03.010

de Oliveira, D.G.G. and Diederichs, M.S. 2016. TBM interaction with soil-rock transitional ground, TAC 2016 Annual Conference, Canadian Tunnelling Association, Ottawa, ON, Canada.

Oliveira, D.G.G., Rocha, H.C., Monteiro, M.D., and Dias, C.C., 2017. Caracterização geotécnica dos maciços ao longo do projeto da Linha 4-Amarela, Fase 3 do Metrô de São Paulo. Proceedings of the 4° CBT – Brazilian Tunnelling Conference, Sao Paulo.

de Oliveira, D.G.G.. 2018. EPB excavation and conditioning of cohesive mixed soils: clogging and flow evaluation. Doctoral thesis, Queen's University, Kinsgton, 282 p.

Oliveira, D.G.G. de Thewes, M., Diederichs, M.S. and Langmaack, L. 2018a. Consistency index and its correlation with EPB excavation of mixed clay-sand soils, Geotechnical and Geological Engineering Journal, Springer Int. Publ., (2018): 1–19.

Oliveira, D.G.G., Thewes, M., Diederichs, M.S., Langmaack, L. 2018b. EPB tunnelling through clay-sand mixed soils: proposed methodology for clogging evaluation. Geomechanics and Tunnelling Journal 4, vol. 11, http://dx.doi.org/10.1002/geot.201800009

Sass, I, and Burbaum, U. 2009. A method for assessing adhesion of clays to tunneling machines. Bull Eng Geol Environ 68: 27–34.

Thewes, M. 1999. Adhäsion von Tonböden beim Tunnelvortrieb mit Flüssigkeitsschilden. Doctoral Thesis. Bericht aus Bodenmechanik und Grundbau, Bergische Universität Wuppertal, Fachbereich Bauingenieurwesen.

Thewes, M. 2004. Schildvortrieb mit Flüssigkeits-oder Erddruckstützung in Bereichen mit gemischter Ortsbrust aus Fels und Lockergestein. Geotechnik 27: 214–219.

Zumsteg, R. and Puzrin, A.M. 2012. Stickiness and adhesion of conditioned clay pastes. Tunnelling Underground Space Technol 31: 86–96.

Tunnels and Underground Cities: Engineering and Innovation meet Archaeology, Architecture and Art, Volume 5: Innovation in underground engineering, materials and equipment - Part 1 – Peila, Viggiani & Celestino (Eds)
© *2020 Taylor & Francis Group, London, ISBN 978-0-367-46870-5*

Hybrid solution with fiber reinforced concrete and glass fiber reinforced polymer rebars for precast tunnel segments

B. De Rivat
BMUS – Bekaert-Maccaferri Underground Solutions, Belgium

N. Giamundo
ATP srl, Italy

A. Meda, Z. Rinaldi & S. Spagnuolo
Tunnelling Engineering Research Centre, University of Rome "Tor Vergata", Italy

ABSTRACT: The interest in using fiber reinforced concrete (FRC) for the production of precast segments in tunnel lining, installed with Tunnel Boring Machines (TBMs), is continuously growing, as witnessed by the studies available in literature and by the actual applications. The possibility of adopting a hybrid solution of FRC tunnel segments with GFRP reinforcement is investigated herein. Full-scale tests were carried out on FRC segments with and without GFRP cage, with a typical geometry of metro tunnels In particular, both flexural and point load full-scale tests were carried out, for the evaluation of the structural performances (both in terms of structural capacity and crack pattern evolution) under bending, and under the TBM thrust. Finally, the obtained results are compared, in order to judge the effectiveness of the proposed technical solution.

1 INTRODUCTION

In the last few years the adoption of fiber reinforced concrete (FRC) in precast tunnel segments, has encountered a great interest, as witnessed by theoretical and experimental studies (Plizzari and Tiberti, 2008; Caratelli et al., 2011; Liao et al., 2015), and actual applications (Kasper et al. 2008, De La Fuente et al., 2012, Caratelli et al., 2012). The solution of FRC elements, without any reinforcement, provides the great advantages, in terms of cost and precast production. Nevertheless, in some part of the tunnel, for particularly loading condition (typically under prevalent bending actions, as in cross-passage or shallow tunnel), the FRC solution could not satisfy the requirement. In this the adoption of a hybrid system, with the addition of rebars, could be a realistic solution.

The possibility of adopting glass fiber reinforced polymers (GFRP) reinforcement in precast tunnel segments in ordinary concrete was investigated (Caratelli et al., 2016; Caratelli et al., 2017; Spagnuolo et al., 2017). GFRP rebars in concrete structures can be proposed as an alternative to the traditional steel rebars, mainly when a high resistance to the environmental attack is required. Indeed, GFRP reinforcement does not suffer corrosion problems and its durability performance is a function of its constituent parts (Micelli and Nanni 2004; Chen et al. 2007). From the mechanical point of view, the GFRP rebars are characterised by an elastic behaviour in tension, and, with respect to the steel ones, present higher tensile capacity, lower elastic modulus, and lower weight (Nanni 1993; Benmokrane et al. 1995; Alsayed et al. 2000). The compression strength is often neglected, due to its low value. GFRP is also electrically and magnetically non-conductive, but sensitive to fatigue and creep rupture (Almusallam and Al-Salloum 2006). Furthermore, the structural effects of the low elastic modulus and bond behavior (Cosenza et al. 1997; Yoo et al. 2015; Coccia et al., 2017) have to be considered. Due to all these aspects, this type of reinforcement is not suitable for all applications, but it appears appropriate for tunnel segments, both for provisional and permanent elements.

In order to evaluate the synergic effect of the above mentioned composite materials, tunnel segments, with a typical metro tunnel geometry, made in FRC with and without GFRP bars were cast and experimentally tested. Both bending and point load tests are carried out, in order to evaluate the structural performances, both in terms of strength and crack width. The obtained results are finally compared and discussed.

2 SEGMENT GEOMETRY AND MATERIALS

Four full-scale fiber reinforced concrete segments were cast in moulds available at the Laboratory of the University of Rome Tor Vergata. The specimens have an external diameter of 6400 mm, thickness of 300 mm, and width of about 1400 mm (Figure 1). Two of the four segments were further reinforced with a perimetric GFRP cage.

Steel fibers Bekaert Dramix 4D 80/60BG were added to the concrete matrix with a content of 40 Kg/m^3. The average compressive strength, measured on 6 cubes having 150 mm side, was equal to 62.35 MPa.

The tensile behavior was characterized through bending tests on eight 150x150x600 mm notched specimens according to the EN 14651. The diagrams of the nominal stress versus the crack mouth opening displacements (CMOD) are plotted in Figure 2.

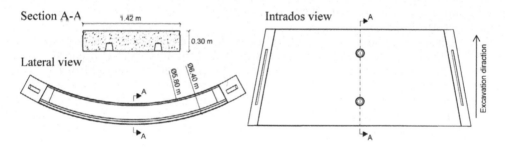

Figure 1. Segment geometry.

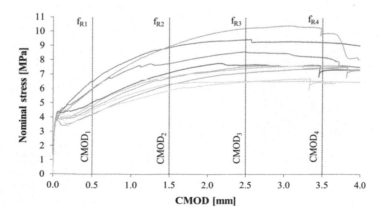

Figure 2. Results of the beam bending tests.

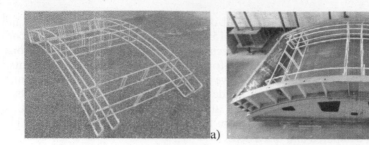

Figure 3.   a) GFRP cage; b) casting phase.

Two fiber reinforced segment, named SFRC-GFRP, were further reinforced with a perimetric Glass Fiber Reinforced Polymeric (GFRP) cage, as shown in Figure 3. The GFRP bars have a nominal diameter of 18 mm, and are characterized by Young's Modulus of about 40 GPa, and ultimate tensile strength equal to 1000 MPa.

For both the segment typologies (SFRC-steel fiber reinforced segments without any reinforcement and SFRC-GFRP), both flexural and point load full-scale tests were carried out, for the evaluation of the structural performances (both in terms of structural capacity and crack pattern evolution) under bending, and under the TBM thrust.

3   BENDING TESTS

The bending tests were performed with the loading set-up illustrated in Figure 4, in displacement control, by adopting a 1000kN electromechanical jacket, with a PID control and by imposing a stroke speed of 10 μm/sec.

The segments were placed on cylindrical support with a span of 2000 mm and the load, applied at the midspan, was transversally distributed be adopting a steel beam as shown in Figure 4. The measure devices, consisting in three potentiometer wires and two LVDTs, are shown in Figure 4.

The behaviour of the segments SFRC and SFRC-GFRP are compared in Figure 5, where the average value of the displacement, measured by the three potentiometer wires, is plotted versus the load. The first cracks appeared for a load value of about 125 kN and 120 kN, for

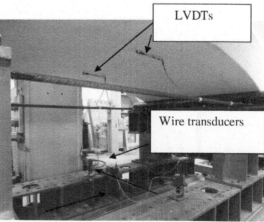

Figure 4.   Bending test set-up and instrumentation.

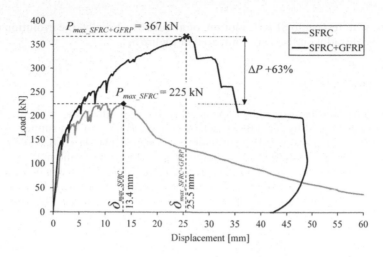

Figure 5. Bending test. Load- average displacement: comparison between SFRC and SFRC-GFRP segments.

Table 1. Maximum crack widths: comparison.

	Loading						
Load [kN]	125	160	180	210	222	250	270
	Crack width						
SFRC	<0.05	0.25	0.35	0.60	1.00	n/a**	n/a**
SFRC-GFRP	<0.05	0.10	0.15	n/a**	0.35	0.45	0.70

n/a*= measure not available since the crack width was not recorded at this load step
n/a**= measure not available since the segment did not reach this load value

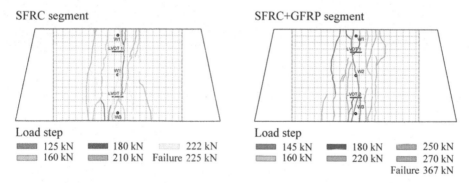

Figure 6. Bending test: crack pattern: a)SFRC segment; b) SFRC-GFRP segment.

the SFRC and SFRC-GFRP segment, respectively. In both the cases the first cracks were opened on the lateral surfaces close to the midspan and propagates on the intrados.

After a first comparable almost elastic response, the SFRC-GFRP segment presented a peak load about 63% higher than the SFRC one (367 kN against 225 kN of the SFRC segment). The maximum crack widths, measured at different load steps, are compared in Table 1. The obtained results clearly show the synergic effects of the two materials in reducing the crack widths, with respect to SFRC solution, of about 60%. Finally, the evolution of the crack patterns, at the intrados surface, are compared in Figure 6.

4 POINT LOAD TESTS

The point load test was performed by applying three-point loads at the segment, and adopting the same steel plates used by the TBM machine (Figure 7). A uniform support is considered, as the segment is placed on a stiff beam suitably designed (Meda et al., 2016). Every jack, having a loading capacity of 2000 kN, is inserted in a close ring frame made with HEM 360 steel beams and 50 mm diameter Dywidag bars (Figure 7). The load was continuously measured by pressure transducers. Six potentiometer transducers (three located at the intrados and three at the extrados) measure the vertical displacements, while two LVDTs transducers are applied between the load pads, to measure the crack openings. (Figure 7).

Two cycles were performed, as shown in Figure 8. The chosen reference load levels equal to 1580 kN and 2670 kN (for each pad) refer to the service load and unblocking thrust of the TBM machine.

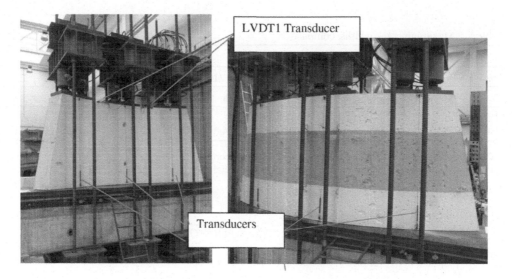

Figure 7. Point load test and instrumentation.

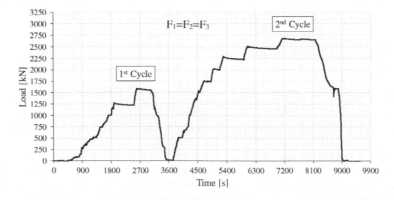

Figure 8. Point load test: Load (single pad) vs Time; a) SFRC; b) SFRC-GFRP segment.

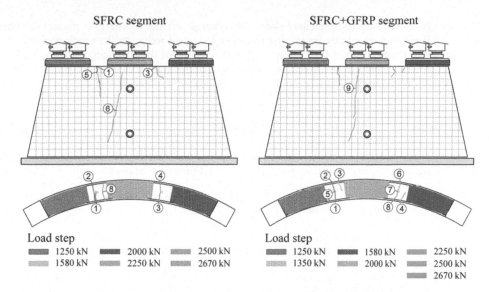

SFRC segment   SFRC+GFRP segment

Load step
▬ 1250 kN   ▬ 2000 kN   ▬ 2500 kN
▬ 1580 kN   ▬ 2250 kN   ▬ 2670 kN

Load step
▬ 1250 kN   ▬ 1580 kN   ▬ 2250 kN
▬ 1350 kN   ▬ 2000 kN   ▬ 2500 kN
                          ▬ 2670 kN

Figure 9.   Point load test; Crack pattern: a) SFRC segment; b) SFRC-GFRP segment.

Table 2.   Maximum crack widths: comparison.

Load		Maximum crack width [mm]	
		SFRC	SFRC -GFRP
1st crack [kN]	1250	0.05	<0.05
Service load [kN]	1580	0.10	0.05
Unblocking thrust*[kN]	2670	0.40	0.25
Unload [kN]	0	0.15	0.10

Note
* For metro tunnel, TBM pushing capacity coincides with unblocking thrust

The final crack pattern after the point load test is shown in Figure 9, for both the segments SFRC and SFRC-GFRP. Similar patterns were registered. The first cracks appeared for a load level of 1250 kN (for each steel pad) between two pads at the top and lateral surfaces (Figure 9a and 9b), in both the cases. Besides the splitting cracks (between the pads), a bursting crack (under the point load), formed in both the cases.

Finally, the crack widths measured for the segments SFRC and SFRC-GFRP, at three significant load steps (related to the first cracking, service load and maximum TBM thrust load), are compared in Table 2.

The addition of the perimetric cage led to halve the crack width under the service load, and to reduce it of about 37.5%, under the unblocking thrust force. Furthermore, a reduction of the crack width of about 33% was measured after the complete unloading.

5   CONCLUSIONS

The experimental results of full-scale tunnel segments, subjected to flexural tests and TBM thrust actions, presented in the paper, allows to draw the main concluding remarks listed in the following.

- The results of bending tests, clearly show the synergic effects of the two materials (fibers and GFRP reinforcement) by increasing the peak load and reducing the crack width.
- The results of the point load test confirm the effectiveness of the solution, since the addition of the perimetric cage led to halve the crack width under the service load, and to reduce it under the unblocking thrust force, and at the complete unloading, respectively.

REFERENCES

Almusallam, T. H., Al-Salloum, Y. A. (2006). "Durability of GFRP Rebars in Concrete Beams under Sustained Loads at Severe Environments." Journal of Composite Materials, Vol. 40, No 7, pp. 623–637.

Alsayed, S.H., Al-Salloum, Y.A., Almusallam, T.H. (2000). "Performance of glass fiber reinforced plastic bars as a reinforcing material for concrete structures." Composite Part B: Engineering, Vol. 31, No 6–7, pp. 555–567.

Benmokrane, B., Chaallal, O., Masmoudi, R. (1995). "Glass fibre reinforced plastic (GFRP) rebars for concrete structure." Construction and Building Materials, Vol. 9, No 6, pp. 353–364.

Caratelli A., Meda A., Rinaldi Z., Romualdi P. (2011) "Structural behaviour of precast tunnel segments in fiber reinforced concrete". Tunnelling and Underground Space Technology 26, pp. 284–291.

Caratelli, A., Meda, A., Rinaldi, Z. (2012). "Design according to MC2010 of a fibre-reinforced concrete tunnel in Monte Lirio, Panama". Structural Concrete, Vol. 13, Issue 3;. pp. 166–173.

Caratelli, A., Meda, A., Rinaldi, Z., Spagnuolo S. (2016). "Precast tunnel segments with GFRP reinforcement". Tunnelling and Underground Space Technology vol. 60 (Nov. 2016) pp. 10–20.

Caratelli, A., Meda, A, Rinaldi, Z., Spagnuolo, S., Giona Maddaluno. (2017). Optimization of GFRP reinforcement in precast segments for metro tunnel lining. Composite Structures 181, pp. 336–346.

Chen, Y., Davalos, J. F., Kim, H-Y. (2007). "Accelerated aging tests for evaluations of durability performance of FRP reinforcing bars for concrete structures." Composite Structures, Vol. 78, March, pp. 101–111.

Coccia, S., Meda, A., Rinaldi, Z. (2015). "On shear verification according to the fib Model Code 2010 in FRC elements without traditional reinforcement". Structural Concrete. Volume 16. Issue 4. Dec. 2015. Pp. 518–523.

Coccia S., Meda, A., Rinaldi, Z., Spagnuolo S. (2017). Influence of GFRP skin reinforcement on the crack evolution in RC ties. Composites Part B. Vol. 119, pp. 90–100.

Cosenza, E., Manfredi, G., and Realfonzo, R. (1997). "Behavior and Modeling of Bond of FRP Rebars to Concrete." Journal of Composites for Construction, Vol. 1, No. 2, May, pp. 40–51.

De la Fuente A, Pujadas P, Blanco A, Aguado A. (2012) Experiences in Barcelona with the use of fibres in segmental linings. Tunnelling and Underground Space Technology 27(1):60–71.

Di Carlo, F., Meda, A., Rinaldi, Z. (2016). "Design procedure of precast fiber reinforced segments for tunnel lining construction". Structural Concrete. Volume 17, Issue 5, 1; pp 747–759.

Ding, Y., Ning, X., Zhang, Y., Pacheco-Torgal, F., Aguiar J.B. (2014). Fibres for enhancing of the bond capacity between GFRP rebar and concrete. Construction and Building Materials 51; pp. 303–312.

fib Model Code for Concrete Structures 2010. (2013). Ernst &Sohn.

Issa, M. S., Metwally, I. M., Elzeiny, S. M. 2011 Influence of fibers on flexural behavior and ductility of concrete beams reinforced with GFRP rebars. Engineering Structures 33: 1754–1763

Kasper, T, Edvardsen C, Wittneben G, Neumann D. (2008) Lining design for the district heating tunnel in Copenhagen with steel fibre reinforced concrete segments. Tunnelling and Underground Space Technology; 23(5):574–587.

Kim, B., Doh J-H., Yi, C. -K., Lee, J.-Y. (2013). Effects of structural fibers on bonding mechanism changes in interface between GFRP bar and concrete, Composites: Part B 45; pp 768–779.

Liao, L., De La Fuente A, Cavalaro S, Aguado A. (2010). "Design of FRC tunnel segments considering the ductility requirements of the Model Code 2010". Tunnelling and Underground Space Technology 2015; 47:200–210.

Meda, A., Rinaldi, Z., Caratelli, A., Cignitti, F. (2016). "Experimental investigation on precast tunnel segments under TBM thrust action" Engineering Structures, Volume 119, 15 July 2016, Pages 174–185.

Micelli, F., Nanni, A. (2004). "Durability of FRP rods for concrete structures." Construction and Building Materials Vol. 18, No 7, September, pp. 491–503.

Nanni, A. (ed.). (1993). "Fiber-Reinforced-Plastic (GFRP) Reinforcement for Concrete Structures: Properties and applications." Elsevier Science. Developments in Civil Engineering, 42, p.450.

Plizzari, GA, Tiberti G. (2008). "Steel fibers as reinforcement for precast tunnel segments. Tunnelling and Underground Space Technology; 21(3–4).

Qin, R., Zhou, A., Lau, D. (2017). Effect of reinforcement ratio on the flexural performance of hybrid FRP reinforced concrete beams. Composites Part B 108; pp. 200–209.

Spagnuolo, S., Meda, A., Rinaldi, Z., Nanni, A. (2017). Precast Concrete Tunnel Segments with GFRP Reinforcement. Journal of Composites for Construction. ASCE.

Spagnuolo, S., Meda, A., Rinaldi, Z., Nanni, A. (2018). Curvilinear GFRP bars for tunnel segments applications. Composites Part B, 141, pp. 137–147.

Wang, H., Belarbi, A. (2011). "Ductility characteristics of fiber-reinforced-concrete beams reinforced with FRP rebars", Construction and Building Materials 25, pp. 2391–2401.

Wang, H., Belarbi, A. (2013). Flexural durability of FRP bars embedded in fiber-reinforced-concrete. Construction and Building Materials 44; pp. 541–550.

Wona, J. P., Park, C. G., Kim, H. H., Lee, S. W., Jang, C. I. (2008). Effect of fibers on the bonds between FRP reinforcing bars and high-strength concrete. Composites: Part B 39, pp. 747–755.

Yang, J.M., Min, K. H., Shih, H. O. Yoon, Y. S. (2012). Effect of steel and synthetic fibers on flexural behavior of high-strength concrete beams reinforced with FRP bars. Composites: Part B 43; pp. 1077–1086.

Yoo, D-Y., Kwon K-Y., Park J-J., Yoon Y-S. (2015). "Local bond-slip response of GFRP rebar in ultra-high-performance fiber-reinforced concrete." Composite Structures Vol. 120, pp. 53–64.

Tunnels and Underground Cities: Engineering and Innovation meet Archaeology,
Architecture and Art, Volume 5: Innovation in underground engineering,
materials and equipment - Part 1 – Peila, Viggiani & Celestino (Eds)
© 2020 Taylor & Francis Group, London, ISBN 978-0-367-46870-5

Twenty years of FRC tunnel final lining: Lessons learnt, design proposal and new development

B. De Rivaz
Bekaert Maccaferri Underground Solutions, Erembodegem-Aalst, Belgium

ABSTRACT: New guideline concerning the use of FRC precast segment has been recently published to provide detail state of the art (ACI, FIB, ITA) and design proposal based on Model Code 2010. This paper will provide an overview of the lessons learnt and proposed design principle. The behaviour of fibre reinforced concrete is more than a simple superposition of the characteristics of the concrete matrix and the fibres. A unique concept of fibre has been developed which combined: high tensile strength + perfect anchorage (double hook) + ductile wire. This unique post cracking behaviour has been used to design very innovative solution. Three projects will be presented in this paper: the inner lining of Lee Tunnel in UK submit to internal pressure, the Jansen Mine shaft in Canada where shaft walls are slip formed and CIP tunnel in Turkey. All the design approach has been supported by a specific testing program.

1 INTRODUCTION

Steel fibres have been used to reinforce concrete since the early 1970s. Employed initially for applications such as industrial flooring, the 1980s saw the start of underground applications, in first shotcrete and then in both precast tunnel lining segments and Cast in Place Final lining.

However, a lack of regulation and standards hampered the spread of fibre-reinforced concrete for final tunnel linings. With the publication of international design guideline, the fib Model Code for Concrete Structures 2010 (Fib Bulletins), this obstacle has been overcome and designers are gaining confidence in working with fibres.

"This publication is targeted to be a source of information for updating existing codes or developing new codes for concrete structures. It specifically addresses non-traditional types of reinforcement, such as steel fibres, that have reached a status of recognition in previous years, with special attention being given to the use of fibre concrete in structural applications. At the same time, the Model Code is also intended to be an operational document for normal design situations and structures".

2 FRC BASIC PRINCIPLE

Steel fibres come in many different sizes, shapes and qualities with each having its own effect on the concrete behaviour and quality. The dosage of fibres needed to meet the desired structural performance will vary, depending on the characteristics of the fibre itself.

The behaviour of fibre reinforced concrete is more than a simple superposition of the characteristics of the concrete matrix and the fibres. To analyse the behaviour of this composite material, the way that the loads transfer between concrete matrix and fibre also must be taken into consideration.

For efficient load transfer, three conditions must be satisfied. There must be sufficient exchange surface, governed by the number of fibres, their length and diameter; the nature of the fibre-matrix interface allows for a proper load transfer; and the mechanical properties of the fibre such as Young's modulus, tensile strength and anchorage strength must allow the forces to be absorbed without breaking or excessively elongating the fibre.

To analyse the performances of fibre concretes, we must consider the combination of fibre and concrete as a composite material, which means integrating the transfer of the concrete matrix charges to the fibres network. This transfer is schematically shown in three distinct steps, identified on the following typical graph corresponding to (Asquapro technical booklet):

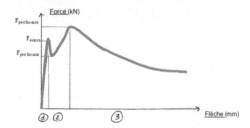

Figure 1. schematically graph explaining the load transfer.

Felmax: Maximum strength corresponding to the elastic limit of the fibre material Fpost fissmin: Minimum force reached after cracking.

Fpost fissmax: Maximum force reached after cracking, thanks to the absorption of forces by the fibres.

In the guideline published by Asquapro the following step are described:

Step 1: Adhesion over the entire length of the fibre during the phase of elastic strain, depending on:

• the specific exchange surface area (number, length, diameter of the fibres),
• the surface quality and thus the quality of the interface fibre/matrix,
• the compactness of the concrete matrix.

Step 2: Mobilization of the fibres right under the micro cracks, depending on:

• the elasticity modulus of the fibres,
• the number of fibres,
• the profile and orientation of the fibres,
• the quality of the interface fibre/matrix (adhesion),
• the compactness of the concrete matrix.

Step 3: Full mobilization of the anchoring, which may prove to be « total » or « sliding » (in support, « sliding » is sought-after):

• the shape of the anchoring,
• the possible sliding of the fibre in its sheath (interface quality and orientation fibre/cracks),
• the compactness of the concrete matrix,
• the number of fibres,
• the tensile strength of the fibre.

This transfer is efficient if the following three points are observed:

1. The exchange surface is adequate (number, length and diameter of fibres). For example, an inadequate number and length of fibres can lead to fragile behaviour even if the characteristics of the interface and of the fibres are satisfactory. Indeed, the loads then not being sufficiently transferred to the fibre network, the crack may bypass the fibres and the break becomes fragile.

2. The quality of the fibre/matrix interface allows a good transfer of loads (anchoring of the fibre in the concrete). Although the number and characteristics of the fibres are satisfactory, an inadequate anchoring of the fibres (slipping, little compact interface) causes a fragile or pseudo-fragile break; the fibres are extracted from their concrete sheath without mechanical stress or with a stress widely below resistance capacity.
3. The intrinsic mechanical properties (Young's modulus and tensile strength) of the fibre allow to take up the stresses without a risk to break or to stretch too much. In contrast, a fibre with insufficient properties causes a fragile behaviour despite a large number of fibres and an effective interface. A fibre with a low Young's modulus leads to a large crack before the fibre take up the stresses. A fibre whose tensile strength is lower than the capacity of the anchoring, will break in its concrete sheath before being extracted.

3 FRC BASIC DESIGN PRINCIPLE

The existing technical guidelines/recommendations/codes provide the structural engineer advice on how to quantify the reinforcing properties of steel fibres based on the measured post crack tensile strength of SFRC.

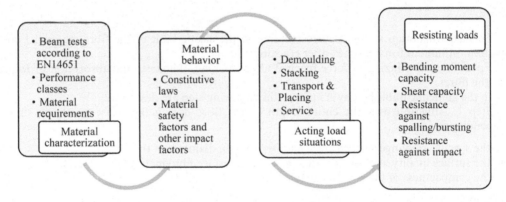

Figure 2.   Design process.

In accordance with fib CEB-FIP Model Code 2010, the structural design of SFRC elements is based on the post-cracking residual tensile strength provided by the steel fibres.

Nominal values of the material properties can be determined by performing a flexural bending test: one of the most common refers to the EN 14651, which is based on a 3-point bending test on a notched beam.

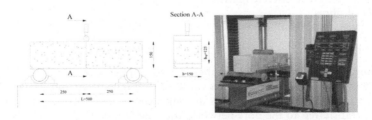

Figure 3.   Flexural bending test set up.

The results can be expressed in terms of force (F) vs. Crack Mouth Opening Displacement (CMOD). In order to obtain reliable statistically results, a minimum of 6 beam tests are recommended. A typical F-CMOD curve for FRC is shown below (figure 4):

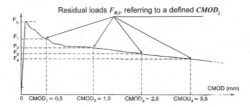

Figure 4.   F-CMOD curve.

Parameters fR,j representing the residual flexural tensile strengths are evaluated from the F-CMOD relationship according to the below equation (simplified linear elastic behaviour is assumed):

$$f_{R,j} = \frac{3F_{R,j}l}{2bh_{sp}^2} \tag{1}$$

Where:
fR,j is the residual flexural tensile strength corresponding to CMOD = CMODj
FR,j is the load measured during the test [kN]
l is the span length (distance between support) = 500 mm
b is the width of the beam= 150 mm
hsp is the distance between the tip of the notch and the top of the beam = 125 mm
From the above residual flexural tensile strengths, the characteristic values can be evaluated as follows:

$$f_{R,jk} = f_{R,jm} - k\,\sigma \tag{2}$$

Where:
k is the Student's factor dependent on the number of the specimens σ is the standard deviation of the test results.
For the classification of the post-cracking strength of FRC, a linear elastic behaviour can be assumed, by considering the characteristic residual flexural strength values that are significant for serviceability (fR,1k) and ultimate (fR,3k) conditions.
In particular two parameters, namely:

• fR,1k Minimum value for SLS
• fR,3k Minimum value for ULS

The designer has to specify these two-minimum value.
Fibre reinforcement can substitute (also partially) conventional reinforcement at ultimate limit state, if the following relationships are fulfilled:
fR,1k/fLk > 0.4
fR,3k/fR,1k > 0.5
In FRC design process a specific constitutive model must be considered.
A typical constitutive model used for FRC, especially at low strain level, is the rigid plastic.

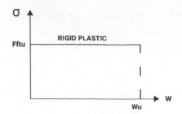

Figure 5. Simplified rigid-plastic constitutive relationship for FRC tensile behaviour.

4 FRC INNOVATIVE SOLUTION

4.1 *Precast segment*

Fibres can be used as reinforcement in precast concrete segment tunnel linings, either "fibre only" or in combination with conventional (bar) reinforcement, "combined solution". Its state of the art is defined by a large number of reference projects, where fibre reinforced concrete (FRC) segments have been used successfully. Projects using FRC segments report the following benefits of its use:

- Excellent durability
- Handling, installing damage and repairs are minimized
- Performance in the relevant ULS and SLS can be reliably demonstrated
- Overall manufacturing costs are lower than for conventionally reinforced concrete

This year there has been a flurry of additional guidance for fibre-reinforced segments: the American Concrete Institute's (ACI) Report of Design and Construction of Fibre-Reinforced Precast Concrete Tunnel Segments (ACI 544.7R-16); the International Tunnelling and Underground Space Association (ITA-AITES)'s Twenty Years of FRC Tunnel Segment Practice: Lessons Learned and Proposed Design Principles; BSI PAS8810:2016 Tunnel Design - Design of concrete segmental tunnel lining – Code of Practice; and ITAtech Guidance for Precast Fibre Reinforced Concrete Segments – Vol. 1 Design Aspects.

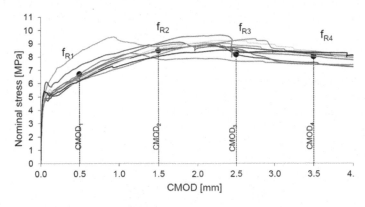

Figure 6. EN 14 651 test result with 4D 80/60BG.

A minimum performance class as C40/50 4c for the FRC as described by Model Code 2010 is recommended.

The use of fibre as Dramix 4D 80/60BG with the following properties are recommended as:

- l/D = 80
- Tensile strength > 1800 Mpa
- Optimum anchorage

- Glued for easy mixing and homogeneous reinforcement

The goal is to guaranty with the lowest dosage as possible a clear hardening post crack behaviour on beam and confirm on test scale 1 realized on segment.

4.2   Permanent spray concrete lining

Fibre reinforcement in the secondary linings of Sprayed Concrete Lining (SCL) designs, such as those used for Crossrail's station and platform tunnels, is also becoming more commonplace. And, as this article demonstrates, designers now have the option of fibre-reinforced concrete for cast-in-situ final linings too.

This application will be also structural reinforcement and could be design using the same process. Shorter fibre length 35mm and 0,55mm are commonly used to ensure with 20kg/m^3 already a network of 10 000ml/m^3. Glued and high tensile strength (> 1800Mpa) are recommended to meet project requirement. The typical Dramix 4D 65/35BG will be used for this application

4.3   Cast in place final lining

Bekaert spent five years developing the 5D fibre, trying many combinations of hook design and wire strengths to optimise the performance and the cost. The goal was to create a fibre that could be used in structural applications such as foundations slabs, rafts, suspended structures – and the final linings of tunnels.

Dramix 3D – the name for the Bekaert Maccaferri's original steel fibre – and the Dramix 4D, which uses higher-strength wire both work in the same way. The kinked ends provide ductility to the concrete by the slowly deforming as the wire is pulled out of the concrete. This is the mechanism that generates concrete ductility and post-crack strength.

The Dramix 5D fibre, with four kinks at either end rather than two or three respectively, actually works in a totally different way to its siblings the 3D and 4D: a revolution rather than evolution. Rather than relying on the pull-out mechanism of the fibres to provide ductility, the 5D does not pull out but remains anchored, lengthening itself - as rebar does - to provide ductility.

The tensile strength of a steel fibre has to increase with the strength of its anchorage, otherwise it would snap causing the concrete to become more brittle. So, the 4D, for example, with its stronger anchorage is made of stronger wire than the 3D. The 5D uses wire with a tensile strength of 2.2N/mm^2 and ductility of more than 6 percent.

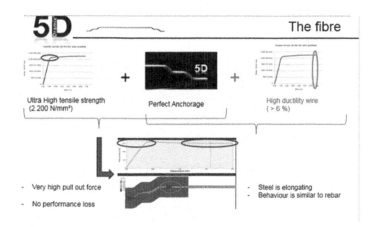

Figure 7.   Dramix 5D a unique combination.

Very soon after the launch of Dramix 5D fibres in 2013, UnPS became one of the first designers to take advantage of the fibres' superior properties on London's multi-million-pound Lee Tunnel (Psomas et al 2014).

The decision to use the fibres was driven by the need to reduce crack widths – although it yielded other The Lee Tunnel in Beckton, London is the first tunnel to be nominated and to win the prestigious UK Concrete Society Award. Built as part of the Thames Tideway CSO project, the tunnel is lined with an innovative Dramix 5D steel fibre reinforced concrete.

Generally inner linings that are reinforced are done so when tension is induced into the lining. For example, hydro and/or sewer tunnels with a high internal head pressure places the inner lining into tension and therefore requires reinforcement to resist the tension in the lining and control cracking.

The design and fabrication of the traditional reinforcement in such a tunnel as the Lee Tunnel would have been a major logistical challenge to the construction team at the jobsite – an estimated 17000 tons of reinforcing bar would have had to have been delivered, stored, transported underground and fixed into position – with iron workers working at height off work platforms of some description in what is technically a confined space and consuming valuable time within the program. One may argue that this operation would have been undertaken as the concrete pours progressed. However, the logistics of moving the high volume of bar reinforcement through the tunnel is a high-risk operation and places additional numbers of operatives and equipment underground and in the immediate vicinity of the concrete works.

MVB JV decided after much research over the last year or so to use the relatively new 5D Dramix steel fibre: this has a double hooked end and a much higher wire tensile strength than the 3D standard Dramix steel fibres previously used. These changes to the fibre geometry and technical characteristics provides the 5D fibre concrete with strain hardening in bending properties that enabled the designer (MS UnPS) of the secondary inner lining to verify the governing design situation for structural crack control.

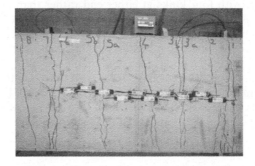

Figure 8.   BRE testing – multi cracking.

Figure 9.   BRE testing – multi cracking over loading span 1.645m.

All hook ended steel fibres, when under loading tend to deform/straighten out by pulling out of the concrete and so a controlled deformation occurs providing the energy absorption that can be measured by beam and plate testing, however the Dramix 5D steel fibre has these double hooked ends and these remain anchored in the concrete, the wire itself is some 2300MPa and this elongates some 7% just as traditional reinforcement behaves and this is what provides the steel fibre concrete element with the bending hardening properties and therefore provide designers with more opportunities to design steel fibre concrete structures both underground and surface structures.

By researching on a large scale seen above and by careful examination and analysis of the results the Designer (MS UnPS) & MVB JV decided to replace the standard traditional reinforcement with the 5D Dramix steel wire fibres – so removing some 17000 tons of bar with >2000 tons of the 5D Dramix steel fibre and eliminating the very large and difficult logistical challenge that would have been placed before the contractors underground team.

However, before work could start underground with the secondary lining, much work was required to establish a suitable concrete mix design that would pump up to 250m and virtually self-compact itself in the enclosed shutter within the tunnel lining. The concrete mix design (and final refinement) has taken almost 6 months to be able to work well and maintain its performance after pumping such distances and while not a great distance compared to some projects it still has to be able to almost self-compact with very little vibration. It isn't designed as a self-compacting concrete – the contractor wished to avoid any risk of the steel fibre segregating, fibre distribution was checked on the early pours and the indications showed there is excellent fibre distribution based on retrieved cores from the first pours in the works that suggest that the two principal orientations are in hoop and longitudinal direction, further analysis is being considered using X-ray photography

The Riva Tunnel is another underground project to use the new Dramix 5D series fibres which can be used to create a material which has a far greater residual strength after cracking than other fibre-reinforced concretes, behaving more like traditionally reinforced concrete. In each of the three cases, designers and constructors faced challenges around constructability or durability – or both – of the final lining which have been solved with the help of this new breed of fibre.

When contractor IC Ictas-Astaldi came to fix the rebar for the secondary lining of the Riva Tunnel in Istanbul, it ran into some problems. The 22m-wide tunnel required reinforcing bars up to 12m long to be fixed; this was proving difficult to do, introducing safety risks and causing damage to the sheet waterproofing membrane which sits between primary and secondary lining.

The solution was a bold and unusual one for Turkey: using fibre-reinforced concrete to create the permanent secondary lining. Following careful evaluation and tests at Istanbul Technical University, designer EMAY International was able to demonstrate that by using high-performance steel fibres a fibre-reinforced concrete lining would be just as good as one reinforced using steel reinforcing bars.

The latest project to make use of 5D's characteristics is the Jansen project, the development of a new potash mine in east-central Saskatchewan, Canada. Contractor DMC Mining is slip-forming the walls of two 1000m-deep shafts, aiming for production rates of 3m a day.

The future will see more use of 5D fibres in the permanent linings of tunnels and shafts. Several projects are already being designed using the fibre and others are considering its use as an alternative to traditional reinforcement for the final lining. Other underground applications have included track slab inside rail tunnels and free-spanning slabs in car parks.

5 CONCLUSION

The possibility of adopting fibre reinforced concrete, without traditional steel reinforcement for precast tunnel segments and cast in place final lining is herein analysed, with reference to test scale 1 and some project realized worldwide.

A minimum performance class or the FRC as described by Model Code 2010 should be defined according to the project requirement.

Fibre materials with a Young's modulus which is significantly affected by:

- time and/or
- thermo-hydrometrical phenomena

are not covered by this Model Code

Steel fibres are suitable reinforcement material for concrete because they possess a thermal expansion coefficient equal to that of concrete, their Young's Modulus is at least 5 times higher than that of concrete and the creep of regular carbon steel fibres can only occur above 370 °C (ISO 13270).

FRC has proven over the years to be a reliable construction material. State of the art concerning steel fibre is complete and validate by the scientific community through many standards and guideline. After 30 years of experience, the first Rilem design guidelines for steel fibre concrete were edited in October 2003 and Model Code 2010 published by fib in 21012 is now recognized as the reference design standard

Nevertheless, in order to obtain the desired results, it is worth noting the necessity to develop an accurate study of the material, i.e. the fibre typology suitable to a peculiar matrix.

In order to optimize the FRC solution some investigation should be conducted at early age of the design process with all actors involved in the project.

There is no good and bad fibre but just the right fibre for the right application using the Relevant testing method to determine engineering properties and for quality control

The use of the finished material should be considered along with the test and performance criteria, post-crack behaviour, match crack widths and deformation in the test to expectations in the project and durability requirement.

REFERENCES

Fib Bulletins 55-56: Model Code 2010 – First complete draft. (2010) Asquapro technical booklet

EN 14651: Test method for metallic fibre concrete. Measuring the flexural tensile strength. (2005)

Psomas et al (2014). SFRC for cast-in-place (CIP) Permanent Linings: Thames Tideway Lee Tun-nel Project in East London, UK. 2nd Eastern European Tunnelling Conference "Tunnelling in a Challenging Environment" 28 September - 01 October 2014, Athens, Greece

(ACI) Report of Design and Construction of Fiber-Reinforced Precast Concrete Tunnel Segments (ACI 544.7R-16);

The International Tunnelling and Underground Space Association (ITA-AITES)'s Twenty Years of FRC Tunnel Segment Practice: Lessons Learned and Proposed Design Principles;

BSI PAS8810: 2016 Tunnel Design - Design of concrete segmental tunnel lining – Code of Prac-tice; ITA-tech Guidance for Precast Fibre Reinforced Concrete Segments – Vol. 1 Design Aspects.

ISO 13 270: Steel fibres for concrete — Definitions and specifications. (First edition 2013-01-15)

Tunnels and Underground Cities: Engineering and Innovation meet Archaeology,
Architecture and Art, Volume 5: Innovation in underground engineering,
materials and equipment - Part 1 – Peila, Viggiani & Celestino (Eds)
© 2020 Taylor & Francis Group, London, ISBN 978-0-367-46870-5

Fibre reinforced spray concrete performance criteria: Comparison between EN14651 and EFNARC three-point bending test on square panel with notch

B. De Rivaz
Bekaert Maccaferri Underground Solutions, Erembodegem-Aalst, Belgium

A. Meda
Roma University, Rome, Italy

J. Kennedy
K&H Geotechnical Service PYLTD, Australia

ABSTRACT: The design of Fibre Reinforced Concrete (FRC) structures is often made adopting the indication of FIB Model Code. The Model code is considered as a reference documents in guidelines for tunnels. To classify and characterize the FRC in tension, FIB Model Code suggest using the EN 14651 bending test. This beam geometry is suitable to cast (poured) concrete. Recently the interest on sprayed fibre reinforced concrete for permanent application is increasing. Design of temporary sprayed FRC lining is usually made considering the energy values given by EN 14488-5 tests. To design permanent structures made with sprayed fibre reinforced concrete, it is necessary to make a material characterization. To fulfil this need, EFNARC prepared a document suggesting the use of bending tests on panel having the same geometry of EN 14488-5 (600x600x100mm) and proposing a correlation between the EFNARC bending test and EN 14651 bending test.

1 INTRODUCTION

The comparison of the results of tests on fibre reinforced concrete adopting the standard EN 14651 and the method proposed on "EFNARC Three Point Bending test on square panel with notch" will be presented in the report.

The analysis is made considered the results of tests performed by Bekaert Asia R&D Center and K&H GEOTECHNICAL LAB.

2 EN14651 BEAM TEST

According to EN14651 requirements, beam tests are performed on beams with nominal square dimensions 150x150 mm and length between 550 and 700 mm. The cast surface was rotated by 90 degrees, and in correspondence to the midspan, a notch with a maximum width of 5 mm and a height of 25 mm is sawn. The notch allows the creating of a weak section where crack can spread easily. In this way in correspondence of the notch, the depth of the section is equal to 125mm.

The beam is tested on 3 points load scheme with a span of 500 mm. The results of the EN 14651 test are the limit of proportionality (LOP) fct,L and the residual tensile strength fR1, fR2, fR3, fR4 evaluated at a crack mouth opening displacement (CMOD) of 0.5, 1.5, 2.5, 3.5 MPa respectively.

Furthermore, since in some laboratory the tests performed measuring the CMOD could complicated, a relationship between CMOD and beam midspan displacement δ is given:

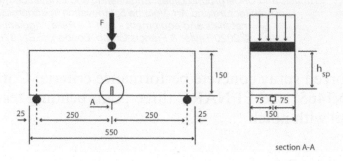

Figure 1. Geometry of the tested specimen.

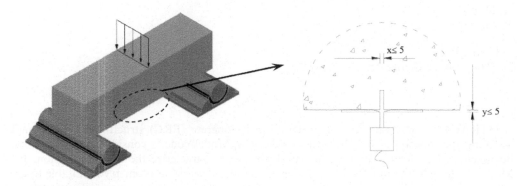

Figure 2. Particular and dimensions of the notch.

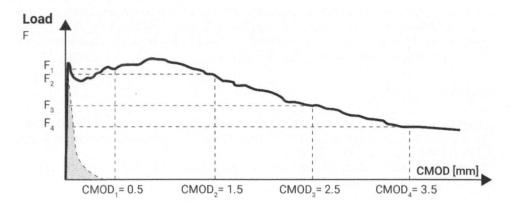

| CMOD | δ |
(mm)	(mm)
0,05	0,06
0,1	0,13
0,2	0,21
0,5	0,47
1,5	1,32
2,5	2,17
3,5	3,02
4,0	3,44

Figure 3. Testing system according to EN14651.

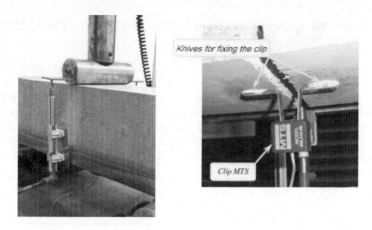

Figure 4.   Measurement devices: LVDT and Clip gauge.

3   EFNARC BENDING TEST ON PANEL

Test is made on square panel 600 x 600 mm with a thickness of 100 mm. The notch has a high of 10 mm.

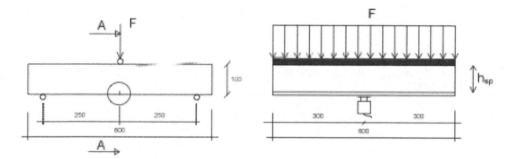

Figure 5.   EFNARC test.

The panels a are placed on a span of 500 mm and tested with a three-point bending configuration.

A relationship between panel tests deformation δ and CMOD and the respective δ and CMOD of EN 14651 are given in the EFNARC guidelines in order to have the same results in term of LOP fct,L and the residual tensile strength fR1, fR2, fR3, fR4.

Table 1.   Correlation between EN 14651 and EFNARC.

	Residual crack strength	CMOD (in mm)	Deflection (in mm)	Crack opening (in mm)
EN14651 beam test	fR.1	0,5	0,454	0,409
	fR.2	1,5	1,364	1,227
	fR.3	2,5	2,273	2,045
	fR.4	3,5	3,182	2,864
3 Point bending test on square panels with notch of 10mm	fR.1	0,5	0,631	0,409
	fR.2	1,5	1,894	1,227
	fR.3	2,5	3,156	2,045
	fR.4	3,5	4,420	2,864

4   COMPARISON OF THE RESULTS OBTAINED WITH THE TWO TESTING SET-UP

In order to compare the results obtained with EN 14651 and EFNARC tests, a experimental campaign was organized.

Tests were performed in 2 different laboratories: Bekaert Asia R&D Center and K&H GEOTECHNICAL LAB. In total 6 different mixes were prepared, mainly considering different fiber content, different concrete matrix, different type of fibers. The adopted mixes are summarized in Table 2 In Bekaert Asia R&D Center tests 6 beams and 6 panels were cast for every mix whereas in K&H GEOTECHNICAL LAB 12 beams and 12 panels were prepared for every mix.

Table 2.   Different adopted mixes.

Mix name	Concrete compressive strength [MPa]	Fibre type	Fibre content [kg/m^3]
MIX BEKAERT 1	55.3	4D 65/35BG	25
MIX BEKAERT 2	55.3	4D 65/35BG	40
MIX BEKAERT 3	86.6	4D 65/35BG	25
MIX K&H 1	46.0	4D 65/35BG	25
MIX K&H 2	43.5	4D 65/35BG	40
MIX K&H 3	42.3	5D 65/60	40

Table 3.   Average results in of fL, fR1, fR2, fR3, fR4 obtained with EN 14651 and EFNARC.

	EN 14651				
	fL	fR1	fR2	fR3	fR4
	MPa				
Mix Bekaert 1	5,1	2,7	3,0	2,9	2,6
Mix Bekaert 2	5,3	3,5	4,0	4,0	3,5
Mix Bekaert 3	7,6	4,4	4,8	4,0	3,2
Mix K&H 1	3,3	1,7	2,1	2,3	2,1
Mix K&H 2	3,4	2,2	3,2	3,7	3,4
Mix K&H 3	2,9	2,3	3,7	4,3	4,2
	EFNARC				
	fL	fR1	fR2	fR3	fR4
	MPa				
Mix Bekaert 1	5,2	3,3	3,7	3,5	3,0
Mix Bekaert 2	5,1	4,2	4,8	4,6	3,9
Mix Bekaert 3	7,9	4,3	4,6	3,8	3,0
Mix K&H 1	3,4	1,7	2,1	2,1	1,9
Mix K&H 2	3,9	2,6	3,7	3,7	3,5
Mix K&H 3	3,8	2,9	4,5	4,5	4,2

Figure 6 show the ratio between EN 14651 and EFNARC average results for fL, fR1, fR2, fR3, fR4.

Figure 7. shows the average value for all the mixes of the results presented in Figure 6. It can be noted that the ratio is very close to 1.0 for all the quantities fL, fR1, fR2, fR3, fR4. This it means that the different between EN 14651 and EFNARC are limited.

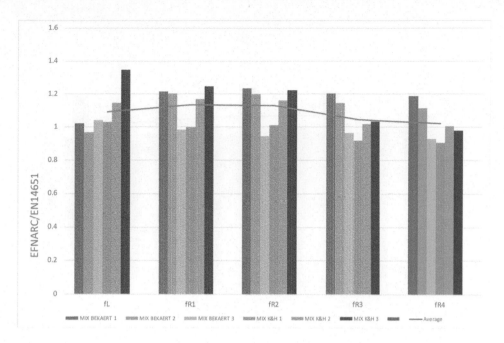

Figure 6.   Ratio between EN 14651 and EFNARC for fL, fR1, fR2, fR3, fR4.

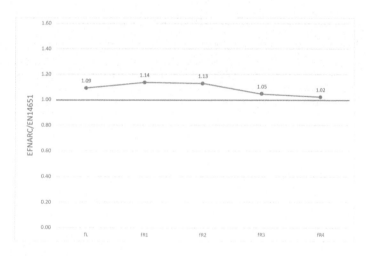

Figure 7.   Average value mixes.

The same analysis id made considering only the three mixes having 25 kg/m3 of fiber. This is a typical content for sprayed concrete. In this case (Figure 8) the different between EN 14651 and EFNARC is very limited.

ENFARC test in general give a characterization of fibre reinforced concrete similar to EN 14651. EFNARC results in terms of fL, fR1, fR2, fR3, fR4 are slightly higher respect to the EN 14651: this is probably due to the higher fracture area (almost 3 times higher in EFNARC respect to EN14651).

The aforementioned statement is confirmed by the analysis of the standard deviation and the coefficient of variation of the results. Table 4 and table 5 show the standard deviations and the coefficients of variation of fL, fR1, fR2, fR3, fR4 obtained with EN 14651 and EFNARC.

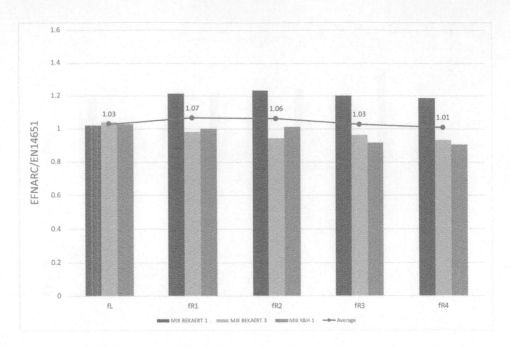

Figure 8.   Mixes with fibre content 25 kg/m3.

Table 4.   Standard deviation.

	EN 14651				
	LOP	CMOD 1	CMOD 2	CMOD 3	CMOD 4
	MPa				
Mix Bekaert 1	0,14	0,54	0,62	0,63	0,50
Mix Bekaert 2	0,24	0,65	0,94	0,93	0,73
Mix Bekaert 3	0,34	0,77	1,00	0,80	0,61
Mix K&H 1	0,91	0,56	0,71	0,76	0,73
Mix K&H 2	0,3	0,7	1,0	1,1	1,0
Mix K&H 3	0,9	0,7	1,1	1,3	1,3
			EFNARC		
	LOP	CMOD 1	CMOD 2	CMOD 3	CMOD 4
	MPa				
Mix Bekaert 1	0,26	0,4	0,48	0,35	0,23
Mix Bekaert 2	0,18	0,50	0,66	0,62	0,50
Mix Bekaert 3	0,46	0,26	0,32	0,29	0,27
Mix K&H 1	0,5	0,3	0,5	0,4	0,4
Mix K&H 2	0,3	0,3	0,5	0,5	0,5
Mix K&H 3	0,3	0,6	0,8	0,8	0,7

Table 5. Coefficient of variation.

	EN 14651				
	LOP	CMOD 1	CMOD 2	CMOD 3	CMOD 4
	MPa				
Mix Bekaert 1	0,03	0,20	0,20	0,22	0,20
Mix Bekaert 2	0,05	0,19	0,23	0,23	0,21
Mix Bekaert 3	0,05	0,18	0,21	0,20	0,19
Mix K&H 1	0,28	0,33	0,34	0,33	0,34
Mix K&H 2	0,10	0,31	0,31	0,29	0,30
Mix K&H 3	0,32	0,32	0,31	0,30	0,30
	AFNARC				
	LOP	CMOD 1	CMOD 2	CMOD 3	CMOD 4
	MPa				
Mix Bekaert 1	0,05	0,12	0,13	0,10	0,08
Mix Bekaert 2	0,04	0,12	0,14	0,14	0,13
Mix Bekaert 3	0,06	0,06	0,07	0,08	0,09
Mix K&H 1	0,16	0,19	0,22	0,21	0,21
Mix K&H 2	0,07	0,13	0,13	0,14	0,14
Mix K&H 3	0,08	0,21	0,19	0,17	0,16

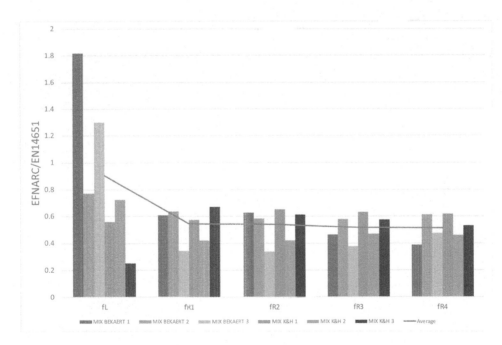

Figure 9.   EN 14651 and EFNARC ratio.

Figure 9 shows the ratio between EN 14651 and EFNARC coefficient of variation results for fL, fR1, fR2, fR3, fR4. It can be noted as the coefficient of variation for fR1, fR2, fR3, fR4 are remarkably lower for EFNARC tests.

5 CONCLUSION

EFNARC test appears to be a suitable for characterizing sprayed fibre reinforced concrete.

The specimen adopted in EFNARC can be easily prepared with spay concrete since the same moulds for EN 14488-5 can be used. It has to remark that beams with the geometry proposed in EN 14651 are difficult to prepare with sprayed concrete and should not be representative of the actual material properties.

The results obtained with EFNARC test can be used for the characterization of sprayed fibre reinforced concrete as demonstrated by the results of the laboratories tests.

We recommend this test for all spray concrete application in order to check the minimum performance required by the Model Code 2010

Fibre reinforcement can substitute (also partially) conventional reinforcement at ultimate limit state if the following relationships are fulfilled:

$fR1k/fLk > 0.4$; $fR3k/fR1k > 0.5$

REFERENCES

fib Model Code for Concrete Structures 2010 (2012), Final Complete Draft, fib bulletins 65 and 66, March 2012-ISBN 978-2-88394-105-2 and April 2012-ISBN 978-2-88394-106-9.

ISO 13 270: Steel fibres for concrete — Definitions and specifications. (First edition 2013 01-15)

Test method for metallic fibre concrete – measuring the flexural tensile strength (limit of proportionality (LOP), residual)". EN14651:2005, April 3rd, 2005. To check with paper (EN14487-1)

Testing sprayed concrete – part 5: Determination of energy absorption capacity of fibre reinforced slab specimens". EN14488-5, February 27, 2006.

EFNARC. "Three-point bending test on square panel with notch".

Tunnels and Underground Cities: Engineering and Innovation meet Archaeology, Architecture and Art, Volume 5: Innovation in underground engineering, materials and equipment - Part 1 – Peila, Viggiani & Celestino (Eds)
© 2020 Taylor & Francis Group, London, ISBN 978-0-367-46870-5

Digital engineering on the Sydney Metro

R. Dickenson & B. Harland
Mott MacDonald, London, United Kingdom

ABSTRACT: Australia has no government mandate for digital projects in the construction industry. The Sydney Metro represents a significant step forward for digital engineering on major projects in the Australasian transport sector.

Programme and cost benefits were achieved by coordination in 3D across all subcontracts and the client: eliminating unnecessary production of drawings and facilitating client review directly in a federated model.

The design programme was integrated within the Common Data Environment, enabling advanced analytics and monitoring. Rigorous model requirements for Uniclass classification and COBie deliverables were achieved, providing a foundation for the rest of Sydney Metro and beyond. The effective deployment of virtual reality gave next-level customer engagement and crucially, design feedback from the public early enough to make a difference.

This paper provides a practical summary of these and other digital engineering elements, and champions the value gained by taking a project digital by default.

1 INTRODUCTION

1.1 *Opportunity: strong foundations in a mixed-up industry*

Australia has had no government mandate for digital projects in the construction industry, and as a result the field of digital engineering has been fractured and diverse. This has led to innovations and the evolution of new solutions among technical experts, but it has hampered uptake of digital ways of working in a broader sense. Digital collaboration in the construction industry remains largely a parliament of many languages: words vary, expectations conflict, coordination on major projects can be slow and often confused.

The Sydney Metro represents a milestone for major project digital engineering in the Australasian transport sector: applying sound information management principles to global standards, a best practice virtual environment, and dramatically improved use of technology across the supply chain from client to subcontractors. By building strong digital engineering requirements into the project, the Sydney Metro team initiated a major step forward in efficient, quality and coordinated project delivery.

Mott MacDonald's role as digital engineering lead on the Underground Stations Design and Technical Services (USDTS) contract was to make that happen, working closely with technical experts from joint venture partner Arcadis, and subcontractors Foster + Partners and Architectus, among others.

An underlying foundation for the success of the project has been to acknowledge that good digital engineering involves a profound change to traditional ways of working; a change that involves every member of the team and can appear challenging and unfamiliar to many. New digital ways of working therefore need to be introduced with care, appropriate change management, and adequate coaching. This paper focuses the lens on how digital engineering can be implemented well in an environment where staff are unfamiliar or nervous. A practical explanation of the implementation of digital engineering on the project is provided, including:

- Programme and cost benefits achieved through full 3D coordination across all subcontracts and the client,
- Measures to improve user-acceptance of model-based design reviews,
- Customer-centred design through virtual reality testing and early design feedback,
- Governance of work in progress model information in a common data environment,
- Automation of project collaboration processes, and
- Exploiting structured data to improve progress monitoring and project controls.

Figure 1.   Concept image of Victoria Cross Station.

2   DISCUSSION

2.1   *Whole supply-chain change management*

In today's industry the majority of project staff are new to the systematic collaboration processes outlined in the new standard on information management using building information modelling from the International Organization for Standardization (ISO, 2018a). Additionally, it is common for staff of all levels to fail to understand the level of competence required to make proper use of a modern collaboration platform. Numerous projects fall into confusion because of attempts to apply old-style 'Windows Explorer' techniques to more comprehensive project information systems, or due to a belief that there is no need for systematic collaboration processes. A whole-team change management approach is required.

Ensuring the supply chain is properly engaged may require careful updates to contracts, diligent communication, persuasion, training and perhaps even provision of software, but it results in strong efficiencies from well-structured and quality-assured information. It is essential that information management requirements are passed carefully to subcontractors, to ensure the digital engineering approach remains a consistent whole.

For Sydney Metro, digital engineering systems and procedures were planned as an integrated way of working across the entire supply chain, from specialist subcontractors up to the client. All 24 organisations in the joint venture were required to use the Common Data Environment to collaborate on design information, and the federated 3D model was the centrepoint for coordination and comment exchange.

Use of multiple disparate systems for generating and managing information is common and arguably necessary on complex transport projects. For this project, Bentley ProjectWise and Autodesk BIM 360 were both used, but with carefully defined roles. To avoid confusion and inefficiencies these software platforms were set-out as a single system of various connected parts, each of which required planning and training. Where exchange of data between different systems required human input, effort was deliberately invested in automation.

Induction, competency management and training on the project were mandatory. Over 400 people were trained in the project's collaboration processes and systems, and staff with digital engineering responsibilities were assessed against a set of required skills. Any claim of 'we don't work in 3D' from organisations without prior experience was met with free software, personalised training sessions and ongoing coaching. One specialist service provider was carrying out virtual reality design reviews within weeks of initially making this claim.

2.2 3D collaboration

Coordination in 3D was central to the success of the station design. All interdisciplinary issues were captured and resolved using 3D processes implemented from the start of the project. Repeatable and scale-able processes were essential to enable tracking and accountability for design issues within the model. The design review process was facilitated by standardising ways of working with off-the shelf industry software.

Autodesk Navisworks has built in mark-up functionality for capturing design issues and saving each markup as a 'view'. The native comment tool itself has limited functionality for tracking comment details (such as comment status), and no provision for custom metadata. To keep the user-experience as simple and repeatable as possible across the supply chain, the project team chose to use this software instead of other alternatives, and to rule out customisation. To enhance coordination effectiveness a comment convention was developed and included in training, along with coaching on how to write clear inter-discipline comments. The comment convention (shown in Figure 2) enabled clear accountability, searchability and automation of analytics, and captured a full 3D audit trail of the design.

Ownership and accountability are crucial for carrying a design forward with integrity. By holding issue owner information in the federated model, it was clear to all whose issues were being addressed and whose were outstanding.

Automated macros were created to correct common errors in the team's use of the comment convention and to push the information into a database for visualisation. This enabled more efficient prioritisation of discussion in coordination meetings and formal design reviews.

2.3 Achieving model acceptance: user-friendly federated models

Client model review was carried out in the federated model, with training, coaching and guidance provided by the design team. User-friendly federated models were critical to the adoption of model-based reviews for both the design team and the client. Unfamiliarity with 3D tools and the new approach to design review were significant obstacles to overcome. The design team smoothed this transition by tailoring the experience to mimic previous methods of working as far as possible within the model review tools. A standard model form was established which contained various easy-to-navigate views for users less familiar with model tools. To replace the traditional stack of plans a set of flat 'drawing' plan views were set up within every federated model. The architectural drawings were spliced into the models, displaying details from the drawings such as room names and vehicle swept paths (Figure 3). Standard views including the drawing details could then be marked up in the same way as a PDF, but the markups fed into the single auditable design history stored within the model. From this

Comment convention:

Field:	Comment from	to	Comment to	-	Action on	-	Risk	-	Comment
Format:	Discipline code		Discipline code		Individual's name in format J.BLOGGS		H/ M/ L		Free text
Example:	ST	to	AT	-	J.BLOGGS	-	M	-	Add door here

Figure 2.   Comment convention used in Autodesk Navisworks.

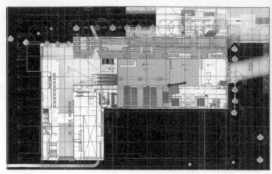

Figure 3. Standard federated model with spliced general arrangement drawing (left), system view (right).

baseline the skills to take advantage of the other benefits from 3D functionality are then more easily developed by reviewers. This introduction to model review proved to be a soft gateway to new ways of working, without enforcing advanced coordination processes in an intimidating manner.

The standard form model also contained key area/room views for quick reference. These were shown with a 3D camera view to give perspective on the room contents. In addition, engineering system views showed the structure in ghost form with each system highlighted, enabling engineers to trace the system through the station in a more intuitive manner. Standardising these views was particularly useful for describing the workings of each station to new team members, or when presenting the designs to client subject matter experts who wanted a quick way to understand the systems and their maintenance or delivery routes.

The elements in the federated model also contained carefully managed system and product information according to the project element metadata schema.

Early in the project these federated models were demonstrated to be suitable for client review at interim design milestones, in fact providing more useful information than the drawings which were required by the services brief. The client agreed to hold model-based intermediate design gates (with a small number of drawings). This eliminated the need for approximately 900 drawings, shortening the programme and saving more than $1m in one stage alone.

2.4 *Governance of work in progress model information*

The federated model is a valuable tool for collaboration, but it can also be a powerful distributor of design errors if the models are not appropriately checked prior to federation. For this reason, industry standards such as ISO 19650 (ISO, 2018a, 2018b) advise careful governance of the work in progress information, including restriction of a file's visibility until it has been checked and approved for sharing. Restricting one discipline's access to another discipline's 'live' working models can appear to stand in the way of progress, because of the need to wait for the next round of checking. However in trials on this project, the advice of the standard was found to be correct, leading to reductions in confusion, misinterpretation and proliferation of errors.

Both approaches were trialed: models linking in other disciplines' work in progress models (using Autodesk Collaboration for Revit), and models linking in other disciplines' checked and approved models (drawn from ProjectWise using Bentley's Revit Advanced Integration tool). In the former case technicians worked in constantly evolving design models, which meant that markups made by other staff often became unusable, and uncertainty was amplified between disciplines. This opportunity for uncertainty created delay, and left technicians to improvise solutions that in some cases differed from the original markup's intentions. While a completely open approach may at times work for smaller co-located project teams, it

484 Users	25 Organisations	5 Stations
4235 Deliverables	36 Disciplines' Documents Issued	

Figure 4.   Statistics from the common data environment.

was not optimal for a project of this scale and complexity. Figure 4 gives a sense of the scale of the design team's common data environment.

2.5   *Customer engagement for design input*

A project's digital models can be efficiently converted for virtual reality testing, putting customers inside the design years before it is built. This level of immersion in customer engagement offers a valuable source of inspiration for excellent station design, and is particularly relevant in underground infrastructure where members of the public more commonly express a sense of unease or anxiety.

Working with customer centred design subcontractor 'Symplicit', the team used Enscape and Unity to bring customers into the design and improve the relevance of each station to its users. Using the digital models in place of physical prototypes not only saved money but also obtained valuable customer insights early enough to make a difference to the design - long before physical prototypes could have been established.

100 members of the public were recruited for testing, deliberately covering a range of personas developed to encompass Sydney Metro's diverse array of customers. The following examples demonstrate feedback which might never have arisen in traditional engineering reviews:

- The arrangement of the walls beyond the gate-line at one station was amended to enhance visibility of lift access, following feedback from people with restricted mobility.
- A teenage boy identified an area in one station where the design needed adjusting to avoid gatherings of school groups which would obstruct pedestrian flow.
- The number of wider ticket gates was increased after feedback from parents.
- Signage was carefully tested and updated regularly to improve wayfinding.

Sadly this level of digitally enabled customer-centred design could still be considered innovative in this industry, because in many cases in-depth customer input is still taken at such a late stage in design that there is limited scope to incorporate the feedback.

2.6   *Automation of project processes*

The arrival of new technologies has come at a cost: the need for more administrative activity, to manipulate data correctly, control its use, coordinate it to higher degrees, validate its outputs, visualise the results, communicate the implications and more. Despite the resulting value, this additional burden is a significant contributor to the resistance which is commonly felt when new technology is introduced to a team. To overcome the administrative burden of increased technological activity, the project allocated time and budget to set up scheduled automation for federation, clash detection, and other model collaboration processes.

Federation and clash detection was automated using iConstruct and Autodesk Navisworks. Automation of these model-handling tasks alone is estimated to have saved the project 40 man hours per week, compared to similar tasks on the previous phase of the Sydney Metro.

Scripts and macros were also developed to:

- Dashboard statistics on deliverables and project progress,
- Manage comments in Navisworks,
- Manage parameters in the Revit models to align with project requirements (Figure 5),
- Validate completeness of non-graphical data in the models.

These scripts enabled design coordinators to focus time on design instead of administration.

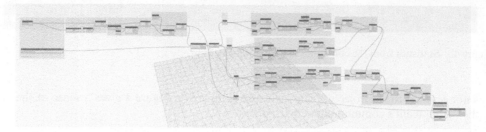

Figure 5. Dynamo script to adjust Uniclass parameters in models ready for COBie exports.

As well as support from project management, a critical resource for introducing and developing these innovations is the availability of team members with the skills and appetite to write scripts and implement automation. Without this skillset and attitude the project would not have been able to achieve the necessary quality of information at the efficiency required.

Good information management is essential to the smooth-running of the project but also for achieving buy-in from the team, especially where processes are new. Cumbersome, repetitive tasks will see team members reject the new working methods in favour of the old.

2.7 *Making the most of structured data to plan and manage deliverables*

To automate effectively, the project's information must be structured systematically. Beneficial automation of the structured data was not limited to just the models.

It is common for projects to expend significant manual effort in maintaining largely manual records of deliverables, trying to keep a spreadsheet up to date with each deliverable's approval status or transmittal status. For this project such manual handling was eliminated by automatically collecting information *about* each deliverable as metadata against the deliverable itself in ProjectWise. In this way the metadata of each file became the prime source of information about each file. This unlocked powerful tracking, data aggregation and dashboarding capabilities.

For this project, the metadata for all deliverables in the common data environment was exported to an external dashboard termed the Master Information Delivery Plan (MIDP), along with corresponding metadata from the client's document management system. The MIDP was the central tool for planning and monitoring deliverables, functioning as a deliverables schedule and statistical report on project progress.

2.8 *Incorporating design programme information*

Programme information such as activity IDs and due dates were associated with each deliverable in the following manner.

1. Each key milestone from the programme had an equivalent 'milestone' file in ProjectWise. This file carried various parameters about the milestone: programme due dates for comment and for approval (imported from Primavera P6), internal project due dates such as inter-disciplinary checks (managed by station design leads), design lead name, etc.
2. All the files due for issue at one milestone were linked to that milestone in ProjectWise and inherit the metadata from the 'milestone' file in ProjectWise.
3. If the programme changes and the milestone's metadata changes, then the update can be pushed out across ProjectWise, and all the files which have been assigned to that milestone update their due dates too.

To efficiently align information in the programme with deliverables in the collaboration platform, care is required to structure the data well from the start of the project. Figure 6 shows how project information is connected across various platforms and ultimately presented in a project controls dashboard.

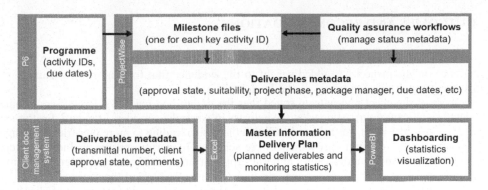

Figure 6.   Flow of project information.

2.9   *Analytics based on deliverables information*

Once information has been structured and aggregated in a consistent manner, useful project progress analytics are easy and can be achieved in any number of ways – Microsoft Excel and Microsoft PowerBI being common processing tools. This project used lookaheads and lookbacks (pivot tables with simple graphs) and formulas to measure 'percentage completeness'. The following parameters were used to estimate how 'complete' a deliverable is:

• Project phase (relative to the phase at which the deliverable is due),
• Approval state (work in progress/shared/published),
• Suitability code (for coordination/for review and comment/for approval and others),
• Delivery status.

2.10   *Other features built into the common data environment*

Table 1 below provides a summary of other smaller project information management features, all of which combine to improve the efficiency of day-to-day working.

Table 1.   Common data environment features developed for the project

Name	Description	Benefits
Quickshare	Approval workflow feature to allow a discipline lead to fast track the approvals workflow for informal document shares at low suitability codes.	Promotes more efficient document sharing. Empowers discipline lead to informally share documents at low suitability codes.
Distinct Approver/ Reviser permissions	Distinction between users able to approve and revise documents at specific workflow states.	Document producers can revise a document and continue working without reliance on senior staff to review and revise.
Incoming information workflow	Approvals workflow for incoming information ensuring proper checking prior to use and prevents working from uninstructed or superseded data.	Provides assurance that the design team are working from checked, relevant information.
Intelligent document numbering	Pre-populated fields and rules to remove repetitive data entry and improve consistency in metadata.	Time saving and improved metadata consistency. More reliable analytics due to increased data accuracy.

3 OUTCOMES AND RECOMMENDATIONS

3.1 *Positive outcomes for the client*

Experience on the project has added weight to the evidence that for major infrastructure projects there is much value in being 'digital by default'.

Members of the client review team stated that the design had reached a level of coordination and detail significantly surpassing what the industry in Australia considers 'standard' for this stage of design. Customer input was achieved earlier in the design leading to more user-friendly designs.

Cost savings and design improvements were achieved by various means, including the elimination of many interim drawings, and the application of early customer engagement to reduce the need for physical prototypes and improve the relevance of the station designs.

Compared to other similar projects this design experienced a smaller number of client comments at design review gates, thanks to enhanced understanding of the ongoing design as a result of model-based reviews.

Sydney Metro is well known in the region as a radically digital project, and as a result TfNSW has strengthened its position of digital engineering leadership in the industry. The client workforce has been upskilled and is now experienced in engaging with design in a virtual environment, with a better understanding of good information management principles.

3.2 *Recommendations for further work*

Development of optimal techniques for information management using building information modelling continues apace. For infrastructure projects with lengthy timescales, the proposed solution for the next project should never be simply a copy of the last. Based on experience on the Sydney Metro, Mott MacDonald recommends the following three areas of focus as fruitful areas for further work.

Model applications for requirements management: Examples have been carried out to show how a federated model can be used in various ways to connect requirements to their design elements, demonstrate compliance with requirements and justify deviations, all to better effect than drawings and diagrams. Further work is recommended to achieve smart linking between requirements and the models. Links may connect requirements to model element metadata, model views, classification schemas, or all three. Legal questions which may arise from reliance upon models in this manner should also be closed out.

Improved model-based coordination tools: Various software vendors are rapidly improving model-based coordination tools, particularly focused on cloud applications for organisations which cannot easily roll out software such as Navisworks or Navigator. Much value has already been identified in tools which can be seamlessly integrated into design review processes without requiring laborious changes to users' computer software.

Automated model validation: The joint venture developed bespoke algorithms which rip data from models and measure it against requirements. Since this work a greater number of third party tools has become available to audit digital models. For improved consistency and efficiency, further work is recommended on streamlining, standardising and making available these automated interrogation and validation processes.

REFERENCES

International Organization for Standardization. 2018a. ISO 19650-1:2018 Organization and digitization of information about buildings and civil engineering works, including building information modelling (BIM) – Information management using building information modelling – Part 1: Concepts and principles. Geneva: ISO.

International Organization for Standardization. 2018b. ISO 19650-2:2018 Organization and digitization of information about buildings and civil engineering works, including building information modelling (BIM) – Information management using building information modelling – Part 2: Delivery phase of the assets. Geneva: ISO.

Tunnels and Underground Cities: Engineering and Innovation meet Archaeology,
Architecture and Art, Volume 5: Innovation in underground engineering,
materials and equipment - Part 1 – Peila, Viggiani & Celestino (Eds)
© 2020 Taylor & Francis Group, London, ISBN 978-0-367-46870-5

Forrestfield Airport Link project – Variable density TBMs to deal with unexpected ground conditions

M. Di Nauta
Salini Impregilo, Milan, Italy

A. Anders & C. Suarez Zapico
SINRW JV, Perth, Western Australia

ABSTRACT: Forrestfield Airport Link Project foresees the extension of the existing urban railway network in Perth through 8.5 km twin bored tunnel crossing the airport land and with three new stations. The new railway will allow a 20 minutes journey into the Perth CBD from Forrestfield Station and improve bus network from the surrounding community. The two TBM's crossing underneath the Airport runways was a delicate point for the Client, reason why it required a Variable Density TBM, supplied by Herrenknecht, capable to operate in Slurry and EPB mode, based on the variable geological conditions. During the machines design phase, Salini-Impregilo decided to not install the conveyor belt and evacuate the spoil material via the slurry pipes system even when excavating in EPB mode. It's the first time that Salini- Impregilo operates this kind of machines and the scope of this article is to illustrate their complexity and operation modes.

1   INTRODUCTION

The Forrestfield-Airport Link (FAL) Project includes construction of a twin bored tunnel which will provide an improved connectivity between Perth's eastern suburbs, Perth International Airport and its business hubs and the Perth Central Business District. The new underground line will connect the existing Midland line, east of the at-grade Bayswater Station, to a new at-grade station in Forrestfield.

Each tunnel will be about 8,5 km long and will run beneath four main critical points: the Swan river, the Tonkin Highway, the Brookfield rail and the Airport land, including runways, taxiways and aprons.

Part of the Contract are also 12 Cross Passages, 3 Emergency Egress Shafts and 3 Stations (Figure 1). The central one, named Consolidated Airport Terminal Station and in the middle of Perth Airport, will serve both domestic and international flights.

At a glance, the main dimensions of the tunnels are:

- Excavation diameter: 7.100 mm
- Segments outside dia: 6.770 mm
- Segments inside dia: 6.170 mm
- Segments length: 1.600 mm
- Segments type: universal
- Segments arrangements: 5+1

The Contract to design and construct the FAL project was awarded by the WA Government Public Transport Authority (PTA) to Salini Impregilo-NRW JV in April 2016.

In August 2017 the first TBM (Grace) was launched from the portal in Forrestfield, where the main logistic area of 170.000 m^2 was set up with all the plants and equipment necessary for the tunneling operations. In October 2017 followed the launch of the second TBM (Sandy).

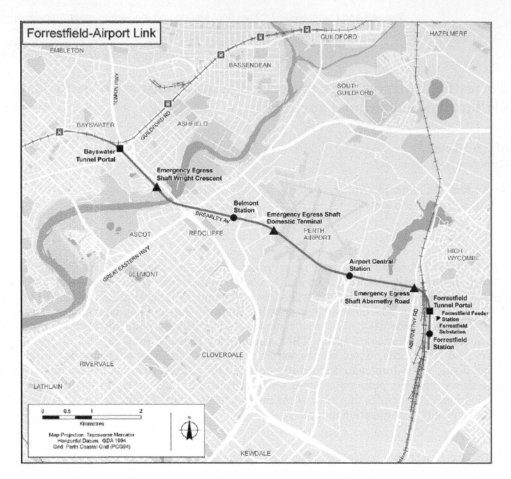

Figure 1.   FAL Alignment.

2   GEOLOGICAL CONTEXT AND CHOICE OF THE TBM'S

The two parallel tunnels run basically through four major geological units:

- Guildford Formation: similar to Perth formation, this unit is comprised of inter-bedded layers of clay and sand dominated material
- Ascot Formation: this material is comprised of a small percentage (13%) of fines with 27% of gravel and the balance sand together with cemented siliceous calcarenite beds.
- Osborne Formation: this unit is made of sandstone (top layer) and mudstone (lower layer) and can be characterized as soft rock with strength of up to 2.0MPa.
- Perth Formation: this material is comprised of both fines dominated material with clays and silt and a sand dominated layers.

For this kind of geology and based on previous experiences of projects in similar conditions, during the negotiations, Salini-Impregilo-NRW JV in its analysis was considering to use EPB machines with bentonite tanks even if the use of Slurry machine was also possible, as showed in the below graph.

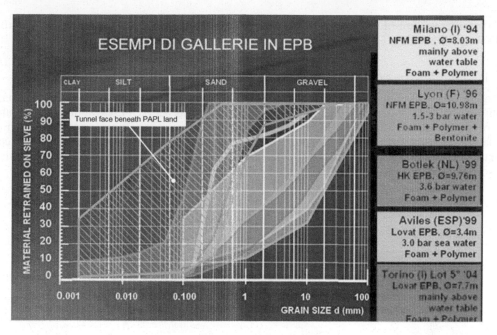

Figure 2. PSD of tunnel face beneath airport land.

Other reasons why the JV was considering EPB machines were:

– Significantly lower costs
– Major schedule advantages
– Less complicated set up of the machine
– Simple and more reliable operation
– Lower maintenance requirements
– Power and water saving (a power comparison between EPB and SPB showed a power saving of almost 50% by using EPB type)
– Less impact on the environment: SPB involves a Slurry Treatment Plant on surface, a series of sedimentation tanks and specific bentonite process
– Improved potential for spoil reuse
– High availability of experienced EPBM personnel
– Saving of site construction areas

However, due to Client, Public Transport Authority (PTA) risk assessment and former studies performed, SI concept of EPB tunnelling has been disregarded due to below main reasons:

– Risk under sensitive airport area where in theory with slurry type machines are better in controlling pressures inside the excavation chamber, excavation parameters which results in stronger control over any potential settlements.
– Variable geology along the alignment which can be consider as both cohesive less materials (Guilford-Ascot) and cohesive material (Osborn-Perth FM).

Thus, the Contract Scope of Works and Technical Criteria (SWTC) required a TBM capable to operate both in Slurry mode and EPB modes with the choice of the mode of operations determined by the JV to best suit the anticipated ground conditions in order to maintain a safe and stable excavated face condition. Having said this, it can be appreciated, that slurry mode should be consider for the first part of tunneling – from Forrestfield to Belmont Station – resulting in all airside land excavated in the slurry mode, which was one of Contract requirements, while the rest of alignment could be excavated in the EPB Mode.

In order to full-fil requirements for slurry tunneling, choice has been made for Variable Density TBM

2.1 Main difference between Slurry TBM and VD TBM

Figure below presents typical configuration of VD TBM. As it can be appreciated, the main difference is screw conveyor and no hole is separation wall between excavation chamber and working chamber. Slurry pressure is applied through communication pipes and extraction of material is done via screw conveyor. Also, no classical jam crasher is installed and instead crusher is located in slurrified box. Some can also appreciate, that regardless of excavation mode, all the material is being transferred by slurry circuit with all its implications.

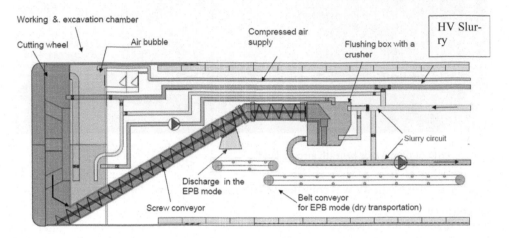

Figure 3.   Variable Density Schematic Overview.

3   TBM USED IN PERTH

As said in the previous paragraph, the choice of a Variable Density machine was dictated by the variable geology at the tunnel face along the alignment, crossing of critical points and specific project requirements for the usage of VD TBMs,

As VD TBMs are new milestone in tunneling, only highly experienced Constructors such as Salini-Impregilo, can cope with complexity and difficulties in introduction of new concept for TBMs. It must be noted, that Perth TBMs are one of the few first VD TBMs worldwide.

After award of Contract, SI-NRW experts together with Herreknecht, have been studying potential solutions and modifications to be applied to TBMs, which than brought to conclusion that TBMs currently used in Perth are unique and there are no TBMs built in that way till date. Major changes resulted in:

– Modification of EPB Mode – due to fact that main part of alignments will be excavated in slurry mode, JV has resigned in introduction of classical EPB resigning from foam generators, belt conveyor and other auxiliary elements for EPB. Instead, it was decided that throughout rotary coupling, water or bentonite will be injected to serve as soil conditioning. In addition, polymers can be added to water to serve as additional lubricant and/or dissolving agent.
– Modification of Slurry Mode – as per above, resignation from classical EPB, changed also the slurry mode. Main change is related to slurryfied box being fix without possibility of removing the box when the TBMs changes the modes from EPB to Slurry.

Due to geology formations that JV has been analyzing during the design of TBM, it has also been observed, that granulometry of soils in general allows for mining with usage of standard slurry without need of introducing High Density Slurry which would result in additional tanks to be added to STP and additional lines for supply. However, it shall be noted, that even without special high-density slurry addition, density of slurry that can be achieved in excavation chamber can be up to 1.5t/m^3 which provides great support medium during excavation. Density is checker and carefully regulated during TBM advancement.

Next figure presents some basic characteristics for the TBMs used in Perth.

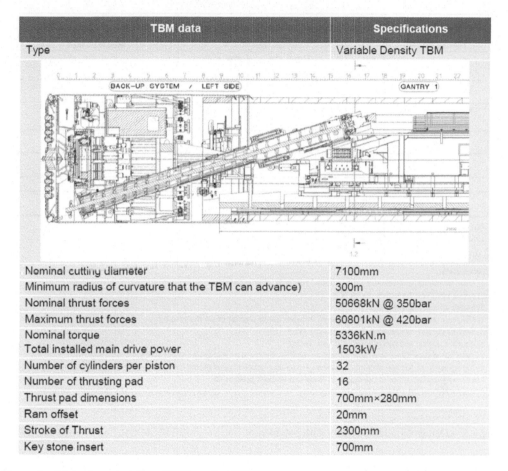

TBM data	Specifications
Type	Variable Density TBM
Nominal cutting diameter	7100mm
Minimum radius of curvature that the TBM can advance)	300m
Nominal thrust forces	50668kN @ 350bar
Maximum thrust forces	60801kN @ 420bar
Nominal torque	5336kN.m
Total installed main drive power	1503kW
Number of cylinders per piston	32
Number of thrusting pad	16
Thrust pad dimensions	700mm×280mm
Ram offset	20mm
Stroke of Thrust	2300mm
Key stone insert	700mm

Figure 4.   Main characteristics of FAL project TBMs.

3.1   *High Viscosity Slurry Injection*

As mentioned in the previous paragraph, TBMs have been modified and no high-density slurry has been planned for usage during TBMs mining. Instead, of HD slurry, SI-NRW JV has studied carefully alternative proposal. After careful studies, it has been decided, that 2 tanks, 5m^3 and 7m^3 will be implemented in the TBM. That tanks, together with additional pump of polymer will create special slurry mix with very high properties which increase slurry properties significantly.

That slurry is called High Viscosity Slurry where increased properties are achieved by mixing driving slurry inside the tanks with addition of special polymer. Dosage of polymer typically is 2l/m^3. Tanks are connected directly with excavation chamber and can be used during mining and/or during TBM stoppage for any reason. Further, JV has also established special method of injection of HV slurry when there are significant drops of slurry levels in working chamber. Two main parameters established for usage of HV slurry are as indicated below:

- Drop of level in working chamber by 0.75m in less than 10 minutes (0.5m^3 of slurry per minute). This could be identified as anomalous geology which would be causing slurry loses through face.
- Drop of face pressure by 0.1bar below Lower Alarm value.

If HV slurry is injected during the mining operation, then the speed of TBM is reduced to 25mm/min allowing for better "sealing" of face with slurry.

Ring No.	TBM1	TBM2	High Viscosity Slurry TBM 2- R896	Fresh Slurry	Active tanks
Primary KPI Test	Test results			Required Range	
Density (g/cm³)	☐Pass ☐Fail	☐Pass ☐Fail	☐Pass ☐Fail 1.11		
Marsh Viscosity (s)	☐Pass ☐Fail	☐Pass ☐Fail	☐Pass ☐Fail 170		
Filter Cake Thickness (mm)	☐Pass ☐Fail	☐Pass ☐Fail	☐Pass ☐Fail > 10		
Filtrate Loss (ml)	☐Pass ☐Fail	☐Pass ☐Fail	☐Pass ☐Fail 16		
pH	☐Pass ☐Fail	☐Pass ☐Fail	☐Pass ☐Fail 7.7		
Secondary KPI Test	Test results			Required Range	
Yield Point (lbs/100ft²)	☐Pass ☐Fail	☐Pass ☐Fail	☐Pass ☐Fail 59		
Gel Strength (lbs/100ft²)	☐Pass ☐Fail 10":___ 10':___	☐Pass ☐Fail 10":___ 10':___	☐Pass ☐Fail 10":___ 10':___		
Plastic Viscosity	☐Pass ☐Fail	☐Pass ☐Fail	☐Pass ☐Fail		

Test Details;
1) Fresh bentonite test frequency: Every 100m³ batched/ when requested
2) Active tanks test frequency: Every Ring
3) Marsh viscosity (s/946 ml)
4) Fluid loss test to be undertaken for 30mins once per day to counter check the reading.
5) Yield Point: 300 rpm reading- Plastic viscosity value
6) Plastic viscosity: 600 rpm reading- 300 rpm reading
7) Tests are in accordance to recommended Appendix 4. Slurry Parameter table (AFTES 2005)
8) Gel strength: At 300 rpm setting, turn knob very slowly (rotation speed should be ~ 3 rpm, controlled maually by Operator) Read maximum deflection of the dial at 10' leave rest and test again at 10"
9) When 'Fail' results are recorded. TBM Engineers/ Yard Manager shall advice replacement rate and changes to be made

Figure 5.   Details of modified slurry with polymer.

During both operations – stoppage and mining – the slurry in the two tanks is always fulfilled with polymer to reproduce the HV slurry to have it available at all the time. During normal procedure, the filling of one tank takes 10–15 min, which is much less than the advancement time. To fulfill both tanks, it takes usually 25–30 min. If the slurry drops in the working chamber are higher, excavation is stopped and all the HV slurry is injected inside the excavation chamber. At this point, the situation will be monitored for one hour to observe the behavior of the machine and bentonite level.To note is that the geotechnical parameters along FAL alignment are not presenting need for constant HV slurry injection. SI-NRW JV has decided, that during passage under critical infrastructure, mix-face conditions or any abnormal situation, HV slurry will be constantly injected to provide robust pressure transfer and reduce any potential slurry loses.

4   SLURRY CONSIDERATION

Choice of bentonite type has been followed by numerous iterations and checks of bentonite prior to selecting the final one. After bentonite and polymer has been decided and implemented into the project, JV decided to perform additional tests of rheological parameters of slurry suspensions. This has been done by Ruhr University of Bochum (RUB), which is world leader in slurry suspensions for tunneling. As known, slurry suspension is key point in transferring face pressure and establishing proper conditions for the safe excavation.

4.1 Basic slurry suspensions

In order to establish the rheological properties of different suspension mixes, bentonite and polymer used by JV has been checked with different concentrations. Tests have been done with following:

– Bentosud 120 T E – this type of bentonite was chosen by JV due to best parameters and cost-effectiveness. Mixes have been established by 30, 40, 50, 60 and 70kg/m^3 of fresh bentonite per 1m^3 of slurry suspension. It is important to notice, that tests with above concentrations have been done after 24h to allow for optimum hydration.
– Sika Foam TBM 900 Bio – as presented above, in order to achieve increased parameters for slurry, JV has decided to use subject polymer for increase of slurry parameters. Main tests have been done by 1l/m^3 and 2l/m^3 of polymer per 1m^3 of slurry suspension.

On that basis, various tests have been done to establish what are the parameters of suspensions that can be achieved by different concentration of solid part of suspension. Table below presents results of bentonite suspensions. As known, for slurry tunneling, main point of interest is static Yield Point and slurry stability meaning filtrate loss. That parameters are key once and other, such as density, pH, viscosity are considered as secondary parameters.

As can be appreciated from the tables and summary graphs, increase of solid parts is changing the properties of bentonite. However, most apparent modification of suspension, can be observed, once polymer is added to fresh slurry. Some would need to appreciate that presented values are based on fresh bentonite (Fresh Slurry) which not necessary can be achieved in the excavation chamber. Also, it must be noted, that those values are based on laboratory density, while during tunneling, the driving bentonite is providing different parameters of suspension.

Properties	Test Method	Tap Water	Bentonite Bentosund 120T E Concentration, (kg/m³)				
			30	40	50	60	70
Density							
Density, ρ [g/cm³]	Pycnometer	1.000	1.019	1.025	1.031	1.037	1.044
Marsh Time							
Marsh time, 946 cm³, t_M [s]	API 13B-1	26	31	32	39	43	52
Marsh time, 1,000 cm³, t_M [s]	DIN 4127	28	32	34	41	46	56
Yield Point			Ball set 3302	Ball set 1312	Ball set 1312	Ball set 1312	Ball set 1312
Ball number	DIN 4127	-	#2	#3	#5	#6	#8
Static yield point, t_F [N/m²]	DIN 4127	-	7.45	10.16	18.62	25.97	42.75
Bingham-yield point, t_B [N/m²]	API 13B-1	0.14	0.430	3.020	5.330	7.100	9.020
Viscosity							
Apparent viscosity, η_a [mPas]	API 13B-1	0.95	2.75	8.35	13.85	16.60	21.00
Plastic viscosity, η_p [mPas]	API 13B-1	0.80	2.30	5.20	8.30	9.20	11.60
Gel Strength							
Gel strength 10s [Pa]	API 13B-1	0	0.00	1.02	4.24	6.85	8.84
Gel strength 10min [Pa]	API 13B-1	0	0.26	2.96	5.62	8.58	15.59
Thixotropy value	API 13B-1	0	0.24	1.824	1.296	1.632	6.336
Stability							
Filtrate water 7.5min [cm³]	DIN 4127	-	13.2	11.2	10.0	8.6	8.0
Filtrate water 7.5min [cm³]	API 13B-1	-	26.0	22.4	19.6	17.4	16.0
Thickness filter cake 30min [mm]	API 13B-1	-	1.0	1.8	2.8	2.0	1.8
Physical Parameters							
Conductivity, [mS]		0.47	1.23	1.67	2.04	2.21	2.50
pH-value		7.8	10.2	9.6	9.8	9.9	10.1

Figure 6.   Rheological parameters of bentonite suspensions made with Bentosund 120T E.

Properties	Test Method	Tap Water	Bentonite Bentosund 120T E Concentration, [kg/m³]					
			50	50	60	60	70	70
			Amount of SIKA Foam TBM 900 Bio, [lt/m³]					
			1.0	2.0	1.0	2.0	1.0	2.0
Density								
Density, ρ [g/cm³]	Pycnometer	1.000	-	-	-	-	-	-
Marsh Time								
Marsh time, 946 cm³, t_M [s]	API 13B-1	26	93	188	413	too thick	too thick	too thick
Marsh time, 1,000 cm³, t_M [s]	DIN 4127	28	101	208	474	too thick	too thick	too thick
Yield Point			Ball set 1312	Ball set 1312	Ball set 1312	Ball set 1312	Ball set 1312	Ball set 309
Ball number	DIN 4127	-	#2	#3	#5	#6	#10	#8
Static yield point, τ_F [N/m²]	DIN 4127	-	22.24	30.25	53.03	62.99	62.92	113.52
Bingham-yield point, τ_B [N/m²]	API 13B-1	0.14	5.86	8.74	14.64	16.46	22.66	25.68
Viscosity								
Apparent viscosity, η_a [mPas]	API 13B-1	0.95	16.2	22.0	23.75	25.35	35.8	38.85
Plastic viscosity, η_p [mPas]	API 13B-1	0.80	10.1	12.9	8.5	8.2	12.2	12.1
Gel Strength								
Gel strength 10s [Pa]	API 13B-1	0	3.68	6.75	10.42	10.37	16.56	22.74
Gel strength 10min [Pa]	API 13B-1	0	8.94	13.9	23.66	16.51	20.64	27.54
Thixotropy value	API 13B-1	0	4.94	6.72	3.648	5.75	3.84	4.512
Stability								
Filtrate water 7.5min [cm³]	DIN 4127	-	7.6	6.0	6.4	5.8	6.2	5.4
Filtrate water 7.5min [cm³]	API 13B-1	-	14.8	12.4	13.0	12.0	12.4	11.0
Thickness filter cake 30min [mm]	API 13B-1	-	3.0	2.5	2.5	2.5	2.5	2.5
Physical Parameters								
Conductivity, [mS]		0.47	1.63	1.76	1.42	1.99	2.25	2.24
pH-value		7.8	10.3	10.3	9.8	9.7	9.9	9.8

Figure 7. Rheological parameters of bentonite suspensions made with Bentosund 120T E and polymer Sika Foam TBM 900 Bio.

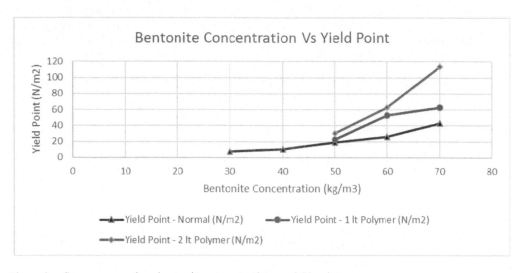

Figure 8. Summary graphs – bentonite concentration vs yield point.

4.2 *Slurry parameters depending on alignment position*

Slurry tests as presented above, have allowed JV to establish the criteria for slurry mixes to be used depending on alignment and geology. It must be mention that such process, is extremely careful and requires involvement of not only JV but also geotechnical consultants, designers and experts. In order to achieve proper suspension, first geology is checked and granulometry of material on given alignment is studied. Once this parameters are known, slurry suspension yield point is calculated. There are two methods which are required for suspension, that are pressure gradient fs0 which represents the stability of entire soil wedge. Second parameter of yield point, is micro-stability, which analyze local transfer of face pressure.

Below is an example of how yield point was established, given excavation in Ascot Formation on certain lot of tunnel.

Item	Properties
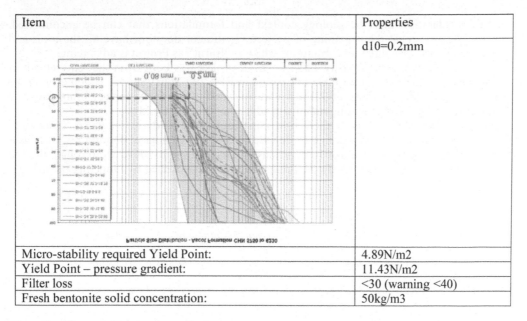	d10=0.2mm
Micro-stability required Yield Point:	4.89N/m2
Yield Point – pressure gradient:	11.43N/m2
Filter loss	<30 (warning <40)
Fresh bentonite solid concentration:	50kg/m3

Figure 9.   Slurry yield point determination.

Paragraph below presents how the mix is established lot by lot of the tunnel. Clearly, one can apricate complexity of entire process and carefulness in establishing parameters of slurry.

4.3   *Slurry adjustments during tunnelling*

One can ask, how the slurry plant is adjusted depending on different geology and required slurry properties. JV has decided that the best method of slurry properties adjustment is dividing plant into two main tanks – driving tank and so called "mother tank". Prior to end of advancement of each ring, series of slurry parameters tests are performed which includes density, filtrate loss, yield point, cake thickness, viscosity and ph. If during the testing, slurry parameters are discovered not to be in line with what prescribed depending on alignment, mother tank is activated and fresh bentonite is added replacing driving bentonite. It can be said that mother tank typically is fulfilled with bentonite of maximum concentration of solids of 70kg/m^3 providing highest possible slurry parameters.

After adding fresh bentonite to the system, next set of tests is performed to see if the slurry is having proper parameters. Mining can only restart once the parameters are reestablished.

4.4   *On-going tunnelling*

As of date of writing this article, TBMs are entering into new formation, Osborn FM. It must be mentioned, that till now there is no relevant experience with tunneling in Osborn. Further next in-coming challenge will be excavation in Perth FM, which is characterized as competent clay with moderate fines content. JV is currently studying different possible options for excavation in EPB mode and transportation of material with slurry circuit with potential usage of very light slurry suspension or even water.

5   CONCLUSION

FAL Project in Perth, presents how important is proper assessment of all aspects of tunneling starting from geology, TBM selection and site restrains. It also must be noted, that base for all the consideration must be geology context and formulations that can be encountered. Sometimes however, at the beginning of project, Contractor does not have all the tools and certain experience with different formations which is "learnt" as Project is advancing.

The choice of a Variable Density TBM for FAL Project was born from the requirement of having a dual mode machine capable to operate in EPB and Slurry mode. Together with the client, during the design phase of TBMs, it was decided to introduce further innovations to already innovative TBMs focusing mainly on adjustments of Slurry Mode.

The aim of this article was to show how important is the study of the slurry suspension characteristics and related mix design to be used in different geological conditions. Establishment of slurry properties have involved various parties lead by SI-NRW JV and had allowed to understand and excavate formations that have never been touched by tunneling before.

REFERENCES

FAL-SINRW-CM-NTE-00088.A.FR – TC. 2018. *Geotechincal Conditions: Ascot Formation West of Airport Central Station*

FAL-SINRW-CM-NTE-00090.C.FR – TC. 2018. *Geotechnical Conditions for Tunneling West of Airport Central Station*

FAL-SINRW-ST-PRC-00002_B. 2018. *Work Procedure – Injection of High Viscosity Slurry*

*Tunnels and Underground Cities: Engineering and Innovation meet Archaeology,
Architecture and Art, Volume 5: Innovation in underground engineering,
materials and equipment - Part 1 – Peila, Viggiani & Celestino (Eds)*
© 2020 Taylor & Francis Group, London, ISBN 978-0-367-46870-5

Durability design for large sewer and drainage tunnels

C. Edvardsen & A. Solgaard
COWI A/S, Lyngby, Denmark

P. Jackson
VINCI Construction Grands Project, Paris, France

ABSTRACT: In the last 10 years there has been a number of major deep gravity sewerage and drainage tunnels constructed. The projects provide a sustainable solution to increase the capacity of large sewer/drainage systems and include the DTSS in Singapore, the STEP in Abu Dhabi, the IDRIS in Doha, the Thames Tideway in UK and the Dubai Stormwater project. They are large tunnels, mostly in excess of 5 m in diameter, constructed with a segmental lining and in most cases a secondary in-situ lining. These tunnels need to have a long service life, thus the durability design of the concrete segmental or in-situ lining to resist the deterioration mechanisms associated with the ground conditions and sewerage is of primary importance. Clients and Consultants have adopted quite different strategies and solutions in order meet the service life requirements. These strategies together with the experience from a number of recent projects is presented.

1 INTRODUCTION

Global urbanization including growth in population of major cities calls for expansion and improvement of crucial infrastructure, such as systems for transport of sewer, groundwater and stormwater. Within the past decade, a large number of construction projects have been initiated worldwide, e.g. construction of networks of pipes collecting sewer, groundwater and the like transporting it to a large diameter (main) tunnel, which is connected to treatment facilities or outfall to the sea.

Considering the return on investment and their importance to the society, major assets for the infrastructure such as tunnels for sewer and drainage are usually designed for extended service life, i.e. 80–120 years. Moreover, as the accessibility to these assets is limited during operation, it is often requested that the design of such assets is based on a need for very limited maintenance.

Traditional design codes, such as Eurocode, provide guidance on the durability design of concrete structures up to 100 year service life. However, the durability design of sewer and/or drainage tunnels often needs particular attention; either because the required service life exceeds that covered by the standards, or because the special exposure conditions of sewer tunnels, i.e. the risk of microbiological microbiologically influenced corrosion (MIC), is not covered by the standards.

Therefore during the initial design phase of such projects, there is a need to assess the additional durability measures required compared to those given by the usual codes and standards. Alternatively, there is a potential risk that durability-concerns are not handled correctly jeopardizing the durability and operation of the asset in question.

This paper focuses on the durability design of segmental (concrete) linings for bored tunnels used for sewer and drainage. This includes presentation of durability design strategies to mitigate the following, selected deterioration mechanisms:

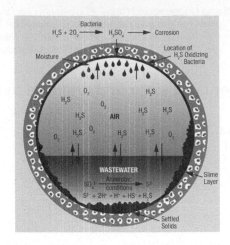

Figure 1.   Sewer tunnel: Schematic illustration of conditions for MIC. After (Pennwell, 2015)

- Chloride-induced reinforcement corrosion,
- Sulphate attack of the concrete, and
- Sulphuric acid attack of the concrete, i.e. microbiologically influenced corrosion (MIC)

Specifically for MIC, Figure 1 presents schematically the issue of sulphuric acid attack of concrete, referred to as "corrosion" in the figure.

2   REFERENCE PROJECTS

The authors of this paper have been the involved in the durability design of a number of sewer and drainage tunnels worldwide for the past decade. This section contains brief descriptions with regard to the durability design of the selected sewer and drainage tunnels, see Figure 2 for an overview of the location of the tunnels described and Table 1 for selected details of the tunnels.

2.1   *Step, Abu Dhabi, Uae*

The Strategic Tunnel Enhancement Programme (STEP) was constructed to improve the waste water system in Abu Dhabi and it is designed to carry 1.7 million m^3 sewage each day. The project was divided into three contracts of which COWI was the designer (structural and durability) for two of the contracts, i.e. STEP 2 and STEP 3 with a combined length of the bored tunnel of 25 km. In addition those two contracts covered other types of structures, e.g. 10 shafts.

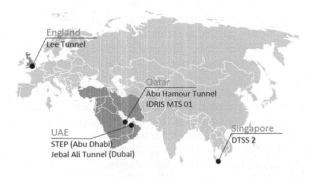

Figure 2.   Map with locations and names of the sewer and drainage tunnels presented herein.

Table 1. Overview of projects and details on bored tunnels

Project	Country	Design service life	Use	Tunnel length geometry	
				Length	Inner diameter
STEP	UAE	80 years	Sewage	41 km	5–8 m
Abu Hamour Tunnel	Qatar	100 years	Stormwater and groundwater	9.5 km	3.7 m
IDRIS MTS 01	Qatar	100 years	Sewage	16 km	3–4 m
Lee Tunnel	England	120 years	Combined Sewage	6.8 km	7.2 m
Thames Tideway Tunnel	England	120 years	Flow and Retained Foul Sewerage	25 km	7.2 m
Jebal Ali Storm and Groundwater Tunnel	UAE	100 years	Stormwater and groundwater	12 km	10 m
Deep Tunnel Sewer System 2 (T-09)	Singapore	100 years	Sewage	7.9 km	6 m

Three options were considered for the design of the segmental lining of the bored tunnel in the initial phase of the project:

- Pre-cast concrete segments with traditional (carbon steel) reinforcement cages
- Pre-cast concrete segments with stainless steel reinforcement cage, and
- Pre-cast segments with steel fibre reinforced concrete (SFRC)

The bored, segmental lining is located in an area with very high chloride (10–12%) and sulphate content (up to 5,000 mg/l) in the soil/groundwater. Moreover, as the tunnel is designed for carrying sewage, special attention was given to the risk of MIC at the internal surface of the tunnel. Considering the very high content of chlorides in the soil/groundwater the first option for the segmental lining, i.e. pre-cast segments solely reinforced with traditional carbon steel reinforcement cages, was ruled out. The reason for that was that, this option would require a very high concrete cover thickness (≥ 80 mm) even when using a highly durable concrete. Such a high concrete cover thickness is impractical for tunnel segments as there is a significant risk that the (un-reinforced) concrete cover of the segments would crack during installation due to splitting stresses caused by the push rams of the TBM.

The second option, i.e. pre-cast segments solely reinforced with stainless steel reinforcement was effectively ruled out due to the large cost with stainless steel reinforcement being approximately ten times the cost of carbon steel. Moreover, the initial structural design revealed that the third option (pre-cast segments solely reinforced with steel fibres, SFRC) was feasible. Hence this option was initially considered for the segmental lining. At a later stage of the design it was noticed that additional reinforcement, i.e. steel bars, was required to cope with splitting forces at the radial joints. The development of the concrete mix design for the pre-cast segments (SFRC) took about one year in order to tune the concrete mix to achieve the durability-related requirements arising from the use of carbon steel reinforcement in combination with 65 mm concrete cover.

Hence, for the first segments produced, the additional reinforcement constituted of stainless steel reinforcement, and once the acceptable concrete quality in terms of resistance to chloride ingress was achieved, traditional (carbon steel) reinforcement bars were used as additional reinforcement, together with the steel fibres.

It is generally accepted that steel fibres embedded in un-cracked concrete has a high intrinsic resistance to chloride-induced corrosion, i.e. a high chloride threshold value, several times higher than that of traditional (carbon steel) bar reinforcement, even when manufactured from the same virgin material, see e.g. (Dauberschmidt, 2006). This is, among other factors, due to their limited size, the casting condition which ensures a dense steel/matrix interface protecting the fibres, and the manufacturing process (cold-drawing) which evens out defects at the surface of the steel fibres prone to initiation of corrosion. Hence, the use of un-cracked SFRC for the pre-cast segments is an obvious choice to ensure the durability of the tunnel. By design, the segments are designed un-cracked (SLS).

The concrete mix design for the pre-cast segments contained a triple blend binder; 50% ordinary Portland Cement (OPC), 20% fly ash (FA), and 30% ground granulated blast-furnace slag (GGBS), all by weight of total binder content. The w/b ratio of the concrete for the pre-cast segments was ~0.33, to ensure a sufficiently dense concrete. In order to protect the additional traditional carbon steel reinforcement bars against chloride-induced corrosion, strict requirements were established for the concrete's resistance to chloride ingress. The requirements to the concrete cover thickness and the chloride migration coefficient were established using the full-probabilistic approach for chloride-induced corrosion presented in (fib, 2012), yielding 65 mm concrete cover and a maximum chloride migration coefficient of 2.4 x 10-12 m²/s when tested in accordance with (Nordtest, 1999). While the selected concrete mix design was advantageous with regard to resistance to chloride ingress, the selected binder combination was also sufficient to mitigate the risk of sulphate attack of the concrete.

It is a well-known fact that concrete subject to sewer is prone to deterioration due to attack of the gaseous sulphuric acid formed above the sewage transported in the tunnel. To achieve the required service life of the tunnel, the Employer's requirements specified the use of a corrosion protection lining (CPL) at the intrados of the segmental lining. The cross section of the bored tunnel including the CPL is presented in Figure 3.

As seen from Figure 3, the CPL at the intrados of the segmental lining consists of a concrete lining, 225 mm thick, lined with a high density polyethylene (HDPE) membrane (2.5 mm thick) at the upper 330° of the cross section. The thicknesses of the concrete lining and the HDPE membrane were pre-defined by the Employer, however in-situ casting of a concrete lining also requires a certain nominal thickness around 225–250mm due to practical reasons. After installation of the segmental lining, the cast-in place concrete lining at the intrados was cast against the HDPE membrane which was placed in the formwork. Installation of the HDPE membrane in one part of the tunnel is shown in Figure 3. The applied system is often referred to as a two-pass lining system, where the first pass is the installation of the pre-cast segments, and the second pass is the construction of the in-situ cast concrete layer with or without the HDPE membrane.

The design of SFRC for the pre-cast segments was carried out in accordance with the German Guideline for SFRC, (DBV, 2001). The SFRC was manufactured with hooked-end steel fibres (length = 47 mm and diameter = 0.8 mm), 40 kg/m³, yielding fibre class F1.4/0.6 in accordance with the aforementioned reference. Prior to production of the pre-cast segments, a strict testing regime was set up in order to ensure that the expected mechanical and durability-related properties of the concrete mix design were achieved. The distribution of steel fibres in the fresh concrete was examined by simple wash-out testing, while the distribution of steel

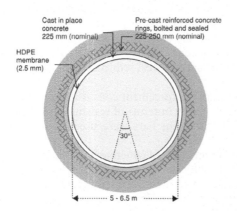

Figure 3. Cross section of bored tunnel incl. CPL, STEP

Figure 4. Installation of HDPE lining, STEP

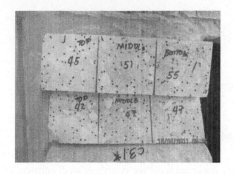

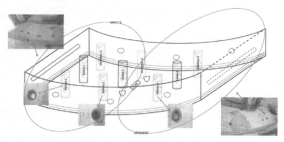

Figure 5.   Fibre count in hardened concrete   Figure 6.   Illustration of segment cored for round robin NT Build 492 testing

fibres throughout the pre-cast segments was examined from saw-cut segments, see Figure 5. To confirm, that the specified chloride migration coefficient was achieved, a segment was cast from the concrete mix design (without steel fibres) and subsequently cores were taken for round robin testing at three independent laboratories in Abu Dhabi and Denmark, Figure 6.

The introduction of pre-cast segments reinforced with steel fibres was, according to the authors' knowledge, the first time this innovative construction material was used for this purpose in the Middle East. Furthermore, testing of the concrete's chloride-migration coefficient in accordance with (Nordtest, 1999) was applied for the first time in the Middle East for tunnels. Since then, the approach has been adopted on all the tunnel projects since in the Gulf such as the Doha Metro, Abu Hamour and on the Dubai Stormwater water project, as a measure during running production for the concrete's durability-performance.

2.2 Abu Hamour Tunnel, Qatar

The Abu Hamour Tunnel in Qatar was constructed for transport of storm and groundwater in Doha, Qatar. As for all underground projects in Qatar, the structures of the Abu Hamour Tunnel project are in contact with soil/groundwater with very high levels of chlorides as well as sulphates. According to the Employer's geotechnical investigations the content of chloride and sulphate in the soil/groundwater is up to 30,000 mg/l and 5,000 mg/l, respectively at the location of the Abu Hamour Tunnel. The tunnel is designed for the same aggressive conditions on the outside as well as the inside.

For this project, the Employer's Requirements, i.e. the contractual requirements, had stipulated tight durability-related requirements to the permanent concrete structures in order to meet the design service life (100 years). The durability design for the segmental lining of the bored tunnel conforming to the Employer's Requirements is presented to the left of Figure 7, referred to as the "Original solution".

As seen from Figure 7, two types of segmental linings are used; one with "Typical segments", reinforced with steel fibres (comprising the vast majority of the segments for the bored tunnel) and "Special segments", reinforced with steel fibres and rebar (used in selected locations where additional structural capacity was required, e.g. close to adits/shafts). While the durability design of the segmental lining, as illustrated in the left sketches of Figure 7, addresses the exposure conditions, the required service life, and the Employer's Requirements, the Designer together with the Contractor decided to value-engineer the durability design of the segmental lining. The motivation for this exercise was to optimize the durability design without compromising the overall project requirements to durability and service life, as a number of the requirements given by the Employer were considered superfluous for a tunnel designed for stormwater. The outcome of the value-engineering is schematically illustrated in the right of Figure 7 showing the "New Solution".

As seen from a comparison of the two solutions presented in Figure 7, the HDPE membrane at the internal surface and the epoxy coating at the external surface were removed while

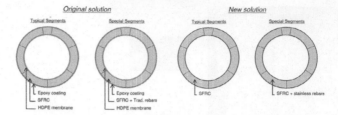

Original solution

New solution

Typical Segments   Special Segments   Typical Segments   Special Segments

Epoxy coating   Epoxy coating   SFRC   SFRC + stainless rebars
SFRC   SFRC + Trad. rebars
HDPE membrane   HDPE membrane

Figure 7.   Schematic illustration of original solution (left) and new solution (right) for the segmental lining

the carbon steel reinforcement of the special segments was replaced by stainless steel reinforcement. The use of an HDPE membrane at the internal surface of a tunnel is generally speaking considered for tunnels transporting sewerage effluent, to protect the permanent concrete lining against sulphuric acid attack, as previously described. However, as the Abu Hamour Tunnel only carries stormwater and groundwater, i.e. no risk of sulphuric acid attack of the concrete, the HDPE lining was considered superfluous.

The very high sulphate content of the soil/groundwater corresponds to exposure class S3 in accordance with (Concrete Society, 2008). To mitigate the risk of sulphate attack of the segments, the concrete mix design for the pre-cast segments contained 30% OPC, 65% GGBS, and 5% micro silica (MS) all by weight of total binder content. Furthermore, the concrete mix design for the pre-cast segments was verified by the recently developed short-term test of the sulphate-resistance, i.e. the German SVA method. That test method allows for a significantly faster evaluation of the concrete's resistance to sulphate attack; the duration of the test method is approximately 3 months, compared to the traditional test method, e.g. (ASTM, 2015), which is usually taking 6 – 18 months. The testing was carried out at the Technical University of Munich. As an illustration, Figures 8 and 9 present a failed and a passed test specimen, respectively, after testing in accordance with the SVA test method.

The concrete mix design passed the SVA tests and the time gained by using the SVA test method, meant that the contractor at a much earlier stage than usually, felt comfortable that their concrete mix design fulfilled the Owner's requirements, reducing his risk and potential, associated delays in the construction phase. In following projects in the Middle East the Designer has convinced Owners to allow this test method instead of more traditional methods (and with longer duration) to the benefit of Contractors as well as Owners.

The "New Solution", which was approved by the Employer's engineer, resulted in substantial savings (cost and materials) for the Employer and the Contractor without compromising the overall requirement to 100 year service life.

The SFRC segments for the segmental lining were designed in accordance with fib Model Code 2010, (fib, 2012), and the post-cracking classification of the SFRC in accordance with the criteria established in that reference was 4c achieved by using 40 kg/m³ hooked-end steel fibres.

2.3   IDRIS MTS 01, Qatar

The IDRIS MTS 01 project in Doha, Qatar is part of the major IDRIS project which since its inception has been reduced in scope, with the MTS 01 project being the only part of the main tunnel to be constructed. It is designed for the transport of sewage from urban areas to sewage treatment facilities and is currently (2018) under construction. The main trunk sewer (MTS) is

Figure 8.   SVA test – failed specimen, Courtesy of VDZ

Figure 9.   SVA test – passed specimen. Courtesy of VDZ

Figure 10. Cross section of segmental lining

Figure 11. Mould for pre-cast segment with cast-in HDPE membrane

constructed as a bored tunnel with segmental lining, with a design service life of 100 years. The length of the bored tunnels for MTS 01 is approximately 16 km with an inner diameter varying in the range 3–4 m.

Considering the use of the tunnel, i.e. transport of sewage, with the risk of MIC of the concrete, the Employer required a double-protective design to mitigate such attack of the segmental lining, i.e. a sacrificial concrete layer (not to be considered as part of the permanent structure) and an HDPE membrane on the internal surface of the tunnel, see Figure 10. The sacrificial concrete layer and the structural part of the segments were cast together with an over thickness segment that allowed for the sacrificial layer within its thickness. For the concrete mix design, a high content of supplementary cementitious materials (GGBS) and a low w/c ratio (< 0.40) was used. The HDPE membrane was attached to the internal surface of the segments during casting, see Figure 11. After installation of the segments in the tunnel, all joints between HDPE membranes, i.e. joints between each ring, and joints between individual segments of a ring were welded on site.

The approach presented above i.e. pre-casting the structural part of the segment together with the sacrificial layer and the HDPE membrane is often referred to as a one-pass lining system. This approach has a number of benefits, of which the major benefit is the omission of in-situ casting of the sacrificial layer and installation of HDPE membrane after ring-build. This was particularly important given the small diameter of the tunnel with a finished inside diameter of 3 m making an in situ secondary lining difficult to construct.

The thickness of the sacrificial layer of concrete was 120 mm, see (Olliver & Lockhart, 2017). The thickness of the sacrificial layer of concrete was determined based on CFD modeling and estimation of the annual loss of concrete, using the Pomeroy approach. For the determination of the thickness of the sacrificial layer of concrete, it was conservatively assumed that the sacrificial layer locally would be exposed to sewer shortly after opening of the tunnel, i.e. the protective effect of the HDPE membrane was not considered in the determination of the thickness of the sacrificial layer of concrete.

The durability of SFRC against chloride-induced corrosion as well as the resistance of the concrete mix design towards sulphate attack considering the prevailing exposure conditions in Doha are already discussed in this paper.

2.4 Thames Tideway and Lee Tunnel, England

The Thames Tideway project and the already constructed Lee Tunnel project are a major improvement scheme for London's waste water system in order to tackle the problem of overflows from the capital's Victorian sewers and they will protect the River Thames from increasing pollution for at least the next 100 years. The Tideway project is 25 km in length and connects with a series of the major sewer combined sewerage overflows along the Thames. The Tideway project links with the Lee Tunnel that takes the effluent from a pumping station at Abbey Mills to Beckton where there is an existing treatment works.

The tunnels and shafts are design for Combined Sewerage Flows, i.e. a combination of raw sewerage and surface water flows together with a number of specific design cases where raw sewerage may be retained in the tunnel for specific periods. The tunnels have an internal diameter of 7.2 m and are designed with a secondary lining (two pass) which has been reinforced with either conventional carbon steel reinforcement or steel fibre reinforcement (SFRC). No inner membrane (HDPE) is used on the Lee Tunnel/Tideway project as the exposure from the sewerage effluent and the consequent deterioration mechanisms such as MIC can be accommodated by the incorporation of a sacrificial layer in the tunnel secondary lining or in the case of the shaft structures by a sacrificial layer in addition to the durability cover to the reinforcement.

The thickness of the sacrificial layer has generally been determined by approaches such as those developed by Pomeroy, and the sacrificial layer thickness varied depending on the particularly environment in each section of the works. The typical sacrificial thicknesses used is approximately 70 mm with additional thickness applied in areas where abrasion was considered to be high.

2.5 Jebal Ali Stormwater and Groundwater Tunnel (DS 233-2), Dubai, UAE

The Jebel Ali Stormwater and Groundwater Tunnel, is located in Dubai, and is currently under design. The main tunnel, having a length of approximately 10.4 km is designed as a bored tunnel with segmental lining and shall, according to the Employer's Requirements, have a design life of 100 years. The main tunnel is the largest stormwater tunnel constructed in the Middle East.

The main tunnel is in close proximity to the Arabian Gulf, and the chloride and sulphate content in the soil/groundwater is very high, with chloride levels up to 55,000 mg/l and sulphate levels up to 6,800 mg/l being present. Hence, from a durability perspective, the use of steel fibres instead of traditional reinforcement cages would be favoured to mitigate the risk of chloride-induced corrosion. However, as the required internal diameter of the tunnel is 10.0 m, a design solution for the segmental lining without traditional reinforcement bars/cages is difficult to verify and there are very few segmental linings worldwide with an internal diameter of 10.0 m or greater, constructed solely from SFRC (ITA, 2016). In case traditional reinforcement bars are required for the pre-cast segments, due to structural capacity, these will need to be epoxy-coated as specified in the Employer's Requirements.

While the tunnel is mainly designed for transport of stormwater and groundwater, the tunnel shall, in accidental cases, also carry treated sewage effluent (TSE), i.e. diluted sewage. TSE is significant less onerous exposure compared to un-treated sewage, and considering the fact, that exposure of the concrete to TSE is very limited, it is not necessary to protect the concrete surfaces to mitigate the risk of sulphuric acid attack such as with a HDPE membrane.

2.6 Deep Tunnel Sewer System 2, Singapore

The Deep Tunnel Sewer System 2 (DTTS 2) project in Singapore is a continuation of DTSS 1 which was completed in 2008. The DTTS 2 project contains a total of 40 km deep tunnels, connected to 60 km of sewer pipes. The design service life is 100 years.

Compared to the Middle East, the exposure conditions from soil/groundwater, i.e. sulphate and chloride exposure, are significantly less onerous in Singapore. However, the design for sulphuric acid attack requires particular attention. The deep tunnel is constructed as a bored tunnel, segmental lining. The cross section of the segmental lining and the connection to a mined adit tunnel is schematically illustrated in Figure 12.

As seen from Figure 12, the segmental lining consists of pre-cast reinforced concrete (RC) segments, i.e. segments reinforced with traditional (carbon) steel reinforcement cages, with an inner lining consisting of in-situ cast SFRC (secondary) lining and a HDPE lining. Hence, a two pass lining system is applied.

For the traditional reinforced segments, a 40 mm nominal cover is proposed by the designer, for the internal and external surfaces, following the Singapore standard. A number of requirements

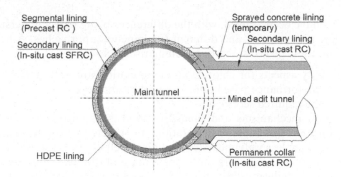

Figure 12. Cross section of segmental lining and connection to mined adit tunnel

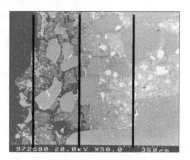

Figure 13. Testing setup for the MPA Performance test (Courtesy of Kiwa Gmbh)

Figure 14 Damage front determined by scanning electron microscopy (Courtesy of Kiwa Gmbh)

have been specified by the Employer with respect to the concrete mix designs and durability properties for the concrete, e.g. high replacement of OPC by GGBS (80% or more) and/or silica fume (15% or more), water absorption of less than 1% (BS 1881: Part 122), a pressure penetration of less than 10 mm (BS EN 12390-8) and a shrinkage limit of less than 0.01% at 28 days.

To account for the risk of sulphuric acid attack, the inner lining is designed to protect the permanent structural segmental lining against such deterioration. The design takes into consideration that the HDPE lining is not there or it has been punctured, and therefore the SFRC forming part of the inner lining needs to resist against MIC. This sets very tight requirements to the SFRC lining in terms of composition of SFRC and thickness of the lining in order to achieve the required 100 year service life. The MIC resistance of the concrete needs to be be tested by an accelerated biogenic sulphuric corrosion test (MPA-Performance test), where the hydrogen sulphide concentration is not less than 100 ppm and the test is to be undertaken for a minimum period of one year. The obtained testing results will be used as the basis to determine the required sacrificial layer thickness for a service life of 100 years.

At the present (initial) stage, it is proposed to use a mix design for the SFRC MIC resistant inner lining with a very low w/c ratio not exceeding 0.35 and a minimum binder content of 400 kg/m^3 in order to achieve a sufficiently dense concrete. Based on necessity, the MPA-Performance test results may be verified by the ODOCO (Odour and Corrosion) test, which real site conditions can be simulated. It should be mentioned that the MPA-Performance test was used, among others, in the Emscher project, which is the largest construction project for sewage systems in Germany in the past 30 years (design life of 120 years).

3 CONCLUSION

As discussed above there are now a number of major drainage and sewer tunnels that have been constructed or are under design and construction, they illustrate that a number of different solutions have been applied, with the differences being partly understood by different project conditions, requirements and different fundamental approaches. A number of lessons can be drawn from these projects, these are summarised below.

- That the durability aspects for chloride and carbonation are well understood and are well covered by the performance based approaches that are now in use, these approaches can reliably deliver the required service life.
- The deterioration mechanisms and predictive models associated with Microbiologically induced deterioration that causes deterioration of concrete by producing acids that degrade concrete appear to be less well understood and less well defined. The solutions to resist this type of deterioration are therefore more uncertain and different solutions are adopted on different projects. Significant differences are also seen in the thickness of a sacrificial concrete layer used between different projects.
- The majority of the projects considered have used a two pass lining, i.e. a segmental lining and a secondary in-situ concrete lining, either unreinforced, or with carbon steel reinforcement or steel fibre reinforcement. A single pass lining has been used on only very few projects such as IDRIS MTS01 and has many benefits particularly for the smaller diameter tunnels.
- The use of a HDPE membrane, such a membrane has been applied on a number of projects, such as the STEP project, where the tunnel is conveying sewerage, this provides a robust solution which is not dependent on the prediction of the deterioration and thickness of the sacrificial layer. It essentially provides a very robust first layer of defence to microbiological induced deterioration with a sacrificial layer providing a second layer. However, not all comparable projects have included such a solution with evidently differing interpretations of how a reliable system can be developed.
- The use of either unreinforced, carbon steel reinforcement or steel fibre reinforcement for the secondary in-situ lining. Again there is evidently differing opinions on these solutions and where possible an unreinforced lining gives many advantages in terms of durability followed by a steel fibre reinforced lining.

It is evident that further development is possible in order to understand the deterioration mechanisms and to optimise the solutions for large sewer tunnels. This includes the deterioration models particularly for Microbiologically induced deterioration and in the developments in progress with one pass lining systems, where the primary, secondary lining are combined with a membrane such as the Combisegments lining system developed by Herrenknecht. Such systems show significant promise in efficiently developing large gravity sewer systems.

REFERENCES

Dauberschmidt. 2006. *Untersuchungen zu den Korrosionsmechanismen von Stahlfasern in chloridhaltigem Beton*. Aachen. Germany.

Fédération internationale du béton (fib). 2012. *Model Code 2010*. Lausanne. Switzerland

NORDTEST. 1999. *NT BUILD 492, Concrete, mortar and cement-based repair materials: Chloride migration coefficient from non-steady-state migration experiments*. Espoo. Finland.

Deutscher Beton- und Bautechnik-Verein (DBV), 2001, *Guideline Stahlfaserbeton*. Berlin, Germany

The Concrete Society. 2008. *CS163 – Guide to the design of concrete structures in the Arabian Peninsula*. Surrey. England

ITA Working Group 2. 2016. *Twenty years of FRC tunnel segments practice: Lessons learnt and proposed design principles*. Avignon. France.

ASTM. 2015. *C1012/C1012M. Standard Test Method for Length Change of Hydraulic-Cement Mortars Exposed to a Sulfate Solution*. PA. USA.

Olliver & Lockhart. Innovative One-pass Lining Solution for Doha's Deep Tunnel Sewer System. *Proceedings, World Tunnel Conference – Surface Challenges – Underground Solutions*:1–10

PennWell. 2015. *Surface Coating Solution Protects Sewer Assets from Corrosion*. Retrieved from Water World. http://www.waterworld.com/articles/print/volume-31/issue-9/special-section/surface-coating-solution-protects-sewer-assets-from-corrosion.html

Tunnels and Underground Cities: Engineering and Innovation meet Archaeology,
Architecture and Art, Volume 5: Innovation in underground engineering,
materials and equipment - Part 1 – Peila, Viggiani & Celestino (Eds)
© 2020 Taylor & Francis Group, London, ISBN 978-0-367-46870-5

Refitting strategies for Italian historical railway tunnels

A.R. Fava, M. Ghidoli, L. Carli & P. Galvanin
Alpina S.p.A., Milan, Italy

ABSTRACT: Historical Tunnels in Italy are normally called the ones built between 1850 and 1940. They were excavated according to different methods, among which the Italian one, and they exhibit a final masonry lining composed by stone blocks, in the bottom part, red bricks at top heading, or, alternatively, red bricks for the whole lining. Most of these tunnels, still in service, are railway tunnels. There are many reasons why railway companies decide to refit historical tunnels: the paper, after a brief historical overview of tunneling methods applied for construction of historical tunnel in Italy, examines the main reasons for their maintenance, proposes a design flow-chart and then presents its practical application to two case histories, showing refitting technical solutions, together with the verification methods compliant to the newest codes of practice in tunnel construction.

1 INTRODUCTION

The opening of Stockton-Darlington railway in 1825, the first commercial line in England designed by George Stephenson - the inventor of the engine locomotive - marked the beginning of railway's age throughout Europe: a flagship project destined to deeply change industry, trades and people's lives, with an impressive rapidity in subsequent decades.

Just few years later in 1839, king Ferdinando II di Borbone, inaugurated the first railway stretch in Italy, from Naples to Granatello di Portici (7,640 km); he also promoted the construction of the first tunnel in the reign, the Orco tunnel having a length of 442 m, opened in 1858, just 11 years later the first railway tunnel completed in Europe, the Schlossberg near Baden (1846-47), 90 m long.

Despite this first successful experience, the spread of railways in Italy was then delayed by the political fragmentation before the unification in 1861 and by objective unfavorable conditions for construction of new lines, due to the complex orography, marked by the presence of Alps in the North and of Appennini, along the whole "boot-shaped" territory.

The mountainous territory deeply affected the Italian railway network, characterized by the presence of many tunnels excavated between 1850 and 1940, which are considered "historical tunnels": a lot of them are still in service nowadays, since - in many cases - the investment costs necessary for their complete reconstruction would be prohibitive. Therefore, the issue of their maintenance is becoming dramatically important over last decades.

2 HYSTORICAL TUNNELS IN ITALY

As aforementioned, a great impulse in railway's construction in Italy was given after the unification in 1861, following the "vision" of the prime minister C.B. Cavour, already anticipated in his article "Des chemins de fer en Italie" written in 1846.

Figure 1.   Railway network in Italy; from left to right: in 1861, in 1870 and in 1955 (source-Wikipedia, the free Encyclopaedia).

The current total length of Italian railway network is approximately 24.500 km - according to RFI database – with roughly 1460 railway tunnels, having a total length of 1480 km. On the basis of data shown in Figure 2, two thirds of the existing network were built before 1930, so approximately 800 km of railway tunnels can be considered historical tunnels.

The territorial complexity, as previously remarked, determined some typical characteristics of Italian railway network and tunnels:

– the high slope of railway lines - up to 35‰ in the Turin-Genoa Line, executed in 1853 - considered a limit also for modern railways;
– the fact that almost all the historical tunnels are one-tracked, rarely double-tracked: this was certainly due to the severe construction constraints at that time, limiting the tunnel excavation width;
– the high length of tunnels: of course, the presence of the Alps determined the presence of very long tunnels playing a crucial role in the railway expansion and tunnelling history in Italy.

Crossing the Alps was the great challenge faced between the half of '800 and the first decades of '900. The oldest Alpine tunnel is the Fréjus-Moncenisio, 13.7 km long, which was inaugurated in September 1871. The construction of the tunnel was initiated by Cavour in 1857, to connect Piemonte to Savoy. In following decades, other long tunnels were built: San Gottardo (1882), Colle di Tenda (1898), Sempione (1906), The Sempione tunnel measures roughly 19′800 meters: at the time of construction, and for the following 76 years, it was the longest railway tunnel in the world. In addition to these striking projects, well studied, monitored and maintained - since the date of their completion - hundreds of minor tunnels were built on the whole Italian network: the attention of National Railways Authorities is increasingly focusing on maintenance and improvement problems of these "minor", but not less important, structures.

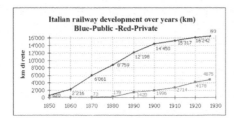
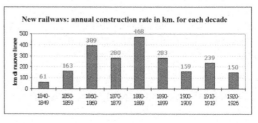

Figure 2.   Extension of Italian railway network up to 1930 and average km per year of new railway, for each decade. A very high value is reached for no less than 30 years, between 1860 and 1890 (source http://www.miol.it/stagniweb/fs101).

Engineers who were engaged in tunnels construction between the end of '800 and first years of '900 had to face a challenge, well beyond the scientific and technological knowledge available at the time: as highlighted by several Authors (G. Barla 2005), they concentrated their efforts on wooden supports to be used while excavating and on final masonry for long term support of the rock mass.

It is worth to note that the elastic theory by Cauchy was set up in 1822, so it was far from being a well-known instrument in the design and construction, especially for minor tunnels.

Conceptual approach to tunneling at that time is well shown by the wooden models preserved in Politecnico di Torino archives, representing the 3 principal excavation methods used for the majority of tunnels built at the time throughout Europe (G. Barla 2005):

– the Austrian-English method, in which a full-face excavation is carried out before the installation of the final revetment;
– the Belgian method: according to this approach, the excavation and lining of the tunnel crown was completed before excavating the benches and the invert;
– the Italian Method, with the so-called "in cunetta" attack, is the opposite of the Belgian: the execution of the invert revetment is completed before the upper part (crown and benches).

The revetment of these historical tunnels is generally composed by stone blocks, in the bottom part along the benches - often obtained from rocks available nearby the construction site - and by red bricks at the crown, or, alternatively, red bricks for the whole lining.

Figure 3. Construction models preserved at Politecnico di Torino; from left to right: Austrian-English method, Belgian method and Italian method (G.Barla 2005).

The thickness of the masonry was often dimensioned according to empirical rules based on experiences and first guidelines of mining engineers. According to these rules (G.Curioni (1877):

– tunnels excavated in rocks were not lined;
– tunnels in rocks with a tendency to degrade, if in contact with air were lined with masonry, having a thickness of 25-40 cm;
– for tunnels excavated in soils, the revetment had to be at minimum 50 cm.

These minimum thicknesses were gradually superseded in railways construction over the years, as observed by the Authors in many tunnels recently refitted (for example the Cucciago or Mombello tunnels on the Milan-Chiasso Line): thicknesses in the range 80-140 cm are not unusual, especially for double track tunnels. Tests with flat jacks on the internal linings of several tunnel demonstrate that stresses on these masonries are generally low, in the range 0÷2.5 MPa, depending on the materials used.

Undoubtedly, great attention was paid - at that time - to the shape of the internal lining to minimize the excavation volume and maximize the arch effect: in this way it was possible to reduce bending moments and to guarantee very low tensile stresses inside the revetment. Generally, the greater part of the rock mass stress was relieved during excavation, due to the lack of an adequate confinement of the excavation profile and of the tunnel face.

Hence, the masonry had the role to counteract the plastic zones and the weight of the rock mass or blocks immediately behind the excavation profile, disturbed while excavating.

The lack of information about construction methods, material characteristics, calculation procedures, geotechnical conditions represents usually a challenging problem when a refitting of these tunnels is required.

3 REFITTING STRATEGIES: A DESGIN FLOW CHART

3.1 *Reasons for refitting of hystorical tunnels*

There are many reasons leading a National Authority (in Italy generally RFI) to plan a refitting intervention:

- structural reasons, when presence of cracks, progressive settlements, masonry deterioration become evident and possible structural damages or serious tunnel deterioration can be feared;
- hydraulic reasons: water leakages, drain clogging, alterations of the hydrogeological and hydrological conditions, induced, for example, by chances at the ground surface caused by new urban transformation, new irrigation channels or sewage systems, or by new/different crops affecting the total amount of water infiltrating into the rock mass surrounding the tunnel;
- adoption of new electrification standards, for example aerial line with consequent reduction of track *gabarit*;
- *gabarit* adjustment to fulfil new national or international standards, STI standards;
- fire protection improvement for safety reason, depending on the presence and distances of buildings and other facilities nearby the tunnel subject to refitting.

Structural reasons, of course, are one of the most important and alarming causes determining the necessity of a refitting intervention. The first step, in this case, is to define what is causing the problem, to identify the correct type of structural improvement. The lack of an accurate analysis of the problem can lead to a not adequate choice; as observed by the Authors, many refit interventions - executed in the '60s -'70s of last century - cover and mask problems, but have a low durability (see Figure 4, where the deterioration of the additional reinforcement layer is well visible on the right).

An important role for stability and durability of structures is often played by hydraulic conditions and their possible changes over time; a correct drainage of the underground water has to be always guaranteed, especially if a new revetment layer is applied on the existing one. Moreover, if existing or new drains are clogged and not periodically maintained, water over-pressure can rise, with subsequent cracks of the existing revetment, deterioration of bricks, instability problems, especially at the benches. This is particularly risky when the surrounding rock mass is prone to quick alteration (for example clays, some types of marls, swelling soft rocks), if in contact with water, as shown in the following case histories.

The electrification of existing railway lines, as well as new *gabarit* standards applied for trains, may cause the tunnel cross section reduction: consequently, lowering the tracks can be necessary, and this can have an impact on the tunnel revetment and on the invert, if present.

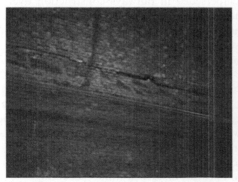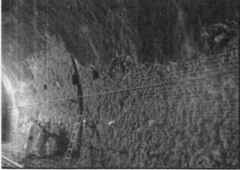

Figure 4. On the left: example of cracking into the revetment, requiring a static analysis. On the right: examples of deteriorated previous structural reinforcement, executed by means of shotcrete (Alpina Archive).

In other cases, a reshaping of the lining at the crown is necessary, ensuing a reduction of the revetment section, to be compensated - in many cases - by additional strengthening interventions to assure the safety conditions.

Other aspects that are becoming quite important in recent years are the firefighting protection together with safety strategies for passengers, in the event of an accident, as well as the seismic refitting. It is worth to note that, quite recently, the Commission Regulation (EU) No 1303/2014 was published. This regulation applies to all tunnels and requires that, in the event of a fire, the integrity of the tunnel lining has to be maintained for a sufficiently long period of time to allow the self-rescue and evacuation of passengers and the intervention of the emergency teams. This period of time must comply with the evacuation scenarios considered in the emergency plan. National stakeholders are called to evaluate risks connected to fire events and necessary/possible countermeasures for existing tunnels. New fire protection guidelines require also the choice and testing of protective materials to enhance the fire resistance of the existing linings, as shown in following paragraphs.

3.2 *The design flow chart*

Since conditions of existing tunnels, problems detected, and type of interventions are specifically related to each single case, it is – generally - not possible to set a unique intervention strategy in advance: design solution are specifically developed for each tunnel, taking into account:

- data collected on site by means of targeted surveys;
- type of problem to be solved (structural, functional improvement, safety strategy);
- service condition of the tunnel: generally, there are different approaches depending on the fact that the railway line is or could be kept out of service, or conversely, has to be maintained in service during refitting works;
- construction exigencies, equipment movable on site, internal dimensions of the tunnel, limiting the equipment to be used.

When an intervention on an existing tunnel is planned, a design flow chart can be summarized in the following 6 main steps:

- *Step 1*: the first step is the understanding of the reasons why the refitting is required: in many cases the original request can be widened after a first analysis: for example, a first required change in the cross section for electrification can become a chance for a more detailed structural check and for a proposal of improvement intervention;
- *Step 2*: the second step is usually the creation of a database containing all the geometrical, structural, geotechnical information the stresses acting on the internal lining, as better explained in next paragraphs; the analysis of data required, and their interpretation, is used to understand the reasons of structural defaults, if present;
- *Step 3*: if structural interventions are required, generally a back analysis of the existing tunnel has to be carried on, thereby it is possible to set up a model, useful to compare *ante-operam* with *post-operam* conditions;
- *Step 4*: in the fourth step, design solutions are developed and analyzed not only from a structural point of view: constructability problems have to be kept well on mind, as well as on site specific constraints, as before mentioned;
- *Step 5*: when a suitable solution is defined, a set of structural analyses is carried out: limits, problems and possible countermeasures are analyzed and discussed. Countermeasures, to manage the refitting risks on site, are usually set, according to the so called "observations method". This method is reliable only if a monitoring plan - during and after works - is conceived in this stage;
- *Step 6*: the construction stage on site can start: the monitoring plan is the key instrument for management of interventions during works and, once the refitting is completed, to follow its effects over time.

After the basic understanding of the design, the second step of the above procedure envisages an accurate survey of the existing structures: the standard preliminary activities - to be carried out before starting the design stage - have generally to include:

– targeted on-site visual inspections;
– 3D laser scanner, to obtain the internal profile of the existing lining, thereby achieving an accurate information, useful above all when *gabarit* checks have to be done to accommodate new standards or new electric devices;
– geognostic surveys outside the tunnel;
– geognostic/structural surveys inside the tunnel, at least at the base of benches, at the tunnel axis, at the crown; when the existing tunnels has not an invert, a key aspect is usually the foundation depth of the benches, above all when the level of tracks has to be lowered to comply with new standards for trains;
– georadar to detect the discontinuities and gaps behind the existing lining as well as the lining thickness for the whole tunnel length;
– flat jacks, single or double, the simplest way to achieve information about the stresses acting on the internal lining, the masonry elastic properties and to compare them with the allowable ones, ascertained with laboratory tests
– laboratory tests on the materials and mortars used for the internal lining (uniaxial tests, shear tests and so on);
– crack meters, optical targets, total and automatic stations, piezometers, extensometers or strain gauges installation inside the tunnel, to measure movements, stress, strains, hydraulic heads and monitor their changings over time.

The back analysis of the revetment can be carried out only if information on the stresses acting on the revetment are known: for these reasons measures with flat jacks, strain gauges, optical targets are all useful to set up these calculation models. They are generally based on analytical solution, on convergence-confinement-methods, or even on more complex FEM models, as shown in the following case histories. A key problem, in this stage, is that the construction procedures followed for tunnel excavation are unknown: so iterative analyses of the rock-mass stress relief is necessary to comply with the stresses measured on site.

Steps 4, 5, 6 of the above flow-chart cannot be standardized for the aforesaid reasons: analysis of case histories can be useful to understand possibilities, problems and countermeasures, while facing a refitting intervention in historical tunnels. These experiences could be a useful guidance for future or similar interventions: the cases of Brichetto and Fey tunnels are presented in next paragraphs, showing the practical application of the design procedure previously described.

4 TWO CASE HISTORIES: BRICHETTO AND FEY TUNNELS

4.1 Brichetto Tunnel

The Asti - Acqui Terme single track railway line, opened on 19 January 1893, runs entirely in the Piemonte region area. Brichetto Tunnel has a total length of about 455 m, a section characterized by an internal width equal to 4.40 m and a height measured from the track level of $5.28 \div 5.37$ m. The tunnel exhibits the typical horseshoe shape with a masonry lining having a thickness equal to 60 cm up to 120 cm on the benches, but it has no invert. The maximum depth from the ground surface is approximately 60 m.

4.1.1 Step 1: understanding of the refitting requirements

After a significant increase in the convergences measured at the tunnel benches up to a value 40 mm, it was decided by RFI-Turin Department to carry out a structural rehabilitation to contain the detected movements. In some stretches the benches of the revetment shown significant deformations and cracks, so serious concerns rose about the stability of the lining.

From a geological point of view the tunnel crosses the so-called formation of the blue-clays, characterized by a friction angle equal to 25-26° and a cohesion 25-30 kPa, without a significant swelling behavior along the examined stretches.

4.1.2 *Step 2: database collection, analysis and interpretation*

Detailed surveys on site, according to the standard procedure described at § 2.3 were carried out along the whole tunnel. It was possible to exclude serious structural deficiencies in the tunnel lining. The observed movements and cracks were related to a local deterioration of masonry

in the areas in contact with water, especially where a local thickness reduction of the masonry was present. These reductions, or the presence of voids behind the lining, may occur frequently in these tunnels, due to construction methods, especially in weak rocks and soils.

The presence of water behind the revetment was related to irrigation changes, occurred for new crops on the surface, affecting the total amount of water infiltrating into the rock mass.

4.1.3 *Step 3: structural back-analysis*

After the completion of stress measures inside the tunnel lining, exhibiting a maximum compression of 0.7 MPa, simple back analyses were carried out, by means of:

– convergence-confinement method for the higher depths;
– structural models – based on the hyperstatic reactions method developed with the FEM code Strauss 7 - for shallow stretches, applying loads according to the Terzaghi's theory.

These first models were useful to check the global behaviour of the lining and to define the *ante-operam* conditions. It was also possible to calculate the possible overpressures acting on the revetment, compatible with the observed displacements.

4.1.4 *Step 4: design solutions for refitting*

The design solutions focused on data collected on site: the bottom part of the revetment seemed no more suitable to withstand the applied loads, moreover, the absence of the invert arch prevented the possibility to counteract the observed displacements of the benches inside the tunnel. The solution was divided into two different stages:

– the first stage envisaged the creation of a concrete invert slab having a "U" shape as shown in figures 5 and 6 during construction: this structure acts as a strut, thereby preventing further movements of the benches and encasing the deteriorated part of the lining up to the center line. Hence, a complete new structural support element - *in lieu* of the damaged portion of the revetment - was created: since the presence of clays (softened by water) didn't assure the lack of settlement over time, the "U" shape structure was underpinned using micropiles placed at an interval of 1 m. Before starting with the excavation stage, the revetment was injected with resins behind the masonry to empty the gaps at the crown detected with the geo radar survey. As shown in figure 6, the "U shape" structure was executed

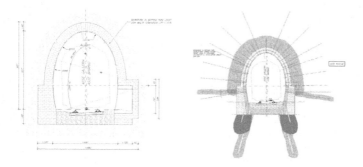

Figure 5. Brichetto Tunnel: on the left: cross section with a rescue-niche; on the right proposed structural interventions (Alpina Archive).

Figure 6. Brichetto Tunnel during rehabilitation works; from left to right: injection stage, execution of the invert; a detail of the "U" shape structure and the sequence of excavation/demolitions concrete casting (Alpina Archive).

following a precise pattern of excavation/demolition and concrete casting, to guarantee a sufficient support to the existing lining during construction.
- The second step included the strengthening of the upper portion of the revetment, to create a full-round new lining, able to withstand overpressures estimated by back analysis models. To achieve this goal, steel ribs were designed and installed within the masonry, previously cut by means of a hydro mill. A final shotcrete revetment was applied to make monolithic the whole intervention.

4.1.5 *Step 5: verifications of design solution and monitoring plan*
The solutions defined in previous step 4, were analysed by means of plane strain FEM models, developed with the FEM code Straus 7: models were able to reproduce the non linearity of materials and restraints and were loaded with pressures estimated by back analysis model previously described.

4.1.6 *Step 6: construction and post-operam monitoring*
No significant problems raised during construction, executed by the Contractor Luigi Notari-S.p.A. Milano because it was possible to keep the tunnel out of service. The intervention was developed during August, a month traditionally not very busy; the challenge was to respect the thought work schedule imposed by the Client. The intervention enabled a significant reduction in the trend of measured displacements; data collected up to now don't show further movements in the refitted zones.

4.2   *Fey Tunnel*

Fey tunnel is part of the Alba-Pra Line in Piemonte: it is a 483 m long tunnel built in 1885, having a masonry of variable thickness: the maximum depth of the tunnel from the existing ground surface is roughly 30 m. The section is a U-shaped one, with safety niches along one side. The geological map of Italy (scale 1:25000), shows that the tunnel crosses an area characterized by an alternation of sand and clays (Astiano and Piacenziano formations).

4.2.1 *Step 1: understanding of the refitting requirements*
The initial reason for refitting was the electrification of the Alba - Bra line: the existing tunnel cross section was enough to guarantee a *gabarit* type PMO n.1, but not enough for the installation of the new contact line at 3kVdc. The required minimum internal height - from the track level - was 4,650 m and this value was not available along the tunnel. To minimize the impacts on on the masonry structure, the solution was to lower the rail level (P.F.) along the entire length of the tunnel. Moreover, the stakeholder RFI, taking into account the presence of numerous buildings along the tunnel alignment, decided to apply the Commission Regulation (EU) No 1303/2014, for fire protection of the existing lining.

Figure 7. Fey Tunnel, from left to right: cross section; internal section after removal of previous interventions by means of hydro-milling; not adequate invert executed in the past (Alpina Archive).

4.2.2 *Step 2: database collection, analysis and interpretation*

No data were available on the existing structures. It was known that, in the past, some interventions were executed at the invert to counteract excessive water pressure. Surveys in the tunnels were based on: georadar investigation, boreholes to calibrate georadar, laser scanner, flat jack tests (single and double), collection of undisturbed and disturbed samples, laboratory geotechnical tests, pull-out tests on swellex anchors. From a geotechnical point of view, soil parameters obtained from the investigations were defined as follow: weight equal to 20-21 kN/m^3; drained cohesion up to 40 kPa; friction angle variable in the range 19-22°.

4.2.3 *Step 3: structural back-analysis*

The back analysis was carried out with the aid of the FDM code FLAC 2D, developed by the ITASCA. This code allows to analyze mechanic problems of continuum, determining the stress-strain conditions, in the two-dimensional or axisymmetric field. A section at the maximum depth (30m) was modeled, and elasto-plastic laws, assuming the "Mohr-Coulomb" were used. The contact areas between the lining and the ground were separated by a so-called interface layer. This interface allows to describe the behavior between the lining and the surrounding rock mass, taking into account possible debonding, relative sliding, and so on.

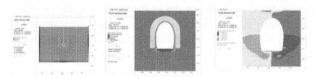

Figure 8. Fey Tunnel: FDM model used for back analysis (Alpina Archive).

4.2.4 *Step 4: design solutions for refitting*

After the analysis of main results obtained by surveys and on site inspections, it was decided that the lowering of tracks would had been possible only by means of a new concrete slab to be placed at the invert, underpinned by micropiles, as done for the Brichetto tunnel. In this case the "U" shape structure was not considered necessary due to the good conditions of the internal lining. It was applied just in some stretches with significant defects - shown in figure 9 - detected in advance during survey stage. The construction stages foresaw:

- removal of rails and the injection behind the existing revetment;
- execution of swellex bolts as temporary reinforcement, to be used while excavating below the foundation, followed by micropiles drilling;
- execution of the invert slab completed by the drainage system to relieve water pressure below the slab;
- fire protection installation, as required by RFI for this tunnel: a protection for 120 minutes assuming a RWS curve according to UNI 11076 and max temperatures 1300° - 1350° was considered.

For fire protection of existing lining, three different options were examined:

- analytical approach: assessment of fire load, fire curve followed by a finite element analysis of the thermal problem;

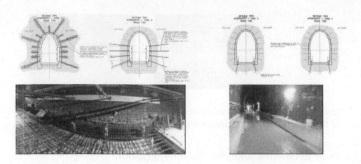

Figure 9. Fey Tunnel: design construction stages (above) and works inside the tunnel; below on the left: defects in the existing lining detected during construction; on the right: the new invert (Alpina Archive).

– experimental approach: evaluation of the fire resistance by taking and testing masonry samples - taken from existing lining - in official laboratories;
– conservative approach: installation of a sacrificial element able to preserve the masonry in case of a fire event.

After analyses and according to RFI-Trurin Department final decision, it was decided to implement the third option using a sacrificial layer specifically developed by Mapei for underground application (called Ignisilex type 4): before spaying the layer, the existing revetment was accurately cleaned by means of an hydro mill, as shown in previous figure 7. Where required by the presence of humidity on the internal masonry, a special support (made by thixotropic mortars and a steel mesh mechanically anchored to the tunnel walls) was prescribed. Laboratory test were done on the prescribed sacrificial thickness to verify the fire-resistance performances.

4.2.5 *Step 5 and 6: verifications of design solution, construction and monitoring*
Design solutions were verified by mans of the FDM models used for back analysis: the numerical analysis was divided into 10 phases equal to the number of the design phases: this procedure allowed to compare final stresses and deformation acting on the lining with the initial ones.

Construction stages (executed also in this case by Luigi Notari S.p.A Milano) were continuously monitored, but no significant problems occurred.

5 CONCLUSIONS

The paper, after a brief historical overview of railway tunneling methods, examines the main reasons for historical tunnels refitting and then presents two case histories showing technical solutions, together with the verification methods compliant to the newest codes of practice in tunnel construction. A design flow chart is proposed and the its practical application to two case histories is discussed. Since conditions of existing tunnels, problems detected, and type of interventions are specifically related to each single case, it is – generally - not possible to set in advance a unique intervention strategy: design solutions are specifically developed for each tunnel, taking into account data collected on site by means of targeted surveys. Analysis of case histories can be useful to understand possibilities, problems and countermeasures, while facing a refitting intervention in historical tunnels. These experiences could be a guidance for future or similar interventions.

REFERENCES

Curioni, G. 1877. *"Sulla determinazione delle grossezze dei rivestimenti delle gallerie in terreni mobili"* - Atti della R. Accademia delle Scienze di Torino, Torino, 1877.
Barla, G. 2005. *"Sviluppi nell'analisi progettuale delle opera in sotterraneo"* - Rivista Italiana di Geotecnica, Volume 3, pages 11–67, 2005.

Tunnels and Underground Cities: Engineering and Innovation meet Archaeology,
Architecture and Art, Volume 5: Innovation in underground engineering,
materials and equipment - Part 1 – Peila, Viggiani & Celestino (Eds)
© 2020 Taylor & Francis Group, London, ISBN 978-0-367-46870-5

BIM – Model-based project management for optimizing project development. Insights into the practical application at the "Albvorland Tunnel" project

W. Fentzloff & J. Classen
Implenia Construction GmbH, Munich, Germany

ABSTRACT: Building Information Modelling (BIM) is a method used to plan and manage projects based on 3D-models. Linking the 3D-models to the dimensions of "time" (4D) and "costs" (5D) generates the so-called process model which gives rise to a graphic project management tool. Tunnelling projects are unique when it comes to time and cost tracking, especially since the works are executed in an unknown underground environment. The anticipated geology is based on an interpretation of geological investigations prior to the start-up of the works. During the execution phase, deviations are commonplace. The way this is to be shown has to be incorporated in the process model in an appropriate way and should be agreed upon by all parties involved. Such methods are being used in the current "Albvorland Tunnel" project in order to gain insights into the future use of BIM on large-scale infrastructure projects.

1 INTRODUCTION

BIM – Building Information Modelling – takes advantage of the continuous use of a digital building model over the individual phases of the life cycle. This means that the structure is planned with the help of 3D models that allow a lifelike interactive presentation. The models consist of components in which relevant information is stored.

In addition to the general information such as geometries, construction materials, etc., the components are also linked to activities from the construction program and to costs from the items of the Bill of Quantities (BoQ), making it possible to make reliable statements on the various completion stages and the associated costs within the scope of project management.

DB Projekt Stuttgart–Ulm GmbH (PSU) has been commissioned with the planning and technical implementation of the highly complex "Stuttgart–Ulm Railway Project" in southern Germany which is currently one of the largest infrastructure projects in all of Europe. It is broken down into two sub-projects "Stuttgart 21" (S21) to rearrange the Stuttgart rail hub and the "New Wendlingen–Ulm Line" (NBS) with high-speed tracks.

One of the most prominent structures of the NBS is the double-tube Albvorland Tunnel running in a west-east direction with a length of around 8.2 km; three further tunnels are shorter than 500 m (see Figure 1).

2 BASIS FOR USING BIM

The German Federal Ministry of Transport and Digital Infrastructure (BMVI), saw the importance of BIM for the near future in terms of digitalisation of the German construction industry and thus released a graduated scheme for the digitalisation in design and construction works in 2015.

In the meantime, the BMVI is performing research into practical application of the method within the scope of four pilot projects.

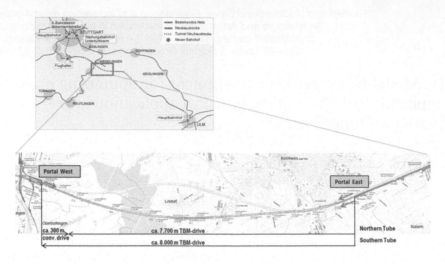

Figure 1. Situation of the "Albvorland Tunnel" project.

In order to gain more experience with **BIM** in regards to the large number of high-value contract awards the PSU took **BIM** into consideration in other parts of the S21/NBS project, such as the bridge over the Neckar River in Stuttgart and specifically in one sector of the Albvorland Tunnel. In the case of the Albvorland Tunnel, the first practical experience will be gained with 5D links (linking costs to the 3D model), i.e. cost budgeting, in addition to 3D planning and 4D links to the scheduling.

3 THE SCOPE OF BIM IN THE TENDER PROCESS

At the beginning of the design phase, the BIM method was not part of the planning process. Therefore, the design planning was prepared conventionally using 2D planning methods. It was not until the tendering stage that specifications were made as to what the employer intended to implement by means of the BIM method. In order to limit the scope involved, the use of BIM in the Albvorland Tunnel project has been restricted to the west modelling area (see Figure 2).

The structures/components integrated into the modelling area cover the complete range of disciplines used in infrastructure projects: tunnelling, civil engineering, special foundations, road and earth works.

The client intends to achieve the following by using BIM:

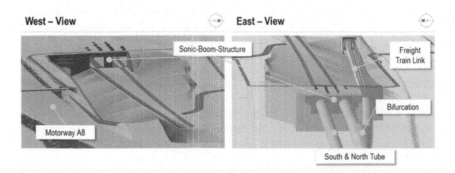

Figure 2.   Modelling area at portal west of the Albvorland Tunnel.

- Building-up expertise in handling BIM in large-scale infrastructure projects;
- Gaining insights into where BIM can generate potential benefits in terms of cost effectiveness;
- Developing the greatest possible system interoperability, specifically taking the Deutsche Bahn's own systems into account;
- Facilitating optimization and quality assurance in planning and implementation;
- Achieving improved communications and exchange of information with project participants;
- More precise and efficient determination of quantities (also in the event of changes).

All the appropriate measures necessary to be successful in the favoured goals, as well as the following defined BIM applications, were set-out in a specific appendix to the tender documents and then became part of the contract. This can be seen as a first draft version of the Employer's Information Requirements (EIR).

3.1 *The defined BIM applications*

The BIM applications defined in the EIR shall be implemented at the west modelling area of the Albvorland Tunnel; these are outlined in the following.

3.1.1 *3D-assisted detailing and coordination for planning purposes*

The application "3D-assisted detailing and coordination for planning purposes" contains the generation of a 3D model of the area concerned with all required geometries and objects including the information and properties assigned to them (see Figure 3).

Geometry is usually described by length, width, height and diameter. In addition, every component is assigned its own identification number (important for later links), the structure number, information on the material and specifics to tunnel construction, e.g. the tunnel meter, length of one round, driving class, block length and even the pertinent item of the BoQ for linking costs is recorded.

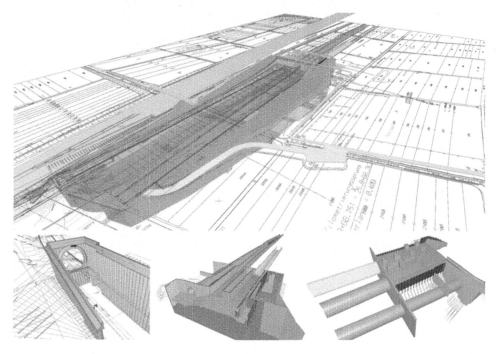

Figure 3.   3D model combined with 2D drawings showing land parcels and detailed views.

The conflicts identified at an early stage relate, for example, to utilities crossing the structure (power supplies, telephone lines) that have to be relocated away from the construction area prior to the start of work or anchors that would have led to ill-defined situations at places where several rows of anchors would cross each other or even penetrate underground structures.

In addition, the 3D model is linked to the 2D technical planning documentation from which the status and overviews of planning packages are prepared according to the target, actual and forecast. Furthermore, the 3D model is used as a basis to determine model-based quantities for plausibility testing of scheduling resources and BoQ-items.

3.1.2 *4D and 5D simulations*

The 4D model of the BIM modelling area in the Albvorland Tunnel is created by linking the 3D model to the construction schedule. This is used as the basis for 4D simulations, visualizations, analyses and optimization work for scheduling feasibility. Moreover, regular review of the 4D model (this means continual adjustment of the construction processes if they change on account of external influences), evaluations of the construction sequences as well as quantities and resources according to target, actual and forecast.

In order to generate the 5D model, the 4D model is linked to the pertinent contractual prices contained in the BoQ and the related items themselves (see Figure 4). If the component-related performance reports are continually integrated into this 5D model it can be used to present and evaluate not only construction progress but also performance levels and costs (target, actual, forecast) for project management purposes in the so-called "process model".

3.1.3 *Reporting*

The third application for the Albvorland Tunnel consists of reporting. In this respect, the 3D, 4D and 5D models are used to generate presentations and evaluations of the key performance indicators relating to schedules, costs, performance and planning. One possible visualization can be seen in Figure 5.

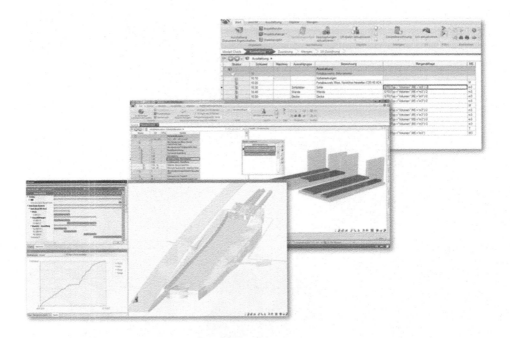

Figure 4.   Examples of visualization of 4D & 5D simulations in the process model.

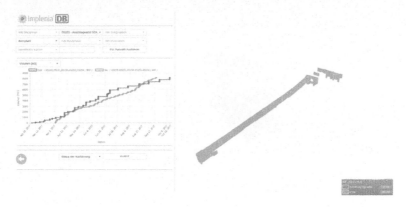

Figure 5. Visualization of construction progress according to target, actual and forecast.

3.2 *Challenges relating to the implementation of BIM*

As BIM is still a recently developed method and as various project participants often do not have sufficient BIM expertise at the beginning of the project or acceptance of the need to use BIM methods has to still be developed in some cases, there is a danger that the BIM method will be pushed to the side during the implementation phase if it is applied simultaneously to conventional planning methods. The active involvement of management and employees alike is vital for the implementation of BIM methods. For this reason, workshops must be held with all participants from the very beginning with the objective of providing support in the various stages of BIM implementation and to present to all individuals involved the potential and advantages of BIM.

Furthermore, it is important to be aware of the possibilities offered by BIM and to specify in the call for tender exactly what is expected at what point in time. Due to differences in the levels of expertise on the part of project participants, there are also different expectations. If this is not laid out unambiguously it will be impossible to ensure that participants work with BIM with as little friction as possible.

4 PRACTICAL IMPLEMENTATION OF THE CALL-FOR-TENDER TERMS

The project initiator took a courageous and trailblazing step in the call for tender by deciding to deploy BIM for the Albvorland Tunnel after the construction project had already been taken to the tendering stage using conventional planning methods. While this only applies to a sub-area at the West portal it does, however, include all of the main works involved in the project: tunnelling works, civil engineering works, special foundation works, earthworks.

This approach makes it possible to process the complex relationships within the scope of preparing a 3D/4D/5D model with the necessary depth of detail in order to gain the desired level of experience.

Revit was the main software used, among other programs, to prepare the 3D models and RIB iTWO for the links to scheduling and costs (4D, 5D).

A few other programs such as Desite, Navisworks, Thekla BIMsite are currently being tested with respect to their effectiveness for BIM on the project.

The following topics were handled during this process:

* Preparation of a BIM implementation plan (BIP);
* Preparation of guidelines and recommendations for modelling;
* Preparatory work on model structure and model coding;
* Specification of the level of development and information to be contained in the partial models (LoDs);

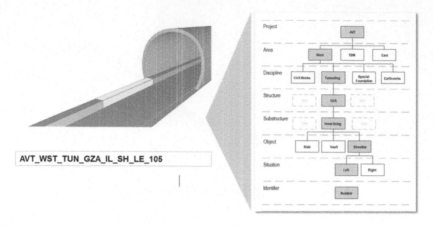

Figure 6. Implementation of the Work Breakdown Structure.

- Use of the shared parameters tool BIM*Q to determine the parameters in Revit families including adjustment on an ongoing basis;
- Preparation of the process models in RIB iTWO to link the construction program to the 3D model;
- Adjusting the structures of the BoQ (items) to achieve a proper link of costs to the model (→ 5D);
- Export of the CPIXML files from Revit using a plug-in to input the data into iTWO.

In this context, make sure that the software used implements an interface for the standardized data format IFC (Industry Foundation Classes) in order to ensure compatibility of the results from the various programs used.

In principle, the challenge coming up in the specific case of the Albvorland Tunnel arose from the fact that the project structure corresponded to the conventional planning process but not to the requirements of the model-based structure. The greatest challenge to be overcome was the lack of a consistent work breakdown structure (WBS) within the project that had to be prepared by the Contractor after the contract had been awarded. The existing structure was adjusted in various steps. Figure 6 shows an example of the WBS adapted to a shoulder of the inner lining of one of the tunnels.

The basic setup is designed by identifying an object using information about it. The information goes from general to specific. Finding proper acronyms for each level in the WBS results in a unique code per object, the so-called ID. In this example, the ID for the shoulder in a particular tunnel on the left side of the inner lining of section #105 is "AVT_WST_TUN_GZA_IL_SH_LE_105". You will find the ID as a property in the 3D model as well as in the 4D model and the 5D model. This system is the crucial backbone in order to be able to generate a realistic simulation of the construction process taking scheduling and costs into account. A further benefit can be achieved for the running phase of the tunnel when RFID tags are implemented into the real objects that again contain the same ID and store the relevant information for facility management purposes (see Figure 7).

5 SPECIAL INSIGHTS GAINED DURING IMPLEMENTATION WITH RESPECT TO THE ALBVORLAND TUNNEL

One of the most important insights gained is the necessity to train all parties involved to attain the same understanding of how BIM is to be applied.

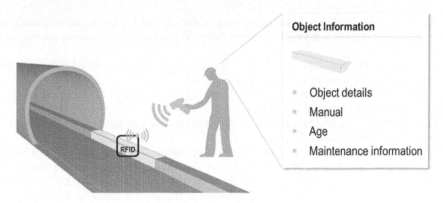

Object Information

- Object details
- Manual
- Age
- Maintenance information

Figure 7. Object information stored in RFID tag for FM purposes.

Another important issue is the need for approval or to mutually agree on a processing status for the 3D model as a basis for meaningful project management. The fact that infrastructure projects in general and tunnelling jobs in particular deal with a defined construction material, specifically the ground, does not easily allow the transfer of BIM applications from the more developed routines of architectural building to tunnel projects. Due to fluctuating ground characteristics, it is difficult to make consistent predictions and results in unforeseen surprises over and over again. Usually there is a preliminary ground investigation of the project area to determine the geology, but a certain risk of ground conditions (water, tensions, fault zones, petrochemical conditions, etc.) that cannot be detected is immanent. Interpretation of the anticipated ground conditions are based on sporadic probe drills and other auxiliary information. This knowledge has already been taken into consideration in the project management of tunnelling projects in the past as well as today. Using BIM with its capability to fix, analyse and evaluate project progress, it must be taken into consideration that the described uncertainties require continuous adjustment of the appointed models with special respect to target and forecast.

There is still a long way to go before 3D planning will be used as the sole basis for model-based project management. 2D planning is currently indispensable if only for the release and approval processes as it is apparent that the technical tools (hardware and software) still do not always fulfil their purpose and further skills will have to be developed among the individuals involved.

Changes in the scope of work during the execution phase necessarily evoke an adjustment of either the 3D model or even the 4D and 5D model according to the respective impact. For the time being, it is a common practice that a long time passes between the occurrence of those changes and the willingness to raise change orders accordingly. The significance of any BIM process model in-use will tremendously suffer from the lack of willingness to agree quickly upon changes and their consequences in time and costs for the project.

6   USE CASES IN ADDITION TO CONTRACTUAL REQUIREMENTS

In addition to the above described contractual use cases the contractor implemented its own model-based applications. Regarding the fact that parts of various tunnel drives run beneath the adjacent motorway A8 there is a distinctive focus on settlements and safety issues (with respect to the traffic on the Autobahn). Two sensitive areas were identified:

- A stretch of approximately 1400 m of the TBM-drive of the main tunnel from the eastern portal
- The underpass of the motorway A8 with the freight train link (GZA-BAB-tunnel) connecting a regional track with the north tube of the main tunnel

In both cases the tunnel drives are monitored at the surface by means of automatic surveying equipment that deliver a myriad of values every 30 seconds from various profiles with different survey points.

The aim was to link the data of the monitoring with tunnelling related data and visualize them in an appropriate way by using a capable viewer.

The results were absolutely satisfying. Real-time information and tracking can be displayed to relevant persons even on mobile devices.

The fact of an existing and efficient 3D-model puts one in the position to link, analyse and visualize any data for required purposes.

In the second situation (underpass of the motorway A8 with the GZA-BAB-tunnel) the overburden is very small. For that reason the tunnel drive is supported by a pipe umbrella to be bored in sections in advance of the excavation of the tunnel profile. Figure 8 shows the minimum overburden between the crown pipes of the pipe umbrella and the bottom edge of the street construction which is less than 20 cm (theoretical value). The motorway-authority decided therefore to close the impacted lane of the motorway (three lanes in each direction) where the boring process is in operation. Simulating this situation, the 3D-model can give and visualize the relevant information (see red-coloured lane in Figure 9).

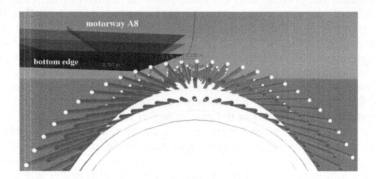

Figure 8.   Situation of pipe umbrella beneath the A8 motorway.

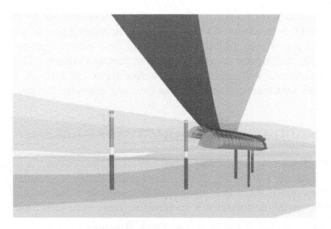

Figure 9.   Visualization of lane closure referring to boring operations beneath the A8 motorway.

7 CONCLUSIONS

The following conclusions can be drawn from the current project progress with regard to BIM:

- The 3D model must be prepared during the drafting stage and become part of the call for tender.
- It is necessary to create a uniform project structure (WBS) with regard to the activities from the construction program and the items of the BoQ.
- The check for conflicts using the 3D model is, if employed correctly, a very suitable instrument for identifying execution problems in a timely manner.
- The 4D and 5D linking, in particular, are challenging issues that require the corresponding systems. In the event that the linking is successful, however, this provides a powerful project management tool.
- A large amount of detail coordination is necessary between the project participants in order to achieve a uniform understanding of the objectives aspired to in the application of BIM. Regular BIM meetings are a suitable means to this end.
- When it comes to cost tracking, it must be taken into consideration that costs are defined differently by the client and the contractor.
- In the event of amended or additional works, it will be necessary to specify within the framework of the construction contracts a non-bureaucratic way to allow for changes.
- The provisions of the intended construction agreement have to meet the needs arising from BIM.
- One issue that has to be clarified is the point at which the project initiator involves the individual "protagonists" who make their important contributions to the project development, execution and running to which contractual model.
- BIM is an important tool that, on the basis of models, makes it possible to compile and process a wide range of data within the framework of the progressing digitalisation of the construction industry.
- The existing structures relating to project handling have to be adapted to the new requirements of BIM in terms of technical, legal and human issues.

The technical side of the structural change has been addressed in this report. To the same extent, however, it will also be urgently necessary to reflect on new contractual models and the willingness of the individuals involved to implement BIM.

Ultimately, BIM is no longer just a project management method or tool, but rather constitutes a radical cultural change in the development and handling of a project. At this point, we are still in the beginning stages but the great potential and the necessary cultural change are the driving force behind this pioneering work.

REFERENCES

Fentzloff, W. (2016): *BIM in Der* Ausschreibungs- *und Ausführungsphase – Erfahrungen vom Albvorlandtunnel* Implenia Construction GmbH, Wiener Gespräche 2016-10-20

Fentzloff, W. (2017): "BIM – Model-Based Project Management, Using The Albvorland Tunnel As A Practical Example" Implenia Construction GmbH, 11th International Tunnelling And Underground Structures Conference 2017, Slovenia, Ljubljana

Tunnels and Underground Cities: Engineering and Innovation meet Archaeology,
Architecture and Art, Volume 5: Innovation in underground engineering,
materials and equipment - Part 1 – Peila, Viggiani & Celestino (Eds)
© 2020 Taylor & Francis Group, London, ISBN 978-0-367-46870-5

Tunnel damage management system: An application for Italian railway tunnels

P. Firmi, F. Iacobini, A. Pranno & C. Matera
Rete Ferroviaria Italiana – Infrastructure Department, Rome, Italy

ABSTRACT: Italy is the European country with the highest number of railway tunnels whose cover about 10% of the entire infrastructure. Rete Ferroviaria Italiana (RFI) is the company in charge of almost all the Italian railway network management. It performs maintenace activities on over 1.500 km of tunnels, which have been built until 150 years ago and are still in service.

In order to perfom the maintenance of this amount of tunnels and achieve effective maintenance plan, the assessment of tunnels conservation status is necessary.

In this respect, RFI developed the "TDMS - Tunnel damage management system", an experimental method to perform the surveys of tunnels and to evaluate the conservation status of these.

The paper will illustrates the criteria underlying the TDMS method, the main features of the technology and how the conservation status of tunnels is defined.

1 INTRODUCTION

Railway tunnels, such as most of structures, require regular inspection and maintenance to assess the structure conditions and extend their life cycle.

However, a prolonged service life of the tunnels with respect to the useful life, which they are designed for, increases the likelihood that degradation phenomena will arise unless appropriate monitoring plan are adopted.

Tipically, cracks, leakage, loss of mortar between the bricks joints, detachments and ejections of the lining are among the most common effects of degradation that may be observed in the oldest tunnels.

That issue becomes of fundamental importance for the Italian railway infrastructure which includes tunnels built until 150 years ago and these are still in service. Furthermore, Italy is the European country with the highest number of tunnels, up to 1681, whose cover about 10% of the entire railway network.

Most of the Italian railway network is managed by RFI which performs routine inspections on tunnels to diagnose lining deterioration, preserve the tunnel's structural integrity and ensure the safety standards of the railway network.

Traditional inspections are conducted by humans during rail traffic interruptions mainly using night slots. A subjective judgement of the tunnel is assigned by the inspector, who, depending on the experience, identifies and carries out direct measures of damages on the lining surface.

This approach may be expensive in terms of time consuming due to the short available traffic interruptions. Moreover an inadequate lighting of the area inspected wouldn't allows to identify all damages, especially the presence of cracks and loss of mortar between the bricks joints. Additionally, the commonly used measuring instruments, such as steel tapes or tape extensometers, are unfit for this purpose and are labour intensive.

In order to overcome the disvantages of human visual surveys of tunnels, several applications of automated inspection systems have been introduced in recent years.

These technologies include impact acoustic methods, laser methods, GRP methods and others and most of them allow to identify the defects that affect tunnels.

In 2015 RFI started a research project for the automation of tunnels damages detection. The project involved part of the railway network tunnels and it will be extended to all national tunnels in the next years.

Tunnels inspection system used for this project is based on high-resolution laser cameras that allows to carry out lining damage evaluation at speeds up to 30 km/h with a millimetric resolution. System is also able to automatically detect most of damages without there being any human intervention.

Following this experience, RFI developed an algorithm, Tunnel Damage Management System (TDMS), to process results provided by the mobile system in order to define the damage index (ID) of the structure and manage potentially hazardous situations. TDMS is also a particularly useful tool on decision making process of planning of structural and maintenance interventions which allows to determine what tunnels needs priorities actions.

This paper will describe the advantages in using a tunnel inspection system to assess the state of conservation of Italian railway tunnels and the key aspects of the Tunnel Damage Management System (TDMS) developed by RFI. Section 2 describes the Italian railway tunnels heritage. Section 3, shows the RFI visual inspection system program. In section 4, the principles of the inspection system and of data processing are described. The criteria of tunnel damage level assessment are illustrated in Section 5. The conclusions of the paper are included in Section 6 and in the section 7 the bibliography is shown.

2 ITALIAN RAILWAY TUNNELS HERITAGE

Italy has one of the largest rail assets in the world with a rail network of about 22,000 km, 16,800 km of which are managed by RFI. Moreover, Italy, is the European country with the highest amount of railway tunnels in number and in length (Figure 1).

Taking into account tunnels longer than 100 m, RFI manages 1681 tunnels whose a length of 1518 km; where percentage between single track and double track tunnels is almost the same.

The type of lining depends on the construction materials available at the time of the construction, bricks and stone were used until 1960, then concrete and segmental linings.

In rock terrain, tunnels are in some cases unlined. However, the rock is more and more covered with shotcrete in order to prevent falling of blocks.

The pie chart illustrates a classification of Italian tunnels by lining materials. At first glance it is clear that common material is the masonry. In particular, a large number of tunnels are made of brick and stone masonry, almost 17% of them is in brick works and only 9% in stone works. In addition, less than a third of tunnels are made in concrete, reinforced concrete and segmental lining. Finally, less than 7% are other types of lining, including the unlined ones (Figure 2).

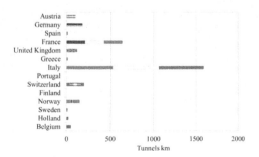

Figure 1.   Railway tunnels heritage in some European countries.

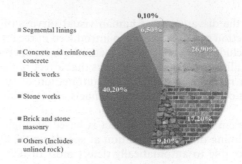

Figure 2.   Classification of Italian tunnels by lining materials.

3   THE RFI VISUAL INSPECTION SYSTEM PROGRAM

The main tool used for tunnel conditions monitoring is the visual survey which is regulate by the "Procedure for the inspections of bridges, tunnels and other structures of the railway network".

It sets out the types of survey relating on the structure, frequency of inspections and how to assign a rating code to the tunnel, which varies from 0010 to 0110. Rating code is related on tunnel structural integrity and is recorded on company database. Moreover, the procedure specifies the hierarchy of responsibilities in the inspection process, personnel equipment and qualifications requested to carry out the surveys and a list of the main damages that may be detected on structures.

The following Table 1 shows types of visual inspections defined by the procedure:

Table 1.   Visual inspections defined by RFI.

Type	Cycles	Description
Ordinary visual inspections	1 year / 6 months	Carried out by specialized personnel with a base certification (MI OC1). Tunnels are regularly checked one time per year. Intermediate visit (6 months) is requested for tunnel with a rank higher than 0060 (see Table 2).
Extraordinary visual inspections	When necessary/exceptional occurrences	Carried out by specialized personnel with an advanced certification (MI OC2) following exceptional occurrences or for critical conditions of tunnels.

The inspection aims to verify the structural elements conditions (portals, lining, platform, track, niches, etc.), not-structural ones (waterproofing, drainage and sidewalks) and the area immediately adjacent to the tunnel, such as slopes and retaining walls.

The inspector, taking into account the lining structure types (unlined, masonry or concrete) shall identify tunnel disorders such as: cracks, geometrical defects, lining surface defects, missing elements, open joints, water inflows, etc.) and their severity.

As conclusion of each tunnel survey the inspector has to record all the information about tunnel lining assessment by a specific tool (Avviso V1) and define a global rating code taking in mind the most critical defects.

Then global rating code has to be associated with one of the following expression for the global judgment:

- The tunnel is efficient with regard to the safety and regularity of railway traffic;
- The tunnel is efficient with regard to the safety and regularity of railway traffic with the following limitations and cautions. . . . . .until the proposed measures are completed;
- The tunnel is efficient with regard to the safety and regularity of railway traffic with the following limitations and cautions. . . . .

Eventually, the global judgment is recorded on Avviso V1 and this latter stored on company database.

4 RESEARCH PROJECT EXPERIENCE

In 2015 RFI started a research project for the automation of tunnels damages detection. The project involved part of the railway network tunnels.

Tunnels inspection system used is based on high-resolution laser cameras that allows to carry out lining damage evaluation at speeds up to 30 km/h with a millimetric resolution. System is also able to automatically detect most of damages without there being any human intervention.

Following this experience, RFI developed an algorithm, Tunnel Damage Management System (TDMS), to process results provided by the mobile system in order to define the damage index (ID) of the structure and manage potentially hazardous situations. TDMS is also a particularly useful tool on decision making process of planning of structural and maintenance interventions which allows to determine what tunnels needs priorities actions.

4.1 The principles of the inspection system

The mobile system is based on the use of high-speed cameras, custom optics and laser line projectors to obtain 2D images and high-resolution 3D profiles of the surveyed tunnel. It can be operated under all types of lighting conditions, providing high-quality data in both illuminated and shaded areas.

The system enables to acquire both 3D and 2D images data with a depth accuracy of 0.5mm, a longitudinal resolution of 1mm and a transverse resolution of 1mm over 12m long arc tunnel sections at survey speeds up to 30km/h. This rapid tunnel condition assessment ability makes it possible to perform long tunnel inspections more frequently on a regular basis.

The high-resolution images lighted up by laser emitters enable the analysis automation for defects detection, so that for instance, cracks, damages, and moisture can be detected and analysed.

Its computer vision-based software for the automatic evaluation of tunnel condition offers the following advantages:

- Tunnel linings: detection of cracks and areas with missing or chipping lining, dampness and running water, areas with poorly assembled segments, protruding edges and poor workmanship, as well as tunnel installation assessment.
- Railway evaluation: includes the assessment of the transverse section by means of 3D geometry and rail flaws detection such as: lack of fixing elements, corrosion, cracks on sleepers or on slab tracks.
- Structure evaluation: 3D reconstruction and clearance analysis.

The mobile system extracts 3D information using a pattern of lighting projected from the laser into the object to be inspected. The pattern is recorded by a digital camera positioned away at a fixed distance, at an oblique angle relative to the projected light. The intersection between the pattern of emitted light and the field of view of the digital camera defines the range of operation of the 3D sensor. The positions of the lighted points on the surface of the object are displayed in the image obtained by the camera. This technique enables high-quality digital images and superimposed 3D information to be obtained in a single capture.

Compared to other equipment based on LiDAR (laser imaging) technology which operates at very low (walking) speeds, this system enables long tunnels to be inspected in short periods of time and with much higher resolutions. Mobile system software also allows the data from two different inspection runs to be rapidly compared, and structural changes and all types of wall lining defects to be assessed.

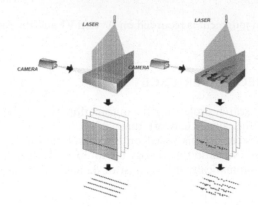

Figure 3. Laser camera operating principle.

4.2 *Data processing*

The data provided comprises 2D images with greyscale intensity information which contain information in a radial direction relative to the direction of the condition's survey. Next Figure 4 shows a 2D image, an intensity image and a profile.

Moreover a global vision of tunnel lining is provided as 2D images maps with different information layers (Figure 5): lining material, defects on 2D images and defects on white background 2D images.

High-quality digital images and a 3D reconstruction of the tunnel lining are also available, as the following figure clearly shows. If necessary to have a better focus of an inspected area, 3D reconstruction of a single laser camera can be obtained from the captured data.

Some algorithms have been developed to detect certain common defects in tunnels with concrete or masonry lining. The defects founded during the post-processing analyses have been catalogued in different principal damages types, as reported below:

- Cracks (longitudinal, transversal and diagonal)
- Loss of mortar between the bricks joints
- Lining falling off
- Damp patches
- Deformation/detachments

As result of this organization, basing on RFI engineers expertise, for each defects type has been founded a geometrical parameter that influence the intensity value of it. For example, the width represents the intensity for the cracks while the depth represents the intensity of some defects like as lining falling off and loss of mortar between the bricks joints.

In function of lining material, different thresholds have been studied to define the intensity of each defects; for example centimetric thresholds have been defined referring at the cracks for masonry lining while millimetric one for the concrete lining. For the lining falling off defects the thresholds are centimetric for both the material, but with different numeric values.

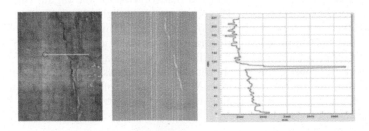

Figure 4. Data with 2D images.

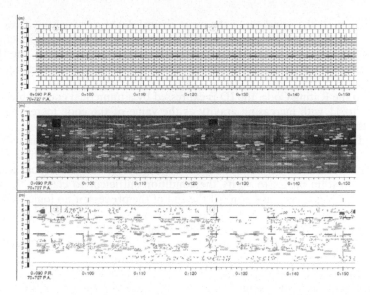

Figure 5. Map with lining material - defects on 2D images - defects on white background 2D images.

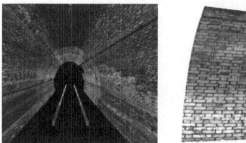

Figure 6. 3D reconstruction of the tunnel lining.

5 TUNNEL DAMAGE LEVEL ASSESSMENT

The data acquired during the surveys by the mobile diagnostic system and processed semi-automatically by a post-processing software have been used, experimentally, to calibrate and validate an algorithm for the definition of synthetic indexes that can describe the state of damage of the tunnel. The following two indices have been defined by RFI:

1) the DAMAGE INDEX (ID) which reproduces the "severity" of the damage in terms of safety of the tunnel;
2) the DIFFUSION INDEX (ID_{IFF}) that represents the "spread" of the damage in the tunnel.

An algorithm was elaborated and was then calibrated on the dimensional values of the damages relating to a sample of about 100 tunnels in the Italian railway, supplied by the mobile diagnostic system.

Through the application of the algorithm, the ID has been defined trying to reproduce the Rating Code assigned by an expert RFI inspector who performs traditional inspections in railway tunnels according to the RFI procedures.

The ID is based on the principle of the most critical damage, which involves identifying a logical condition sufficient for the membership of the tunnel in question to a condition class. Therefore, considering each sector of the tunnel and each damages present, the ID is assigned to the entire tunnel on the basis of the intensity and extension of the dimensional damage of the relevant sectors.

In analogy to the Rating Code, the ID is expressed by a numerical code, which identifies the entity, from 1 to 11 with increasing gravity as the value of the code increases. So there is a scale of 11 classes that characterize the damage magnitude.

The method experimentally developed for the evaluation of the ID consists in verifying if there is a sufficient logical condition to be able to classify the tunnel in the highest expected class, that is 11. In this case, the ID will assume the value 11. If this logical condition is not satisfied, the algorithm passes to verify the condition sufficient to classify the tunnel in the immediately lower class (10) and so on. At the end of the process, if no logical condition exists to classify the tunnel in class 2, it will be automatically assigned an ID equal to 1 (Figure 10). The logical conditions are defined for each damage family.

The calculation of the ID, assessed on the whole tunnel, is performed for each sector and is equal to the maximum value assigned to the individual sectors.

The alignment between the ID, comes from algorithm, and the Rating Code expressed by the RFI inspectors, was made using the Pearson Index as correlation coefficient, which takes values between -1 and 1. For the whole sample of the 100 tunnels the correlation value was 63%. Non-aligned cases essentially depend on:

– the impossibility of the mobile diagnostic system to detect and classified some category of damages (for example: Damage of the inverted arch, Corrosion/exposure of the armature, etc.);
– the inevitable subjectivity of the judgment of the RFI inspector who carries out the inspections in the tunnels.

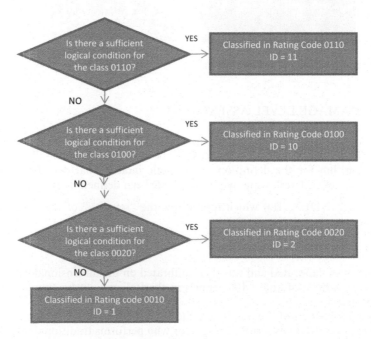

Figure 7.   Logical conditions of the algorithm.

Table 2. ID classes.

			Damage family		
		Intensity	slight	medium	severe
0	< Ext [m] <	S1	1	1	1
S1	< Ext [m] <	S2	1	1	2
S2	< Ext [m] <	S3	1	1	3
S3	< Ext [m] <	S4	1	1	4
S4	< Ext [m] <	S5	1	1	5
S5	< Ext [m] <	S6	1	1	6
S6	< Ext [m] <	S7	1	2	7
S7	< Ext [m] <	S8	1	2	8
S8	< Ext [m] <	S9	1	2	9
S9	< Ext [m] <	S10	2	2	10
S10	< Ext [m] <	/	2	2	11

The information acquired by the mobile system is stored in a database and the post-processed data is provided in the form of a report. Each report is specific to a tunnel and reports the extension of the damage divided into intensity classes for each damage family. For example, for the cracks families, the intensity classes depend on the opening of the slots, while for the lining falling off the classes depend on the depth of the gap. Information on damage is further subdivided along the tunnel's development longitudinally in sectors of 25 meters. The mobile system therefore provides an objective and quantitative photograph of the damage to the tunnel.

For each of the 7 families of damages described above, identified and geometrically classified, have been defined the extension thresholds (S1–S10) and the limits corresponding to the different ID classes (with values from 1 to 11), depending on the extension and intensity of each detected damage.

The following table shows the ID values to be assigned to the entire tunnel. Applying the logic of the method, the ID value derives from the value of the most critical damage identified in a tunnel sector, according to the extension and intensity of the damage itself.

In the Table 2 the term "Ext" refers to the total extension (measured in meters) of a specific damage family (cracks, mortar loss, lining falling off) in a sector of 25 meters; the terms "S1"...... "S10" refer to the extension limit value for the specific damage family assessed in a sector of 25 meters.

In addition to the ID, the "Diffusion Index" (IDIFF) has been defined to represent the spread of damage within the tunnel. The objective in defining the Diffusion Index is to express, with a single number or code, the conservation status of a tunnel and to evaluate how far the tunnel is from the design condition. The difference between IDIFF and ID is that IDIFF not reproduce the Rating Code expressed by the inspectors and is therefore not linked to the most critical damage, but is related to the amount of damage present in each sector of the tunnel.

For the definition of the IDIFF RFI has elaborated some logical conditions that, with the same maximum ID, lead to the determination of the greater severity of a tunnel compared to another and therefore to the assignment of the priority of intervention to the tunnel that has the ID maximum in the remaining sectors.

The logical conditions underlying the processing of the IDIFF are as follows:

tunnels with at least one sector with the highest ID are judged to be more damaged and therefore with higher priority;
with the same maximum ID (between the ID assigned to the individual sectors), priority is given to the tunnel that has the maximum ID in the remaining sectors;

Table 3. Tunnel priority.

| Tunnel | Rating Code | | | | Priority |
	Sector 1	Sector 2	Sector 3	Sector 4	
A	7	1	1	1	Higher
B	6	6	6	6	Minor

Table 4. Tunnel priority.

| Tunnel | Rating Code | | | | Priority |
	Sector 1	Sector 2	Sector 3	Sector 4	
A	7	7	1	1	Higher
B	7	6	6	6	Minor

Table 5. Tunnel priority.

| Tunnel | Rating Code | | | | Priority |
	Sector 1	Sector 2	Sector 3	Sector 4	
A	7	7	7	1	Higher
B	7	7	X	X	Minor

if two tunnels have different lengths, priority is given to the tunnel which has the largest number of sectors in the highest ID. The IDIFF is therefore independent of the length of the tunnel and the percentage of sectors with maximum ID.

The algorithm for the ID_{IFF}, which simulates the mechanism described by the logical conditions listed above, is as follows:

– application of the algorithm for the evaluation of the ID at sector level, so as to realize a damage mapping at the sectorial level along the tunnel;
– calculation of the number of sectors that fall within the classes defined in the ID, from 1 to 11 indicated with nID;
– calculation of the IDIFF (Equation 1):

$$IDIFF = \log_{Nmax} \sum_{ID=1}^{11} n_{ID} N_{max}{}^{ID} \qquad (1)$$

where Nmax is the maximum number of sectors with the same ID (maximum or immediately lower ID), the value of which was set at 100. In the case in which the same tunnel had more than 100 sectors with maximum ID, the ID would automatically switch to the upper class.

This value corresponds to a real number between 0 and 12, where:

– the whole number indicates the ID associated with the sector with the largest ID;
– the mantissa represents the spread of the ID of the sector and is greater as higher is the number of sectors with high ID.

6 CONCLUSION

The objective of the Tunnel Damage Management System research project is to provide a decision support system to RFI who, starting from the information deriving from tunnel inspection activities through mobile diagnostic systems, allows to identify the most suitable maintenance activities and to optimize then the interventions on the inspected structures.

Information on damage to tunnels' linings allows an assessment of the most detailed and objective damage status of the tunnel. In this way the evaluation is not influenced by the subjectivity of the operator who carries out the inspection and is not conditioned by operational situations that are not always easy.

The TDMS is a method, applied experimentally by RFI, which made it possible to calculate the damage level of a sample of railway tunnels in a rapid manner by assigning an ID damage index to each tunnel. Depending on the value of the assigned ID, it is also possible to make a first selection and a subsequent classification of the tunnels to be monitored, identifying those with the highest number of sectors with an advanced degree of damage. In this way, RFI, based on a priority scale that takes into account the value of the ID, can easily program targeted and more detailed inspections in order to confirm the assessment of the tunnel damage and plan specific restoration and maintenance interventions of the tunnel. In this way it would be possible to guarantee also an optimization of the economic resources available for maintenance, aiming at the tunnels that most need restoration or for which an evolution of the significant degradation is expected in the short term.

REFERENCES

RFI Procedure for the inspections of bridges, tunnels and other structures of the railway network.

Gavilán, M., Sánchez, F, Ramosand, J.A., Marcos, O. 2013. Mobile inspection system for high-resolution assessment of tunnels. The 6th International Conference on Structural Health Monitoring of Intelligent Infrastructure. Hong Kong

Montero, R., Victores, J.G., Martínez, S., Jardón, A., Balaguer, C. 2015. Past, present and future of robotic tunnel inspection. Automation in Construction Volume 59/2015: Pages 99–112

Tunnels and Underground Cities: Engineering and Innovation meet Archaeology,
Architecture and Art, Volume 5: Innovation in underground engineering,
materials and equipment - Part 1 – Peila, Viggiani & Celestino (Eds)
© 2020 Taylor & Francis Group, London, ISBN 978-0-367-46870-5

Tunnels widening and renewal using precast lining: The Swiss Rhaetian railway construction method applied to tunnel Mistail

D. Fortunato & G. Barbieri
AF TOSCANO AG, Zürich, Switzerland

ABSTRACT: The Rhaetian Railways AG (RhB) manages about 384km of rail network in the Canton Graubünden in Switzerland. Due to the mountainous topography, many civil engineering structures had to be built along the network, 115 of which are tunnels realized between 1901 and 1914. 75 tunnels with an overall length of 26km have damages due to water income, therefore representing a serviceability issue. The challenge and the purpose of the RhB is the rehabilitation of these tunnels in the next 50 years defining an expected service life of 100 years. To reach this purpose, a standard method was studied called "Standard Tunnel Construction Method", which allows to replace the masonry ring with a close ring of prefabricated concrete elements, to widen the cross-section, to excavate the invert and place prefabricated concrete slabs completed with poured concrete. Application of this construction method to the tunnel Mistail will hereby shown and described.

1 INTRODUCTION

1.1 *The Rhaetian Railways (RhB)*

The Rhaetian Railways manages 384km of rail network in the Canton Grisons in the east of Switzerland. These 384km of narrow-gauge railway tracks run through 617 bridges and 115 tunnels in the heart of the Swiss Alpine mountains. The company operates in the transport of goods and in the tourist and commuter traffic.

The Albula and Bernina routes are part of the UNESCO World Heritage since the beginning of July 2008 and with their panoramic trains, the GLACIER and the BERNINA EXPRESS, they are worldwide famous and offer unforgettable travel experiences throughout the territory of the Grisons.

In 1889 the pioneers of the Graubünden railway, on the initiative of the Dutch Willem-Jan Hol-sboer, inaugurated the Landquart – Klosters line. The first steam trains from Landquart to Davos were already in circulation in 1890, then the routes to St. Moritz, Disentis and Scuol-Tarasp were added.

Thanks to the merger with the Arosa and Bernina railway lines, the network expanded, and only 25 years later almost the entire Rhaetian Railway network was already completed. In 1999 the 19-km long tunnel of the Vereina was inaugurated as the last extension of the track. Over the years, the Rhaetian Railway developed continuously, remaining faithful to its exclusive peculiarity: linking the most beautiful mountain resorts with equally striking lines.

1.2 *RhB Tunnels and their conservation status*

There are 115 tunnels on the Rhaetian Railway (RhB) railway network, with a total length of 58.7 km. Most of these tunnels were built between 1901 and 1914. These tunnels have all the same standard: single track, inner ring covered with masonry or sometimes (where it was possible) without an inner ring, with rock at sight. The analysis of the results of periodic

Figure 1.   The GLACIER EXPRESS crosses the world famous Landwasser Viaduct on the Albula line.

inspections highlighted that 67% of the tunnels are in classes 3 (damaged condition) and 4 (bad condition), meaning that rehabilitation works are needed in the next 25–35 years (class 3) and in the next 5–10 years (class 4).

Since the tunnels were all built with the same features, even the detected damages are similar:

– damages to the masonry due to water infiltration and frost
– damages to the abutments, due to the erosion of the masonry or to an excessive horizontal load (deformation of the masonry due to water infiltration and frost)
– for the majority, a missing or malfunctioning evacuation of the water

These damages are mainly due to water infiltrations, which create a problem of serviceability. The structural safety is minimally reduced or even still guaranteed. From this point of view, the time horizon for the rehabilitation of the tunnels in class 3 (75 tunnels for a total length of about 26 km) could be extended up to 50 years.

From the point of view of compliance to the current Swiss regulations, the inner dimensions of the tunnels, built for steam trains, are insufficient to fullfill requirements on the free space between tunnel lining and train standard profile, moreover, the distance between the tunnel wall and the wagons is not enough to provide a safe escape route when the train it is stuck in the tunnel. In addition, security systems are missing.

The interventions therefore envisaged, as well as the number of tunnels in which they are necessary, are of considerable entity, implying the almost complete rehabilitation of the tunnels and the widening of their cross-section. This involves a considerable financial effort for the next decades, accompanied by complex intervention plans aimed to avoid services interruptions that would entail a further financial burden resulting from lower revenues.

Given the repetitiveness of the planned interventions, the idea of the Rhaetian Railway to face this challenge is to develope a standardized method both for planning and execution of the interventions, in order to have well-defined processes, facilitating time and financial planning and allowing a reduction of time and costs.

This new rehabilitation system also takes account of the appearance and the conservation of architectural heritage. Two of the Rhaetian Railway lines, the Albula line and the Bernina line, have been crowned as UNESCO heritage sites. One of the peculiarities of these lines is the characteristic portals of its tunnels: all in masonry, they have in common even the shape and the dimensions. The abutments, straight and slightly inclined to the outside, have a height of 2.55m and support the circular crown, which has a span of 2.15m. The width of the bottom slab is 4.04m. The rehabilitation project has to maintain the characteristics and proportions described above.

2 PROJECT

The tunnel Mistail is the third tunnel of the Rhaetian Railway that has been restored with the new standardised system.

This tunnel is located on the Albula line, approximately at 2.2 km from the Tiefencastel station, in the Canton Grisons. Commissioned in 1903, it has a total length of 300 m and a longitudinal slope of 6.94‰. The tunnel cross-section has a horseshoe shape with a total height of 4.70m and a width of 4.04m. The net area is of 17.9m^2. The internal ring is covered in masonry.

Figure 2.   Existing masonry Portal (Solis side).

2.1   *Current state and geology*

Based on the defects found during the last periodic inspection in 2015, the tunnel was classified in damaged conditions (class 3), on an evaluation scale from 1 (without damages) to 5 (in alarming condition). The main defects are due to water infiltration.

During the tunnel's life, temporary systems were adopted to reduce water infiltration. The masonry ling was covered with a layer of sprayed concrete as to create an impermeable sheath. This system is effective, but has a short lifespan. Over time, the concrete cracks allowing the water to infiltrate, with the consequence of having drips in the tunnel, limestone formations and sometimes even concrete detachments. The areas of the masonry lining not cover by shotcrete show runoff of the mortar in the joints.

The Mistail tunnel is located mainly in a rocky mass composed of schist limestone (40–60%), fillites (30–50%) and sandy limestone (0–20%).

The rock is classified as fractured; there are risks of falling rock blocks, detachment of rock layers and formation of fractured bodies. The three types of rock formations alternate frequently in the rock mass and show very marked folds. Depending on the intensity of the folds, detachments of packages of even large dimensions can occur. The area is covered with a layer of loose material (Albula moraines and ground debris) of varying thickness.

2.2 Underground section

The current section has a horseshoe shape with straight abutments inclined outwards.

The new section maintains the horseshoe shape, but the abutments are curved, to better withstand horizontal loads. The new lining consists of precast concrete rings 1.5 m long, each one made of 5 precast elements.

The bottom slab is composed by a central prefabricated element of 2.80m width, completed on the right and on the left by cast in situ concrete. On the precast central element, the provisional tracks are mounted to allow the circulation of the trains during the day, whereas the excavation and assembly of the other precast elements take place at night, during breaks of the railway traffic. At the end of the works, the temporary tracks will be disassembled and replaced with the final ones after the ballast laying.

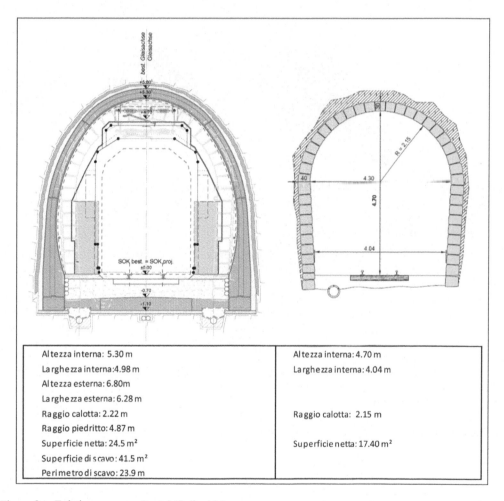

Altezza interna: 5.30 m	Altezza interna: 4.70 m
Larghezza interna: 4.98 m	Larghezza interna: 4.04 m
Altezza esterna: 6.80m	
Larghezza esterna: 6.28 m	
Raggio calotta: 2.22 m	Raggio calotta: 2.15 m
Raggio piedritto: 4.87 m	
Superficie netta: 24.5 m²	Superficie netta: 17.40 m²
Superficie di scavo: 41.5 m²	
Perimetro di scavo: 23.9 m	

Figure 3. Existing masonry Portal (Solis side).

The section does not have the classic waterproofing sheet, since the precast elements are waterproof and the joints between them are provided with special gaskets: radial gaskets between adjacent rings and longitudinal gaskets between the elements of the same ring.

Water evacuation takes place with a mixed system. The space between the rings and the rock is filled with round gravel, creating in this way a draining layer which allows the mountain water to flow to the foots of the internal ring and, through the opening at the abutments foot, to enter the water evacuation pipe (HDPE) which is placed on both sides of the bottom slab.

Even the water that could be present on the bottom slab can flows into the longitudinal evacuation pipes, through openings of 1.50m length and 0.80m width in the slab. These openings are placed at both sides of the slab, but alternatively each 9m and they are filled with draining gravel. Since there is no space for inspection wells, vertical pipes are located every two openings to allow inspection of the evacuation system and to wash longitudinal pipes.

Water is then collected at the east portal in a well with proper systems to collect mud and it is lead into the external receptor.

At the feet of the abutments, 4 cable pipes (100/112) are mounted on each side for wiring of electrical systems.

The ballast, on which tracks are laid, has a minimum thickness of 30cm and a minimum width of 3.20m.

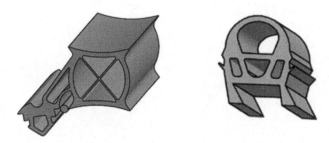

Figure 4.   Horizontal gasket (on the left) and radial gasket (on the right).

2.3   *Precast elements*

Each ring consists of 5 precast elements: 2 abutments, 2 lateral crowns and a keystone segment. The abutments lay on a footing, which is also prefabricated. The thickness of the segments ranges between 30 cm (in the 3 crown elements) till 65 cm at the foots.

The concrete used is type C30/37 (XC4, XD3, XF2), with high resistance against corrosion due to carbonation induced by chlorides and against damages to the cement matrix due to freezing and thawing cycles. With these characteristics, the concrete is more compact, less porous and more resistant to water infiltration. To make the edges more resistant to impact, poly-propylene fibres are added.

The weight of the precast elements ranges depending on the considered element: abutment elements are the heaviest, with a weight of 5.8t; lateral crown elements weight 2.8t, whereas the keystone weights 4.8t.

In the precast elements, there are openings which allow the insertion of the round gravel through blowing. These openings have retention valves that prevent the gravel from leaking into the tunnel.

The front face of the precast elements has an indentation aimed to properly position the elements to each other, thanks to a "male-female" system.

The rings of precast elements are of three types: straight ring, ring for right-hand bend and ring for left-hand bend. The last two types are distinguished from the straight one by a slight taper of the elements. The route of the tunnel is a sequence of these elements, forming a broken route as close as possible to the track layout with curves and clotoids (Construction tollerance 2.5 mm).

2.4   *Portals*

Although the tunnel is completely rehabilitated internally, the portals shall maintain their original characteristic. As the tunnel section is larger than the current one, even the portal

Figure 5.   Existing masonry Portal (Tiefencastel side).

dimensions have to be larger, whereas maintaining their original proportion between the elements and masonry portal features.

The abutments remain straight and inclined outwards, supporting the round crown which has a radius of 2.50m.

The shape of the stone blocks as well as the joints layout is the same as in the original structure. Where possible, the blocks of existing portals, integrated with new ones, are reused. For instance, on both portals, there are four blocks which form a cross right above the tunnel crown: these stones are stored and reused for the new portals, recreating the same crosses. The arch of the portal consists of 21 blocks of stone. Since the dimensions of the new arch are larger, these blocks will be replaced with new blocks. The internal walls of the portals are also masonry.

2.5   Operating and safety systems

The tunnel rehabilitation foresee also the upgrade of the operating and safety systems. The escape route is located on both sides of the tracks and has a width $B \geq 0.90m$. At the abutments, at an height of 1.10 m, an integrated handrail with emergency lighting with LED lamps is mounted. The escape route signal is located on the abutments at a height of 0.60m. Given the reduced length, the escape route leads directly to the tunnel portals.

3   EXECUTION

Construction was divided in two phases: the first, for preliminary works and the latter for the main works.

In the first phase the existing bottom slab was demolished and lowered up to a height of 1.20 from the level of the existing tracks. In detail, the following activities are foreseen: disassembly of the tracks, removal of the ballast, lowering of the foundation level to the desired height, laying of the precast elements of the bottom slab and installation of the provisional tracks on the slab. For the execution of these works, the total close of the railway line is foreseen for 20 days, with working times of 24 hours per day, 7 days per week.

In the second phase the following activities are foreseen: demolition of the existing cladding, widening of the tunnel cross-section and securing it, assembly of the precast elements, laying of the pipelines, demolition and rebuilding of the portals, arrangement of the quays and assembly of electromechanical systems. The execution of the main works takes place at night, during the night interruption of the railway traffic, from 8:50 PM to 05:45 AM. During the day, the trains travel at reduced speed on the sections affected by the construction site.

3.1 *Excavation method*

The enlargement of the tunnel takes place conventionally through blasting.

Rock reinforcement and temporary support, aimed for the construction phase till precast elements installation, is made of fibre-reinforced shotcrete, reinforcement steel meshs and rock bolts, depending on the local rock conditions.

The safety of the railway is fundamental during the rehabilitation works as every morning the tunnel is reopened to the railway traffic: before the daily reopening of the tunnel to the railway traffic, the temporary supports applied in the tunnel section excavated during the night. The critical zone is the one near the excavation face, between the already widened tunnel and the area which is still with the masonry lining, for two reasons: first, depending on when shotcrete was sprayed, it could be still not hardened enough; moroever, vibrations and loads induced on existing lining during by excavation operation may also damage the masonry near to the excavation area. In order to cover these risks and to guarantee the safety of the trains passage in the interested area, it is envisaged to build a protection shield, called "protection tunnel".

The "protection tunnel" consists of a surface of rigid metal elements which protect the free profile in the area in front of the excavation zone from accidental falls of stones blocks.

Its dimensions are slightly smaller of the internal profile of the existing tunnel and they allow the train passage. Its length is equal to 8m and it is composed by two elements of 4m each.

It is moved on rails fixed laterally to the precast central element of the bottom slab and positioned in front of the excavation area, so that it is centred between the last excavated section and the existing tunnel. The half along the existing tunnel is equipped with inflatable cushions that allow a better adaptation to the masonry lining and, thus, favours the support of any loose masonry elements. At night, during the works, the "protection tunnel" is moved away from the excavation area.

In order to reach the necessary height of the new section, the bottom slab must be lowered to 1.20m from the tracks level. The lowering of bottom slab was originally foreseen by drill and blast, but during construction the Contractor proposed and adopted a cutting head.

Figure 6.   Protection shield moved forward to let excavation procedures to widen the tunnel.

To allow the continuity of the railway traffic, the catenary line was temporary replaced by a power line installed along the lining crown. The guides of the power line are divided into 6m elements and can be easily dismantled. In this way, in the area affected by the excavation, the guides can be dismantled at the beginning of the shift and reassembled before the tunnel is reopened.

3.2 *Assembly of the elements*

For the system to assemble precast elements, the Rhaetian Railways didn't impose any limit to the companies: in the technical bid, each competitor should offer its solution. Since it was a new rehabilitation system, the equipment for precast elements assembly was a prototype developed specifically for this purpose.

Equipment precision during precast elements handling is fundamental to avoid damages to precast elements themselves during moving and position them. The footing elements are laid at the foot of the excavation area and positioned according to the project coordinates. Then it is fixed to the ground with anchors and their bases filled with mortar.

Each ring of 1.50m should be completely assembled before beginning the assembly of the next one.

The segments are laid from the bottom upwards on both sides. On the footing elements, the abutments elements are mounted on both sides, then lateral crown elements; finally, the keystone element is positioned which, with its conical shape, is inserted between the lateral crown elements through specific guiding rods. The precast elements are then fixed to each other through bolts.

Once the ring is finished, the radial gasket has to be installed before starting mounting the next ring, by fixing it in the groove located on the back face of the precast elements. Longitudinal seals are instead directly installed on the moulds of the precast elements before casting them and, hence, are already fix to the concrete.

The rings are temporarily fixed to each other with bolts, until the space between the rock and the elements is filled with the round gravel.

For the Mistail tunnel, the company has developed a machine to assemble precast rings starting from a common excavator. The precast elements are transported from the storage area to the assembly point in the tunnel using a mobile platform on rails, also designed by the company itself.

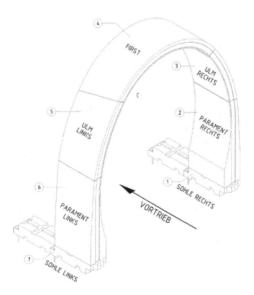

Figure 7.   Trimensional model of a ring of the precast lining.

4 CONCLUSIONS

The Mistail tunnel is the third tunnel to be rehabilitated with the standardized tunnel system of Raetian Railway with the railway in operation. This concept, conceived by the Rhaetian Railway, represents an important step in the field of tunnel rehabilitation. The use of predefined work methods allows to reduce the processing times and consequently also the costs of the project both in the design phase and in the execution phase. Moreover, the possibility of keeping the rail traffic in operation is an important added value, as the construction site does not create inconvenience to users and the Rhaetian Railway does not have to face the costs of interrupting railway traffic.

5 ACKNOWLEDGEMENTS

the authors gratefully acknowledge the Rhaetian Railway and in particular the project manager Urs Tanner for giving us the opportunity to submit an interesting and innovative project in the field of rehabilitation of railway tunnels.

REFERENCES

AF Toscano AG 2018. *Tunnel Mistail, rehabilitation – Detailed design*
Rhaetian Railway AG 2014. *Standardized tunnel rehabilitation model, description of the concept*

*Tunnels and Underground Cities: Engineering and Innovation meet Archaeology,
Architecture and Art, Volume 5: Innovation in underground engineering,
materials and equipment - Part 1 – Peila, Viggiani & Celestino (Eds)
© 2020 Taylor & Francis Group, London, ISBN 978-0-367-46870-5*

Experimental investigation of the flow behaviour of conditioned soils for EPB tunnelling in closed mode

S. Freimann, M. Schröer & M. Thewes

Institute for Tunnelling and Construction Management, Ruhr University Bochum, Germany

ABSTRACT: The application range of EPB shields can be extended using conditioning agents. The excavated material serves as support medium during advance in closed mode and must have suitable properties, notably workability and permeability. In this research, the workability of soil-foam mixtures is investigated to elaborate a more precise material description. Up to now, the workability of soil-foam mixtures is typically tested with the Slump Test. However, this test does not provide detailed information on rheological parameters.

Therefore, soil-foam mixtures with varying characteristics were investigated using a Ball Measuring System. The test results provided information on the rheological parameters of the samples, so that the findings could be linked to the results of the slump tests. Since the Slump Test is not suitable for examining coarse-grained gravelly soils or purely cohesive soils, a new penetration test was developed to examine the workability of all mixtures of soils and conditioning agents.

1 INTRODUCTION

The earth pressure balance shield (EPB shield) is the machine type mainly used for tunnelling in cohesive soils but also in sand and gravel. A support muck from the excavated soil is used to transmit the support pressure at the tunnel face during advance. Since the soil usually does not have the appropriate properties for optimum transfer of the support pressure, a suitable support medium can be temporarily produced by the use of conditioning agents (e.g. foam or other project-specific conditioning agents). The addition of conditioning agents aims at generating an adequate workability (consistency), a decrease of permeability of the support muck and to achieve sufficient compressibility to ensure the smooth advance. Accordingly, the application area of EPB shields has extended into coarse grained soils (Budach & Thewes 2015).

For tunnelling in unstable ground below the groundwater level, the EPB machine is operating in closed mode. Only in this mode the machine can create a support pressure to the water and earth pressure. The tunnel face is supported to minimize settlements on the surface. In EPB closed mode the excavation chamber of the machine is filled completely with excavated soil. This way, a support pressure can created by means of the interaction between excavation advance, by volume flow of conditioning and by the speed of the screw conveyor.

The properties of conditioned soils can be investigated by laboratory tests, which is recommended before tunnelling advance starts (Budach 2012, Martinelli et al. 2015, Budach & Thewes 2015, Galli 2016). A concept for conditioning is developed where the workability is one of the first investigated characteristics. Numerous project-related works have shown that laboratory tests prove to be a helpful tool for the planning of EPB tunnel advance with soil conditioning in non-cohesive soils. The results of laboratory tests provide a fundamental understanding of the practicability, processes and interactions that occur in soil conditioning, but are only used as a practical guide.

1.1 Workability of conditioned soils in closed mode

The workability or flow behaviour of the conditioned soil has a big influence on the material flow in the excavation chamber (Maidl 1995). In 2015, Budach & Thewes presented recommendations for standardized and reproducible soil conditioning tests. Based on these results, Peila et al. (2013) as well as Galli & Thewes (2014) developed the specific challenges in soil conditioning for both overconsolidated, cohesive soils and highly permeable, non-cohesive loose and solid rocks.

In the laboratory as well as on the construction site the Slump Test is an adequate test to investigate the workability of soil-foam mixtures. A slump between 10 and 20 cm is recommended by several researchers (Vinai 2006, Budach 2012). However, the Slump Test is not useful for every type of soil. Cohesive soils, e.g. pure clay or silt cannot be investigated in the Slump Test, just as very coarse-grained soils with high gravel content. For fine grained cohesive soils, a widely used guideline to define the ideal state of the support medium is the consistency index. Therefore Oliveira et al. (2018) performed laboratory tests using a flow table to investigate the flow behaviour of artificially mixed clay-sand soils. The results of the investigations show that the flow table is a better alternative to determining the flow behaviour of conditioned and unconditioned cohesive soils. Peila et al. (2016) concentrated on conditioning investigations of clayey soils. The results of the tests have led to some important considerations regarding conditioning with water, foam and polymer solutions. Further laboratory tests regarding shear strength were carried with cohesive soils out by Zumsteg et al. (2012). Based on a further developed test setup of the wing shear test, initial findings on the influence of various combinations of conditioning agents on the shear strength were obtained both under atmospheric pressure conditions and under pressure.

During advance of the EPB shield, a direct evaluation of the support medium inside the chamber is not possible. Therefore Galli (2016) used the Ball Measuring System (BMS), based on the work of Tyrach (2000) and Schatzmann (2005), in the small laboratory scale to investigate the rheological properties of foam conditioned fine sand and sand and correlates the results to numerous Slump Tests as a first step. Based on these results, the spectrum of examined soils was extended to all types of sand by Freimann et al. (2017a, 2017b). The aim of the investigations with the BMS is to obtain and analyze data from laboratory tests to install a BMS in the excavation chamber of an EPB shield.

Another possibility to investigate the workability of soil-foam mixtures is the Penetration Test newly developed at Ruhr University Bochum for EPB tunnelling. The Penetration Test is based on the Kelly Ball Test, which was developed in the 1950s as an alternative to the Slump Test to investigate the consistency of fresh concrete. Abd Elaty & Ghazy (2014) tested various concrete mixes with a Penetration Test with conical indenter and showed that this test is well suited to investigate the properties of fresh concretes regarding workability.

1.2 Rheological properties of conditioned soils

Depending on shear load, temperature and time, the flow behaviour of a liquid is determined and can be classified by its load-dependent flow behaviour. It is characterized by its shear rate shear stress interaction and its viscosity. Ideal viscous liquids are called Newtonian fluids. Non-Newtonian fluids have shear-thickening or shear-thinning properties due to a non-proportional relationship between shear rate and shear stress. In addition, the flow behaviour of fluids with and without yield point can occur. The yield point must first be exceeded to trigger a fluid flow.

The shear stress can be determined based on the shear rate by rheometer tests. To perform a model function by approximating the data points, torque measurements and shear rates must be transformed into rheological parameters. To describe the different flow patterns, there are different flow models (e.g. Bingham, Herschel-Bulkley).

With reference to newton's parallel plate model, the basic parameters of rheology are considered below. In the parallel plate model, a fluid is between the two plates with defined shear area. One of the two plates, the base plate, is fixed and the upper plate is stressed by the force

F so that the fluid in between is sheared. The resulting velocity v is measured or vice versa. The requirement for this model is the assumption that the fluid exhibits wall adhesion on both plates. Another requirement is the assumption that the flow conditions in the shear gap are laminar, resulting in a layer flow. The shear stress τ is defined as the quotient of the acting force F and the shear area A of the movable upper plate (Equation 1).

$$\tau = \frac{F}{A} \tag{1}$$

where τ = shear stress (Pa); F = force to stress the upper plate (N); and A = shear area of the upper plate (m^2).

In Equation (2) the shear rate $\dot{\gamma}$ is defined as the quotient of the velocity v of the movable upper plate and the height of the fluid h in the shear gap. It describes the speed for deforming the fluid. If wall adhesion and laminar flow conditions exist, the necessary conditions for the parallel plate model are fulfilled and the shear rate is constant within the entire shear gap. It describes the speed difference between all fluid layers and their distance from one another.

$$\dot{\gamma} = \frac{v}{h} \tag{2}$$

where $\dot{\gamma}$ = shear rate (1/s); v = velocity (m/s); and h = gap distance (m).

The viscosity η of a liquid describes the internal frictional forces. These will be created by the flow resistance of a liquid in relation to external agitation. It can be calculated with Equation (3).

$$\eta = \frac{\tau}{\dot{\gamma}} \tag{3}$$

where η = (shear) viscosity (Pa·s).

2 MATERIALS AND METHODS

The used testing materials must reproduce realistic conditions in EPB shield tunnelling practice and allow for unrestricted reproducibility.

2.1 Investigated Soils

The soils examined were selected based on the investigations carried out in Budach (2012). The grain size distribution curves cover the entire spectrum for sand (0.063-2.0 mm). The soils investigated for the Slump Tests and the rheological investigations with the BMS and the Slump Test are fine sand, fine sand middle sand, middle sand, sand, middle sand coarse sand and coarse sand. So far, the Penetration Test has only been carried out with fine sand. The soil samples are produced with special washed and dry laboratory sands with a well-defined grain size distribution to ensure high reproducibility. The water content of the investigated soils was varied to get an overview of the influence of the water content on the conditioning behaviour of the different soils. For Slump Tests and BMS tests the water content was between 2 and 12 wt%, varied in steps of 2 wt%. For the Penetration Test, the water content was selected from 4 to 12 wt% by steps of 4 wt%.

2.2 Foam and foam production

For foam production, a real-scale original TBM foam generator was used. The setting options of the foam generator enable a practical, reproducible foam production. It has three different foam guns, which differ in design and their foaming behaviour. Two of the foam guns are constructed in the same way as foam guns used in practice on EPB shields and therefore are dimensioned for large foam flow rates (Q_F = 150-550 l/min). For the investigations of the present research, a self-developed foam gun was used. This foam gun was specially developed for laboratory usage and the production of reduced flow rates Q_f < 100 l/min. Low volume flows are particularly suitable for laboratory tests. In Equation (4), the foam expansion ratio (FER) is defined. The FER described the ratio between the foam and the liquid phase of the foam, the liquid (water + surfactant). The higher the FER, the "drier" the foam. Usually, values of the FER are between 5 and 25. For the present study, the foam was produced with a FER of 15.

$$FER = \frac{Q_{foam}}{Q_{liquid}} \qquad (4)$$

where FER = Foam Expansion Ratio (–); Q_{foam} = volume flow of the foam (l/min); Q_{liquid} = volume flow of the liquid (l/min)

The amount of foam injected into the soil is described by the percentage ratio between the foam volume and the soil volume (Equation 5). Depending on the geotechnical boundary conditions, the foam injection ratio (FIR) can vary widely. For fine-grained or cohesive soils, a high FIR is required, while for coarse-grained soils lower FIR values are usually injected into the soil.

$$FIR = \frac{V_{foam}}{V_{soil}} \cdot 100\% \qquad (5)$$

where FIR = Foam Injection Ratio (vol%); V_{foam} = volume of the foam (l); V_{soil} = volume of the soil (l)

The surfactant concentration c_f is defined as follows (Equation 6).

$$c_f = \frac{m_{surfactant}}{m_{liquid}} \cdot 100\% \qquad (6)$$

where c_f = surfactant concentration (vol%); $m_{surfactant}$ = surfactant mass (kg); m_{liquid} = liquid mass (kg)

2.3 Setup and testing procedure

To determine the conditioning behaviour of the different soils, it was necessary to first produce the desired grain size distribution using the various grain fractions of the laboratory soils. After that, different amounts of water for the specific water contents and foam were added to the soil in a gravity tumbler to generate a soil foam mixture. After production of the conditioned soil mixtures, the samples were investigated in the tests presented in the following chapters to determine its workability and flow behaviour. These include the Ball Measuring System, the Slump Test according to DIN EN 12350-2 and the newly developed Penetration Test.

2.3.1 Investigation with Ball Measuring System (BMS)

The Ball Measuring System (BMS) was developed for rheological investigations on suspensions with larger particles. Müller et al. (1999) and Schatzmann (2005) have already carried out flow curve tests with the BMS on cement-bound pastes or debris-flow material. However, they have chosen different approaches to determine the system parameters in rheological parameters. First, the theoretical background will be explained before the application of the BMS to the investigations with soil-foam mixtures are described.

The test setup of the ball rheometer consists of a cylindrical sample container with a volume of approximately 500 ml and an eccentrically arranged ball (see *Figure 1*). The ball can have

Figure 1. Rheometer with Ball Measuring System (BMS) before the tests (left), during the tests with soil-foam mixture (middle) and System Sketch of the BMS according to Tyrach (2000).

different diameters (d_{BMS08} = 8 mm, d_{BMS12} = 12 mm, d_{BMS15} = 15 mm). The eccentricity depends on the selected ball diameter (L_{BMS08} = 38 mm, L_{BMS12} = 37 mm, L_{BMS15} = 35 mm). The immersion depth of the sphere was selected so that it corresponds to the average height of the sample container. Based on the work of Tyrach (2000) and Schatzmann (2005), the test software of the rheometer offers a standard test routine and was adapted for the investigation of soil-foam mixtures by Galli (2016) and Galli & Thewes (2018). The shear rate is logarithmically increased over three decades of the shear rate within one revolution. The shear rate range is 0.1 to 100 1/s. An essential point for the revision of the measurement profile was, on the one hand, that a test run should allow several rounds and, on the other hand, that the area in which the sphere dips into the sample should be excluded from data acquisition on a large scale. By immersing the ball in the sample, the material structure of the medium to be examined is disturbed, so that this zone was defined to a ball diameter before and after the immersion point. By repeatedly spinning the sample with the ball, the development of the yield point from uncured material to sheared material is to be investigated. A total measurement consists of six rounds. The sample material is produced as already described in chapter 2.3 and first filled into the sample container before it is placed with the sample on a Haegermann table and compacted with 15 strokes. The surface is then pulled smooth with the aid of a ruler and positioned in the rheometer so that the ball dips into the sample. After the ball has been immersed, the measurement starts with a delay of 60 seconds so that the sample material can completely enclose the ball by reflow.

2.3.2 *Investigation with Slump Test*

In contrast to the Ball Measuring System, the Slump Test typically is not used directly to determine the rheological properties (although there are theoretical approaches describe by Galli (2016)). The Slump Test is certainly suitable for coarse-grained soils, while the BMS is only suitable for soil with a maximum grain size of 2 mm. Furthermore, the next larger scale for investigating the flow behaviour of soil-foam mixtures after the BMS is the Slump Test. Various researchers previously employed the Slump Test for the investigation of the workability of the soil-foam mixtures (Peila et al. 2008, Peila 2014, Budach 2012). Previous research has shown that for EPB tunnelling the range for the slump value should be between 10 and 20 cm (Vinai 2006).

First, the cone shape and the base plate of the test setup are moistened with water to reduce the friction between the inner surface of the cone and the sample. At least 12 kg soil-foam mixture must be prepared for each test. The soil foam mixture is filled in in three layers of approx. 10 cm each. Each layer is compacted in 25 thrusts with a steel rod. The filling process ends when the excess material can be levelled off with a steel ruler at the upper edge of the cone. Adhering material is removed from the test setup. After the cone has been raised quickly, the settling size of the soil sample can be determined (Figure 2, II). The vertical difference between the initial sample height (30 cm) and the maximum height after lifting is determined. Furthermore, the slump-flow of the sample is determined by the average values of the diameter. To make the lifting of the cone more reproducible and to minimize the manual influence, the Slump Test devise at Ruhr University Bochum was equipped with guide rails (Figure 2, I).

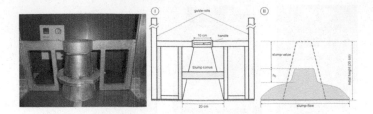

Figure 2. I: Slump Test & System sketch of the Slump Test with guide rail extension; II: Recorded values.

2.3.3 *Development of a Penetration Test for the investigation of a wider range of conditioned soils*

For a continuous and safe support of the tunnel face, the workability of a conditioned soil is of highest importance for the operation of an EPB shield in closed mode. Furthermore, the soil conditioning counteracts possible sticking of the soil in the excavation chamber and the screw conveyor. The Slump Test is well suited for granular non-cohesive and foam-conditioned soils but it does not work well in in fine-grained cohesive soils, in conditioned changeable solid rock chips (e.g. clay-stones) and in extremely coarse-grained particle mixtures such as rock or concrete chips. (Galli & Thewes 2014).

For this reason, the Penetration Test was developed based on the Kelly Ball test which is suitable for determining the workability of all soil types (Abd Elaty & Ghazy 2017, Juradin 2012). Originally, this test was developed for the consistency testing of fresh concrete. The test was modified to quantify the workability and determination of rheological properties, see Figure 3.

In the present research, a correlation between the penetration resistance of the rounded indenter and the degree of conditioning of different sands is investigated. The penetration depth depends on the specified weight force of the indenter (total weight: 4.503 kg). Depending on the respective FIR and density of the soil foam mixture, the indenter penetrates into the soil sample. The indenter displaces the soil-foam mixture and the level of the soil-foam mixture rises slightly. One further aim of the tests is to validate the new test device for determining the rheological properties of a soil-foam mixture in the laboratory and on the construction site on the basis of rheological investigations with BMS.

Initially, 12 kg of soil-foam mixture with the corresponding parameters (water content, FER, FIR) are produced. Then the soil-foam mixture is filled into the penetration bucket.

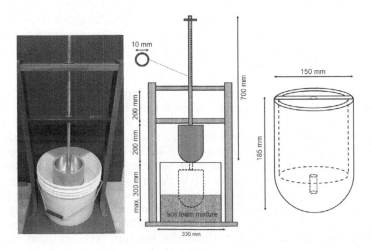

Figure 3. Penetration Test device & System sketch of the Penetration Test device and indenter.

During installation, care must be taken to ensure that the soil-foam mixture is installed in three layers of 3-4 cm (maximum total height 13 cm ± 1 cm). Each layer is evenly to consolidate with 25 impacts of the steel bar. If the installation height is the same at all points, the bucket will be centered under the indenter. Then, the indenter is slowly deposited on the soil-foam mixture and height is marked on the rod and the rod is released. Then the penetration depth of the indenter is measured on a scale at the fixing rod.

3   CORRELATION OF SLUMP TEST AND RHEOLOGICAL TESTS WITH BMS

The flow behaviour of the conditioned soil in the excavation chamber is influenced by various rheological parameters. Two essential parameters are viscosity and yield stress. Stators on the pressure wall and rotors on the back of the cutting wheel are to enable the mixing and kneading of the soil material within the excavation chamber (Düllmann et al. 2014).

Taking into account the results of the Slump Tests, the corresponding results of the rheological tests with BMS can be reflected. Since there are generally accepted limits for workability for the Slump Test with a slump between 10 and 20 cm, the results of the rheological investigations were examined with the corresponding slump results to determine a yield stress range for the soils examined.

Figure 4 shows an example of the yield stress range for fine sand, considering the ball diameters used. The results of the Slump Tests were only evaluated for a slump of 10 and 20 cm without consideration of the conditioning parameters (w, FER and FIR). The horizontal lines show the mean value of the corresponding BMS yield stresses for these slumps independent of the ball diameter. The average values are 250 Pa for a slump of 10 cm and 130 Pa for slump 20 cm and indicate a range of yield stresses for optimum conditioning of the fine sand.

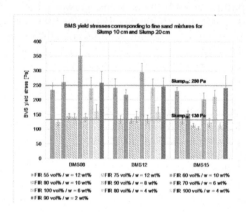

Figure 4.   Average BMS yield stresses and standard deviations corresponding to the Slump results of fine sand-foam mixtures (Galli 2016).

Table 1.   Average yield stresses for slump range d_{BMS12} with $c_f = 3\%$.

Soil	yield stress Slump$_{10}$	yield stress Slump$_{20}$
Fine Sand	250 Pa	130 Pa
Fine sandy middle sand	310 Pa	190 Pa
Middle sand	270 Pa	190 Pa
Sand	250 Pa	130 Pa
Middle sandy coarse sand	–	130 Pa
Coarse sand	–	100 Pa

The results for the other soils show that the yield stress range can vary depending on the soil, whereby medium sand and coarse sand could not be conditioned with foam alone for a slump of 10 cm. The values in Table 1 represent the mean value of the yield stresses for all examined conditioned soils with foam, which represent the desired slump.

4 PENETRATION TESTS

This chapter presents the results of Penetration Tests of foam-conditioned fine sand with different water contents. Three test series were performed, whereby each test series is characterized by a constant water content. Each individual test was repeated three times. *Figure 5* (left) shows the influence of FIR on the penetration depth. The penetration depth is given here as a percentage to ensure comparability of the results because of the slightly varying soil installation heights. It is also clearly recognizable that the penetration depth increases with increasing water content. Furthermore, linear trends of the penetration depth depending on the foam injection rate can be identified for the individual test series. With a few exceptions, the individual data points for a degree of conditioning are relatively close together, which suggests good reproducibility of the tests.

Considering the depth of penetration in relation to the density of the soil-foam mixtures examined, it is evident that the depth of penetration decreases with increasing density (*Figure 5*, right). Here also a linear trend can be observed for the different test series. Individual values of the test series show a density of less than 1.0 g/cm^3 (minimum = 0.93 g/cm^3). This can be explained by the high FIR of the samples, so that excessive conditioning by foam has taken place, which also results to slump values of more than 20 cm. The maximum measured density value of a sample is 1.4 g/cm^3 (see Figure 5, right).

When correlating the Penetration Test results with the corresponding Slump Test results, it is evident that the penetration depth increases with increasing slump value, regardless of the water content or foam parameters. A linear trend between the penetration depth and the results of the Slump Tests is not discernible (see Figure 6, left) because the penetration depth increases significantly from a slump of 20 cm. In addition, the results show a certain scatter, so that for example for a slump of about 23 cm a penetration of approx. 65% to 100% was measured. Based on the results presented, a clear correlation between the Slump Test and the penetration test can be established, but some uncertainty remains requiring further research.

In addition to the Slump Tests and Penetration Tests performed, the results obtained were correlated with the results from previous rheological investigations with the BMS for fine sand (see Figure 6, right). It is evident that the higher the penetration depth, the lower the yield stress. For the limit values of the yield stress of 130-250 Pa investigated by Galli (2016), the corresponding values of penetration are between approx. 65% and 100% independent of the water contents. It should be noted that the yield stresses determined from previous test series may show slightly different Slump Test results. If a regression analysis is performed over the entire number of tests and a curve fitting is carried out, the regression curves shown in Figure 6 are obtained. A good correlation coefficient is achieved. Based on the data

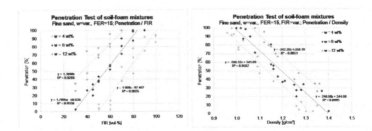

Figure 5. left: Penetration depth depending on the FIR for different fine sand-foam mixtures; right: Penetration depth depending on the density for different fine sand-foam mixtures.

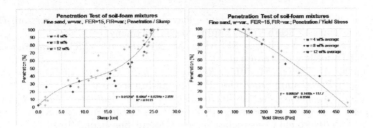

Figure 6. left: Correlation of the Penetration depth and Slump values with borderlines for a Slump of 10 and 20 cm; right: Correlation of the Penetration depth and Yield stresses with borderlines according to Galli (2016) with 130 and 250 Pa.

obtained, the Penetration Test is suitable for determining the workability of a soil-foam mixture. This test is particularly well-suited for the use directly on EPB TBMs to test the material extracted from the screw conveyor.

5 CONCLUSION

To obtain information from the excavation chamber about the flow behaviour and workability of the soil-foam mixture during the tunnelling of an EPB shield in closed mode, it is necessary to generate a suitable test procedure and basics from small-scale laboratory tests.

The test results from the Slump Tests and rheological investigations presented in this paper provide initial information on the rheological parameters and flow behaviour of foam-conditioned sands. In future investigations, cohesive soils will also be investigated with the BMS. For the investigation of coarse-grained soils with a gravelly content, the test scale must be extended. Research work at Ruhr University Bochum is continued in this direction. For the new Penetration Test of soil-foam mixtures, the spectrum of investigated soils has to be extended to the full application range of EPB shields in further test series. The rheological properties of cohesive soils and highly permeable coarse-grained soils will also be examined in the new Penetration Test. For this purpose, the evaluation of the tests should not only be carried out regarding the penetration depth, but bearing stresses are be calculated and evaluated. Furthermore, the geometry of the indenter will be varied to recognize its influence on the penetration depth.

In summary, the soil-foam mixtures investigated in all three test methods behave as expected. However, further research is needed to carry out a comprehensive and detailed correlation of the different test methods.

6 ACKNOWLEDGEMENT

This paper presents the results from the subproject A4 "Model Development for the Conditioned Soil used as Face Support Muck of Earth-Pressure-Balance-Shields". Subproject A4 is a part of "Collaborative Research Centre - SFB 837" at Ruhr University Bochum in Germany funded by DFG (Deutsche Forschungsgemeinschaft).

REFERENCES

Abd Elaty, M. A. A. & Ghazy, M. F. 2017. *Evaluation of consistency properties of freshly mixed concrete by cone penetration test*. In HBRC journal 12. Nr. 1. 1–12.

Budach, C. 2012. *Untersuchungen zum erweiterten Einsatz von Erddruckschilden in grobkörnigem Lockergestein*. Ph.D. Thesis. Ruhr University Bochum, Germany.

Budach, C. & Thewes, M. 2015. *Application ranges of EPB shields in coarse ground based on laboratory research*. In Tunnelling and Underground Space Technology. Vol. 50. 296–304.

DIN EN 12350-2 2009. Testing fresh concrete - Part 2: Slump-test, German version. DIN Deutsches Institut für Normung e.V. Beuth Verlag GmbH. Berlin, Germany.

Düllmann, J.; Hollmann, F. S.; Thewes, M.; Alber, M. 2014. *Practical TBM excavation data processing.* In Tunnels and Tunnelling International. 53–66.

Freimann, S.; Galli, M.; Thewes, M. 2017a. *Rheological Characterization of Foam-conditioned Sands in EPB Tunneling.* In Proceedings Euro: Tun. 241–248 Innsbruck, Austria.

Freimann, S.; Galli, M.; Thewes, M. 2017b. *Rheology of Foam-conditioned Sands: Transferring Results from Laboratory to Real-world Tunneling.* In Proceedings 9th International Symposium on Geotechnical Aspects of Underground Construction in Soft Ground. Sao Paulo, Brazil.

Galli, M. & Thewes, M. 2014. *Investigations for the application of EPB shields in difficult grounds.* In Geomechanics and Tunnelling 7. No. 1. 31–44.

Galli, M.; Thewes, M. 2018. *Rheological Characterisation of Foam-conditioned Sands in EPB Tunneling.* International Journal of Civil Engineering. 1–16. DOI: 10.1007/s40999-018-0316-x

Galli, M. 2016. *Rheological Characterisation of Earth-Pressure-Balance (EPB) Support Medium composed of non-cohesive Soils and Foam.* Ph.D. Thesis. Ruhr University Bochum, Germany.

Juradin, S. 2012. *Determination of Rheological Properties of Fresh Concrete and Similar Materials in a Vibration Rheometer.* In Materials Research 15. 103–113

Maidl, U. 1995. *Erweiterung der Einsatzbereiche der Erddruckschilde durch Bodenkonditionierung mit Schaum.* Dissertation. Ph.D. Thesis. Ruhr University Bochum, Germany

Martinelli, M.; Peila, D.; Campa, E. 2015. *Feasibility study of tar sands conditioning for earth pressure balance tunnelling.* In Journal of Rock Mechanics and Geotechnical Engineering. No. 7. 684–690.

Müller, M.; Tyrach, J.; Brunn P. O. 1999. *Rheological characterization of machineapplied plasters.* In ZKG International 52. No. 5. 252–258.

Oliveira, D. G. G.; Thewes, M.; Diederichs, M.; Langmaack, L. 2018. *Consistency Index and its Correlation with EPB Excavation of Mixed Clay-Sand Soils.* In Geotechnical and Geological Engineering. https://doi.org/10.1007/s10706-018-0612-x

Peila, D. 2014. *Soil conditioning for EPB shield tunnelling.* In KSCE Journal of Civil Engineering. No. 18 (3). 831–836.

Peila D.; Picchio A.; Chieregato A. 2013. *Earth pressure balance tunnelling in rock masses: Laboratory feasibility study of the conditioning process.* In Tunnelling and Underground Space Technology. No 35. 55–66. Torino, Italy.

Peila, D.; Oggeri, C.; Borio, L. 2008. Influence of Granulometry, time and temperature on soil conditioning for EPBS applications. In Proceedings of the World Tunnel Congress-2008, *Facilities for Better Environment and Safety.* Kanjlia, V. K. (Ed.). Central Board of Irrigation & Power 2008. 881–891. Agra, India.

Peila, D.; Picchio, A.; Martinelli, D.; Dal Negro, E. 2016. *Laboratory tests on soil conditioning of clayey soil. In Acta Geotechnica.* No. 11. 1061–1074.

Schatzmann, M. 2005. *Rheometry for large particle fluids and debris flows.* Ph.D. Thesis. Eidgenössische Technische Hochschule. ETH Zurich, Switzerland.

Tyrach, J. 2000. *Rheologische Charakterisierung von zementären Baustoffsystemen.* Ph.d. Thesis. Universität Nürnberg-Erlangen, Germany.

Vinai, R. 2006. *A contribution to the study of soil conditioning techniques for EPB TBM applications in cohesionless soils.* Ph.D. Thesis. Politecnico di Torino, Italy.

Zumsteg, R.; Plötze, M.; Puzrin, A. M. 2012. *Effect of Soil Conditioners on the Pressure and Rate-Dependent Shear Strength of Different Clays.* In Journal of Geotechnical and Geoenvironmental Engineering 138. No. 9. 1138–1146. DOI: 10.1061/(ASCE)GT.1943-5606.0000681

Tunnels and Underground Cities: Engineering and Innovation meet Archaeology,
Architecture and Art, Volume 5: Innovation in underground engineering,
materials and equipment - Part 1 – Peila, Viggiani & Celestino (Eds)
© 2020 Taylor & Francis Group, London, ISBN 978-0-367-46870-5

The role of mechanized shaft sinking in international tunnelling projects

S. Frey & P. Schmaeh
Herrenknecht AG, Germany

ABSTRACT: Tunnelling projects require shafts, either as start and reception shafts for the tunnelling process or for inspection, ventilation and rescue purpose. Inner-city shaft structures demand safe working principles. The Vertical Shaft Sinking Machine method procures safety for the surrounding environment and for all personnel. As the water level in the shaft equals the groundwater level outside the shaft, there is no water flow which can cause ground movement. All installations including the lining erection are completed from the surface. No personnel have to enter the shaft until it has reached the final depth and is fully secured. The lining consists of either precast segments or cast in place concrete. As the lining installation is completed on the surface, a high quality installation can be reached. This paper will describe the role of shaft sinking in tunnelling projects along with all necessary safety and planning aspects.

1 INTRODUCTION

Almost all tunnelling projects require shafts, either as start and reception shafts for the tunnelling process or for inspection, ventilation and rescue purposes. Also, a current trend towards infrastructure installations in growing depths can be observed. It is driven, among other things, by deep sewer construction projects that aim to avoid pumping stations as well as the need to build new installations below existing infrastructure. Our paper discusses the benefits of the mechanized shaft sinking technology and presents a selection of worldwide references from a variety of tunnelling projects.

The Vertical Shaft Sinking Machine (VSM) was originally developed by Herrenknecht for the mechanized construction of deep launch and reception shafts for microtunnelling. After starting design and testing in early 2004, the first Herrenknecht VSM equipment went into operation in Kuwait and Saudi Arabia in 2006. The machine concept, fully remote-controlled from the surface, as well as its implementation on site proved to be an efficient solution right from the start for the safe and fast realization of shafts especially in difficult, inner-city environments without lowering the groundwater table. To date, approx. 75 shafts have been successfully installed worldwide with the Herrenknecht VSM technology, reaching depths of up to 85m. They serve today, for example, as ventilation shafts for metro systems, maintenance or collector shafts for sewage, or as temporary microtunnelling shafts (Figure 1).

2 BENEFITS OF VSM TECHNOLOGY

The VSM technology masters the main challenges associated with shaft sinking: inner-city shaft structures demand safe working conditions for surrounding buildings and the environment, especially regarding potential ground settlement. There is an increased requirement to avoid the lowering of the groundwater during the construction of shafts in order to avoid the

Figure1. Overview of VSM applications, from left to right: ventilation/emergency shaft, microtunnelling shaft, sewage collector shaft, U-Park® shaft.

associated settlement, which can affect a wide area. Deep shaft construction companies often encounter difficult geological conditions such as high groundwater pressure combined with layers of hard and soft material. In addition, deep shafts need special attention for the safety of the operating personnel.

2.1 Cost and time benefits

Simultaneous excavation and ring building facilitate high advance rates and shortened overall project duration. At the same time, continuous performance ensures overall high planning reliability for all stakeholders.

The lining of a VSM shaft consists of either precast segments or cast-in-situ concrete. As the lining installation is completed on the surface, high quality installation can be accomplished, leading to greater accuracy of the shaft structure. In most cases, a secondary, time-consuming lining is not required, resulting in a reduced wall thickness of the shaft and, thus, less soil excavation.

Furthermore, each VSM type is very flexible as its excavation diameter can be adjusted within a specific range. A VSM10000, for example, can cover an inner shaft diameter range from 5.5m to 10.0m and is, therefore, a one-time investment for multiple use.

2.2 Construction and occupational safety benefits

As the water level in the shaft is maintained close to the groundwater level outside the shaft, water flow is prevented which otherwise could cause ground movement and lead to a high risk of settlement. The Herrenknecht VSM can be applied below groundwater with a hydrostatic pressure of up to 10 bar and in heterogeneous soil and hard rock of up to 140MPa compressive strength.

All installations, including the lining erection, are remotely controlled from the surface. No personnel have to enter the shaft until it has reached the final depth and is fully secured. In general, mechanized shaft sinking requires less personnel and machinery on site, which leads to minimized risk exposure.

2.3 Environmental benefits

Measures for groundwater lowering are not necessary, as the VSM machine concept is designed for operation under groundwater. As the VSM technology applies a high degree of accuracy of shaft construction, the shaft lining thickness can be reduced to a minimum, which reduces the amount of excavated soil.

3 VSM MACHINE COMPONENTS

The VSM consists of two main components (Figure 2): the excavation unit and the lowering unit. The excavation unit systematically cuts and excavates the soil and consists of a cutting drum attached to a telescopic boom that allows excavation of a determined overcut. The

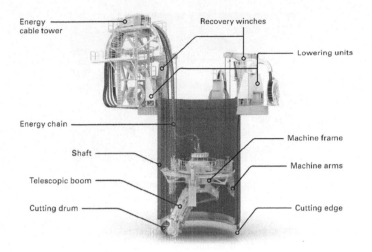

Figure 2. VSM components.

lowering unit on the surface stabilizes the entire shaft construction against uncontrolled sinking by holding the total shaft weight with steel strands and hydraulic jacks. When one excavation cycle is completed, the complete lining can be lowered uniformly and precisely.

A slurry discharge system removes the excavated soil and a submerged slurry pump is located directly on the cutting drum casing. It transports the water and soil mixture through a slurry line to a separation plant on the surface. The whole operation takes place from the surface and is controlled by the operator from the control container on the surface. All machine functions are remote-controlled without the necessity to view the shaft bottom or the machine. Power supply for the submerged VSM is secured by the energy chain. After reaching its final depth, the VSM is lifted out of the shaft by the recovery winches and the jobsite crane.

4 VSM JOBSITE PREPARATION

4.1 Jobsite layout

Depending on the space conditions on site, the VSM components can be positioned flexibly to suit local circumstances (Figure 3). As most obsites are located in heavily built-up urban

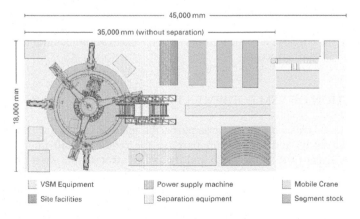

Figure 3. Exemplary layout of a VSM site.

Figure 4. Completely installed start section with attachment brackets.

areas, the access for logistics, e.g. trucks, ring segment stock or soil disposal, is limited. Special concepts to relocate components such as the separation plant already exist for this purpose, and can be discussed if required.

4.2 *Ring foundation*

After preparation of the required site surface, a concrete ring foundation has to be installed in a pre-excavated pit. This foundation bears the loads of the VSM and serves as a support for the lowering units, which guide and hold the shaft at all times. The size of the ring foundation depends on the ground conditions and the size of the shaft. Connection bolts for various VSM components such as lowering units, recovery winches and energy winch tower are also integrated into the ring foundation.

4.3 *Cutting edge*

After pre-excavation and ring foundation, the installation of the cutting edge and the first segment rings are the next steps. The cutting edge is designed to cut the shaft profile in soft and loose soil conditions. Its design depends on the shaft diameter and wall thickness. It can be integrated as the first concrete segment ring or welded onto the shaft lining as a separate steel ring.

4.4 *Start section and machine attachment*

The first five meters of the shaft lining constitute the so-called start section (Figure 4). The start section has a stronger steel reinforcement to be able to take the loads and reaction forces of the machine during excavation and shaft sinking. Furthermore, the shaft lining is equipped with cast-in steel plates, onto which the brackets for the machine arms of the VSM are welded. As the VSM employs a sequential partial face excavation technique there is no torque transmitted into the shaft structure.

5   INSTALLATION OF VSM EQUIPMENT

The excavation unit arrives on site in three parts: the telescopic boom with the cutting drum, the machine main body, and the adapter parts to the required shaft internal diameter. The lowering unit consists of the strand jacks and the coiled steel strands on a drum. The number of strands depends on the total predicted weight of the shaft including the machine weight and the estimated buoyancy and friction forces. The strand jacks are bolted to the ring foundation by anchor bolts. Coming from the strand drum, the strands are fed through the strand jack, lowered through the outer annulus of the shaft wall and connected to the cutting edge. When all strand jacks are installed and connected to the cutting edge the strands can be tensioned and carry the loads.

Now, the preassembled excavation unit can be lifted into the start section. The VSM is secured by hydraulically activated locking bolts. When the VSM is in place, the recovery winches are installed and connected to the three arms of the machine. The recovery winches are used to recover the VSM for required maintenance or for final machine recovery.

Next, the energy chain tower is installed and the excavation unit is connected to the hydraulic and electrical supply as well as to the feed and discharge lines. The energy chain tower with its winch has bolted connections for easy assembly.

As a final step, all the electrical and hydraulic connections are done, and the equipment is now ready to operate. Before starting the excavation, a calibration of the VSM in reference to the projected alignment of the shaft is required to ensure the accurate action of the cutting boom.

6 VSM OPERATION – SHAFT SINKING PROCEDURE

Excavation is completely remote-controlled from the operator cabin at the surface. Stored data, together with the position of the cutting boom, is shown on a graphic display, giving the operator full control of the excavation and sinking process. The excavation unit can be operated in three different overcut options, which requires the installation height measured from the level of the cutting edge to be adjustable. In its highest position, the excavation unit is not able to create an overcut under the cutting edge, which is important in soft or unstable soil to maintain surrounding stability. When an overcut is required, e.g. in stable or cohesive soil, the excavation unit works in its lowest position. In this case, the annulus should be stabilized by a bentonite-water suspension. In addition, each segment can be equipped with bentonite nozzles for lubrication, which can also later be used to grout the annulus. The standard installation is to connect the segments in ring number 3 and 5 with the bentonite mixing unit right from the beginning and to lubricate from these two segment rings at the shaft bottom during the sinking. The stabilization of the annulus together with the controlled sinking of the shaft by the strand jacks minimizes the risks of settlement.

During the excavation and sinking process the shaft is kept full of water to balance the level of the groundwater table in the surrounding geology (Figure 5). The cutting drum cuts and crushes the material to a granular size that can be handled by the pumps (pump capacity: 200-400m³/h). A slurry circuit transports the excavated material from the shaft to a separation plant on the surface.

Figure 5. VSM excavation unit working below groundwater.

The telescopic boom allows varying diameters: flexible use is possible up to 12m and up to 14m with a reinforced frame structure. Diameters of up to 18m are realized through an enlarged concept. Excavation is possible even in water depths up to 85 m. The cutting arm moves radially from the center to the outside of the shaft with an additional telescopic extension of 1m. With a rotation of +/-190° the cutting boom covers the whole cross section of the shaft. The cutting speed and the movement of the boom can be varied to achieve the best excavation rate.

6.1 *Shaft lining*

In most cases, the shaft lining consists of precast concrete segments installed at the surface. This so-called ring building is comparable to segmental lining in tunnelling. The ring is built at the surface by crane. The number of segments depends on the shaft diameter. Ring building work includes the proper connection of the rings by anchors and bolts, which can be handled from outside the shaft. The excavation process of the VSM is not affected by the ring building process. This increases the shaft sinking performance significantly.

Alternatively, in-situ concrete casting of the shaft walls is another solution, especially for larger shaft diameters where segment handling becomes more difficult. In this case, the progress of shaft construction works is slowed down by the necessary time to build the formwork and the setting time of the concrete structure. The benefit of in-situ casing is the "continuous" structure without joints and the possibility to integrate entire entry and exit structures, e.g. for microtunnelling activities in the shaft walls.

7 COMPLETION OF THE SHAFT

After reaching the final depth, the bottom plug has to be installed. Usually, the VSM is used to excavate the required overcut. When this final excavation is done, the VSM can be disconnected, recovered by the recovery winches and lifted up to the surface and out of the shaft by a crane. The bottom plug is cast with underwater concrete. In a next step, the shaft annulus is grouted through the lubrication lines to stabilize and anchor the shaft to the surrounding ground. Finally, the shaft water can be pumped out and the shaft is completed (Figure 6). Personnel access is now possible.

8 DEEP SHAFT APPLICATIONS AND REFERENCES

8.1 *Ventilation and emergency shafts*

Girona and Barcelona, Spain
For the high-speed rail link from Barcelona to the French border, a total of four shafts were built in Girona by a Herrenknecht VSM as ground stabilization shafts before tunnelling

Figure 6. VSM reaching final depth.

works (4 shafts, ID 5,250mm, depth 20m). The same VSM built one additional shaft in Barcelona for ventilation and as an emergency exit (ID 9,200mm, depth 47m).

The major challenge for the construction contractors were the extremely confined working conditions. For example, one of the shafts in Girona was located between two rows of houses with a spacing of only 12m (Figure 7). Here, the VSM's ability to work under limited space conditions in inner cities proved to be a major benefit.

Due to the lack of ground stability in the center of the city of Girona, only small, lightweight cranes could be used and this led, in turn, to a complete reorganization of assembly logistics: The main components of the VSM10000 were delivered just in time and assembled directly in the shaft start section. The average daily performance was 3.0m.

Naples, Italy

A total of 13 ventilation and emergency shafts (ID 4500/5500, depth up to 45m) for the subway line were sunk in Naples, Italy, by a Herrenknecht VSM. The site was located in a densely built-up area in the inner city with high traffic (Figure 8). The required jobsite footprint was approx. 300m² in the narrow streets of Naples. Noise exposure for the residents had to be kept at a low level. Because the excavation of the shafts and their lining with precast concrete segments could be realized simultaneously, the production of the shafts could be finished quickly with performance rates of up to 5m per day. Due to its modular setup, the VSM was rapidly disassembled and transported to the next site after completing one shaft.

Upcoming: Grand Paris Express

In August 2018, a Herrenknecht VSM will be installed on a shaft sinking site in France for the first time. In the context of Grand Paris Express, currently the largest infrastructure

Figure 7.   Confined space conditions in Girona.

Figure 8.   VSM site in Naples.

Figure 9.   VSM site in Honolulu.

project in Europe (the large-scale extension of Paris's metro network), a VSM12000 will sink emergency and ventilation shafts for the Line 15 tunnels excavated by Herrenknecht tunnel boring machines. Four shafts are to be constructed with inner diameters of 8,300mm, 10,300mm and 11,900mm and depths of up to 53m.

8.2   Microtunnelling shafts

Hawaii, USA

Two large shafts with a 10m inner diameter were sunk in Honolulu, Hawaii. These 36m deep shafts were to be used as launch shafts for a pipe jacking project (Figure 9). Cast-in-situ was the preferred lining method in order to handle the necessary thrust forces in the shaft wall when launching the pipe jacking machine. Moreover, fiberglass reinforcement simplified the launch process for the TBM. In Hawaii, the VSM successfully handled a challenging geology comprised of hard basalt as well as coral that would have been problematic for conventional methods. Best daily performance with the Herrenknecht VSM was 2.3m.

8.3   Sewage collector shafts

St. Petersburg, Russia

The deepest VSM shaft under groundwater to date was sunk in St. Petersburg, Russia, where a total of four shafts were realized to depths ranging from 65 to 83m. In St. Petersburg, the Herrenknecht VSM technology proved to be especially efficient in the face of tight time schedules. Together with the customer, Herrenknecht assembled the machine in nine days following site preparation, and successfully finished the first shaft of 83m depth and an internal diameter of 7.7m in 50 working days (Figure 10).

8.4   Launch shaft and sewage collector shafts

Upcoming: DTSS Phase 2 Singapore

In the autumn of 2018, the first VSM project in Asia will see the use of VSM technology for the Deep Tunnel Sewer System (DTSS) Phase 2 in Singapore with a total of approx. 100km of main sewer tunnels and link sewers. Seven shafts with inner diameters of 10 and 12m will be sunk down to depths of up to 60m. A project-specific feature will be the combination of segmental lining in the upper section for fast construction progress and in-situ concrete casting in the lower section for the connection of the tunnel. The shafts will be used first as launch shafts for the tunnelling operation and later as collector shafts.

Figure 10. Completed sewage collector shaft in St. Petersburg.

Figure 11. Shaft sizes used for U-Park®

8.5 *Outlook: U-Park® shafts*

Especially in large cities, new parking concepts have to be developed because space above ground is extremely built-up and expensive. Therefore, new parking solutions are being designed that make use of underground space. One of them, called U-Park®, was conceptualized as a combination of VSM technology for creating shafts and automatic parking systems, which are accommodated in these shafts (Figure 11). The number of parking lots per system depends on the diameter and depth of the shaft.

9 CONCLUSIONS

The current and worldwide trend to construct more and more infrastructure underground, e.g. metro, road and railway as well as a large variety of utility lines, promotes a growing demand for the construction of shafts.

With increasing depth and groundwater levels, conventional shaft construction methods reach their technical and economical limits. Herrenknecht has developed the solution: the Vertical Shaft Sinking Machine (VSM). Its efficiency and benefits in terms of budget, construction time and occupational safety have led to a total of approx. 75 VSM projects with a total depth of 3,8km where shafts have been successfully sunk e.g. in inner-city environments with tight space contraints and a requirement to avoid all settlement. Since its first design in 2004 and its first deployment in 2006, Herrenknecht has continuously developed the VSM machine design to a proven technology with a growing range of applications.

Tunnels and Underground Cities: Engineering and Innovation meet Archaeology,
Architecture and Art, Volume 5: Innovation in underground engineering,
materials and equipment - Part 1 – Peila, Viggiani & Celestino (Eds)
© 2020 Taylor & Francis Group, London, ISBN 978-0-367-46870-5

Risk-based tunnel design utilizing probabilistic two-dimensional finite element analyses

A. Gakis & C. Kalogeraki
Dr. Sauer & Partners, London, United Kingdom

P. Spyridis
TU Dortmund University, Dortmund, Germany

ABSTRACT: The present paper investigates the impact of geotechnical and structural parameters on the robustness of the tunnel lining capacity, using probabilistic, 2D finite element analysis with the software Abaqus. Seven cross-sections of a mined tunnel are investigated, with similar cross-sectional areas, but varying dimensions and curvatures. Normal and log-normal distributions are assumed, and a Monte-Carlo simulation is used to generate a 1000-point sample of input parameters. The force distribution along the circumference of the tunnel and the lining capacity are assessed on the basis of a sensitivity/parametric analysis. The paper offers a better insight into the design and shape optimization procedure with potential benefits in increased structural reliability, as well as cost savings resulting from improved stress flow in the lining and reduced support requirements due to more informed selections in light of the specific geotechnical circumstances.

1 INTRODUCTION

Reliability-based optimization tools and methods aided by the advancements in computational engineering, allow for improved solutions in the design of structures. The nature of tunnels is such that the geometry of their cross-section is one of the most influential factors affecting their efficiency and robustness. The size and shape of a tunnel is primarily designed to satisfy spaceproofing requirements but additionally may depend on the in-situ stress-state, type of waterproofing, health & safety criteria, constructability and construction methods.

Geotechnical conditions play a primary role in the design of tunnels. Their inherent uncertainty and spatial variability call for risk-based approaches which, however, are not the common engineering practice due to limitations in precise quantifications of the impacts of multiple parameters on the robustness of the tunnel lining capacity.

In the present study, the impact of geotechnical and structural parameters on the capacity of the tunnel lining is investigated using probabilistic, 2D non-linear finite element (FE) analysis utilizing the general-purpose FE software Abaqus. The finite element mesh for an indicative tunnel section is shown in Figure 1. Seven mined tunnel cross-sections with equivalent areas are investigated. Assuming normal and lognormal distributions, a Monte-Carlo simulation is deployed to generate a 1000-point sample of input parameters for the FE analyses. A parametric/sensitivity analysis of the force distribution along the circumference of the tunnel and the lining capacity is assessed. The paper provides an insight in-to the design and shape optimization procedure of mined tunnel linings, with potential benefits in increased structural reliability and cost savings.

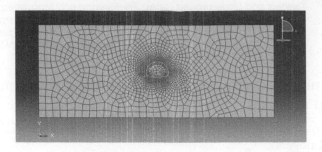

Figure 1.   Finite element meshing of Section Wide 3.

2   PROBABILISTIC FINITE ELEMENT

2.1   *Cross section geometries*

The seven tunnel cross sections that were selected for the presented study as shown in Figure 2, have a cross-sectional area of A ≈ 50.0 m². Section 0 is a circular tunnel and serves as the base case/control model. The set is completed by three sections with height/width ratio < 1, referred to as Section Wide 1, 2 and 3 and three sections with height/width ratio > 1 referred to as Section Tall 1, 2 and 3. The maximum/minimum dimension ratios are 1.2, 1.3 and 1.4 for sections Wide/Tall 1, 2 and 3 respectively. The overburden measured from the tunnel crown for all seven sections is 16 m.

2.2   *Soil-structure interaction models*

The numerical models implement non-linear Mohr-Coulomb plasticity with a graded mesh of 6-noded triangular solid material elements to model the ground and Timoshenko-beam elements to model the concrete liners. The concrete liners are 300 mm thick and were modelled with an elastic concrete material model with Young's modulus of E_c = 13 GPa (John & Mattle 2003), and Poisson's ration v = 0.2. The geological model and the range of parameters used are selected based on an extensive survey of available geotechnical information in relevant publications (Bond & Harris 2008, Phoon 2008, Spyridis et al 2016) and they are anticipated to be representative of a typical shallow tunneling project in urban environments. An undrained total stress analysis is carried out using the soil self-weight and a lateral earth pressure coefficient at rest, in order to simulate the soil stress pattern. The analysis assumes a full-face excavation and a relaxation factor (β) is applied in the soil through the stiffness reduction method (Möller 2006 and Potts & Zdravkovic 2001) in order to simulate the soil deformation prior to advance and installation of the lining at each excavation step.

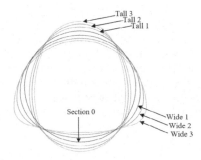

Figure 2.   Overview of the analyzed cross section geometries.

Table 1.   Geotechnical input parameters.

Soil property	Units	Mean	Std. dev.	Distribution
Material density, γ	[kN/m^3]	20	deterministic	
Poisson's ratio, v	[-]	0.45	deterministic	
Relaxation factor, β	[-]	0.5	deterministic	
Undrained Young's modulus, Eu	[MPa]	200	50	Normal*
Undrained shear strength, Su	[kPa]	200	50	Normal*
Lateral stress coefficient, Ko	[-]	1.00	0.25	Normal*

* truncated distribution with lower limit at 1% and upper limit at 99% of the normal distribution

Table 2.   Tunnel support input parameters.

Concrete property	Units	Mean	Std. dev.	Distribution
Material density, γ	[kN/m^3]	25	deterministic	
Poisson's ratio, v	[-]	0.2	deterministic	
Young's modulus, Ec	[GPa]	2.5*	0.25*	Log-Normal

* parameters refer to Log-Normal distribution

2.3   Probabilistic models

A 1000 – point sample of input parameters is populated using a Monte – Carlo simulation in order to assess the variability of the lining section forces due to the uncertainty of the geotechnical and concrete parameters. Each set of parameters is then used as stochastic input to a simulation run. The selected stochastic parameters (see 2.4) are assumed to follow a normal distribution except for the concrete's Young's modulus which is assumed to follow a log – normal distribution. The same 1000 – point sample is used in the FE analyses of all 7 cross sections.

2.4   Input variables

The soil's undrained Young's modulus E_u, the undrained shear strength S_u, the coefficient of lateral earth pressure K_0, and the concrete's Young's modulus E_c are the stochastic parameters, while all other parameters are kept as deterministic. Table 1 and Table 2 provide an overview of the geotechnical and tunnel support input parameters respectively. The soil Young's modulus is found to be in strong correlation with the undrained shear strength and a correlation was assumed where $E_u = 1000 \times S_u$. Additionally, the Young's modulus beneath the tunnel springlines was assigned a value of $3 \times E_u$ to account for a stiffer response during unloading.

3   RESULTS AND DISCUSSION

An overview of the section-force results is presented in the graphs of Figure 3 to Figure 8. The deterministic analyses results are indicated with solid lines and the upper (95%) and lower bound (5%) fractile with dashed lines (half sections presented due to symmetry).

The wide cross sections develop higher axial forces at the tunnel arch (upper half of the section) due to the vertical in-situ stress components. These accordingly decrease along the inverts as the latter become flatter, and thus bear less of the horizontal in-situ stress components. Accordingly, the tall sections show an opposite trend, a more pronounced axial-load bearing behavior and narrower dispersion. This is reflected in the presented results, since (a) the axial forces in the crown through to the springlines of the wide sections are of the same

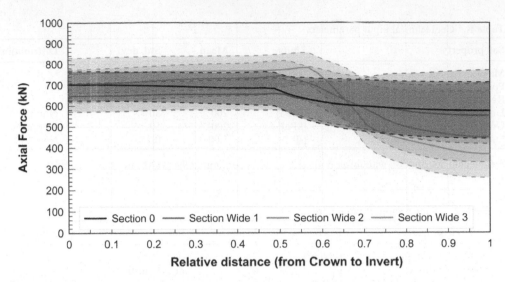

Figure 3. Axial force deterministic values and 5% - 95% percentiles, for the wide sections and for the development of the axial force from the tunnel crown (left) to the invert (right).

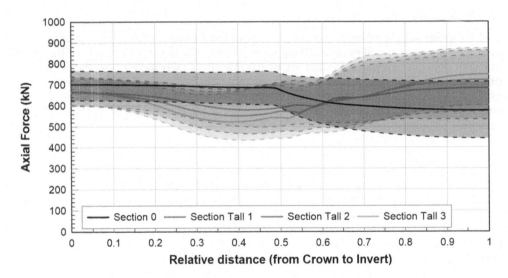

Figure 4. Axial force deterministic values and 5% - 95% percentiles, for the tall sections and for the development of the axial force from the tunnel crown (left) to the invert (right).

order of magnitude both in terms of mean value (700–800 kN) and standard deviations (40–50 kN), whereas in the inverts the axial force mean values ranges from 600 to 350 kN with the standard deviation gradually increasing to approx. 130kN and (b) the mean value and standard deviation are fairly constant along the inverts of the tall sections (approx. 700 and 90 kN respectively).

The bending moments and shear forces in the linings are shown in the graphs of Figure 5 to Figure 8. The efficiency of each cross-section design with respect to the assumed variabilities becomes more noticeable. As the inverts become flatter, and consequently flexure becomes more pronounced, a higher dispersion of bending moment values becomes evident. Respectively, higher shear forces and dispersions develop as the invert/arch transition areas for the

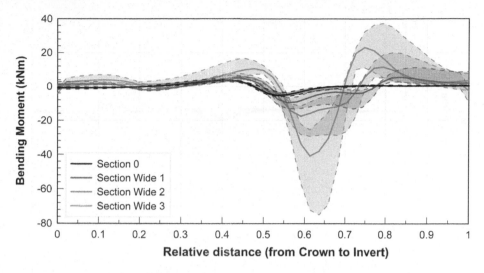

Figure 5. Bending moment deterministic values and 5% - 95% percentiles, for the wide sections and for the development of the bending moment from the tunnel crown (left) to the invert (right).

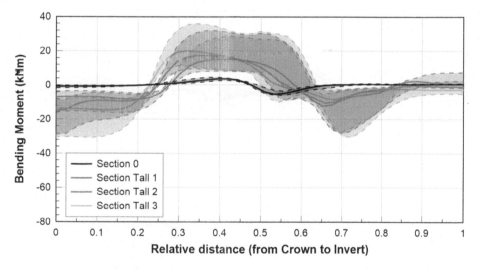

Figure 6. Bending moment deterministic values and 5% - 95% percentiles, for the tall sections and for the development of the bending moment from the tunnel crown (left) to the invert (right).

cross sections are realized with smaller local curvatures. In the circular section (base case, Section 0) the moment and shear mean values are negligible and an optimum force flow develops within the lining; some variations in bending are observed at the crown, invert, and springlines of the lining, mainly due to the variations in the lateral earth pressure coefficient K_0. Bending moment and shear patterns are recognized at locations of strong curvature, at the transitions from the bottom of the arch of the lining toward the invert. In these instances, both the mean values as well as the standard deviations increase as the inverts get flatter and the change in curvature more acute (from Sections 1 to Sections 3), yet the overall variation coefficients remain in the same order of magnitude for each location of the tunnel perimeter.

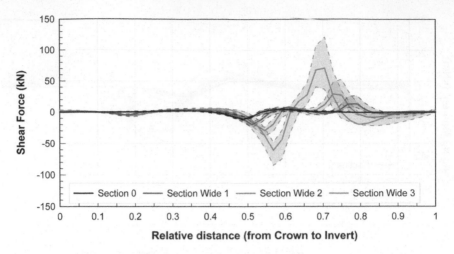

Figure 7. Shear force deterministic values and 5% - 95% percentiles, for the wide sections and for the development of the shear force from the tunnel crown (left) to the invert (right).

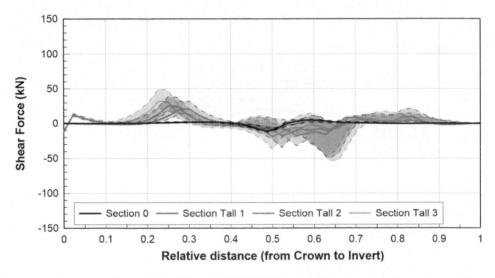

Figure 8. Shear force deterministic values and 5% - 95% percentiles, for the tall sections and for the development of the shear force from the tunnel crown (left) to the invert (right).

The results are also aggregated on the basis of histograms, with best-fit probability density functions. The probabilistic concept is mainly based on Ang & Tang 2007, Bergmeister et al 2009, Strauss et al 2009 and Nasekhian et al 2012. Figure 9 demonstrates the axial force variation for the widened cross sections at each sections springline (location at the end of the upper arch on the side of the tunnel).

The results in Figure 10 are presented graphically based on interaction diagrams (or also called capacity limit curves – CLC) which pose a very useful assessment tool in the case of shells with uniform thickness and reinforcement ratio, under axial and bending loads (Sauer et al 1994. For the development of these graphs, one can rely on national and international codes and standards. Herein, an unfactored, deterministic resistance of an unreinforced lining, with a C30/35 concrete is implemented, in accordance with Eurocode 2 (CEN EN1992-1-1).

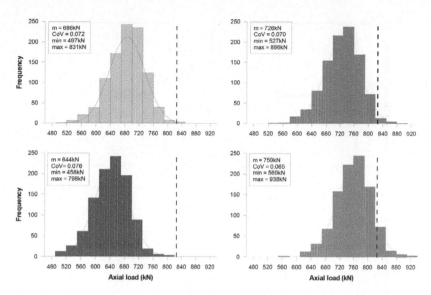

Figure 9. Histograms of the axial force at the springline of the analyzed wide cross sections. The 99.9% percentile of the control section (0) is shown with a dashed line.

All design combinations of axial forces and bending moments relative to the envelope of the cross-section's design capacity are presented, providing a comprehensive graphical and numerical structural verification. The reliability and robustness of the design can then be assessed based on the distance of the design point from the capacity curve; this distance is essentially the design's Limit State Function (LSF). Based on this concept, a comparison of the performance and reliability for various tunnel lining designs is accommodated, which leads to a consistent design optimization. A tunnel lining design concept typically utilizes the compression induced by the soil/structure interaction to exhibit flexural resistance (an alternative of active pre-stressing in standard concrete engineering). Therefore, the minimum failure probability appears for a centrically compressed section well above zero-stress, or in other words well beyond the threshold of tension stresses in the cross-section of the shell. Structural reliability in the context of this study is identified as the load situation, which is less prone to failure, and – counter-intuitively – it is not the unloaded tunnel structure.

The outcomes of the analyses reveal a significant dependence of the section forces dispersion on the cross-section shape. In particular, axial force flow in the invert is characterized by a higher variability for sections with higher invert curvature, while bending moments and shear forces in certain areas of the tunnel perimeter exhibit a very distinct variability for cross-sections with more acute curvature changes and flatter inverts. The main outtake of these comparisons is that the shape of the structure (i.e. how optimized it is against the imposed loads, the soil-structure interactions, and the involved uncertainties) strongly defines the structure's reliability level. Furthermore, it also characterizes the associated redundancy and robustness against further uncertainties in the geotechnical and structural properties, the tunnel idealization and modelling, as well as the as-built shape.

With regards to surface settlements, the magnitude at the projection above the tunnel was recorded and plotted in Figure 11 for all the analyses. The tall sections result in smaller settlements compared to the wide ones, with increasing spread as the width/height ratio increases.

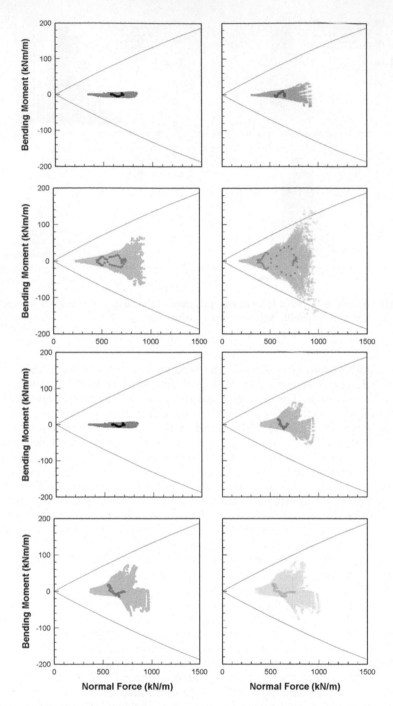

Figure 10. Interaction diagrams for the analyzed wide and tall cross sections; black for Sections 0 (base case), blue, red, green, for Sections Wide 1, 2, and 3 and brown, magenta, light blue for Sections Tall 1, 2, and 3 respectively.

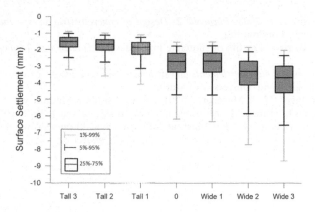

Figure 11. Surface settlements above the tunnel. The caps or whiskers at the end of each box indicate the extreme values as a percentage of the data, the box is defined by the lower (Q1, lowest 25% of data) and upper (Q3, highest 75% of data) quartiles, and the line in the center of the box is the median.

4 CONCLUSIONS

In the present study, probabilistic finite element analyses are utilized to evaluate the impact of geotechnical and structural uncertainties on tunnel design against an optimization of the tunnel shape. The case study is a shallow tunnel excavated in uniform, undrained soft soil. The soil's undrained Young's modulus Eu, the undrained shear strength Su, the coefficient of lateral earth pressure K_0, and the concrete's Young's modulus Ec, are the varying input parameters (stochastic variables).

The findings are based on a specific set of section shapes and geotechnical conditions. In the context of a broader investigation, a quantifiable tunnel shape optimization classification can be established, in relation to various influencing parameters and their variabilities, including geotechnical and hydrological conditions, tunnel space functionality and architecture, construction logistics and costs.

A clear dependence pattern between cross-section curvatures and geotechnical variabilities is disclosed. The study concludes that a cross-section shape optimization can undoubtedly lead to substantial savings in terms of lining design (e.g. thickness dimensioning or reinforcement), but also to an enhancement on the overall structural safety of the tunnel lining elements. Conversely, cross-sections with a lower degree of optimization (e.g. flat inverts, or acute curvature changes in soft soil), should be treated in the design with additional reliability and robustness considerations, as for example by use of appropriate safety factors, or probabilistic analysis and design tools.

The assessment is completed with the variation of the induced surface settlements for the various tunnel shapes. Such an approach can be potentially used for risk-based impact assessments. The results show that the taller cross-sections produced significantly smaller spreads than the wide sections. Additionally, the deviation from the circular tunnel assumption which is simplistically used in volume loss assessments is significant.

REFERENCES

Ang, A.H.S. & Tang, W.H. 2008. Probability Concepts in Engineering: *Emphasis on Applications to Civil and Environmental Engineering 2nd Edition.* New York: Wiley.
Bergmeister, K., Novak, D. Pukl, R. & Červenka, V. 2009. Structural assessment and reliability analysis for existing engineering structures, theoretical background. *Structure and Infrastructure Engineering* 5 (4): 267–275.
Bond, A. & Harris, A. 2008. *Decoding Eurocode 7.* London: Taylor & Francis.

British Standards Institution. 2008. *Eurocode 2: Design of concrete structures – Part 1-1: General rules and rules for buildings*. London: BSi.

John, M. & Mattle, B. 2003. Factors of shotcrete lining design. *Tunnels and Tunnelling International* 35 (10): 42–44.

Möller, S.C. 2006. *Tunnel induced settlements and structural forces in linings*. Mitteilung 54 of the Institute of Geotechnical Engineering, University of Stuttgart.

Nasekhian, A., Schweiger, H. & Marcher, T. 2012. Point Estimate Method vs. Random Set – A case study in Tunnelling. In Huber, M., Moormann, C. & Proske, D. (eds), *Proc. 10th International Probabilistic Workshop*. Mitteilung 67 of the Institute of Geotechnical Engineering, University of Stuttgart.

Phoon, K. K. 2008. *Reliability-based design in geotechnical engineering: computations and applications*. London: Taylor and Francis.

Potts, D. M. & Zdravkovic, L. 2001. *Finite element analysis in geotechnical engineering: theory & application*. London: Thomas Telford Publishing.

Sauer, G., Gall, V., Bauer, E. & Dietmaier, P. 1994. Design of tunnel concrete linings using limit capacity curves. In Siriwardane & Zaman (eds), *Computer methods and advances in geomechanics; Proc. intern. conf., Morgantown, W. Va, USA, May 1994*. Rotterdam: Balkema.

Spyridis, P., Konstantis, S., & Gakis, A. 2016. Performance indicator of tunnel linings under geotechnical uncertainty. *Geomechanics and Tunnelling* 9(2): 158–164.

Strauss, A., Hoffmann, S., Wender, R. & Bergmeister, K. 2009. Structural assessment and reliability analysis for existing engineering structures, applications for real structures. *Structure and Infrastructure Engineering* 5(4): 277–286.

Tunnels and Underground Cities: Engineering and Innovation meet Archaeology, Architecture and Art, Volume 5: Innovation in underground engineering, materials and equipment - Part 1 – Peila, Viggiani & Celestino (Eds)
© 2020 Taylor & Francis Group, London, ISBN 978-0-367-46870-5

Soil conditioning for EPB tunnelling: Relevant foam properties and characterisation

M. Galli
PORR Deutschland GmbH, Düsseldorf, Germany

M. Thewes & S. Freimann
Institute for Tunnelling and Construction Management, Ruhr University Bochum, Germany

ABSTRACT: Soil conditioning is a key factor for an effective, economical and safe construction of tunnels using EPB shield machines in closed mode. Depending on the geotechnical conditions there are different requirements regarding the soil conditioning process to be considered. Main goals among others are enhancements of face pressure control and soil flow/transport. The most frequently applied conditioning agent in EPB tunnelling is foam. Foam consists of time- and pressure dependent properties, which makes the design of the conditioning process challenging. In this publication, the authors will provide fundamental knowledge on the relevant foam properties for EPB tunnelling and methods of characterisation. Besides investigations on the time- and pressure dependent behaviour of foam, significant influences from the production process on the foam and approaches on the rheology of foam will be presented. Therefrom, indications will be given for soil conditioning in practice.

1 INTRODUCTION

Tunnelling with EPB shields is typically conducted in soft ground with soft consistency (IC = 0.40–0.75) consisting of at least 30% of clays and silts (Figure 1). If the ground does not possess the required properties when excavated, the TBM drive undergoes significant impacts on the excavation performance and the TBM availability. Therefore, conditioning agents have to be used, which are being added into the excavation chamber in order to temporarily form a material which complies with the demanded properties.

In closed mode operation with active face support, the key properties of the excavated material are:

- Suitable consistency and flow behaviour under different pressure loads for
  - sufficient face pressure transfer,
  - low clogging potential,
  - effective material transportation through the excavation chamber,
  - and inexpensive material disposal,
- Low inner friction to avoid significant abrasion to the TBM
- Low hydraulic conductivity to avoid destabilizing forces from seepage flow at the tunnel face

In clayey and silty soils, usually water or foam are used to adjust the material properties above to the particular tunnelling situation. In coarse soils, like sands or gravels, foams, polymers, or slurries of fines are applied, in order to replace the missing fines content. The fines content steers the hydrologic budget of the ground decisively and thus the material consistency. The fields of application of EPB TBMs in soft ground are summarized in Figure 1.

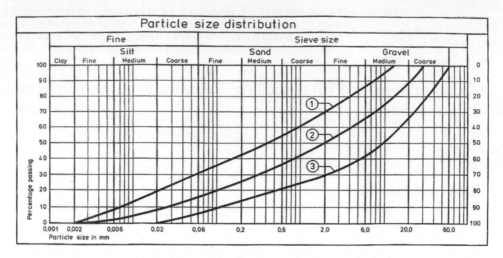

Figure 1. Application ranges for EPB shields according to Maidl et al. (2012). Area 1 with IC = 0.40 - 0.75, Area 2 with kf < 10^{-5} m/s and water pressure < 2 bar, Area 3 with kf < 10^{-4} m/s and without water pressure.

However, it is not only the amount of conditioning agent to be added which has to be evaluated carefully (Thewes, Budach & Galli 2010, Galli & Thewes 2014), but it is also the agent itself that has to be prepared cautiously. This is why in the subsequent chapters main requirements of the conditioning agents for soil conditioning are presented. It is followed by a short description of recommended tests, which can be easily conducted even on construction site. Since surfactant-based foams represent the most frequently and most common soil conditioning agent, the authors put emphasis on foams in this study. The paper closes with indications for the tunnelling practice.

2 REQUIREMENTS ON FOAMS FOR SOIL CONDITIONING

Foam is a three-phase medium composed of water, air and surfactant. Foam evolves from the interaction of air and water, when the interfacial surface is covered by a surfactant. Surfactants (= surface-active agents) reduce the interfacial tension significantly so that water and air can mix. They consist of chains of hydrocarbon molecules with one hydrophilic and one hydrophobic end. Each end adsorbs the respective phase.

The morphology and the stability of foams may be best described by the laws of Laplace and Plateau. The law of Laplace treats the pressure difference in interfaces between two phases. It is responsible for the bending of the interfaces. Thus, a stable equilibrium state is achieved expressed by the bubbly structure. Plateau established a geometrical construct defined by tetrahedral angles. The geometrical structure consists of lamellae (bubble membranes), three lamellae form a channel, and three channels end in a knot. The form of this system is influenced by the liquid fraction of the foam. When the foam is very dry, the structure is distorted (polyhedral foam), while a high liquid content forms a spherical foam structure.

However, foams possess a limited life-span only. The stability of the foam is time-dependent and of dynamic nature. Due to gravity and differences in pressure between lamellae and channels, the liquid fraction drains from the foam. When lamellae become too thin, they crack and lead to bubble bursts and structural redistributions (coarsening). The deformation processes can be measured by evaluating bubble sizes over time.

Furthermore, the structure of the foam is influenced by the surrounding pressure conditions. Due to the air content, foam is a compressible fluid. Therefore, the compressible fraction follows the rule of Boyle-Mariotte. Foams experience different pressure conditions from production to injection. While produced at high pressures and high velocities, foam is compressed again when entering the conveying system towards the injection point. It is expanded again when the pressure decreases over the conveying length and certain pressure impacts exist in bends or at cross-sectional changes. Finally, when injected into the soil, face pressure varies throughout the excavation chamber depending on the shield diameter. Hence, foams need a robust design to face the different pressure conditions with the least possible influence on the drainage and deformation behaviour.

The foam expansion ratio (FER) describes the expansion behaviour of a foaming liquid compared to its unexpanded volume, see eq. 1. The FER is a pressure-dependent factor, which needs to be defined with respect to the conditions at the location to be supplied to. Since the parameters for the production of foam are derived from the FER value, it has to be considered that in the excavation chamber or in a specific testing environment the pressure conditions differ from the conditions at the point of production.

$$FER = \frac{Q_{Foam}}{Q_{Liquid}} = \frac{Q_{Liquid} + Q_{Air} \cdot (1 + p_{supp})}{Q_{Liquid}} \quad (1)$$

with Q_{Foam} the flow rate of foam [l/min], Q_{Liquid} the flowrate of liquid (water + surfactant) [l/min], Q_{Air} the flow rate of air [l/min], and p_{supp} the supplementary pressure above atmospheric pressure [bar].

Typical FER values range from 5 to 20 depending on the geotechnical situation. As a rule to remember: the lower the FER the wetter the foam.

3 STATE OF THE-ART OF FOAMING TESTS FOR EPB TUNNELLING

Investigations on soil conditioning are being carried out all over the world. Since face pressure control and muck disposal are key factors in safety and costs (especially with increasing diameter) conditioning concepts become indispensable. With regard to the muck disposal, the chemical additives should be well chosen (Langmaack 2017). Recent research is summarized e. g. by Galli (2016). However, focus in these investigations is mainly put on the conditioning behaviour of the construction ground. The preparation guidelines of the foam itself usually come either from the manufacturer of the conditioning agent or the TBM manufacturer.

The European Federation for Specialist Construction Chemicals and Concrete Systems (EFNARC) developed guidelines for soft ground tunnelling (EFNARC 2003), which consist of recommendations for testing of foams. Besides others, it describes a test setup and procedure in order to determine the foam stability based on its half-life. Furthermore, some indicative values for soil conditioning in typical construction grounds can be found therein, too.

Maidl (1995) researched on soil conditioning with different types of foams. For an evaluation of the foam quality, he investigated the flow and foaming behaviour. Quebaud et al. (1998) looked into the influence of FER and liquid concentration on the foam consistency, the foam half-life and the compression behaviour of foams. It was pointed out, that the generation process is of great importance regarding the foam quality. Psomas (2001) investigated the bubble structure. Using microscopy, he determined the bubble size between 0.1 and 1.0 mm. Furthermore, he examined the drainage behaviour. Depending on the FER, the foam half-life span ranged between 15 and 25 minutes. Merrit (2004) used drainage tests on foams according the British Defense Standard 42–40 for fire extinguishing foams, which is similar to the EFNARC test.

Mooney et al. (2016) and Wu et al. (2018) focused on foam conditioning and foam properties under pressure. Therefore, they compared foam structure and foam stability both under atmospheric pressure and under supplementary pressure. It was found, that generation against counter pressures can have positive/retarding effects on foam degeneration.

All tests have in common, that finally comparative analyses were possible. However, a standardisation of test methods and procedures does not exist and makes global comparison of results without closer evaluation of the testing conditions impossible.

4 TESTING OF FOAMS

4.1 *Setup and procedure*

In the present study, testing was conducted using two real-scale foam guns from the tunnelling industry (FG1 and FG2) as well as a laboratory-scale foam gun (FG3) at the Institute for Tunnelling and Construction Management at Ruhr University Bochum (Figure 2).

Foam gun 1 (FG1) contains a sieve carrier installed perpendicular to the direction of flow by which air and liquid are vortexed. Foam gun 2 (FG2) consists of glass pearls between two sieves, through which liquid and air are forced to mix. Foam gun 3 (FG3) is a micro-version of FG2 for reduced flow rates, which can be even varied in its assembly (pipe length, pipe diameter, filling).

The whole foam unit provides variable production conditions for all foam guns. Besides flow rates and supply pressures (water, air), the conveying system can be altered in length and diameter (1" and 2"). The foaming liquid is premixed with the predefined concentration. Testing was carried out using the surfactant Condat CLB F4/TM.

A parametric study was carried out, altering only one production factor at a time. The variation parameters were: c_f (concentration of the foaming liquid), FER, flow rate, inlet pressure (air, water), foam gun, flowrate, conveying length, pipe diameter, counter pressure, and sample age. All foams produced were examined in order to evaluate the foam quality, the foam stability and the foam structure. Therefore, the following test methods were applied:

A. Foam quality (foaming behaviour, density measure / evaluation of FER)

Determining the density of a foam sample that was taken in a hermetical sealed container gives information on the production quality of the foam with the distinct parameter set and of the foaming ability of the surfactant. The actual FER enables a calibration to the target FER on the foam unit. This test can be easily conducted on the jobsite. Sampling has to take place immediately after foam production. After sampling, the time of measuring is irrelevant, since the sample should be sealed and its weight is only dependent of the liquid content.

The foam quality is a measure for the production quality. If the actual FER and target FER differ, the foam generation process needs to be revised. At Ruhr University Bochum for instance, deviations in actual FER are allowed within ± 1.0 of the target FER.

B. Foam stability (drainage tests)

The drainage behaviour of the liquid phase is the most important indicator of the degeneration effects taking place within the foam. Therefore, drainage tests were conducted according to the test proposed by EFNARC.

Figure 2.   Foam generation unit at Ruhr University Bochum (Budach & Thewes 2015).

A foam sample with 80 g ± 1 g is inserted into a glass suction filter with defined porosity of the filter plate (Figure 3). Testing was initiated immediately after sampling. 1 minute prior to sampling, the filter plate was moistened with water for 1 minute. The drainage time was evaluated each 10 g. The representative value is the sample half-life at 40 g passage of drained water.

The description by EFNARC does not define geometry and porosity of the filter funnel. In pre-tests, different geometries and filter porosities were investigated. Both parameters influence the test results significantly. The size of the contact zone between filter and foam sample steers the drainage velocity of the foaming liquid: the bigger the contact area, the faster the drainage process. The same applies for the filter porosity. These facts need to be considered in a comparative analysis. Wu et al. (2018) have drawn similar conclusions.

In order to evaluate the drainage behaviour under pressure, sampling was also conducted into a pressurized container (Figure 4). Considering the initial filling volume, the drainage

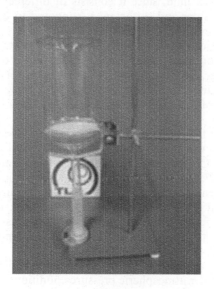

Figure 3. Drainage test for foams according to EFNARC (Galli 2009).

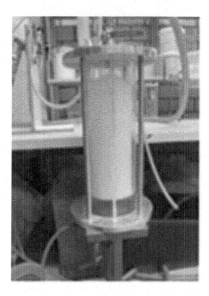

Figure 4. Pressurized drainage test for foams (Galli 2009).

behaviour (and the actual FER) could be determined over time. In addition, compression and relaxation test could be conducted. It provides information on the mechanical impacts during conveyance through the foam system.

C. Foam structure (evaluation of bubble-sizes)

In order to evaluate the degeneration of foam, the foam structure was observed over a period of 30 minutes. The samples were illuminated to enhance contrast for measurements. In this way, the lamellae in the foam structure can be recognized more clearly, which simplifies the measurement of the bubble diameters. Maximum bubble-sizes were recorded at predefined sample ages.

D. Flow behaviour of foams (rheology)

Once generated, the foam is transported through the conveying system towards the injection nozzle. On its way, the foam undergoes different pressure and flow conditions. Foam is characterised as a complex fluid, since it consists of different phases with different fluid behaviours. The mix of water and surfactant is supposed to behave Newtonian. The gaseous phase exhibits compressible non-Newtonian behaviour.

Rheological investigations shall provide information on the flow behaviour of foam. Usually, the flow behaviour of the pure foam (without soil) is of minor interest for the tunnelling industry. However, information on the foam flow is very relevant for understanding foam production and foam transportation as well as for design purposes of the conveying system. The complexity of foam flow and foam production is well highlighted by Mooney et al. (2017).

In order to determine first flow patterns of tunnelling foam, a rheometer (Anton Paar Rheolab QC) was used, which was equipped with a coaxial cylinders measuring system (CCMS; Figure 5). The CCMS used was of the Searle type and consisted of the cylinder "CC27" (diameter of 26.7 mm) and the corresponding cup (diameter 28.9 mm). Hence, the present gap width between the sample container and the measuring system was 1.1 mm.

The foam investigated possessed the following properties: FER=15, cf=3%. Additionally, a sample of synthetic foam (Gillette Shaving Foam) was examined, too. In this way, comparison with results from micro-scale experiments was possible, see Galli (2016) for details.

Testing was executed applying a logarithmic ramp profile with shear rate increases from 0.01 to 1,000 1/s under atmospheric pressure conditions. Therefrom, the flow behavior should be determined applying suitable flow models to the test results. The reference temperature was defined to 20±0.1°C using a Peltier cooling system for temperature regulation.

4.2 *Findings*

Detailed test results from the parametric study mentioned above can be found in Galli (2009), Thewes & Budach (2010), and Budach (2012). In the following, main findings are summarized.

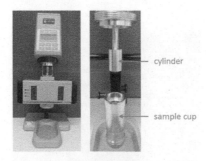

Figure 5. Rheometer Anton Paar Rheolab QC with Peltier cooling system; right: concentric cylinder system CC27 (Galli 2016).

A. Foam quality

The foam quality is very much dependent on the production unit and the conveying system. The different types of foam guns exhibit specific areas of application. FG1 necessitates high production rates in order to generate the desired foam density/FER. On the other hand, the pressure gradient over FG1 is not as high as over FG2. It facilitates applications even at very high production rates.

At very low injection rates, both industry foam guns lack in production quality. Vortexing velocity is not sufficient to mix the phases. Then, a greater conveying length or smaller pipe diameters improve foam quality. Wall friction and higher conveying velocities lead to further foaming effects.

Foam quality increases with increasing liquid concentration up to about 3%. Higher concentrations did not show a significant enhancement but lead to higher costs.

B. Foam stability

Foam stability increases with increasing concentration of the foaming liquid. However, the foaming product (surfactant) had a significant impact on all results. This is why it is recommended to evaluate the products prior to application with respect to the particular tunnelling conditions. Therefore, consultation of an expert is advised. Drainage times ranged between 5 and 10 minutes (FG1) and between 10 and 15 minutes (FG2) for foams with FER=15.

The target FER of course determines the half-life of the foam. The FER should be selected according to the excavation ground. Usually, in coarse grounds under the groundwater table a higher FER is chosen, while in cohesive or dry grounds a low FER should be considered.

The necessary production rate of foam is dependent on the advance speed of the TBM and ground volume being excavated. Therefrom, the flowrates of liquid and air can be calculated. The impact of flowrates on foam stability depends on the foam gun. Generally, the foam stability increases with increasing flowrate.

Testing under pressure showed, compression (and relaxation) of foam accelerates drainage of the liquid phase. The compression behaviour of the foam followed precisely the law of Boyle-Mariotte.

C. Foam structure

Generally, the relationship of foam structure and foam stability according to the model of Plateau could be verified in all tests. The foam structure differs depending on the type of foam gun used. FG1 produces foam with significant bigger bubbles than FG2. Bubble-sizes increase over time. According to the mentioned rules of Laplace and Plateau, this has to do with degeneration and drainage effects going on in the foam. The principle structure of the foam depends on the FER. A low FER leads to a more spherical shape of the bubbles, while a high FER leads to "honeycomb"-type foam. The higher the concentration of the foaming liquid the smaller the bubble-size and the more stable the structure over time. Smaller bubbles possess a higher energy level and thus greater binding forces.

Maximum bubble diameters ranged between 1.1 and 1.3 mm (FG1) and between 0.4 and 0.6 mm (FG2) after production.

D. Flow behaviour of foams

The results of the rheological investigations are depicted in Figure 6. The flow curves of the pure foams (tunnelling foam, shaving foam) show a similar flow behaviour, although at same shear rates pure shaving foam evokes higher shear stresses than the pure tunnelling foam. This difference certainly results from the basic constitutions of the foams. Compared to tunnelling foam, shaving foam consists of a considerably higher share of stabilisers. Cells and lamellae are also much smaller, which certainly retards destabilising effects.

Curve fitting was possible using the established non-Newtonian model functions, here: Herschel-Bulkley. The application of the Papanastasiou-Herschel-Bulkley model as proposed by Özarmut & Steeb (2015) did not lead to a significant melioration in fitting the flow curve data in comparison to Herschel-Bulkley. Curve fitting for the shaving foam resulted in a zero yield-stress condition, while for the tunnelling foam, an inclusion of a small yield stress value enhanced the data fit slightly.

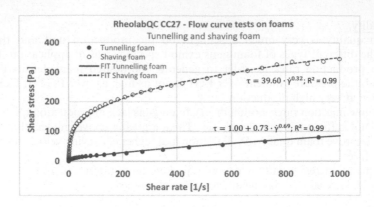

Figure 6. Flow curve data of shaving foam and tunnelling foam using a concentric cylinder configuration (Galli 2016).

5 EXPERIENCE FROM TESTING

5.1 Indications for future research

Based on the findings from the study presented above, some open questions remain as well as experience the authors would like to share.

- In the tests described before, no significant difference in foam stability and foam structure was detected between testing under atmospheric pressure and testing under pressure, when the drop in pressure between production and testing environment was similar. However, testing must be extended for verification. Wu et al. (2018) found out that producing foam against higher counter pressures can retard foam degeneration.
- The setup of FER on the foam generator must be different to the FER at the point of application. Otherwise the actual FER of the sample under different pressure conditions is not the same and a comparison in results not feasible.
- The temperature of the foaming liquid was controlled in all tests and maintained lower than 30°C. It appeared that with higher liquid temperatures, the foaming behaviour and the foam properties were affected. Further investigations on temperature-dependence did not take place, yet.
- The flow behaviour of tunnelling foam was determined to behave shear-thinning under atmospheric condition and could be well described by the flow model of Herschel-Bulkley. With respect to tunnelling practice, testing needs to be extended to different ambient pressure conditions in order to simulate realistic circumstance during transportation to the excavation chamber.

5.2 Scaling of foam production

Testing in the lab usually requires small amounts of conditioning agent. The preparation of such small amounts often differs from the actual generation in practice. With regard to foam conditioning, scaling of the foam production is difficult. Scaling here means generation of small sample volumes with the same foam quality and without affecting foam stability and structure.

In Bochum, the foam generator was designed to feature a foam gun (FG3) with smaller dimensions and variability in length, diameter and filling as well as different pipes with different diameters. Through alteration of these design parameters, foam production with about 1/6 of the volume could be achieved. This of course reduces the demand in resources and disposal significantly, too.

Overall, the results in foam quality, stability and structure were thereby similar to FG2. Variation of the design of the foam gun led to significant information:

- Drainage stability increases with increasing length of the foam gun. This can be explained by the longer stretch over which air and liquid are swirled.
- An increased length however, increases the pressure gradient, which limits at some point the applicability of the foam gun or the production rate.
- The same applies for smaller fillings to generate the turbulent flow within the foam gun.
- A bigger diameter requires higher volumetric flow rates in order to properly fill the cross-section of the foam gun.

6 INDICATIONS FOR THE TUNNELLING INDUSTRY

The results from parametric studies as presented here have to be evaluated carefully, when transferring results to practice. First, results from different laboratories cannot be compared so easily, since test methods and procedures might differ from each other. Second, the target objective of testing needs to be defined in advance. The aim of an investigation solely of the conditioning agent is to ensure a qualitatively proper and stable foam at the point of injection. Nonetheless, some advice is provided in the following to be considered in practice:

- Besides ground properties, production parameters, the design of the foam unit and the foaming product have a decisive influence on the soil conditioning process.
    - The foaming unit should be designed considering the conditioning concept. It steers the demand for maximum injection rates. Thus, pipe diameters and the number of foam guns and injection points can be optimized.
    - The surfactant to use during advance should be investigated prior and parallel to advance. Ground conditions can change unexpectedly and need therefore new experiences regarding the conditioning behaviour.
- The actual FER should be checked regularly. It is an indicator for the production unit's serviceability. The foam quality is a measure for the production quality. If the actual FER and the target FER differ too much, the foam generation process needs to be revised. Pollution and biodegradability of the surfactant reduce the foam quality significantly.
- The recommendations for correct storing of the surfactant have to be regarded.
- Depending on the parameter set for production, foams for soil conditioning in EPB tunnelling should possess a half-life time of at least 5 minutes to cover the time span for conveyance to the injection nozzle. Once injected, the material behaviour changes significantly in combination with the excavated ground. Through mixing and rotation in the excavation chamber, re-foaming occurs as well as is drainage impeded by the soil skeleton.
- Tests under pressure have been successfully executed on TBMs using pressure cylinders. It could be considered to install similar devices for routine checks.
- Apart from the conditioning process, the design of the conditioning concept should also consider the disposal of conditioned soil.
- In case the soil conditioning design is accompanied by laboratory investigations, the experimental results need to be analysed with respect to transferability to practice. One key factor is the scaling of foam production. If foam production in the laboratory is conducted with a small-scale foaming unit, the foaming behaviour should be compared to the foaming behaviour on the particular machine.

ACKNOWLEDGEMENTS

Financial support was provided by the German Science Foundation (DFG) in the framework of project A4 of the Collaborative Research Center SFB 837. This support is gratefully acknowledged.

REFERENCES

Budach, C. 2012. *Untersuchungen zum erweiterten Einsatz von Erddruckschilden in grobkörnigem Locker-gestein*. Aachen: Shaker.

Budach, C. & Thewes, M. 2015. Application ranges of EPB shields in coarse ground based on laboratory research. *Tunnelling and Underground Space Technology* 50: 296–304.

European Federation of National Associations Representing for Conrete (EFNARC). 2003. *Specification and Guidelines for the use of specialist products for Soft Ground Tunnelling*. Surrey: Farnham.

Galli, M. 2009. *Untersuchungen zur Nutzung von Tensidschäumen für die Konditionierung von kohäsionslo-sen Lockergesteinsböden bei EPB-Vortrieben unter realitätsnahen Randbedingungen*. Ruhr-Universität Bochum.

Galli, M. & Thewes, M. 2014. Investigations for the application of EPB shields in difficult grounds. *Geo-mechanics and Tunnelling* 7(1): 31–44.

Galli, M. 2016. *Rheological Characterisation of Earth-Pressure-Balance (EPB) Support Medium com-posed of non-cohesive Soils and Foam*. Aachen: Shaker.

Langmaack, L. 2017. How to protect our Nature – Chemistry in TBM tunnelling. From the laboratory to on-site use and muck disposal. *Proceedings World Tunnel Congress 2017, Bergen, Norway*

Maidl, U. 1995. *Erweiterung der Einsatzbereiche der Erddruckschilde durch Bodenkonditionierung mit Schaum*. Ruhr-Universität Bochum.

Maidl, B., Herrenknecht, M., Maidl, U. & Wehrmeyer, G. 2012. *Mechanised shield tunnelling*. Berlin: Ernst & Sohn.

Merrit, A. 2004. *Conditioning of clay soils for tunnelling machine screw conveyors*. University of Cambridge.

Mooney, M.A., Wu, Y., Mori, L., Bearce, R. & Cha, M. 2016. Earth pressure balance TBM soil condi-tioning: it's about the pressure. *Proceedings World Tunnel Congress, San Francisco, CA, April 22 –28*.

Mooney, M.A., Tilton, N., Parikh, D. & Wu, Y., 2017. EPB TBM foam generation. *Proceedings Rapid Excavation and Tunneling Conference, San Diego, CA, June 4 –7*.

Özarmut, A.Ö. & Steeb, H. 2015. Rheological Properties of Liquid and Particle Stabilized Foam. *Journal of Physics: Conference Series* 602:1–6.

Psomas, S. 2001. *Properties of foam/sand mixtures for tunnelling applications*. University of Oxford.

Quebaud, S., Sibai, M. & Henry, J.P. 1998. Use of chemical foam for improvements in drilling by earth-pressure balanced shields in granular soils. *Tunnelling and Underground Space Technology* 13(2): 173–180.

Thewes, M., Budach, C. & Galli, M. 2010. Laboratory Tests with various conditioned soils for tunnelling with Earth Pressure Balance shield machines. *Tunnel* 2010(6): 21–30.

Wu, Y., Mooney, M.A. & Cha, M. 2018. An experimental examination of foam stability under pressure for EPB TBM tunneling. *Tunnelling and Underground Space Technology* 77: 80–93.

Geotechnical impact on fiber reinforced segmental liner experiencing large deformation

Brierley Associates Corporation, Denver, CO, USA

ABSTRACT: A project involved the construction of 2845 mm internal diameter liner using 200 mm thick fiber reinforced segmental liner. The contractor successfully installed segmental liner rings in the very stiff to hard clay with slickensides. For the successfully installed rings the contractor was able to inject grout behind the segments without difficulties using a grouting pressure of 0.40 MPa. In the tunnel stretch within interglacial deposits, the contractor experienced significant difficulties trying to inject grout. According to the GBR, these deposits consist of interbedded layers of cohesive and granular materials. The granular material was described as medium dense to very dense sand. Shortly after installing the first 4 rings in the granular material, the rings experienced excessive deformation with excessive diametric deformation as high as 3%. The paper discusses the causes of that large deformations in relation to the ground condition and the superior response of the FRC segmental liner under these ground conditions.

1 INTRODUCTION

A CSO project in Seattle (WA) involved sliplining approximately 1,300 feet to replace the existing 36-inch ID wood stave siphons with 30-inch OD HDPE pipe, and construction of approximately 1,980 feet of EPB tunnel. The tunnel is comprised of 200 mm (8") thick and 2850 mm (112.2") ID segmental FRC liner and each ring is 1000 mm long. The tunnel originates from a launch shaft located on the south side of Salmon Bay and terminates at the retrieval shaft near the existing Ballard Regulator on the north side of Salmon Bay. The depth from the ground surface to the tunnel invert is approximately 130 feet at the tunnel portal and approximately 83 feet at the retrieval shaft.

2 THE GEOLOGICAL SETTING

The majority of soils in the project area were deposited during the most recent glacial advance in the Vashon period of the Fraser Glaciation, approximately 14,000 years ago, with a surficial ice layer thickness atop the soil that that was approximately 3,000 feet thick. The thick glacier preloading caused significant densification, and the Vashon till, Vashon advance outwash, and older geologic units are typically dense/very stiff to very dense/hard due to overconsolidation by the weight of the ice. Figure 1 shows the variation of the stratigraphy along the tunnel alignment.

The general geology of the project area consists of Holocene fill (Hf) which consists of loose, sandy, silty fine to coarse gravel and medium stiff, sandy clay with wood debris. The post-glacial sediments (PG) encountered in the borings consist of very loose sand with varying amounts of silt and clay to very soft to stiff clay. Vashon till (Ovt) consists of medium dense to very dense, clayey to silty fine sand with gravel. Vashon outwash deposits (Ova) were emplaced by meltwater streams flowing from the front of an advancing glacier. The Vashon

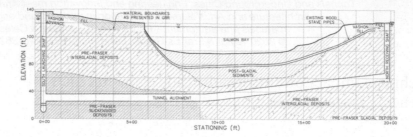

Figure 1. Stratigraphy along tunnel alignment.

outwash materials encountered in the borings consists of medium dense, silty fine to medium sand with fine gravel.

Pre-Fraser Glaciation age (Qpf) are broken out into three separate units:

Pre-Fraser Interglacial Deposits (Qpfi) consists of interbedded layers of cohesive and granular materials. The cohesive material is very stiff to hard silt and clay. The granular material is medium dense to very dense sand with varying amounts of silt and clay.

Pre-Fraser Slickensided Deposits (Qpfs) are geologically similar to the finer-grained portion of the interglacial deposits. The slickensided deposits encountered in the borings consist of very stiff to hard clay with slickensides and sand filled fractures. Slickensides are abundant but very small, less than one centimeter in width. Both low and high plasticity clays are contained within this deposit.

Pre-Fraser Glacial Deposits (Qpfg)

The Pre-Fraser glacial deposits encountered in the borings consist of very dense sand with clay and trace gravel to very dense, very clayey fine sand. These deposits do not contain slickensides. This unit was encountered below the proposed tunnel alignment.

The physical and engineering properties of the different soil units are presented in Table 1.

3   SEGMENTAL LINER DESIGN

The use of pre-cast segments as a tunnel liner facilitates the control of the construction phases and enhances the quality of the finished work. Often, the overall cost is lower than that of the conventional cast in place method. For this project, a 1000 mm long ring comprised of 6 gasketed segments was used to support the tunnel excavation. The fiber reinforced liner was 200-mm thick with fiber dosage of 65 lbs per cubic yard (0.49% by volume). The minimum unconfined compression strength of the liner concrete was 6000 psi. For the 2850 mm internal diameter tunnel, the loads presented in Table 2 were provided in the specifications and these loads include the action from both of the earth and water pressures on the liner. It is worth noting

Table 1. Physical and engineering properties of different soil units.

SOIL Deposit	Density (kN/m^3)	ϕ	C' (MPa)	K_o	E (MPa)	Poisson's ratio
Holocene fill (Hf)	18–20	32	0	0.47	41	0.3
Post-glacial sediments (PG)	17–18	25	0	0.58	55	0.3
Vashon till (Qvt)	20–145	40	0	0.36	69	0.3
Vashon outwash (Qva)	110–125	34	0	0.44	55	0.3
Pre-Fraser interglacial* (Qpfi)	17–135	35	0	1.1	38	0.4
Pre-Fraser slickensided (Qpfs)	17–20	24	2.1	1.5	38	0.4
Pre-Fraser glacial (Qpfg)	17–20	34	0	0.95	38	0.4

* Pre-fraser interglacial (Qpfs) has undrained shear strength of 13.8 MPa.

that the ratio of the lateral pressure to the vertical pressure is about 1.5, reflecting the impact of locked insitu horizontal stress due to glaciation. However, the lack of shallow bedrock beneath the Qpf's units and the initial introduction of an annulus around the liner, the actual K_o value was expected to be significantly less than the specified 1.5. The forces applied to the liner are presented in the Table 2 below:

Table 2. Pressure distribution around the tunnel.

Crown (vertical)	Springline (horizontal)	Invert (vertical)
0.71 Mpa	1.11 Mpa	1.09 Mpa

Successful liner design optimization considers the system flexibility based on the interaction of the liner and the surrounding media. The magnitudes of the thrusts and moments are dependent upon the stiffness of the lining relative to that of the surrounding medium (soil/ rock) under given loading condition. Flexible liners permit deformation that causes a nearly uniform pressure distribution resulting in minimal moments developing within the liner. Flexibility ratio is the parameter used to make this determination. A liner with flexibility ratio of 10 or greater is considered flexible. For circular conduit, the flexibility ratio is given by the following equation:

$$F = 0.25 \frac{E}{E_L} \frac{D^3}{t^3} \frac{(1 - \gamma_L^2)}{(1 + \gamma)}$$ (1)

Where,

F = flexibility ratio;
E = Elastic modulus of the ground;
E_L = Elastic modulus of the liner;
D = Mid diameter (measured to the centerline of the liner);
t = Liner Thickness;
v = Ground Poisson's ratio;
v_L = Liner Poisson's ratio

For the liner used in this project, the flexibility ratio was determined to range from 1.1 to 1.5 and considering the 6 joints along the perimeter of the liner, the flexibility ratio value was slightly higher but did not exceed 3. This indicates very rigid liner compared to the surrounding soil. Several closed form elastic solutions were used to confirm the FE modeling results. These closed form elastic solutions include the solution presented by Peck et al. (1972) who modified the original work by Burns & Richard (1969) to determine the forces acting on a tunnel liner. Another approach is based on the work of Muir Wood (1975) who introduced a similar approach to that of Peck et al. but was based on Airy stress function, and finally the work of Curtis (1976) which considers the shear stress at the interface between the liner and the medium.

The flexure values obtained from the closed form solution and the finite element ranged from 14.7 kN.m/m to 29 kN.m/m, while the thrust value ranged from 222 Kip/ft to 276 Kip/ft. The thrust-moment diagram is presented in Figure 2 with the values of thrust and moments obtained from both the elastic closed form solution and the finite elements are plotted.

The thrust moment capacity of the 200-mm concrete liner was determined by assuming linear strains across the liner. Related stresses were estimated across the liner with stress-strain relationships using the more accurate method of slices. The compression/tension stress-strain relationships were determined using methods cited in Barros (2004) that are based on beam tests, which take into account the post-cracking behavior of the steel fiber reinforced concrete. RILEM TC 162-TDF 2002 and 2003 papers were also used to understand the behavior of SFRC in flexure.

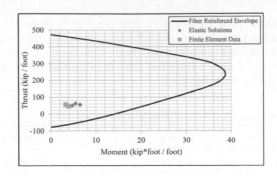

Figure 2.   Thrust-Moment envelope for 200 mm thick FRC liner.

4   RESPONSE OF LINER UNDER EXCESSIVE DEFORMATION

The diametric deformation, defined as the percentage of the maximum changed in diameter length to the original diameter, ranged from 0.08% using finite element analysis to about 0.18% using closed form elastic solution. These values were determined for the stretch of the tunnel situated in the Qpfi unit assuming a medium dense to dense formation if in cohesionless formations, and very stiff to hard in cohesive formations.

Peck (1969) summarized diametric deformation data that were collected from several projects around the world and presented in Table VII in one his papers that he presented in the 7th international soil mechanics and foundation engineering conference. The typical diametric deformation values ranged from less than 0.1% to 0.67%. Based on the stability of these liners, Peck recommended a diametric deformation value of 0.5% for design of concrete liners which he tied to a flexible response of the liner. The fiber reinforced concrete liner used in this project was classified as rigid liner and the calculated diametric deformation was significantly less than the 0.5% typically associated with liners that are more flexible.

The observed diametric deformation for this project ranged from about 0.1 to about 0.2% with the exception of few stretches where the diametric deformation approached 0.3%. After advancing the tunnel about 285 m away from the launching shaft, the excavation and advancing the ring took much shorter time compared to the rings installed closer to the launching shaft. However, the contractor was not able to inject grout to fill the annulus between the shield and the segments even at pressures as high as 0.47 MPa. At this location the tunnel was about 38 m below the Salmon Bay. The TBM operator continued to advance the tunnel and indicated the difficulty of keeping the face stable. After adding 4 more rings, excessive deformation and squatting for all 4 rings occurred. One of the rings experienced 3% of diametric deformation and eventually cracked, and at the time upon the initiation of the first crack the deformation ceased. Figure 3 below shows the ovaling and cracking of that ring. At this point, the contractor was able to come back and inject grout behind the deformed rings.

Figure 3.   A photo showing one of the excessively deformed rings.

The excessive ovaling of the rings and the initial inability to inject grout are all good indications that the tunnel at this stretch was not within the medium dense to dense Qpfi unit but rather in the more recent post glacial deposits which are described in the GBR as very loose sand or very soft clay to stiff clay. The presence of seashells discovered from random soil samples taken from spoil near the deformed rings further confirmed that the tunnel was likely within the PG unit.

In fact, the contractor was able to push a standard 25 mm probe rod a distance of 750 mm with minimal resistance, which indicated that the tunnel was not within the medium dense to dense Qfpi unit.

The inability to initially inject grout, but the ability to do so later during the tunnel advancement is another indication of the loose nature of the surrounding soil. The only explanation for this important observation is that when the highly permeable and loose sand was exposed to the shear forces induced by mining at the face, the soil skeleton collapsed and caused a temporary spike in porewater pressure. The additional pore water pressure increment triggered further reduction in the shear strength due to the reduction of the effective stress. The reduction in effective stress negatively affected the lateral passive resistance from the soil medium, thus allowing the ring to squat. The additional porewater pressure increment dissipated in relatively short time due to the porous and cohesionless nature of the loose sand, allowing for easier grouting with time passage. If the material is dense to very dense as indicated in the GBR, the dense to very dense sand would have dilated upon exposure to shear forces. The dilation causes the pore water pressure to drop below static value or to remain unchanged depending on the degree of dilation which is dependent on the density of the sand. In the latter case, the Contractor would have been able to grout immediately after excavation, with grouting likely becoming a little more difficult with the passage of time.

The fiber reinforced segmental liner endured significant amount of deformation with one of the rings cracking. Paul et al. (1983) carried out an extensive testing program in which large scale monolithic and segmental liners, with flexibility ratios greater than 100, were loaded up to failure. Paul et al. showed that segmental liner that were reinforced using smooth steel bars, endured larger deformation than regularly reinforced monolithic liners with a maximum diametric deformation value of 1.86%. The initial cracking in the segmental liner occurred at 90% of the peak capacity of the liner while the initial cracking in the monolithic liner occurred at about 32 to 42% of the peak capacity. The observation in this project showed that the fiber reinforced segmental liners endured much larger diametric deformation of 3%.

The liner used in the project was rigid with a calculated flexibility ratio of less than two to three, yet it endured 3% deformation without cracking except at some locations at dowel inserts. This demonstrates the superior ductile ability of steel fiber reinforced segment to adsorb distortion energy without cracking.

For the ring that was cracked, it was clear that flexure cracking was arrested. This can be further explained be examining Figure 4 below. The figure shows the difference between linear and nonlinear analyses for unreinforced lining section. In the figure, the moment-thrust paths are plotted for two different conditions of flexibility ratio.

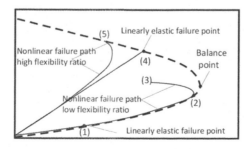

Figure 4.   Moment-Thrust paths for an unreinforced concrete lining.

The nonlinear and linear paths, which intersect the interaction diagram below the balance point, pertain to a flexibility ratio less than that for the paths that intersect the same interaction diagram above the balance point.

When linear analyses are performed, the material stress-strain response must follow a linear relationship which yields a triangular stress distribution on the liner section, thus yielding a liner capacity less than actual capacity of the liner. For non-linear analysis of cracked section, the cracking causes the eccentricity to decrease with the neutral axis of the section shifting away from the crack. The neutral axis shift results in a higher value of thrust and essentially increasing the axial capacity of the section due to the non-linear stress distribution, while the moment will drop off as shown by point 3 on the plot. Above the balance point, the thrust capacity calculated by nonlinear analysis will be closer to that calculated by linear analysis, as evidenced by comparing the percentage difference between points 4 and 5 with that of points 1 and 3.

The nonlinear analysis demonstrates that a concrete tunnel lining does not fail by excessive moment, but instead will fail by excessive compressive thrust, which is affected only indirectly by moment.

5 CONCLUSION

Encountering an unexpected loose sand caused significant difficulties at one stretch of the tunnel within what is supposed to be medium dense to dense PreFraser interglacial deposits. These difficulties included the inability to inject grout, even at pressures, as high as 0.47 MPa, to fill the annulus behind the installed rings and the excessive deformation of 4 rings with one ring experiencing 3% of diametric deformation. It was proven that the difficulties were caused by collapsible structure of loose sand. The spike in porewater pressure caused grout injection difficulties and the inability to fill the annulus between the ring and the soil medium. The loss of lateral passive support caused significant ring deformation.

The fiber reinforced segmental liner endured significant amount of deformation with only one of the ring cracking at dowel insertion location. Generally, regularly reinforce flexible segmental liner endures larger deformation than regularly reinforced flexible monolithic liner with a reported maximum diametric deformation of the cracked segmental liner section of 1.86% at peak strength. The fiber reinforced segmental liner used in the project was rigid with a calculated flexibility ratio less than 3, yet it endured 3% deformation without cracking except at some locations where dowel inserts existed. This proves the superior ability of steel fiber to adsorb distortion energy, thus increasing safety during tunneling even under extreme conditions.

The nonlinear analysis proves that a concrete tunnel lining does not fail by excessive moment but will fail by excessive compressive thrust which is affected only indirectly by moment.

REFERENCES

Burns, J. Q., & Richard, R. M. (1964). Attenuation of Stresses for Buried Cylinders. Proceedings Symposium on Soil-Structure Interaction: 378–392. Tucson, AZ.

Cruz, J. M. S., & Barros, J. A. O. (2004). Bond Between Near-Surface Mounted Carbon-Fiber-Reinforced Polymer Laminate Strips and Concrete. Journal of composite for Construction, 8(6): 519–527.

Curtis, D.J. (1976). Discussion on the Circular Tunnel in Elastic Ground, Geotechnique, 26(1): 231–237.

Muir Wood, A. M. (1975). The Circular Tunnel in Elastic Ground. Geotechnique, 25(1): 115–127.

Paul, S. L., Hendron, A. J., Cording, E. J., Sgouros, G. E., & Saha, P. K. (1983). Design Recommendation for Concrete Tunnel Linings – Volume 2. US Department of Transportation – Urban Mass Transportation Administrations.

Peck, R. B., Hendron, A. J., & Mohraz, B. (1972). State of the Art of Soft-Ground Tunneling. *International Proceedings of the North American Rapid Excavation and Tunneling Conference* 1(1): 259–286. Chicago, IL.

Peck, R. B. (1969). Deep Excavations and Tunneling in Soft Ground. Proceedings 7th International Conference on Soil Mechanics – State-of-the-art volume: 225–290. Mexico City.

Tunnels and Underground Cities: Engineering and Innovation meet Archaeology,
Architecture and Art, Volume 5: Innovation in underground engineering,
materials and equipment - Part 1 – Peila, Viggiani & Celestino (Eds)
© 2020 Taylor & Francis Group, London, ISBN 978-0-367-46870-5

Excavating through hard rock formations with TBM: Technical challenges of the Follo Line Project

G. Giacomin, A. Mariotti & M. Ortu
Ghella S.p.A., Rome, Italy

ABSTRACT: The Follo Line Project is Norway's largest infrastructure project and Scandinavia's longest railway tunnel. Excavation is done by four Double Shield Tunnel Boring Machines (DS TBMs) all with a diameter of 9.96 m. It is the first tunnel in Norway built with a precast segment lining. Excavation in the Norwegian extreme hard rock conditions with ground water has defined the main challenges of the Project. The experience gained so far, at a progress of 83% on the TBM drives, has been extremely positive. TBM penetration, cutter wear and TBM advance rates are well within the expectations. The management of ground water has been more of a challenge than expected. To control water inflow, pre-excavation grouting has been successfully performed although compromising the TBM advance rate but with satisfying results.

1 INTRODUCTION

The Follo Line Project is currently Norway's largest infrastructure project and Scandinavia's longest railway tunnel. It consists of a 22 km new double track railway line between Oslo Central Station and Ski, south of Oslo, where a new station is built. Of the total length, 18km are excavated by 4 Double Shield TBMs, for overall 36km of single track tunnels. This largest portion of the project is part of the EPC TBM contract which Bane NOR, the Norwegian National Railway Administration, has awarded to the Acciona-Ghella Joint Venture (AGJV). The scope includes the design and construction of the tunneling works, the slab track, and the installation and commissioning of the railway systems.

The main challenges of the Project are the impact on excavation due to the extreme Norwegian hard rock conditions; the logistic required to operate 4 TBMs from one same access point; the ground water management while boring through compact rock with interconnected horizontal and vertical fracture systems and weakness zones.

2 SITE LOGISTICS

All supplies to the 4 TBMs are managed from a single access point, in the middle of the tunnel alignment, where the entire site installation is located. The location is well served with direct access from the motorway. The site includes the precast plant, tunneling logistics base and spoil management facilities as well as the site management offices, the camp and canteen facilities to host 480 tunnel and precast workers coming from all over Europe (see Figure 1). This set-up guarantees high logistical efficiency with minimum impact on the local community at the same time.

A total of 40.000 m^2 of shed structures have been installed to cover precast factories, workshops, warehouses, water treatment and filter press plants, spoil discharge and handling facilities, crushing plants, aggregate distribution facilities, concrete and grout batching plants.

Figure 1. Aerial photo of the site set-up.

A total of 300.000 m^3 of rock has been excavated with the drill and blastto obtain the access tunnels, logistical tunnels and the TBM assembly caverns. The 4 TBMs have started excavation each one month apart from the other, two towards Ski and two towards Oslo from the same central area which also serves as the unique access point for all the coordination.

Road based transports for the material supply to the TBMs have been reduced to a minimum. Mucking is transferred to conveyor systems: the material excavated from each TBM is collected onto four different conveyors then shifted onto two independent conveyor lines which move the muck to the surface. Once outside, the material is discharged inside a closed, noise and dust proof shed structure then loaded onto dumpers and towards the final deposit.

Grout for the annular gap backfill is transported then pumped via pipelines directly from the grout plant to the TBM backups.

Custom made multi service vehicles (MSVs), composed by four synchronized units, with a length of 45 m and a load capacity of 125 T, have been procured. They each transport up to 2 precast rings at a time. To cope with the adit tunnels inclination of 10%, special attention has been paid to the breaking systems provided on the MSVs.

Segment handling on surface is managed with 3 gantry cranes with 40T capacity and 30m span.

A tailormade conveyor system has been installed for the automatic distribution of aggregates to the 3 concrete batch plants.

3   PRECAST SEGMENTS

The single-shell, watertight precast concrete segmental lining consists of 6 trapezoidal and rhombic segments plus one key stone. The segment ring has an internal diameter of 8.75 m, a thickness of 40 cm and a length of 1.8 m (see Figure 2).

To assist in a quick ring erection with minimum lips and steps the segment design includes shear connectors in the transversal joint and guiding rods in the longitudinal joint.

Figure 2. Precast segments on the Follo Line.

The waterproof EPDM gasket has been engineered and tested for a total pressure of 33 bar to be able to withstand the design water load over the required 100-year design life. AGJV has chosen to utilize an anchored gasket solution and is so far very satisfied with the result.

The segment reinforcement is with conventional steel bars combined with steel fibers. PP fibers are added to improve the spalling behavior in case of a tunnel fire.

The elements are produced inside three independent precast factories which have been built within the construction area, each equipped with: a carousel plants for 6 sets of molds, including the invert segment; a dedicated concrete batch plant; a curing chamber and a covered evacuation line for 36 hours of controlled curing following the demolding activities.

4 BACKFILL GROUTING

The watertight single-shell precast lining has originally been developed for tunnels in soft ground excavated with pressurized shield TBMs and with lower levels of ground water pressure. Its application in the open Double Shield TBMs with high overburden and water pressure sets special requirements for the backfill grouting. A quick stabilization of the grout after injection needs to be achieved to provide the erected ring with the necessary support to avoid deformation. At the same time, the grout needs to be sufficiently fluent to guarantee a complete and void-less filling of the annular gap. This combination can be achieved with a two-component grout. Cement, bentonite, retarder, and water are mixed as the A-component while sodium and silicate form the B-component. The two components remain separate and combine only at the injection nozzle.

Special testing requirements focus on the bleeding time, the gel time and the strength development.

The backfill grout is injected in two stages. Primary grouting is performed via the tailskin injection lines during the regrip of the TBM gripper shield with the aim to fill the lower 85% of the gap. Secondary grouting of the remaining 15% of the section in the tunnel crown is performed through the segments via cast-in grout ports (see Figure 3).

Figure 3.   Primary and secondary grouting.

Complete filling of the annular gap is systematically controlled by volume and pressure monitoring and by sporadic scanning with non-destructive impact echo equipment.

5   TBM TUNNELING

5.1   *TBM comeback in Norway*

Norway is generally considered to provide one of the toughest hard rock challenges in the world. Tunnels, mainly related to hydro power development works, have been bored with TBMs in Norway in the 70s and 80s with Main Beam Gripper TBMs mostly within a diameter range of 3 to 4 m.

The end of the hydro power construction boom in the early 90s, low infrastructure development in the cities in the following decades, the proven efficiency of conventional tunneling in Norway and the challenges experienced with TBM boring through the extreme hard rock conditions have caused infrastructure growth to change strategy, and completely excluded TBMs. Tunnels have since exclusively been excavated with conventional drill and blast methods.

The permanent lining of Norwegian traffic tunnels generally consists of permanent rock support in form of rock bolts and only locally applied shotcrete, thanks to the typically excellent geological conditions. The installation of a watertight membrane supported by concrete lining is rarely found. The required water tightness is mainly achieved through pre-grouting of the fissures in the rock mass in front of the tunnel face. This tunneling practice is the main element characterizing what is referred to as the Norwegian tunneling method.

Examples of tunnels built in the past have shown that in Norwegian rock conditions the groundwater level can be irrecoverably lowered: lakes have disappeared and surface settlements produced following tunneling activities where pre-grouting was not applied successfully.

In 2014, after over 20 years of absence, a Main Beam Gripper TBM had been picked to bore the Røssaga headrace tunnel. The tough hard rock conditions led to a main bearing damage suffered along the TBM drive but the tunnel works were nevertheless completed with success.

Within two years, another Main Beam Gripper TBM with conventional rock support was utilized for the Ulriken Railway Tunnel in Bergen.

On the 3rd of September 2016, the first of four Double Shield TBMs of the Follo Line project in Oslo started the excavation. The other three have followed with a monthly gap between one another so that the fourth TBM was launched just before Santa Barbara. For the first time in Norway, a tunnel is built with precast segment lining.

5.2 *Geological and hydrogeological conditions along the tunnel alignment*

The average overburden of the tunnel has been approximately 80 m, with variations between 20 to 170 m. The rocks in the area consist predominantly of Precambrian gneisses with banding and lenses of amphibolite and pegmatite. In addition, several generations of intrusions have occurred. Generally, the rock mass has been quite homogenous and competent with moderate jointing. Two main joint sets, one oriented NNW-SSE and one oriented E-W, have been consistent along the tunnel alignment.

Laboratory tests have shown that the rock mass is tough hard and abrasive. The expected UCS varied in a range between 100 MPa and 300 MPa.

Various groups of sub-vertical fractured or weakness zones, oriented in semi-parallel to the tunnel alignment with a thickness between 1 and 5 m, seldom more than 10 m, have intersected the tunnel.

The maximum ground water level is close to the surface, which means about 160 m above the tunnel level in some areas. Ground water flow in the rock mass has mostly been restricted to fissures and weakness zones. The hydraulic conductivity has been low. This has not applied to the weakness zones which were predicted to act as "drainage channels".

Sub-horizontal fractures and weakness zones have been encountered an all four tunnel drives. Such geological features have increased significantly the hydraulic conductivity in the area.

5.3 *The TBMs*

AGJV has procured the four Ø 9.96 m Double Shield TBMs from Herrenknecht (see Figure 4).

Figure 4.   Follo Line TBM.

with particular attention to their capacity to face the Norwegian hard rock constantly over the 9 km of excavation. The TBM shield and cutterhead have thus been designed with a weight of 265T resultig on the very heavy range compared to other hard rock TBMs of this size.

The main bearing diameter of 6.6 m and the ratio of 0.66 compared to the TBM diameter is higher than on any other hard rock TBM of this size, built by HK. A larger main bearing reduces the impact of eccentrical cutter head loads on the main bearing life time. The cutter-head is fitted with 71 x 19" cutters rated 315 kN.

The TBM is further equipped with sealing systems for telescopic shield, gripper windows and muck ring which can be activated in case of emergency to stop uncontrollable water ingress and prevent environmental and structural damages.

5.4 *Daily probe drilling and mapping*

The contract requires close geological monitoring to be performed from the TBMs. This includes probe drilling, face mapping, televiewing, 3D photographing and core drillings which AGJV performs systematically on a daily basis during the maintenance shift.

The probe holes are normally drilled up to around 40 m in length with the help of two drill rigs which are installed on the erector ring of the TBM (see Figure 5). Depending on the actual geological situation and taking into consideration the sensitiveness of the area, usually between 1 to 3 probe holes are drilled. Probing can be done in 38 fixed positions evenly distributed along the circumference of the front shield.

Systematic televiewing is performed on at least one probe hole. The instrument for televiewing is equipped with compass and gyro to keep track of the borehole orientation. The high resolution televiewer picture is used to determine fracture sets, spacing and fracture classes (see Figure 6).

Geological face mapping serves to observe rock types, fracture plane roughness, infilling or aperture and confirm the televiewer results at the tunnel face. 3D photographing of the tunnel face is performed with an autonomous imaging unit installed on the inspection hole of the TBM

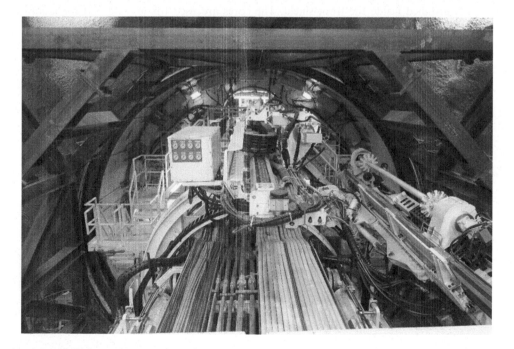

Figure 5. Drill rigs of the Follo Line TBMs.

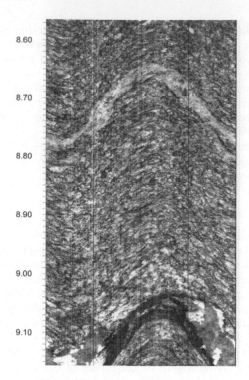

Figure 6.   Extract of televiewer picture.

cutterhead which is turned while the device captures a video. Advanced software generates scaled and oriented 3D images from measurements taken (3GSM) from which it is possible to identify and measure overbreaks (see Figure 7). Core drilling at the face is performed with 2 drill units installed inside the front shield. The cores are used for logging and for laboratory testing to determine DRI, CLI and mineral content.

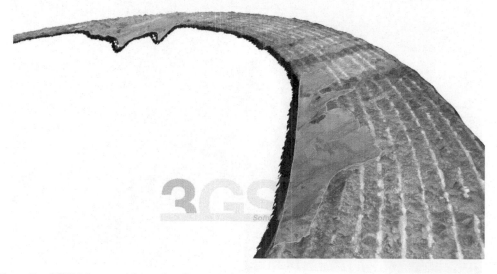

Figure 7.   3GSM image with overbreak measurement.

5.5 Compensation on basis of the NTNU prediction model

The contract foresees additional compensation when the TBM advance rate and cutter wear is negatively affected by the encountered geological conditions. The assessment is made on NTNU prediction model basis, which is used for the evaluation of the geological data and the calculation of the theoretical TBM advance rate and cutter wear, which are then compared to the contractual baseline.

5.6 Actual penetration, cutter wear and advance rates

The TBMs have so far excavated 95% of the total tunnel length (status end of November 2018). During this period, the TBMs have been excavating with an average penetration of 6 to 7 mm/rev.

The average cutter lifetime has so far settled on an overall average of 220–250 m^3/cutter. Mixed face in hard rock conditions, resulting from the unexpected breaking mechanism of the amphibolite veins combined with hard and compact gneiss, encountered along 40% of the tunnel drives, led to significant reduction of the cutter bearing life time and increased cutter refurbishment cost.

The TBMs operate 6 days per week. So far they have achieved a best 4-week production of 583 m on the northern drives and 518 m on the southern drives, with a best daily production of 31 m achieved on both driving directions.

Best weekly productions are 155 m in the north and 145 m in the south.

Average weekly productivities have to date settled on 90.0 m/week along the northern drives and 78 m/week along the southern drives. The TBM drives have been affected by water ingress which have slown down the production, as further explained under section 5.7.

5.7 Ground water management

Draining of the ground water can lead to the lowering of the ground water table and loss of pore pressure in clay filled zones on the surface. As a result, there might therefore be environmental damages and surface settlements, producing damages to the structures in the affected area. Such situations need to be avoided.

For the Follo Line TBM tunnels the following elements are applied in the overall groundwater management:

• installation of a water tight precast concrete segment lining,
• backfill grouting with two-component cementitious grout,
• probe drilling and monitoring of water ingress to in the TBMs,
• pre-excavation grouting,
• operation of infiltration wells.

The permanent water tightness of the tunnel is guaranteed by the segmental lining. However, the success in achieving a leakage free segmental lining in hard rock with high overburden and ground water pressure depends on the hydraulic conductivity of the rock mass, the accuracy of the ring building, the quality of the gasket sealing, the efficiency of the backfill grouting and the sealing of the grout ports used for secondary grouting. Any problem, imperfection or damage which results in a leakage need to be treated with additional repair injections etc., which can delay the process in achieving a complete watertight segmental lining.

In Double Shield TBM tunneling, the watertight lining is placed approximately 20 m far from the tunnel face. Furthermore, significant ground water flow is mainly expected during the excavation of weakness zones with limited extension. Ground water inflow to the TBMs and the subsequent impact on ground water levels and pore pressure should either be negligible or of a duration of 1–2 days only.

To control the amount of water inflow, especially in weakness zones, pre-excavation grouting need to be performed where necessary. Triggers are the measured inflow of water in probe holes and in the TBM tunnel face and shield area.

The preparation of a pre-grouting round requires a minimum of 20 grout holes to be drilled ahead of the tunnel face with approximately 40m length. After the installation of hydraulic packers, the bore holes are grouted over the full length with cementitious grout at grouting pressures between 40 and 60 bar. AGJV has made good experience with the use of microfine cements which provides a quick setting time.

Aim of this treatment is to reduce the water inflow to a sustainable level during the time until the segmental lining is in place. It is not required to completely seal off the surrounding rock, how it is usually done in Norwegian D&B tunnelling. Such an approach requires a much tighter drill pattern than it can be achieved on the TBMs with reasonable effort.

Pre-excavation grouting is an efficient method from a technical point of view. It provides sealing of the fissures and fractures in the rock and a significant reduction of the water flow. The negative aspect is the significant impact on the TBM advance rate. The contract therefore foresees additional compensation for grouted TBM tunnel sections once the contractual baseline quantities have been exceeded.

To date, pre-excavation grouting has been performed on 17% of the tunnel length, 4 times more than initially expected.

While tunnelling through sub-horizontal weakness zones in areas where pre-excavation grouting has not been performed, water inflow has been continuous and values up to 30 l/s have been registered. Due to the increased hydraulic conductivity in the area piezometers have been affected at more than 1 km distance.

At such flow ratios, the two-component grout risks to be washed out towards the open TBM shield. In such cases, additional treatment needs to be performed, for example by controlled injection of two-component polyurethane into the annular gap in order to cut-off the water flow and enable the backfill grouting to be executed properly.

To manage the pore pressure in such situations, infiltration wells are installed at selected locations and operated during the passing of the TBMs. Such infiltration wells inject water into the transition zone between rock and soil in the area of fault zones in order to compensate the temporary loss of ground water and maintain the pore pressure in the soil fill.

6 CONCLUSION

The Follo Line Project is a challenging and exciting large scale TBM project, which has introduced Hard Rock TBM Shield Tunnelling with precast segmental lining to the Norwegian environment. The future of the TBM technology in Norway will mainly depend on whether the 4 Follo Line TBMs will successfully complete tunnelling on time.

The experience achieved with excavating through the tough hard rock conditions is satisfactory. TBM penetration, cutter wear and TBM advance rates achieved so far are well within expectations in the encountered conditions and highlight the advantages that mechanized tunneling offers over conventional methods, especially for such long tunnels.

Ground water management is more complicated than expected. The situation can be well managed with the pre-grouting approach, but TBM advance rates are affected significantly.

REFERENCES

Bruland, Amund. 1998. Hard Rock Tunnel Boring. Doctoral Thesis. Trondheim University of Science and Technology, 1998:81.

Grasbakken, Elisabeth et. al. 2017. Geological monitoring of Follo Line Project in Norway. Proceedings of the World Tunnel Congress 2017 in Bergen.

Kalager, Anne Kathrine. 2016. The Follo Line Project – A large scale project that includes a complex excavation of the longest railway tunnel in Norway. Proceedings of the World Tunnel Congress 2016 in Iguazu.

Norwegian Tunnelling Society. The principles of Norwegian Tunnelling. Publication No. 26.

Tunnels and Underground Cities: Engineering and Innovation meet Archaeology,
Architecture and Art, Volume 5: Innovation in underground engineering,
materials and equipment - Part 1 – Peila, Viggiani & Celestino (Eds)
© 2020 Taylor & Francis Group, London, ISBN 978-0-367-46870-5

Fiber reinforced concrete segmental lining: Evolution and technical viability assessment

J. Gollegger
Technical Manager EPC TBM, Follo Line Project, Bane Nor, Norway

L.M. Pinillos Lorenzana
Mott MacDonald, UK

S. Cavalaro
Reader in Infrastructure Systems, Loughborough University, Leicestershire, UK

T. Kanstad
Professor, Department of Structural Engineering, NTNU, Trondheim, Norway

ABSTRACT: This paper focuses on the assessment of the technical viability of fiber reinforced concrete (FRC) with structural responsibility in tunnel segmental linings. An analysis of historical information about tunnels built with FRC shows the typical geometry and reinforcement strategy used. An extensive parametric study with several tunnel geometries and boundary conditions was conducted using two different design approaches: beam model and numerical continuum model. The results show the differences between these approaches and support the proposal of a graphic representation of normal and bending actions, delimiting zones that favor the application of fibers either in partial or full substitution of traditional reinforcement. The graphic representation allows a rapid assessment of the viability of using FRC in segmental lining with positive repercussion on the future application of the material and the confidence of designers, contractors and project owners.

1 INTRODUCTION

In the last decades, fiber reinforced concrete (FRC) has risen as an alternative to conventional reinforced concrete in segmental lining, bringing simplifications to the construction process and savings. Despite the abundant literature and design guidelines available, is in some countries this alternative still hindered by doubts regarding the structural contribution of FRC and the lack of knowledge about the long history of successful applications.

The objective of this paper is to provide a graphical assessment of the technical viability of using FRC with structural responsibility in segmental lining depending on the load combination and interaction with the ground. An extensive literature review was conducted to compile historical information that was analyzed to identify patterns in terms of typical geometry and reinforcement strategy in tunnels with FRC. This information and the recent trends in design of FRC tunnel segments were combined in an extensive parametric study using two different design approaches. One is based on beam models and the other one on numerical continuum models. Possible differences of resulting internal forces are discussed.

The analysis of the results underpins the proposal of an indicative graphic that relates non-dimensional normal and bending actions, delimiting zones that would favor the application of fibers either in partial or full substitution of traditional reinforcement. The paper provides a

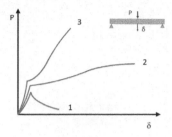
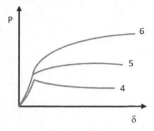

Figure 1.   Illustration of design principles related to ductility, (a) ordinary reinforced concrete (left), (b) fiber reinforced concrete (right).

reference for the decision-making process on whether or not the application of FRC is feasible.

2   PRINCIPLE OF REINFORCED CONCRETE AND FRC DESIGN

In many projects during startup phase, and especially when fiber concrete is considered, the design principles are heavily discussed.

Failure in ordinary reinforced concrete is commonly grouped into the three following cases, as seen in Figure 1a:

1. Brittle failure occurring in ordinary reinforced concrete structures when the minimum reinforcement requirements are not fulfilled. This situation should be avoided in load-carrying structures and in all cases where large crack-widths must be avoided.
2. Ordinary reinforced cross section where the reinforcement yields before the concrete crushes. After the first cracks occur the loads may increase, and a distributed crack pattern occurs. This is the preferred situation used in a large majority of structures since we achieve both large ductility (ULS) and limited crack widths (SLS).
3. Heavily reinforced section where the concrete crushes before the reinforcement yields. This situation is unwanted due to the brittle nature of the failure and can be avoided by use of compressive reinforcement.

For fiber reinforced concrete, a set of cases may be considered as illustrated in Figure 1b:

4. Lightly reinforced fiber reinforced cross sections with bending softening behavior. This is typical for structures with steel fiber contents less than about 45 kg/m3 (dependent on the fiber properties). In structures without static redundancies, brittle failures will occur.
5. Concrete with larger amounts of fibers (>45 kg/m3) bending hardening behavior may occur. The failures may be considered as ductile, but the ductility is still considerably lower than for sections with ordinary reinforcement fulfilling the minimum requirements.
6. Combined reinforced section with good ductility. This solution can be used in all sorts of load carrying structures, and all requirements in SLS and ULS may be fulfilled.

The principles are illustrated in Figure 1 for simple bending of beams and one-way slabs, but the discussion is still relevant and can be extended to other failure types as those resulting from combined bending and axial forces, and the various types of shear failure, which are typical load/stress conditions for tunnel linings.

3   HISTORICAL OVERVIEW

A literature review was conducted to compile cases of tunnels constructed or under construction with TBMs and FRC segmental linings in which fibers are added for structural purposes (not exclusively to increase fire resistance). Figure 2a shows the accumulated number of

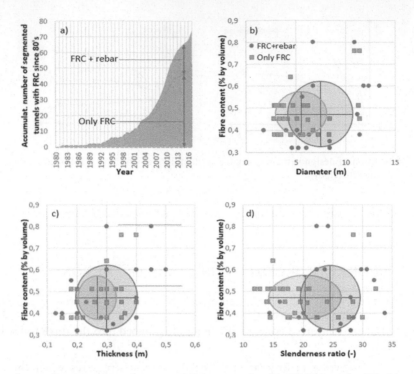

Figure 2. Accumulated number of segmental lining tunnels with FRC since the 80's (a) and relation between fiber content and inner diameter (b), thickness (c) and dimeter/thickness ratio (d) for 57 segmented tunnels built with FRC.

tunnels reported since the 80's in the literature that meet these criteria. Notice that a significant increase in the number of tunnels with FRC was observed in the last decade despite unfavourable economic factors. Such increase is mainly attributed to the increase of know-how enshrined in the publication of codes and guidelines about the structural design of FRC elements.

A discontinuous line indicates the proportion of tunnels constructed only with fiber and with a combination of FRC and rebar. From the total of 74 tunnels analysed, 22 have a combination of fiber and traditional reinforcement (30%) and 52 were reinforced only with fibers (70%). The significant number of tunnels with FRC reveal the high acceptance of the material and the confidence in its structural contribution considering the typical loading conditions found in projects. Steel fibers predominate in the majority of tunnel with plastic fiber being reported in only 3% of the tunnels.

An analysis was conducted to assess the relation between the fiber content and geometrical parameters of each tunnel (lining inner diameter, segment thickness and slenderness ratio). The analysis was conducted separately for cases only with FRC or with a combination of FRC and rebar. Results are presented in Figures 2b, 2c and 2d. Notice that each point in the graph represents a project. An area with the same colour as the points indicates the typical interval of values obtained for a range equal to ±1 standard deviation. The centroid of the area represents the average values for each series. The average fiber content for tunnels only with FRC and with the combination of FRC and rebar are nearly the same, equating to 0.42% by volume (equivalent to approximately 37 kg of steel fiber per cubic meter of concrete). Conversely, the average inner diameter of tunnels constructed only with FRC is 26% smaller than that of tunnels with a combination of FRC and rebar (see Figure 1b). Considering a confidence interval of ±1 standard deviation, the inner diameter of the former ranges from 3.0 m to 8.0 m, whereas the inner diameter of the latter ranges from 4.2 m to 10.7 m. These observations are explained by the lower resistant capacity of segments only with FRC in comparison with segments with FRC and rebar for similar fiber contents. Figure 2c shows that the thickness of tunnels constructed with FRC and rebar ranges from 0.19 m to 0.40 m, with an average of 0.30. Tunnels only with FRC show thicknesses ranging from 0.21 m to 0.33 m, with an average of 0.27. The smaller thickness of tunnels only with FRC is possibly a consequence of the smaller diameter mentioned in the previous paragraph. This suggests that linings only with FRC are generally associated with smaller diameter and thickness. A more accurate appreciation of the typical characteristics of tunnels only with FRC is obtained in the analysis of the slenderness ratio - calculated by dividing the tunnel inner diameter by the segment thickness (see Figure 2d). The slenderness ratio of these tunnels usually goes from 14.2 to 26.4, with an average of 20.3. Conversely, in the case of tunnels that also include conventional reinforcement, the typical range goes from 19.5 to 29.6, with an average of 24.5. As expected, the use of rebar in combination with FRC enables bigger slenderness ratios owed to the ability to reach bigger resistant capacity. Interestingly, however, the use of segments only with fiber is compatible with the slenderness ratio found in the majority of segmented linings.

4 STRUCTURAL ANALYSIS

4.1 *Parametric study with continuum model*

A parametric study was conducted with typical tunnel geometries and ground conditions. Table 1 summarizes the ranges of lining diameter, lining thickness, concrete characteristic compressive strength, depth of the tunnel, and the elasticity of the ground that was considered. In each case, the tunnel and the surrounding masses were simulated in the software RS2 from Rocscience. A preliminary study was conducted to determine the boundary conditions and size of the model needed to avoid influence of the external constraints on the results.

Figure 3 shows the model geometry and mesh used. Note that the total size of the simulated area varies with the internal diameter (D) of the tunnel. The water level was placed 1 m below the surface in every case and K0 was chosen as 1 for different situations to reduce the number of simulated cases. For simplification purposes, only 3 different calculation steps were performed for each model: initial situation, 40% relaxation and final configuration with segmental lining. A homogeneous isotropic numerical model was assumed, which might imply a certain underestimation of the forces applied to the lining.

The critical combination of bending moment (M) and axil force (N) was assessed after each simulation. Since the geotechnical model uses directly the given geotechnical parameters and the results are unfactored, a Factor of Safety of 1.35 according to EC-7 for the unfavorable situation was used to obtain pairs of M-N that were later compared with the resistant capacity of the lining.

Such resistant capacity was assessed through sectional analysis considering the geometry of the lining and its material properties. The constitutive behaviour of concrete in compression followed the parabolic curve proposed in EC-2, whereas the behaviour in tension followed the rigid-plastic model described in fib Model Code 2010. M-N diagrams were obtained for all considered linings.

Table 1. Variables of the parametric study.

Concrete f_{ck} (MPa)	D -Internal Diameter (m)	t -Lining Thickness (mm)	Tunnel Depth*	Elastic modulus of the soil E_m (MPa)
	7 m	300 mm	1D - 2D - 5D - 10D	50 - 100 - 500 - 1000 - 10000
		400 mm	1D - 2D - 5D - 10D	51 - 100 - 500 - 1000 - 10000
	10 m	300 mm	1D - 2D - 5D - 10D	52 - 100 - 500 - 1000 - 10000
45, 55 and 65		400 mm	1D - 2D - 5D - 10D	53 - 100 - 500 - 1000 - 10000
		500 mm	1D - 2D - 5D - 10D	54 - 100 - 500 - 1000 - 10000
	13 m	400 mm	1D - 2D - 5D - 10D	55 - 100 - 500 - 1000 - 10000
		500 mm	1D - 2D - 5D - 10D	56 - 100 - 500 - 1000 - 10000

* Depths are given in relation to the internal diameter (D) of the tunnel

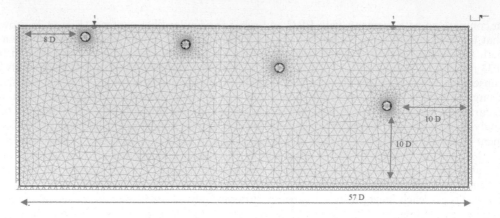

Figure 3. Model geometry and mesh.

Table 2. Overview of boundary conditions for tunnel diameter, lining thickness, concrete strength, contact width and longitudinal joint as well as calculated limits for splitting reinforcement.

Internal Diameter (m)	7		10			13	
Characteristic compressive cylinder strength (MPa)	From 45 MPa to 65 MPa		From 45 MPa to 65 MPa			From 45 MPa to 65 MPa	
Lining thickness (mm)	300	400	300	400	500	400	500
Contact width (mm)	140	220	140	220	310	220	310
Splitting reinforcement limit at longitudinal joint (kN)	2700	3200	2700	3200	3700	3300	3800
Splitting reinforcement limit at circumferential joint (kN)	2200	2800	2400	3000	3700	3400	4000

A critical point for the design of many segment linings is the verification of the capacity of joints, particularly against bursting effects. From the experience of the authors and using different verifications for each of the contact widths and a maximum eccentricity fixed of 30 mm. The verifications are based on a German Guideline (DAUB, 2013) and ITA (ITA WG No.2, 2000) recommendations. The permittable forces for the splitting reinforcement limit at longitudinal and circumferential joints are presented in Table 2.

Note that the resistant capacity of the longitudinal joints increases by approximately 500 kN for every 100 mm increase of lining thickness, regardless of the internal diameter and characteristic compressive strength of concrete. This is a simple way to understand the influence of the thickness in the assessment without changing the tensile capacity given by the increased quality of the concrete in each case.

In total, 140 scenarios were simulated. Figure 4 summarises the results obtained in the continuum model. There is a clear influence of the tunnel diameter in the results. For 10 m diameter tunnels, values of axil force (N) below 2700 kN are found for all types of grounds, if the depth (at the crown level or overburden) of the tunnel is smaller than 5D, or if the ground has an elastic modulus over 5000 MPa.

The bending moments for the applied boundary conditions are unrealistically low. With these results the authors want to illustrate the consequences of too many simplifications and possibly wrong assumption in the model. Asymmetric stress conditions or/and load situations shall be modelled as well.

The red line in the graph establish the limit explained above regarding the joints splitting capacity for the segmental lining design. The red line is related to a certain segmental lining joint geometry, eccentricity of load considered and contact width available. The values are 2700 kN

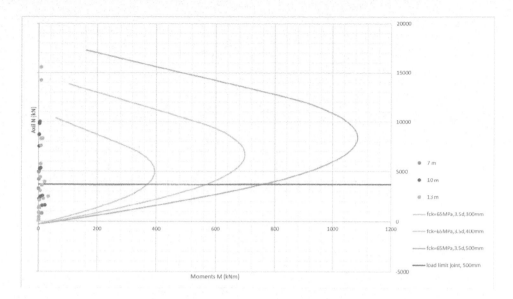

Figure 4.   Results of continuum model for every scenario and M-N diagrams.

(for 300 mm lining thickness) up to 3800 kN (for 500 mm lining thickness) for the splitting of the longitudinal joint in the permanent situation and 2200 kN to 4000 kN for the splitting of the circumferential joint in accidental situation, commonly from ram-shoes forces worked out for typical dimensions. The grey area is the permitted area for segmental lining design without any rebar reinforcement for the different types of concrete, given by the different 3 sets of curves.

4.2   Parametric study with the beam model

The parametric study was repeated using beam models to represent the ground-lining inter-action. The analysis was conducted in Staad Pro, a structural engineering software used for different types of structural design. The beam-spring models represent the tunnel lining with beams and the ground modelled by radial springs. Only compressive stresses can be transferred between the ground and the lining. The stiffness of the springs was calculated considering the modulus of elasticity of the ground surrounding the tunnel.

Typically, the segmental lining is loaded from the ground and water loads in the permanent situation (Table 3). Load scenarios from the production and handling are not considered for

Table 3.   Overview of load cases and load combinations.

Load cases	Self-weight	Wedge load	Symmetric ground load	Asymmetric ground load	Water load
	q_c=25kN/m³	q_j=60kN/m²	q_s=250kN/m²	q_s=250kN/m²	qw=1200kN/m²
Load combinations		q_c+q_j $q_c+q_j+q_w$	q_c+q_s q_c+q_s+qw	q_c+q_a $q_c+q_a+q_w$	q_c+q_w

2171

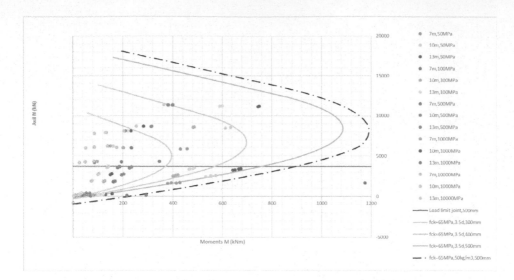

Figure 5.   Results of beam model for every scenario and M-N diagrams.

this paper, although they are for several projects the most unfavourable situation. A set of M-N interaction diagrams with different concrete properties and fiber performances using ITA recommendation (ITA WG No.2, 2000) as well as M-N interactions for conventional steel bar reinforcement were assessed. The purpose of this, is to define the limits for different segmental lining thickness with various amount and type of reinforcement.

Figure 5 shows the results obtained in the beam model. Results reveal the axil forces and bending moments increase with the tunnel diameter. This is consistent with the results previously obtained with the continuum model. The high water load (1200 kPa which is equivalent to 120 m of water table over the crown of the tunnel) led to very high axil and bending moments.

The self-weight as result from lining thickness and the water load have maximum effect on axial forces. For bending moment evaluation of asymmetric loads has the maximum effect and self-weight has the minimum effect as in general symmetric loads, like water load, reduce the bending moments.

As illustrated in figure 5, the load bearing capacity of conventional reinforced concrete is not significantly bigger than those of concrete reinforced only with fibers. The high limitation induced by the splitting of the joints is also clearly demonstrated. Apart from the limitation at the joints does FRC provide for many cases sufficient load bearing capacity. But this consideration is not sufficient for a final evaluation since it does not provide indications of possible failure development, reference is made to the design principles described in chapter 2.

5   DESIGN PROPOSAL

5.1   *Conclusion of performed calculation*

The comparison of results from continuum models and beam models reveal significant differences. The former shows low bending moments compared to the latter. None of the models can reproduce the actual loading situation. This isotropic continuum model is only able to reproduce homogeneous symmetric load cases, and the beam model is highly dependent on the chosen load case, which in many cases is defined as a result of the experience of the geotechnical engineers. For rock masses, discontinuum models are required, at least for the definition of load cases, if for the structural design beam models are used.

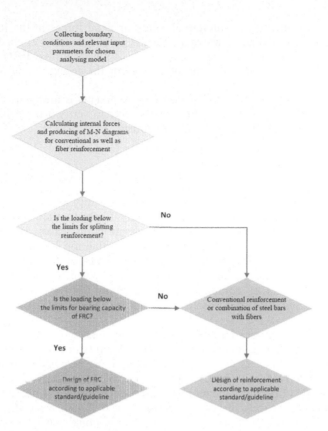

Figure 6. Flow chart for design process.

The resistant capacity at the joints for a certain segmental lining thickness is given by the contact area, eccentricity of the load and geometry of the joints (flat or curved). Such resistant capacity is not significantly influenced by the concrete strength, which affect mainly the capacity of the segment to resist bending moment. Considering the limitation induced by the splitting at the joints, the use of segments reinforced only with fibers is recommended in cases with axial forces below 2700 kN, 3200 kN and 3700 kN for lining thicknesses of 300 mm, 400 mm and 500 mm, respectively. For tunnels with internal diameter of 7 m, 10 m and 13 m and cover smaller than respectively 5D, 3D and 2D, a valid recommendation is always to consider the water table close to the surface. This general recommendation is valid from the lowest values of isotropic grounds from $E_m = 50$ MPa up to hard rock masses. Special boundary conditions and load situations should be analysed in detail since these general recommendations do not consider the presence of faults, anisotropy, karst, caverns, among others.

The failure mechanism should also be verified. A ductile failure characterized by a maximum load bearing higher than the first crack load is recommended. All these studies are only relevant if the calculated internal forces are correct. Therefore, engineers must reproduce the ground-structure behaviour as realistically as possible. Isotropic homogeneous 2D models might be sufficient for tunnels in soft grounds, while anisotropic 3D models are highly recommended for tunnels in hard rock.

5.2 Principle design process

Figure 6 illustrates the decision-making process about the use of only fibers or a combination of fibers with conventional reinforcement. A big limitation of the FRC application is the requirement for splitting reinforcement, which applies to many tunnel projects. Either high

2173

thrust forces pressed on the circumferential joints or axial forces in the longitudinal joints introduced by high ground and/or water load result in high loading applied the limited area for load transfer of these joints. Nevertheless, does this not mean that FRC is not possible, it only means no standards or guidelines are available for the design of splitting reinforcement using FRC.

A verification through test with full-scale tunnel segments is always possible, which in is strongly recommended by the authors. These tests serve to verify if the behaviour of the segments satisfy the design principles described previously.

6 CONCLUSION

Despite the increased application of FRC and availability of standards and guidelines for the design, the decision about the use of only fibers (without traditional reinforcement) is not always clear. The use of fibers is in many cases driven by advantages during construction, rather than by clear technical evaluation based on design principles. This paper has focused on design principles but does not discuss practical issues during the construction phase. The workability of the concrete, which might be challenging with higher amounts of fibers or combination of different fiber types (e.g. steel fibers for structural purpose and PP-fibers for fire scenarios) is certainly an important matter to be considered.

Current standards, recommendations and test procedures enable the adequate use of FRC in tunnel segmental linings. Lining geometry, type of joints (and the evaluation of bursting and splitting), tunnel diameters and type of concrete must be considered in addition to the simple M-N interaction diagrams for the study of possible load combinations.

It is well known that fibers provide more advantages regarding fire protection and local failure during handling operations, this is justifying the use of fibers. The decision to use concrete reinforced only with fiber or in combination with traditional reinforcement should be the result of an analysis of the load combination acting for the permanent and transient situations as well as also on other parameters in the design.

The application of FRC shall be evaluated for each project separately based on the established design principles. The trend shows that tunnels with smaller diameter and tunnels with water loads are more suitable, if the load in the joints can be transferred without splitting reinforcement. If the design is done according to the flow chart given in this paper, a technically transparent approach is secured. Special care shall be taken regarding the design approach and verifications tools chosen. This paper shows that internal forces may vary significantly depending on the chosen design approach.

Further research and development of standards contemplating various design aspects is still needed to increase the acceptance of FRC application in tunnels. The implementation of design guideline of simple structures like slabs, beams and walls reinforced with fibers in the Eurocodes is an important step forward. Currently, many projects require intensive verification, which are time consuming, expensive and may be difficult to perform, especially during an ongoing project. Tests on real tunnel segments are recommended to verify the behaviour under certain load conditions. For the quality control during production, the traditional small-scale bending test (EN 14651) or the Barcelona test (Blanco et al, 2014) can be used to verify the mechanical properties in tension and the inductivity tests (Cavalaro et al, 2015) can be used control the steel fiber content.

ACKNOWLEDGEMENTS

The authors would like to thank the companies Bane Nor, Mott MacDonald and the UPC (Polytechnic University of Barcelona) for the source of data and help during the preparation of this paper.

REFERENCES

Blanco, A. & Pujadas, P. & Cavalaro, S.H.P. & A. de la Fuente & Aguado, A. 2014. Constitutive model for fibre reinforced concrete based on the Barcelona test. Cement and Concrete Composites 53, 327–340. DOI 10.1016/j.cemconcomp.2014.07.017

Cavalaro, S.H.P. & López, R. & Torrents, J.M. & Aguado, A. 2015. Improved assessment of fibre content and orientation with inductive method in SFRC. Mater Struct 48: 1859. https://doi.org/10.1617/s11527-014-0279-6

DAUB (German Tunnelling Committee) Working group "Lining Segment Design". 2013. "Recommendations for the design, production and installation of segmental rings", Deutscher Ausschuss für unterirdisches Bauen e. V. (DAUB), ITA-AITES.

ITA-AITES, Working Group No. 2, Official report of International tunnelling association. 2000. "Guidelines for the Design of Shield Tunnelling Lining", Tunnelling and Underground Space Technology, Vol. 15, No. 3, pp. 303–331, Elsevier Science Ltd.

Tunnels and Underground Cities: Engineering and Innovation meet Archaeology,
Architecture and Art, Volume 5: Innovation in underground engineering,
materials and equipment - Part 1 – Peila, Viggiani & Celestino (Eds)
© 2020 Taylor & Francis Group, London, ISBN 978-0-367-46870-5

Sprayed concrete composite tunnel lining design: Analytical study on its behaviour at the bond interface

C. Green, H. Jung & A. Pillai
Arup, London, United Kingdom

ABSTRACT: Traditionally, no bond has been assumed across waterproof membranes installed between tunnel primary and secondary linings. Spray-applied membranes offer the opportunity to take into account some bond between the two linings, producing a composite action. The composite action has been verified through a significant laboratory testing programme from recent research from Pillai et.al. (2017). In addition, Jung et.al. (2017) have proposed a design chart for the determination of the secondary lining thickness which will allow designers to reduce the total lining thickness when composite action is considered. This paper presents a microscopic interpretation of the bond interface behaviour in composite lining that will allow further understanding of the mechanism of load transfer between the outer and inner linings, and allows the authors to provide a practical commentary on the application of the bond layer. Also, this paper provides the results of the analytical work done to investigate the effects of membrane thickness, tunnel diameter and shear strength parameters of the interface.

1 INTRODUCTION

There has been a significant increase in published material relating to sprayed concrete composite shell lining design in recent years. There are a number of concepts which have been presented to the industry to challenge the traditional sprayed fibre reinforced concrete lining design methods, where the primary sprayed lining capacity has been ignored in the long-term design of the total lining. Both 'single pass' and composite shell lining solutions have been proposed, the latter based on the theory that sprayed waterproofing membranes produce a bond between the primary and secondary linings, providing a composite structural action.

Jung et.al. (2017) proposed a design chart concept for quantifying the secondary lining thickness reduction which may be possible when composite action is considered in design. This concept was based on an extensive bespoke laboratory testing programme and a finite element model calibration exercise for sprayed concrete composite shell linings; this work is presented by Pillai et.al. (2017) and this includes test-calibrated tunnel section analysis modelling results.

This paper will present more detail on the tunnel section analysis, and will build on the previously presented results by providing a microscopic interpretation of the bond interface, for London Clay. The design chart proposed by Jung et.al. (2017) is also developed within this paper, with a parametric analysis presented to consider the effect of tunnel diameter, interface thickness and shear parameters on the proposed application of laminate glass theory to composite shell tunnel linings.

2 IMPORTANCE OF UNDERSTANDING THE MICRO SCALE BEHAVIOUR AT THE BOND INTERFAC

The bond interface behaviour in a composite shell tunnel lining has been presented on a macro scale by Pillai et.al. (2017), based on laboratory test calibrated numerical modelling. The load sharing behaviour of the composite lining was presented in comparison to full slip and fully connected tunnel lining cases. It is important, however, to fully understand the behaviour of the bond interface on a micro scale, to allow designers to simplify modelling of the bond interface and choose the right representative structural element types in a structural analysis model. In order to do this, understanding behaviour such as the relative displacement together with the mechanism of load transfer between the outer and inner linings at the bond interface is essential.

Pillai et.al. (2017) noted that numerical analyses results indicated that the maximum shear stress developed in the membrane interface is much smaller than the shear capacity of the membrane interface. This was to determine that the bond interface should not reach the failure mode and therefore should not be damaged over the design life of the tunnel, ensuring that the composite action, at least structurally, is ensured for the design period.

This paper will investigate the interface layer displacement to check whether the membrane interface is taking compression or tension or shear. Also, the hoop thrust change along the section of the composite lining will be investigated to give designers further understanding of the stress conditions at the interface.

3 COMPOSITE LINING BOND INTERFACE MOVEMENT ON THE MICRO SCALE

3.1 Ground models

The analysis was carried out on a tunnel model in London Clay ground conditions with 10m diameter tunnel. Figure 1 shows the tunnel analysis model used for this study. The primary and secondary lining thickness are 260mm and 280mm respectively, and the sprayed waterproof membrane thickness is modelled as 5mm. The sprayed waterproof membrane layer was modelled as Mohr-Coulomb strain softening elements. A cohesion value of 1.56MPa, friction angle of 12.12°, and a shear stiffness of 7MPa were inputted for the membrane interface.

For this study, the tunnel excavation and analysis stages were simplified to help straight forward interpretation of the results with respect to load sharing between the primary and secondary linings. Full-face excavation was assumed, and the lining was modelled with 28-day design strength and stiffness parameters, rather than considering the strength/stiffness change over time. This was done to aim to eliminate complexity, which could make the interpretation difficult.

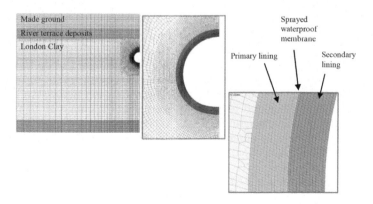

Figure 1.   Numerical analysis model used for this study (Pillai et.al. 2017).

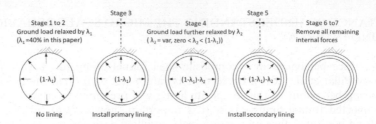

Figure 2.   Numerical analysis stages used for this study (Pillai et.al. 2017).

Table 1.   Analysis cases for micro scale assessment of tunnel section analysis.

Analysis case	Ground conditions	λ_1	λ_2
1	London Clay	40%	0%
2	London Clay	40%	20%
3	London Clay	40%	40%

The construction stages were modelled as shown in Figure 2 and Table 1. Analysis cases for micro scale assessment of tunnel section analysis.

As per the previous modelling work carried out by Pillai et al (2017), the tunnel model was modelled with a single full-face excavation stage, to simplify the problem for analysis to focus on composite behaviour. As such, there were limitations to the modelling in terms of achieving a realistic degree of displacement around the tunnel ring. High immediate stiffness of the sprayed concrete lining had to be modelled to support the full-face excavation in the model, therefore the displacement generated in the model is low compared to that which would be expected in a real tunnel in London Clay. To address this limitation, a surcharge was applied to the model ground surface at the final stage, to increase tunnel deformation to a level that allows the microscopic interpretation to be practically feasible. In this study, the surcharge was applied until the tunnel diameter change reached approximately 10mm.

3.2   Hoop thrust distribution

Figure 3 shows a contour plot of the hoop thrust along the section of the tunnel lining. It can clearly be seen that the hoop thrust has a sudden change at the bond interface location (for example, note at the axis). To give better illustration of the stress condition at the interface layer, the hoop thrust distribution along the section of the lining was outputted for crown, axis, knee and invert locations. These are presented in Figures 4 to 7.

Figures 4 to 7 show that the sprayed waterproof membrane interface layer creates a step change of hoop thrust distribution, due to the existence of a relatively thin softer layer (waterproof membrane layer). The hoop thrust is unable to be completely transferred to the adjacent layer at the interface location, and so there is a certain drop of the hoop thrust – but not to zero – through the interface, then the hoop thrust develops again within the other layer. This mechanism can be observed from all the four locations shown in Figures 4 to 7.

The magnitude of the developed hoop thrust step is in the range of 1000kN to 3500kN. Although the interface shear strength parameter is constant along the perimeter of the interface, the hoop thrust step size is different, based on the location around the tunnel and from the level of ground load relaxation. From this study, the maximum hoop thrust step is observed at the axis.

At a certain location around the tunnel lining, the inner and outer layers are shown to be bending to the same direction i.e. if the outer shall is in sagging then inner shall is also in sagging. However, the level of hoop thrust and its distribution slope is different between the inner

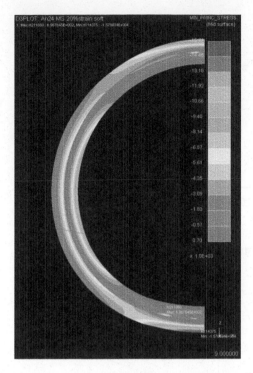

Figure 3. Hoop thrust distribution on the composite lining at the final analysis stage.

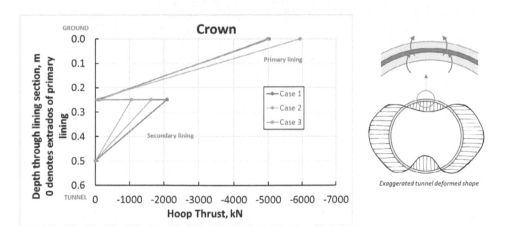

Figure 4. Hoop thrust distribution through the section of lining at crown.

and outer layer. This indicates that the primary and secondary lining do not necessarily share the load evenly, and each layer is developing its own hoop/bending subject to the design condition.

3.3 *Relative displacement at the bond interface*

The relative displacement of the primary and secondary linings is presented in this section from the tunnel section analysis models. Using the numerical models, the initial positions of the nodes are determined for the inner and outer edges of the interface layer, before

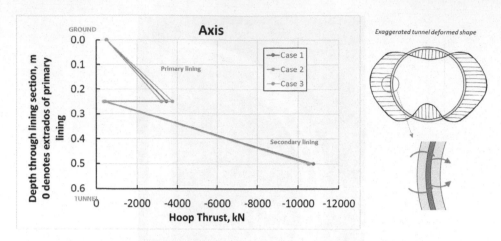

Figure 5. Hoop thrust distribution through the section of lining at axis.

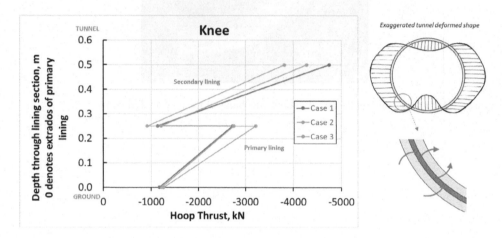

Figure 6. Hoop thrust distribution through the section of lining at knee.

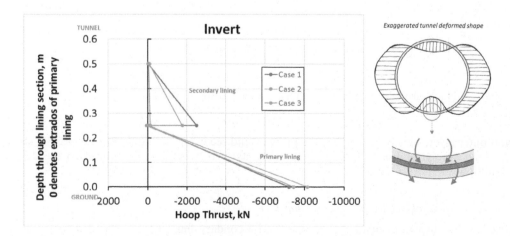

Figure 7. Hoop thrust distribution through the section of lining at invert.

deformation of the tunnel lining (illustrated in Figure 8 as the 'a' positions). The final positions of the nodes following deformation under loading are also determined (illustrated in Figure 8 as the 'b' positions). Since the interface layer is receiving both axial and shear displacement, simple measurement of the distance between the node points after the loading - i.e. distance between the b1 and b2 points from Figure 8 – was found not to indicate whether the layer is in compression shear or in tension shear because both cases can increase the node distance from the original position. To resolve this issue, firstly one of the 'b' points was manually shifted to the location where the relevant 'a' point is, and the remaining 'b' point was shifted by the same amount. Once one of the a and b point is overlaid (e.g. a1 and b1' from Figure 8), then the relative movement of the other nodes (a2 and b2') is calculated. For this study, Va is defined as the original thickness of the interface layer, Vb is defined as an interface layer thickness after the load, and Hb is defined as a shear relative displacement between the inner and outer layer (see Figure 8). It is considered that the interface layer is in compression when Vb > Va, in tension when Vb < Va.

Table 2 summarises the study results. Quite interestingly, the result showed that the interface layer thickness decreases at the axis, so the membrane is in compression, but the knee/

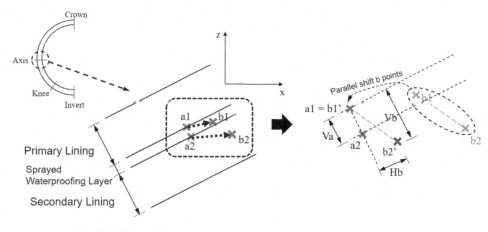

a1, a2 : initial position
b1, b2 : deformed position
b1', b2' : parallel shifted position of b points
Va : initial thickness of sprayed waterproofing layer
Vb : deformed thickness of sprayed waterproofing layer
Hb : shear displacement of waterproofing layer

If Vb>Va, interface layer in tension. Vb<Va, interface layer in compression

Figure 8.   Sprayed waterproofing layer relative displacement - microscopic interpretation.

Table 2.   Interface layer thickness change and shear displacement summary.

Location	Va (mm)	Vb (mm)	Membrane thickness change, Vb – Va (mm)	Comp./Ten.	Hb (mm)
Crown	5.6	5.6	0	-	-
Axis	7.08	6.92	–0.16	Compression (strain = 2.2%)	0.02
Knee	5.15	5.17	0.02	Tension (strain = 0.38%)	0.20
Invert	4.54	4.56	0.02	Tension (strain = 0.44%)	-

invert area showed slight tension. The crown showed neutral change of thickness. At the axis, the membrane is compressed by 0.16mm (=2.2% axial strain), with shear displacement of 0.02mm. At the knee and invert, the membrane is in slight tension at 0.02mm (=0.38% to 0.48% axial strain) with shear displacement of 0.2mm.

When the interface is in pure compression, there are less concerns about the bond failure of the interface, but when tension and/or shear is generated at the interface, the loss of interface bond could be a concern as this will damage the composite action.

Figure 9 is a typical pull out test curve of the double bond interface layer. This graph was produced using a lab test record carried out by a manufacturer of the sprayed waterproofing material, BASF. Figure 9 indicates that to mobilise 1MPa tensile stress at the bond interface, the bond interface needs to be stretched by approximately 1.5mm to 2.0mm. From Table 2, the observed tensile displacement at the interface layer is 0.02mm, which is much less than the peak value.

Figure 10 shows a typical shear stress-deformation curve extracted from other research projects. This test was carried out with the use of 20cm x 20cm sample. The peak shear stress is

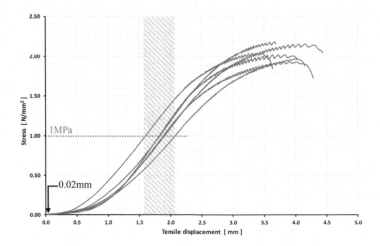

Figure 9.   Pull out test of interface layer (from BASF internal lab test record).

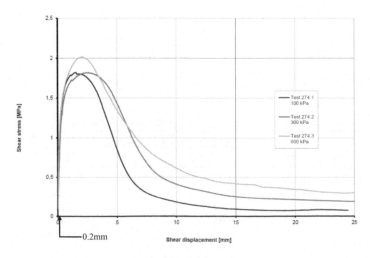

Figure 10.   Direct shear test of interface layer (TU Graz, 2014).

2182

reached at a shear displacement level well above 1mm. Note that the maximum shear displacement observed from the analysis model is 0.2mm, from Table 2.

This indicates that the level of axial and shear deformation observed from the analysis is well below the interface failure point. However, further analysis and sample testing will be required to verify; a) whether the level of tensile/shear deformation will remain low when the tunnel ovalisation level approaches the limit design value; and b) whether the membrane layer saturation level would influence the interface stress-strain behaviour and its peak strength value.

4 EXPANDING THE LAMINATE GLASS THEORY APPLICATION BY PARAMETRIC STUDY

Jung et.al. (2017) presented the chart shown in Figure 11. The black line was determined based on the equations presented by Bennison (2008) for equivalent monolithic thickness for laminated glass beams. This was shown by Jung et.al. to be applicable for composite shell tunnel linings, using laboratory test-calibrated finite element modelling, shown by the red markers in Figure 11.

Further modelling work has been carried out for a parametric study to expand this design chart; this is to determine its practical applicability for first-pass type analysis for larger tunnel diameters, different interface layer thicknesses and for varied shear modulus values, for analyses in London Clay.

4.1 Parameters for analysis

For a parametric study of the composite lining, a separate analysis model was used from what was used for Section 3 of this paper. The model was developed with a tunnel diameter of 6m with no ground around the lining. For the details of the analysis model, read Jung et.al. (2017). The 6m diameter tunnel was considered an appropriately representative tunnel diameter for tunnels where sprayed concrete construction methods would be employed. However, sprayed concrete tunnel lining methods can and have be applied to tunnels up to 10m in diameter. The laminate theory and the LS-DYNA model were used to determine the applicability of the laminate glass theory for larger tunnel diameters by considering a 10m diameter tunnel. For this larger diameter tunnel, the primary lining thickness was increased from 250mm to 350mm.

The interface layer was considered for the initial design chart as 3mm thick; in reality this can vary considerably and is a key parameter which governs the buildability of the composite

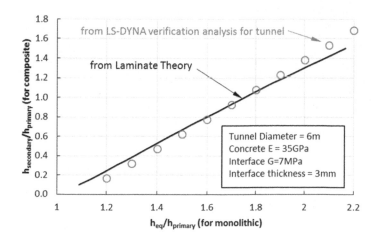

Figure 11.   Design chart theory verification (Jung et.al. 2017).

shell lining. If variation of the waterproofing membrane thickness within a few millimetres governs this design concept, then it could restrict the use of the design tool, since ensuring that level of control on site would be challenging. An interface layer thickness of 5mm was used to understand the effect of increased thickness on the application of the laminate theory.

A shear modulus value of 7MPa was derived from laboratory test data, by determining an average of the lower, more regularly occurring shear modulus values from the numerous tested core samples. A shear modulus of 15MPa has been selected as a larger value to determine its effect on the composite behaviour.

4.2 Expanded design chart

The variations in the tunnel diameter, interface layer thickness and interface shear modulus were assessed with the laminate theory equations and the LS DYNA finite element model, which has shear strength parameters determined from laboratory testing. The results of these parametric analyses are shown in Figure 12. All of the parametric studies show a good correlation between the LS DYNA modelling and the laminate glass theory equations. It appears that the laminate glass theory equations can be used for different tunnel configurations to capture the same behaviour as a numerical model calibrated to test data. It is encouraging that something relatively straightforward has been shown to represent the same results as a detailed model and that this can be used to determine an initial secondary lining reduction potential for consideration in detailed design.

From Figure 12 it can be seen that there is a slight gradient difference between the laminate theory predicted results and the numerical modelling result. The effect of this is that when $h_{eq}/h_{primary}$ <1.75 approximately, the laminate theory is shown to slightly over predict the $h_{secondary}/h_{primary}$ value, and when $h_{eq}/h_{primary}$ >1.75, the laminate theory is shown to under predict the $h_{secondary}/h_{primary}$ value. The difference in predicted values for $h_{eq}/h_{primary}$ >1.75 is not large and this method is intended for a first pass approximation of secondary lining thickness requirements only. Nevertheless, the authors would advise caution in application of the design chart method for tunnel lining configurations where the monolithic permanent lining thickness is >1.75 times the primary lining short-term support thickness.

When the shear modulus is modelled as 15MPa, the line drops down from where it was for G = 7MPa, meaning that for the same h_{eq} and $h_{primary}$ values, a lower $h_{secondary}/h_{primary}$ value is found. This is also the case for an increased tunnel diameter of 10m, giving a lower $h_{secondary}/h_{primary}$ value for the same h_{eq} and $h_{primary}$ values. When the interface layer

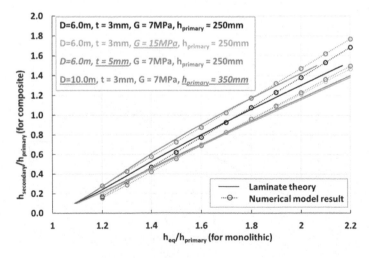

Figure 12. Expanded composite lining design chart verification from parametric study.

thickness is considered as 5mm thick, a higher $h_{secondary}/h_{primary}$ value is found for the same h_{eq} and $h_{primary}$ values.

While there is a good correlation between the two analyses for all four cases, Figure 12 does show that the results are dependent on the varied parameters. The assumed value for the interface shear modulus has an impact on the $h_{secondary}/h_{primary}$ ratio; a larger value to give an improved secondary lining reduction would need to be justified. During the laboratory experiment programme, there were core tests which had lower shear modulus values than 7MPa.

In a real tunnel, the interface layer can vary between 3mm–5mm over a small area, due to the challenges of spraying an accurate depth at all times. The difference of 3mm to 5mm for, for example, a $h_{eq}/h_{primary}$ of 1.5 is approximately 0.1. If a primary lining of 225mm is considered, the secondary would be found as 146.25mm for a 3mm thick interface layer and 168.75mm for a 5mm thick interface layer. This is 15% change for a small difference in interface layer thickness, so care should be taken in assuming an interface thickness value.

5 CONCLUSION

A microscopic review of the numerical analysis model showed that the sprayed waterproof layer can be either in compression or in tension when the tunnel is deformed. The interface layer is also found to experience a certain level of shear displacement. For the ovalisation level of 0.1%, the numerical analysis model showed that the interface layer experiences 0.16mm compression at the axis, and 0.02mm tension at the knee and invert. Shear displacement observed was maximum 0.2mm at the knee level, and 0.02mm at the axis level. The observed tension and shear displacement level from the numerical analysis is well below the peak tensile/shear stress from laboratory tests, but further analysis and test results are required to expand this observation to the full extent, to consider tunnel deformation level and the level of saturation at the interface.

Good agreement was found between the laminate theory and numerical modelling results for tunnel section analysis when varying the interface layer thickness, shear modulus and tunnel diameter. The application of the laminate theory equations by Bennison (2008) has been shown to be applicable to various tunnel lining configurations for determining initial secondary lining reduction potential. Care should be taken over assumptions for interface layer thickness and shear modulus values as they impact the $h_{secondary}/h_{primary}$ ratio. Care should also be taken in application of the design chart method for tunnel lining configurations where the monolithic permanent lining thickness is >1.75 times the primary lining short term support thickness.

ACKNOWLEDGEMENTS

The authors would like to acknowledge the significant support from Frank Clement and Diletta Traldi of BASF.

REFERENCES

Bennison S.J., Qin M.HX, Davies P.S., 2008, High-performance laminated glass for structurally efficient glazing. Innovative light-weight structures and sustainable facades, Hong Kong

Jung H., Clement F., Pillai A., Wilson C., Traldi D. 2017. Composite tunnel linings allowing a more cost effective and sustainable tunnel design. WTC 2017, Bergen, Norway.

Pillai A., Jung H., Clement F., Wilson C., Traldi D., 2017, Sprayed concrete composite tunnel lining – load sharing between the primary and secondary lining, and its benefit in reducing the structural thickness of the lining, WTC 2017

TU Graz, 2014, Direct Shear Test Results BASF (unpublished report)

Tunnels and Underground Cities: Engineering and Innovation meet Archaeology,
Architecture and Art, Volume 5: Innovation in underground engineering,
materials and equipment - Part 1 – Peila, Viggiani & Celestino (Eds)
© 2020 Taylor & Francis Group, London, ISBN 978-0-367-46870-5

Development of a non-radioactive density meter in tunneling

H. Greve & K. Zych
Royal IHC, Sliedrecht, The Netherlands

ABSTRACT: The management of the slurry system in tunneling projects plays an important part in the safety and efficiency of the tunneling operation when using a slurry shield TBM. The bentonite slurry is used for maintaining face stability and transporting the cuttings to surface. This requires continuous monitoring and control of flow rates and density. The current method for measurement of density is by means of a radio-active source and a receiver. The disadvantages of using a radioactive source are numerous, in particular with regard to safe handling and storage of the device. This paper will describe the development of an alternative, non-radioactive method for measuring the slurry density, by using radio frequency waves. The design, system architecture and testing methodology will be presented, followed by assessment of applicability of the meter to the monitoring of slurry density in a slurry shield TBM.

1 INTRODUCTION

The management of the slurry system in tunnelling projects plays an important part in the safety and efficiency of tunnelling projects. Whether the slurry is used for maintaining face stability or transport of the cuttings to surface, the overall mass balance of solids contained in the slurry going in and out of the tunnel has to be monitored and controlled. Ideally the excavated amount of solids should match the increased density of the discharge flow. If this is not the case, it could be an indication that too much soil is excavated in relation to the theoretical advance rate, which may lead to settlements on surface. In addition it is important to monitor the density of the recirculated slurry, as part of the fines of the soil particles will not be separated in the separation plant. The density and quality of the feed slurry has to be corrected to maintain the required slurry characteristics. The monitoring of density and flow rates is achieved by a combination of measuring flow (Q, m^3/s) and density of the mixture (ρ, kg/m^3). The combination of the two provides the instantaneous production in kg/s (Duhme et al. 2015).

Measurement of the density is currently done by a radioactive source and a receiver (RA meter). The level of absorption of radiation is an indicator of the number of particles between source and receiver. Application of this equipment ranges in pipe diameter from 300 mm up to 1000 mm or even more. The method is well-proven in dredging and tunnelling, but has it disadvantages:

- Handling of a radioactive source requires special permits
- Transportation of the equipment must be done with special precautions
- Only certified personnel is authorized to work with the equipment
- Certified personnel has to be present on the site of installation of the equipment

Using technology and extensive experience from the dredging industry, IHC has developed a density meter using harmless radio waves instead of radioactive material as source. This new technology can be used as standard equipment, it requires no special permits and is completely safe.

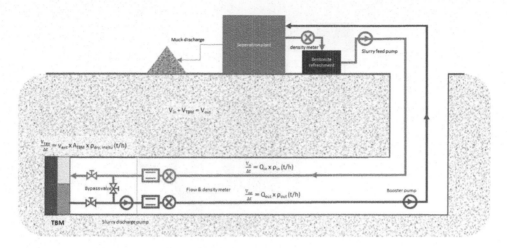

Figure 1. Overview of soil management in a slurry TBM.

2 DEVELOPMENT OF RF DENSITY METER

Rationale for the RF density meter can be found in the dredging industry. Slurry density is one of the crucial parameters to be monitored on board a dredger for efficient process. To measure mixture density, RA gauge, an ionising radiation based measurement is commonly used and considered the golden standard in the dredging industry. Despite the radioactive source is contained in a protective casing, providing the highest level of radiation shielding, the system is still subject to very strict radiation safety regulations. Permits for operation, as well as separate ones for transportation and storage need to be issued by a national radiation safety authority. This causes additional cost of ownership and extra bureaucratic and logistic attention.

Royal IHC devoted many years of research and development efforts to a system which would provide a radioactivity-free density measurement for industries utilizing a slurry transportation systems (Zych & Osnabrugge 2017). We aimed to find an apparatus which can closely match the gamma density meter in terms of accuracy and robustness. In this section, we present the development of a patented density meter which uses radio waves to sense the volume ratio of solids present in the carrier fluid.

2.1 Measurement principle

The method is based on a fact that water has a very high dielectric relative permittivity (ε_r=79, which quantifies the resistivity of a material to an external electric field) compared to typical soil constituents, which exhibit relatively low values ($\varepsilon_r \approx 4$ for quartz sand). This order of magnitude contrast in permittivity, results in the permittivity of water-based mixture to be a strong function of amount of solid particles present in it. Inversely, the volumetric solids concentration can be calculated from permittivity of the mixture, obtained through a carefully designed measurement.

For application in slurry pipeline, a radio-frequency transmission measurement geometry was considered. The measurement section of the slurry pipeline is fitted with a pair of collinear antennas, see Figure 2. The transmitting antenna (T_x) creates an electromagnetic field, which propagates across the pipe and is sensed by the receiving antenna (R_x). The received signal amplitude and phase is measured by an electronic circuit.

The signal phase is governed mainly by the permittivity of the mixture, whereas amplitude represents the energy losses the signal encounters. The major energy loss mechanism is related to the absorption caused by ions in the mixture. Since certain received signal strength is

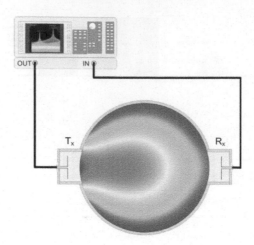

Figure 2. Schematic view of the measurement principle in the slurry pipe. The colour map shows the intensity of the EM field emerging from the transmitter.

required for a proper phase detection, a highly conductive water will limit the operation of the system. We set the design requirement that the system must be operational in brackish water, up to 1.5 S/m.

2.2 R&D process

The development process focused on converting the principle of operation into an industrial density measurement for slurry.

Preliminary research was to verify our assumptions regarding the electrical nature of sand, water, and their mixtures, together with varying amount of sea salt content. For that reason, commonly a dielectric characterization cell is used (Keysight Technologies 2017). Since the off the shelf systems are a fragile, desktop instruments, not very suitable to handle sand and corrosion-inducing water. we designed our own version of a coaxial measurement cell, with measurement capacity of 6 liters of mixture. With this simple instrument the practical values of mixture permittivity, conductivity, as well as bounds for the transmitted signal were found.

The hardware engineering included development of a special type of spool piece which can house the transmit and receive antenna. Secondly, the high frequency electronics had to be designed from scratch.

To really tackle the requirements related to the slurry, we adopted an experiment-based-design approach. The proposed measurement principle and the developed hardware had to be validated by series of laboratory trials. We opted for full scale tests with a flowing slurry. We thus developed a custom-made and independent circuit in a IHC laboratory, containing a 500 mm diameter RF measurement pipe (Figure 3).

The central piece was a vertical water column, 500 mm in diameter, containing the RF prototype. A high-speed water jet from a small centrifugal pump was fed into the column. Downward, laminar fluid flow was created inside the prototype pipe. Sand could be added to the circuit manually, in portions, to gradually increase the density. Sand once added, circulated continuously in the circuit, providing a stable condition for measurement.

The RF system was fitted with the radioactive gauge on the same measurement pipe, as can be seen in Figure 2. Thus both measurements intend to measure exactly the same mixture. In addition, both the RF density and the reference RA density values are computed by the same signal conditioner unit. Thus both spatial (location) and temporal (synchronization) consistency between the two measurements was achieved. This circuit was used over a period of 3 years, with 4 subsequent prototype measurement pipes. Eventually, the measurement

Figure 3. The small scale test circuit with two stacked RF measurement pipes (orange and brown) are visible.

concept was validated, the hardware adjusted and the whole system thoroughly tested, covering the full range of densities and water conductivities a dredging vessel may encounter.

The final step in the development was to have the RF system validated in a real life dredging operation. Together with our dredging partner, the RF system (still supplemented with the same radioactive gauge) was installed on board a river dredger. For a period of over a year, the ship operated in various locations along the Merwede and Waal rivers in The Netherlands. At each location, the operation agreed with the laboratory model. No significant deviations between radioactive gauge and RF indications were observed. The ship worked with several sand grades, but we didn't notice dredging any other material than quartz sand. In no case the type of sand was causing deviation.

The field trials successfully concluded the 9 year-long effort to develop a radio-wave density measurement system suitable for operation in slurry pipeline.

2.3 System architecture

The current day RF density meter architecture is schematically shown in Figure 4. The two main components of the system are the measurement pipe and the RF-enabled signal conditioner.

As shown in Figure 2, the principle of operation requires two openings in the pipe wall, placed collinearly on the opposite sides of the pipe. That will allow the electromagnetic wave to couple into and from the pipe volume. The opening is backed by a cavity, which holds the antennas. The implementation of this structure is patented by Royal IHC and took the form of rectangular flanges, oriented axially along the pipe. To provide protection of the antennas, the cavity is filled with the high endurance irathane polymer, well known in dredging industry for its wear resistance. The surface of the irathane inside the pipe has the same curvature as the pipe wall, resulting in a flush, protrusion-free pipe cross section, as shown in Figure 5.

The signal conditioner is a robust, custom-built electronic circuit to handle the high-frequency measurement signal. The function of the circuit is to provide the electromagnetic stimulus signal for the transmit antenna and to detect, filter and amplify the faint signal from

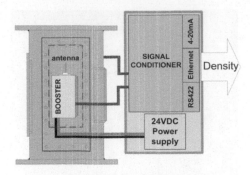

Figure 4. Architecture of the RF density measurement system.

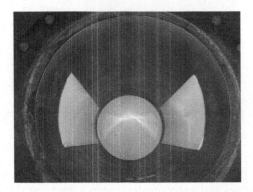

Figure 5. Inside look of the RF density measurement pipe. The irathane windows in the pipe wall are visible.

the receiving antenna. Besides the radio front-end, the conditioner contains the digital signal microprocessor required for density calculation and supplemented with typical industrial automation communication protocols- analog, serial and Ethernet.

The transmitting antenna panel contains a RF booster amplifier. It is connected between the output of the signal conditioner and the transmitting antenna, to amplify the transmitting signal to a power level of several dozens of watts.

Concerning safety, the RF density system is totally free from ionising radiation and only uses harmless level radio waves. The highest power signals, with a maximum power of 75W, only exist after the power booster on the pipe. The steel structure of the pipeline provides 100% screening from any electromagnetic waves. Thus, under normal operation, the electromagnetic waves are entirely contained in the pipeline. Even without that, in case the measurement pipe is taken out of the pipeline without powering it down, the resulting EM emission emerging from an open flange of the measurement pipe would be lower than the emission from a regular WiFi router.

3 ADOPTION FOR TUNNELING

In principle, suction dredging and slurry TBM both belong to the class of hydraulic transportation techniques, there are several specific differences between those two, resulting from excavation process in TBM and the operation of a separation plant. The features relevant for the operation of the RF density meter are: usage of bentonite; slurry, additives and slurry recycle.

In this section we will discuss the feasibility of RF density meter for slurry TBM, regarding the aforementioned aspects.

3.1 *Bentonite*

Bentonite is a common name for a class of sodium kaolinites, essentially a powdered clay, which is pre-mixed with water and used as a thickening agent for slurry. What that means, the TBM slurry is not water-based, but bentonite-based. As discussed in section 2.1, the electrical properties of carrier fluid are crucial for proper and accurate operation of the RF density measurement. The reported experimental results from available literature (Fam 1996) suggest the permittivity of aqueous bentonite solution does not deviate significantly from permittivity of fresh water. Which means the high permittivity contrast, essential for accurate operation is maintained.

To verify this data in our application, we again used the dielectric measurement cell. We performed several measurements with measurement cell filled up with water and followed by a pre-mixed bentonite mixtures with 10g, 15g and 25g of bentonite powder per dm^3 of water. All three mixtures didn't exhibited a significant difference referred to water. We can conclude the bentonite solution has the same permittivity as water and therefore will not hinder the operation of RF meter.

3.2 *Additives*

Besides bentonite, additional substances are used to modify the properties of TBM slurry. In order to help with excavation, polymers, foams or mineral oils are added to the slurry in front of the shield. The slurry processing plant also uses its own set of chemicals, mainly to induce flocculation and sediment of particles in the sedimentation pits. Among others, limewater and iron chloride are the most common flocculants. The former is of special attention to us, as it is a strong electrolyte. Thus it introduce free ions in water and rises conductivity of water. Which brings a risk of the radio signal being blocked.

The solubility of lime in water is about 9g per dm^3 at room temperature. We evaluated the resulting conductivity to be 1.1 S/m. Which is below the operational limit for the RF meter. However, this amount of lime would flocculate the bentonite suspension at the cutting shield, and hence would hinder the whole operation. Therefore it is in TBM operator interest not to use extensive dosage of flocculants or polluted processed water. We conclude that this self-regulating mechanism will ensure the level of additives will not cause excessive radio wave absorption in slurry.

3.3 *Fine clays*

Contrary to dredging, where mixture passes the hydraulic system only once, the TBM slurry is continuously cycled between the TBM and the separation plant. The separation plant is responsible for removal of coarse phase of excavated soil, refill any water + bentonite loss and feed the processed slurry back to the TBM. Inefficiency in separation of the finest solid phases, leads to a gradual build-up of density of the recycled slurry. When the density reaches $1.3 t/m^3$, the whole slurry is discarded, and fresh batch of bentonite mixture is prepared. The interaction of fines with bentonite and it net effect on the measured signal must be verified experimentally.

3.4 *Test program*

In order to verify the above theoretical discussion, we applied our experiment-based-design once again. The slurry circuit, described in section 2.2, can be implemented for TBM slurry test program. For that purpose we built in a bentonite pre-mixer. It can prepare a batch of

150 litres of thick bentonite slurry, which is afterwards injected into the main, water running circuit. We therefore ensure a homogenous, properly mixed bentonite base slurry is created.

With the test circuit prepared in such way, we are able to recreate the typical slurry TBM process in the laboratory and verify the usefulness of RF density meter. Two test scenarios are particularly interesting:

- Excavation: sand and gravel is added to a circulating bentonite mixture;
- Build-up of density in a process plant: clay is added to already circulating bentonite mixture.

3.5 *Results*

To test the excavation scenario, we prepared the bentonite mixture. We used the TUNNEL-GEL™ PLUS, from Baroid. The TUNNEL-GEL was loaded into the mixer, and pumped with water for 25 minutes, until homogenous and well hydrated bentonite solution was obtained (Figure 6). After transferring to the main measurement circuit, a density of 1.015 t/m^3 was obtained.

Subsequently, portions of sand were gradually added to the mixture, until density of 1.115 t/m^3 was reached. Results are presented in Figure 7. When working with low densities in the circuit, it is possible to stop the pump and observe the dynamics of solids sedimenting in the circuit. By turning the pump on again, we can observe how the RF meter reacts to a rapid rise of density. Those two actions are visible in Figure 7: switching the pump off and on at minutes `22, `35 and `29 respectively.

The mixture created in Figure 7 was a starting point for a second test run. Again material was being added in batches, this time using sand and gravel (diameter 3–5mm), and reaching higher density. The result is presented in Figure 8. There is a discrepancy between RF and RA indications, visible after the first batch of sand was added. We believe sand was too dry and adding in introduced extra bubbles of air, which caused the overestimation of the RF meter.

To test the build-up of fine materials in the circuit, we used a kaolinite based clay Mahlon FT-204 manufactured by Sibelco Deutschland. Material was mixed with water and added in

Figure 6.   The bentonite pre-mixer added to the slurry test circuit.

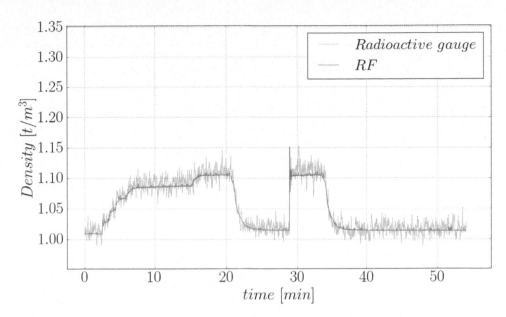

Figure 7. The first test run of bentonite slurry.

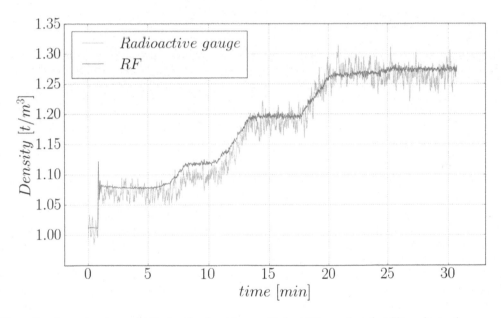

Figure 8. Second test run. At 7' a batch of sand was added, at 12' gravel, and at 18' again sand.

small portions to the circuit, which was running with bentonite and sand mixture. Unfortunately, the clay exhibited high shear strength and tend to stuck in the circuit, thus limited amount was added. Result is presented in Figure 9. In the graph, the RF indication rises faster than the RA meter. That is due to the fact that the RF meter measures the volumetric ratio of solids in the carrier water. To convert the volumetric ratio into density, the meter assumes all the solid material in the mixture is sand with density 2.65 kg/dm^3. Whereas the density of the clay used was 2.35 kg/dm^3. Hence the RF meter calculated higher density.

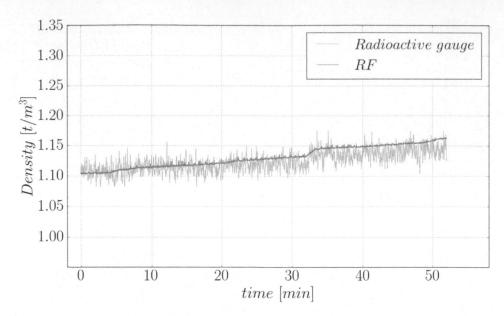

Figure 9. Test run simulating the build-up of density due to fine particles.

4 CONCLUSIONS AND OUTLOOK

It has been confirmed experimentally that the RF system can detect solids concentration in bentonite-based slurries, with an accuracy comparable with the conventional radioactive gauge. Density of slurry containing gravel, coarse sand, fine sand as well as clay can be measured by the system. It can be concluded that the RF density measurement system is suitable for monitoring the density in slurry shield TBM, in both the feed line and the return line. Furthermore, from the previous, long-term tests (Zych & Osnabrugge 2017) in a dredging environment it was learned:

– RF system is independent of the type of soil pumped;
– has lower measurement noise compared to radioactive gauge and no measurement drift;
– is maintenance free.

Having proven the effectiveness and accuracy of the RF meter in the laboratory, there is a high potential to use non-radio-active equipment on tunnelling jobsites. This will increase the safety on the jobsite, reduce potential environmental hazards and moreover it will make the handling, storage and transport of density meters much more easy.

The next step is the install the meter on an actual tunnelling job. This is planned to be done by the end of 2018 on a slurry shield TBM operating in Egypt. The meter will work in series with the standard RA meter to establish the performance. The comparison of the two meters will be based on the theoretical density that can be calculated and validated by taking samples of the slurry and actually measure the density by weight. This will be the reference for both the RA and RF meter and will allow to make a data based comparison of the two methods. Not only the incidental density will be monitored, but also the reactivity to changes in density, stability of the output signal and influence of larger blocks in the pipeline will be assessed.

REFERENCES

Duhme, R., Rasanavaneethan, R., Pkianathan, L., Herud, A. 2015. Theoretical basis of slurry shield excavation management system, *International Conference on Tunnel Boring Machines in difficult grounds*. Singapore.

Fam, M. A. 1996. Study of clay-cement slurries with mechanical and electromagnetic waves. *Journal of Geotechnical Engineering.* 122(5):365–373.

Keysight Technologies. 2017. Basics of measuring the dielectric properties of materials. *Application Note.* 5989-2589EN.

Zych K. & Osnabrugge J. 2017. Non-radioactive slurry density measurement for inland dredgers. *CEDA Dredging Days 2017.* Rotterdam.

Tunnels and Underground Cities: Engineering and Innovation meet Archaeology,
Architecture and Art, Volume 5: Innovation in underground engineering,
materials and equipment - Part 1 – Peila, Viggiani & Celestino (Eds)
© 2020 Taylor & Francis Group, London, ISBN 978-0-367-46870-5

Artificial intelligence support for tunnel design in urban areas

M. Hafner, D. Rajšter & M. Žibert
Elea iC, Ljubljana, Slovenia

T. Tušar, B. Ženko & M. Žnidaršič
Jožef Stefan Institute, Ljubljana, Slovenia

F. Fuart & D. Vladušič
XLAB, Ljubljana, Slovenia

ABSTRACT: The article describes a comprehensive integrated project delivery approach based on digital transformation of the classic BIM (Building Information Model/Modelling) workflow and integration of project stakeholders by means of a single cloud-based IT platform. The topic falls into BIM technology and associated processes to support design, construction and operation of underground traffic connections based on maximal utilization of underground space in urban areas. The step forward is made by putting Information in central place of BIM. Through this approach the requirements are set for state-of-the-art analytic methodologies - artificial intelligence utilization in civil engineering as seen in other engineering branches. Most important are state of the art support services for optimization and decisions, better communication between the client, experts and public, interdisciplinary collaboration between disciplines by multi-criterial decision support process.

1 INTRODUCTION

Research results in the fields of AI (Artificial Intelligence) and cloud-based ICT (Information Communication Technology) solutions are becoming part of our everyday life through a rapid growth of available applications, services and products based on those technologies. They have triggered the digital transformation of many human activities to the point where traditional business models have been disrupted by the new digital economy. However, the adoption of these approaches within the construction practice for tunnel design and infrastructure design in general is still relatively slow and limited.

In recent years, big steps were made in infrastructure design using the practically established BIM approach by market available software, which enables greater parametric control of the infrastructure and structural elements' data. But there are some drawbacks in the current BIM process, which are very clear:

- The purpose of BIM in its current state is not oriented towards the analysis of infrastructure effects and consequences of different design solutions. An important step forward was achieved by enhanced control of the input data, which further enabled parametric variation of design variables and control over them;
- Its inability to share data between models implemented with software tools offered by different suppliers (software companies);
- The traditional, linear design workflow is divided into numerous distinct steps from formalizing the client's idea to defining the architecture, geotechnics, structural design, etc., until forming of the final solution is confirmed. This approach is limited due to its linear workflow performed as a sequence of distinct steps (Figure 1), which seriously hinders the

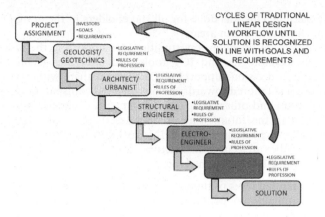

Figure 1. Traditional design process workflow.

possibilities for controlled consideration of many interconnected infrastructure characteristics and effects.

The consequences of these drawbacks include the possible loss of some important information within the workflow and the inability to efficiently control the workflow itself. In addition, the overall solution optimization and decision-making tends to be very limited, expensive and highly sensitive to human errors and obstacles due to the hierarchical nature of the workflow. This is even more problematic in the cases with complex, mutually connected interdisciplinary problems that include clashes with existing infrastructure and assets in urban areas where cost/benefit and overall performance of the transport solution is very hard to achieve. To better address this challenging task, the traditional design approach was transformed into a data interconnected model, called the Information Model (IM), and a newly developed, cloud-based workflow. The IM workflow offers an important opportunity for tunnel design since it enables a clearly defined mapping of infrastructure variables (infrastructure information, element data, etc.) into infrastructure effects and consequences (see Section 2), which in turn are vital for a thorough and transparent examination and analysis

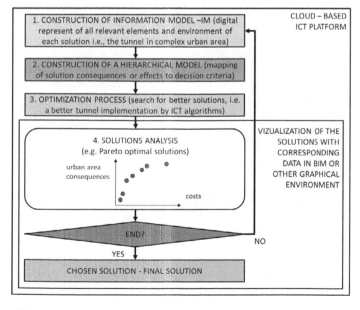

Figure 2. Steps of the proposed infrastructure design process.

needed for decision making. This was done for all stakeholders in the design process to fulfill interoperability requirements in urban areas.

From the described, we are opening opportunities for supporting the design process with the state-of-the-art AI technologies and ICT tools that integrates technologies for multi-objective optimization and decision support, and implement them with efficient cloud-based technologies. The process is tailored towards the optimal and transparent design of tunnels in the most challenging urban and other complex environments experienced in our construction practice and anticipated in the future. The process described in this article is particularly tailored to the tunnel alignment design, but the IM workflow platform is quite general and is applicable also to many other areas of infrastructural design:

The above described workflow is designed for:

- Detailed and transparent analysis of tunnel alignment in complex urban areas;
- Interactive, interdisciplinary cooperation between the client, experts and non-experts, which is vital for building infrastructure in urban areas;
- Public participation and interactive cooperation of non-experts through a collaborative web-based platform;
- The solution offered as SaaS (Software as a Service). SaaS is a software licensing and delivery model in which software is licensed on a subscription basis and is centrally hosted (https://en.wikipedia.org/wiki/Software_as_a_service).

2 INFORMATION MODEL

2.1 *Description*

To fulfill the requirements of AI to support the tunnel design, an enhanced information model as well as Level 2 BIM technology and associated processes are required. To achieve this, we need to consider the following tunnel design features:

- The nature of the tunnel design workflow. Each expert or non-expert participant (designer) needs to be included in the model concerning its responsibility and autonomy. This is done by partial models that communicate with other partial models via shared data space (input and output variables).
- Model requirements. The IM merges the project assignment and the model's partial data from the designers and stakeholders. This means that the data exchange is needed between traditionally used drawings, tables, numeric values, BIM, etc. This is achieved by standardizing input and output variables and storing them in a centralized, cloud-based, data repository.
- Data management requirements. The partial models and the project assignment need to be digitized in the form capable of representation of solutions by variables. Numerous solutions are expected to be explored by the optimization algorithms; therefore, digitalization is required regardless of the level of expertise of the designers involved.

To fulfill those, the IM is only one step in the workflow to acheive one-to-one mapping from decision space to effects/consequences of each generated solution (see Figure 3).

Each solution in the IM is constituted by many variables:

- Decision space (or search space) variables. Only few variables are relevant for optimization, e.g. the tunnel alignment axis, the tunnel cross-section, etc. To find the best solution according to the given criteria, the decision space is explored during the optimization phase.
- Other variables are defined by the digital project assignment (client) and partial models (by expert or non-expert designers).

To achieve the proper functionality of the IM to reflect the problem, exact data flows need to be established. This can be described by the following example: Figure 4 shows an example

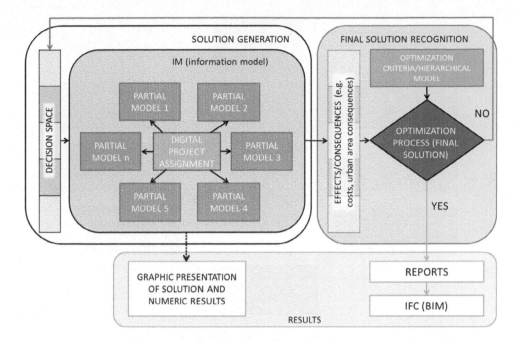

Figure 3. The IM in the workflow.

of the data flow between the partial models and the digital project assignment in the IM. Each designer (expert or non-expert) operates with its own specific data (e.g. rules of profession, legislative boundaries, etc.), which are represented by the corresponding partial model. The new approach of data sharing between partial models works as follows:

- The project assignment is the same for all participants in the IM, therefore it is digitized and shared with all partial models (designers and stakeholders);
- The decision space (DS in Figure 4) must be the same for all partial models for the generation of each solution;
- Different participants need different information or must provide it to specific partial models;
- Generated solutions are/need to be admissible. Since partial models include specific legislative or profession-based requirements and limitations, the final solution of the IM includes all these features, which lead to admissible solutions for the optimization process;
- The approach enables equal involvement of non-expert stakeholders for which effects or consequences can be evaluated;
- Each partial model outputs its own effects/consequences (E/C in Figure 4) that are then gathered by the hierarchical model and used in the optimization process.

2.2 *Relation to BIM*

By the solutions proposed for the IM we are making a step from the effort focused on upgrades of traditional Level 2 BIM practices into open BIM or Level 3 BIM (as defined in Sacks et. al, 2018). This is done by the data-based approach – approach where the information occupies the central role of the BIM process. Drawbacks of Level 2 BIM described in chapter 1 are alleviated by the collaboration between relevant disciplines through IM which connect separate parametric models or parametric objects. This means that traditional BIM tools remain as a constituting part of the central information model (IM) e.g. Revit, ARCHICAD, Grasshopper for Rhinoceros etc. This way a new workflow and process can offer the following benefits:

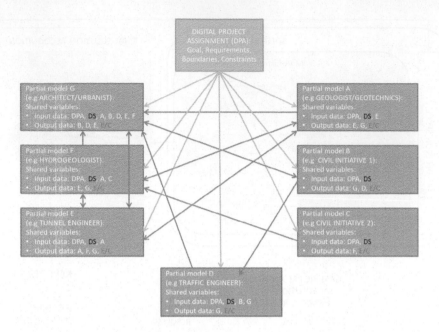

Figure 4. Shared variables between partial models of the IM data flow (an example).

- Full collaboration between disciplines on a single model;
- All stakeholders can access and modify the same model in accordance with their level of authorization rules. IM is designed for this task;
- Model variables are stored in a centralized, cloud-based, data repository;
- Risk of conflicting data is minimized.

2.3 Test model

The IM approach was tested for the case of road tunnel alignment optimization with the focus on mapping from the decision space to the effects/consequences for each generated solution, which is a prerequisite for being able to use the optimization algorithm. A simple example of determining the optimal alignment solution is shown in Figure 5.

For this purpose, we have constructed an interdisciplinary parametric model in Dynamo plug-in for Revit (http://dynamobim.org/). The test model supports automatic generation of the necessary infrastructural elements (tunnels, excavations, embankments, bridges, etc.) on the generated route axis. The route axis is defined by control points which define decision space of the analysis:

With such a model, the calculation of the corresponding effects/consequences for each solution can be done automatically. Figure 6 shows the four results obtained when one or more effects/consequences were selected as the sole criterion for the optimization:

This test has shown that the model is adequate for the proposed optimization purposes:

- Decision space variables or infrastructural attributes are values which can be connected to the digital project assignment or shared by other models;
- Effects'/consequences' values for each solution are obtained automatically as predefined decision criteria.

The example shown in figure 6 shows only the result of optimization on the basis of one criterion – one criterion for each subfigure. In case of multiple criteria, the optimal solutions are not so obvious. This is even more important when the criteria are mutually exclusive. The approach for such cases is explained in Sections 3.3 and 3.4.

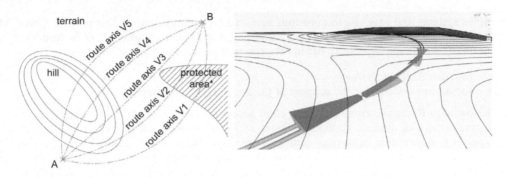

Figure 5. IM implementation example for automatic generation of the infrastructural elements using the parametric model.

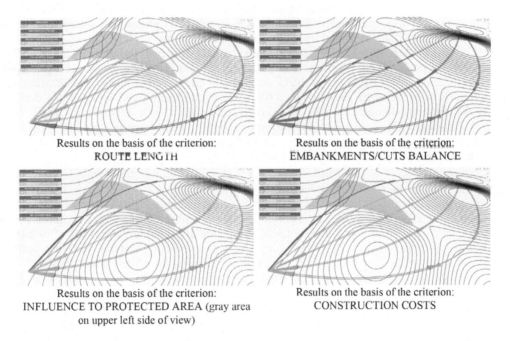

Results on the basis of the criterion:
ROUTE LENGTH

Results on the basis of the criterion:
EMBANKMENTS/CUTS BALANCE

Results on the basis of the criterion:
INFLUENCE TO PROTECTED AREA (gray area
on upper left side of view)

Results on the basis of the criterion:
CONSTRUCTION COSTS

Figure 6. Results of the optimization of the test model for different optimization criteria (criteria 1 - 4).

3 AI TECHNOLOGIES

Steps defined in previous chapters are prerequisites for implementation of AI technologies:

3.1 *Artificial intelligence*

Artificial intelligence (AI) is a science and research field investigating methods and technologies that enable machines (computers) to perform tasks that typically require human intelligence (a quite loose definition, a more detailed discussion on different definitions can be seen, e.g. in (*Russell & Norvig, 2014*)). Although the development of AI started already after the Second World War, within the last decade we are witnessing the appearance of many AI-based applications and services offering support, as well as previously unimaginable solutions in different areas of human endeavor, including science, engineering and everyday life. Currently, the most prominent area of AI is machine learning, which is typically used for data-driven modeling of complex systems and gave rise to notorious applications such as chess and go playing. Other AI disciplines include decision support systems and evolutionary computation. It is the technologies from these two disciplines that we use in the proposed optimization workflow:

3.1.1 *Decision Support Systems*
Are information systems for the support of the approach in the form of:

- Knowledge Representation is an important aspect of decision models, which serve also as a formalization of domain knowledge about various decision factors and their relations. By applying algorithmic reasoning and analysis capabilities on structured and formalized human-provided domain knowledge, we are combining the best capacities of humans and computers. Decision models therefore serve as knowledge representations and mechanisms for transparent and elaborate reasoning and simulations. This relates well with decision making in tunnel design in urban areas, which is influenced by many goals, opinions and interests, which are hard to grasp for a human to reason about, but also impossible for a computer to efficiently learn from empirical data. Even by using the models for knowledge representation only, one can recognize a limited set of good options from a large number of acceptable ones that are gained from optimization algorithms.
- Multi-Criteria Decision Modelling (MCDM) is aimed at formal (mathematical) modelling of decision problems that consider a multitude of criteria. This kind of model represents a formalization of the specified problem and usually enables assessment, visualization and comparison of decision alternatives. There are several established and mature decision modelling methodologies like *MAUT (Keeney and Raiffa, 1993), AHP (Saaty 2008), Electre (Figueira, 2005) and DEX (Bohanec et al., 2013)*. The last one is especially well suited for problems that are hard to quantify and influenced by many goals, opinions and interests, such as the problems related to Smart cities.

3.1.2 *Multi-Objective Optimization by Evolutionary Algorithms.*
Evolutionary algorithms are iterative methods that improve on solutions using principles that mimic the natural evolution such as selection, crossover and mutation (*Eiben & Smith, 2003*). Because they operate on populations of solutions, they are especially appropriate to handle multi-objective optimization problems, where, due to conflicting criteria, multiple optimal solutions exist—each representing a different trade-off among the criteria (*Deb, 2001*).

The use of AI technologies in optimization and decision support helps automate and speed up the tunnel design process, while the final decision is still made by humans (the decision-making team). Additional benefits of using AI technologies in the proposed process are:

- Effective exploration of the decision space that enables us to find a large number of admissible solutions, some of which could not be conceived by the designers.
- The objectiveness (given the provided inputs and rules) and increased transparency of the decision-making process reduces the risk of subjective judgements of stakeholders, omission of important consequences, lack of design goals and violation of valid constraints.

On the other hand, using AI technologies in such application has its limitations:

- If the decision space is large and the partial models very complex, there might not be enough time for the optimization algorithms to find all the optimal solutions. In such cases, the final decision has to be made on near-optimal solutions.
- Although AI techniques speed up and automate the design procedure, they are only as good as their input. If the problem is not defined well, the returned solutions might not be appropriate or even admissible.

3.2 *Construction of a hierarchical model*

The construction of a decision model is a collaborative process in which domain experts provide necessary domain knowledge, such as relevant variables, target concepts and the rules

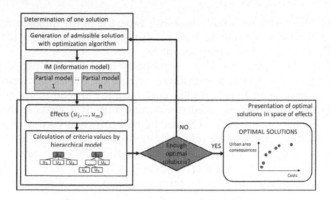

Figure 7.   Multi-objective optimization in the proposed process.

and constraints that need to be considered. After this is done, a decision analyst takes care of proper and effective formalization. In our proposal, the hierarchical MCDM methodology is used to define the aggregation transformations of raw data inputs (measurements) into higher level concepts (criteria), which are afterwards used in the optimization process (see Figure 7, bottom left).

3.3   *Multi-objective optimization*

Figure 7 shows how multi-objective optimization is used to find optimal solutions in the tunnel alignment problem. The two considered criteria are the costs (to be minimized) and the benefits to the urban area (to be maximized). These two criteria are conflicting, i.e. achieving favorable urban area consequences usually comes at a high cost, while low-cost solutions can bring unwanted consequences to the urban area. To solve such a problem, multi-objective optimization algorithms search among all admissible solutions from the decision space so that both criteria are optimized. The optimization process ends when enough (optimal) solutions have been found. The result is a set of solutions.

The recognition of many optimal solutions is the most important improvement over the traditional design practice since all these solutions are non-dominated – they all lie on the Pareto front (the graph on the right side of Figure 7). This means that no solution dominates any other and none of them can be improved in one criterion without degrading the value of the other criterion. The final solution is chosen after the solution analysis step (Section 3.4).

3.4   *Solution analysis*

The criteria used in optimization already have their hierarchical structure defined, as well as the rules of how the criteria are calculated from input values. This allows for a subsequent (post-hoc) analysis of the solutions proposed by the optimization methods, as (particularly *DEX*) MCDM allows for easy and transparent comparison of alternatives, analyses of trade-offs among them and sensitivity analysis.

4   A CLOUD-BASED TUNNEL DESIGN SOLUTION

BIM and other infrastructure design tools provide a practical platform to manipulate data and visualize building elements. The IM involves this toolset as described in previous sections. In addition to these technical needs, with the development of cloud-based solutions we are further digitally revolutionizing the tunnel and infrastructure design industry, introducing new business models and gaining a competitive advantage over other actors. The implementation of the cloud-based IM workflow is built upon our expertise in the introduction of cloud-computing, big data, cybersecurity and analytics solutions across a range of industries, from healthcare to IoT (Internet of Things) applications.

From the stakeholders' point of view, a properly implemented proposed workflow as Cloud Computing solution provides several advantages over classical approaches. It allows actors in the design process to focus on their core business instead of spending resources on the maintenance of own ICT infrastructure. Consequently, ICT costs decrease significantly, as the upfront investment is minimal. The used cloud infrastructure is charged on demand (pay-as-you-go), meaning that during tunnel design optimization/simulation phases there is a surge in use of computational resources, while most of the time those resources are not needed, thus not charged to the end user. Maintaining own servers for this purpose means that effectively, the infrastructure is oversized to support sporadic needs for lots of computational power.

Collaboration, data sharing, secure access and multitenancy are all concepts that are introduced by design from the very beginning of the development process. The cloud-native workflow has been designed and implemented to support those needs. This approach allows effective introduction of the SaaS (Software as a Service) business model in tunnel design. The workflow is licensed on subscription basis to clients/stakeholders, while the system is cloud-hosted. This model allows efficient ICT management and support, allowing clients to obtain access to the system from the moment they need it. There is no need to purchase and deploy additional ICT equipment. The data is stored securely, which includes strictly controlled access rights as well as guarantees of continuous backups and data-accessibility.

5 CONCLUSION

The wave of digitalization (known as BIM-Building information models/modelling) introduced a new design technology, processes and also vision to the architecture, engineering and construction industry. Most important is the digital parametric representation of objects and efficient data management to describe the characteristics, state or behavior of the object through concerned time period. Despite these achievements we see that the state-of-the-art analytic methodologies and ICT in civil engineering are not on the level as seen in other engineering branches.

Based on years of practical experience and knowledge acquired through cooperation with high-tech ICT companies and scientific research institutes, we decided to combine our knowledge and experiences to:

- Enhance traditional design approach based on BIM by putting information (data) in central place of modeling process. The proposed Information Model (IM) enables coherent consideration of non-expert disciplines in tunnel design, construction and operation e.g.: civil initiatives, public participation and other relevant societal groups;
- Develop a practical approach for utilization of state-of-the-art ICT to offer new solutions to tunnel design problems. The support is on the highest ICT level compared to other engineering disciplines.

The results are new workflows and procedures for decision making and optimization support for underground traffic connections based on maximal utilization of underground space in urban areas. This topic is becoming very important due to fast urbanization of society, migrations and other societal changes/challenges of the future.

ACKNOWLEDGMENTS

We acknowledge the financial support from the Slovenian Research Agency (research core funding No. P2-0103 and research project No. Z2-8177).

REFERENCES

Bohanec, M., Žnidaršič, M., Rajkovič, V., Bratko, I. & Zupan, B. 2013. DEX methodology: three decades of qualitative multi-attribute modeling. Informatica 37: 49–54.

Deb, K. 2001. Multi-Objective Optimization using Evolutionary Algorithms, John Wiley & Sons.

Eiben, A.E. & Smith, J.E. 2003. Introduction to Evolutionary Computing, Springer-Verlag Berlin Heidelberg.

Figueira, J., Mousseau, V. & Roy, B. 2005. ELECTRE methods. Multiple criteria decision analysis: State of the art surveys: 133–153.

Keeney, R.L. & Raiffa, H. 1993. Decision with Multiple Objectives: Preference and Value Tradeoffs. Cambridge University Press, New York.

Russell, Stuart J & Peter Norvig. 2014. Artificial Intelligence: A Modern Approach, Pearson.

Saaty, T.L. 2008. Decision making with the analytic hierarchy process. International journal of services sciences 1: 83–98.

Sacks, R., Eastman, C., Lee, G. & Teicholz, P. 2018. BIM Handbook: A Guide to Building Information Modeling for Owners, Designers, Engineers, Contractors, and Facility Managers. John Wiley & Sons.

Tunnels and Underground Cities: Engineering and Innovation meet Archaeology,
Architecture and Art, Volume 5: Innovation in underground engineering,
materials and equipment - Part 1 – Peila, Viggiani & Celestino (Eds)
© 2020 Taylor & Francis Group, London, ISBN 978-0-367-46870-5

Key technologies of double shield TBM for urban metro tunnel with small radius curve

F. He, X.J. Zhuo & L.H. Jia
China Railway Engineering Equipment Group Co., Ltd., Zhengzhou, Henan, China

ABSTRACT: With reference to the construction conditions and features of metro tunnel, this paper analyzes the design features of double shield TBM and proposes key issues to be considered and settled when double shield TBM is used for metro tunnel construction, including small curve excavation, cutter head's rock-breaking capacity, selection of backfill grouting technology and excavated material removal method, which all have a direct effect on the geological adaptability, tunnel lining quality and tunnelling performance of double shield TBM. This paper also studies the specific design and optimization scheme which includes the design of cutter head thick steel plate, stepped shield and monorail hoist. The successful performance of double shield TBM(φ6.5m) in Shenzhen metro project well proved its remarkable geological adaptability and advantages in mechanized construction(Max. daily advance 24m). Besides, this paper puts forward advice on the application of double shield TBM in the future engineering projects.

1 INTRODUCTION

Double shield TBM, one type of full-face hard rock tunnel boring machine, was firstly applied to Diverting Datong River into Qinwangchuan Irrigation Project in Gansu Province in 1991 (Qiao 2015; Gong 2014). Then this type of TBM was used in succession in headrace tunnels such as Yunnan Zhangjiuhe Water Conservancy Project, Xinjiang Bashiyidaban Diversion Tunnel Project, the Main Channel of Drawing Water from Datong River into Huangshui River Project in Qinghai Province, Wanjiazhai Yellow River Diversion Project in Shanxi Province, Water Diversion Project of Transferring Water from Hongyan River to Shitou River in Shaanxi Province etc. Moreover, double shield TBM was firstly applied to urban metro project in China in 2014, and by far it is mainly used in Qingdao and Shenzhen City. Unlike mountain tunnel, metro tunnel has its own characteristics, such as short tunnelling section, small radius curve, frequent station passing, restricted construction site, and startup from launching shaft. Consequently, the design of double shield TBM in accordance with the mountain tunnel no longer has universal applicability and needs to consider the actual construction circumstances and characteristics. Taking Qingdao Metro Line 2 as an object of study, reference(Huang 2016; Lin 2014) analyze the geological factors affecting tunnelling efficiency and the geological adaptability of the double shield TBM for Qingdao metro tunnel, and point out that Qingdao metro tunnel is more suitable for using double shield TBM on the basis of compressive strength, rock integrity and abrasion resistance etc. According to the construction method of station, reference(Tang 2016) proposes compounding lining method (anchoring and shotcreting+mold lining) in consideration of the difference between several support forms used in different drives, so as to address the problems of repeated disassembly, assembly, and the impact of commissioning on the station. Reference(Li 2006; Guan 2003; Chen 2014; Xia 2014) expound on the technical features and geological adaptation range of double shield TBM, and analyze the strong and weak points of double shield TBM construction. Reference(Sun 2016) analyzes the ultimate bearing capacity of segment lining based on some power station diversion tunnel and points out that the maximum bearable external water pressure of the segment has a linear relation with the gap width, concrete strength and segment thickness, and a quadratic curve relation with rock modulus of deformation. Reference(Wen 2008) analyzes the risk of TBM jamming based on some tunnel project and points out that when the TBM encounters difficult geological conditions such as faults and fracture zones, rock reinforcement measures should be taken in advance. Pre-grouting and proper adjustment of tunnelling parameters could reduce the risk of TBM jamming as well.

Double shield TBM technology has been matured as early as last century, and merely German Wirth has manufactured over 200 double shield TBMs with a variety of diameters, mostly applied for mountain tunnels. However, the use of double shield TBM in metro tunnel is still at the preliminary stage with a very few application cases. Besides, up to now in China the double shield TBM is mainly used in Qingdao metro and Shenzhen metro. The author has participated in the TBM selection, design, manufacture and application of some double shield TBM projects and summarized some experience for reference.

2 THE FEATURES OF DOUBLE SHIELD TBM FOR METRO TUNNEL

Double shield TBM has two excavation modes (double shield mode and single shield mode). Under the double shield mode, the thrust of TBM is not transferred to segment by auxiliary cylinders, but to the tunnel wall via the grippers. The double shield TBM demonstrates remarkable advantages particularly when enormous thrust is required under extreme hard rock ground condition, because it allows for segment erection parallel to tunneling and has high excavation performance. On the contrary, in weak rock and fractured ground, the support provided by the tunnel wall is limited. Therefore, the single shield mode is used, which relies on the segment ring to provide thrust. Because of its high efficiency in hard rock condition and reliability in weak rock ground condition, the double shield TBM is increasingly used for excavating metro tunnel where the ground is dominated by hard rock. However, the metro tunnel has its own characteristics, which requires scientific selection of the double shield TBM in order to meet the requirements of the metro tunnel project.

(1) Since major cities in China have different geological conditions, the cities using double shield TBM are mainly located in the eastern or southeastern coastal area, such as Shenzhen Metro Project. The tunnel alignment of the project mainly goes through slightly to moderately weathered granite whose quartz content is high and uniaxial compressive strength largely falls between 80MPa - 180MPa. Thus, under the hard rock condition efficient rock-breaking capacity is required.

(2) The distance between two MRT stations is generally 500m-3000m, which means that the distance of continuous excavation is short and the TBM has to pass station frequently. In addition, due to the restricted construction space, TBM is usually assembled in the launching shaft. Therefore, the length of TBM main machine, single piece and the whole machine should be as short as possible, and the structure and size of the TBM components should be suitable for easy disassembly, transportation, lifting and assembly, so as to assemble the TBM within the shaft and shorten the whole assembly time.

(3) Since the metro tunnel alignment is subject to conditions such as buildings and existing city layout, the curvature should be properly designed. Contrary to the straight alignment or big radius of curve of diversion tunnels, metro tunnel's radius of curve is generally small between R250m and R400m. In that sense parameters such as the length/diameter ratio of the TBM main machine, annular gap between the shield and the tunnel wall, and segment hoist track should be specially designed, so the machine could perform small curve excavation.

(4) According to the Code for Design of Metro: GB 50157 and Code for Construction and Acceptance of Shield Tunnelling Method: GB 50446, the design life of metro tunnel's major structure is mostly 100 years. Thus, it has stringent requirement for segment erection quality especially for the control of altitude difference among segments and segment damage in the erection process. Apart from that, it is generally required that TBM should meet the demand of all erection points and pay special attention to the back-grouting

technology and stability of the segment. Different from pressure-balanced shield, double shield TBM is under open mode, and uses pea gravel instead of synchronous grouting for back-fill. So, the grouting technology and its quality shall have a direct impact on the segment lining quality.

(5) When passing through faults and fracture zones, the TBM might run into the risks of unstable tunnel face, water and mud ingress, and jamming because the rock is unstable and the machine does not have the pressure-balancing function. Double shield TBM can be transformed into single shield mode, but the machine still faces the risk of entrapment owing to the long shield body. For this reason, double shield TBM should be able to prevent jamming and release itself once got stuck.

3 SPECIFIC DESIGNS FOR DOUBLE SHIELD TBM

According to the characteristics of metro tunnel excavation, the design of double shield TBM used for metro tunnel needs to attach importance to the following aspects and be carried out accordingly: efficient rock-breaking capacity, jamming prevention and breakout function, small radius curve as well as backfill grouting.

Figure 1. Typical TBM double shield composition.

Figure 2. Segment geometry.

3.1 Specific design for construction on small curve

The small curve adaptability of double shield TBM mainly relies on the articulation of the shield, the length and width of the biggest inseparable components and the capacity of the track walking mechanism etc.

(1) By enlarging the initial eccentricity of TBM cutter head, the gap between the shield and rock is relatively big and the gap between telescopic shields is also enlarged. Theoretical simulation should be carried out based on actual alignment radius. While making sure the TBM could successfully pass through, the erection quality of the segment should also be guaranteed. When necessary the cutter head excavation could be expanded on the construction site in order to enlarge safety margin.

(2) The travelling wheel of the back-up trains should well match the track. On the given radius of curve, the rim of the front and back wheels should be able to walk on the track with some margin left. The articulation between trains should also meet the curve requirement.

(3) The travelling of the segment hoist should match the track. The segment hoist uses monorail rack engagement structure which meets the curve requirement at the articulation position, so as to guarantee no rack or chain drops off.

(4) The main machine belt conveyor is usually long (about 20m). When the TBM turns, it tends to interfere with nearby structures. Therefore, the safety gap should be enlarged, and proper safety margin should be left between the rims of belt conveyor driving wheels and track.

In actual construction, attention should be paid to the influence of TBM excavation axial deviation, segment erection error and TBM shield partial deformation. The following measures should also be taken:

①Use segments with relatively small width (such as 1200mm);
②Expanding excavation should exceed 50mm (radial direction);

3.2 Specific design for faults and fracture zones

Throughout the alignment difficult conditions such as faults and fracture zones will be encountered due to the geological structure. When passing through these ground conditions, the TBM is faced with the risk of cutter head jamming, machine jamming and unstable tunnel face. Hence, while obtaining full understanding of the geological conditions along the tunnel alignment, special design and emergency measures for the above risks should be considered in advance.

(1) In order to reduce the risk of unstable tunnel face and collapse, the contact area beyond the cutter head periphery should be as small as possible, so as to reduce the disturbance to the surrounding rock during cutter head rotation. Besides the speed of the cutter head rotation should also be reduced, so that the TBM can pass by slowly.

(2) The excavated material is removed through the muck scoop scattered around the periphery of the cutter head. Therefore, the design of the muck scoop directly determines the performance of excavated material removal. In the fracture zones, the rock chips caused by unnormal rock-breaking such as rock collapse could cause the stuck of the belt conveyor. So attention should be paid to the size of the muck scoop especially in fracture zones. If necessary, part of the muck scoop could be blocked so as to reduce the excavated material removal volume and maintain the stability of the rock.

(3) The cutter head is equipped with variable frequency drive and has a breakout torque which is 1.5 times of the rated torque or even bigger breakout torque. When the cutter head gets stuck, the big breakout torque is activated in order to reduce the risk of TBM jamming. Besides the thrust system is designed with high-pressure breakout mode. Once the thrust resistance increases, the big thrust mode is on use, which could avoid stoppage and TBM jamming caused by insufficient thrust.

(4) Tapered shield design. Because the double shield TBM main machine is relatively long and its length/diameter ratio is big, it faces a high risk of jamming, especially in soft converging ground. Stepped shield design enables the successive decrease of the shield diameter, which

therefore reduces the risk of TBM jamming. Besides proper gap is left at the crown area of the front shield, which also helps to reduce the risk of TBM jamming.

(5) TBM expanding excavation design. The most direct and easy way for TBM expanding excavation is to install cushion blocks at the corresponding gauge disc-cutter support or thickening the C-shaped block, so as to enlarge the excavation diameter of the cutter head and realize the mechanical expanding excavation. In order to make sure the gap between the cutter head bottom and the shield stays unchanged and to avoid the fall-down of the main machine due to overcut at the bottom, the cutter head is lifted by the main drive's lifting function. Therefore, the expanding excavation volume is confined within the crown area, which reduces the risk of TBM jamming.

Apart from that, the TBM shield should be equipped with proper emergency windows for the convenience of manual emergency measures when necessary. At the same time, it should also leave some space for the installation and operation of probe drill for the convenience of grouting reinforcement.

3.3 *Specific design for cutter head with efficient rock-breaking capacity*

Efficient rock-breaking is the first issue to be considered for the TBM, because it directly determines the geological adaptability of the TBM cutter head. Efficient rock-breaking depends on the reasonable design and reliable quality of the cutter head and cutters, which include the following aspects:

(1) High strength design of cutter head plate. Currently there are two types of cutter head plate. One is tailor-welded with thin steel plate (about 90mm). This type of welding requires too much workload, produces too many welds and leaves limited space for the cutters. Although it is cost-effective and time-saving, its fatigue resistance capacity is insufficient which leads to high risk of cracking and low reliability. The other type is solid or local forging thick steel plate (about 300mm). The tool support holes are machined out of the plate itself. Therefore, the total number of welds and stress concentration are reduced, and the cutters could be evenly spaced. Besides the reliability of the cutter head plate is enhanced as well. In comparison of these two types, they both have their respective advantages. But in hard rock or extreme hard rock, the reliability of the cutter head plate is the priority to be considered.

(2) Proper cutter spacing design. The rock-breaking mechanism of hard rock TBM is different from that of the pressure-balanced shield. TBM is equipped with heavy duty disc-cutters, which break the rock by their rolling force. Under sufficient thrust, proper cutter spacing could achieve crack coalescence and fragmentation with one-time rolling, which spares the secondary rock-breaking and significantly reduces extra abrasion. In consideration of the rock strength of Shenzhen metro tunnel, the TBM cutter spacing is designed around 80mm, as shown in figure 3. With proper cutter spacing design, rock chips can be broken out of the rock with smaller penetration force. While ensuring good rock-breaking performance, small cutter spacing can also effectively reduce the vibration of the cutter head and prolong the service life of the cutters.

(3) Non-linear cutter arrangement. There are several cutter arrangement patterns. In consideration of the cutter head plate structure, some cutter heads (tailor-welded thin steel plate) can only adopt linear arrangement (figure 4) whose cutters are densely spaced. The thick steel plate cutter head on the contrary could make the best use of its surface and use non-linear cutter arrangement (figure 5). Therefore, the cutters can be evenly spaced and its dynamic balance is more stable. Besides, this design could avoid high stress concentration and therefore prolong the service life of the cutter head.

Figure 3.   Cutter spacing.

Figure 4. Linear arrangement type cutter head.

Figure 5. Non-linear arrangement type cutter head.

3.4 *Specific design for backfill grouting*

The ring distribution (outer diameter 6m) of metro tunnel usually adopts "3 standard segments+2 adjacent segments +1 key segment" pattern, of which the lifting holes are the corresponding backfill grouting holes. The back-grouting technology of double shield TBM is different from that of the regular pressure-balanced shield and domestic engineering projects often use different back-grouting technologies. It is recommended to use the combination of pea gravel backfill and cement slurry grouting based on the comparison of various technologies and their performance.

(1) Pea gravel backfill

When the TBM advances, the erected segments gradually emerge from the rear shield. When the central holes of the bottom segments leave the grout excluder, the pea gravel back-fill of the bottom segments could begin. The operation starts from the bottom two segments, then the segments above the tunnel centerline and at last the top segments. This kind of backfill can guarantee the backfill quality and effectively prevent altitude difference among segments. In addition, pea gravel backfill should be completed within 5 rings from the rear shield.

(2) Cement slurry grouting

The conventional cement slurry grouting has two ways: one is to prepare cement slurry inside the tunnel. Namely the cement powder is mixed with water according to a certain proportion on the back-up gantries and then the finished cement slurry will be pumped to

Figure 6. Linear arrangement type cutter head.

grout; the other is to prepare cement slurry outside the tunnel. Namely the cement mixed outside of the tunnel will be transported into the tunnel by tankers with mixing function, so as to prevent slurry from solidifying on the way.

The TBM used for metro tunnel mostly launches in the working shaft where uses portal crane for vertical lifting. Regarding the fact that TBM excavation drive is often short, it is recommended to prepare the cement slurry outside of the tunnel and use pipeline to feed the slurry into the tunnel (figure 6). During the transmission process, solidification of the cement slurry inside the steel pipe can be effectively prevented by controlling the flow rate and pressure of the cement slurry. Besides to rinse the pipeline after each transmission also helps.

There are two circumstances for cement slurry grouting: (1) Under normal condition, it is operated in back-up gantry where it is easy to grout continuously to achieve efficient grouting with excellent quality, because all grouting holes are exposed in the tunnel. (2) When necessary, the segments located four rings ahead and five rings behind of the total station should be grouted in advance to guarantee the segment stability near the installation position of the total station. This grouting area is usually located in the connection bridge where is suitable for full-ring grouting.

4 APPLICATION CASE OF DOUBLE SHIELD TBM

4.1 *Project profile*

The double line tunnel from Maling station to Yabao station of Shenzhen metro line 10 uses two double shield TBMs which were independently designed and manufactured by China Railway Engineering Equipment Group. The total length of the section is 3.8km including 2.96km-long shield tunnel (figure 7). The tunnel mainly goes through the slightly weathered granite (accounting for 95%) and partially moderately weathered granite. The maximum rock strength is over 120Mpa, and the average is around 100Mpa. Along the tunnel alignment there exist F1 and F2 faults and local structurally fractured zones with 0.6–9.9m underground water level which mainly consists of pore phreatic water and bedrock fissure water. The excavation diameter isφ6.5m, and the deepest buried depth of the tunnel is 232m. The advancing direction of TBM is one way uphill, among which the maximum longitudinal slope is 27‰, the maximal longitudinal length of 112m and the minimum radius of plane curve of 350m. This project adopts the ring distribution of "3 standard segments+2 adjacent segments+1 key segment" pattern and the outer diameter, thickness and width of the segment are 6200mm, 400mm and 1500mm respectively. Besides, pea gravels and cement slurry grouting are used for backfilling.

4.2 *Technical parameters and TBM performance*

Main technical parameters of double shield TBM
TBM performance
Subject to the site space restriction, the project uses the launching shaft and portal crane to vertically lift the material and rock chips. The launching shaft is 44m deep and the vertical

Figure 7.   Layout plan between Maling station and Yabao station.

Figure 8.   Longitudinal profile.

Figure 9.   The photo of launching shaft.

Table 1.   Main technical parameters of TBM.

Item	
Excavation diameter(mm)	φ6500
Length of main machine(m)	about 11.7
Overall length(m)	about 135
Radius of curve(m)	≯300
Rock chips removal capacity(t/h)	1020
Cutter head power(kW)	2100
Cutter head rotation speed (rpm)	0-4.96-10.28
Rated torque(kNm)	4040
Breakout torque(kNm)	6318
Thrust speed(mm/min)	0-120

Figure 10. The layout of launching shaft.

Figure 11. The photo of launching shaft.

Figure 12. TBM launching photo.

Figure 13. Completion photo.

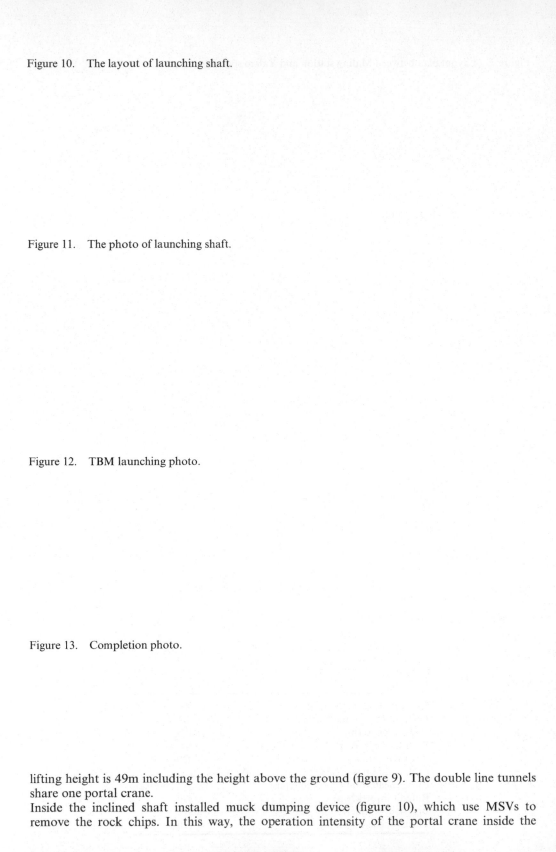

lifting height is 49m including the height above the ground (figure 9). The double line tunnels share one portal crane.

Inside the inclined shaft installed muck dumping device (figure 10), which use MSVs to remove the rock chips. In this way, the operation intensity of the portal crane inside the launching shaft could be enormously reduced, which helps to save excavated material removal time and increase excavation efficiency.

In order to guarantee the back-grouting quality of segment, rounded natural pea gravels were used (figure 11), which have great fluidity and exert low abrasion on the pipeline.

The TBM started excavation in March 2017, and its right line achieved breakthrough on the 22nd December 2017 (figure 12 and 13). The accumulated excavation length is 2840m, excavation speed is 20–60mm/min, thrust is 7000–12000KN, maximum daily and monthly advance are 24m and 468m respectively.

5 CONCLUSION AND RECOMMENDATION

In accordance with scientific selection and reasonable specific design, the application of double shield TBM allows the machine to achieve highly efficient rock-breaking capacity, to construct on small curves, to pass through faults and fracture zones and to guarantee the high erection quality for metro tunnels in hard rock. Together with installations such as muck dumping device inside the tunnel, the double shield TBM could give the best play of its high excavation efficiency and shorten the construction time. However, since the metro tunnel removes excavated materials via shaft instead of continuous belt conveyor, its tunneling efficiency is affected to some extent. It is suggested that new construction methods should be developed to improve the effectiveness of rock chips transportation for metro tunnel such as adopting vertical lifting belt conveyor. In the above-mentioned project, the muck dumping device using inclined shaft to remove rock chips is a positive and effective try. It is believed that double shield TBM will have wider application space in the excavation of metro tunnel by reasonably improving the removal method of excavated material.

REFERENCES

Chen, Y. 2014. *Research on TBM selecting and key construction technology.* Shijiazhuang: Shijiazhuang Tiedao University.

China Railway Eryuan Engineering Group Co., Ltd. 2017. Southwest Jiaotong University, China Railway First SurveyDesign Group Co., Ltd. Comprehensive technical research of double-shield TBM project of Qingdao Metro: Key technology of design research. Chengdu: Southwest Jiaotong University.

Code for design of Metro: GB 50157-2013. 2013. Beijing: China ArchitectureBuilding Press.

Code for construction and acceptance of shield tunnelling method: GB 50446-2008. 2008. Beijing: China ArchitectureBuilding Press.

Gong, Q.M. 2014. *Summary of tunnel boring machine.* Beijing: Science Press.

Guan, B.S. 2003. *Key points of tunnel engineering design.* Beijing: China Communications Press:11.

Huang, J. 2016. On the geological adaptability of the double shield TBM for the Qingdao Metro Tunnel. *Modern Tunnelling Technology* 53 (3):42–46.

Li, J.B. 2006. Geology adaptability of double shield TBMs and relevant calculations. *Tunnel Construction* 26(2):76–78, 86.

Lin, G.Li, D.C.Zhang, J.X. 2014. Application research of improved shield-type TBM in urban rail transit projet. Proceedings of Symposium on Urban Rail Transit Management and Technology Innovation of 2014 in Qingdao:185.

Qiao, S.S.Mao, C.J.Liu, C, et al. 2005. *Tunnel boring machine.* Beijing: Petroleum Industry Press.

Sun, B.Tang, B.H.Liu, Y, et al. 2016. Ultimate bearing capacity analysis for segment lining of double shield TBM. *Chinese Journal of Underground Space and Engineering* 60(12): 689–695.

Tang, Z.Q. 2016. Key technologies of double-shield TBM applied to urban rail transit. *Railway Standard Design* 60(11):81–89.

Technical code of urban rail transit: GB 50490-2009. 2009. Beijing: China ArchitectureBuilding Press.

Wen, S.Xu, W.Y. 2008. Risk analysis on TBM jamming in deep buried tunnel. *Journal of Yangtze River Scientific Research Institute.* (5):135–138.

Xia, Y.M.Wu, Y.Guo, J.C., et al. 2014. Numerical simulation of rock-breaking mechanism by gage disc cutter of TBM. *Journal of China Coal Society* 39(1):172–178.

Tunnels and Underground Cities: Engineering and Innovation meet Archaeology,
Architecture and Art, Volume 5: Innovation in underground engineering,
materials and equipment - Part 1 – Peila, Viggiani & Celestino (Eds)
© 2020 Taylor & Francis Group, London, ISBN 978-0-367-46870-5

The subgrade reaction modulus method in tunneling

M. Hofmann, T. Cordes & K. Bergmeister
Brenner Base Tunnel BBT SE, Innsbruck, Austria

ABSTRACT: In tunnelling, a method based on the modulus of subgrade reaction is often used to design tubbing elements and the (cast-in-place) concrete inner lining. To this end, the modulus of subgrade reaction is an important parameter. As early as 1980, the "Recommendations for the calculation of tunnels in soft rock" were published by the German society for earthworks and foundation engineering, in which standard approaches for the modulus of subgrade reaction were specified and which are still used widely today. At the time, it was said: "With numerical calculation methods non-linear material laws for the soil and lining can also be taken into account". However almost 40 years later, the subgrade reaction modulus method is frequently used in favor to more recent numerical methods. This article questions whether the assumptions and principles of the subgrade reaction modulus method are still up-to-date.

1 INTRODUCTION

The history of the subgrade reaction modulus method began in 1867: Emil Winkler introduced it to model railway tracks. Starting in 1921, it was used to calculate steel-reinforced concrete foundations (Kurrer, E. 2002).

Several decades later the method was used in structural analysis in tunnelling (Schulze, H. 1964), where it is still often used to design cast-in-place inner linings and tubbing rings. The recommended formulas for the modulus of subgrade reaction were summarized in 1980 in the "Recommendations for the calculation of tunnels in loose rock" (Working Group Tunnelling, German society for earthworks and foundation engineering, 1980) and subsequent literature (Ahrens, H. 1982).

The advantage of the subgrade reaction modulus method as compared to the numeric methods is to be found in its simplicity. Loads can be more easily superimposed and calculations become clearer and can be more easily checked.

This contribution comprises a motivation to the topic in chapter 2, in chapters 3 and 4 the subgrade reaction modulus method for rock mass and for the combined rock mass-backfilling-internal lining systems. Subsequently the question is discussed if the assumptions and principles of the subgrade reaction modulus method in tunnel construction are still up-to-date. As regards (Ahrens, H. 1982), the results apply to tunnels with overburdens over three times the tunnel diameter.

2 MOTIVATION: SENSITIVITY OF THE LOAD-BEARING CAPACITY IN DEPENDENCE OF THE MODULUS OF SUBGRADE REACTION

To display the influence of the modulus of subgrade reaction in the tunnel design a sensitivity analysis with nonlinear 2d load-bearing simulations of a build cross-section of the BBT rescue tunnel is performed.

2.1 Simulation

According to the BBT guide design (Insam R. 2019) the inner lining has to be designed to support totally the rock loads from the outer lining. Hence for the permanent double lining design the load-bearing capacity of the outer lining is not considered. For the analyzed cross-section the present modulus of subgrade reaction was determined by back-analysis of the primary lining to 100 MN/m³ (back-analysis according to Marcher, T. 2019).

For a design of an unreinforced inner lining in this section the load-bearing capacity of the unreinforced lining is simulated with nonlinear analysis with Atena Science 5.4. (Červenka J. 2016). The 30 cm thick, C25/30 concrete lining with an outer radius of 4.35m is radial supported with a modulus of subgrade reaction of 100 MN/m³. For the sensitivity analysis the modulus of subgrade reaction is varied to 10, 50, 100, 500, 1000 MN/m³. The lining is loaded first by dead load (24 kN/m³), second the lining shrinks (-0.18‰), third it is cooled down (-14 K cooling over 55 h) and last it is loaded by the rock load. The rock load is applied by a projected vertical and horizontal load to the lining and is linearly increased until failure. These loading levels of the rock loads are displayed relatively to the loading of the present lining, determined with the modulus of subgrade reaction of 100 MN/m³.

2.2 Result

According to Figure 1 the load-bearing capacity increase strongly with increasing radial modulus of subgrade reaction. Compared to a high sensitiveness at a low modulus of subgrade reaction (10–100 MN/m³) the simulation show marginal differences for high modulus of subgrade reaction (500–1000 MN/m³). This nonlinear increase of the load-bearing capacity results out of different failure modes and different crack patterns.

Figure 2 displays the deformation of the left sides of the inner lining just before failure for different moduli of subgrade reaction a) 10 MN/m³, b) 50 MN/m³, c) 100 MN/m³, d) 500 MN/m³, e) 1000 MN/m³. The cracks in black display the overloaded sections in compression, shear or in tension and illustrate the occurring failure mode.

For the lowest radial modulus of subgrade reaction of 10 MN/m³ (Figure 2a) a compressive failure in the inside of the side-walls is dominating the behavior and the typical flexural crack of unreinforced linings in the crown is propagating to great crack width. Increasing the modulus of subgrade reaction to 50 MN/m³ (Figure 2b) doubles the load-bearing capacity of the lining and reducing the horizontal deformations from 11 cm to 7 cm in the upper sidewalls. Hence the support of the upper arch is improved, firstly by the higher position and secondly by the higher rigidity. From the first it results a reduction of the span width of the upper load-bearing shell and the latter reduces the horizontal deformations and decreases the flexural bending in the crown. The failure mode of Figure 2c improves the load-bearing capacity again analogously. The crushing takes now place already at the top of the side-walls. The 50 % higher normal stress state results in an earlier compressive failure of the uncracked outer

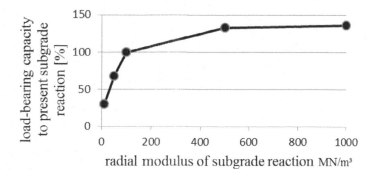

Figure 1.   Relative load-bearing capacity in dependence of the radial modulus of subgrade reaction.

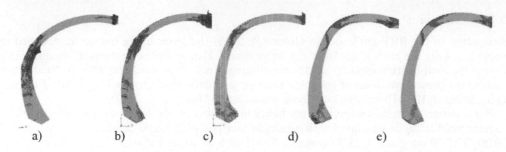

Figure 2. Deformations immediately before failure with crack developments in black of the left secondary lining. From left to right different radial modulus of subgrade reaction a) 10 MN/m³, b) 50 MN/m³, c) 100 MN/m³, d) 500 MN/m³, e) 1000 MN/m³ and different maximum load-bearing capacities of a) 30 %, b) 68%, c) 100 %, d) 132 %, e) 136%. Horizontal deformations of the upper side-wall are a) 11cm, b) 7cm, c) 6 cm, d) 5 cm, e) 4.5 cm.

surface of the crown. Further increasing of the radial modulus of subgrade reaction improves the load-bearing capacity only slightly without changing the mode of failure.

3 MODULUS OF SUBGRADE REACTION

For design calculation of tunnels up to a overburden of two tunnel diameters, a partially embedded system (no bedding in the area of the crown) with a modulus of subgrade reaction in radial direction k

$$k = \frac{E_S}{r_a} = \frac{2G}{r_a} \frac{1 - v}{(1 - 2v)} \tag{1}$$

is recommended in (Working Group Tunnelling, German society for earthworks and foundation engineering, 1980) and (Ahrens, H. 1982). In Equation 1 E_S is the stiffness modulus, G the shear modulus, v Poisson's ratio and and r_a the outer radius of the lining. This approach for the modulus of subgrade reaction has become established in practice and is still commonly used and recommended (see e.g. DAUB 2007 and 2013, Winselmann, D. 2000 and ITA 2000).

The subgrade reaction modulus according to Equation 1 is also used for deep tunnels frequently, although in (Working Group Tunnelling, German society for earthworks and foundation engineering, 1980) for tunnels with a overburden of more than three times the tunnel diameter the reduced modulus of subgrade reaction

$$k = 0,5 \frac{E_S}{r_a} = \frac{2G}{r_a} \frac{1 - v}{(1 - 2v)} \tag{2}$$

was recommended. In either case, no tangential bedding has to be applied according to (Ahrens, H. 1982). These recommended subgrade reaction moduli are based on studies by (Ahrens, H. 1982). Equation 1 follows from numerical comparative calculations for shallow tunnels and Equation 2 from an analytical solution of a continuum model with elastic material behavior.

For simple load patterns and initial stress states, circular tunnels and elastic material behavior, further analytical solutions of the continuum model ("slab with a hole") can be found in the literature (e.g. Carranza-Torres, C. 2013, El Naggar, H. 2008, Zhang, D. 2014). For the calculation of the subgrade reaction modulus for a constant vertical load p_v and a horizontal load $p_h = K_0 p_v$ it is advantageous to split this load into the hydrostatic (rotationally symmetrical) component

$$p_r = p_\vartheta = \frac{1 + K_0}{2} p_v$$

and the deviatoric part

$$p_v = \frac{1 - K_0}{2} \cos 2\vartheta\, p_v \quad p_\vartheta = \frac{1 - K_0}{2} \cos 2\vartheta\, p_v \quad p_\vartheta = \frac{1 - K_0}{2} \sin 2\vartheta\, p_v$$

(see e.g. Timoshenko, S. 1951)

For the hydrostatic part, the subgrade reaction modulus is (Ahrens, H. 1982, Sonntag, S. 1976)

$$k = \left.\frac{\sigma_r}{u_r}\right|_{r=r_a} = \frac{E}{(1+v)r_a} = \frac{2G}{r_a} \tag{3}$$

For the deviatoric part, a distinction must be made whether a full (radial and tangential) or only a radial bond is applied between rock mass and support.

For a full bond, the modulus of subgrade reaction for the deviatoric part of the load of the rock results from the stresses and displacements in (Carranza-Torres, C. 2013)

$$k = \frac{2G}{r_a} \frac{1}{3 - 4v} \tag{4}$$

In this case is the modulus of subgrade reaction in tangential direction is equal to that in radial direction.

With only a radial bond between subsoil and support, the modulus of subgrade reaction in radial direction follows from (Carranza-Torres, C. 2013) to

$$k = \frac{2G}{r_a} \frac{3}{5 - 6v} \tag{5}$$

For $v = 1/3$ Equation 5 equals Equation 3, i.e. the modulus of subgrade reaction in radial direction $k = 2G/r_a$ is the same for the hydrostatic and deviatoric load component.

Strictly speaking, the same modulus of subgrade reaction may only be used in this case for both load components. The recommendation for the modulus of subgrade reaction according to Equation 2 in (Working Group Tunnelling, German society for earthworks and foundation engineering, 1980) for deep tunnels was derived from Equation 5, where agreement only for $v = 1/3$ exists.

The subgrade reaction moduli according to Equations 3–5 apply only to hydrostatic or deviatoric load components. For full bond and constant vertical and horizontal loads, the modulus of subgrade reaction in radial direction follows from (Carranza-Torrec, C. 2013) to

$$k = \frac{2G}{r_a} \frac{(1 + K_0) - (1 - K_0)\cos 2\vartheta}{(1 + K_0) - (1 + K_0)(3 - 4v)\cos 2\vartheta} \tag{6}$$

In Equation 6 the dependence of the modulus of subgrade reaction on the load and the polar angle ϑ ($\vartheta = 90°$ at the crown) can be clearly seen. With full bond a constant modulus of subgrade reaction $k = 2G/r_a$ only if $v = 1/2$. In Figure 3 the dependence of the subgrade reaction modulus on the polar angle according to Equation 6 and a comparative calculation with FEM for $v = 0.2$ and $v = 0.4$ is shown.

The different values for the modulus of subgrade reaction according to Equations 1–5 and the non-constant modulus of subgrade reaction show the difficulties of determining the modulus of subgrade reaction from a theoretical point of view.

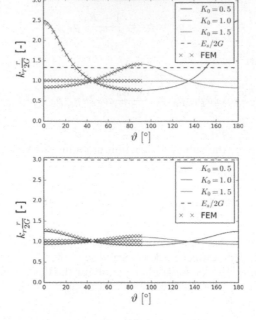

Figure 3.   Curve of subgrade modulus according to Equation 6 for $\nu = 0,2$ (left) and $\nu = 0,4$ (right).

The moduli according to Equations 2–6 follow from analytical solutions of the system "slab with hole" with elastic material parameters. In practice, springs with a tension cut off are often used. This assumption does not correspond to the requirements for deriving the sub-grade reaction modulus. It is also not plausible that if there is a rock pressure that the springs are inactive. The bedding reaction should only be deactivated when the resulting force (or stress) from load and bedding reaction is tensile.

4   SUBGRADE REACTION MODULUS OF THE SYSTEM ANNULAR GAP - ROCK MASS

For the sake of simplicity, only rotationally symmetric states are considered below.

4.1   *Common Approach*

In practice, for the determination of the modulus of subgrade reaction k of the combined system rock mass-annular gap, an approach of serial springs is used. The modulus of subgrade reaction k_R for annular gap filling

$$k_R = \frac{E_S}{d} = \frac{2G_R}{d}\frac{1 - \nu_R}{(1 - 2\nu_R)} \tag{7}$$

with the thickness $d = r_a - r_i$ of the annular gap is used (index „R"for annular gap).
For serial springs, the subgrade reaction modulus k of the system rock mass - annular gap filling (index "G" for rock mass):

$$\frac{1}{k} = \frac{1}{k_R} + \frac{1}{k_G}\, bzw. k = \frac{k_R k_G}{k_R + k_G} \tag{8}$$

With $\beta = \frac{r_a}{r_i}$, and the ratio of the shear modulus of the rock mass and the annular gap $k = \frac{G_G}{G_R}$, it follows for the ratio of the subgrade reaction modulus of the overall system k to the modulus of subgrade reaction of the rock k_G:

$$\frac{k}{k_G} = k\frac{r_a}{2G_G} = \frac{\beta\frac{1-v_R}{1-2v_R}}{\beta\frac{1-v_R}{1-2v_R} + K(\beta - 1)} \tag{9}$$

For serial springs, the replacement spring is softer than the individual springs, i.e $k < k_G$. and $k < k_R$, Thus the modulus of subgrade reaction for the combined system is smaller than the modulus of subgrade reaction for the rock mass and annular gap filling respectively. This result is implausible because the bedding has to improve in relation to the bedding of the rock alone if the annular gap filling is stiffer than the rock mass.

In the following, the modulus of subgrade reaction is derived from the continuum-mechanical equations for elastic material behavior.

4.2 Consistent Subgrade Reaction Modulus of the System Annular Gap - Rock Mass

The basic equations in a cylindrical coordinate system are for rotational symmetry and plane strain conditions:

4.2.1 Kinematic relations

$$\varepsilon_r = \frac{du}{dr} \quad \varepsilon_\theta = \frac{u}{r} \quad \varepsilon_z = 0 \tag{10}$$

4.2.2 Equilibrium equations

$$\frac{d\sigma_r}{r} + \frac{\sigma_r - \sigma_\theta}{r} = 0 \tag{11}$$

4.2.3 Boundary conditions

$$\sigma_r(r_i) = -p \ \text{and} \ \sigma_r(r_a) = -k_G u(r_a) \tag{12}$$

4.2.4 Constitutive relationships of annular gap filling (elastic - Hooke's law)

$$\sigma_r = E_{s,R}(\varepsilon_r + \frac{v_R}{1-v_R}\varepsilon_\theta) \quad \sigma_\theta = E_{s,R}(\varepsilon_\theta + \frac{v_R}{1-v_R}\varepsilon_r) \quad \sigma_z = v_r(\sigma_r + \sigma_\theta) \tag{13}$$

Inserting the kinematic relations Equation 10 in Equation 13 and the result obtained from this into the equilibrium conditions Equation 11 gives the homogeneous Euler's differential equation

$$r^2 u'' + ru' - u = 0$$

From the solution of this differential equation, taking into account the boundary conditions Equation 12 follows the modulus of subgrade reaction of the tunnel lining for elastic material behavior of annular gap filling

$$k\frac{r_a}{2G_G} = \frac{\frac{\beta}{K}}{2c_1(1-v_R) + 1} \tag{14}$$

with the dimensionless auxiliary value

$$c_1 = \frac{1 - K}{\beta^2 + (\beta^2(1 - 2v_R) + 1) - 1}$$

The scaled subgrade reaction modulus $k\frac{r_a}{2G_G}$ thus depends on the ratio of the radii β, the ratio of the shear modulus K and the Poisson's ratio of the annular gap.

5 COMPARISON WITH NUMERICAL CALCULATIONS

In this section the moduli of subgrade reaction according to Equation 9 and Equation 14 are compared with FE calculations. The FE model used was an axially symmetrical model with 50 elements over the thickness of the annular gap. The bedding of the rock was modeled with interface elements. The inner and outer diameters of the annular gap are $r_i = 5$m and $r_a = 5,2$m $(\beta = 1,04)$ respectively.

5.1 Stiff Rock - Soft Annular Gap ($\kappa > 1$)

For this comparison, typical values for a stiff rock (intact rock) and a soft annular gap filling (e.g. gravel) are used. The used values of the rock are $E = 10000$ MPa and $v = 0.2$, thus $k_G = 1602$ MN/m^3. The modulus of elasticity of the annular gap is $E = 100$MPa. For $\kappa = \frac{G_G}{G_R} > 1$ it can be seen that the Poison ration of the annular gap has a large influence on the modulus of subgrade reaction. Therefore, $v = 0,4/0,25/0,1$ was used.

The analytical and numerical results for the modulus of subgrade reaction are practically identical, the modulus of subgrade reaction according to the model "serial springs" deviates slightly

5.2 Soft Rock - Stiff Annular Gap ($\kappa < 1$)

For this comparison, representative values for soil (loose rock) and a stiff annular gap filling (e.g. mortar) are used. The used values of the soil are $E = 100$ MPa and $v = 0,2$, thus

Table 1. Comparison analytical – numerical calculation of subgrade modulus for $\kappa > 1$.

	Unit	comparison 1	comparison 2	comparison 3
Annular gap				
v	-	0.40	0.25	0.10
κ -		117	104	92
Modulus of subgrade reaction				
k acc. (Eq. 14)	MN/m^3	636.73	440.58	395.43
k acc. (Eq. 9)	MN/m^3	642.12	436.55	387.66
k acc. FEM	MN/m^3	636.73	440.58	395.43

Table 2. Comparison analytical – numerical calculation of subgrade modulus for $\kappa < 1$.

	Unit	comparison 1
Annular gap		
v	-	0.20
κ	-	0.01
Modulus of subgrade reaction		
k acc. (Eq. 14)	MN/m^3	97.19
k acc. (Eq. 9)	MN/m^3	16.02
k acc. FEM	MN/m^3	97.19

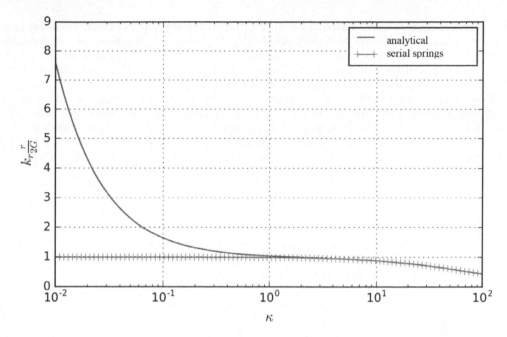

Figure 4. Comparison of modulus of subgrade reaction according Equation 9 and Equation 14.

$k_G = 16,02\,\mathrm{MN/m^3}$. The modulus of elasticity of the annular gap is $E = 10000\,\mathrm{MPa}$. The investigation showed, that for $\kappa = \frac{GG}{GR} < 1$ the Poison ration of the annular gap has no great influence on the modulus of subgrade reaction. Therefore $\nu = 0,2$ and thus $\kappa = 0,01$ was used in all cases.

Again the analytical and numerical results for the modulus of subgrade reaction are practically identical, but the modulus of subgrade reaction according to the model "serial springs" is only about 20% of it.

5.3 *Interpretation Der Ergebnisse*

In Figure 4, the modulus of subgrade reaction for the combined system of annular gap-rock according to Equation 9 (serial spring) and the analytical approach according to Equation 14 are compared.

The parameters used are $\beta = 1,05$ and $\nu_R = 0,3$.

From the diagram it follows that for $\kappa = \frac{GG}{GR} < 1$, i.e. in relation to the annular gap filling of soft subsoil (e.g. tunnel in loose rock and annular gap mortar), the use of Equation 14 for the calculation of the modulus of subgrade reaction is recommended, with the approach "serial springs" the modulus of subgrade reaction is underestimated.

6 CONCLUSIONS

The following conclusions can be drawn from the investigations in this contribution:

The commonly used modulus of subgrade reaction of the rock $k_r = E_s/r$ is only justified for shallow tunnels (overburden smaller than two tunnel diameters) by comparisons with numerical calculations. For deep tunnels the modulus of subgrade reaction is lower.

The modulus of subgrade reaction depends on the geometry and the load. The given subgrade reaction moduli apply approximately to circular tunnels, constant vertical and horizontal loads and linear elastic material behavior. If the conditions deviate from this (e.g. open profiles), the applicability of the method of subgrade reaction modulus must be checked.

The model "serial springs" leads to implausible results for the system annular gap - rock with relatively soft rock (loose rock with mortared annular-gap). For the calculation of the subgrade reaction modulus Equation 14 is recommended.

For the application of the subgrade reaction modulus method it is important to know the simplifying assumptions and to assess whether the approximations are permissible. For more detailed investigations of tunnels, numerical calculations (e.g. FEM, FDM) with more sophisticated material models are preferable. In the literature (e.g. Schwarz, G. 2010) it was also pointed out for other geotechnical applications (e.g. slab foundations) that the modulus of subgrade reaction method can lead to implausible results.

DEDICATION

This contribution is dedicated to Walter Hofmann, who gave valuable suggestions in discussions on this topic. As civil engineer, he designed numerous bridge structures and cut-and-cover tunnels. Walter Hofmann died on 2017 at the age of 74.

REFERENCES

Ahrens, H., Lindner, E., Lux, K. 1982. Zur Dimensionierung von Tunnelausbauten nach den "Empfehlungen zur Berechnung von Tunneln im Lockergestein, 1980". *Bautechnik Bd. 59, Nr. 8*

Bergmeister K. 2011. Brenner Basistunnel - Brenner Base Tunnel - Galleria di Base del Brennero. *Tappeiner Verlag*

Carranza-Torres, C., Rysdahl, B., Kasim, M. 2013. On the elastic analysis of a circular lined tunnel considering the delayed installation of the support. *International Journal of Rock Mechanics and Mining Sciences* (61), 57–85

Červenka J. 2016. ATENA Program Documentation, ATENA Science – GiD Tutorial. Červenka Consulting. Prague, Czech Republic.

El Naggar, H., Hinchberger, S., LO, K. 2008. A closed-form solution for composite tunnel linings in a homogeneous infinite isotropic elastic medium. *Canadian Geotechnical Journal* (4), 266–287.

German Tunnelling Committee (DAUB). 2007. Empfehlungen für statische Berechnungen von Schildvortriebsmaschinen (Recommendations for structural anaylsis of shield tunnelling machines). *Tunnel* (07)

German Tunnelling Committee (DAUB). 2013. Recommendations for the design, production and installation of segmental rings

Insam, R., Eckbauer, W., Matt, K., Gangkofner, T. 2019. Brenner Base Tunnel Guide Design, *ITA-AITES World Tunnel Congress*, Naples, Italy

International Tunnelling Association (ITA), Working Group No. 2. 2000. Guidelines for the Design of Shield Tunnel Lining

Kurrer, K. E. 2002. Geschichte der Baustatik. *John Wiley & Sons*

Marcher, T., Cordes, T., Bergmeister, K., 2019. SEM - Single Shell Lining Application for the Brenner Base Tunnel. *ITA-AITES World Tunnel Congress*, Naples, Italy

Schulze, H & Duddeck, H. 1964. Spannungen in schildvorgetriebenen Tunneln. *Beton-und Stahlbetonbau* (59), 169–175

Schwarz, G. 2010. Technisch und wirtschaftlicher Vergleich von Bodenplatten mit unterschiedlichen Gründungssystemen, Thesis, TU Graz

Sonntag, G & Fleck, H. 1976. Zum Bettungsmodul bei Tunnelauskleidungen. *Ingenieur-Archiv* (45), 269–273

Timoshenko S. & Goodier, J.N. 1951. Theory of Elasticity, *McGraw-Hill book Company*

Winselmann, D., Städing, A., Babendererde, L., Holzhäuser, J. 2000. Aktuelle Berechnungsmethoden für Tunnelauskleidungen mit Tübbingen und deren verfahrenstechnische Voraussetzungen. *DGGT, Baugrundtagung Hannover*

Working Group Tunnelling, German society for earthworks and foundation engineering, 1980. Empfehlungen zur Berechnung von Tunneln im Lockergestein (Recommendations for the calculation of tunnels in soft rock) *Bautechnik* Bd. (57), 349–356

Working Group Tunnelling, German society for earthworks and foundation engineering, 1980. Empfehlungen zur Berechnung von Tunneln im Lockergestein (Recommendations for the calculation of tunnels in soft rock) *Bautechnik* (57), 349–356

Zhang, D., Huang, H., Phoon, K., Kwang, H. 2014. A modified solution of radial subgrade modulus for a circular tunnel in elastic ground. *Soils and Foundations* (54), 225–232

*Tunnels and Underground Cities: Engineering and Innovation meet Archaeology,
Architecture and Art, Volume 5: Innovation in underground engineering,
materials and equipment - Part 1 – Peila, Viggiani & Celestino (Eds)*
© 2020 Taylor & Francis Group, London, ISBN 978-0-367-46870-5

Permanent sprayed concrete tunnel linings waterproofed with bonded membranes. A review of the current state-of-the-art for hard rock conditions

K.G. Holter
Norwegian University of Science and Technology NTNU, Trondheim, Norway

ABSTRACT: Permanent sprayed concrete tunnel linings waterproofed with bonded membranes have been used at a number of important projects. The layout of this lining gives fundamentally different system properties compared to the traditional lining systems. The current understanding of the function and properties of such lining structures is presented based on review of recent research carried out in Norway, as well as field observations and monitoring carried over several years. The influence of the water exposure on the final condition of the concrete and membrane materials has proven to be of vital importance for proper material testing and acceptance, assessments of the mechanical contribution of the bonded membrane, as well as assessments of the longterm durability of such linings. Finally, some recent results from ongoing research on such linings, particularly the hydraulic response of the rock mass and the long term behavior of the concrete and membrane materials are presented.

1 INTRODUCTION

1.1 *Final lining and waterproofing in hard rock*

Final tunnel linings based on sprayed concrete and a spray-applied waterproofing membrane was initially developed as a response to excessive technical solutions for such as cast-in-place concrete when the ground conditions don't require a heavy lining structure. Ground conditions which can be supported permanently with a system based on sprayed concrete with a limited thickness pose opportunities to design a permanent and stable linings without using high concrete thicknesses. Furthermore, in soft rock conditions, the primary lining can be given a permanent function, which allows for a reduction of the thickness of the secondary lining and hence a lower total lining thickness.

In Norway the prevailing hard rock conditions in many cases allow a design with this lining system. Two railroad tunnels and several test sections were constructed in the period 2009 – 2014 with this lining system.

This paper focuses on the experiences and research results from the construction of these hard rock tunnels and test sections in Norway. Findings from in-situ investigations covering monitoring of temperatures, measurements of moisture content and tensile bonding strength, as well as laboratory investigations of important material parameters such as vapor and hydraulic conductivities, water absorption properties and mechanical properties. The field investigation results and material property test results were used for numerical simulations of the water transport and moisture condition of the lining structure.

A model for the function and properties of this lining system, as well as the key material performance parameters are presented. Finally, an outlook to the further development of the permanent sprayed linings is discussed.

2 TECHNICAL BACKGROUND

2.1 *Experience with permanent sprayed concrete linings*

Sprayed concrete linings in rail and road tunnels in the Scandinavian countries have been used for the permanent rock support. Mix design and requirements for this sprayed concrete have been laid out for the required mechanical strength and durability for a permanent rock support purpose. The required functionality for the final inner lining which includes waterproofing and the aesthetic design was not feasible to achieve with this rock support lining.

The use of sprayed concrete for a final inner lining with required functionality and durability is not a new approach. One of the first major successes with this method was the construction of the 19 km long Vereina railroad tunnel in eastern Switzerland (Röthlisberger, 1993). In addition to the low investment cost for this tunnel, 20 years of operation and maintenance has been very favorable. Several minor projects constructed over the last decade in the Czech Republic, Switzerland, Italy and Austria have shown favorable experiences with such a permanent lining approach.

A typical obstacle to implement the design of the final inner lining based on sprayed concrete are mainly two-fold: the durability of the sprayed concrete and the functionality and durability of the waterproofing.

2.2 *Main boundary conditions for tunnel linings in hard rock and cold climate*

The main geomechanical boundary condition posed by hard rock ground is that the rock support lining is not exposed to a global radial loading from the ground. In most cases the rock support lining is subjected to local or punctual loads caused by instable blocks or weakness zones. This leaves the sprayed concrete primary support lining a passive structure, apart from local or singular spot loads. The explanation to this is the high Young's moduli of the rock material compared to that of concrete. (Holter, 2015)

Traffic tunnels in hard rock are mostly constructed with a drained waterproofing structure and a drained invert without any waterproofing. Therefore, there are no hydrostatic loads on the waterproofing. The rock support sprayed concrete surface allows water to seep into the tunnel through cracks and discontinuities. In drill-and-blast excavated tunnels the hydrostatic loading of the rock mass in the immediate vicinity of the lining surface has proven to be negligible. Hydrostatic loads have only caused instabilities in extreme cases, such as permeable weakness zones. Design of rock support in hard rock, not even in subsea tunnel construction, does not account for any hydrostatic loading.

2.3 *Currently used systems for final lining, water proofing and thermal insulation*

Rail and road tunnels are currently being constructed with a separate drainage and insultation lining system which is designed as a separate structure. The two commonly used systems are shown graphically in Figure 1.

2.4 *Trends*

The trend in the construction of Norwegian railroad tunnels is to adopt the traditional central European method for final linings with cast-in-place concrete. A main issue is to achieve a service lifetime of 100 years in combination with low maintenance costs and foreseeable downtime during operation. In hard rock this implies the construction a primary rock support lining during excavation, followed by an additional smoothening layer of sprayed concrete to facilitate the installation of the sheet waterproofing membrane on a surface with suitable evenness. The technical solution with cast-in-place concrete linings for modern road tunnels is shown in Figure 2. A similar technical solution is adopted for rail tunnels (Bane NOR, 2018) which has been realized on on one major project, shown in Figure 3, right. For modern road tunnels there has been a further development of the drainage and frost insulation lining system which meet the functional requirements for highways. The recent version of this lining system is shown in Figure 3, left

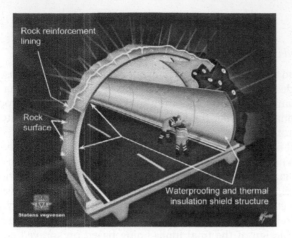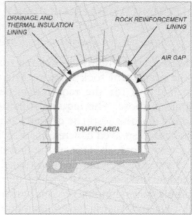

Figure 1. Commonly used final inner lining systems which meet the functional requirements for waterproofing, thermal insulation and aesthetic design. Left: layout for a road tunnel (NPRA, 2012). Right: Layout for a single-track rail tunnel (Holter, 2015).

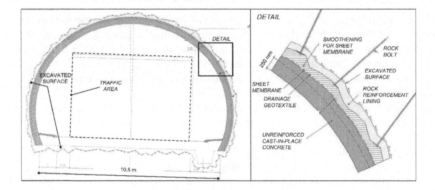

Figure 2. Principal design drawings of cast-in-place concrete lining in hard rock with drained invert (NPRA, 2012)

Figure 3. Trend for future final tunnel linings. Left: The Eisdvoll road tunnel in Norway completed 2012 with concrete segments with layout as shown in Figure 1 (courtesy Ådne Homleid). Right: the Ulvin rail tunnel in Norway completed in 2014 with cast-in-place concrete lining

3 TECHNICAL SOLUTION BASED ON SPRAYED CONCRETE AND SPRAY APPLIED WATERPROOFING

3.1 *Layout of lining system*

The technical solution of making a sprayed concrete lining waterproof described here is the use of a double-bonded membrane which is spray applied onto the sprayed concrete rock support lining, and subsequently covered with a covering layer of sprayed concrete. In this way a continuously bonded multilayered structure is created. The main geometrical feature of this lining is that it follows the excavated rock surface and is multilayered and monolithic. Hence, a final lining with a thickness only slightly higher than the primary support lining can be achieved. In hard rock ground, total lining thicknesses in the range of 100 to 150 mm, with 200 mm as the maximum thickness under exceptional conditions can be achieved.

In its basic form the lining does not contain any draining or separating features. The layout of the lining and the main boundary conditions are illustrated in Figure 4.

3.2 *Material properties, system properties and function of the lining system*

Several material properties of the concrete and membrane materials are important for the system properties and function and function of the lining. The sprayed concrete is the main constituent material in the lining. Investigations of hardened sprayed concrete samples, obtained in-situ in tunnels at ages ranging from 300 to 400 days. The mechanical strength properties are subject to contractual specifications. Properties such as porosities, capillary sorptivity, hygroscopic sorptivity, vapor permeability and hydraulic conductivity have proven to be essential for the understanding of the function and properties of this lining system.

Sprayed concrete exhibits high capillary porosity, extremely low water permeability, relatively high vapor permeability. The illustrations in Figure 5 show polished surfaces of sprayed concrete colored with black ink with enhanced contrast of the pore structure obtained with barium sulphate powder. The macro pore structure is shown as the white fields, whereas the solid components remain black. The capillary pore structure is not visible. The images show the range in size and shape of the pores in sprayed concrete, and hence demonstrate the importance of an application technology which ensures even and good compaction during the concrete spraying.

The membrane materials investigated in this study are from two different commercial suppliers (BASF and MINOVA) and are both of the category ethylene-vinyl-acetate copolymer (EVA),

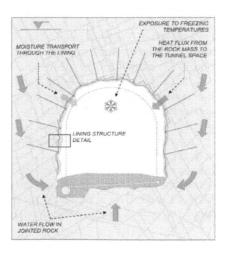
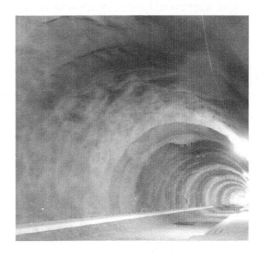

Figure 4. Left: Illustration of the main boundary conditions and processes affecting the function and properties of a continuously bonded sprayed concrete lining with a drained invert in hard rock (Holter, 2015). Right: Completed waterproof lining before the final installation of rails and infrastructure.

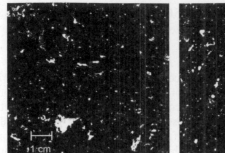

Figure 5. Images of sprayed concrete obtained by an enhanced contrast procedure, showing the air voids (macro porosity), which account for approximately 5% volume. Both images are to scale. Left: examples of coarse pores and voids. Right: example of the most even pore distribution which could be observed (Holter et.al, 2016).

Table 1. Some properties of hardened sprayed concrete, wet mix, age 400 days, cured in-situ in tunnel.

Measured property	Value	Remark
Binder quantity	480 - 520	1.18"
Water/binder ratio	0,43–0,45	Accounting for all sources of water in the concrete
Uniaxial compressive strength [MPa]	3.0	1.18"
Young's modulus [GPa]	25 - 26	
Macro porosity (air void volume) [%]	4 - 6	
Suction porosity [%]	19 - 21	
Hydraulic conductivity [m/s]	0.0	< 10 -14 m/s
Vapor permeability [kg/m·s·Pa]	7 – 22	Measured at relative humidity contrast of 50 to 94%

* 470–500 kg/m^3: CEM II A-V 42.5 (17–18% flyash content)
20–25 kg/m^3: micro silica fume

which is a hydrophilic material. Extensive testing was carried out which comprised in-situ field measurements of moisture content and tensile bond strengths, as well as laboratory testing of in-situ produced and cured samples. Furthermore, large scale physical simulations of freeze-thaw exposure were carried out in order to obtain realistic data of the performance of the lining when exposed to freezing.

Properties of the concrete and membrane materials are shown in Tables 1 and 2.

The important consistent observation which was made during the field investigations was the high tensile bond strengths and the relatively low moisture content of the membrane. One might expect that an undrained watertight lining structure would create a complete water saturation of the concrete behind the membrane, as well as a significant moistening of the membrane material from the bonded interface to the concrete. The explanation to the field measurements was found in the measured hygroscopic sorptivities of the sprayed concrete and sprayed membrane materials. This material property expresses how much water a material can retain at a certain relative humidity. This is shown in Figure 7.

The noteworthy observation is the large contrast in hygroscopic sorptivity between the concrete and membrane materials at high relative humidities. From the graph Figure 6 one can read that at RH 95%, which corresponds to the in-situ humidity of the concrete in the lining, the membrane will exhibit at water content which is approximately 35% of the maximum water uptake.

3.3 *System properties and function of the lining structure*

A synthesis of the laboratory and field investigations was made and a model for the properties and function of the lining system could be established. In principle, the lining structure is

Table 2. Some properties of cured membrane

Measured property	Value	Remark
Membrane thickness [mm]	3 - 4	Measured thickness after curing
Maximum water uptake at immersion	30 - 44	% of dry weight, measured after 7 days
Water content in-situ in tunnel lining	12–15	% of maximum water uptake at immersion
Tensile bond strength [MPa]	1.1–1.5	at 7 °C in situ tunnel, no freezing exposure
Tensile bond strength [MPa]	1.0–1.2	after cyclic freeze-thaw exposure to -3 °C
Shear strength, dry samples [MPa]	2 - 3	Specimens stored at norm conditions RH 60%
Shear strength, moist samples [MPa]	0.7-0.8	Specimens from in-situ tunnel lining, RH 90-95%
Shear modulus, dry samples [MPa]	7 - 15	Specimens stored at norm conditions RH 60%
Shear modulus, moist samples [MPa]	2 - 3	Specimens from in-situ tunnel lining, RH 90-95%
Crack bridging capacity at 20 °C	4 mm	
Crack bridging capacity at -3 °C	1,5 mm	
Watertightness		Watertight according to EN-12390
Vapor permeability [kg/m·s·Pa]	9 - 11	Measured at contrast of RH from 50 to 94%

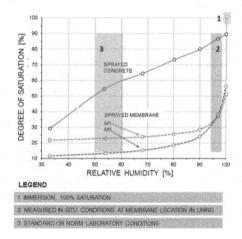

Figure 6. Hygroscopic sorptivities of sprayed concrete and sprayed membrane materials represented as desorption isotherms at 20 °C (Holter, 2015).

vapor permeable and watertight. The precise waterproofing function of the membrane is to seal and bridge the cracks in the concrete in order to prevent water to flow towards the lining surface. The membrane and concrete materials exhibit vapor permeabilities in the same magnitude. Hence the membrane allows vapor to pass without posing a barrier to vapor transport through the lining. The low hygroscopic sorptivity of the membrane compared to concrete keeps the membrane at low moisture contents and hence provides the in-situ moisture conditions which gives a high tensile bonding strength.

The function of the lining, the main processes and boundary conditions are illustrated in a graphic with a section of the lining structure in Figure 7.

The vapor transport feature of this lining system has an important consequence for the moisture condition. Moisture contents were measured in a large number of drill core samples from tunnel linings. The results were compiled in moisture content profiles across the lining structure from rock mass to lining surface. Based on these measurements the model for moisture condition and water transport mechanisms could be established. An important finding of this study was the type of water transport mechanism, as well as a calculation of the amounts water which can pass through the lining in the form of capillary and vapor

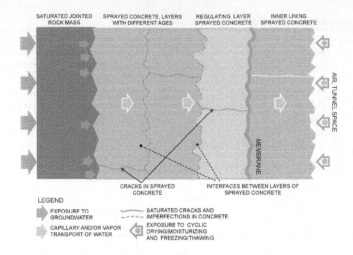

Figure 7. Section of the lining structure from rock mass to lining surface showing the function and properties of the lining.

transport. The principles which apply to these calculations are well known from building physics. A numerical simulation was carried in the software WUFI, using measured data for the boundary conditions and the material properties. The simulations could reproduce the measured in-situ moisture contents in the investigated tunnel linings (Holter & Geving, 2016). The model for moisture condition and transport mechanisms is shown in Figure 8. The important feature is that when the water cannot flow on the cracks and imperfections in the concrete, water transport takes place in the form of vapor transport. This leads to a gradient in moisture content in the concrete lining, with gradually less water in the concrete closer to the lining surface. With realistic boundary conditions as in the investigated tunnels, the amounts of water vapor which can pass through a 300 mm thick sprayed concrete lining is in the range of $2 - 3$ milliliters per m^2 lining surface per day (Holter & Geving, 2016).

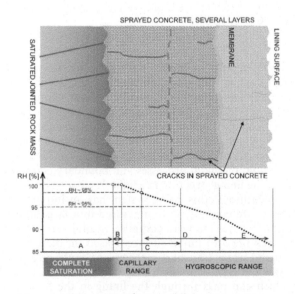

Figure 8. Model for moisture condition across the tunnel lining structure.

The moisture condition and water transport mechanisms have been used in assessments of freezing damage and other degrading mechanisms such as leaching of sprayed concrete, feeding of aggressive ions from the groundwater and build-up of complete saturation in the concrete material and hence a risk of hydraulic pressure acting on the membrane interface.

3.4 Impact on groundwater pressure near the tunnel lining, importance of the excavation damaged zone (EDZ)

A bonded waterproof lining structure is in principle undrained and therefore poses a barrier to water flow through lining as well as along the lining. Therefore, a situation with increased water pressure in the rock mass in the immediate vicinity of the lining might occur. Field investigations with monitoring of the groundwater pressure were undertaken at several test sites with full scale constructed tunnel lining sections, all of them with drained inverts and waterproof undrained lining structures in the walls and the crown. The results of these investigations show that there is a water pressure drop in the rock mass behind the lining to a level significantly below the theoretical hydrostatic pressure. The measurements show a slight pressure drop over a distance of 2-8 m from the lining surface, where the hydrostatic exposure would imply a slight increase of the pressure. Furthermore, the pressure drop is the largest closest to the lining, occurring within approximately 1 m distance from the lining. The interpretation of this is that there is a waterflow in the rock mass towards the invert and which created a certain water pressure reduction compared to the hydrostatic case. The pressure drop closest to the lining surface is attributed to the EDZ with its increased jointing and hence increased hydraulic conductivity.

The hydraulic properties of the EDZ are very likely the key feature to be understood in order to make precise calculations of the hydrostatic loading of undrained lining structures. A detailed study carried out at the Swedish Nuclear Fuel and Waste Company SKB (Ericsson et al. 2015) suggests that there is a significant increase in the hydraulic transmissivities of the

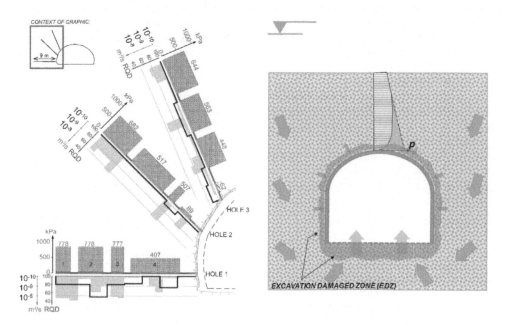

Figure 9.   Effect of the increased hydraulic transmissivities on rock joints in the Excavation Damaged Zone on rock joint water pressure p in the immediate vicinity of the tunnel lining. Left: results from in-situ measurements of groundwater pressure behind an undrained lining structure with a drained invert (Holter, 2014). Right: Basic model for groundwater pressure around an undrained lining structure with a drained invert in hard rock.

rock joints at decreasing distance to the rock surface. Further detailed measurements of the hydraulic properties of the EDZ are currently being undertaken at NTNU in order to obtain more data covering a larger range of geological conditions including in-situ rock stress conditions.

Some results from the monitoring of the groundwater pressure in the rock mass around an undrained waterproof lining are shown in Figure 9.

4 OUTLOOK, FURTHER DEVELOPMENT

Waterproof sprayed concrete based linings which meet modern requirements regarding functionality and durability represent one of the largest opportunities for investment cost savings and reduction of the carbon dioxide footprint. For this reason further development of this tunnel lining approach is taking place. Specific research projects are being initiated in Switzerland and Norway in order to improve the material technology for sprayed concrete, as well as carrying out field investigations in order to improve the understanding of the precise behavior of the lining and the groundwater response. An important goal of the further research work will be to obtain an even better understanding as to under which ground conditions and with which limits one can adopt this lining method.

Currently (2018) there are a number of companies in the supplier industry which are working on improving the sprayed waterproofing technology. Even better performing material technology on sprayed waterproofing can therefore be expected.

5 CONCLUSIONS

Sprayed concrete linings which are designed with sprayed waterproofing and given a permanent function in modern rails tunnels has been successfully attempted in large scale on several projects. Favorable experiences have been obtained regarding the extent of required maintenance and low downtime.The undrained layout of the lining structure results in several favorable modes of which water can expose the lining, most notable static water which prevents leaching and feeding of aggressive ions. The field investigations of constructed linings show consistent high strength performance and constant low water content, hence suggesting long term durable linings.

The limitations as of today are related to constructability issues in situations with water seepages in the tunnel during construction, durability of sprayed concrete and legislation regarding the use of sprayed concrete in permanent lining structure, as well as the precise hydrostatic loads when designing.

The further development of this lining method will be oriented towards the sprayed concrete as well as the sprayed membrane in order to achieve improved in-situ material properties and application methods which create more homogenously constructed materials.

REFERENCES

Bane NOR (2018). Technical Specifications (in Norwegian, Teknisk regelverk), www.banenor.no/tekniskregelverk.

Ericsson, L. O., Christiansson, R., Hansson, K.; Butron, C., Lehtmaki, T., Sigurdsson, O.; Thörn, J.; & Kinnbom, K. 2015. A demonstration project on controlling and verifying the excavation-damaged zone. Experience from the Äspö Hard Rock Laboratory. Swedish Nuclear Fuel and Waste Company. Report no R-14-30

Holter, K. G. 2014. Loads on sprayed waterproof tunnel linings in jointed hard rock

Holter, K. G. 2015. Properties of waterproof sprayed concrete tunnel linings. PhD thesis, Norwegian University of Technology and Science, Trondheim, ISBN: 978-82-326-1050-1

Holter, K. G., Geving, S. 2016. Moisture transport through sprayed concrete tunnel linings. Rock Mechanics and Rock Engineering, Vol 49/1, pp 243–272

Holter, K. G., Smeplass, S., Jacobsen S. 2016. Freeze-thaw resistance of sprayed concrete in tunnel linings. Materials and Structures, Vol 49, pp 3075–3093

Holter, K. G. 2016. Performance of EVA-based membranes for SCL in hard rock. Rock Mechanics and Rock Engineering, Vol 49, pp 1329–1358

NPRA Norwegian Public Roads Administration. 2012 Report No 127, Major research and development projects, modern road tunnels. Norwegian Public Roads Administration, Oslo

Röthlisberger, B. (1993) Sprayed Concrete – the wet mix method considering quality, rebound and dust (in German). Proceedings of the Sprayed Concrete Conference, Alpbach Austria

Tunnels and Underground Cities: Engineering and Innovation meet Archaeology, Architecture and Art, Volume 5: Innovation in underground engineering, materials and equipment - Part 1 – Peila, Viggiani & Celestino (Eds)
© 2020 Taylor & Francis Group, London, ISBN 978-0-367-46870-5

A fast detection method for tunnel surface defects based on video processing

M. Hu, X.W. Zhou & X.W. Gao
SHU-SUCG Research Centre for Building Industrialization, Shanghai, China

ABSTRACT: Identification of tunnel defects is a main method of tunnel inspection, however, collection of images will be affected by uneven illumination, wide-angle distortion, inefficient collection. Compared with taking photos, video data is recorded more efficiently. Fixed distance and angle meet the requirements of efficient and standardized inspection. However, there is data redundancy, obstruction of facilities and difficulty in offline positioning. An algorithm named Fast detection algorithm of tunnel defects based on video data (FDA-TDV) is proposed in this paper. FDA-TDV maximizes image correlate rate to identify key frames. Harris angle points were set as features to correct distortion. Canny operator identified ring edges to realize offline positioning. SIFT was used to extract defect features which were clustered to build feature dictionary, which built BoW model including defects and noises. Semantic segmentation of defects was carried out to classify defects. The inspection method mentioned has been tested with higher accuracy.

1 INTRODUCTION

With the continuous growth of the amount of construction and maintenance. The operation and maintenance of urban highway tunnels became difficult, due to complex geological conditions, climate change, man-made sabotage and lack of stable and effective inspection (Xue & Li, 2018). Highway tunnels in operation are prone to surface defects and structural damage. And common surface defects include cracks, leakage, flaking and corrosion (Koch et al. 2015), while structural damages include structural distortion and slab staggering.

Recently, the image processing and detection technology that adopts computer vision has made remarkable progress (Huang et al.2017 & Zhang et al. 2014). These detection methods use a large amount of image data of tunnel structure for defect recognition. However, collecting image data in the tunnel for processing will be affected by the following factors:

1. Uneven illumination. Uneven illumination produces light spots and shadows, which reduce the accuracy of defect recognition. Existing image dodging methods focus on how to obtain the brightness change trend of images, and accordingly compensate different parts of images (Sinha & Fieguth, 2006 & Zhang et al. 2014 & Yeung & Liu, 1995 & Zhao et al. 2000). Currently, the image dodging algorithm based on mask is commonly used. The algorithm can simulate the brightness distribution of the images well to serve as the background image, which is subtracted with original one to obtain a more ideal dodging effect. However, the selection of filter is complicated when adopting this method to obtain background image. And there is no suitable solution for determining the optimal filter size adaptively.
2. Wide-angle distortion of images. To ensure the integrity of the information, wide-angle approach is used when recording the video. Therefore, serious optical distortion often exist in video stream images, which will seriously affect the recognition of defect.
   Generally, lens distortions include radial distortion, centrifugal distortion and thin prism distortion (Wang et al. 2005). And the main factor of lens distortion is radial distortion.

Zhang & Flexible (2005) proposed that the first term of distortion function has a decisive effect on distortion, among which radial distortion is more important. When modeling lens distortion, considering too many other factors will reduce the stability of calculation instead of optimization. Hideaki et al. (1995) adopted grid template to fit the relationship between ideal and distorted images, while Devernay (2001) proposed polynomial method for distortion correction. Besides, Hughes et al. (2009) studied the application of wide-angle lens in vehicle monitoring system.

3. Inefficient and nonstandard collection of images. Due to the limitations of the collection ways, the efficiency is relatively low. The shooting angle, shooting distance and sharpness of images are also not consistent.Video data can be collected automatically for processing. Compared with taking photos, recording video data can not only solve the problem of recording distance and angle, but also be more efficiently, which is suitable for automatic inspection. However, recording video still have problems:

1) Redundancy of video data. The recording of video can quickly obtain image information by frame. Generally, a 1.5-meter-long ring surface has at least 72 redundant frames of image. However, the redundant information cannot be deleted due to inconstant collecting speed. So key frames need to be selected from the video stream as images to be processed.

   The previous studies of extracting key frames mainly focused on the use of color features. However, in the tunnel inspection, the information of the images is highly repetitive. Moreover, the spatial motility is not obvious and the speed is required. Therefore, lens-based method is more suitable, which includes the classical frame average method and histogram average method. Zhang (1997) proposed an effective method that extracts key frames based on color histogram, while Yeung (1995) proposed a method calculating the maximum distance of feature space. Sheena (2015) et al. proposed a method which based on continuous frame histogram of video data absolute difference thresholding and experiments were carried out on KTH action database, which verifies the high accuracy of the method.

2) Obstruction from facility structure. Due to fixed angle, the image collection will be disturbed by the tunnel structure. The pipeline, bolt hole, grouting hole, repair material, fire door, stairwell and other facilities will affect defect recognition.

   Therefore, it is necessary to identify these noises in the process of defect recognition. Image features are often needed to be extracted in the digital image-based pattern recognition. The extraction and research of image features can provide a statistical expression of image content. Generally, image features can be divided into area features, line features, local regional features, point features, invariant point features and scale invariant features. Liu et al. (2002) used SVM to recognize defects in the image block. Weimer (2013) proposed a machine vision system, and it uses basic plaque statistics from original images and combines two-layer neural network to detect surface defects on arbitrary texture and weakly labeled image data. Sykora (2014) et al. compared SIFT with SURF, which are two common methods of feature extraction. Then the effectiveness of the two methods for fingerprint recognition was tested on a set of depth images. SVM was used for classification, and the results showed that SVM prediction is of high accuracy in image selection.

   However, these methods based on pattern recognition rely heavily on training data to build robust classifiers. The training and validation data are usually manually marked (supervised learning), which is prone to errors during marking. The wrong training samples will reduce accuracy of these algorithms in practice.

3) Difficulty in offline positioning. Most tunnels are lack of network, so positioning with only video images is difficult.

   As there are gaps in each ring edge of collected images, the offline positioning can be achieved through edge detection. The classical edge detection method is to build an edge detection operator for a small neighborhood of pixels in the original images. Several commonly used edge detection operators include Prewitt, Roberts, Sobel, LOG, Laplace and Canny edge detection operator. Abdel Qader et al. (2003) compared crack detection performance of fast Haar transform, Fourier transform, Sobel filter and Canny filter in 25 defective concrete images and 25 healthy concrete images. And Fast Haar transform is the most accurate method, with a total accuracy rate of 86%,followed by Canny filter (76%), Sobel filter (68%) and Fourier transform (64%). However, the processing time is not considered in the analysis, and the error criteria in the binary image are not clear.

A method of Fast Detection Algorithm of Tunnel Defects based on Video data (FDA-TDV) is proposed in this study. Based on the video data collected from tunnels, key frame extraction, wide-angle distortion correction, off-line positioning and other preprocessing operations are carried out. Then, a fast classification and recognition algorithm based on image features is used to detect defects in the images (only leakage is identified in this paper). Finally, a set of video data is processed by this method and other feature extraction methods respectively. Results of the method are compared to verify the performance of the algorithm.

In this paper, section 2 introduces a video-based image preprocessing method and the experimental result. Then the fast detection method of surface defects and the experimental result are described in section 3. In section 4, applications of this method in engineering are presented and the accuracy of defect recognition is also computed which are compared with other methods. Section 5 is the conclusion of this paper.

2 AN IMAGE PREPROCESSING METHOD BASED ON VIDEO STREAM

2.1 Description of image preprocessing method

To reduce the redundant data collected from videos, key frames should be captured from the video stream as the images to be processed. Then these images are corrected of distortion to recovery information. Finally, gray the images and the use Mask method for image dodging. Canny operator is then used to capture ring edges, which are used to count the rings, to realize offline positioning.

2.2 Key frame extraction method of video images

To solve the problem of data redundancy in the video stream collected in the tunnel, the key frame extraction of the video stream needs to be completed by two steps.

Step 1. Select the first standard frame of the video stream as the standard graph. Canny operator is then used to extract the feature points of ring surface after graying the standard graph. Gray level of feature points is arranged according to positions to form a new set of feature points. Finally record position of each point.

Step 2. Set the maximum and minimum step length. Calculated from the first video image, select the image between the minimum and the maximum step length. Obtain verification points from every image and calculate gray level of all points. Pearson correlation analysis was performed on grayed verification points and standard points. Select the picture with the maximum correlation as the key frame of this step length. Take this picture as the new standard picture within the next minimum and maximum step length and choose pictures from the next step length. Repeat Step 2 until all pictures have been calculated with Pearson correlation and compared with the standard picture.

2.3 Image distortion correction

There are many models for describing lens distortion. If radial distortion is considered alone, the model can be simplified and only radial distortion coefficient should be considered. In this paper, a more comprehensive model including radial distortion and thin prism distortion is adopted. The description model is:

$$\begin{cases} \lambda_x = x - u = (k_0 x_d + k_1)(x_d^2 + y_d^2) + k_3 x_d^2 + k_4 x_d y_d \\ \lambda_y = y - v = (k_0 y_d + k_2)(x_d^2 + y_d^2) + k_3 x_d y_d + k_4 y_d^2 \end{cases} \tag{1}$$

$$\begin{cases} X_d = (u - c_x)/F_x \\ Y_d = (v - c_y)/F_y \end{cases} \tag{2}$$

$$\begin{cases} F_x = f/d_x \\ F_y = f/d_y \end{cases} \tag{3}$$

Where, $k_0 \sim k_4$ is distortion parameters; x and y are actual image coordinates. u and v are ideal image coordinates. c_x and c_y are main coordinates of images. f is camera focus length. d_x and d_y are the horizontal and vertical sizes of each pixel.

Furthermore, use the standard calibration template as a reference and photo this template with the camera need to be calibrated. Then, obtain grid points in the pictures by Harris point recognition and calculate coordinate difference of these grid points in calibration image and distortion image. With certain control points selected, distortion parameters in formula 1 were solved. In that case, the distortion change relationship was identified.

X coordinates and Y coordinates of a point in normal images are all integers, while the position on a distorted image is basically not an integer after the distortion. It is necessary to use bilinear interpolation for processing.

2.4 Edge identification method with image dodging

In this paper, Canny operator, which is improved by Mask dodging algorithm, was used to identify ring edges so as to realize off-line image positioning in the tunnel. Illumination is a major problem that interferes with machine vision methods, especially in feature extraction based on gray scale.

Then the image was grayed and Canny operator was used to extract the ring edges. Finally, make sure non-maxima suppression. The maximum value of gradient intensity on each pixel was retained and other values were deleted. For each pixel, do the following steps:

a) Gradient direction was approximated as one of the following values (0,45,90,135,180,225, 270,315) (That is, the horizontal, vertical and 45 degreed direction).
b) Compare the gradient strength of the pixel with pixel on its plus or minus gradient direction.
c) Keep the pixel if its gradient strength of the pixel is at its maximum, otherwise, suppress it (delete and set to 0)

2.5 Experiment results of image preprocessing

2.5.1 Key frame extraction.
Through key frame extraction method proposed in this paper, part of key frames were extracted in Figure 1. The maximum step length is set as 48 frames and the minimum step length is set as 12 frames. Through the calculation of the correlation coefficient of feature points, set threshold value θ. Result of key frame extraction meets the expectation. Total number of rings is set as M and the number of identified rings as m. The accuracy rate $\alpha(\alpha = M/m)$ of this method is 95.23%. While partial rings cannot be effectively identified due to the existence of fireproof doors, stairwells and other structures.

2.5.2 Distortion correction.
After key frames being extracted, distortion correction of each image was conducted. The standard calibration template was photographed by Gopro 5 Silver camera, which was used

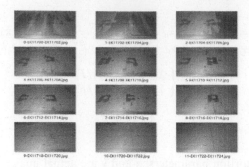

Figure 1.   Effect of Key frame extraction.

Figure 2.   Effect of Wide-angle distortion correction.

for data collection and distortion parameters were calculated by capturing Harris Angle points. Result of distortion parameters were as follows: $k_0 \sim k_4$ were -0.28043, 0.13491, -0.00034, 0.00020, -0.04355; x and y were 0.287 and -2.392. u and v were 58.666 and 58.661. Figure 2. shows the image after distortion correction.

2.5.3 *Off-line positioning based on edge identification*

In this paper, mask dodging algorithm was used to remove the shadow caused by uneven illumination in the image. The effect was shown in Figure 3. In order to clearly reflect the processing effect, binary images were used for comparison. Figure on the left shows the area of leakage after binarization, while Figure on the right shows the area of binary leakage after image dodging. It shows that the area affected by shadow at the corners is reduced. Therefore, the shadow treatment effect is significant. Since the shadow region is difficult to be defined and marked, the accuracy of this algorithm is not further studied in this paper.

Figure 4 shows the result of the edge identification of the tunnel rings by Canny operator with image dodging. The red border in Figure on the left is the result of the Isosceles trapezoid fitting of the edge identified by of Canny operator. Figure on the right shows the image processed by uniform light, which is the original image in this part.

Figure 3.   Effect of Mask dodging method.

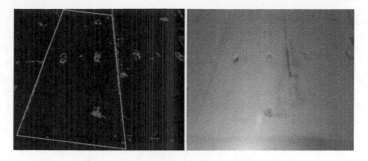

Figure 4. Effect of Edge identification by Canny operator.

3 A FAST DETECTION METHOD OF SURFACE DEFECTS

3.1 Description of fast detection method of defects

Firstly, SIFT feature points were extracted from the preprocessed images to determine the main direction of each feature. Then, 128 dimensions of these feature points were extracted as feature vectors to carry out k-means classification, forming a vocabulary containing K visual words. Next, we counted the occurrence of these words (including defects and noises) in the images, and weaved the image into a k-dimensional vector. Finally, SVM classifier was used to classify the vector to determine the existence of defects or noises.

3.2 Extraction of SIFT features

Compared with other feature extraction algorithms, SIFT searches feature points in different scale spaces and calculates the direction of key points. Therefore, it can adapt to problems with illumination or scale.

Firstly, gaussian pyramid was used to detect the extreme points of the image. Through sub-pixel interpolation method, continuous spatial extreme points were obtained based on the discrete spatial points interpolation. To improve the stability of key points, the Taylor expansion of DoG function in scale space should be used as interpolation function.

The local features of the images are essential to allocate a reference direction to each key point. The stable direction of local structure is calculated by means of image gradient. The gradient histogram divides the range of direction from 0 to 360 degrees into 36 bins, each of which is 10 degrees. In the histogram, the peak represents the feature point is just in the direction of neighborhood gradient, and the maximum is taken as the main direction of the key point. The peak value of the directional histogram represents the direction of neighborhood gradient at the feature point, and the maximum value of the histogram is taken as the main direction of the key point.

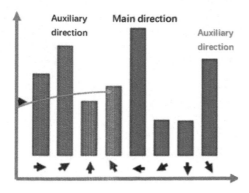

Figure 5. Histogram of gradient and direction.

By following these steps, each key point has three types of information: location, scale, and direction. The next step is to create a descriptor for each key point, which will not change with various changes, such as illumination, viewing angle and so on. Moreover, the descriptors should be highly unique to improve the probability of correctly matching with feature points.

Then, the area of feature points was divided into 4 × 4 grids as feature angles and angle index numbers, and each cell generated 8-unit-high columns to represent the gradient magnitude. Then the gradient of eight directions was calculated by means of tri-linear interpolation. Finally, the normalization process is carried according to formula:

$$L_i = \frac{h_i}{\sqrt{\sum_{j=1}^{128} h_j^2}}, i = 1, 2, 3, \dots, 128 \tag{4}$$

3.3 Building Bag of Words through feature clustering

In this study, k-means algorithm is adopted to cluster the features. 128 dimensions of all feature points extracted by SIFT served as feature vectors, and Euclidean distance was used as classification distance. We set the constant K as the number of categories to be clustered. The algorithm process is shown as below:

(1) Randomly select K points from N feature points as the centroid;
(2) Measure the distance from each of the remaining feature points to each centroid, and classify it into the class of the nearest centroid;
(3) Recalculate the centroid of each new class;
(4) Iterate step (2) ~ (3) until the new centroid is equal to or less than the specified threshold value, and the algorithm ends.

The k-means algorithm is used to build a vocabulary containing K visual words. Next, we counted the occurrence of each word in the images, and weaved the image into a k-dimensional vector. In the image-based defect detection method, the vocabulary may include defects or fixed noises, such as bolt holes, grouting holes and so on.

3.4 Defect recognition by means of SVM classifier

This paper used one versus rest method. We built K two-class classifiers (K categories), in which the i-th classifier divides congruence class of the i-th category. In the training, the i-th classifier takes the i-th class in the training set as positive category and the rest points as the negative category. In the discrimination, the input signal goes through k classifiers and we can obtain k output values. If there is only one +1 appears, then the corresponding category is the input signal category. In practice, there are errors in the decision function we built. If the number of +1 in the output is more than one (more than one class claims it belongs to itself) or less than one (no class claims it belongs to itself), then the output value is compared and the input category belongs to the category corresponding to the maximum. The advantage of this method is that only K two-class SVM need to be trained for K-class problems. The number of classification functions obtained is less, which results in faster speed of classification.

4 APPLICATION IN ENGINEERING

4.1 The source of data

In this research, we used automatic inspection equipment to collect the experimental data in two cross-river tunnels in Shanghai. And the figure of the equipment at worksite is shown as Figure 6.

Figure 6.   The equipment at worksite.

The video collected was greater than 700G in total, and more than 5,500 images of torus were captured. In the experiments, 200 images were randomly selected for training, while another 800 images were selected for verification.

4.2   *The experimental results*

4.2.1   *The comparison of fallout ratio*
This research compared some common fast defect detection methods with the methods proposed in this paper, including OTSU, Watershed and Region growing.

Through comparison, it can be found that OTSU was seriously affected by pipelines, joints and bolt holes, and was difficult to effectively distinguish noises from defects. Watershed had good performance in defect recognition of small local region, but was easily disturbed by illumination, shadow, bolt hole and so on. Region growing relied heavily on manual selection of initial points, which will determine the accuracy of recognition. In addition, Region growing was apt to detect the connected pipelines, and miss the disconnected small defects. While the FDA-TDV proposed in this paper can effectively identify the defect and noise area with high accuracy.

In order to quantify the advantages and disadvantages of each detection algorithm, this paper used fallout ratio as the evaluation index, which refers to the ratio of pixel count of fault detection to the total pixel count of images and its calculation formula is:

$$P_i = \frac{N_i}{N_{total}} \times 100\% \tag{5}$$

Where, P_i is the error detection rate of algorithm i and i refer to the FDA-TDV, OTSU, Watershed and Region growing. N_i is the pixel number incorrectly detected of algorithm i. N_{total} is number of total pixels in the images.

Compared with the other three methods, fault detection rate of FDA-TDV is lower than that of the other three methods.

Table 1.   Comparison of different defect detection methods.

Origin	OTSU	Watershed	Region growing

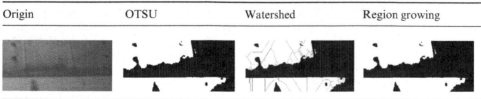

Table 2. Comparison of fault detection rates for various disease detection algorithms.

Category	FDA-TDV	OTSU	Watershed
1	0.08734	0.400721	0.464391
2	0.08585	0.364464	0.484923
3	0.06341	0.143225	0.323888
4	0.08991	0.488706	0.542315

Defect No.	Defect Type	Photo Direction	Signed Seg.	Evalution Seg.	Area(mm²)	Criticality	Photo No.
20180315004	Leakage	Ring surface	10001	101	22875	b	1
20180315005	Leakage	Ring surface	10002	101	37508	b	2
20180315006	Leakage	Ring surface	10003	101	2693	a	3
20180315007	Leakage	Ring surface	10004	101	44582	c	4
20180315008	Leakage	Ring surface	10005	102	17743	b	5
20180315009	Leakage	Ring surface	10006	102	30165	c	6
20180315010	Leakage	Ring surface	10007	102	19923	b	7
20180315011	Leakage	Ring surface	10008	102	31678	c	8
20180315012	Leakage	Ring surface	10009	102	43674	c	9
20180315013	Leakage	Ring surface	10010	102	23056	b	10

Figure 7. The test results.

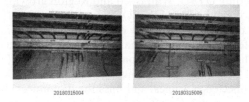

Figure 8. Effect of leakage identification.

4.2.2 *Results report*

Figure 7 is a part of the test results report. In order to systematize defect detection, the defect number is managed in the system mentioned in this paper. The whole tunnel is divided into marking segments and evaluation segments according to the fixed length for statistical analysis. The result of the algorithm described in this paper is a method of detecting leakage water defect in the system. Through research and supplement, it can be extended to multi-defect recognition system.

Figure 8 is an effect diagram of leaking water disease identification. In the diagram, the leaking area is leaked out with a red frame coil and the confidence rate of detection is marked in the blue area in the upper left corner.

5 CONCLUSION

In this paper, an algorithm named Fast Detection Algorithm of Tunnel Defects based on Video data (FDA-TDV) is proposed. This algorithm calculates and maximizes the characteristic correlation coefficient of images, in order to capture valid frames. And the Harris Angle Point is used as the feature to calibrate the feature point and correct the distortion. While the Canny operator grabs ring surface as a counter to realize offline positioning, which is combined with the correction of special position. Then the SIFT method is used to extract features of defects of the ring surface. Next, these features are clustered to build feature dictionary and Bag of Words including defects and noises. And the Bag of Words is used to train the SVM classifier, which is adopted to carry out the semantic segmentation in the area classified as defects.

By collecting data in 2 cross-river tunnels in Shanghai, this paper compares FDA-TDV with three other defect detection methods (OTSU, Watershed and Region growing methods), which are used to process images and identify leakage defects. And the error detection rate of each method is calculated respectively. Among them, FDA-TDV is the fastest one and can effectively identify the defect and noise area with low error rate, which is suitable for applications in engineering.

REFERENCES

Abdel-Qader, I. & Abudayyeh, P. & Kelly, M.E. 2003. Analysis of edge-detection techniques for crack identification in bridges. *Journal of Computing in Civil Engineering* 17(4): 255–263.

Devernay, F. & Faugeras, O. 2001. Straight lines have to be straight: automatic calibration and removal of distortion from scenes of structured environments. *Machine Vision & Applications* 13(1):14–24.

Haneishi, H. & Yagihashi, Y. & Miyake, Y. 1995. A new method for distortion correction of electronic endoscope images. *IEEE Transactions on Medical Imaging* 14(3):548–555.

Huang, H. & Sun, Y. & Xue, Y. 2017. Inspection equipment study for subway tunnel defects by greyscale image processing. *Advanced Engineering Informatics* 32: 188–201.

Hughes, C. & Glavin, M. & Jones, E. & Denny, P. 2009. Wide-angle camera technology for automotive applications: a review. *Iet Intelligent Transport Systems* 3(1):19–31.

Hunt, M. A. 2002. Tunnel crack detection and classification system based on image processing. *Proceedings of SPIE - The International Society for Optical Engineering* 4664(11):495–499.

Koch, C. & Georgieva, K. & Kasireddy, V. & Akinci, B. & Fieguth, P. 2015. A review on computer vision-based defect detection and condition assessment of concrete and asphalt civil infrastructure. *Advanced Engineering Informatics* 29(2): 196–210.

Sheena, C V & Narayanan, N.K. 2015. Key-frame Extraction by Analysis of Histograms of Video Frames Using Statistical Methods. *Procedia Computer Science* 70: 36–40.

Sinha, S.K. & Fieguth, P.W. 2006. Neuro-fuzzy network for the classification of buried pipe defects. *Automation in Construction* 15(1): 73–83.

Sykora, P. & Kamencay, P. & Hudec, R. 2014. Comparison of SIFT and SURF Methods for Use on Hand Gesture Recognition based on Depth Map. *Aasri Procedia* 9:19–24.

Wang, J. & Shi, F. & Zhang, J. & Liu, Y. 2011. A New Calibration Model and Method of Camera Lens Distortion. *International Conference on Intelligent Robots and Systems* 2007:5713–5718.

Weimer, D. & Thamer, H. & Scholz-Reiter, B. 2013. Learning Defect Classifiers for Textured Surfaces Using Neural Networks and Statistical Feature Representations. *Procedia CIRP* 7(12): 347–352.

Xue, Y.D. & Li, Y.C. 2018. A Fast Detection Method via Region-Based Fully Convolutional Neural Net- works for Shield Tunnel Lining Defects. *Computer-Aided Civil and Infrastructure Engineering* 33 (8): 638–654.

Yeung, M.M. & Liu, B. 1995. Efficient Matching and Clustering of Video Shots. *Proceedings of IEEE ICIP* 1: 338–341.

Zhang H.J. & Wu, J. & Zhong, D. & Smoliar, S.W. 1997. An Integrated System for Content Based Video Retrieval and Browsing. *Pattern Recognition* 30(4):643–658.

Zhang, W. & Zhang, Z. & Qi, D. 2014. Automatic crack detection and classification method for subway tunnel safety monitoring. *Sensors* 14(10): 19307–19328.

Zhang, Z. & Flexible, A. 2005. New Technique for Camera Calibration. *IEEE Trans on Pattern Analysis & Machine Intelligence* 22(11):1330–1334.

Zhao, L. & Qi, W. & Li, S.Z. & Zhang, H.J. 2000. Key-frame extraction and shot retrieval using nearest feature line. *Proceedings of ACM Workshop on Multimedia* 2000:217–220.

Tunnels and Underground Cities: Engineering and Innovation meet Archaeology,
Architecture and Art, Volume 5: Innovation in underground engineering,
materials and equipment - Part 1 – Peila, Viggiani & Celestino (Eds)
© 2020 Taylor & Francis Group, London, ISBN 978-0-367-46870-5

Research on applicability of limit state method in lining design of railway tunnel

W. Hu & X. Tan
China railway eryuan engineering group Co. Ltd, Chengdu, China

ABSTRACT: In order to explore the applicability of the limit state method in tunnel lining design, the limit state method and the damage stage method are used to test the design of railway tunnel lining in this paper, then the test results of the two methods are compared and analyzed. Conclusions are as follows: (1) for concrete lining, the calculation thickness obtained from limit state method is higher than that obtained from damage stage method, and the maximum difference reaches 20cm that means it's necessary to get the cracking resistance coefficient (γ_d) in the limit state equation recalibrated, and the final value is 1.55; (2) for reinforced concrete lining structure and open-cut tunnel structure, the design results base on the limit state method and damage stage method are basically in agreement which means the value of partial coefficients and adjustment coefficients in the limit state equation of reinforced concrete lining are reasonable; (3) the project investment based on the limit state method when the value of γ_d equals 1.55 and damage stage method are basically the same. In terms of all projects that selected, the investment based on limit state method is 2370 thousand and 1% more than that based on damage stage method.

1 GENERAL INSTRUCTIONS

For a long time, the damage state method has been adopted in the design of railway tunnel in China (National Railway Administration of the People's Republic of China, 2017, Zhao & Yu 2014, Zhao et al. 2015), and the limit state method has been widely used as a tunnel design method internationally (Ben-Haim & Elishakoff 1989, Oreste 2005, Goh & Kulhawy 2003). In order to improve the scientific nature of tunnel design and integrate with international standards, the former Ministry of Railways and the China Railway successively carried out the basic research work on the transition of the design standard of tunnel since 1990s (Zhang & Wang 1994, Bian & Huang 2005, Qi 2014, Zhao et al. 2015), and published the Interim Code for Limit State Design of Railway Tunnel in 2015(China Railway Corporation, 2015).

Through the reliability calibration of the general reference map of the tunnel, the target reliability indexes of the ultimate bearing capacity of the tunnel lining are put forward referring to the target reliability value of the various industry structures. Based on the target reliability value, the limit state equations of lining structure are proposed in the code. However, the partial factors and adjustment coefficients in code are obtained through the calibration of the general reference map, and not all tunnel sections are designed through the general reference map. So the applicability of limit state method in tunnel design especially for some special cases needs further study. In this paper, the tunnel linings are designed by using limit state method based on interim code and using damage method based on current design code, and the difference of the design results are compared and analyzed. The paper verifies the rationality of limit state method for tunnel lining design and discovers some deficiencies of the interim code which provide reference for its revision.

2 RESEARCH ROUTE

At present, the standard design method is adopted under general conditions and standard drawings are used for lining design. The design parameters of standard drawings are obtained based on the numerical calculation combined with engineering experience which leads to the value of the parameters are often safer than the numerical results. The premise of the design verification is that the design method in current code is reasonable and reliable, and the basis for judging the rationality of the limit state method is whether the design result of limit state method is consistent with the results of the damage stage method or not. Therefore, the research route can be summarized as follows:

(i) Reducing the thickness of original secondary lining 0.25cm every time.

(ii) Calculating the internal force of secondary lining using finite element software based on limit state method and damage stage method respectively.

(iii) Calculating the control index and obtaining the minimum required thickness and quantity of steel bar of secondary lining when the control index reaches the target value.

(iv) Comparing and analyzing the difference of design results of limit state method and damage state method and optimizing the design method.

The control index is bearing capacity when the lining is concrete and the maximum crack width or reinforcement design basis when the lining is reinforced concrete. The checking calculation equations of concrete and reinforced concrete based on limit state method and damage state method can be found in interim code and current code respectively.

3 CALCULATION CONDITIONS

The calculation conditions are listed in Table 1. The selected tunnels contain different types of composite lining and open cut tunnel lining

4 CALCULATION MODELS AND PARAMETERS

The widely used finite element software ANSYS is adopted in the calculation based on the load- structure model. The lining in the model is simulated by the element type of BEAM3, and contact affection between the lining and surrounding rock is simulated by the type of LINK10. The calculation model of singe line tunnel and open cut tunnel are shown as Figure 1. The parameters of concrete, steel bar and surrounding rock can be found in Section 3.2, 5.2, 5.4 of the interim code and Section 4.3, 6.2, 6.4 of the current code.

Table 1. The calculation conditions.

Tunnel	Yongshou-liang tunnel	Bijiashan tunnel	Qishan tunnel	Houshi-shan tunnel	Shangye-tian tunnel	Mawei-shan tunnel	Xinbada-ling tunnel
Rock grade	III~V	III~V	IV~V	III~V	III~V	III~V	II~V
Composite lining	4	11	3	13	8	6	8
Open cut tunnel lining	2	2	2	2	2	2	2
Single/double line	Double	Single	Single	Double	Double	Double	Double
Design speed	350km/h	160km/h	120km/h	200km/h	250km/h	350km/h	250km/h

(a) Secondary lining   (b) Bias pressure open cut tunnel   (c) single pressure open cut tunnel

Figure 1.   Calculation models.

5   ANALYSIS OF DESIGN VERIFICATION

5.1   *Concrete lining*

According to the research route illustrated in Section 2, all of the concrete linings for selected conditions are redesigned. In consideration of the limited space, only the design results of Bijiashan tunnel are listed in Table 2, and the laws of design results of other concrete linings are the same as the Bijiashan tunnel.

Just as shown in Table 2, the two methods have the same control position and mode, but the minimum required thickness of limit state method is higher than that of damage state method for one condition, which means the adjustment coefficient γ_d=2.35 in the checking formula of anti-cracking in interim code is too conservative. Therefore, the adjustment coefficient should be modified.

The γ_d is gradually decreased from 2.35 by 0.05, and when the minimum required thickness of the limit state method is the same as that calculated with the damage stage method, the γ_d can be considered as a suitable value. Taking the concrete lining located in IV-class surrounding rock as an example, the relation between the lining thickness and the safety factor of the anchor crown section is shown in Figure 2.

Just as shown in Figure 2, when using the damage stage method, the thickness of lining h=30cm, the lining passes checking, and the thickness of lining h=25cm, the lining doesn't pass the checking. When using the limit state method, the relation between γ_d and R-S(R is the resistance and S is the action effect) when h=25cm and h=30cm is as shown in Figure 3.

In Table 3, the h_{LM} and h_{DS} mean the calculation thickness of lining when using limit state method and damage state method respectively. According to the above calculation results, it is suggested that the γ_d can be adjusted from 2.35 to 1.55, which can ensure the results of the two methods are the same at most conditions and close in a few cases.

Table 2.   Calculation results of concrete linings in Bijiashan tunnel.

Lining type	Interim code		Current code	
	Arch wall thickness	Control position/ mode	Arch wall thickness	Control position/ mode
III$_a$	10cm	Arch crown/anti-cracking	5cm	Arch crown/anti-cracking
IV$_b$	50cm	Arch crown/anti-cracking	30cm	Arch crown/anti-cracking
IV$_a$	45cm	Arch crown/anti-cracking	30cm	Arch crown/anti-cracking

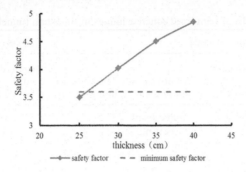

Figure 2. Relation curve of thickness and safety factor.

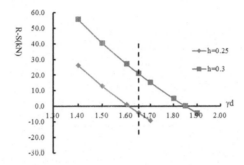

Figure 3. Relation curve of γd and R-S.

Table 3. Calibration Results of anti-cracking adjustment coefficient.

Calculation results	coefficient	140km/h single line Va	160km/h single line IIIa	160km/h single line IVb	160km/h single line IVa	200km/h double line II	200km/h double line III
$h_{LM}=h_{DS}$	γ_d	1.4~1.6	1.4~1.45	1.5~1.65	1.5~1.65	1.45~1.7	1.5~1.6
$h_{LM}>h_{DS}$	γ_d	1.65~1.95	1.5~1.8	1.7~1.85	1.7~1.85	1.75	1.65

5.2 Reinforced concrete lining

The design results of Bijiashan tunnel are listed in Table 4, and the laws of design results of other reinforced concrete linings are the same as the Bijiashan tunnel.

According to Table 4, the design results of the two methods are basically the same. Specifically, the control position and control mode are the same. Besides, the minimum required thicknesses of the two methods are close and the minimum required quantities of steel bars are almost the same, so the final design quantities of reinforcement are completely equal.

In conclusion, the design results of the two methods are basically the same for reinforced concrete lining. Moreover, the design results of open cut tunnel are almost the same as the reinforced concrete lining because the open cut tunnel lining is also made of reinforced concrete.

Table 4. Calculation results of reinforced concrete linings in Bijiashan tunnel.

Lining type	Interim code			Current code		
	Thickness of lining/ cm	Required quantity of rebar/cm^2	Control position/mode	Thickness of lining/cm	Required quantity of rebar/cm^2	Control position/ mode
IV	30	5.24	arch crown/ bearing capacity	32.5	5.73	arch crown/ bearing capacity
V$_B$	37.5	6.23	arch crown/ bearing capacity	37.5	6.6	arch crown/ bearing capacity
V$_a$	35	4.55	arch crown/ bearing capacity	35	4.61	arch crown/ bearing capacity
V$_b$	40	6.82	arch crown/ bearing capacity	40	7.13	arch crown/ bearing capacity
V$_c$	40	6.91	arch waist / bearing capacity	40	5.22	arch waist / bearing capacity
V$_d$	40	7.71	arch crown/ bearing capacity	40	6.96	arch crown/ bearing capacity

Table 5. Investment and comparison of main building materials in Bijiashan tunnel.

Structure	Total investment of tunnel					Investment per meter		
	Length /m	Damage stage method /million	Limit state method /million	Increased /million	Increased proportion	Damage state method /thousand	Limit state method /thousand	Increased /thousand
Concrete lining	2956	12.341	12.341	0	0%	4.18	4.18	0
Reinforced concrete lining	1178	8.403	8.249	−0.154	−1.8%	7.13	7.00	−0.13
Open cut tunnel	8	0.229	0.229	0	0	28.57	28.57	0
Overall length	4142	20.97	20.82	−0.154	−0.73%	50.6	20.3	−0.03

6 ECONOMIC ANALYSIS

According to the results of the above design verification, the investment amount of the main building materials of secondary lining is calculated and the difference of the investment between the two methods is compared in order to analyze the economics of the limit state method. Taking the Bijiashan tunnel as an example, the investment is as shown in Table 5.

Just as shown in Table 5, the investment of Bijiashan tunnel using limit state method is almost the same with that using damage state method. Considering all the conditions, the total investments of the two methods are basically the same, the investment based on limit state method is 2370 thousand and just 1% more than that based on damage stage method.

7 CONCLUSIONS

In this paper, the different types of lining are redesigned by adopting the limit state method and damage stage method respectively. The research results not only verify the operability of limit state method in tunnel design, but also provide the basis for the revision of the interim code.

When the lining structures are made of concrete, the required thickness of lining is different for these two methods. In view of this, the adjustment coefficient of anti-cracking is adjusted from 2.35 to 1.55 which can ensure the design results of two methods keep consistent in most cases. For reinforcement concrete lining, the required thickness of lining and quantity of rebar are almost the same. Therefore, the factors in limit state equations of reinforcement concrete are reasonable. From the economic analysis result, the tunnel total investments of two methods are almost the same when adopting the calibrated adjustment coefficient of concrete anti-cracking. In summary, the design results of two methods are basically the same after correcting the adjustment coefficient of anti-cracking.

REFERENCES

Ben-Haim Y. & Elishakoff I. 1989. Non-probabilistic models of uncertainty in the nonlinear bucking of shells with general imperfections: theoretical estimates of the knockdown factor. *Journal of applied mechanics* 56(2): 403–410.

China railway corporation. 2015. Interim code for limit state design of railway tunnel. Beijing: China railway publishing house.

Dongping Zhao. & Yu Yu. 2014. Comparative research between China and Britain on design approach of railway tunnel with drilling-and-blasting-method, *Railway standard design* 58(5): 99–103.

Dongping Zhao. & Yu Yu. 2015. Research on the target reliability index of railway tunnel lining. *Journal of railway engineering society* 6: 51–56.

Goh A T C. & Kulhawy F H. 2003. Neural network approach to model the limit state surface for reliability analysis. *Canadian geotechnical journal* 40(6):1235–1244.

Jianxu Qi. 2014. Research on calculation model and structure reliability in railway composite lining of tunnel, M.S. Dissertation of shijiazhuang tiedao university.

National railway administration of the people's republic of China. 2017. Code for design on tunnel of railway. Beijing: China railway publishing house.

Oreste P. 2005. A probabilistic design approach for tunnel supports. *Computers and geotechnics* 32(7): 520–534.

Qing Zhang. & Dongyuan Wang. 1994. Reliability analysis of railway tunnel structure. *Chinese journal of rock mechanics and engineering* 13(3): 209–218.

Wanqiang Zhao. & Yuxiang Song. 2015. Reliability analysis of reinforced concrete lining structure for railway tunnel. *Journal of railway engineering society* 8: 81–86.

Yihai Bian. & Hongwei Huang. 2005. Application of reliability theory in determining reasonable parameters of railway tunnel lining. *Chinese journal of underground space and engineering* 1(1): 129–132.

Tunnels and Underground Cities: Engineering and Innovation meet Archaeology,
Architecture and Art, Volume 5: Innovation in underground engineering,
materials and equipment - Part 1 – Peila, Viggiani & Celestino (Eds)
© *2020 Taylor & Francis Group, London, ISBN 978-0-367-46870-5*

A complex variable solution for the excavation of a shallow circular hole in gravitational elastic ground

X. Huang & Y. Xiang
School of Civil Engineering, Beijing Jiaotong University, Beijing, China

ABSTRACT: This paper presents a complex variable solution of the displacement and stress induced by the excavation of a shallow circular hole in a ground modelled as a gravitational elastic half-plane. Assuming that the diameter of the hole is adequately small so that the initial stress on the perimeter of the hole can be taken to be equal to the initial stress of the center of the hole, the complex integral of the released initial tractions on the perimeter of the hole is derived analytically, first on the half-plane and then transformed via a holomorphic function onto a mapped annulus. The complex potential functions are expressed as Laurent series, with their coefficients determined by using the stress release condition at the perimeter of the hole as well as the traction free condition of the surface of the half-plane. The comparison with numerical solutions for hypothetical problems verifies the analytical solution.

1 INTRODUCTION

In order to predict the ground displacement and stress induced by the excavation of a shallow horizontal circular hole/tunnel, both empirical methods, such as the Peck formula and its likes (Attewell et al., 1986; Mair & Taylor, 1999), and analytical methods, such as the virtual image methods, the random medium methods, the Airy stress function methods, and the complex variable methods, have been developed.

With regard to the analytical methods, the basic assumption in the virtual image methods, as in Sagaseta (1987) and in Verruijt and Booker (1996), that the ground loss induced by the excavation of the hole can be treated as a point source located at the centre of the hole, renders the solution inaccurate or incorrect in the vicinity of the hole. In addition, the virtual image methods cannot explicitly consider the stress release at the perimeter of the hole. The random medium methods, as in Liu (1993), also need the deformation of the perimeter of the hole be prescribed and cannot explicitly consider the stress release. One of the major shortcomings of the analytical solutions in Matsumoto and Nishioka (1991) and in Bobet (2001) using the Airy stress function methods is the precondition that the hole must be located sufficiently deep below the surface of the ground. The analytical solutions in Verruijt (1997, 1998) using the complex variable methods have considered neither the gravity nor the corresponding stress release at the perimeter of the hole. Strack and Verruijt (2002) and Verruijt and Strack (2008) considered the gravity related tunnel buoyancy but not the stress release.

In this paper, a complex variable analytical solution is developed for the displacement and stress induced by the excavation of a shallow circular hole in a ground modelled as a gravitational elastic half-plane. In what follows next, the complex integral of the released initial tractions on the perimeter of the hole is derived analytically, first on the physical half-plane and then transformed via a holomorphic function onto a mapped annulus. Then the complex potential functions are expressed as Laurent series, with their coefficients determined by using the stress release condition at the perimeter of the hole as well as the traction free condition at the surface of the half-plane. Finally, comparisons are made with numerical solutions for hypothetical problems, to serve as a verification of the analytical solution.

2 THE ANALYTICAL SOLUTION

2.1 *Expression of the Boundary Condition of the Hole*

Refer to Figure 1, consider the excavation of a shallow circular hole in a ground that behaves as an elastic half-plane under gravity, let γ and μ be the unit weight and the at-rest lateral pressure coefficient of the ground, respectively; (x_c, y_c) denotes the (x, y) coordinates of the center and a the radius of the hole, respectively; t'_x and t'_y are the x-direction and y-direction tractions of the initial stress on the perimeter of the hole, respectively.

Assume that the radius of the hole is sufficiently smaller than the depth of the hole such that the initial stress on the perimeter of the hole may be taken approximately equal to the initial stress at the center of the hole, then let the stress be positive when tensile and negative when compressive, the tractions corresponding to the stress release on the perimeter of the hole due to the excavation can be expressed as:

$$\begin{cases} t_x = -t'_x = -\mu\gamma y_c \cos\theta \\ t_y = -t'_y = -\gamma y_c \sin\theta \end{cases} \tag{1}$$

The resultant force of these tractions can be expressed by a complex integral as:

$$F(\tau)\big|_{|z-\bar{b}-\iota h|=a} = i \int_{s_0}^{s} (t_x + i t_y) ds = i \int_{s_0}^{s} (-\mu\gamma y_c \cos\theta - i\gamma y_c \sin\theta) ds \tag{2}$$

where, $z = x + iy$ represents the physical plane, with $i = \sqrt{-1}$ being the unit imaginary number; s_0 and s denote a prescribed starting point and an arbitrary point on the perimeter of the hole, respectively; θ is the counterclockwise angle of the radial direction of an arbitrary point with regard to the x axis.

Since, on the perimeter of the hole, $ae^{i\theta} = (z - x_c - iy_c)$, $ae^{-i\theta} = (\bar{z} - x_c + iy_c)$, and $ds = ad\theta$, Equation (2) can be developed into:

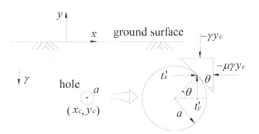

Figure 1. Shallow circular hole in elastic half-plane under gravity and its initial boundary tractions.

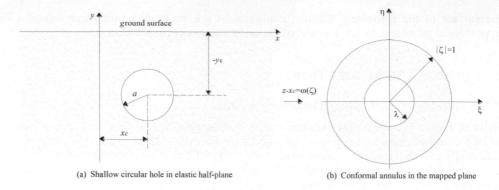

(a) Shallow circular hole in elastic half-plane   (b) Conformal annulus in the mapped plane

Figure 2.   Correspondence between the physical and the mapped planes.

$$
\begin{aligned}
F(z)|_{|z-x_c-iy_c|=a} &= i\int_0^\theta (-\mu\gamma y_c\cos\theta - i\gamma y_c\sin\theta)a\,d\theta \\
&= -ay_c\gamma(i\mu\sin\theta + \cos\theta - 1) \\
&= -ay_c\gamma\left(\frac{\mu+1}{2}\frac{z-x_c-iy_c}{a} + \frac{-\mu+1}{2}\frac{\bar{z}-x_c+iy_c}{a} - 1\right) \\
&= -\frac{y_c\gamma}{2}(\mu+1)(z-x_c-iy_c) - \frac{y_c\gamma}{2}(-\mu+1)(\bar{z}-x_c+iy_c) + ay_c\gamma
\end{aligned}
\tag{3}
$$

The half-plane in z may be transformed into an annulus in ζ, as shown in Figure 2, by using the mapping function:

$$
z - x_c = \omega(\zeta) = iy_c\frac{1-\lambda^2}{1+\lambda^2}\frac{1+\zeta}{1-\zeta}
\tag{4}
$$

where, $\zeta = \xi + i\eta$, $\lambda = \left(y_c - \sqrt{y_c{}^2 - a^2}\right)/a$.

Substitution of equation (4) into Equation (3), noting $\zeta = \lambda e^{i\beta} = \lambda X$ with β being the counterclockwise angle of the radial direction of an arbitrary point with regard to the ξ axis in ζ, yields

$$
\begin{aligned}
L^*(\zeta)|_{\zeta=\lambda X} &= (1-\lambda X)F(\omega(\zeta))|_{\zeta=\lambda X} \\
&= (1-\lambda X)[-\frac{y_c\gamma}{2}(\mu+1)\left(iy_c\frac{1-\lambda^2}{1+\lambda^2}\frac{1+\lambda X}{1-\lambda X} - iy_c\right) \\
&\quad -\frac{y_c\gamma}{2}(-\mu+1)\left(-iy_c\frac{1-\lambda^2}{1+\lambda^2}\frac{1+\lambda X^{-1}}{1-\lambda X^{-1}} + iy_c\right) + ay_c\gamma] \\
&= (1-\lambda X)[i(\mu+1)\frac{y_c{}^2\gamma\lambda}{1+\lambda^2}\frac{\lambda-X}{1-\lambda X} - i(-\mu+1)\frac{y_c{}^2\gamma\lambda}{1+\lambda^2}\frac{\lambda-X^{-1}}{1-\lambda X^{-1}} + ay_c\gamma]
\end{aligned}
$$

Let $K_1 = (\mu+1)\dfrac{y_c{}^2\gamma\lambda}{1+\lambda^2}$, $K_2 = -(1-\mu)\dfrac{y_c{}^2\gamma\lambda}{1+\lambda^2}$, $K_3 = ay_c\gamma$, then

$$
\begin{aligned}
L^*(\zeta)|_{\zeta=\lambda X} &= [iK_1\frac{\lambda-X}{1-\lambda X} + iK_2\frac{\lambda-X^{-1}}{1-\lambda X^{-1}} + K_3](1-\lambda X) \\
&= iK_1(\lambda-X) + iK_2\frac{\lambda-X^{-1}}{1-\lambda X^{-1}}(1-\lambda X) + K_3(1-\lambda X)
\end{aligned}
$$

2254

in which, by using Taylor series expansion

$$\frac{1}{X - \lambda} = \frac{X^{-1}}{1 - \lambda X^{-1}} = X^{-1} \sum_{k=0}^{\infty} \lambda^k X^{-k} = \sum_{k=0}^{\infty} \lambda^k X^{-k-1}$$

the second term can be written as

$$iK_2 \frac{\lambda - X^{-1}}{1 - \lambda X^{-1}} (1 - \lambda X) = iK_2 \frac{\lambda X - 1}{X - \lambda} (1 - \lambda X)$$

$$= iK_2 \frac{-1 + 2\lambda X - \lambda^2 X^2}{X - \lambda} = iK_2 \left(-1 + 2\lambda X - \lambda^2 X^2 \right) \sum_{k=0}^{\infty} \lambda^k X^{-k-1}$$

thus

$$L^*(\zeta)|_{\zeta = \lambda X} = iK_1 (\lambda - X) + iK_2 \frac{\lambda + X^{-1}}{1 - \lambda X^{-1}} (1 - \lambda X) + K_3 (1 - \lambda X)$$

$$= iK_1 (\lambda - X) + iK_2 \left(-1 + 2\lambda X - \lambda^2 X^2 \right) \sum_{k=0}^{\infty} \lambda^k X^{-k-1} + K_3 (1 - \lambda X)$$

$$= iK_1 (\lambda - X) + iK_2 \left[2\lambda - \lambda^2 X - \lambda^3 - \left(1 - \lambda^2 \right)^2 \sum_{k=1}^{\infty} \lambda^{k-1} X^{-k} \right] + K_3 (1 - \lambda X)$$

$$= iK_1 \lambda + iK_2 \left(2\lambda \quad \lambda^3 \right) + K_3 + \left(-iK_1 - iK_2 \lambda^2 - K_3 \lambda \right) X - iK_2 \left(1 - \lambda^2 \right)^2 \sum_{k=1}^{\infty} \lambda^{k-1} X^{-k}$$

$$\tag{5}$$

2.2 Series Expressions of the Complex Potential Functions

Let the complex potential functions in ζ be expressed as Lauren series

$$\Phi(\zeta) = a_0 + \sum_{k=1}^{\infty} a_k \zeta^k + \sum_{k=1}^{\infty} b_k \zeta^{-k}, \ \lambda \le |\zeta| \le 1 \tag{6a}$$

$$\Psi(\zeta) = c_0 + \sum_{k=1}^{\infty} c_k \zeta^k + \sum_{k=1}^{\infty} d_k \zeta^{-k}, \ \lambda \le |\zeta| \le 1 \tag{6b}$$

where, $a_0, c_0, a_k, b_k, c_k, d_k \ (k = 1, 2, \cdots)$ are complex coefficients to be determined by using the prescribed boundary conditions.

2.3 Determination of the Complex Coefficients

The boundary condition of the hole can be expressed in ζ as

$$(1 - \lambda X) \left[\Phi(\zeta) + \frac{\omega(\zeta)}{\overline{\omega'(\zeta)}} \overline{\Phi'(\zeta)} + \overline{\Psi(\zeta)} \right]_{\zeta = \lambda X} = L^*(\zeta)|_{\zeta = \lambda X} + (1 - \lambda X) C \tag{7}$$

where, C is a constant to be determined, $L^*(\zeta)|_{\zeta = \lambda X} = (1 - \lambda X) F(\omega(\zeta))|_{\zeta = \lambda X}$, which in general can be expanded into a Fourier series of the form

$$L^*(\zeta)\big|_{\zeta=\lambda X} = \sum_{k=-\infty}^{k=\infty} A_k X^k = \sum_{k=1}^{k=\infty} A_{-k} X^{-k} + \sum_{k=0}^{k=\infty} A_k X^k \tag{8}$$

while for Equation (5) the non-zero coefficients are

$$\begin{cases} A_{-k} = -iK_2\left(1-\lambda^2\right)^2\lambda^{k-1}, k=1,2,3,\dots \\ A_0 = K_3 + i\left[K_1\lambda + K_2\left(2\lambda - \lambda^3\right)\right] \\ A_1 = -K_3\lambda + i\left(-K_1 - K_2\lambda^2\right) \end{cases} \tag{9}$$

By comparing the two sides of Equation (7) and taking similar treatment of the traction free condition at the surface of the half-plane, just as the procedure given in Verruijt (1998), the coefficients $a_0, c_0, a_k, b_k, c_k, d_k, (k=1,2,\cdots)$ can be determined.

2.4 Calculation of the Displacement and Stress

Assuming plane strain condition, the displacement and the stress induced by the stress release at the perimeter of the hole can be calculated by using the following so called Kolosov-Muskhelishavili equations

$$u_x + iu_y = \frac{1}{2G}\left[(3-4\nu)\Phi(\zeta) - \frac{\omega(\zeta)}{\overline{\omega'(\zeta)}}\overline{\Phi'(\zeta)} - \overline{\Psi(\zeta)}\right] \tag{10}$$

$$\sigma_{xx} + \sigma_{yy} = 4\mathrm{Re}\left[\frac{\Phi'(\zeta)}{\omega'(\zeta)}\right] \tag{11a}$$

$$\sigma_{yy} - \sigma_{xx} + 2i\sigma_{xy} = \frac{2}{\omega'(\zeta)}\left\{\overline{\omega(\zeta)}\left[\frac{\Phi'(\zeta)}{\omega'(\zeta)}\right]' + \Psi'(\zeta)\right\} \tag{11b}$$

where, u_x and u_y are the x-direction and y-direction components of the displacement, σ_{xx}, σ_{yy} and σ_{xy} are the x-direction normal stress, y-direction normal stress and xy-plane shear stress, respectively, and G is the shear modulus of the material of the half-plane.

The procedure for determining the coefficients of the complex potential functions and for calculating the displacement and stress are implemented using the mathematical software MATLAB.

It is worthwhile to emphasize that the displacement and the stress calculated here are relative to the initial state of the half-plane, i.e. they must be added to the corresponding initial displacement and initial stress, respectively, to arrive at the absolute (total) displacement and stress.

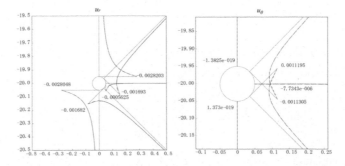

Figure 3.   Induced displacements around a hole with a center depth of 20m and a radius of 0.05m (unit: meter).

2256

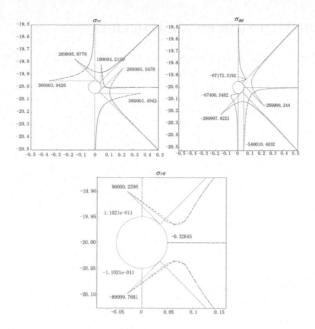

Figure 4. Induced stresses around a hole with a center depth of 20m and a radius of 0.05m (unit: kPa).

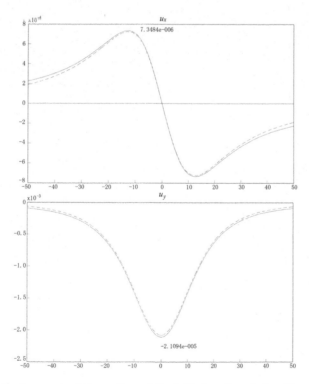

Figure 5. Induced displacements at the surface of the half-plane for a hole with a center depth of 20m and a radius of 0.05m (unit: meter).

3 COMPARISON WITH NUMERICAL MODELING

Consider a gravitational elastic half-plane with the unit weight, the at-rest lateral pressure coefficient, the Young's modulus, and the Poison's ratio being 18kN/m^3, 0.5, 104kN/m^2, and

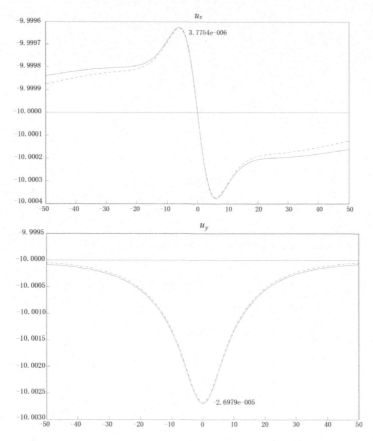

Figure 6.   Induced displacements at one half of the center depth of a hole with a center depth of 20m and a radius of 0.05m (unit: meter).

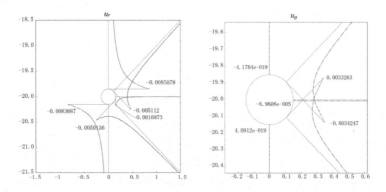

Figure 7.   Induced displacements around a hole with a center depth of 20m and a radius of 0.15m (unit: meter).

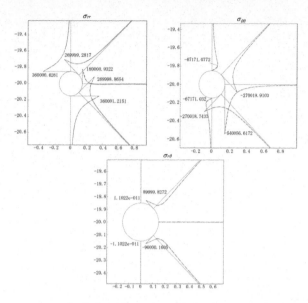

Figure 8. Induced stresses around a hole with a center depth of 20m and a radius of 0.15m (unit: kPa).

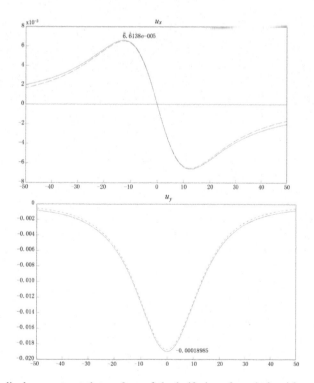

Figure 9. Induced displacements at the surface of the half-plane for a hole with a center depth of 20m and a radius of 0.15m (unit: meter).

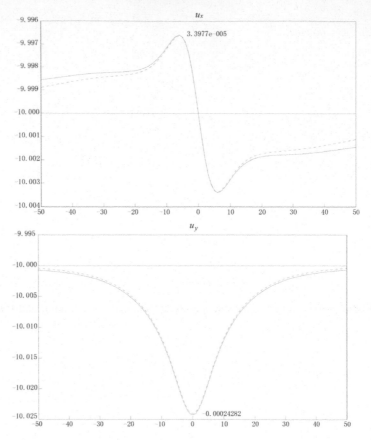

Figure 10. Induced displacements at one half of the center depth of a hole with a center depth of 20m and a radius of 0.15m (unit: meter).

0.25, respectively. The displacements and stresses are calculated using the analytical solution and the finite element software MIDAS. A system of $r\theta$ polar coordinates, with the origin at the center of the hole and the angular ordinate θ measured counterclockwise from the x axis, is employed to illustrate the results around the hole. The results for a hole with a center depth of 20m and a radius of 0.05m are shown in Figures 3 through 6, and for a hole with a center depth of 20m and a radius of 0.15m in Figures 7 through 10, with the solid lines for the analytical solutions and the dashed lines for the numerical solutions.

4 SUMMARY

A complex variable solution is developed for the displacement and stress induced by the excavation of a shallow circular hole in a ground assumed as a gravitational elastic half-plane, and the comparisons with numerical solutions for hypothetical problems are presented as a verification of the analytical solution.

It is worthy to note is that, in order to get the analytical solution, the radius of the hole has been assumed to be sufficiently smaller than the depth of the hole so that the initial stress on the perimeter of the hole be taken approximately equal to the initial stress at the center of the hole. Obviously, the analytical solution may be used for large diameter holes though the results would become less accurate or even incorrect with the increase of the diameter of the hole.

ACKNOWLEDGEMENT

This work is supported by the National Science Foundation of China under grant 50978019, and by the China National Key Basic Research and Development Project (973 Project) under Grant 2015CB057800.

REFERENCES

Attewell, P.B. & Yeates, J. & Selby, A.R. 1986. Soil Movements Induced by Tunnelling and Their Effects on Pipelines and Structures. Chapman and Hall.

Bobet, A. 2001. Analytical solutions for shallow tunnels in saturated ground. Journal of Engineering Mechanics, 127(12):1258–1266.

Liu, B. 1993. Ground surface movements due to underground excavation in the People's Republic of China. In: Hudson. Comprehensive Rock Engineering. Pergamon Press, 4:781–817.

Matsumoto, Y. & Nishioka, T. 1991. Theoretical Tunnel Mechanics. University of Tokyo Press: 162–167.

Mair, R.J. & Taylor,R.N. 1999. Theme lecture: Bored tunneling in the urban environment. (Post Conference) Proceedings of the 14th International Conference on Soil Mechanics and Foundation Engineering (Hamburg, 1997), Balkema, Rotterdam, 2353–2385.

Sagaseta, C. 1987. Analysis of undrained soil deformation due to ground loss. Géotechnique, 37(3): 301–320.

Strack, O.E. & Verruijt, A. 2002. A complex variable solution for a deforming buoyant tunnel in a heavy elastic half-plane. International Journal for Numerical and Analytical Methods in Geomechnics, 26: 1235–1252.

Verruijt, A. & Booker, J.R. 1996. Surface settlements due to deformation of a tunnel in an elastic half plane. Géotechnique, 46(4): 753–756,

Verruijt, A. 1998. Deformations of an elastic half plane with a circular cavity. International Journal of Solids and Structures, 35(21): 2795–2804

Verruijt, A. & Strack, O.E. 2008. Buoyancy of tunnels in soft soils. Géotechnique, 58(6): 513–515.

Tunnels and Underground Cities: Engineering and Innovation meet Archaeology,
Architecture and Art, Volume 5: Innovation in underground engineering,
materials and equipment - Part 1 – Peila, Viggiani & Celestino (Eds)
© 2020 Taylor & Francis Group, London, ISBN 978-0-367-46870-5

Polymer Rubber Gel technology high performance waterproofing for shotcrete and blindside applications

J. Huh
RE-Systems Group Americas, Minneapolis, MN, USA

T.S. Lee & GE
WSP, San Francisco, CA, USA

ABSTRACT: Polymer Rubber Gel (PRG) waterproofing technology is effective and proven for the challenges of underground construction. Key characteristics for PRG waterproofing systems: (1) superior substrate adhesion, (2) responsiveness to substrate movement, (3) non-curing, (4) self-healing, (5) chemical resistance, and (6) environmentally friendly. PRG technology combined with durable, flexible, fleece, reinforced HDPE creates a dynamically responsive waterproofing assembly. This blindside system comprises the following (1) 20-mil HDPE fleece sheet installed against the lagging, (2) PRG is spray applied at a thickness of 100 to 115-mil, (3) 20-mil HDPE fleece sheet is embedded into the PRG. Shotcrete or concrete is installed against the assembly. This system is a high-performance membrane providing the benefits of fluid and rigid protection. This technology is successfully used for waterproofing large-scale subway stations. This paper describes (1) PRG's characteristics of a high-performance waterproofing system, (2) PRG providing effective solutions for shotcrete and blindside applications.

1 INTRODUCTION

Selection of a high-performance waterproofing system is essential to the success of any tunnel or below-grade structural waterproofing application. Waterproofing poses distinct challenges in performance, design, and application. One of the key determining factors of a waterproofing system is based on the method of planned construction. Two distinct methods of application are used for waterproofing structures; Positive Side and Blindside application. Positive side application is commonly used when the application of the waterproofing membrane installed directly to the outside surface of a concrete structure. Blindside application is commonly used when the application of waterproofing is installed prior to placing the concrete or shotcrete lining. The substrate for application is usually to the soil support of excavations i.e. sheet pile walls, secant/tangent pile walls, CDSM, etc. For blindside applications, the final lining or structural wall is formed against the waterproofing membrane.

The preferred method for waterproofing has been the direct application of waterproofing to the exposed concrete substrate - positive side application. This method is generally preferred because it allows the installer to see the substrate that is receiving the waterproofing and to ensure that proper membrane detailing and proper adhesion of the membrane to the substrate is achieved. However, over excavation of the structure may be impractical due to adjacent lot lines and is typically costlier than utilizing a soil support of excavation method. For this condition, the use of a blindside waterproofing assembly is preferred. Utilizing a blindside waterproofing assembly reduces the amount of excavation necessary for site construction this reduces costs. Traditionally, blindside waterproofing of structures has been accomplished utilizing bentonite clay panels or more recently, composite moisture activated adhesive panels

Table 1. Waterproofing application methods.

Waterproofing Method	Description
Positive Side (Figure 1)	Waterproofing is installed to the existing structure directly to the outer concrete substrate
Blindside (Figure 2)	Waterproofing is installed to the support of excavation or lagging wall, prior to structural concrete or a spraying shotcrete lining.

with sheet laminate. Bentonite requires hydration and compaction for effective waterproofing ability. Care must also be taken to protect the exposed membrane prior to concreting from pre-hydration caused by environmental conditions such as rain, site runoff, or sunlight,. Within the last decade, preformed HDPE laminate pressure sensitive adhesive membranes have also been used to varying degrees of success for blindside applications. These preformed blindside membranes do not require compaction or hydration. Various challenges are inherent for both systems, including adhesion, flexibility, durability, and environmental conditions.

Through innovative technological advancements in waterproofing materials, a state-of-the-art solution called Polymer Rubber Gel (PRG) has been utilized for both positive side and blindside membranes for construction. With the introduction of PRG, new hybrid composite waterproofing systems have been developed that attain superior waterproofing performance for underground structures. As an innovative concept in waterproofing, a PRG composite waterproofing system effectively wraps the underground structure in a monolithic layer of flexible, self-healing, adhesive, non-curing gel combined with vapor and chemical resistant HDPE sheets. PRG's unique physical characteristics were specially formulated to effectively retain the integrity of the waterproofing envelope through exceptional engineering properties such as being adhesive, chemical-resistant, environmentally friendly, continuous flexibility, and self-healing.

Figure 1. Positive side Waterproofing.

Figure 2. Blindside Waterproofing.

2 GUIDELINES FOR EFFECTIVE WATERPROOFING

2.1 *General design guidelines and waterproofing system selection*

Site factors must be considered for effective waterproofing of underground structures. First, a set of watertightness criteria must be identified specific for the structure. A waterproofing system selection process should then be based on the watertightness criteria considering both physical site conditions and type of construction chosen for the underground structure. The substrate for which the blindside waterproofing system is to be applied should be considered during the soil support of excavation design process. A relatively smooth surface with sufficient rigidity should be specified to prevent the risk of cavitation, tears or punctures in the waterproofing system subsequent to concrete or shotcrete placement. For example, with sheet pile walls, a protection board with some sort of structural fill behind the board in the flutes is necessary to prevent blowouts in the waterproofing system during concrete or shot-crete placement. For CDSM or secant/tangent pile walls, the substrate should either be shotcrete to a relatively smooth surface or the face of the CDSM/secant/tangent pile walls should be shaved flush with the face of the soldier beams. Protrusion or cavitation in the excavated CDSM/secant/tangent pile walls will need to be smoothed to create an acceptable substrate for the application of the blindside waterproofing system. All waterproofing systems require careful consideration of the type of substrate prior to final design.

Effective waterproofing of underground structures requires that the entire waterproofing system must be continuous throughout the building envelope. Proper detailing for tie-backs to other structural systems, penetrations, protrusions (such as tiebacks in secant/tangent pile walls), transitions, terminations and seams within the waterproofing system are essential for maintaining the integrity of the continuous waterproofing envelope. Maintaining the same waterproofing system throughout the structure is preferred to limit any issues of either incompatibility or difficult tie-in details.

2.2 *Site conditions, constructability, and installation*

Site conditions and constructability play a strong role in the waterproofing system design process. Oftentimes, waterproofing systems used on large scale applications will be exposed to the elements for extended periods of time prior to placement of concrete. In addition, other trades must work in direct proximity such as the mud slab on top of the installed waterproofing system. The environmental and physical durability of the exposed waterproofing membrane is important to prevent potential damage to the waterproofing system prior to completion. Pre-construction meetings with the waterproofing applicator and project general contractor should be conducted to ensure proper work staging to limit the exposure to possible damage. Waterproofing systems that are easily damaged or require extensive protection can cause project delays due to the necessity for repairs and/or complicated protection schemes. Manufacturer approved applicators, skilled and experienced in the installation of the specified waterproofing systems, are essential to the positive outcome of any waterproofing installation. Onsite QA/QC for the waterproofing work should also be provided to document and help ensure that the waterproofing system is installed per design specifications and plans.

2.3 *Key physical guidelines of the waterproofing system*

After successful installation and proper detailing, a waterproofing system must perform based on the physical attributes of the product. The following is a list of physical attributes that are important to the long term watertightness of a cut and cover waterproofing system.

2.4 *Additional water mitigating components to the waterproofing system*

In addition to an underground structure's primary waterproofing system, additional consideration must be given to accessory waterproofing components such as prefabricated drainage

Table 2. Important performance criteria of the waterproofing system.

Performance Criteria	Benefit to waterproofing
Adhesion to concrete	Adhesive bond of the waterproofing to the substrate which it is protecting ensures no path for water migration.
Responsiveness to substrate movement (non-curing)	Ability to retain a waterproof seal during seismic events or joints that experience constant movements.
Elongation	Ability to bridge cracks in concrete and construction joints without debonding ensures water tightness.
Hydrostatic pressure resistance	The system must have the durability to withstand continuous hydrostatic pressure without rupture.
Self-healing/sealing capability	Mitigates failure of the system with the ability for it to self-heal if punctured or penetrated.
Chemical resistance	Prevents degradation of the waterproofing system from soil contaminates.
Environmentally Friendly	Made from non-toxic and recycled materials. No significant VOCs during application. Low odor.

composites and various types of waterstops. For underground structures that are not below the water table or where additional water control measures are desired, prefabricated drainage composites may be a suitable addition to the waterproofing system. A drainage system removes direct hydrostatic pressure from the waterproofing membrane. It is also advisable to utilize waterstops on critical construction joints as a last means of defense against water inflow. Regroutable injection tubes should also be considered for critical interfaces, such as between subway stations, structures, and tunnels, and construction of a thick base slab in stages. There are many various forms of prefabricated drainboards, waterstops and reroutable injection tubes to help with additional unforeseen water intrusions.

3 POLYMER RUBBER GEL WATERPROOFING SYSTEM

3.1 Introduction

PRG waterproofing systems have excelled in meeting the requirements of challenging underground waterproofing applications. Developed specifically for the waterproofing industry, PRG is composed of a polymer modified rubberized asphalt emulsion. However, unlike typical rubberized asphalt materials, PRG's polymers never completely cross-link. This retains the gel always in a semi-cured state. This innovation enables PRG to act as an exceptionally flexible, adhesive, never-cured, continuously self-healing membrane. As a proven concept in waterproofing, composite waterproofing systems utilizing a PRG component exhibit superior elongation properties, adhesion and self-healing ability. A PRG composite waterproofing system consists of a layer of polymer rubber gel at minimum thickness of 2.5 mm +/- .5 mm combined with a sheet membrane of laminate fleece reinforced HDPE. PRG with varying manufacturer-produced viscosities permits different delivery methods including (1) spray applied, (2) trowel applied, and (3) preformed waterproofing sheet applied. The flexible, non-curing, highly adhesive PRG combined with a durable, chemical resistant, hydrostatic pressure resistant HDPE sheet creates a dynamically responsive high-performance waterproofing system for demanding conditions of underground structures. Application of a PRG system is effective, efficient, and economical.

3.2 Unique physical characteristics of Polymer Rubber Gel (PRG)

PRG exhibits many unique physical characteristics that make it an ideal component to a dynamic waterproofing system. The physical characteristics that are unique to PRG are

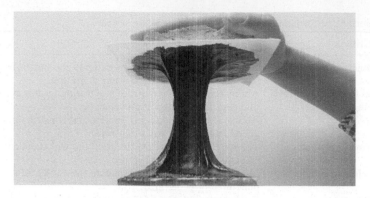

Figure 3.   Polymer rubber gel's exceptional elongation property.

principally the gel's ability to remain in an uncured state and its extreme cohesion and adhesion attributes. PRG's elongation to break is greater than 350% (ASTM C1135). PRG's adhesion to concrete is rated one (1) for excellent (ASTM D412-98). PRG's self-healing ability has been tested to 3.0 bar of direct hydrostatic head (2 mm thickness of PRG membrane). (Figure 3)

3.3   *Advantages of polymer rubber gel waterproofing systems*

Sheer force of the waterproofing membrane against the concrete substrate caused by either seismic activity, foundation settlement, vibration, thermal expansion and contraction or shrinkage cracks in concrete can cause traditional waterproofing membranes to de-bond from the substrate and fail. Since PRG is a non-cured flexible gel, it effectively creates a ball bearing effect that allows it to dynamically respond to the movement of two substrates moving independently of one another. This non-cured, flexible bond retains the integrity of the waterproofing envelope better than traditionally fully adhered, cured waterproofing systems. (Figure 4)

Due to PRG's non-curing characteristic, it has the unique ability to repeatedly self-heal under direct hydrostatic pressure. This ability helps mitigate some common pre-construction waterproofing system damage, such as; accidental formwork penetrations, construction site debris (nails, fasteners, etc.) or applicator installation mistakes. This allows for a greater "margin of error" in the waterproofing system resulting in a system that achieves a higher level of predictable performance. (Figure 5)

PRG also does not require substrate primers or conditioners. It can be applied to freshly poured or cast concrete, eliminating a 28-day cure time prior to waterproofing application. Since the PRG never completely cures, there is no cure wait time for application of other components to the waterproofing system and concreting can be completed immediately after application of the waterproofing system. Easy surface preparation and simple application procedures of the PRG help shorten time for project completion, thus saving construction costs. (Figure 6)

Figure 4.   Polymer rubber gel's ball bearing effect.

Figure 5. Testing PRG's self-healing ability by slicing through the PRG with a knife.

Figure 6. Field application of polymer rubber gel composite waterproofing system.

Spray application of the PRG allows for fast application, because PRG uses thermal heat and not a chemical reaction to reduce viscosities for application. Prior to application, PRG is heated to 87 degrees Celsius in a specialized spray pump kettle. The low heated temperature is ideal for enclosed areas, allows for safe application and has low VOC's. The spray application allows for a monolithic coverage of the material allowing the PRG to be applied on many types of substrates in a timely manner. Contactors have improved scheduling and sequencing of waterproofing installations with these properties. This spray system allows for waterproofing application to be installed in temperatures below zero centigrade.

4 PRG COMPOSITE SHOTCRETE AND BLINDISDE WATERPROOFING

4.1 *Introduction: the challenges of shotcrete assembly and blindside assembly*

For underground construction, shotcrete and blindside waterproofing systems play a significant role to protect from water intrusion. The challenge in this type of construction is that the waterproofing must endure the exposure to adverse environments, survive and withstand the concrete pour of shotcrete pressure. In addition, rebar must be supported, the waterproofing will have numerous penetrations from tie-backs and rebar making these areas prone to water leakage. Most importantly, it is critical that concrete or shotcrete must bond to the waterproofing after placement. This will ensure that water does not migrate between the membrane and the concrete. Inspection and monitoring during application is critical since the waterproofing will be inaccessible once concrete is in place.

4.2 *Polymer Rubber Gel Shotcrete or Blindside assembly*

Each assembly is composed of a composite system of two layers of 20-mil HDPE fleece reinforced sheets sandwiched with a 100-mil to 110-mil thick PRG in the center. The durability and additional chemical resistance of the sheet combined with the flexibility of the gel creates the dynamically responsive waterproofing system. The final layer of 20-mil HDPE is applied to the negative face of the system with the fleece facing the installer to protect the waterproofing system from job site contamination, weather, or damage. The fleece layer forms a mechanical bond to the concrete. This 3 layered PRG waterproofing system allows greater flexibility for timing of pours during construction. (Figure 7)

Preparation of the soil support or excavation substrate for blindside application of the PRG system may require the application of a plywood protection board or application of a shotcrete smoothing layer to create a sufficiently rigid and smooth substrate for the mechanical attachment of the waterproofing sheet. Typical applications where this would be necessary are for sheet pile/secant/tangent pile walls or some types of deep soil mix walls. Care must be taken to prevent the possibility of protrusions from the wall or cavities that could damage the waterproofing assembly either during assembly or at the time of concrete pour. (Figure 8, 9)

The principal design concept with a PRG waterproofing system is to achieve a complete monolithic building envelope of the gel system. This requires proper detailing of the transitions from base slab to walls and walls to ceiling. Once the first layer of 20-mil HDPE is installed and penetrations and transitional details are in place, a layer of PRG is sprayed creating a monolithic non-curing, self-healing membrane. (Figure 10)

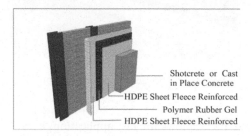

Figure 7. Polymer Rubber Gel Blindside Assembly.

Figure 8. Installation of the outer fleece reinforced HDPE sheet.

Figure 9. Spray application of PRG on fleece reinforced HDPE.

Figure 10.   Spray application of PRG on fleece reinforced HDPE.

Once the PRG spray layer is in place, a final layer of fleece reinforced HDPE is installed with the fleece layer facing the rebar. This HDPE layer bonds to the PRG and creates a mechanical bond with the shotcrete or cast concrete. Due to the seal-healing aspect of the PRG waterproofing assembly, the risk of damage during rebar installation is minimized and mitigated.

5   PRG CASE STUDY CENTRAL SUBWAY SAN FRANSISCO, CA

5.1   Introduction

The Central Subway Project will improve public transportation in San Francisco by extending the Muni Metro T Third Line through SoMa, Union Square and Chinatown. By providing a direct, rapid transit link between downtown and the existing T Third Line route on 3rd Street, the Central Subway will greatly improve transportation to and from some of the city's busiest, most densely populated areas. When the Central Subway is completed, T Third Line trains will travel mostly underground from the 4th Street Caltrain Station to Chinatown, bypassing heavy traffic on congested 4th Street, Union Square, and Stockton Street. Four new stations will be built along the 1.7-mile alignment: 4th and Brannan Station; Yerba Buena/Moscone Station; Union Square/Market Street Station; Chinatown Station.

The Central Subway Project is funded by the Federal Transit Administration (FTA), the State of California, the Metropolitan Transportation Commission, the San Francisco County Transportation Authority and the City and County of San Francisco. (Figure 11)

5.2   Polymer rubber gel waterproofing selection

A PRG waterproofing system was specified for use on all 4 subway stations within the scope of work. Project design calls for the construction of four subway stations, constructed utilizing the top down constructions methods. In conjunction with SFMTA, the stations were designed

Figure 11.   Central Subway under construction.

Figure 12. Central Subway top down construction.

by WSP/Parsons Brinckerhoff. The design consisted of different but unique box structures, with twin bore TBM tunnels, connecting into the stations. Total square footage for all installed PRG waterproofing is greater than 1,000,000 sq. ft. Because of the significance of this project and the high-profile use of the PRG product on the Subway System, a pre-evaluation and mock-up were required prior to installation. Based on the evaluation and mock-up results, PRG was the chosen waterproofing system by the SFMTA. Many waterproofing systems were evaluated for specification for the Central Subway project including typical bentonite panel and pressure adhesive systems. However, known limitations with these systems in seismic areas precluded their use. Structural engineers were particularly concerned with improving seismic performance of the construction. PRG's superior flexibility, non-curing, and self-healing characteristics help ensure that the tunnel and station waterproofing system can better withstand seismic events. In addition, the owner was given a watertight performance guarantee that covers both labor and materials for 15 years after completion of the project.

5.3 *Polymer rubber gel waterproofing system application*

The concrete box with embedded steel beams provided the support of excavation for the station. The face of the walls was shaved flush with the beam flange and a 3" smoothing layer of shotcrete was applied. Construction required a blindside PRG waterproofing system as the structure was not over-excavated. The structural walls of the stations were formed directly against the shotcrete walls. As the stations were expected to withstand constant hydrostatic pressure, a prefabricated drainage composite was applied to assist and control water in the shotcrete walls. The blindside PRG assembly was applied directly to the prefabricated drainboard composite. A shotcrete application to the PRG system was then installed and mechanically fastening to the fleece on the HDPE sheet. This creates a dry and protected final lining in the underground structure. (Figure 12)

6 CONCLUSION

Polymer rubber gel technology for shotcrete and blindside applications provide a high-performance waterproofing solution for underground construction. PRG waterproofing has been successfully installed internationally since 2005. Notable North-American large-scale infrastructure applications utilizing PRG waterproofing technology include the following projects: WSX, Bay Area Rapid Transit, Presidio Parkway, Caltrans and Downsview Park/Shepard West, Toronto Transit Commission. The unique physical characteristics of PRG composite waterproofing assemblies ensure peace of mind and performance against water intrusion. Specification of composite PRG waterproofing assemblies provides effective solutions to construct and protect underground structures.

REFERENCES

CalTrans. 2018. Presidio Parkway. www.presidioparkway.org/construction_info.

Mackenzie, R. 2017. A Look Inside the TTC's Downsview Park Station. http://urbantoronto.ca/news/2017/01/look-inside-ttcs-downsview-park-station.

SFMTA. 2018. SFMTA Central Subway. www.sfmta.com/projects/central-subway-project.

BART. 2018. Warm Spring Extension Project Overview. www.bart.gov/about/projects/wsx.

Tunnels and Underground Cities: Engineering and Innovation meet Archaeology,
Architecture and Art, Volume 5: Innovation in underground engineering,
materials and equipment - Part 1 – Peila, Viggiani & Celestino (Eds)
© 2020 Taylor & Francis Group, London, ISBN 978-0-367-46870-5

New monitoring technology to notify the displacement situation of the tunnel face

A. Ichikawa & H. Sato
Takenaka Civil Engineering & Construction Co., Ltd., Tokyo, Japan

S. Akutagawa
Kobe University, Kobe, Hyogo, Japan
Enzan Koubou Co., Ltd., Kyoto, Japan

Y. Hashimura
Keisokugiken Co., Ltd., Amagasaki, Hyogo, Japan

ABSTRACT: It has been confirmed that the behavior of tunnel face extrusion of a fragile ground can be captured by the measurement of a cutting face using a laser rangefinder, and there is a possibility to anticipate the rock falling events or collapse of the cutting face. Using this prediction technology, the authors have developed a new system named "Face Condition Viewer", which is a tunnel face visualization system, that uses augmented reality (AR) technology to generate warnings by directly marking on the tunnel face through monitoring of the extrusion behavior in real-time. It also displays real-time images and displacements of the tunnel face on a terminal equipment such as PCs and wearable terminals. By grasping the real-time displacement state of the tunnel face, it is possible to prompt evacuation instructions or warn the workers about any possible danger present in the vicinity of the tunnel face, thereby improving their safety.

1 INTRODUCTION

In the mountain tunneling works, in addition to the excavation work performed by the heavy machineries, there are many occasions when workers have to approach the face of the tunnel such as while loading the blast holes with explosives, installing the steel frame to support the rock, rock bolting work after shotcrete etc. Often occupational accidents occur due to tunnel face collapse and the rock fall events as the workers have to approach the tunnel face multiple number of times. In such rock fall events, 6% of the cases have resulted in the death of the workers, 42% of the sites were shut down for more than one month. It is alarming to know that the severity of such occurrences is on the rise. Table 1 shows the rock fall prevention countermeasures described in the recently revised guideline "Guidelines for Preventing Rock Fall Disaster in the face of Mountain Tunnel Construction" MHLW Guideline (2018). There exists several countermeasure techniques being implemented to reduce the risk of rock fall events near the tunnel face and there is a continuous effort to improve those existing technologies and develop new countermeasures.

Usually, the degree of stability of the tunnel face is visually confirmed by the on-site staffs or the workers. However, while confirming by the visual observation, there are many occasions which requires the knowledge of the experience site staffs. There is a risk of over sighting the unstable areas of the tunnel face leading to rock fall events or even collapse of the face. Therefore, in the fragile grounds such as, road tunnel natural ground grade D, E or railroad tunnel natural ground grade II, I, there are high possibilities of the rock fall events or the

Table 1. Selection of rock fall preventive measures.

Measures to prevent rock fall	Appropriateness of measures against rock falling prevention				Effect as spring water counterme asure	Constructabili ty (Ease of construction)	Others	
	IV、B	III、C	II、D	I、E			Compatibility when conducting deformation observation	High level of human body protection
Shotcrete	△	○	◎	◎	○*	◎	◎	△
Bolting	△	△	○	◎	○	△	×	△
Blasting	◎	◎	◎	△	◎	◎	△	△
Drain/Counterbore	○	○	◎	◎	◎	○	×	×
Displacement measurement	×	△	◎	◎	×	○	◎	×
Facility protective measures	△	△	△	△	△	△	△	○

Caution: ◎: Best、 ○: Good、 △: Possible、 ×: Inappropriate
○*: Good by combining water drainage measures.

tunnel face collapse after the extrusion of the face due to time-dependent displacement. Thus, in such cases, displacement measurement of the tunnel face is considered effective. As for the displacement measurement of the tunnel face, there exist measurement techniques such as using the laser rangefinder (Kumagaya (2011), Yokota (2012)) or make use of the image processing technology Fujioka (2018) etc. The development of such technologies to capture the extrusion behavior of the tunnel face is advancing. Therefore, it is important to promptly and directly inform about the unstable areas of the tunnel face so as to evacuate the people working in the immediate vicinity of the tunnel face.

In this paper, a case study to grasp the extrusion behavior of the tunnel face of the mountain tunnel construction using a laser rangefinder along with the development of the "Face Condition Viewer" (patent pending), which is a tunnel face deformation visualization system that enables all the workers involved in tunnel construction, such as those working in the immediate vicinity of the face and on-site staff, to grasp the degree of the safety condition of the tunnel face in real-time are presented.

2 CASE STUDY OF THE EXTRUSION BEHAVIOR OF THE TUNNEL FACE

In the vicinity of the face of the mountain tunnel, the deformation of tunnel face was measured using the laser rangefinder. Figure 1 shows the measurement condition of the face and Figure 2 shows the measurement points by the laser rangefinder. In this laser rangefinder,

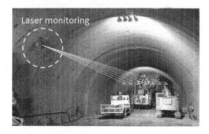 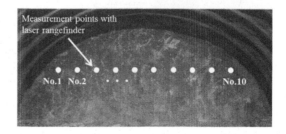

Figure 1. Measurement state of the tunnel face by a laser rangefinder. Figure 2. Measurement points with a laser rangefinder.

laser light is irradiated in the horizontal direction, and the distance to the irradiation point is measured by a non-prism at plurality points in intervals of about 1.0m. By measuring the tunnel face in this way, the extrusion amount of the face, based on the displacement of the real-time measurement data is determined. In this tunnel, machine excavation was used as the geology of the tunnel mainly consists of the weak mudstone with soft tuff rock sandwiched between. Figure 3 show the measurement results before and after the time of the rock fall. It also shows an example in which the extrusion speed of No. 3 abruptly increased to about 24.1 mm/h and the extrusion amount was 45.5 mm at maximum, thus confirming the rock fall event.

Figure 4 show the measurement results and image of the tunnel face at the time of collapse. Immediately after the excavation (shown by green), the cutting face was extruded with time, and the maximum extrusion amount at collapse was about 100 mm. Thus, as described above, at the time of fragile ground excavation having time-dependent extrusion of the tunnel face, there is a possibility that the tunnel face becomes unstable, and it is possible to quantitatively grasp the behavior in such conditions and take necessary measures in advance to reduce the risk against such unstable phenomenon. Based on these measurement results, it was confirmed that the extrusion behavior can be captured adequately in fragile grounds, and controlled standard values (extrusion amount/extrusion rate) was formulated, and the possibility of predicting the rock fall events and collapse of the tunnel face in advance. Therefore, it was considered that the safety of the tunnel face work could be improved by using this prediction technology and transmitting the displacement state of the tunnel face (unstable areas of the tunnel face) to all the people involved in the tunnel construction in real-time.

3 DEVELOPMENT OF THE SAFETY IMPROVEMENT OF TUNNEL FACE WORKS

Using the ability to predict rock fall events or collapses of the tunnel face in advance, from the measurements of the face using a laser rangefinder, the authors have developed a new system, Face Condition Viewer, which is a face visualization system that visualizes

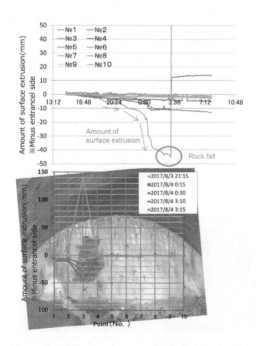

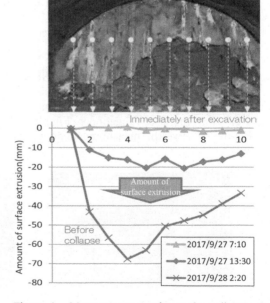

Figure 3.   Measurement results before and after the rock fall and photo of the tunnel face.

Figure 4.   Measurement results at the collapse of the face and photo of the tunnel face.

real-time displacements of the tunnel face (unstable areas of the tunnel face) by sending direct warnings through marking the tunnel face and visualizing them, while displaying them on a wearable terminal (scouter type or transmissive type glasses) using the AR-technology. The outline of the Face Condition Viewer is shown in Figure 5, the underground equipment arrangement diagram is shown in Figure 6, the equipment layout at the time of on-site application is shown in Figure 7, and the equipment enlarged photographs are shown in Figure 8. The equipment used consists of a laser rangefinder for tunnel face measurement and a control PC, a green laser for generating warnings by marking, a camera for imaging the tunnel face, and a tunnel excavation system for controlling these. The green laser is used to project a shape on a tunnel face by irradiating green light in a high-speed rotation. Figure 9 shows the unstable areas clearly indicated by the green marking. The amount of extrusion of the tunnel face is measured continuously by a laser rangefinder, and the periphery of the measurement point exceeding the control level is marked directly on the tunnel face by a green laser, and the unstable areas of the tunnel face is visualized by all the workers working in the immediate vicinity of the tunnel face.

Furthermore, the video projected on the wearable terminal in Figure 10, the actual situation showing the tunnel face monitoring and wearable terminal usage state in Figure 11, and the

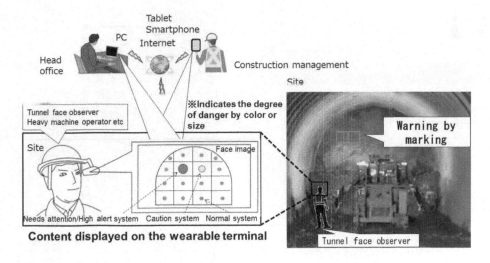

Figure 5.   Outline of Face Condition Viewer (tunnel face visualization system).

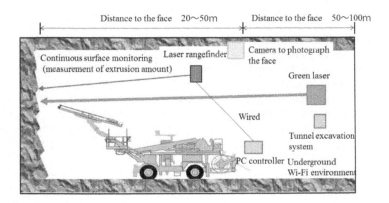

Figure 6.   Equipment setup inside the underground.

Figure 7. Equipment placement situation at the time of on-site application.

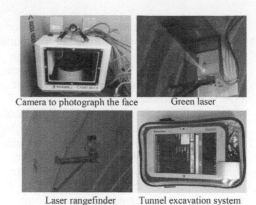

Camera to photograph the face   Green laser

Laser rangefinder   Tunnel excavation system

Figure 8. Enlarged photos of the equipment.

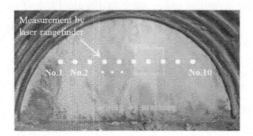

Figure 9. Showing the state of unstable areas by marking.

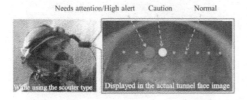

Figure 10. Image displayed on wearable terminal.

Figure 11. Monitoring situation of the face and usage of a wearable terminal.

general concept JRA Guideline (2009) at the control level in Figure 12, are shown respectively. The displacement data measured by a laser rangefinder was classified into three stages: normal: green, caution: yellow, needs attention/high alert system: red, based on the general idea of the management level. In addition, the wearable terminal (scouter type or transmissive type glasses type) visualized the actual tunnel face image captured by the face-capturing camera and the real-time displacement state of the tunnel face (the unstable area of the tunnel face) according to the management level. It should be noted that the present system is also capable of sending lights, sounds, and vibrations as warnings. Further, this terminal was selected so as not to interfere with the visibility and work of the user. This technology can be used not only at the site by the workers and tunnel face inspectors, but also at offices outside the underground, such as at headquarters and branches, using the terminal equipment such as smartphone, tablet, PC etc. Therefore, the on-site staffs who cannot always be present in the tunnel face work and all the people concerned with the tunnel construction would be capable

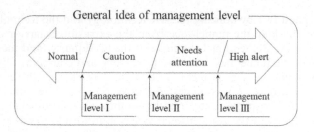

Figure 12. General idea of management level.

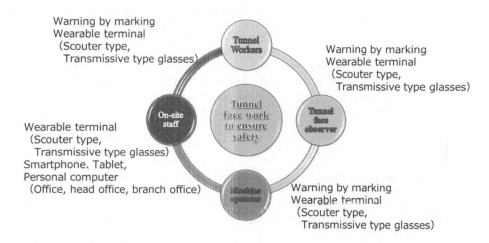

Figure 13. Structure aimed for this development.

of grasping the real-time condition of displacement of tunnel face (the unstable area of the tunnel face) using this technology.

Figure 13 shows the outline of the structure aimed for the current development. By grasping the state of displacement of the tunnel face (the unstable area of the tunnel face) in real-time, tunnel engineers at headquarter and branch offices can judge the state of stability of the face and take preventive and conservative measures to stabilize the tunnel face. Further, by mounting such wearable terminal by on-site workers, face inspectors, etc., and monitoring the tunnel face by using the present system in combination with the conventional visual monitoring system, it is possible to prompt the people working in the immediate vicinity of the tunnel face to give an instruction or to call for attention, thereby improving the safety against rock fall events or collapse of the tunnel face. However, this technology should be applied in consideration of the ground grade, underground environment, etc., and should be used in combination with the conventional monitoring system in the supplementary role.

4 SUMMARY AND FUTURE PROSPECTS

It was confirmed that the behavior of the tunnel face extrusion in a fragile ground was captured by the measurement of the tunnel face using the laser rangefinder and the standard reference value (extrusion amount and extrusion velocity) was formulated. It was also confirmed that it is possible to predict the rock fall events or the collapse of the tunnel face in advance. In addition, we have developed a system to visualize the actual face image and real-time displacement of the tunnel face (unstable areas of the tunnel face) on terminal equipment such as wearable terminals and PCs by utilizing the AR technology and the technology for generating direct warnings by marking the tunnel face from the extrusion behavior of the face in real-time. In this technology, all the tunnel engineers at the headquarter and the branch offices can judge the stability of the tunnel face and take preventive and conservative measures to stabilize the face by grasping the displacement state of the face (the unstable areas of the tunnel face) in real time. In addition, it was suggested that such wearable terminals could be worn by site workers, tunnel face inspectors, etc., and by monitoring the tunnel face using the present system in combination with the conventional visual monitoring, it was possible to urge the person who is working in the immediate vicinity of the tunnel face to give an instruction or call for attention, thereby improving the safety of the tunnel face against rock fall events and collapse of the tunnel face.

Furthermore, this system can be used in combination with the technology for capturing the tunnel face behavior such as a laser rangefinder, a 3D laser scanner, and a video camera, and it is expected that the safety of the tunnel face work will be broadly improved.

The effectiveness of this technology has already been verified at the tunnel site in Fukui Prefecture, Japan, and the application of this technology to the vulnerable grounds with extrusion properties, is being considered in the future. In addition, using the visualization technology, the authors intend to contribute not only to the mountainous tunnels, but also to the various different sites for the improvement of safety of the workers.

REFERENCES

Ministry of Health, Labour & Welfare 2018. *Guidelines on measures to prevent labor accident due to fall of loosen rocks in cutting of mountain tunnel works* (in Japanese)

Kumagaya, A. Terashima, Y. & Yoshida, R. 2011. Development and Test Apply of the Harbor Safety Monitoring System in Poor Areas, *66th Annual Scientific Lecture of the Society of Civil Engineers* 6-327: 653–654 (in Japanese)

Yokota, Y. & Yamamoto, T. 2012. Measurement of tunnel face displacement by a multi-point high precision laser rangefinder, *Geotechnical Engineering Research* (in Japanese)

Fujioka, D., Nakamura T., Nishiyama T. & Nakaoka, K. 2018. Development of real-time face monitoring system using image treatment technology, *45th Symposium on rock dynamics* (in Japanese)

Japan Road Association 2009. Road tunnel observation and measurement guidelines (in Japanese)

Tunnels and Underground Cities: Engineering and Innovation meet Archaeology, Architecture and Art, Volume 5: Innovation in underground engineering, materials and equipment - Part 1 – Peila, Viggiani & Celestino (Eds)
© 2020 Taylor & Francis Group, London, ISBN 978-0-367-46870-5

Smart Office; application of a unified analytics center in tunneling construction

K. Jahan Bakhsh, J. Kabat, R. Bono, S. Tabrizi & G. Pini
Salini-Impregilo Healy JV, USA

ABSTRACT: While tunneling customers are becoming increasingly sophisticated, the tunneling construction sector remains severely under-digitized. Mechanized tunneling construction is among the most innovative areas in the construction industry, however, this industry depends on expert skills and experience to deliver the project successfully. Building a unified analytics center to gather, integrate, and analyze data from disparate databases on a construction site (e.g., TBM itself, conveyor system, inventory, schedule, and daily activity reports) and then contextualize this information into a visual, meaningful representation provides a robust tool to enhance instant, and far-off decision-making for all levels of site and office personnel. In this paper, the application of a unified analytics center in tunneling construction is introduced. Dugway Storage Tunnel project is considered as a case study. Throughout the construction period of the tunnel, vast quantities of data from TBM sensors is pulled out and streamed directly from TBM's PLC. Data is then pushed into Power BI, which is an analytics service provided by Microsoft. This system connects the project personnel (e.g., Project manager, Construction manager, Project engineer, and Site Engineers) to a broad range of data via easy-to-use dashboards, interactive reports, and meaningful interactive visualizations that bring data to life.

1 INTRODUCTION

There is a proverb saying that necessity is the mother of innovation. While early tunnels constructed to be for the illegal purposes of smuggling (e.g., Salem tunnel), what drives today's societies for developing tunnels is its value to develop and shape the modern world. Continual development and improvements have changed the shape of the tunneling industry from highly labor intensive, with several hundred workers engaging in daily activities, to a highly mechanized cost-effective industry. Today's Tunnel Boring Machine is equipped with hardware (e.g., mechanical components and cutters) and software (e.g., sensor networks and computer-supported applications) technologies that enable us to build tunnels with bigger diameters in ever more challenging ground conditions.

Although TBM creates a lot of data throughout the construction period of the tunnel, except for a small amount of data which is consumed by TBM operator, much of that data is siloed and filed away once a project is completed. In addition to TBM data, as technology has changed, tunneling construction firms are capturing more data than ever before through other information sensing devices (e.g., job site sensors, smartphones, and heavy equipment tracking devices). Transforming these heterogeneous, unrelated sources of data into coherent, visually immersive, and interactive insights will enhance every aspect of the project execution.

The use of data-driven modeling processes (e.g., BIM, CIM) become more popular in the tunnel engineering construction. The Crossrail tunnel in London (Heikkila, R. & Makkonen, T. 2014), State Route 99 tunnel in Seattle (Lensing, R. 2016, Trimble, 2011), Hallandsas tunnel in Sweden (Smith, C. 2014), Hangzhou Zizhi tunnel in China (Wang, J. et al., 2015), and Mikusa tunnel in Japan (Sugiura, S. 2015) are all examples of projects that executed by utilizing data-driven methods (figure 1). Despite the several advantages of BIM models in the design field, due to the disruptive nature of this system, there are several challenges of implementation in the construction firm (Davidson, A. 2009). For instance, BIM doesn't help to facilitate communication in the construction field, and it should be implemented with the full capacity to be effective.

Tunneling construction firms have already realized the power of data. However, just sporadic attempts have been made to utilize this power to make better decisions, to increase productivity, to control the quality of materials, and to improve safety in job sites. Companies such as Babenderede Engineering and Tunnelware have developed data management and visualization software compatible for tunneling construction market. For example, Babenderede Engineering has developed a modular software to support tunneling construction process by equipping contractor's TBM with an external data acquisition system that can pull all relevant data from the machine PLC. They also developed software exclusively for tracking segmental lining through the lifecycle of the project (Cicinelli, L. et al., 2017). Tunnelware which is still at the developing stage tries to consolidate data from several sources (e.g., TBM, site personnel reports, and job site sensors). According to the Tunnelware experts, the software will provide a robust tool for the constructors and will increase their productivity by letting them to visualize and analyze the processes on a 5D format. There are several other features that Tunnelware specialists are developing (e.g., cutter tool management, and virtual meeting rooms).

Figure 1. Example of BIM model. The Crossrail tunnel in London (top left), State Route 99 tunnel in Seattle (top right), Hallandsas tunnel in Sweden (bottom left), and Mikusa tunnel in Japan (bottom right).

Although available commercial services, i.e., software have several benefits, there are few drawbacks that make contractors hesitate to utilize these services in their projects. For instance, the way data is visualized is old-fashioned and dissociated. Indeed, it is very crucial to prepare an easy-to-use interactive visualization platform that can facilitate construction process for the project personnel; otherwise, it will be abandoned throughout the construction period. In addition, since every tunnel project is unique, existing software in the market are not flexible enough to meet the uniqueness of the project.

In this paper the concept of unified analytics center compatible for tunnel construction is presented. First, Dugway Storage Tunnel project which is considered as a case study for testing our model is introduced. Later The structure of the conceptual model, data collecting, and required hardware and software is explained in the Afterward, application of the proposed model throughout the construction period of the tunnel is partially examined. The advantages of implementing such a data management system as a tool for project managers, engineers, safety superintendents, etc. are also addressed.

2 DUGWAY STORAGE TUNNEL

2.1 General description of the project

The Dugway Storage Tunnel (DST) is a deep tunnel for combined sewer overflow which will provide an additional storage to combined sewer flows during events of wet weather. This will reduce the number of the combined sewers overflows which discharge into the environment without proper treatment. The tunnel alignment is approximately 4.5 km in length with seven curves of variable radius, excavated using a single shield hard rock TBM with an 8.22 m excavation diameter The finished internal tunnel lining diameter is 7.30 m using concrete segments of 30 cm thickness. Depths of the tunnel invert is ranged from 55 to 70 m below ground surface. The project includes a total of six deep shafts along the path of the tunnel with an internal lined diameter between 5 m to 15 m and four Adit connections between these shafts and tunnel of variable length between 15 and 300 meters (figure 2). The 14m diameter shaft known as DST-1 is the TBM launch shaft where all the main conveyor systems will be installed. Part of the shafts was constructed through soft ground and part also encountered chagrin shale bedrock. The project includes the construction of an additional structure as Diversion Structures, Gate Structures, Control Vaults, Ventilation Vaults, Drop Manholes and modifications to existing regulatory structures. The project area is mainly older residential, i.e., pre-1950's interspersed with commercial properties and urban parks.

2.2 General description of the utilized TBM

The Tunnel Boring Machine used to excavate the 8.22 m diameter tunnel through the Chagrin Shale was a hard rock single shield Herrenknecht machine type S-684. The machine was reconditioned on site by the contractor after excavating the first phase of the tunnel i.e., Euclid Creek Tunnel. The TBM was partially assembled for the launch with only three of the six gantries in the starter tunnel which was previously excavated by employing drill and blast method. At this stage, materials were hauled out using locomotives and muck boxes. After the first 100 meters of tunnel has been excavated, the TBM was assembled in its final setup with six gantries. The tunnel conveyor system was comprised 5 sections transporting materials from the tunnel, to the vertical conveyor to overland belt and finally to the stacker. The total rings installed at the completion of the tunnel were 2911 for a total length of 4,438 meters. The production average was 17 rings per day. The average of the excavation parameters were as follows: 1) Penetration 10.3 mm/rotation, 2) Cutter-head rotation 6 rpm 3) Advance of the TBM 66.75 mm/min 4) Thrust force = 10,000 KN 5) Torque = 2,500 KNm. There was no presence of water in the material, but the quantity of methane trapped between the layers was sometimes relevant.

Figure 2.   Project overview.

3   UNIFIED ANALYTICS CENTER

3.1   *Conceptual model*

Unlike traditional ways of managing construction process, nowadays with an extensive amount of data available from projects, managing these data and using it as a management tool to make better decisions during the construction period is of utmost importance. The purpose of the conceptual model of our unified analytics center is to provide a robust tool to enhance instant and far-off decision-making for all levels of site and office personnel. The conceptual model is comprised of 1) a physical smart office equipped with digital display screens 2) sensors, equipment, and digital tools to gather data from project site and 3) a platform to integrate, analyze, and contextualize this information into a visual, meaningful representation. Figure 3 shows the Data Flow Model (DFM) of the proposed conceptual model. Data captured from TBM, conveyors system, equipment operating, inventory, material used, site personnel reports, labor hours, project documents, and even online weather reports are structured and pushed into a cloud-based collaboration and sharing system. Microsoft Power BI which is a suite of analytics tools is utilized to create interactive visualization and dashboards and share insight with individuals across the project and company. Depends on the end-user, model can provide customized dashboard that can get updated either in real-time, near real-time or over a longer period. For instance, real-time monitoring of critical parameters of the TBM (e.g., Thrust force, torque) can be visualized in form of interactive dashboards to assist optimizing excavation process. As another example, historical data that have collected over the years on the project can be presented to the headquarter to give them an overall insight on project progress and assists them to gain a competitive advantage when estimating and bidding on the new project with similar characteristics.

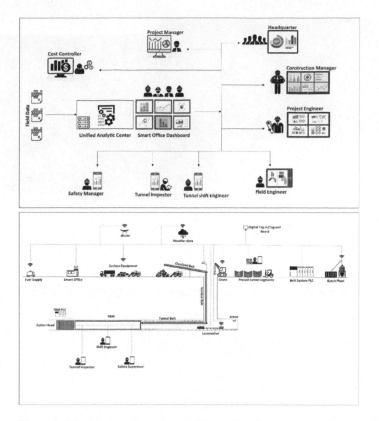

Figure 3. Data Flow Model of the conceptual model. Data flow from various sensors and human inter-action to the unified analytics center (top), and flow of processed data from unified analytics center to end-user personnel (bottom).

3.2 Application of the model in construction of the Dugway Storage Tunnel

Since the idea of a unified analytics center is introduced in the middle of the project execution, there wasn't enough time and resources to implement the model completely. Therefore, through-out the excavation phase of the tunnel, just data from TBM PLC is pulled out and pushed into the Power BI platform. Nevertheless, this partially implemented system allowed us to connect the project personnel (e.g., project manager, construction manager, project engineer, tunnel shift engineers, and TBM operator) to a broad range of data via easy-to-use dashboards, inter-active reports, and meaningful interactive visualizations. In fact, customized dashboards eased up the decision-making process for everyone that involved directly and indirectly in the project.

Figure 4, 5, and 6 show examples of crafted dashboards. For instance, as shown in the figure 4, the real-time interactive map of the TBM location and data recorded from geotechnical instrumentations can give a better insight into the behavior of the ground subjected to excava-tion. Figure 5 represents the historical graph of the volume of the accelerator admixture injected from each port for every single installed ring. With this data available in the hand of the engin-eer or TBM operator, in case of any problem related to accelerator injection, they can easily explore more data to solve the problem rather than just relying on the last ring data. As another example, say you are trying to improve your daily production by adjusting working shifts hours. Having access to historical data of shift production through an easy-to-use dashboard will give you a holistic view of your ongoing progress and accordingly an insight to take a proper decision (figure 6).

Above examples are just few out of many informative dashboards that can be generated to get the right information to the right person at the right time. In order to have a better picture

Figure 4.   Real-time location of the TBM on satellite map.

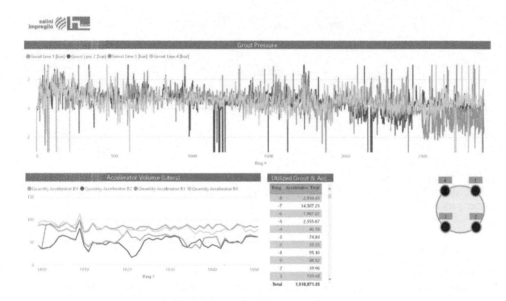

Figure 5.   Real-time dashboard of accelerator admixture parameters.

of the benefits of employing such a data-driven system, it is useful to address the model's advantages as follows:

3.2.1 *As a tool for engineers*

Wi-Fi system in the tunnel allowed the engineering team to have permanent access to the real-time dashboard of all critical parameters of excavation, grout utilization, and navigation. As shown in figure 7 grout utilization parameters (e.g., grout and accelerator admixture pressure and volume) are integrated into a customized easy-to-use dashboard which used as robust tool by engineers to monitor grout utilization per each installed ring. As another example, we conjoined daily production with push and ring build time to get better real-time picture of TBM utilization per day throughout the excavation phase (figure 8). We also formulated several other dashboards to facilitate monitoring TBM cylinders pressure and extensions, gas infiltration location and values, ring installation effect on navigation, and the variability of tendencies during the advance and many other parameters and behaviors.

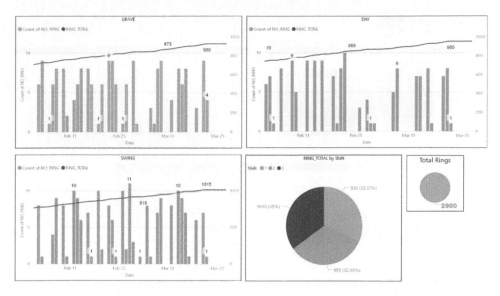

Figure 6. Near real-time of the tunnel production progress based on three-shift schedule.

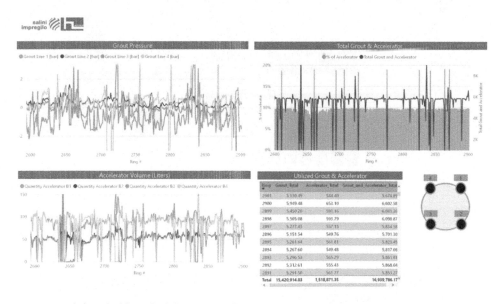

Figure 7. Real-time dashboard of the grout utilization parameter.

3.2.2 *As a tool for project managers*

The physical smart office, i.e., meeting room, where people can join to discuss on going tunnel construction process while all critical data modeled in shape of interactive dashboards provided a unique tool in the hand of project managers to hover over all aspects of the project from different viewpoint. For example, a dashboard is generated to illustrate the production progress on shift, daily, weekly, and monthly bases. As shown in figure 9, monitoring the project progress through this dashboard is far more convenient than exploring archived hard-copies of Personnel report.

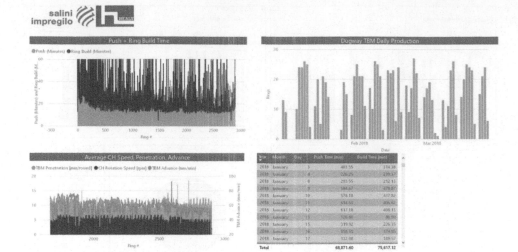

Figure 8.   TBM utilization dashboard.

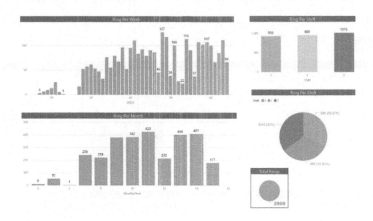

Figure 9.   Production progress dashboard.

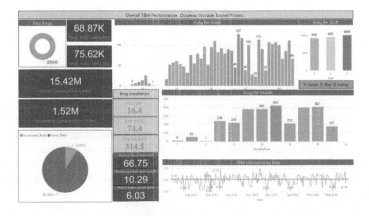

Figure 10.   Overall TBM performance dashboard.

2286

3.2.3 *As a tool for reporting*

Unified analytics center allowed us to generate digital custom reports for construction process based on different time scales (e.g., shift, daily, weekly, monthly, full project) automatically. This platform also let us share automatic reports on predefined timeframes with co-workers and any other recipients outside of the company just by including the list of email addresses in the system. For instance, a dashboard comprised of production progress, grout utilization, TBM utilization, and the average of TBM excavation parameters can be an informative means for program managers in headquarter and estimators when estimating and bidding on the new project with similar characteristics (figure 10).

4   CONCLUSIONS

In this paper, we introduced the principles of unified analytics center as a robust tool for contractors. Moreover, we examined the application of this data modeling system in the construction of the Dugway Storage Tunnel. Although, data management in mechanize tunneling construction is not a novelty, there are several features that make our proposed model distinct. Interpreting data into meaningful easy-to-use, real-time, visual dashboards that gives everyone involved in the process an insight into the project is the foremost advantage of the system. In addition, Microsoft Power BI as a backbone of the model is an analytics service entrenched in the Microsoft stack. Indeed, unlike other inflexible pre-defined data management and visualization platforms, Power BI compatibility increases adoption of our data modeling for contractors. Even though, we implemented the system just by using TBM data and started in the middle of the project execution, due to the flexible characteristic of the model we could take advantage of the system in our decision-making process. The proposed system also allows to automatically generating consistent reports on shift, daily, weekly, monthly scales. As a future work, we have planned to examine the system as a whole by inducing data from other sources (e.g., job site sensors, smartphones, and heavy equipment tracking devices). The all-inclusive system will allow improving decision-making and accordingly increasing efficiency and reducing construction costs.

REFERENCES

Cicinelli, V., Stahl, F., & Gronbach, T. 2017. Copenhagen Cityringen Project: Big Data to Manage Quality Control in Megaprojects. *RETC San Diego, June 4- 7,2017* page 190–202.

Davidson, A. 2009. A Study of the Deployment and Impact of Building Information Modelling Software in the Construction Industry, *England University of Leeds, 2009.* Available: http://www.engineering.leeds.ac.uk/eengineering/documents/AndrewDavidson.pdf

Heikkilä, R., Kaaranka, A., & Makkonen, T. 2014. Information Modelling based Tunnel Design and Construction Process. *The 31st International Symposium on Automation and Robotics in Construction and Mining (ISARC Proceedings2014). Sydney, Australia, July 2014.*

Lensing, R. 2016. BIM and construction process data in mechanized tunnel construction; Milestone control for tunnel construction sites using automatically created process data in comparison with 4D BIM. *Master of Science thesis (Geographical Information Science & Systems). MSc(GIS)" Heidelberg. 20.12.2016*

Smith, C. 2014. BIM at Sweden's Hallandsås Tunnel: Planning pioneer [Online]. New Civil Engineer. Available: http://www.newcivilengineer.com/features/geotechnical/bim-atswedenshallandss-tunnel-planning-pioneer/8663146.article

Sugiura, S. 2015. First Application of CIM to Tunnel Construction in Japan. *Proceedings of International Conference on Civil and Building Engineering Informatics (ICCBEI 2015),* 82. Tokyo, Japan, 2015.

TRIMBLE. 2011. Seattle's Massive Tunnel Makes travel safer – with Tekla BIMsight [Online]. Available: https://www.teklabimsight.com/references/seattles-massive-tunnel-makes-travel-safer [Accessed 24.07.20162016].

Wang, J., Hao, X., & Gao, X. 2015. The Application of BIM Technology in the Construction of Hangzhou Zizhi Tunnel. *3rd International Conference on Mechatronics, Robotics and Automation (ICMRA 2015). ISBN 978-94-62520-76-9.*

Tunnels and Underground Cities: Engineering and Innovation meet Archaeology, Architecture and Art, Volume 5: Innovation in underground engineering, materials and equipment - Part 1 – Peila, Viggiani & Celestino (Eds)
© 2020 Taylor & Francis Group, London, ISBN 978-0-367-46870-5

Key technologies and engineering application of horseshoe-shaped TBM

L.H. Jia & S.H. Tan
China Railway Engineering Equipment Group Co., Ltd., Zhengzhou, Henan, China

ABSTRACT: In the past, mining method or open-cut method were widely used which proved inefficient and unsafe. Therefore prompt solution is urgently needed in order to realize mechanized construction. Based on the project of Baicheng Tunnel of Menghua Railway, this paper briefly describes the background and project overview of the horseshoe-shaped shield machine, expounds the main problems in its design, and explores its key technology which includes low disturbance multi-cutterhead and multi-drive cooperative excavation technology, multi-curvature segment erection technology, efficient muck discharge system, horseshoe-shaped shell design, anti-roll attitude control technology, etc. Besides, it also summarizes the engineering application of the shield machine, analyzes the problems occurred during engineering test, and carries out targeted optimization design to provide reference for future horseshoe-shaped shield machines.

1 INTRODUCTION

At present, circular tunnels are widely adopted for urban subway tunnel, diversion tunnel, underground utilities and other projects because of its favorable load-carrying structure, high level of construction automation, safety, low pollution, and cost-effectiveness(Qian 2002; Wang 2014). However, at the same time circular tunnels also have some weaknesses. For example, after the completion of the tunnel, the bottom of the circular tunnel has to be paved with prefabricated inverts for the sake of vehicle operation, which will inevitably take up some excavation space(Ryunosuke 2013; Zhang 2010). Although rectangular shield machine has gone through rapid development in recent years with its remarkable space utilization, its structure form results to limited overburden-carrying capability and can hardly meet the requirement of big cross-section and deep-buried tunnels(Jia 2016; Kazunari 2002). On the contrary, the horseshoe-shaped tunnel has better load-carrying structure(compared with rectangular tunnels) and higher cross-section utilization(15% higher compared with circular tunnels). Combining the technological advantages of both rectangular and circular tunnels, the horseshoe-shaped tunnel could meet the application requirement of double lane railway tunnels with soft rock and deep-buried ground conditions, and exhibits evident technical advantages.

Besides, mining method or open-cut method are usually used for the construction of horseshoe-shaped tunnels at home and abroad which have the problem of poor safety, unfavorable construction conditions and low excavation efficiency which is one third of the shield method (Sun 2017). Thus, in order to overcome the above-mentioned problems, this paper conducts studies on the horseshoe-shaped shield machine, explores low disturbance multi-cutterhead and multi-drive cooperative excavation technology, multi-curvature segment erector technology, efficient muck discharge system, horseshoe-shaped shell design, anti-roll attitude control technology, etc. Moreover, it also analyzes the problems occurred during the engineering application of the horseshoe-shaped shield machine, proposes targeted optimization methods, and verifies machine performance after modification.

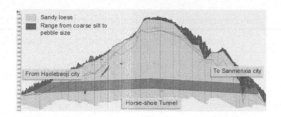

Figure 1. Profile view of the Baicheng tunnel.

2 PROJECT PROFILE

Menghua railway Baicheng tunnel project is located at Jingbian County, Shaanxi Province. This 3345m-long tunnel is a 120km double lane electrified railway tunnel, whose cross section profile can be seen in Figure 1. All the tunnel alignment is on the straight section with the deepest overburden being 81m. Its longitudinal slope is herringbone slope, whose gradient and length are 4.5‰/1935m, -3‰/900m, and -11‰/510m respectively. Ground conditions within the tunnel mainly consist of silts, fine sand and sandy new loess. The tunnel passes through Baomao highway, Jinghuang road, natural gas pipeline, and Baichengzi water supply pipeline, all of which have strict requirement on the settlement.

Its biggest longitudinal height is 9589mm, and the largest transverse width is 10540mm. Distance between centers of two tracks is 4m, and its tunnel clearance is 2B with overhead rigid suspension catenary and ballastless track bed.

3 KEY TECHNOLOGIES OF HORSESHOE-SHAPED SHIELD MACHINE

Horseshoe-shaped shield machine refers to the kind of high-end intelligent equipment which combines mechanics, electrics, hydraulics, sensors and communication, and is related to multiple discipline fields. In terms of its research at home and abroad, only Japan has few concept designs. With the aim to break technical bottleneck, this paper studies on its key technologies including:

3.1 *Low disturbance multi-cutterhead and multi-drive cooperative excavation technology*

Although at present conventional full-face shield machines mostly have one single cutterhead with circular cross section both at home and abroad, for super large horseshoe-shaped shield machine, it is very difficult to achieve effective excavation with one single circular cutterhead. Therefore, in reference to rectangular shield machine cutterhead, and taking into consideration of the advantages and disadvantages(Liu 2017) of combined rotating cutterhead and eccentric wagging cutterhead (Figure 2), this project uses overlapping cooperative exaction method by means of multiple cutterheads with different diameters, which has the advantages of little disturbance to the ground, better settlement control, and high excavation efficiency. The comparison of two types of cutterhead can be seen in Table 1. The overlapping design

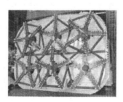 

Figure 2. Combined rotating cutterhead(left) and eccentric wagging cutterhead(right).

Table 1. Excavation form comparison of two commonly used cutterheads.

Cutterhead type	Combined rotating cutterhead	Eccentric wagging cutterhead
Excavation ratio	over 90%	100%
Muck mixing	big	small
Mixing torque	big	small
Ground disturbance	small	big
Reliability	high	low

could guarantee excavation area to the most extent, reduce blind areas, and ensure muck-mixing of the cutterhead at the same time.

Horseshoe-shaped shield machine has nine parallel shaft cutterheads, three at front, six at back, which could realize over 90% excavation coverage. The structure of the cutterhead and serial number can be seen in Figure 3. No.2–7 big cutterheads have five spokes, and No.1, 8 and 9 cutterheads have three spokes.

The excavation feature of combined rotating multi-cutterhead leads to excavation blind areas. By means of installing high pressure water flushing device, soil conditioning passage at the blind areas and other auxiliary measures, the muck at the blind areas could be handled. In addition, cutters around the periphery of the shield skin could ensure the size of the excavation section.

3.2 *Horseshoe-shaped multi-curvature segment erection technology*

As for the horseshoe-shaped cross section, segments have different surface curvatures, various shapes, biased center of gravity, and large weight, which all cause the complex erecting path and high erecting precision. Figure 4 shows segment distribution of horseshoe-shaped tunnel (a), diagram of stagger-jointed segment erection(b), and contrast diagram of circular segment and horseshoe-shaped segment(c).

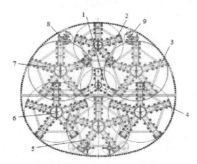

Figure 3. The composition and number of the cutterheads.

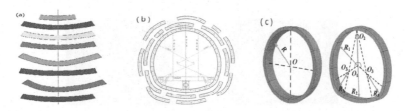

Figure 4. (a), (b) and (c).

2290

(1) Multi-degree of freedom segment erection system

Multi-degree of freedom segment erection system (Figure 5) is designed to reduce the deviation of segment erection and improve the stability of grasping from multi-points as shown in Figure 5. The main structure includes rotating ring, longitudinal telescopic mechanism and pick-up unit. The rotating ring moves around the Z axis, and the longitudinal telescopic mechanism realizes the extension and retraction function by means of cylinders.

The mechanical pick-up unit (Figure 6) mounted on the longitudinal telescopic mechanism can grasp and retain segments for the entire duration of the positioning operations with three independent movements such as swinging, pitching and rolling. The extension cylinder controls the movement along the Z' axis and the swinging cylinder could completely revolve around Z' axis. Besides, the rotation around the X' and Y' axis are separately driven by pitch cylinder and deflecting cylinder. Consequently, multi-degree of freedom segment erection system ultimately completes the complex trajectories of segment erection.

(2) The electro-hydraulic control technology of erection system

The electro-hydraulic control technology adopted by the erector can reduce the system impulsion, continuously variable transmission and differential pressure compensation to overcome the speed influence produced by changeable load. The ramp time is implanted in the electro-hydraulic integrated control system, so as to eliminate the rigid impact and enhance the stability of the system. A~H shows the segment erection process in Figure 7.

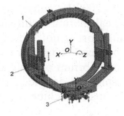

1. rotating ring
2. telescopic mechanism
3. pick-up unit

Figure 5.   Segment erector.

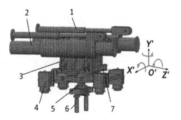

1. translation cylinder
2. translation guidepost
3. pitch cylinder
4. locked cylinder
5. lift cylinder
6. jack screw
7. twist cylinder

Figure 6.   Erector pick-up unit.

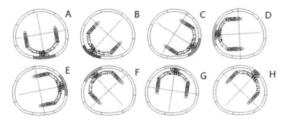

Figure 7.   Erection order of the segment.

Figure 8. Structure of double screw conveyors.

3.3 Double screw conveyor mucks discharge system

Super-large horseshoe-shaped shield machine is horizontally wide, which means its excavation volume is enormous. Therefore, excavation chamber might be clogged, and it is difficult to control the earth pressure at the right and left side of the excavation chamber. Thus, the muck-discharging capability and equilibrium of muck discharge system face some new demands. In consideration of the above concerns, the horseshoe-shaped shield machine is equipped with two screw conveyors (Figure 8).

Since two screw conveyors are simultaneously discharging muck, both of their speed shall have an effect on the balance of the excavation chamber. FLUENT software is used to conduct analysis on the flow field pressure of the excavated muck(conditioned soil density: 1580kg/m; viscosity: 670Pa•S; static shear stress: 12000Pa; consistency index: 1; power law index: 1.1; cutterhead rotating speed: 1rpm). The calculation result shows that the two screw conveyors' muck discharging speed at the loading port is not the same, and the cutterhead rotation direction can cause pressure unbalance between two screw conveyors' muck discharging ports. In order to ensure full-face pressure equilibrium of the excavation chamber, real-time monitoring on the excavation chamber pressure and double screw conveyors pressure is carried out, and the result is transmitted to the Human Machine Interface simultaneously, so as to eliminate the pressure coupling effect of the double screw conveyors.

3.4 Horseshoe-shaped shell design and anti-roll attitude control

The upper part of the horseshoe-shaped shield machine is circular arch which can create natural relieving arch during excavation, and its circumferential load distribution is similar to that of the conventional circular shield machine(Wang 2014). Different from conventional single circular cutterhead, horseshoe-shaped shield machine has multiple cutterheads and therefore its load distribution is quite different at axial direction. As a result, on the basis of meeting the requirement of low-cost, transportation, strength and inflexibility, the weight of the machine shall remain as light as possible.

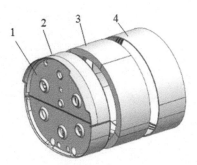

Figure 9. Horseshoe-shaped shell.

When conventional circular shield machine rolls, it does not affect the quality of the completed tunnel, because correction can be done by cutterhead reverse rotation. But for the horseshoe-shaped shield machine, because the tunnel is non-centrosymmetrical, unbalanced earth pressure and changing geology during advancing can cause deflection at horizontal axis direction or rolling. If the machine rolls, the shape of the tunnel shall be affected. Therefore the segment erection accuracy must stay very high, and the attitude control must be very strict. So various measures are used at the same time in order to guarantee the attitude control accuracy (Chen 2011).

Deviation rectification measures include:

1. Horizontal inclinometer on the shell can monitor the attitude in real-time and has warning system. The rotating speed and direction of every cutterhead is adjustable so as to rectify roll deviation. The screw conveyors have continuously various speed, and could help rectify deviation as well by controlling the earth pressure at the left and right side of the excavation chamber.
2. Earth-hitting pump: the grout ports at the periphery of the front shield can help steer the shield machine.
3. Balancing weight: the balancing weight inside of the machine can help with roll-correction.

4 ENGINEERING APPLICATION AND OPTIMIZATION

4.1 Application

The horseshoe-shaped shield machine was completed in July 2016 and started excavation on 11th November 2016. On 26th January 2018 it achieved breakthrough and completed the construction of the Menghua railway. In general, its performance throughout the testing section was satisfactory and met the requirement of safe, fast and environmental-friendly construction, with best daily advance rate 19.2m(12 rings) and best monthly advance rate 308.8m.

During construction, the segment erection system performed very well and the completed tunnel quality was relatively high with altitude difference within 3mm. Every segment ring erection cost 40 min in average and realized safe and efficient operation.

4.2 Problems and machine optimization

4.2.1 Muck deposition and solution

During engineering test, the muck discharge was not smooth, and screw conveyors were idling with average daily advance rate 1–2rings. The earth pressure within the excavation chamber fluctuated significantly, which affected construction efficiency and settlement control. Down-time inspection found that along the tunnel alignment existed collapsible loess which could easily collapse after encountering water. The overlapping cutterhead design could avoid interference, but could also cause mixing blind area at the bottom of the excavation chamber, which resulted to hard soil arch and affected muck movement and muck-discharging (Figure 10).

In consideration of the special geology of the project, corresponding measures are as follows:

(1) Use high pressure water jet to flush muck. In order to ensure the flushing area of the nozzle as large as possible and flushing speed, the relationship between two nozzle sizes and flushing effect is simulated by using FLUENT numerical analysis (Figure 11).

With the use of GAMBIT and FLUENT, the water jet speed can be seen in Figure 12 and Figure 13. Two nozzles' velocity distribution is similar. Namely, the water jet speed increases rapidly and reaches its peak at the moment of leaving the nozzle. Right after the water comes out of the nozzle, the speed decreases quickly. Therefore, under the same water jet pressure, the faster the flow speed at the nozzle is, and the faster the dynamic pressure falls, the bigger the flushing area is. In order to enlarge flushing area, it is decided to increase nozzle opening and water pump pressure with 170°opening angle and 10mm opening width. With the help of telescopic device, water jet can have the effect of water jet scalpel.

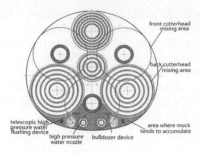

Figure 10.   Anti-muck deposition measures.

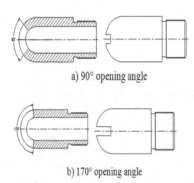

a) 90° opening angle

b) 170° opening angle

Figure 11.   Nozzle sketch.

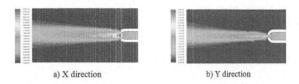

a) X direction   b) Y direction

Figure 12.   Velocity distribution of the nozzle with 90 degree opening angle.

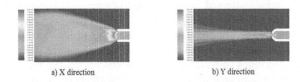

a) X direction   b) Y direction

Figure 13.   Velocity distribution of the nozzle with 170 degree opening angle.

Mixing blind areas are installed with several high pressure water jet nozzles (Figure 10), and telescopic high pressure water jet flushing system at the bottom. There are two telescopic high pressure water jet flushing devices (Figure 14). Since the flushing area of telescopic high pressure water flushing device is quite big, the muck at the bottom of the excavation chamber can be removed so as to increase muck fluidity.

(2) Bulldozer device is installed at the upper part of screw conveyor in order to avoid soil arch inside the excavation chamber and realize smooth muck removal.

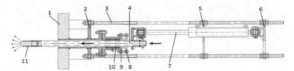

1.bulkhead, 2.grout pipe mounting base, 3.pull rod, 4.nozzle jet, 5.support plate, 6.back plate,
7.cylinder, 8.gland, 9.sealing socket, 10. sealing assembly, 11.high pressure nozzle

Figure 14. Telescopic high pressure water jet flushing system.

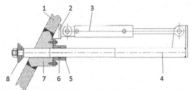

1.bulkhead, 2.lug, 3.cylinder, 4.soil-pushing rod, 5.sealing socket, 6.sealing assembly,
7.deflecting cylinder, 8.soil-pushing board

Figure 15. Bulldozer device.

There are two bulldozer devices (Figure 15). Under the pushing force of the soil-pushing device, the muck could fall into the muck-discharging area of the screw conveyors or the mixing area of the mixing devices, which could destroy the soil arch.

After the above modification, the muck-discharging system is improved obviously and went smoothly.

4.2.2 *Excessive thrust force and attitude control*

During the engineering test, it is found that the torque of the bottom three big cutterheads were almost zero and thrust force reached over 90000kN at No.1045 segment ring. The shield machine was lifted and steering became difficult. Inspection showed that the reason was that the geology started changing. The bottom was sticky old loess mixed with loess-doll whose unconfined compressive strength is 2MPa. The middle part is sticky loess-doll and the ground condition is upper-soft-bottom-hard.

The reason for excessive thrust force and lifted attitude includes: because of the changed ground condition, soft rock caused great squeezing force to the shield machine, especially the rock at the bottom, which caused big frontal and friction resistance force. Liu Quansheng (2013) studied the biggest radial displacement u_r^M in soft rock ground condition without support based on Hoek-Brown principle.

It is known that because wall rock could cause radial squeezing force, when wall rock deformation excels overcutting volume, the friction resistance shall increase. Therefore, by increasing the excavation diameter and the overcut volume at the bottom of cutterhead, the frontal resistance caused by scrapers can be reduced. As a result, thrust force reduces as well and the machine attitude could be adjusted.

The general plan can be seen in Figure 16. The diameter(Φ4900mm) of No.4, 5, and 6 cutterheads is increased with singular side overcutting 300mm, which could decrease shield machine friction resistance while thrusting and blind area, and assist with downward steering. At the same time, the diameter(Φ2700mm) of No.1 cutterhead is decreased because of interference. In order to avoid ground settlement due to overcutting, the synchronous grouting volume is increased so as to reduce soil disturbance.

The spokes of No.4, 5 and 6 cutterheads are enlarged and welded with cutters. Besides, the slant spokes of No.5 cutterhead are removed and new spokes are welded to avoid interference with No.4 and 6 cutterheads. And analysis is carried out on the modified cutterhead and modification effect.

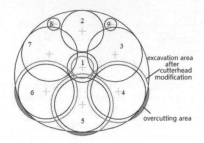

Figure 16. Cutterhead modification diagram.

(1) Finite element analysis

In order to confirm the strength and inflexibility of the modified cutterhead, statics simulation is carried out for the modified No.5 cutterhead with the use of ANSYS Workbench.

Simulation result comparison can be seen in Table 2. The inflexibility of the modified structure decreases, while deformation increases. The biggest stress is smaller than material's permissible stress 295MPa. Therefore modified cutterhead could meet the requirement of inflexibility.

(2) Comparison of thrust before and after modification

Before No.1045 segment ring, the average thrust force is 72000kN and after it the thrust force increases to 85000kN with maximum over 90000kN (Figure 17). After cutterhead optimization, the thrust force decreases to 6000kN, which decreases 2500kN compared to the thrust force before optimization. With the enlarged cutterhead, the cutting length decreases 10340mm and the excavation blind areas decrease as well. The optimization effect is satisfactory in terms of ground settlement and attitude control.

Because of the increased excavation area and changing geology, cutterhead torque after modification increases. The torque of No.5 cutterhead increases from 650kNm to 1250kNm (Figure 18). Although its rated torque is 1250kNm, it still could meet the requirement.

Table 2. Strength and inflexibility comparison of Φ4900 cutterhead.

	Before modification	After modification
Maximium Stress(MPa)	152.74	265.88
Maximum deformation(mm)	3.92	10.16
Deformation rate	0.08%	0.18%

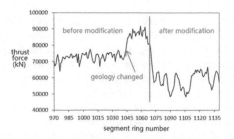

Figure 17. Thrust curve before and after cutterhead modification.

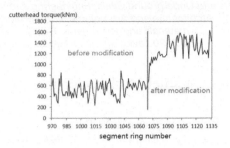

Figure 18.   Torque curve before and after No.5 cutterhead modification.

5   CONCLUSIONS

Based on the Menghua railway of Baicheng tunnel, this paper studies on the overall technology of the horseshoe-shaped shield machine. The design and application of horseshoe-shaped shield machine provides a solution for several key technological problems of non-circular shield machine. The Menghua railway is the first mountainous double lane tunnel which was excavated by horseshoe-shaped machine. The first successful application of horseshoe-shaped shield machine proves that it is possible to realize trenchless excavation of horseshoe-shaped tunnel and safe, fast, and environmental-friendly construction. In addition, it sets an example for large scale mountainous double lane tunnel construction as well.

Research on full-face hard-rock or mix ground horseshoe-shaped shield machine, high automation and intelligence shall be important future research direction.

REFERENCES

Chen, Y. 2011. Study on deflection mechanism and rectification technology of DOT shield. *Urban Mass Transit* (8): 108–114.
Ge, C.H. 2011. *Design and construction of pipe jacking engineering*. Beijing: China Building Industry Press.
Jia, L.H. 2016. Application of rectangular pipe jacking machine to urban underground space development and its prospects. *Tunnel Construction* 36(10): 1269–1276.
Kazunari, K. & Takashi, M. 2002. Development of rectangular shield, *KOMATSU Technical Report* 3, 47(148): 46–54.
Liu, J. 2017. Discussion on design of cutterhead of horseshoe-shaped tunnel boring machine. *Tunnel Construction* 37(suppl. 1): 204–211.
Liu, Q.H. & Huang, X. & Shi, K, et al. 2013. Jamming mechanism of full face tunnel boring machine in over thousand-meter depths. *Journal of China Coal Society* (1): 78–84.
Qian, Q.H. 2002. Application situation and outlook of TBM in underground project in china. *Construction Machinery* 34(4): 28–35.
Ryunosuke, K. & Takuya, N. & Toru, G. 2013. Application of double-O-tube TBM in soft ground, XIAN, *The Seventh China-Japan Conference on Shield Tunneling proceedings*: 533–543.
Sun, T.L. 2017. The research status and engineering application of non-circular shield method. *Journal of Railway Science and Engineering* 14(9): 1959–1966.
Wang, M.S. 2014. Tunneling by TBM/shield in China: state-of-art, problems and proposals. *Tunnel Construction* 34(3): 179–187.
Wang, X.Y. & Shao, Z.S. 2014. Force of the primary support in a horseshoe tunnel. *Modern Tunnelling Technology* 51(6): 83–88.
Zhang, F.Q. 2010. Importance of technological rehabilitation of tunnel boring machine(TBM). *Tunnel Construction* 30(1): 84–90.

Tunnels and Underground Cities: Engineering and Innovation meet Archaeology,
Architecture and Art, Volume 5: Innovation in underground engineering,
materials and equipment - Part 1 – Peila, Viggiani & Celestino (Eds)
© 2020 Taylor & Francis Group, London, ISBN 978-0-367-46870-5

Seal performance test of shield tail under high water pressure

X. Jiang, W. Chen, D.L. Xu, X.H. Zhang & Y. Bai
Department of Geotechnical Engineering, Tongji University, Shanghai, China

ABSTRACT: Shield method has been used for more tunnels projects due to its safety and high efficiency. However, the shield method is facing more unprecedented technical challenges because of the increasingly complex conditions such as high water pressure and mixed ground formation. Shield tail seal plays an important role in the tunneling process, which is greatly affected by high water pressure, soil pressure or other factors that might endanger the construction safety. The injection of grease is of fundamental importance for shield tail seal, which can prevent water and soil from flowing into the shield machine. As a result, the sealed grease must be studied in details. This paper introduces an experiment on shield tail grease used in tunneling. The experiment includes testing in static pressure and propelling state, which can fully reflect the performance of shield tail grease in shield tunneling. The pressure change of grease during pipeline transportation and in the shield tail gap are both studied in this paper. The research results of this paper provide the technical reference for the safety control of shield tail seal during the tunneling process.

1 GENERAL INSTRUCTIONS

After more than half a century's development, the shield method has become one of the leading methods of tunnel construction because of its high safety and automation level compared with traditional mining methods. During the construction of a shield tunnel, the outer steel shield shell is used to support the stratum, and under the protection of the steel shield shell, the assembly and advancement of the segment are realized. The shield shell can isolate the outer formation and has a high sealing property so that water and soil are stopped from flowing into the inside of the shield machine. In general, the shield shell is divided into three parts: the front shield, the middle shield, and the shield tail. The seal between the shield tail and the segment is the key to ensuring the isolation of the surrounding soil. For the structure, a wire brush is installed on the inner side of the shield tail. During shield propulsion, the wire brush is in contact with the segment and is sealed by injecting grease into the gap. Shield tail seal grease is an important material to ensure that the seal of the shield tail is intact, and it also has the function of preventing soil and muddy water outside a shield from infiltrating into the shield, which can guarantee the smooth construction of the shield. There are many accidents caused by the failure of shield tail seal in the world. For example, on February 7, 2018, Foshan (China) Metro Line 2 construction collapsed. Initial investigate by industry experts have identified that it was the shield tail leakage that caused external mud water to flood into the shield machine and tunnel, leading to huge property loss and casualties. This is a wakeup call for the safety of tunnel construction in subways.

In current shield construction, grease used for shield tail seal is important mainly due to its role in shield tail sealing, and some scholars have conducted several researches. For example, Wang (2013) has studied the formulation and properties of seal tail grease, focusing on the domestic product comparison of the sealability and pumpability of the sealed grease. Hanyong et al. (2016) studied the sealing performance of a rectangular shield tail, which provided a technical basis for the safe construction of new shield types. Zhang et al. (2014)

studied the theory and control method of sealed grease injection in the shield tail and proposed the advantage of automatic control in engineering application. Song et al. (2017) studied the method for testing the performance of shield tail grease, which provided a test basis for the efficient and safe operation for shield tunnel engineering.

However, there are no systematic standards or specifications for the testing of shield tail grease at present. For example, the injection amount of grease in the shield tail or on the brushes depends more on the experience of engineers. The rationality of the *experience* has to be verified. Moreover, the researches mainly focused on measuring the properties of shield tail grease as a viscous material, rarely considering the actual workability. Because the viscous grease is extremely complicated in nature, performance test and numerical simulation cannot fully consider the actual working status of the grease, while a specific shield tail seal test can better simulate the state of grease at the shield tail. It can also simulate the sealing performance of the grease under the relative movement of the shield tail and segment. Shield tunnels are developing in such a direction that they will have larger sections, run a longer distance, and go deeper in depth. The shield tail seal will also face the challenge of dealing with high water pressure. For example, the Qiongzhou Strait Tunnel and the Taiwan Strait Tunnel are at the planning stage and the water pressures they face will be above 1.0Mpa. The high water pressure demands a higher level of skills and techniques in shield tunnel construction, and the issue of shield tail seal has become an important topic that we must address.

Due to the superiority of the shield tail seal test relative to other research methods, it is of great significance to study the reliability of shield tail grease under complex construction conditions. Through the waterproof sealing test of the shield tail of a high water pressure shield tunnel, a reasonable detection method is proposed to study the waterproof sealing performance, the flow performance of the shield tail grease, and to set the industry standards as well as testing practice for shield tail sealing of high water pressure shield tunnels. This paper introduces the technical characteristics of the shield tail comprehensive test platform and the shield tail test process. The results are also analyzed.

2 TEST PURPOSES

A circular shielded tube is used to study the following:

(1) Simulate shield tail sealing performance when the central axis is under static pressure state;
(2) Simulate shield tail sealing performance when the central axis is under propulsion state;
(3) Compare the sealing properties of four groups of B, C, E, and G.

This test uses French Condat grease and grease made in China.

3 TEST EQUIPMENT AND MATERIALS

This test uses the 863 large shield simulation test platform (Figures 1 & 2). The main test device consists of a steel simulated shield tail outer tube section and a steel simulated tube with a shield tail clearance of 50mm. Three shield tail brushes are set in the tail of the shield. The distance between each brush is 455mm. The shield tail grease is pumped by hand. The external outer tube section of the simulated shield tail is connected with the pressure gauge, which can react in real time to simulate the water pressure in the outer tube section of the shield tail. The simulation tube is propelled by the jacking device test platform, and the test data such as propulsion speed, propulsion stroke, and cylinder pressure can be recorded by the platform monitoring system.

The propulsion device is mainly composed of a rear leaning device, a jacking device, a tie rod block, and a top iron. Among them, the jacking device is composed of three pairs of cylinders, and the reaction force of the cylinder is carried by a tie rod. In the test, the propulsion cylinder is not in direct contact with the steel simulating tube piece, but the force is transmitted by the middle top block and the top iron (as shown in Figure 3).

Figure 1.   863 shield test platform.

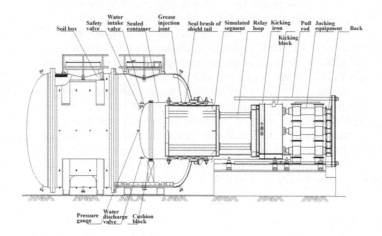

Figure 2.   Schematic diagram of 863 shield test platform.

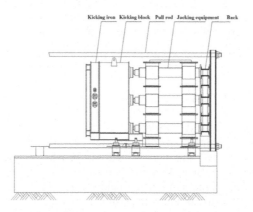

Figure 3.   Plan view of the propulsion device of the 863 shield test platform.

As shown in Figure 4, the shield tail seal test device is mainly composed of a steel shield tail, a steel tube and a relay ring. The front part of the simulated shield tail has a semi-ellipsoid head pipe joint. In the test, the main function of the head pipe joint is to store water, maintain and control the water pressure. The bottom of the head pipe joint is equipped with a water

2300

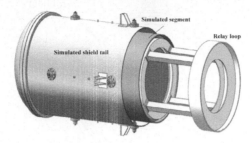

Figure 4.   A three-dimensional diagram of the shield tail seal test device.

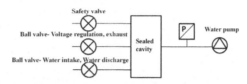

Figure 5.   Water injection schematic.

inlet hole, an overflow valve hole and a water pressure gauge hole from top to bottom, as shown in the sectional view of the device. The specific test principle is based on the water injection principle shown in Figure 5. There are eight screw holes around the simulated shield tail pipe body, and the position of the simulation pipe piece is controlled by the screws during the test. The three-way wire brush has a grease injection hole outside the pie body, where the shield tail grease is injected. The principle of the shield tail grease injection is shown in Figure 6.

The inner diameter of the test steel tailpipe section is 2000mm, and the inner diameter of the test steel simulating tube piece is 1836mm (see Figure 7). The test steel simulated shield tail and steel simulation tube is all scaled models, but the shield tail gap is similar to the actual site construction. As shown in the shield tail gap side view, the simulated shield tail and the simulated tube gap is 82mm, the installation thickness of the shield tail brush is 32mm, and the shield tail clearance is 50mm. Therefore, this is a shield tail seal performance test of a shield tail gap built to a scale.

There are four adjusting screws on the front and back of the test steel simulated shield tail to adjust the attitude of the steel simulation tube inside (see Figure 8). The four adjusting screws at the rear are located at the sealing cavity formed by the oil seal of the analog tube and the head tube. When water is added and pressurized, the four adjusting screws are bound to receive the same water pressure as the sealed chamber. Therefore, the four screws also need to be properly sealed and tested beforehand to avoid the risk of sealing failure, affecting or destroying the test result. The screw mounting member is composed of a 60mm diameter screw, a fixing member and a sealing gasket (see Figure 9).

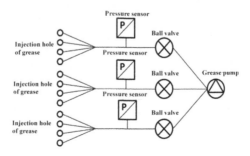

Figure 6.   Shield tail grease injection schematic.

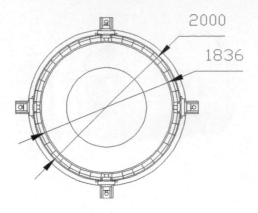

Figure 7. Schematic diagram of the shield tail gap.

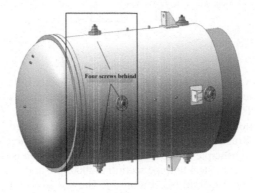

Figure 8. Four adjusting screws.

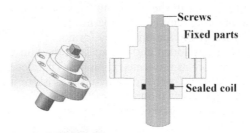

Figure 9. Screw structure.

4 TEST PROCEDURE

The test flow chart is shown in Figure 10.

The shield tail seal test is carried out under static pressure (propulsion speed = 0 cm/min) (see Figure 11) and propulsion state (propulsion speed > 0 cm/min) (see Figure 12).

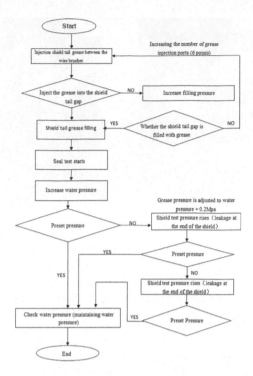

Figure 10. Test procedure.

Figure 11. Shield tail seal static pressure test.

4.1 Static pressure state

(1) Place the test equipment in place as shown in Figure 11, apply the seal grease in the simulated shield tail until fully embed;

(2) Push the simulated lining into the position shown at the end of the simulated shield, leaving a thrust gap of 900mm in front. The simulated lining (brake device) is then fixed;

(3) Fill the propulsion gap with clean water, increase the water pressure, and observe the leakage of the shield tail;

(4) The test ends without leakage.

Figure 12.   Shield tail seal dynamic pressure test.

4.2   *Propulsion state*

(1) Take the same steps in the static pressure state (3), and then open the back of the cylinder to insert the simulation lining. The cylinder jacking speed can refer to requirements set for relevant working condition;

(2) Observe the leakage of the shield tail, and control the water pressure in the gap. When the water pressure reduces, replenish water timely. When the water pressure rises, release the water through the pressure relief pipe until the pressure drops to a level suitable for the working condition;

(3) If the tail leaks, it can be sealed in time by injecting grease. The amount of grease required will depend on the leakage condition;

(4) Repeat the process until the predetermined propulsion distance is completed.

5   TEST PROCESS AND ANALYSIS

5.1   *Group B grease test*

Based on the actual working condition, the shield tail seal test of the sealant grease of various groups are carried out by using the 863 platform. Their waterproof sealing performance under different pressure and different propulsion speeds are verified.

5.1.1   *Static pressure test of group B grease*

Firstly, the static pressure test of Group B grease is carried out on the platform. When the external water pressure is 0 MPa, the shield tunneling machine is in the initial state. The water is injected into the platform cavity by a water pump to adjust the pressure inside the cavity. Then, the sealing grease is injected into the shield tail to seal the shield tail, maintaining the pressure balance between the shield tail and the platform cavity. As the pressure in the chamber gradually increases, the amount of shield tail grease will also increase (see Figure 13), causing the end grease pressure to increase.

The data obtained from the field test is shown in Table 1:

Table 1. Group B seal grease tail seal test under static pressure.

Working condition	Water and soil pressure MPa	End grease pressure MPa	Grease injection amount L	Leakage
1	0	13.1	242.2	No
2	0.2	18.9	251.8	No
3	0.4	40.6	263.8	No
4	0.6	82.9	270.9	No
5	0.8	181.1	293.1	No

5.1.2 *Group B grease dynamic pressure test*

The dynamic pressure test is carried out under two conditions, one at the propulsion speed of 2cm/min, and the other at 6cm/min.

First, the propulsion speed of the platform propulsion cylinder is set to 2 cm/min, the lower speed. When the external water pressure is 0 MPa, the end grease pressure and the amount of grease injection are measured. Water is then injected into the platform cavity by a water pump to adjust the pressure in the chamber. When the water pressure is 0.2 MPa, the cylinder is pushed forward, and the grease pump injects sealing grease into the tail of the shield while maintaining the pressure balance between the shield tail and the platform cavity. The end grease pressure and grease injection amount are recorded and leakage, if any, is monitored. After that, the cylinder is withdrawn. Water is injected into the platform cavity again, with the pressure inside the chamber being adjusted constantly. When the pressure reaches 0.4 MPa,

Figure 13. The pressure of water injection in the platform cavity is increased from 0.2MPa to 0.8MPa.

the end grease pressure and the amount of grease injected are recorded to observe whether any leakage has occurred.

It is observed that when the pressure reaches 0.8 MPa at a propulsion speed of 2 cm/min, water seepage begins to appear (see Figure 14) and it continues until all strokes are completed. No other serious water leakage is observed. The specific data is shown in Table 2.

High-speed propulsion is adopted following the previous test. When the pressure reaches 0.2 MPa at a propulsion speed of 6 cm/min, a big water leakage begins to occur. The greater the water and earth pressure, the more likely it is to leak. Therefore, under the water pressure of 0.4–0.8 MPa in high-speed propulsion state, a large amount of water leakage can be predicted. The data is shown in Table 3.

5.2   Grease test results of group C, E and G

The grease test of Groups C, E and G are also carried out by using the test procedure as Group B grease. Through the shield tail seal test of the sealed grease, the data shows that the sealing performance is G<B<C<E for the four groups. Furthermore, under static pressure, the four groups of sealing grease used in this test can withstand the pressure of 0.80 MPa. When the slow propulsion test is carried out, the four groups of greases, except for the group B, does not lead to leakage, which means the four groups of grease can resist pressure up to

Figure 14.   Water seepage with Group B grease at low speed (2cm/min) when pressure reaches 0.8MPa.

Table 2.   B group seal grease tail seal test at the propulsion speed of 2cm/min.

Working condition	Water and soil pressure (MPa)	End grease pressure (MPa)	Grease injection amount (L)	Leakage
1	0	8.3	249.6	No
2	0.2	15.6	263.5	No
3	0.4	29.8	295.5	No
4	0.6	62.4	326.8	No
5	0.8	96.7	402.8	Yes

Table 3.   B group seal grease tail seal test at propulsion speed of 6cm/min.

Working condition	Water and soil pressure (MPa)	End grease pressure (MPa)	Grease injection amount (L)	leakage
1	0	7.6	276.3	No
2	0.2	8.3	376.7	Large amount of leakage
3	0.4	/	/	/
4	0.6	/	/	/
5	0.8	/	/	/

0.80MPa at low-speed propulsion. When the rapid propulsion test is carried out, however, the pressure resistance of group B is less than 0.20MPa, the compressive capacity of group C is greater than 0.40MPa but less than 0.60MPa, and the compressive capacity of group E is greater than 0.60MPa but less than 0.80MPa, and the pressure resistance of Group F is less than 0.20MPa.

6 CONCLUSION

The shield tail grease test shows that the shield tail seal plays a decisive role in the construction process of a shield tunnel. If the shield tail leaks because of sealing grease, it will threaten the project safety. The following conclusions and recommendations can be drawn from this test:

(1) The test shows that the E group grease (the WR89 grease) produced by the French company Condat, has the best sealing performance, indicating that Chinese sealing grease has room for improvement compared with foreign products;
(2) The quality of the shield tail brush itself and the quality of the grease play an important role in the sealing of the shield tail. However, there is no relevant industry standard for quality control at this stage. Such industry standards should be established;
(3) At present, the shield tail grease used is mostly non-degradable due to the need to control costs. In the future, the use of green and environmentally degradable shield tail seal grease should be promoted to reduce environmental problems caused by shield construction.

This test provides the reference for industry's related standards and specifications and also has engineering and scientific significance.

REFERENCES

Deqian, W. 2013. Formulation Research and Property Characterization of a Kind of Shield Tail Sealing Grease. *Tunnel Construction*, 4, p.007.
Fangjia, S. et al. 2017. 'Test methods for grease of shield tail seal in shield engineering', 143(Iceep), pp. 955–962.
Hanyong, X. et al. 2016. Performance Test for the Quasi-Rectangular Shield Tail Sealing. *Modern Tunnelling Technology*,53(S1): 51–55.
Hepei Z. et al. 2014. Research on shield tail seal oil injection theory and control method. *Shield Equipment & Project*, 09, 68–71.

Tunnels and Underground Cities: Engineering and Innovation meet Archaeology,
Architecture and Art, Volume 5: Innovation in underground engineering,
materials and equipment - Part 1 – Peila, Viggiani & Celestino (Eds)
© 2020 Taylor & Francis Group, London, ISBN 978-0-367-46870-5

Rescuing and rebuilding TBMs in adverse ground conditions

D. Jordan
The Robbins Company, Germany

B. Willis
iPS, The Netherlands

ABSTRACT: Modern TBMs deliver high performance with availability rates that are beyond 90%. The TBM design concepts make the machines highly versatile for employment in varying soil and ground conditions. Machines can now withstand extreme loads and impacts in rough underground environments because of the components made for longtime use. Regular maintenance and planned service is the vital element in prolonging a machine's life and for high performance and availability. A well-serviced machine provides excellent performance as well as active project safety. Proper operation in variable conditions is also key. For instance, a hard rock TBM may run into zones of swelling rock. The most appropriate method to overcoming the swelling rock is to keep going, avoiding any unnecessary stops. Worn disc cutters that have not been maintained in due course are a prominent example of such avoidable stops, which may result in long downtimes and severe damage to the machine. However with modern and advanced techniques to underground tunneling, rescuing and rebuilding TBMs is possible to save the project This paper will discuss methods and tools for modern TBM service and maintenance using present case studies about TBM rebuilds in extreme project conditions.

1 INTRODUCTION: THE TBM LIFE CYCLE

It is essential to consider the total life cycle of a machine, even in its beginning design stages—this is by far the most economical and sustainable way of thinking. Designing machines with ease of rebuilding in mind ensures that the manufacturer does not have to start from scratch every time a machine needs work. It also results in time, cost, and energy savings when the time does come to rebuild a machine, which is then passed on to the customer.

Perhaps even more important than that is the way a TBM is maintained during its project. It is important to remember that the basic structure of a TBM is metal—as long as the structure is intact, one can then check on the bearings, conveyor, hydraulics, and other components. Particular attention should be paid to components that are hard to reach. The main bearing is one of those parts that is difficult to replace during tunneling.

When developing a maintenance plan, it is critical that TBM crews are properly trained on how to operate the machine in the entire gamut of ground conditions that may be encountered on a given tunnel project. Plans must be in place to deal with a wide range of ground conditions as well (e.g., fault zones, water inflows), with protocols as to how the machine should be operated in such conditions. Once the machine has been launched, regularly scheduled maintenance based on tunnel length and geological conditions is also essential. While there are no special guidelines for long-distance tunnels, crews must be diligent and conduct more detailed inspections the longer a TBM is in operation.

Planned cutter inspections are a regular part of maintenance, which is recommended daily. Checking of oil levels, and all fluids, greases and hydraulics, is also of primary importance. Daily logs are recommended for monitoring of all major systems on the TBM. A daily maintenance regime typically involves routine checks without TBM downtime. Protocols for more in-depth monthly, semi-annual, and annual checks of systems should also be in place. These full checks of various systems do require downtime but are all the more critical when tunneling over a long distance or in variable conditions. These checks are also typically based on the rigors of the project schedule—in hard rock, a week is assumed to be equivalent to 100 m of advance while a month is assumed to be equivalent to 500 m as a baseline.

Depending on the tunnel length, some maintenance may be done beyond what is considered normal. Gearboxes, for example, may be designed for long tunnels but if it is known that the tunnel length will exceed the life of the gearboxes then planned refurbishment should occur during tunneling. This procedure has been done on several tunnels including India's AMR tunnel—what will be the longest tunnel without intermediate access at 43.5 km once complete.

Maintenance while storing the TBM between projects can also maximize equipment life—such as storing components indoors, coating the equipment with anticorrosive spray, and making sure the main bearing is filled with oil. Owning and using a new TBM has added hidden benefits including familiarity of machine operation and proven performance for that particular piece of equipment.

1.1 *Maintenance in the Digital Age*

Modern TBMs are making maintenance and replacement of consumables more efficient with monitoring technology. A typical TBM will include sensors and detectors for all manner of functions and will activate an alarm or other type of notification that can shut down the operation or parts of the operation if a critical threshold is reached.

Even disc cutters can be monitored to determine when cutter changes are needed. Sensors can deliver information wirelessly about cutter RPM, vibration, and temperature, indicating when cutter changes are necessary and allowing the operator to avoid cascading cutter failures known as wipeouts (see Figure 1, Mosavat 2017).

However, a visual inspection of the cutterhead is still ideal. Cutter rings are not the only thing that need inspection in this high-wear area of the TBM. The cutter assembly bolts, the seals, the surrounding structure, the wear coating – all this needs to be inspected by trained and experienced cutter technicians.

Data reading and logging systems provide operators further clues as to when maintenance is required. Real-time data loggers can transmit to the surface where crew members can interpret the data, or the data can even be stored on a website where the TBM supplier can monitor the machine's behavior and consult the crew. An example of this would be a high pressure reading at the main thrust rams combined with low torque at the cutterhead—this can be an

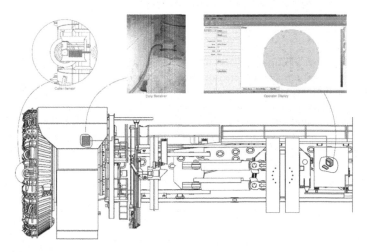

Figure 1.   Remote monitoring system for disc cutters.

Figure 2.   TBM data screens in operator's cabin.

indication of worn gauge disc cutters that in conditions like swelling rock could result in the machine becoming jammed (see Figure 2).

2   TBM DESIGN FOR MULTIPLE PROJECTS

Over the years, Robbins has built a quality assurance system that ensures when it delivers a rebuilt machine, either to the original configuration or a modified one, it still adheres to a design life of 10,000 hours. This standard also includes checks to make sure that all the components are in a functional condition of 'as new' or 'new'.

In order to guarantee the same design life and same warranties on a rebuilt machine, the initial design of the TBM will need to consider that the TBM will be used on several projects. This means that the major structures will need to be strong enough to survive even the toughest conditions and that worn parts can easily be replaced. If the machine is not properly designed for multiple projects, there will be a need to do major work to get the TBM in a working condition, either in its original or modified configuration.

One can argue that project owners typically only have one project and that the condition of the TBM and the suitability of its rebuild is therefore not essential. This is something that is also reflected in many of today's tunneling projects, where the commercial consideration is often given far more attention than the technical one. We would argue, however, that an initially sturdy and robust design of the TBM will give the project more uptime, higher production rates and better flexibility if unexpected conditions are encountered, making it a good and effective insurance against many types of obstacles. Some examples of design aspects that enable longer TBM design life are given below (Khalighi, 2015).

2.1   *Robust Cutterhead and Machine Structure*

A machine designed with multiple projects in mind relies on a heavy steel structure that can stand up to the harsh environments often encountered underground. Designs that take into account high abrasivity of the excavated material or the possibility of high abrasivity are even more robust. Ideally, the cutterhead should be designed with regular cutter inspections and changes in mind. It must also be built to last: this can be difficult with a back-loading cutterhead design, which is full of holes not unlike Swiss cheese. In order to build up the structure, much of the strengthening occurs during the manufacturing process. Full penetration welds are recommended for the cutterhead structure to battle fatigue loading and vibration. Rigorous weld inspections and FEA stress analysis checks can then be made for vulnerabilities in the cutterhead structure.

2.2   *Main Bearing and Seals*

Large diameter 3-axis main bearings, with the largest possible bearing to tunnel diameter ratio have larger dynamic capacity, and therefore are capable of withstanding more load impacts and giving longer bearing life. It is important to retain as high a ratio as possible (see Figure 3).

The bearing and ring gear are in a difficult-to-access spot on the TBM, so they must be designed for longevity, with a super robust structure and high safety factor. Safety factor is defined as any surplus capacity over the design factor of a given element, and overbuilding such structures is of necessity in long distance tunneling.

Robust seal design is also essential. The Robbins Company provides a proven seal design using hardened wear bands. Many other manufacturers don't use wear bands, and so as the TBM operates, it wears a groove into the seal lip contact zone. Robbins sacrificial wear bands can be switched out or replaced, making repairs easier. The abrasion-resistant wear bands, made of Stelite™, can be changed in the tunnel in the unlikely event of excessive wear, or can be relocated on the carrier to ensure that damage is not done to the TBM structure itself on long drives.

In addition to the seal design, other elements of the main bearing such as the internal fasteners must be designed to be durable and of high reliability, as these fasteners are difficult to access and are not easily replaceable. The studs connecting the cutterhead to the main bearing seal assembly must also be closely analyzed for strength, deflection, and adequate fastening/clamping force, and protection against abrasive muck must be provided for the fasteners.

2.3 Lubrication

Dry sump lubrication is a critical way of keeping the main bearing cavity clean by filtering and recycling the oil at a constant rate. Any contamination is cleaned from the cavity, prolonging bearing life. The system also has an added benefit: The oil can be monitored and analyzed for any indications of distress in the main bearing or gears. This monitoring has the potential to allow for correction or intervening maintenance of critical structures/components before a failure occurs.

2.4 Drive System

The right drive system is also important for heavy TBM usage. Variable Frequency Drives (VFDs) and planetary gear reducers allow for infinitely adjustable torque and speed control

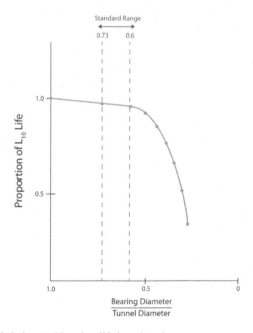

Figure 3. As the ratio falls below 0.6 bearing life is reduced.

based on the encountered ground, which optimizes the TBM advance rate and reduces damage to machine components. This is in comparison to older style drives: In older model TBMs, often the drive system was single speed or 2-speed. If a machine bored into a fault zone, for example, there would be no way to slow down the cutterhead. Such drives would often result in undue wear to the TBM, or even damage to structural components.

Drive motors must also be designed to withstand high vibration as a result of excavating through hard rock conditions. Cantilevered motors must be able to withstand the high g-forces applied to them by violent machine vibration, which is induced by the rock cutting action.

2.5 *Load Path*

A uniform load path, from cutterhead to main bearing to cutterhead support, is always desirable. However, for long distance tunnelling or for multiple uses, the load path can be crucial as high stresses occur wherever the load path shifts. A cutterhead with a cone-shaped rear section can help with this problem by evenly distributing the load across the circumference of the main bearing. In general, everything must be designed in a more robust fashion, and the loads generated by the cutterhead must also translate into a heavier overall structure of the machine.

3 KEY FACTORS IN REBUILDING TBMS

The rebuilding of TBMs—both, the process and the standardization of rebuilds–has become a focus for the industry as more projects with multiple machine requirements and short time frames are being proposed. The focus has been further highlighted by the ITAtech, a technology-focused committee for the International Tunneling Association (ITA-AITES) that produced guidelines on rebuilds of machinery for mechanized tunnel excavation in 2015. While the guidelines are relatively new, Robbins has a long history of delivering robust machines, many of which are rebuilt (many are also 100% new).

In general, Robbins' experience with rebuilding machines has yielded some key insights. As long as the TBM is well-maintained, there will be jobs it can bore economically. Optimal TBM refurbishment on a used machine requires a broad knowledge of the project conditions, and there are some limitations:

- Machine diameter can be decreased within the limits set by free movement of the grippers and side/roof supports
- Machine diameter can be increased subject to the structural integrity of the machine and the power/thrust capabilities
- Propel force can be increased only to the level supported by the grippers' thrust reaction force
- Cutterhead power must be adequate to sustain the propel force in the given rock, but cannot be increased beyond the capacity of the final drive ring gear and pinions
- Cutterhead speed increases must not exceed the centrifugal limits of muck handling or the maximum rotational speed of the gage cutters

Increasing the power of the TBM is one way to make the design more robust for a longer equipment life. Strong designs have been developed in recent years, including Robbins High Performance (HP) TBMs, and have been used on a number of record-setting hard rock tunnels. The HP TBM is designed with a greater strength of core structure and final drive components. They can be used over a much wider range of diameters, whereas older machines (from the 1970s and 80s) are typically limited to a range of less than 1 m of diameter change plus or minus their original size.

HP TBMs have the capability of operating over a broad range. For example, a 4.9 m TBM can be refurbished between 4.3 m and 7.2 m diameters—a range of 2.9 m. Main bearing designs have allowed for greater flexibility, evolving from a 2-row tapered roller bearing to the 3-axis, 3-row cylindrical roller bearing used today. This configuration gives a much higher axial thrust capacity for the same bearing diameter and far greater life in terms of operating hours or revolutions.

Overall, what determines how long a TBM will last is a function of the fundamental design, such as the thrust and gripper load path through the machine and the robustness of the core structure. On older model TBMs, the ring gear and pinions can be strengthened, and larger motors can be added. With sufficient core structure strength, it is also possible to increase the thrust capacity. The limitation is the capacity of the gripper cylinder to handle the increased power and thrust. Once replacement of the gripper cylinder and carrier are required, TBM modification costs are generally considered uneconomic.

4 TBM REBUILDS IN ADVERSE GEOLOGICAL CONDITIONS

Service, maintenance and operation of a TBM can be dramatically impacted by the ground strata, which may appear in the form of rock bursting, swelling and squeezing rock, face collapse, mixed face conditions, or other challenges. In some cases, the TBM operation must be interrupted to allow the machine to undergo substantial reconditioning.

4.1 Case Study: Kargi HEPP, Turkey

The Kargi Kizilirmak Hydroelectric Project involved the excavation of an 11.80 km headrace tunnel at 10.00 m bore diameter. Due to expected variations in geology, a Robbins Hard Rock Double Shield TBM was selected with specific features to bore the tunnel in mixed ground formations.

As expected, the machine ran into adverse ground conditions, but these were much worse than initially predicted during site investigations. The crew employed various well-proven techniques to keep the machine from becoming stuck, including boring half strokes to keep the gripper and rear shield from sitting idle for too long. This worked well for a time, but the machine came to stop when a face collapse occurred followed by a crown cavity. In total, eight bypass tunnels were needed in the first 2 km of boring.

It was determined the machine needed to be reconditioned to continue the drive, and it also needed to be furnished with supplementary features and equipment to enable tunneling in highly variable geologic conditions. Some of the most significant features are listed below (Clark 2015).

4.1.1 Supplementary means for ground support installation
The possibility of installing ground support such as fore-poles or a pipe roof canopy ahead of the tunnel face was investigated as a means of supporting loose and fractured ground. A custom designed canopy drill was delivered to site and installed in the forward shield for installation of a tube canopy. The space in the forward shield area was limited; hence, the extension section of each tube was only 1.0 m in length. However, the advantages of drilling closer to the tunnel face more than compensated for the time spent adding extensions to the tube length. The location of the canopy drill reduced the length of each canopy tube by more than 3 meters when compared to installation using the main TBM probe drills (see Figure 4).

4.1.2 Modification of Cutterhead Drive Characteristics
To further mitigate the effects of squeezing ground or collapses, custom-made gear reducers were ordered and retrofitted to the cutterhead motors. They were installed between the drive motor and the primary two-stage planetary gearboxes. During standard boring operations, the gear reducers operate at a ratio of 1:1, offering no additional reduction and allowing the cutterhead to reach design speeds for hard rock boring. When the machine encounters loose or squeezing ground the reducers are engaged, which results in a reduction in cutterhead

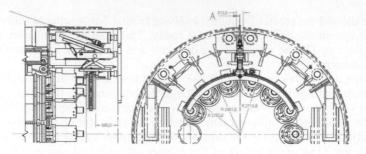

Figure 4.   Custom canopy drill in the forward shield.

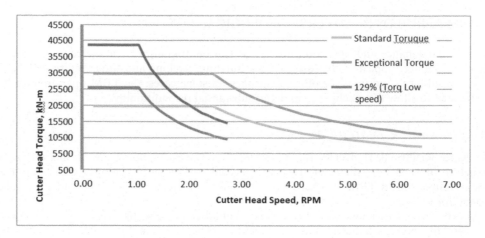

Figure 5.   Cutterhead torque curves after modification.

speed but the available torque is increased. Figure 5 shows the torque curves for both standard and reduced gearing. After the installation of the canopy drill and the increase in available cutterhead torque, the TBM traversed several sections of adverse geology including stretches of severe convergence without becoming trapped.

4.2   Case Study: Namma Metro Phase 1 Project

The Namma Metro Phase 1 Project in the Indian City of Bangalore included dozens of TBMs, many operating in mixed face conditions. One of the sections—a 1.0 km drive from Mantri Square to the central hub station of Majestic Gowda, as part of the North-South Corridor—utilized a 6.23 m diameter European-manufactured EPB. The TBM employed experienced severe wear at the cutterhead with the result that the TBM could no longer excavate a tunnel of sufficient diameter for the shield to pass through. In general, the TBM excavated in weathered granite having approximate compressive strength of 130 MPa with a high content of quartz and feldspars (see Figure 6).

The initial plan was to determine how to dismantle the TBM underground to allow another TBM to excavate the tunnel from the opposite direction. After an intense survey and inspection, it was found that the entire machine was in good enough condition to be refurbished in-situ, permitting completion of tunnel excavation (a further 630 m) with the same machine. The cutterhead and screw conveyor were completely worn out and needed changing (Willis 2017).

To enable the reconditioning effort, a rescue shaft was built to gain access to the front of the machine. While this happened, TBM parts and components were refurbished and repaired.

Figure 6.   Rock sample taken from the face.

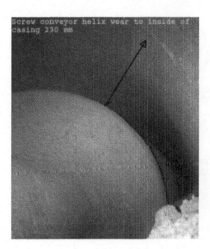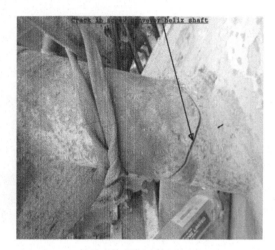

Figure 7.   Extreme wear of helix measured at 230 mm to inside of casing & cracks at screw helix shaft.

The plan was to inspect the cutterhead and drive systems once the shaft was finished and the front of the machine could be accessed. Other components such as the screw conveyor were removed inside the tunnel for subsequent replacement with a combination of new and refurbished parts (see Figure 7).

Once in the shaft, the cutterhead of the machine showed evidence of the extremely abrasive rock encountered. Disc housings, buckets and drag cutter mounts were completely gone. This damage resulted after only 370 m of tunnelling (see Figure 8).

Repair of the cutterhead was not an option anymore, so it had to be replaced in full by a new one. The basic configuration was kept. The new head was dressed with extended wear coating and protection and flown in for installation to make the tight schedule.

Full refurbishment began in January 2015 with a programme of dismantling components such as the screw conveyor, hydraulic system, PLC, and checking the main drive. These actions uncovered even further signs of damage than earlier seen. A completely new drive system for the screw conveyor as well as the top helix section of the screw was ordered from Italy. The refurbishment was completed and re-commissioning began in August 2015.

The TBM was re-launched in late September 2015. Though various challenges were faced, tunnel excavation was completed on 19th April 2016. The remaining 630 m drive was bored in especially challenging geology and completed in seven months, with minimal cover in many places (down to less than 2 metres even in back-filled areas). Significant stretches of mixed face conditions were also encountered where cutter ring damage was frequent along with hub

Figure 8. Cutterhead condition when machine was pushed in the shaft.

Figure 9. Condition of the new cutterhead after breakthrough into the final receiving shaft.

seal failure—filling the hubs with an oil/grease mixture helped overcome this problem. However, a disciplined programme of proactive service and maintenance kept the machine in good shape and at high availability. The service company in charge of the machine's repair actively supported the contractor on site, demonstrating that even in the worst cases when machines are heavily damaged and have suffered extreme wear, they can be rescued and repaired in-situ (see Figure 9).

5 CONCLUSIONS

Regular service, good housekeeping and efficient organization of maintenance periods on site are essential to maximize a TBM's performance, its availability and safe employment on a project.

Suppliers can help and provide guidance and support – in paper and in person. Modern tools and data communication can support maintenance service, but they don't replace intervention by experienced crews on site. TBMs can be reused economically for multiple projects given that the machine design is robust and the equipment is operated and maintained according to requirements.

Machines operating in adverse geological conditions can run into stoppages and need reconditioning or in-situ modification. Suppliers have engineering knowledge and the technical know-how to adjust the machine to the new requirements and bring the TBM back on route.

REFERENCES

Clark J. 2015. Extreme Excavation in Fault Zones and Squeezing Ground at the Kargi HEPP in Turkey. *Proceedings of the ITA-AITES World Tunnel Congress 2015, Dubrovnik.*

Khalighi, B 2015. TBM Design for Long Distance Tunnels: How to Keep Hard Rock TBMs boring for 15 km or More. *Proceedings of the ITA-AITES World Tunnel Congress 2015, Dubrovnik.*

Mosavat K. 2017. A Smart Disc Cutter Monitoring System using Cutter Instrumentation Technology. *Proceedings of the RETC 2017, San Diego*: 109–118.

Willis B. 2017. Rescue and Refurbishment Underground of a Damaged Tunnel Boring Machine. *Proceedings of Tunnel Turkey, Istanbul*: 103–110.

Tunnels and Underground Cities: Engineering and Innovation meet Archaeology,
Architecture and Art, Volume 5: Innovation in underground engineering,
materials and equipment - Part 1 – Peila, Viggiani & Celestino (Eds)
© 2020 Taylor & Francis Group, London, ISBN 978-0-367-46870-5

The effect of testing age on the performance of fibre reinforced concrete

K.P. Juhasz & P. Schaul
JKP Static ltd., Budapest, Hungary

R. Winterberg
BarChip Inc., Okayama, Japan

ABSTRACT: The use of Fibre Reinforced Concrete (FRC) is widely accepted in the tunnelling industry. The generally accepted method to determine the material parameters of FRC is the standard 3-point beam test. The effect of age can be relevant for tunnels, because these structures are usually designed for life cycles over 100 years. In this paper the test results of FRC specimens with different fibres using different testing methods (beam and square panel) at different ages will be presented. FRC changes its properties over time in case of both fibre types. The post crack capacity of steel FRC increases in beam tests, while it decreases in panel tests, which yields the conclusion that steel fibres work better at smaller crack width. This only changes a small amount over time. The energy absorption measured from panel tests reduces in case of steel fibre, but stays at a constant level for synthetic fibre.

1   INTRODUCTION

Fibre reinforced concrete has been widely used over recent decades in the tunnelling industry, in infrastructural structures like tramlines, railway track slabs and in industrial floors for instance, primarily due to its main advantage of increasing the ductility of the quasi-brittle concrete while providing a post-cracking strength to the composite material (Juhasz, 2014). There are several raw material sources for these fibres, but the two main types are steel and macro synthetic fibres. Steel fibres usually have hooked ends and macro synthetic fibres typically have a fully embossed surface. The interaction between these fibres and the concrete matrix can define the properties of the fibre reinforced concrete material. Under loading the addition of fibres in the concrete can bridge the crack sides, but after reaching a critical load level they will fail either in rupture or in pull out (Zollo, 1997). The best residual strength capacity can be obtained when the bond strength of the fibres is high and where the fibres are not rupturing. To reach this performance, highly engineered fibres are needed where both the tensile strength of the fibres and the pull-out resistance of the fibres are high. To increase the pull-out resistance multiple hooks can be used in the case of steel fibres, while in the case of macro synthetic fibres the embossing on the surface plays a key role in pull-out resistance However, the chemical treatment of the surface (sizing) is also a very important factor. Over time the bond between the fibres and the concrete matrix can change, which can lead to an unfavourable behaviour of the composite material. It can happen that the bond strength of the matrix becomes too strong over time and the composite exhibits a brittle fibre rupture instead of a ductile pull-out behaviour. This phenomenon is called embrittlement in recent literature (Bernard, 2008). Over time the bond strength can also decrease, which in turn leads to a similar loss of post-crack performance.

In this paper the results of experimental investigations on panel and beam specimens reinforced with steel and macro synthetic fibre at different test ages will be presented and discussed.

2 CORRELATION BETWEEN THE BOND STRENGTH AND THOUGHNESS

The pull-out phenomenon of fibres in fibre reinforced concretes was investigated by Bartos (1980, 1981). According to his study the failure mode of the composite depends on the length and the tensile strength of the fibres. With regard to the fibre length the failure mode can be a sudden de-bonding or a progressive de-bonding. Regarding the strength of the fibres, the failure mode can be pull-out or rupturing. The fibre is optimal if it can maximally increase the ductility of the composite material. According to Kelly (1973), maximum ductility can be achieved when the bond strength of the fibres is equal to τc, which is the critical bond value. If the bond strength is less than this value the fibres will pull out from the concrete matrix, if it is larger the fibres will rupture and the ductility will decrease (Figure 1).

The most common way to examine fibre reinforced concrete properties is the three point bending beam test according to the European harmonised standard EN 14651:2005 by measuring the applied load versus the constantly increasing Crack Mouth Opening Displacement (CMOD). The maximum value of the CMOD is 3.5 mm in these tests, i.e. the fibres are not necessarily pulling-out from the concrete matrix and thus their effect on the ductility cannot be fully investigated.

Figure 2 shows the Load-CMOD diagrams of two fibre reinforced concrete specimens out to a displacement of 30 mm. One of the fibres has a high pull-out resistance, which is limited to small crack widths only, while the other one has a lower initial pull-out resistance, but it is not significantly decreasing even at wider cracks. The 3.5 mm CMOD value is marked in the figure, representing the standard limit of the beam test.

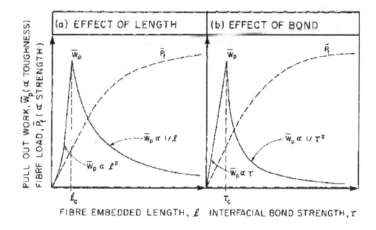

Figure 1. Pull-out work and interfacial bond strength (Kelly, 1973).

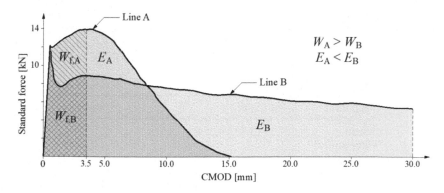

Figure 2. Load – CMOD curves of FRC specimens.

2319

These diagrams represent well the characteristic behaviour of steel fibre and macro synthetic fibre reinforced concrete specimens. While the steel fibres are working well at small crack widths, they lose efficiency dramatically with increasing crack width (line A). On the contrary, macro synthetic fibres exhibit constant pull-out resistance, even at much wider cracks, due to the different energy absorption characteristic (line B) (Juhasz, 2013; Bernard, 2009). This difference can be clearly seen in those tests where the maximum crack width or deflection values are larger than the mentioned 3.5 mm, for example in case of square panel tests according to EN 14488-5:2006. In such a testing configuration, the difference is much larger and in addition, the absorbed energy characteristic will change, i.e. the energy absorption capacity of the macro synthetic fibres will be greater than the capacity provided by steel fibres.

The area under the Load-CMOD diagram of a beam test is referred to as "fibre-work" (Toth, Juhasz, Pluzsik, 2017), where the area under the Load-Displacement curve of a square panel test is referred to as "energy absorption" (see Figure 5). These two parameters represent very well the ductility and the post crack performance of fibre reinforced concrete specimens.

The interaction between the fibres and the concrete matrix is very important, but due to ageing of the surrounding concrete several chemical-physical effects can occur, which have an influence on the bond performance. In "figure 1" it can be seen that the bond strength is a highly sensitive element in the performance of fibre reinforced concrete, whether increasing or decreasing, it can cause a significant change in the ductility of the composite.

3 EXPERIMENTAL INVESTIGATIONS

Steel and macro synthetic fibre reinforced concrete beam and square panel specimens were cast and tested at the Czakó Adolf Laboratory of the Budapest University of Technology and Economics. The objective of this research was to investigate the performance of different fibre reinforced concrete specimen types with increasing age. Specimens were tested at 28 days, 90 days and one year of age.

The concrete mix design was identical in all cases, which was a typical shotcrete mix according to Table 1

To investigate the performance of different fibres one steel fibre and three types of synthetic fibres were used in the test with different dosages. The types of the fibres and their main parameters can be seen in Table 2.

In all cases three square panels and seven beams were cast. To measure the compressive strength of the concrete with every fibre dosage three cubes were also cast. The compressive strength of the concrete was C50/60 according to Eurocode 2. The concrete was mixed with a Collomix –XM2 – 650 professional mixer. The slump of the concrete was 210 mm and the air content was 6% (Figure 3).

In all cases the fibre mixing was good with the stated dosages, the fresh fibre reinforced concrete was homogeneous and well compacted. The square panels were poured into the formwork without any vibration, while the beams were compacted on a high frequency shaking table. All specimens were stored underwater at room temperature until tested.

Table 1.   Concrete mix design.

Component	Type	[kg/m3]
Cement	CEM I 42.5 R	480
Water content		216
Aggregate 0/8		1620
Superplasticizer	Mapei Dynamon SXN	4
Fibres	Macro synthetic, Steel	6 & 8, 55
w/c	0.45	

Table 2.   Fibre types and properties.

Reference	Fibre name	Fibre type	Fibre dosage	Fibre length	surface/end
BC48-6kg	BarChip48	Macro synthetic	6 kg/m3	48 mm	embossed
BC48-8kg	BarChip48	Macro synthetic	8 kg/m3	48 mm	embossed
BC54-6kg	BarChip54	Macro synthetic	6 kg/m3	54 mm	embossed
BC54-8kg	BarChip54	Macro synthetic	8 kg/m3	54 mm	embossed
DT57-6kg	DucTil57	Macro synthetic	6 kg/m3	57 mm	embossed
DT57-8kg	DucTil57	Macro synthetic	8 kg/m3	57 mm	embossed
SF35-55kg	SF35	Steel	55 kg/m3	35 mm	hooked

Figure 3.   Air content and slump test,

Table 3.   Test matrix.

Reference	28 days		90 days		1 year	
	Beams	Panels	Beams	Panels	Beams	Panels
BC48-6kg	7	3	-	-	-	-
BC48-8kg	7	3	7	-	7	-
BC54-6kg	7	3	-	-	-	-
BC54-8kg	7	3	7	3	7	3
DT57-6kg	7	3	-	-	-	-
DT57-8kg	7	3	-	3	7	3
SF35-55kg	7	3	7	3	7	3

The full test matrix with the testing ages can be seen in Table 3.

The beam tests were conducted according to EN 14651:2005 and the panel tests according to EN 14488-5:2006. In both cases the testing machine was a deflection controlled universal testing machine type Zwick Z150.

In case of the beam tests the speed of the crack mouth opening displacement was 0.05 mm/min until CMOD = 0.1 mm, after that the speed was 0.2 mm/min. In case of the square panel tests the speed of the centre displacement was 1 mm/min. The Load-Displacement curve was recorded and the test was continued until a deflection of at least 30 mm was achieved at the centre point of the slab. The supports and loading devices of the beams and panels can be seen in Figure 4.

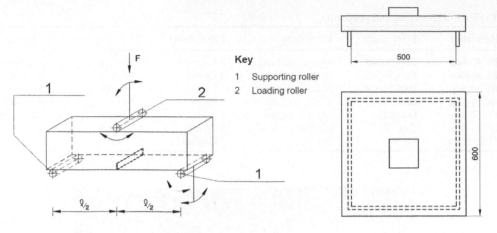

Key
1   Supporting roller
2   Loading roller

Figure 4.   Test supports and arrangement.

4   TEST RESULTS

The "fibre work" of the beams and the "energy absorption" of the square panels can be seen in Figures 5a and 5b.

In the case of steel fibres the ductility decreased continuously with increasing testing age. According to Kelly (1973) (Figure 1) the bond strength is unknown but the decreasing ductility can be due to two reasons: the bond strength decreased and thus the pull-out resistance also decreased or the bond strength exceeds the critical value and the energy absorption of the

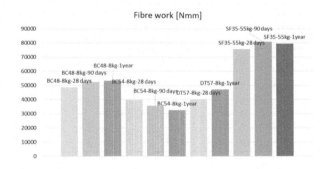

Figure 5a.   Fibre work of the beam specimens.

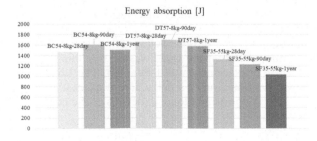

Figure 5b.   Energy absorption of the square elements.

2322

fibre changed or the fibres rupture instead of pulling-out. On checking the cracked surfaces, it was found that none of the fibres were ruptured. The loss of ductility is a well-researched topic in case of steel fibre reinforced concrete (Bernard) who relates this embrittlement effect with the rupture of the fibres due to the increased chemical-physical bond. According to the results it can be seen that the decrease in the bond strength can also decrease the ductility of the fibre reinforced concrete specimens.

In case of synthetic fibre reinforced specimens, the ductility increased or remained constant over the test period. According to Kelly this can be because of the increasing or the decreasing of the bond strength. The number of the ruptured fibres on the cracked cross section was not higher after one year than after 28 days which means that the bond strength increased during the ageing process but it did not reach the critical value.

From the beam test results, it can be seen that the beam test does not represent very well the real fibre performance during the ageing process due to the low level of the deformation and crack widths. Thus, it is questionable whether testing of beams at 28 days of age alone yields sufficient information with regard to the design values obtained hereof and with regard to the design life of the structure.

5 CONCLUSION

By adding fibres to plain concrete, the ductility and the post crack performance increases. The bond interaction between the concrete and the fibres is a very important parameter in the case of ductility but it highly depends on the age of the material. With an optimal anchorage length the bond strength can change during the ageing process which can lead to both an increase and a decrease in ductility.

Steel and synthetic fibre reinforced concrete beams and square panel tests were carried out at different ages. While the performance of steel fibre was better in case of beam tests, the energy absorption was superior with synthetic fibres. During the test period the ductility of steel fibres decreased while most of the synthetic fibres' ductility increased or remained constant. On the cracked surface of the steel fibre reinforced concrete specimens none of the fibres were ruptured, which means that the bond strength decreased during the test period. In the case of synthetic fibres, the number of the ruptured fibres was not larger over the test period, the bond strength in those cases increased (Figure 6).

Panel tests better represent the ductility of fibre reinforced concrete specimens, the beam test does not represent the real fibre concrete composite performance during the ageing process due to the low level of displacement and crack width.

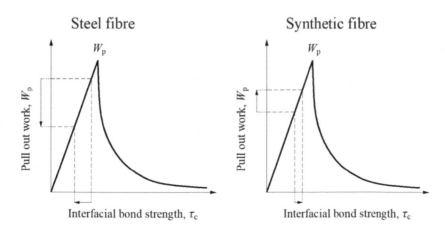

Figure 6.   Interfacial bond during ages.

REFERENCES

Bartos, P. (1981). Review paper: bond in fibre reinforced cements and concretes. *International Journal of Cement Composites & Lightweight Concrete* 3, pp. 159–177.

Bartos, P. (1980). Analysis of pull-out tests on fibres embedded in brittle matrices. *Journal of Material Science* 15, pp. 3122–3128.

Bernard, E. S. (2008). Embrittlement of fiber-reinforced shotcrete. *Shotcrete*, 10 (3), pp. 16–21.

Bernard, E. S. (2009). Design of fibre reinforced shotcrete linings with macro-synthetic fibres. In: *Shotcrete for Underground Support XI, ECI Symposium Series*, P11. (Accessed 15 June, 2013) Retrieved from: http://dc.engconfintl.org/shotcrete/14

Juhász, K. P. (2013). Modified fracture energy method for fibre reinforced concrete. In: A. Kohoutková et al., (eds.) *Proceedings of Fibre Concrete 2013: Technology, Design, Application*. Prague: Czech Republic, pp. 89–90.

Juhász, K. P. (2014). Szintetikus makro szálerősítésű betonok (Synthetic macro fibre reinforced concrete, in Hungarian). Postgraduate Degree in Concrete Technology, master thesis. Budapest University of Technology and Economics, Budapest.

Kelly, A. (1973). Strong Solids. Oxford University Press: Oxford.

Tóth, M., Juhász, K. P., Pluzsik, A. (2017). Effect of mixed fibers on the ductility of concrete. *Journal of Materials in Civil Engineering*, Vol. 29, No. 9.

Zollo, R.F. (1997). Fibre-reinforced concrete: an overview after 30 years of development. *Cement Concrete Composites*, 19 (2), pp. 107–122.

Tunnels and Underground Cities: Engineering and Innovation meet Archaeology, Architecture and Art, Volume 5: Innovation in underground engineering, materials and equipment - Part 1 – Peila, Viggiani & Celestino (Eds)
© 2020 Taylor & Francis Group, London, ISBN 978-0-367-46870-5

Study of the image photographing of the tunnel lining as an alternative method to proximity visual inspection

S. Kaise, K. Maegawa & T. Ito
NEXCO Research Institute Japan, Machida-city, Japan.

H. Yagi
Central Nippon Expressway Co., Ltd., Nagoya-city, Japan.

Y. Shigeta & K. Maeda
Pacific Consultants Co., Ltd., Chiyoda-ku, Japan.

M. Shinji
Yamaguchi University, Ube-city, Japan

ABSTRACT: The detailed lining inspection system of Nippon Expressway Companies (collectively referred to as "NEXCO") includes extraction of priority inspection points for each span according to records of lining development image photographing, assessment of health of the lining, and subsequent on-site proximity visual inspection of all lines. In this study, for a purpose of higher accuracy and advanced application of the lining inspections, we conducted trial tests to examine potentials of the lining image photographing record, which has been employed for assessment of health of the linings as an alternative method for visual inspection. As a result, we found that the desktop inspection with the lining development image photographing can be employed in place of visual inspection.

1 INTRODUCTION

As shown in Figure 1, East Nippon Expressway Company Limited, Central Nippon Expressway Company Limited, West Nippon Expressway Company Limited (hereinafter collectively referred to

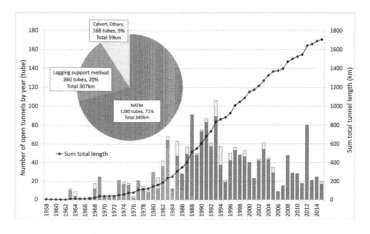

Figure 1. Number of Tunnels under NEXCO Management (as of March 2016)

as "NEXCO") administer 1,808 tubes of expressway road tunnels, which extend in total length of 1,705 km. This time, for a purpose of uniformly and comprehensively maintaining the tunnel assets with high accuracy, we studied applicability of the lining image photographing, which has been employed for assessment of lining soundness, etc., as an alternative method to visual inspection. In addition, we also conducted a trial test regarding inspection of an actual tunnel and verified applicability of the method as an alternative from comparison and analysis of those inspection results.

2 CONVENTIONAL TUNNEL LINING INSPECTION METHOD

As illustrated in Figure 2, the tunnel inspection procedure of NEXCO, which is an operator of the expressways in Japan, are based on lining image photographing record by a lining surface image acquisition system using lasers, CCD cameras, line sensor cameras, etc. According to the NEXCO inspection procedure, detailed tunnel inspection consists of desktop inspection

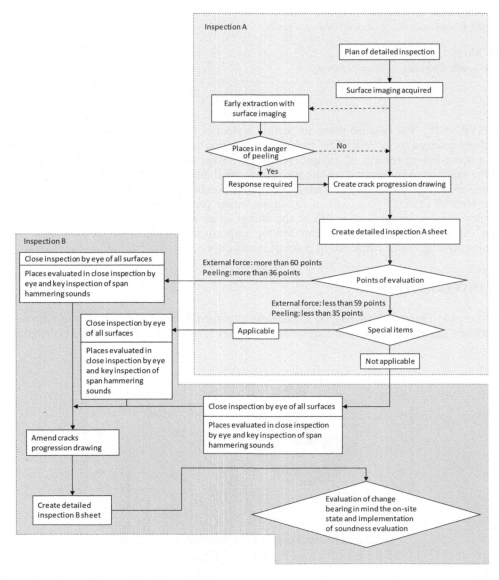

Figure 2. Flow of Tunnel Detail Inspection

2326

(Inspection A) and onsite inspection (Inspection B). In Inspection A, points and spans, which are considered as problem, are extracted from the lining image photographing record as priority inspection points, using thresholds and special notes shown in Figure 2. Thereafter, in Inspection B, based on the determination results in Inspection A, the onsite inspection is conducted at those priority inspection points such as proximity visual inspection and hammer inspection.

3 APPLICABILITY OF PROXIMITY VISUAL INSPECTION BY LINING IMAGE PHOTOGRAPHING AS AN ALTERNATIVE METHOD

3.1 *Purpose of trial test*

In order to study applicability of a lining development by mobile image photographing method as an alternative to visual inspection, we conducted a trial test on an actual tunnel. Figure 3 shows concept of accuracy verification in this test.

3.2 *Purpose of trial test*

Figure 4 shows a flow of the trial test procedure. In the trial test, we conducted three types of inspections, "inspection with a lining development by mobile image photographing", "regular proximity visual inspection", and "detailed proximity visual inspection". Thereafter, we compared and verified accuracy of the defect developments prepared in the respective inspections.

3.3 *Inspections conducted in the trial test*

Details of the three types of inspections conducted in the trial test are as follows:

3.3.1 *Inspection using a lining development prepared by mobile imaging technology*
In the inspection with a lining development by mobile image photographing, a lining development by mobile image photographing was used. In this technology, images for the lining development image photographing are taken from a vehicle, which meets the current NEXCO specifications and does not require restriction of the traffic. In this test, we conducted the inspection using MIMM-R shown in Figure 5, which can perform lining image photographing with precision of 1.5 to 2.0 mm/pixel while driving at 70 to 80 km/hour. Using the obtained lining development images, we detected defects such cracks, blowups/peels, and water seepage and make defect development.

Here, we call "image development(s)" the defect developments prepared from the lining development images, which were taken by the mobile lining development image measurement.

3.3.2 *Regular Proximity Visual Inspection*
"Regular proximity visual inspection" referred herein means proximity visual inspection, which is supposed to typically conduct once every 5 years. With MLIT (the Ministry of Land, Infrastructure, Transport and Tourism) standards for quantity per unit work and past inspection results, we calculated time required per span from the crack density of the tunnel where the test

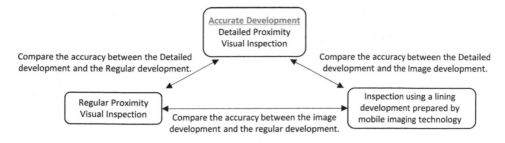

Figure 3. Concept of Accuracy Verification

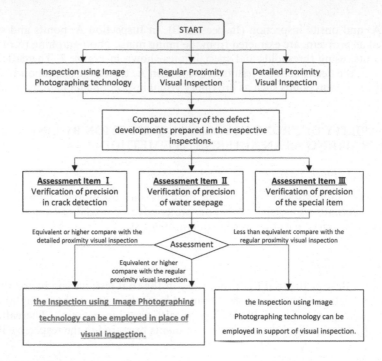

Figure 4.   Trial Test Flow

Figure 5.   Mimm-R

Figure 6.   Regular Proximity Visual Inspection

was conducted. Using the calculation results as reference, we conducted the inspection. Figure 6 shows the implementation status of the regular proximity visual inspection.

Upon test inspection, we did not use inspection data of past years, defect development prepared from the lining images, etc., but an engineer sketched from the beginning to prepare one span of lining defect development. Hereinafter, we refer the defect development prepared from the regular visual inspection as "regular development".

3.3.3 Detailed Proximity Visual Inspection

"Detailed proximity visual inspection" referred herein means detailed inspection, in which a defect development is carefully prepared taking longer inspection time than that in the regular proximity visual inspection.

Upon test inspection, using an image development in which the defect location information is accurately captured, we collected detailed information such as crack widths and lengths, and conducted a full-surface hammer test, etc. so as to prepare more accurate defect development than that obtained by the regular proximity visual inspection. Hereinafter, we refer the defect development prepared through the detailed proximity visual inspection as "detailed development".

In this accuracy verification, we used the detailed development as an accurate development, and compared and examined the accuracy between the image development and the regular development.

Table 1 shows differences of the inspection standards among the three types of inspections. Regular development Image development, Detailed development

3.4 Results of trial proximity visual inspection

3.4.1 Summary of trial test
- Summary of Trial Test Tunnels

A Tunnel, cross-section: two lanes, one direction, Extension: L=3,069 m

Construction Method: NATM, Year/Month of Use: Since March 1984, Test span: three spans (7, 60, and 67)
- Conditions for trial inspection

To have the same inspection conditions, we split the span into two, slow lane and fast lane (passing lane), and conducted inspection the respective lanes in the three spans on the same day.
- Inspector team and duration of inspection

As shown in Table 2, the inspection team consists of four inspectors, which include three for inspection of arches and one for a driver of a high-place working vehicle.

In addition, we compared the time required for the inspection with reference to MLIT's Cost Estimate Standards for Regular Inspections of Road Tunnels (provisional version). The results are shown in Table 3. According to the results, it can be understood that the regular proximity visual inspection can be done in an amount of time generally similar to that calculated from the MLIT standards for labor per unit work.

Table 1.　Differences of Inspection Standard

Inspection types (Development)	Inspection by photogrammetry using a driving vehicle (Image development)	Regular proximity visual inspection (Regular development)	Detailed proximity visual inspection (Detailed development)
Crack width reading precision	• 0.3mm • 0.5mm • 1.0mm • 2.0mm • 3.0mm • 3.0mm or greater	• 0.3mm • 0.5mm • 1.0mm • 2.0mm • 3.0mm • 3.0mm or greater	
Crack width measurement	Includes change[*1]	Max. width	Includes change[*1]
Sketching accuracy	From photographed images	Manual	Based on image development.
Hammering test	Not able to conduct.	• Along traverse joint horizontal joint • along cracks • Priority inspection points	Whole lining surfaces
Defect photo	Whole lining surfaces	Points with abnormalities	Points with abnormalities
Time restriction	None	100 m/day	30 m/day

*1) Crack Measurement

Regular development　　　　　　　　　　　　Image development, Detailed development

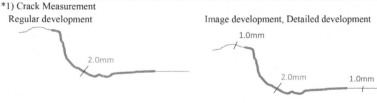

Table 2. Inspection Team

Inspectors	Job Type	Age	Experience	Remarks
Inspector 1	Engineer B	55	8	Arch inspector
Inspector 2	Engineer C	26	7	Arch inspector
Inspector 3	Technician	33	4	Arch inspector
Inspector 4	Technician	65	2	High-altitude working vehicle driver

Table 3. Comparison of Inspection Time

	Regular Proximity Visual Inspection			Detailed Proximity Visual Inspection		
	Slow lane	Fast lane	Total	Slow lane	Fast lane	Total
MLIT's standards for labor per unit work (0.2 < C ≦ 0.3)	13 min.	13 min.	26 min.	—	—	—
Span 07 (C = 0.262 m/m²)	8 min.	8 min.	16 min.	40 min.	35 min.	75 min.
Span 60 (C = 0.262 m/m²)	15 min.	10 min.	25 min.	30 min.	30 min.	60 min.
Span 67 (C = 0.246 m/m²)	15 min.	10 min.	25 min.	25 min.	25 min.	50 min.

Legend				
	Crack width 0.3mm or less (≦0.3mm)	2.0	Crack width about 2.0mm (1.7mm<t≦2.2mm)
0.5	Crack width about 0.5mm (0.3mm<t≦0.7mm)	2.5	Crack width about 2.5mm (2.2mm<t≦2.7mm)
1.0	Crack width about 1.0mm (0.7mm<t≦1.2mm)	3.0	Crack width about 3.0mm (2.7mm<t≦3.2mm)
1.5	Crack width about 1.5mm (1.2mm<t≦1.7mm)	3.5	Crack width about 3.5mm or more (3.2mm<t)

Figure 7. Legends in the defect development

3.4.2 *Investigation Results*

Using the legends indicated in the development in Figure 7 and lining images obtained in the trial test in Figure 8 as the results of this test inspection, we examined the results. Among the cracks, blowups/peels, etc. of the Spans of the inspection, we specifically show comparison defect developments of cracks among the image developments, regular developments, and detailed developments. In addition, we show the detection comparison results of the respective assessment items in Tables 4, 5 and 6.

3.5 *Verification of precision in crack detection (Assessment Item I)*

Upon verification of precision in each inspection, we set the following conditions for comparison and verification of crack widths:

1) Correct values of widths and lengths of the cracks are determined in the detailed development.

2) The widths and lengths of cracks in the range where the cracks occurred are calculated in the image development and the regular development.

3) The crack widths in the image development and the regular development are those that account for the greatest part of the crack lengths. For example, of the crack length of 5 m, if length of 0.5 mm crack is 1 m, the one of 1.0 mm is 3 m, and the one of 1.5 mm is 1 m, the crack width shall be determined as 1.0 mm.

4) In the assessment, the crack widths are assumed to be the same value and are to be assessed separately from the crack lengths.

2330

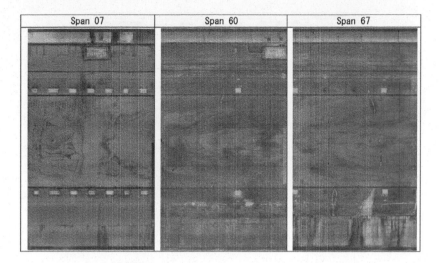

| Span 07 | Span 60 | Span 67 |

Figure 8.   Lining images used in the verification

Table 4.   Detection comparison results related to crack detection

Crack		Image development			Detailed development			Regular development		
		detection	width concordance	length concordance (m)	detection	width concordance	length concordance (m)	detection	width concordance	length concordance (m)
Span 7	Number of detections	35	29	34.98	43	43	56.78	30	30	29.7
	Concordance rate	81.4%	67.4%	61.6%	100.0%	100.0%	100.0%	69.8%	69.8%	52.3%
Span 60	Number of detections	21	14	26.1	26	26	44.3	18	11	36.1
	Concordance rate	80.8%	53.8%	58.9%	100.0%	100.0%	100.0%	69.2%	42.3%	81.5%
Span 67	Number of detections	24	15	29.08	26	26	49.34	22	11	27.6
	Concordance rate	92.3%	57.7%	58.9%	100.0%	100.0%	100.0%	84.6%	42.3%	55.9%
Total	Number of detections	80	58	90.16	95	95	150.42	70	52	93.4
	Concordance rate	84.2%	61.1%	59.9%	100.0%	100.0%	100.0%	73.7%	54.7%	62.1%

3.6   *Results of verification of precision in crack detection*

Upon verification of precision in the crack detection, we verified three items, the number, widths, and lengths of cracks. Upon the verification, the crack widths to be used in calculation of the crack detection rates included those of 0.3 mm or less for the comparison and verification, including the MLIT's Regular Inspection Manual. For the verification data, using the detailed inspection data as correct values, and the regular developments that were under consideration of replacement with the driving-type lining development image technology as standards, we calculated the detection rates of the number of cracks and the concordance rate in the crack widths and crack lengths between the image development and detailed development.

Table 4 and Figure 16 show the comparison and verification results of crack detection between the image development and the detailed development when the precision of the detailed development was assumed as 100%. Hereunder, we will describe the respective assessment items.

1) Assessment of crack detection rate

The crack detection rate in the image development was 84.2 %, which was lower than that of the detailed development, but it can be understood that it detected at higher precision than the regular development.

2) Assessment of the crack width concordance rate

The crack width concordance rate was 61.1 % in the image development, which was lower than that of the detailed development, but it can be understood that it detected at higher precision than the regular development.

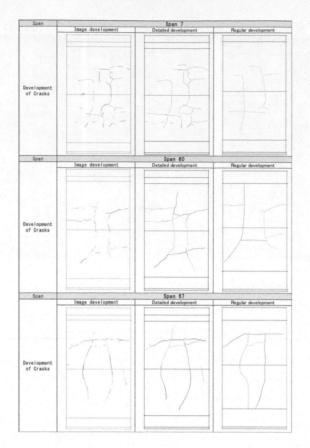

Figure 9. Developments of Cracks

3) Crack length concordance rate

The crack length concordance rate was 59.9 % in the image development, which was lower than the detailed development, and it was detected at slightly lower precision than the regular development. This may be since one crack was taken as a plurality of cracks upon preparation of the image development and lengths between cracks were not detected. In the regular inspection, when an inspector thought that cracks were present at several points of one line based on his/her experience, these cracks tended to be shortened. Therefore, the result seems to reflect that tendency of the procedure.

Furthermore, while the positions of cracks are generally identical between the image development and the detailed development, the crack positions in the regular development are quite different from those two developments. The possible reasons for this could be that it is difficult to understand the positions of the cracks in the regular development, which does not use the image development, and those positions can be accurately obtained in the image development and the detailed development. This is considered very important upon assessment of progression of cracks, which is a representative defect.

3.7   *Water seepages (Assessment Item II), Special Items (Assessment Item III)*

Tables 5 and 6 show detection precision of water seepage and leached matters and that of blowups/peel-offs and repair works, which are not directly related to crack assessment among the special item(s), respectively. The detection of those items in the image development was generally similar to that in the detailed development, and it can be understood that it can detect at higher precision than that in the regular development.

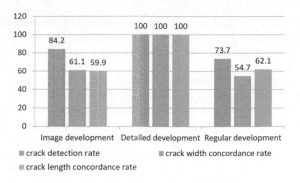

Figure 10. Verification and comparison of precision of crack detection with the regular development

Table 5. Comparison of detection of water seepage and leached matters

detection of water seepage and leached matters		Image development		Detailed development		Regular development	
		All detection	Area (m²)	All detection	Area (m²)	All detection	Area (m²)
Span 7	Number of detections	5	1.4399	7	2.0773	3	2.9535
	Concordance rate	71.4%	69.3%	100.0%	100.0%	42.9%	142.2%
Span 60	Number of detections	4	1.3139	4	1.3139	1	0.0352
	Concordance rate	100.0%	100.0%	100.0%	100.0%	25.0%	2.7%
Span 67	Number of detections	4	3.0729	4	3.0729	1	3.4035
	Concordance rate	100.0%	100.0%	100.0%	100.0%	25.0%	110.8%
Total	Number of detections	13	5.8267	15	6.4641	5	6.3922
	Concordance rate	86.7%	90.1%	100.0%	100.0%	33.3%	98.9%

Table 6. Comparison of detection of blowups, peel-offs, and repair works.

detection of blowups, peel-offs, and repair works.		Image development		Detailed development		Regular development	
		All detection	blowups, peel-offs	All detection	blowups, peel-offs	All detection	blowups, peel-offs
Span 7	Number of detections	4	1	5	2	3	1
	Concordance rate	80.0%	50.0%	100.0%	100.0%	60.0%	50.0%
Span 60	Number of detections	3	0	2	0	1	0
	Concordance rate	150.0%	–	100.0%	–	50.0%	–
Span 67	Number of detections	5	1	5	1	1	0
	Concordance rate	100.0%	–	100.0%	–	20.0%	–
Total	Number of detections	12	2	12	3	5	1
	Concordance rate	100.0%	66.7%	100.0%	100.0%	41.7%	33.3%

However, since the blowups in the Span 07 were internal defects that accompanied by defects such as cracks, it was difficult to distinguish from other points that had no problem from the photographed image of the lining surface, and it was impossible to detect from the image photographing development, in which hammer inspection was not performed to find internal defects. This indicates that the image photographing cannot be a complete alternative and still requires internal defect detection such as hammer inspection.

4 CONCLUSION

Using trial test data of expressway tunnel, we conducted analysis and assessments, and verified validity of introduction of lining image development imaging technology by mobile image photographing as an alternative of proximity visual inspection for tunnel lining inspection method done currently by NEXCO. In the trial test, we conducted three types of inspection accuracy verifications, and verified accuracy of the inspection by mobile image photographing (image development), regular proximity visual inspection (regular development), and detailed proximity visual inspection (detailed development) which takes longer time than the regular proximity visual inspection

In the respective items, i.e., the number of cracks, crack widths, and safety assessment of crack width, detection rate of cracks of 0.5 mm or greater and reproducibility of the crack positions, the detection rate and concordance rate of the respective items were greater in the order of detailed development > image development > regular development. Among the six assessment items, the crack lengths were greater in the order of detail development > regular development > image development, but as we mentioned in 3.6–3) concordance rate of crack length, it seemed to be influence from that the cracks were shortened when an inspector determined as several cracks are on one line based on his/her experience.

In future, the inspection by mobile lining image photographing technology will be done second. More specifically, based on the development made by Inspection B, in which onsite proximity visual inspection and hammer inspection are done first, Inspection A by image photographing will be performed. Therefore, we believe there would be no issue in detection of crack lengths in view of those inspection status, or it will improve. However, we need to consider upon use, such as dividing the contents of the detailed inspection by the number of implementations, including collecting data of internal defects such as blowups, as will be described below.

From comprehensive examination of those assessment items, the accuracy of the inspection itself by the mobile lining development imaging technology was lower than that of the detailed proximity visual inspection. However, this could be because the lining surface of A tunnel was darkened with soot and also because there was a lot of free lime along cracks.

When we compare with the regular proximity visual inspection, the detection precision of the inspection by the mobile lining development imaging technology was considered to be equivalent or higher, and it was also revealed that the reproducibility of the crack positions was also very high. Therefore, with the method, we can expect advancement of inspection in quantification of health assessment of the tunnel lining, progression of cracks, etc.

Accordingly, it can be considered to be reasonable to employ inspection by the mobile lining development imaging technology as an alternative of the regular proximity visual inspection. Moreover, with those inspection methods, we can expect about 20% reduction in the days for inspection and about 6 % reduction in inspection expense in comparison with the conventional method (in which only Inspection B is conducted), according to estimate assuming extraction of points to conduct Inspection B by Inspection A with the mobile lining development image photographing.

In addition, it was found that the blowups (non-clear sound in hammer test), which are internal defects that do not accompany defects, cannot be detected by the image photographing of the lining surface, and the image photographing cannot substitute direct hammering inspection, and we still need to conduct hammer inspection. In future, in order to find out inner defects, we conduct "full-surface proximity visual inspection + full-surface hammer inspection" as well as image photographing in the first detailed inspection, so as to find internal defects such as blowups, which occur at points where there was no defects that cannot be found in the images. By doing this, in the second or later detailed inspection, we can employ the image photographing for proximity visual inspection as an alternative. In addition, for hammer inspection, instead of the full-surface hammer inspection, we can make changes the contents of the detailed inspection depending on the number of past implementations, such as changing to check whether there is any progress in the defect by hammering the points of inner defects found in the first detailed inspection. As a result, we can limit the frequency of the hammer inspection.

In future, for a purpose of limiting hammer inspection, we will analyze inspection data, and verify the points where blowups occur and their tendency to limit the necessity of the hammer inspection.

Tunnels and Underground Cities: Engineering and Innovation meet Archaeology, Architecture and Art, Volume 5: Innovation in underground engineering, materials and equipment - Part 1 – Peila, Viggiani & Celestino (Eds)
© 2020 Taylor & Francis Group, London, ISBN 978-0-367-46870-5

Conventional tunneling method using ICT

K. Kakimi & F. Kusumoto
Shimizu Corporation, Tokyo, Japan

R. Okawa & K. Yamabe
Central Nippon Expressway Company Limited, Nagoya, Japan

ABSTRACT: The Takatoriyama Tunnel in Japan is a twin tunnel approximately 3.9 km long with two lanes in each direction. The boring system is a full-face excavation with the use of a computer-controlled drilling jumbo, blasting and an automated shotcreting machine. The geometry for the surface bored through blasting without support and the geometry of the surface of the shotcreted side wall are measured from the cutting face with a laser scanner. The blasting pattern will be corrected with reference to the difference from the planned excavation face, and the difference determined from the planned shotcrete surface, then shotcrete will be placed on overbreaks to smooth the concrete surface. The face will be a hemisphere, an excellent face shape, and efficient in forming a sturdy ground arch. For weak ground, the support structure will be ring shaped. By closing the full face earlier, the project will establish mechanical stability.

1   INTRODUCTION

This project aims to design an arch support structure with consideration for elastic ground so that a tunnel may be independently stabilized with a ground arch formation. The construction method we will use is a blasting boring of full-face excavation with a spherical face, which is known to be excellent in handling at the cutting face as well as in forming a ground arch. However, for low-strength plastic ground, we designed a ring support structure, using early closure of full face by adopting a curved face with stable geometry to ensure mechanical stability and bore the tunnel reliably. A laser scanner will be used to measure the geometries of the bored surface and the shotcrete surface. By using these data, we can detect excessive overbreaks, correct the blasting pattern with reference to the deviation from the planned excavation face as well as the computer jumbo drilling angle. Shotcrete will be placed using a sprayer over overbreaks, which are identified as the differences from the planned shotcrete surface, and the shotcrete surface will be finished smooth to increase the potential for the tunnel to be independently stabilized.

This paper reports a mountain tunnel Information and Communication Technology construction able to cope with all kinds of geology, which was adopted for the Takatoriyama Tunnel.

2   OVERVIEW OF TUNNEL CONSTRUCTION

The Shin Tomei Expressway is a national expressway from Ebina City, Kanagawa Prefecture to Toyoda City, Aichi Prefecture, passing through Shizuoka City. This expressway route is expected to serve as a main traffic artery linking Tokyo, Nagoya and Osaka. The Takatoriyama tunnel, located in Kanagawa Prefecture, has twin tunnels of about 3.9 kilometers in total length with two lanes in each direction (Figure 1). The inner cross section of the tunnel is around 62 m^2. Construction of the tunnel is divided into two segments: the west segment is

Figure 1. Location of the Takatoriyama tunnel.

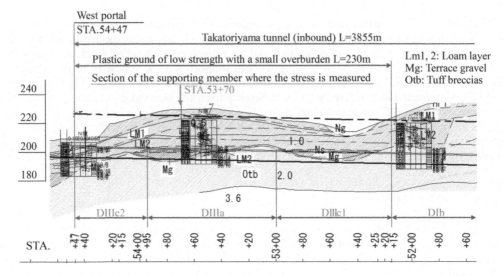

Figure 2. Geological longitudinal profile (west portal of the Takatoriyama tunnel, inbound).

composed of two lanes, inbound 1,573 meters and outbound 1,609 meters, and the tunnel is bored in downgrade

with a longitudinal gradient of 2%.

The geology encompasses spreads of diorite, porphyrite, etc. The maximum overburden is 390 meters in depth. The 230 meters segment of the west portal is geologically composed of terrace gravel deposits in an unconsolidated aquifer and loam layer. The overburden is in a range of 16 to 31 meters and the competence factor is less than 2. From these facts, the site is rated as plastic ground of low strength with a small overburden (Figure 2). For ground of low strength, we designed a ring support structure which is made mainly of shotcrete and steel supports and will construct the tunnel with an early closure of full face.

3 SUMMARY OF THE TUNNEL STRUCTURE

As a tunnel structure in elastic ground, the standard support pattern in elastic ground is typically designed according to the ground class, basically resorting to a ground arch which is formed in the neighboring ground and with a combination of shotcrete and members of steel arch supports (Table 1, Figure 3).

Table 1. Specifications of the tunnel.

Tunnel support structure		Arch structure (Standard pattern)			Ring structure (Early closure pattern)	
Support pattern		CIIb	DIb	DIIIa	DIIIc1	DIIIc2
Competence factor Cf (-)		$2 \leqq$	$2 \leqq$	< 2	$0.5 \sim 2$	$0.5 \sim 2$
Cycle length (m)		1.2	1.0	1.0	1.0	1.0
Deformation allowance (cm)		0	0	0	0	0
Tunnel support structure	Thickness of shotcrete	7cm	10cm	20cm	20cm	20cm
	Compressive strength (28 days)	$36N/mm^2$	$36N/mm^2$	$36N/mm^2$	$36N/mm^2$	$36N/mm^2$
	Arch steel supports	HH-100	HH-100	HH-154	HH-154	HH-154
	Rock bolts	L=3m,290kN		L=4m,170kN	L=4m,170kN	
Early closure structure	Thickness of shotcrete	—	—	—	20cm	20cm
	Ratio of structure radius (r3/r1)	—	—	—	2.69	2.67
	Early closure distance Lf (m)	—	—	—	5	8
Lining thickness (cm)	Arch, side walls	30	30	35	35	40
	Invert	—	45	50	50	50

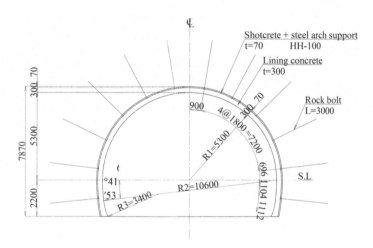

Figure 3. Arch structure CIIb.

In plastic ground of low strength, based upon DIIIa as the standard support pattern, we designed the strategy of an early closure pattern (Figure 4) of DIIIc1 which forms a ring structure by shotcreting the ground side of the invert concrete. For the section of the road on the surface, we use DIIIc2 pattern considering the lining thickness.

4 CONSTRUCTION IN PLASTIC GROUND

4.1 Measurement system of as-built profiles

The digital measurement system of as-built profiles consists of a total station, a Laser scanner and a PC system (Figure 5). The data of point groups about the cavity surface, which is

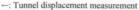

←: Tunnel displacement measurement
○: Measurement of stress of shotcrete, measurement of stress of steel support
△: Measurement of axial force for rock bolts (0.5, 1.0, 1.5, 2.5, 3.5 m)

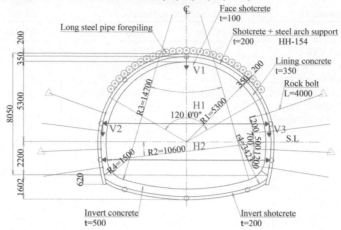

Figure 4.   Early closure DIIIc1 pattern and measurement points.

obtained by the Laser scanner, is converted and plotted in a special coordinate system by the use of the 3D coordinates of the two points (A, B) which are provided on the cavity surface. Table 2 shows the performance of the Laser scanner.

4.2   *Evaluation of as-built profiles*

During the construction cycle, digital measurement is made once before and after shotcreting of the cutting face, twice in total, in which the following three elements, (1) geometry of the cutting face, (2) geometry of the tunnel boring face and (3) geometry of the shotcrete surface, are measured as point group data, and converted to 3 D coordinates for digitalization. The time needed for each measurement is 3 to 5 minutes.

Using this point group data, cutting face shotcrete thickness (4) are measured from the difference between cutting face geometry and shotcrete surface geometry. An excessive overbreak region (5) is determined from the difference between design overbreak and shotcrete surface data. The shotcrete thickness (6) is determined from the difference between the tunnel boring surface geometry and surface data on shotcrete of the previous cycle length (Tables 3-4).

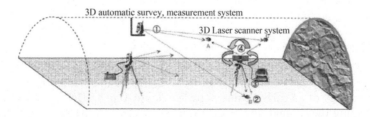

Figure 5.   Summary of digitally measuring as-built profiles.

Table 2.   Performance of Laser scanner (ScanStation P40).

Manufacturer	Leica
Scanning speed	1,000,000 points per second, max.
Laser class	Class 1
Measurement accuracy (after numerical treatment)	+/-2mm (+/-5mm)

Table 3. Measurement results and measures to be taken for the next boring step (tunnel boring).

Measurement timing	Cutting face	Tunnel boring face
Before shotcreting	(1) Cutting face geometry: Large difference from the goal geometry → Correct the blasting pattern	(2) Tunnel boring surface geometry: Excessive overbreak → Correct the blasting pattern
After shotcreting	(4) Cutting face shotcrete thickness, graduation display: Less than 5 cm than the reference value → Additional shotcrete	—

Table 4. Measurement results and measures to be taken for the next boring step (shotcrete).

Measurement timing	Shotcrete surface	Shotcrete surface of the previous cycle length
After shotcreting	(3) Shotcrete surface geometry: Cavity side of the planned shotcrete surface → Scrape off the shotcrete surface placed there	—
	(5) Excessive overbreak regions, graduation display: Large overbreak region exceeding the designed value → Shotcrete the overbreak	(6) Shotcrete thickness, display in graduation: Less than the designed value → Additional shotcrete or rework

Excessive overbeak regions and shotcrete thickness which appear between as-built profiles and design profiles are displayed by means of color toning and graduation on the PC system screen. Based upon this data, the as-built profiles in the construction cycle are confirmed, evaluated and corrected. With an ICT process for confirmation of as-built profiles, rational tunneling can be ensured.

4.3 Tunneling method

The tunneling method in elastic ground used for this project is a blasting boring of full face. Considering the geometry of the cutting face, we use a semi-spherical cutting face which is advantageous in forming an arch, and increases the tunnel quality, safety of cutting face work and tunneling performance (Figure 6) (Kusumoto et al. 2013). In order to create the target shape of a spherical cutting face, we provide a center of a sphere on the plane of the tunnel center SL, and the face excavation length is set at Ls=2L+0.3 m, that is, two times one cycle length L to excavate in the boring direction (Figure 7).

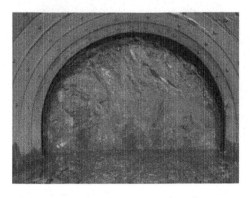

Figure 6. Overview of the spherical face (CIIb, Ls=2L).

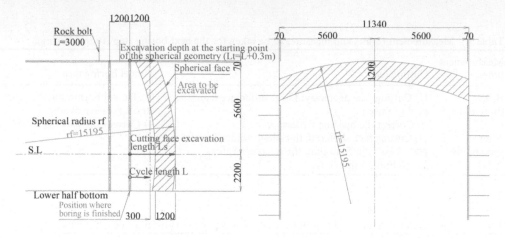

Figure 7. Geometry of the spherical face (CIIb, Ls=2L).

Table 5. Main construction machines (per tunnel).

Type of work	Name	Performance, output	Number
Drilling	Hydraulic computer Jambo	Wheel type, 3 booms, drifter drill of weight 220 kg class	1
Chopping	Large breaker	Hydraulic type 1300 kg class; the base machine is a back hoe of $0.8m^3$ class	1
Muck loader	Wheel loader	Side dumping type, class 3.0 m^3	1
Crasher	Jaw crasher	Capacity 300 tf/h	1
Muck hauling	Continuous belt conveyer	Capacity 150 m^3/h	1
Erection of steel arch support	Elector jambo	Class 1 ton, 2 booms	1
Shotcreting	Shotcreting machine	Wet type, spraying robot, 20 m^3/h	1
	Truck mixer	$4.5m^3$	2
Rock bolt placement	Hydraulic drill jambo	Also used for drilling	(1)

The drilling machine used is a wheel-type computer jumbo with three booms which are able to control penetration angles on the external circumference (Table 5). Based upon the evaluation results of as-built profiles for the cutting face and boring surface, the blasting pattern is evaluated and corrected to control overbreaks (Figure 8).

Figure 8. Cutting face and tunnel boring surface geometry (Table 3, (1), (2)).

Figure 9.   Shotcrete surface geometry (Table 4, (3)).

Figure 10.   Shotcrete thickness (Table 4, (6)).

Figure 11.   Situation of an early closure of the full face (DIIIc1, Ls=6m).

   As a shotcreting machine, we use basically a shotcreting robot which is operated by a worker controlling the nozzle (Table 5). From the evaluation of as-built profiles of shotcrete, shotcrete is placed over excessive outbreaks. When there are portions of shotcrete hanging towards the cavity from the planned shotcrete surface, they are chopped off to ensure a planned as-built profile, and to smoothly finish the shotcrete surface (Figure 9). At the cutting face, the shotcrete that was placed is confirmed to have a planned thickness (Figure 10).

2341

Muck produced in the tunneling route is hauled by a continuous belt conveyer system with a belt width of 610 mm (Table 5). Then, the muck is crushed to a size less than 200 mm, with the use of a self-propelling crusher installed 50–80 meters from the cutting face, then it is loaded on the belt via the tail piece of the continuous belt conveyer before being taken out of the tunnel.

5 CONSTRUCTION IN LOW-STRENGTH PLASTIC GROUND

5.1 *Summary of measurements*

To determine the characteristics of deformation behavior, mechanical stability of the tunnel, tunnel displacement and stress of support members are measured and evaluated. Reference sections for displacement measurement are provided at intervals of 10 meters in the tunnel boring direction. Deformation is measured automatically at intervals of 6 to 12 hours by means of a 3D automatic measurement system (Figure 4). Crown settlement V is set at -50 mm, and convergence H is at -100 mm. The reference section to measure the stress of support members is provided at a point of STA.53+70 under the maximum overburden with a distance of 230 meters.

5.2 *Construction method*

Since the tunnel site is plastic ground of low strength whose competence factor is less than 2, we will use an early closure pattern DIIIc to ensure safety, adapting a curved face with stable geometry (Kimura et al, 2019). The construction unit with early closure is set at Lc=2–3 meters. The distance of early closure is LF=5–8 meters. The unit for early closure is Lc=2 to 5 meters, and the distance of early closure is Lf=5–8 meters. We use the following two different methods alternatively: full cross section excavation with one cycle length = 1 meter, and early closure at each boring unit of 2 to 3 meters (Figure 11). With the target geometry of the cutting face in mind, we use a cutting face with Ls=2L, where the face excavation length Ls is at two times the cycle length. As an auxiliary method to consolidate the rock ground, an injection-type long steel pipe forepiling is used at the same time.

5.3 *Selection of support pattern*

In the project, the unconfined compressive strength qu of the ground was determined by needle penetration test and converted to a competence factor cf. By using cf and tunnel displacement, a boring pattern was selected to determine Lf, early closure distance. At a distance of about 230 meters in the Phase I inbound route, with cf=1–2, there was an aquifer with 70 to 100 liters/min of water inflow. Construction was started with early enclosure pattern DIIIc2. As the ground condition changed, the pattern was changed to DIIIa. Since the displacement of tunnel tended to increase, and cf became nearly 1, the pattern was changed to DIIIc1 (Figure 12).

5.4 *Tunnel deformation characteristics*

In the case of the arch structure DIIIa pattern, the settlement of the crown is -12 mm at maximum. The lower half convergence H2 is -10.7 mm as squeezing. V1 of DIIIc2 pattern for the ring structure with Lf=5 m is -5.5 mm as settlement, and H2 is -2.7 mm as squeezing. In the plastic ground of low strength, the use of an early closure of full face and the shortening of Lf work to control tunnel displacement (Figure 13).

5.5 *Stability of the arch structure tunnel*

With a DIII a pattern of arch structure in plastic ground of low strength whose competence factor is 1.8 to 2.1 on the tunnel lateral side, it is estimated that a ground arch is formed at depths of 1 to 2.5 meters (Figure 14). The axial stress of the shotcrete that has been placed has a compression of 0.9 to 5.8 N/mm^2, less than 1/6 of the standard design strength, continuously

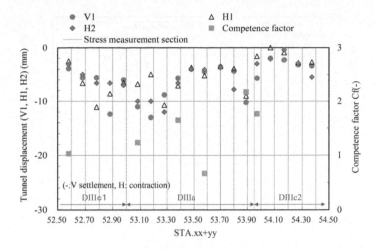

Figure 12.   Support patterns and competence factor.

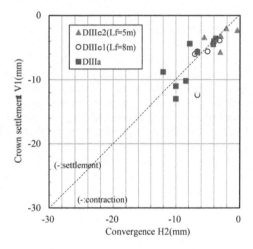

Figure 13.   Lower half convergence H2 and crown settlement V1.

distributed in the axial direction. The axial force of shotcrete against the axial force of support member is 51% to 68%, from which it is proven that the tunnel of arch structure is mechanically stabilized (Figure 15).

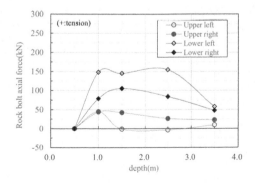

Figure 14.   Distribution of rock bolt axial forces (DIIIa).

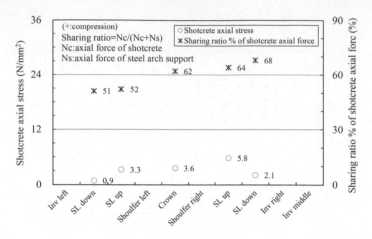

Figure 15.   Shotcrete axial force and axial force sharing (DIIIa).

6   SUMMARY

In elastic ground which is stabilized independently by forming a ground arch, the project utilized an arch structure for support pattern, and used a full face boring method with a spherical cutting face, contributing to boring performance, work safety at cutting face and tunnel quality. In plastic ground with low strength, we adopted an early closure of full face using a curved cutting face and guaranteed the mechanical stability to facilitate tunneling further. At the reference cutting face in the construction cycle, the boring face and the surface of the support structure were measured digitally with a laser scanner to graphically represent the difference between the panned values and the as-built profiles to confirm construction performance. This approach has been proven to be effective and rational for mountain tunnel ITC construction, and sufficiently worthy of being employed in future projects.

As discussed above, both mountain tunnel ICT construction and ICT construction management are demonstrated to have wide applicability, and to be effective and rational in ensuring construction progress, improving tunnel quality and tunnel durability over a long span of time.

7   CONCLUSION

In the Shin Tomei expressway construction project, the segment of 52 km from Gotenba JCT to Atsugi Minami IC is under construction to be opened partially in 2022. For the west segment of the Takatoriyama Tunnel, excavation was started in September 2017 and is currently underway to be completed by August 2020, barring accident or incident. The mountain tunnel ICT construction and ICT construction management which are used for this tunneling project are expected to contribute to development of mountain tunnel construction technology. We will report the details of such efforts on another occasion.

REFERENCES

Kusumoto, F., Tanimura, K. & Sato, J. 2013. The effect of hemi-spherical tunnel face on the stability of mountain tunnels; *World Tunnel Congress 2013 Geneva.*
Kimura, A., Nobunaga, H., Yamanaka, S., Yamamoto, S. & Kusumoto, F. 2019. Enlargement of in-service evacuation tunnel and automated lining placement; *World Tunnel Congress 2019 Naples.*

Tunnels and Underground Cities: Engineering and Innovation meet Archaeology,
Architecture and Art, Volume 5: Innovation in underground engineering,
materials and equipment - Part 1 – Peila, Viggiani & Celestino (Eds)
© 2020 Taylor & Francis Group, London, ISBN 978-0-367-46870-5

Study on applicability of self-compacting concrete with low cement content to tunnel lining

T. Kato
Chubu Regional Development Bureau, MLIT, Nagoya, Aichi, Japan

N. Kurokawa, H. Nishiura, K. Sakurai, Y. Okazaki & S. Hagino
OBAYASHI CORPORATION, Minato-ku, Tokyo, Japan

ABSTRACT: Application of self-compacting concrete which does not require compaction work is effective for improving the productivity of lining work. However, conventional self-compacting concrete requires greatly increased unit cement content compared with the ordinary concrete used for tunnel lining. Therefore, by using a new special thickener and a general-purpose high-performance AE water reducing agent, we have developed low-cement concrete with high fluidity and self-filling ability without increasing the unit cement content. The newly developed concrete was applied to concrete lining of a road tunnel, which realized homogeneous tunnel lining without performing compaction work.

1 INTRODUCTION

In many countries, self-compacting concrete is commonly applied to tunnel lining. On the other hand, in Japan, commonly used is not self-compacting concrete but ordinary concrete with slump value of 15cm. One reason for this is that general self-compacting concrete requires larger amount of unit cement in order to ensure material segregation resistance, which causes thermal cracks, more mixing time and higher cost. Therefore, in Japan, it has rarely been applied to tunnel lining except for the case with a dense reinforcing bar and so on.

On the other hand, in the tunnel lining work, the aging of skilled workers and the shortage of young workers are becoming normal. Also, lining concrete is placed with compaction in the closed space, and the working environment is not necessarily good. Thus, improvement of the productivity and the environment of lining work is strongly desired these days in Japan. In order to realize this, it is effective to apply self-compacting concrete which doesn't require compaction work.

Thus, by using a special thickener and a general-purpose high-performance AE water reducing agent, we have developed low-cement-content self-compacting concrete with almost no increase in cement content compared with conventional lining concrete (Figure 1).

In this paper, firstly the results of various quality tests and model experiments of lateral pressure and fluidity are shown for applying the developed self-compacting concrete to the whole line of a road tunnel. The results that the self-compacting concrete was actually applied to the tunnel lining are subsequently shown.

2 TARGET QUALITY OF THE SELF-COMPACTING CONCRETE

Table 1 shows the target quality to determine the proportion of the self-compacting concrete. The target quality was set with reference to the STANDARD SPECIFICATIONS FOR CONCRETE STRUCTURES-2017 of Japan Society of Civil Engineers. 2017. and the

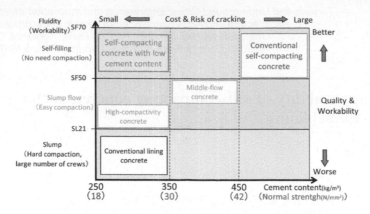

Figure 1. Position of self-compacting concrete with low cement content.

Table 1. Target quality.

Test item	Target quality	Testing method
Filling hight(Rank 3)	Not less than 30cm	JSCE-F511
Slump flow	60±10cm	JIS A 1150
Air content	4.5±1.5%	JIS A 1128
Bleeeding	Less than the conventional lining concrete	JIS A 1123
Pressurized bleeding	In the range of "good pumping" suggested in the Recommendations	JSCE-F502
Compressive strength	Not less than design strength and Not less than the conventional lining concrete	JIS A 1108
Depth of neutralization	Not more than the conventional lining concrete	JIS A 1152 JIS A 1153
Freezing and thawing Resistance	Not less than the conventional lining concrete	JIS A 1148

Recommendations for Mix Design and Construction of Self-Compacting Concrete-2012. of Japan Society of Civil Engineers. 2012b. Since most of the tunnel lining is unreinforced structure, the self-filling rank is rank 3 and the target slump flow value is 60 ± 10 cm.

Bleeding tests and pressurized bleeding tests were also conducted to prevent the occurrence of back cavity and blockage during pumping. Furthermore, various kinds of strength and durability tests were also conducted.

3 DETERMINATION OF PROPORTION AND CONFIRMATION OF QUALITY BY INDOOR TESTS

3.1 Proportion of self-compacting concrete

Table 2 shows the materials used, and Table 3 shows the proportion of the self-compacting concrete determined by the test mixing. Coarse aggregate with the same maximum size 40 mm as used in conventional tunnel lining was used to prevent an increase in unit water content and cement amount. The main ingredient of the special thickener is cellulose ether. As shown in Figure 2, it is a fine powder with an average particle diameter of 80 μm, and the amount used is as small as 30g/m^3. In addition, the high-performance AE water reducing agent is a commercially available. For these reasons, this self-compacting concrete can be easily manufactured in ready-mixed concrete plants all over Japan.

3.2 Quality of fresh concrete

The quality test results of the fresh self-compacting concrete are shown in Table 3. The self-compacting concrete satisfied the target fluidity and self-filling ability. Also, when observing the sample after the slump flow test, segregation of mortar and coarse aggregate or lifting of bleeding water couldn't be observed (Figure 3).

Table 2. Materials used.

Material	Symbol	Physical property
Cement	C	Ordinary portland cement, Density3.16g/cm³
Water	W	Ground water
Fine aggregate	S1	River sand, Surface-dry density2.62g/cm³
	S2	Crushed sand, Surface-dry density2.65g/cm³
Coarse aggregate	G1	Broken stone1505, Surface-dry density 2.66g/cm³
	G2	Broken stone2010, Surface-dry density 2.66g/cm³
	G3	Broken stone4020, Surface-dry density 2.66g/cm³
Admixture	WR	AE water reducing agent (High performance type), General-purpose product, Used in conventional lining concrete
	SP	High performance AE water reducing agent (Policarboxylic acid type), General-purpose product, Used in the self-compacting concrete
	VMA	Special thickener(Powder type, Cellulose ether type, Used in the self-compacting concrete

Figure 2. Special thickener.

The bleeding and pressurized bleeding test results are shown in Figure 4. It was confirmed that the bleeding rate of self-compacting concrete with low cement content was reduced to about 1/4 of the conventional lining concrete and material segregation or back cavity are unlikely to occur at the time of construction. The pressurized bleeding test results of conventional lining concrete deviates from the range where good pumping can be performed suggested in the Recommendations for Concrete Pumping-2012. of Japan Society of Civil Engineers. 2012a., which indicates blockage during pumping tends to occur. On the other hand, that of the self-compacting concrete with low cement content was in the target range, which indicates excellent pumping performance.

3.3 Strength development and durability

The compressive strength test results are shown in Figure 5. It was confirmed that the strength development of self-compacting concrete with low cement content at the early age was equal

Table 3. Proportion of the self-compacting concrete with low cement content and the results of quality test of fresh condition.

Concrete type	Self-filling rank	Target slump flow (cm)	Maximum size of coarse aggregate (cm)	W/C (%)	s/a (%)	Unit content (kg/m³)							Admixture		Quality of fresh concrete			
						W	C	S		G			Water reducing agent (C*%)	Special thickener (g/m³)	Slump flow (cm)	Air content (%)	Filling hight (cm)	Bleeding ratio (%)
								S1	S2	G1	G2	G3						
Conventional lining concrete	–	Slump 15±2.5	40	55.0	48.2	165	300	615	267	287	287	383	HWR 1.0	–	Slump 16.5	5.8	17.9	8.8
Self-compacting concrete with low cement conctent	Rank 3	60±10	40	48.5	47.2	165	340	422	427	335	335	287	SP 0.90	30	55.0	5.5	33.1	2.4

Figure 3. Conventional lining concrete (Left) and self-compacting concrete (Right).

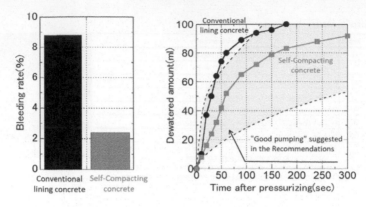

Figure 4.   Results of bleeding (Left) and pressurized bleeding (Right) test.

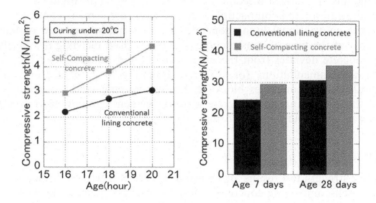

Figure 5.   Results of compressive strength (Left: Early age, Right: Ordinary age).

to or higher than that of the conventional lining concrete and that the lining work using the self-compacting concrete can be performed in the same cycle as the conventional concrete.

The durability test results are shown in Figure 5. It was confirmed that the resistance to neutralization and the resistance to freezing and thawing were equal to or higher than those of conventional lining concrete.

From the above results, it was confirmed that the self-compacting concrete with low cement content has the same strength and durability as the conventional lining concrete.

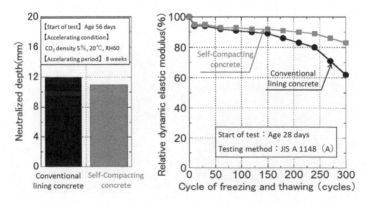

Figure 6.   Results of durability tests (Left: Neutralization, Right: Freezing and thawing).

4 LATERAL PRESSURE EXPERIMENT USING COLUMN MODEL

4.1 *Purpose of experiment*

In general, when using self-compacting concrete, it is necessary to design the lining form as hydraulic pressure acts on the formwork. The design of lining form considering hydraulic pressure requires uneconomically large reinforcement because of the construction height. Therefore, we conducted a lateral pressure experiment using a column model in order to estimate the side pressure acting on the formwork during construction, and reflect it in the design of the lining form.

4.2 *Outline of experiment*

The column model used in the experiment is shown in Figure 7. The self-compacting concrete was placed at 30 cm/layer and divided into seven layers in total at intervals of 40 minutes (placing rate of 0.45 m/h). The pressure gauge was installed at the center of the bottom layer, and side pressure during concrete placement was measured.

The proportion determined in the previous chapter was used. The concrete was manufactured at intervals of 40 minutes in accordance with the above-mentioned placing speed, and was placed after confirming the required quality by the tests. The experiment was conducted indoors controlled at a temperature of 20 °C.

4.3 *Experimental Results and Discussion*

Figure 8 shows the measurement results of the side pressure in relation to the placing height. The lateral pressure was almost equal to the fluid pressure until the completion of the third layer placement (until 80 minutes after the first layer placement).

After that, the increase in lateral pressure suddenly became gentle and reached the maximum value after completion of the fifth layer (after 160 minutes). After that, it gradually declined. The maximum value of the lateral pressure acting on the formwork was about 0.018 N/mm^2, that was about 50% of 0.035 N/mm^2, the case when hydraulic pressure acted at the same point (after 160 minutes).

In accordance with the placement of the first layer, the sample was packed in a slump cone and was allowed to stand, then slump tests were conducted at 40 minute intervals (Figure 9). Until 80 minutes after standing, the specimen flowed in the lateral direction, which inferred that pressure acted on the formwork. On the other hand, at 120 minutes after standing, the sample collapsed without moving in the lateral direction and self-sustained after 160 minutes, which proved that the lateral pressure reached the peak.

Figure 7. Column model used in the lateral pressure experiment.

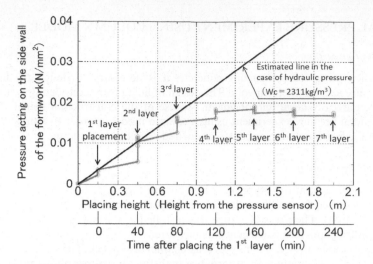

Figure 8. Placing height and the lateral pressure acting on the formwork.

Figure 9. Results of slump tests (The time in each figure indicates the standing duration).

The placing speed of standard lining work is about 1.5 m/h. From the test results, assuming that 50% of the pressure corresponding to the hydraulic pressure acts 160 minutes after placement, it was estimated that side pressure of about 0.046 N/mm^2 acts in actual construction.

The designed load of the lining form used for actual construction was set 0.06 N/mm^2 in anticipation of 30% safety factor because of the deformation of the lining form during lining concrete placement and the possibility of the side pressure increases for a long time depending on the temperature condition.

5 FLOW EXPERIMENT USING WALL-LIKE MODEL

5.1 Outline of experiment

As the placing point of tunnel lining concrete is restricted, the flow distance of the concrete becomes long, and there is a concern of material segregation and deterioration of homogeneity. Therefore, a flow experiment was conducted using a full-size wall-like model (length 10.5 m × width 0.3 m × height 0.9 m, volume 2.2 m^3) simulating a maximum flow distance of 10.5 m in actual construction.

Concrete was produced (2 m^3 × 2 batches) with the actual mixer of a ready-mixed concrete plant. After transporting over 30 minutes to the construction site with an agitator car, it was placed from the end of the model by the shoot of the agitator car. The concrete was placed at intervals of about 5 minutes simulating the unloading speed in actual construction. The

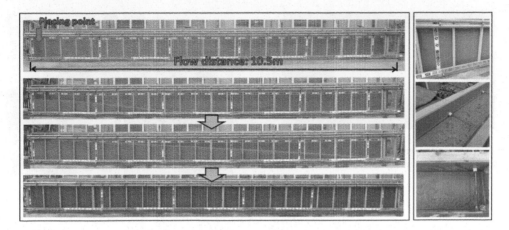

Figure 10. Flowing condition (Left) and the appearance of side and top surface after the experiment (Right).

experiment was conducted in autumn and the concrete temperature during the experiment was about 15 ° C.

5.2 *Experimental Results and Discussion*

Figure 10 shows the flow condition of the self-compacting concrete. Due to the high fluidity of the concrete itself, it easily flowed to the opposite end. In addition, occurrence of limewash or advancement of cement paste couldn't be observed. After the experiment, the height of the concrete at the placing point and the tip point of flowing was measured. As a result, the difference was 6 cm (about 1/175 flow gradient) which implies the concrete flew almost flatly.

The amount of coarse aggregate contained at the tip was measured by taking sample and washing it. As a result, the amount of particle size 4020 was 103% and that of particle size 2005 was 102% with respect to the determined proportion. It was confirmed that the concrete flew in a homogeneous state.

From the above results, it was estimated that the self-compacting concrete can fill every corner of the form uniformly without compaction.

6 CONSTRUCTION OF LINING WITH SELF-COMPACTING CONCRETE

6.1 *Manufacture of self-compacting concrete*

The self-compacting concrete was made at 2 m³/batch, manufactured in 2 batches, in the same method as shown in Chapter 5, and transported to the construction site with an agitator car. As shown in Figure 11 the special thickener was prepared in advance with the used amount (60 g) per batch packed in water-soluble paper, and was introduced from the automatic dropping apparatus shown in Figure 12. This device can be operated by the button in the operation room of the plant, and the thickener can be automatically put in accordance with charging of other materials such as cement and aggregate.

6.2 *Quality test results*

In the construction using self-compacting concrete, since the compaction work is not carried out, the quality at the time of fresh concrete is directly related to the quality of the lining concrete. For this reason, it is extremely important to continuously manufacture and ship the self-compacting concrete with stable quality.

Therefore, at the beginning of construction, the quality test of concrete was carried out by increasing the frequency. Table 4 shows the list of quality test results at the time of unloading as of the end of July 2018.

Figure 11. Appearance of the special thickener packed in water-soluble paper.

Figure 12. Appearance of the automatic dropping apparatus for the special thickener.

In each kind of test, the target quality was satisfied, and it was confirmed that the self-compacting concrete can be stably manufactured and shipped from the plant in town.

6.3 *Method of placement and the flow condition*

Figure 13 shows the outline of the concrete placing method. In the same way as the conventional lining concrete construction work, the side wall part was placed from the left and right placing port near the center of the span and the crown part was placed from the blow-up port at the existing side. At the side wall, placing port was sequentially switched to the upper in accordance with placing the concrete. Also, so that the placing height per layer is 50 cm or less, each half of the agitator car concrete (4 m^3) was placed to the left and right respectively. A hose was connected to the placing port to reduce the free fall height to 1.5 m or less. The placing rate was 1.5 m/h, which is the standard construction speed. Since the concrete was self-compacting, compaction work was not done.

Figure 14 shows the flow conditions in the form of the self-compacting concrete. Although there were some sections with reinforcing bars, the concrete easily passed through the reinforcing bar's clearance, and it flew to every corner of the formwork without material segregation. Almost no rise of limewash or cement paste was observed.

6.4 *Side pressure acting on the form and filling of the crown part*

A pressure gauge was installed at the lower end of the form and the lateral pressure acting on the form was measured along with the placement of the concrete as shown in Figure 15. Although the lateral pressure increased with the placement, there was almost no increase after reaching the peak at about 60 minutes after placement. The maximum value of the lateral pressure was about 0.04 N/mm^2, which was about the same as the estimated value (0.046 N/mm^2) from the experimental results using the column model shown in Chapter 4. On the other hand, Sakurai et al. 2018. show that the lateral pressure of self-compacting concrete with low cement content using the maximum size of coarse aggregate of 20 mm was about 0.055 N/mm^2. The self-compacting concrete used this time didn't show the increase of side pressure that is generally said to occur. It may be due to the influence of using coarse aggregate with larger particle size and so on. Although it is necessary to accumulate data in the future, when using the newly-developed self-compacting concrete with low cement content, it is thought to be unnecessary to provide such a large-scale reinforcement for the conventional lining form.

The crown of the lining concrete is a part which is particularly difficult to fill because it is a blow-up construction from the end part. Therefore, we installed a pressure gauge at 3 top locations of the lining form and verified that the concrete could be filled appropriately. The measurement result is shown in Figure 16. It was confirmed that about 5 times the pressure corresponding to the lining thickness acted at any measurement point and the self-compacting concrete could be densely packed without compaction.

Table 4. Quality test results at the time of unloading.

Test item	Target quality	Unit	Number of measurement times	Test results		
				Average	Maximum	Minimum
Slump flow	60±10	cm	28	57.6	66.0	51.5
Reaching time of 500mm flow	3~15	s	28	8.4	12.7	5.6
Air content	4.5±1.5	%	28	4.5	5.5	3.5
Filling height	Not less than 30	cm	11	33.5	35.5	31.0
Unit water content	165±15	kg/m³	7	161.3	171.4	151.9

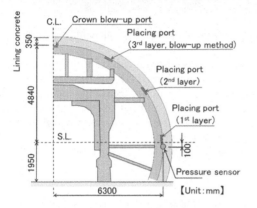

Figure 13. Method to place the concrete.

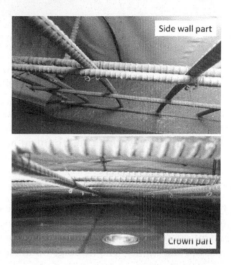

Figure 14. Flowing condition of the concrete.

6.5 Finished condition

Figure 17 shows the appearance of the lining constructed using the self-compacting concrete with low cement content. There was no filling failure or cracking at all. In addition, it was confirmed that the surface bubbles of the side walls, color unevenness at the crown and so on, which are likely to occur in the case of using the conventional lining concrete, were greatly reduced. The finished condition can be regarded as good enough.

According to Sakurai et al. 2018., when the lining of the road tunnel similar to this time was constructed with ordinary concrete, it was shown that the total compaction time by the bar-shaped vibrator was 960 minutes (16 hours) in total. By eliminating compaction work using self-compacting concrete, there can be a possibility of reducing the number of crews by about 2. In addition, since compaction work with bar-shaped vibrators causes large noise and

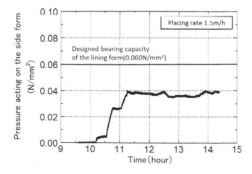

Figure 15. Measured lateral pressure acting on the form.

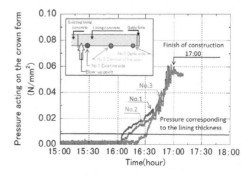

Figure 16. Measurement results to verify the filling condition of the crown part.

Figure 17. Appearance of the finish condition after construction using the self-compacting concrete.

vibration, it can be said that elimination of compaction work can contribute to improvement of work environment. In this construction work, the work performed by the crews during the concrete placement is switching of piping, closing inspection windows, cleaning, and so on. It can be said that concrete placement could be done with smaller number of crews than in the past. In addition, only the pumping sound of the concrete pump is generated in the form, which implies the working environment has been greatly improved.

7 SUMMARY

By using a special thickener, we have developed the self-compacting concrete with low cement content that can ensure high fluidity and self-filling ability almost without increasing unit cement content with respect to conventional lining concrete. It was applied to the lining concrete of road tunnel. The findings obtained by a series of studies are shown below.

(1) The newly-developed self-compacting concrete with low cement content can greatly reduce bleeding and ensure good pumping performance. In addition, strength development and various durability are equal to or higher than those of conventional lining concrete.
(2) From the result of the flow experiment simulating actual construction, the self-compacting concrete with low cement content can easily flow and fill a length of 10.5 m in homogeneous state without material segregation.
(3) The self-compacting concrete with stable quality can be manufactured and shipped from the ready-mixed concrete plant in town.
(4) Even with the reinforcing bars, the self-compacting concrete with low cement content can easily pass through the reinforcing bar's clearance and flow to every corner of the formwork.
(5) When the self-compacting concrete with low cement content was applied at standard placing speed, the side pressure acting on the form was about 0.04 N/mm². Although it is necessary to further accumulate data in the future, in the case of self-compacting concrete using coarse aggregate with a small amount of cement and large particle size, it is considered that such a large-scale reinforcement is unnecessary to the lining form used for conventional lining concrete.
(6) By using self-compacting concrete with low cement content, it is possible to construct tunnel lining with good finish without performing compaction work.

REFERENCES

Japan Society of Civil Engineers. 2012a. Recommendations for Concrete Pumping-2012. Concrete Library 135.
Japan Society of Civil Engineers. 2012b. Recommendations for Mix Design and Construction of Self-Compacting Concrete-2012. Concrete Library 136.
Japan Society of Civil Engineers. 2017. STANDARD SPECIFICATIONS FOR CONCRETE STRUCTURES-2017.
Sakurai, K., Sensui, D., Yamakawa, T. & Ishida, T. 2018. Development of self-compacting concrete with low cement content using a new special thickener and its application to actual structures. *Proceedings of the Japan Concrete Institute*, Vol. 40, No.1: 1161–1166.

Tunnels and Underground Cities: Engineering and Innovation meet Archaeology,
Architecture and Art, Volume 5: Innovation in underground engineering,
materials and equipment - Part 1 – Peila, Viggiani & Celestino (Eds)
© 2020 Taylor & Francis Group, London, ISBN 978-0-367-46870-5

Soil abrasion and penetration test for the evaluation of soft ground TBMs' excavation performance and cutter life

D.Y. Kim, H.B. Kang, Y.J. Shin & J.H. Jung
Hyundai Engineering and Construction, Seoul, S. Korea

E. Farrokh
Amirkabir University of Technology, Tehran, Iran

ABSTRACT: TBM tunneling in soft ground requires careful planning and execution in terms of excavation performance and cutting tool life. It is because these are affected by many parameters including ground conditions, TBM operational parameters, and soil conditioning. Soil abrasion and penetration test (SAPT) is a newly developed device to simulate EPB TBM excavation process in a small scale to evaluate the excavation performance and the abrasion of cutting tools considering the soil conditioning under chamber pressure. Major issues discussed in this paper include the introduction to the new test, its working procedure, parametric study regarding the influential parameters, and finally the draft correlation among the test results and the operational parameters.

1 GENERAL INSTRUCTIONS

Soft ground TBMs are generally categorized into Earth Pressure Balanced (EPB) and Slurry shielded TBM according to the supporting method of the tunnel face. EPB TBMs use excavated materials inside the chamber to control the chamber pressure by means of the discharge speed of screw conveyor. EPB TBMs are known to be used for soft ground applications, but recently with the aid of newly developed additives (e.g. polymers) they are also applied to hard rock cases. Nowadays, the application of EPB TBMs is increasing as compared to slurry TBMs due to its simplicity to use, cheaper cost, and simpler site layout design.

While NTNU and CSM models are widely used for the prediction and the evaluation of TBM performance and cutter wear, their applications are limited to hard rock TBMs as these models are mainly developed based on the data from hard rock tunneling projects. For the application of EPB TBMs in soft ground, there is no implicit reference, particularly for the performance and cutter wear evaluation as these are influenced by various parameters of TBM operation and ground condition including chamber pressure, screw conveyor speed, mixing ratio of TBM additives, water content, soil distribution and quartz content. Due to the involved complexity in the relationships among these parameters, a thorough study has not been conducted so far. In recent years, some researchers from NTNU, Penn State, and Turin Universities (Salazar et al., 2018; Bosio et al., 2018; Jakobsen et al., 2013; Rostami et al., 2012; Barbero et al., 2012; Peila et al., 2012; Gharahbagh et al., 2011) started to look further into this area by developing specific soil abrasion testers, but just focusing on the cutter wear. In addition, detailed field data of TBM tunneling projects were not included in their data analyses. As the testers in these studies do completely resemble the actual EPB TBM excavation cutterhead and its chamber, a definitive conclusion to link the results of these testers and the actual cutting tool wear is yet to be investigated. Nowadays, TBM excavation performance and cutter head intervention interval length are mostly evaluated on the basis of TBM manufacturers' recommendations or site experience, which usually yield large over or under estimations. More importantly, the impacts of additives and chamber pressure on the TBM performance has never been considered. In a regular practice on a jobsite, slump test and vane shear test are conducted using foam mixed soil under atmospheric condition, which do not reflect the effect of chamber pressure.

Table 1. Comparison between the design parameters of SAPT and 3 other testers developed recently.

	SAPT	NTNU tester (SGAT) (Jakobsen et al., 2013)	Penn State Tester (Gharahbagh et al., 2011)	Turin University Tester (Salazar et al., 2018)
Tool shape	5 small rippers *	4 steel spokes	3 propeller blades	Impeller
Material	Standard steel with various hardness and aluminum can be used	Standard steel with HRC of 23	Standard steel with HRC of 17, 31, 43, 51, 60	cemented carbides and conventional steel
Penetration depth (mm)	Up to 1000	Up to 200	Fixed position for the blade	Up to 90
Penetration rate (mm/min)	0~200	0–200	Not applicable	78
Thrust force (N)	0~30,000	0–3,000	Not applicable	-
Chamber pressure (bar)	True chamber pressure up to 7 bars	Pressurized air up to 6 bars	Pressurized air up to 10 bars	-
Grain size (mm)	no certain limit on the soil size (actual soil sample can be tested)	Up to 10 mm	d50 of up to 7 mm has been reported	d60 of 6.5 mm has been reported
Soil compaction	Possible with automatic proctor rammer **	Possible with manual proctor rammer	No soil compaction	No soil compaction
Foam injection	Via injection ports on the excavation tool during the test. A foam injection device controls required FIR and FER during the test.	Via injection ports on the spokes during the test	Premix and Via injection ports on the chamber during the test	Soil sample is mixed with foam prior to the test
Soil extraction	Excavated soil is discharged via a screw conveyor	No soil extraction. The overburden pressure on the blade changes during the test	No soil extraction	No soil extraction
RPM	1–200	1–100	60–180	160
Abrasion measure	Weight loss per pass length	Weight loss	Weight loss	Specific volume loss per pass length

* These small rippers resemble actual rippers on an actual TBM cutterhead with different installation radius to cover the whole excavation face. With this method, it is possible to normalize the tool weight loss based on the travel length of a ripper

** A major issue for a long soil column compaction is how to compact it consistently through its depth. To resolve this issue, soil is compacted with an automatic proctor rammer in small cha2mbers of 20 cm height. Then, a number of compacted soil chambers (up to 5) can be assembled on top of each other. With this method we can generate a pretty consistent compaction throughout the soil column depth

In this study, a new soil abrasion and penetration test (SAPT) developed by Hyundai Engineering and Construction is introduced. SAPT is developed to reasonably predict the TBM performance and cutter life by simulating an EPB TBM excavation process. Table 1 shows a comparison between the design parameters of SAPT and 3 other testers developed recently in 3 universities. As seen in this table, SAPT is carefully designed to mimic the actual excavation process of an EPB TBM. The major goal for the development of this tester was to simulate the excavation process of an EPB TBM as close as possible to actual real size machines in a small scale and to find relationships for the wear and penetration prediction and to link them to the actual TBM performance data through field data investigation. Several tests were carried out in the initial phase to determine the main influential factors. Using the test results of this phase, a preliminary correlation is proposed. This correlation is subject to further revision as additional tests are performed in the future.

2 INTRODUCTION OF SAPT

Soil abrasion and penetration test (SAPT) is a newly developed device to measure the excavation parameters in soil such as thrust, torque and ripper wear and to control the foam injection rate according to the varying penetration rate. In addition, vane shear test device is installed inside the chamber for this device to measure shear strength of foam mixed soil directly under chamber pressure.

SAPT is composed of excavation chamber, a specimen box, driving motors, and a control panel. The excavation chamber is equipped with a vane shear test device, a blade, and a screw conveyor. Also, packing seal is attached to the exterior of the chamber to maintain its pressure. An internal slope is set up inside the chamber to ease the process of soil discharge. The specimen box is composed of a couple of smaller cylindrical boxes, which are completely sealed with packing seal to avoid any possible pressure loss. The multiple compartment specimen-box provides an opportunity to compact each box separately and to simulate a mixed ground condition. The small blade within the chamber is equipped with 5 replaceable aluminum rippers that have different rotation paths. With this configuration, the wear of each ripper can be measured separately. During the excavation, foam and other additives can be injected simultaneously from the blade.

Vane shear tester is mounted on the upper side of the blade and it works independent of the blade. Also it is possible to operate it during excavation or standstill. Screw conveyor can control its revolution; hence the chamber pressure. It can be also modified to conduct experiments with various pitch intervals. Finally, monitoring and control panel can monitor and control thrust, torque, foam injection rate, and penetration rate in real time. Figure 1 shows SAPT

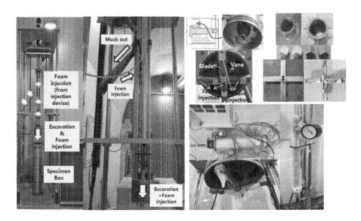

Figure 1. SAPT device and main parts(chamber, blade, ripper, vane shear tester and foam generator).

device and its main parts. The chamber diameter of SAPT device is 200 mm, which is much smaller than a real TBM, so a device that can generate a small amount of foam was required. This foam generator generates the foam using the same principles of a real TBM foam generation system. As such, FIR (Foam Injection Ratio) and FER (Foam Expansion Ratio), which are the main controlling parameters of the foam injection, are adjusted accurately when the penetration rate of the SAPT device is changed.

3 INFLUENTIAL PARAMETERS FOR THE PREDICTION OF TBM PERFORMANCE AND CUTTING TOOL WEAR

3.1 Soil particle size distribution

To simulate an EPB TBM excavation, the widely used soil particle size distribution in general EPB TBM applications, which is known to be from 0.001mm to 2.0mm (Lovat, 2007), is chosen. The soil specimen is artificially made up of 70% silica sand and 30% illite clay.

3.2 Testing parameters and cases

Penetration per revolution (PRev in mm/rev), foam mixing ratio parameters and foam concentration are selected for the testing parameters of TBM excavation process. In particular, in the case of foam-related variables such as FIR, FER and CF, slump tests were performed as preliminary tests to decide the optimal foam-mixing ratio to facilitate the excavation and discharge process. The slump test is widely used to design the parameters of TBM additives (Vinai et al, 2006). The ideal slump value is known to be approximately from 10 to 20 cm (Langmaack, 2000 and Vinai et al, 2008, Peila et al, 2009 and Budach et al, 2015). A variety of slump tests were carried out within the general range of FIR (10%~80%), FER (5~30), CF (0.5%~5.0%) (EFNARC, 2005). When the slump value was low, an additional test was conducted by increasing the FIR value. Table 2 shows the slump test results.

Table 2.   Slump test results.

No.	WC (%)	FIR (%)	FER	CF (%)	Slump (cm)	Volume of Water (L)
1	9	80	5	3.0	25.5	1.95
2	9	80	10	3.0	8.0	1.51
3	9	80	20	3.0	1.0	1.30
4	9	120	10	3.0	16.0	1.73
5	9	120	15	3.0	19.0	1.51
6	9	80	10	1.0	1.0	1.51
7	9	120	10	1.0	5.0	1.73
8	9	160	10	1.0	25.5	1.95
9	9	80	-	0.0	0.0	1.51
10	9	120	-	0.0	0.0	1.73
11	9	160	-	0.0	3.0	1.95
12	12	40	10	3.0	0.0	1.66
13	12	60	15	3.0	2.0	1.66
14	12	80	5	3.0	24.0	2.31
15	12	80	10	3.0	16.0	1.87
16	12	80	15	3.0	8.0	1.73
17	12	80	20	3.0	6.0	1.66
18	12	40	10	3.0	1.5	1.66
19	12	80	10	3.0	11.0	1.87
20	12	120	10	3.0	23.0	2.09
21	12	40	-	0.0	0.0	1.66
22	12	80	-	0.0	2.0	1.87
23	12	120	-	0.0	13.0	2.09

As seen in Table 2, the higher the water contents, FIR and CF values, the higher the slump values. In contrast, the lower the FER, the higher the slump value and the lower the CF, the lower the slump vale in spite of low coefficient of correlation (Figure 2). Results of slump tests show that at 9% water content, optimum excavation performance (slump value between 10 cm and 20 cm) can be obtained when FIR is around 120% and FER is 10 to 15. The results also show at 12% water content, optimum excavation performance can be obtained when FIR is 60% to 100% and FER is around 10 or below (Figure 3).

Input parameters of SAPT and test cases are shown in Table 3 and 4. Experiments carried out in displacement control mode (constant excavation speed). In the case of 15% of water content, first a soil sample with 10% water content is used, and then additional water is injected during the test to reach the 15% of water content.

3.3 Test result of excavation performance

Test results and graphs are shown in Table 5 and Figure 4–7.

The test results show that the thrust force generally increases as the depth of the penetration per revolution increases (cases 1 to 3 in Table 4 and Figure 4). It is also shown that the blade torque and vane torque are slightly increased as PRev increases. In the case of chamber pressure, the results show this pressure is mainly controlled by the screw conveyor speed (Figure 5).

The test results about varying FER are reflected in Table 4 and 5 for the similar cases of 1–4, 2–5, 6–9 and 7–10 (Figure 6). For the majority of these cases, both thrust and torque are

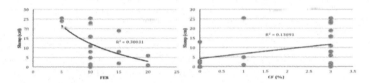

Figure 2.   FER and CF versus Slump value.

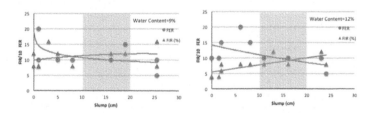

Figure 3.   FIR and FER at optimum excavation state.

Table 3.   Input parameters of SAPT.

Input Parameters	Value	Input Parameters	Value
PRev (mm/rev)	3.0/4.5/ 6.0	FER	10/20
Blade RPM (RPM)	10	Screw conveyor RPM (RPM)	Estimated for constant chamber pressure
CF (%)	3.0/1.0	Water contents (%)	9.0/13.0/15.0
FIR (%)	30/40/80	Displacement (mm)	650±20

Note: PRev is for penetration per revolution

Table 4. SAPT cases.

Case	PR (mm/min)	CF (%)	FIR (%)	FER	Screw RPM (RPM)	Water content (%)	Water Injection (specimen, L)	Water Injection (Foam sol, L)
1	30	3.0	80	10	35	13	5.87	1.63
2	45	3.0	80	10	53	13	5.87	1.63
3	60	3.0	80	10	71	13	5.87	1.63
4	30	3.0	80	20	35	13	5.87	0.82
5	45	3.0	80	20	53	13	5.87	0.82
6	30	1.0	80	10	35	13	5.87	1.63
7	45	1.0	80	10	53	13	5.87	1.63
8	60	1.0	80	10	71	13	5.87	1.63
9	30	1.0	80	20	35	13	5.87	0.82
10	45	1.0	80	20	53	13	5.87	0.82
11	45	3.0	40	15	120	15	6.77	0.54
12	45	3.0	40	15	60	15	6.77	0.54
13	45	3.0	30	15	120	15	6.77	0.41
14	45	1.0	30	15	120	15	6.77	0.41
15	30	3.0	80	10	35	9	4.06	1.63

Table 5. SAPT results.

Case	Ave. Thrust (kgf)	Ave. Torque (N-m)	Ave. Chamber Pressure (bar)	Max Vane Torque (N-cm)	Slump (cm)
1	148.74	20.17	0.68	16.04	16.0
2	199.96	22.69	0.69	29.16	16.5
3	309.70	29.38	0.71	30.10	16.0
4	138.86	30.70	0.66	31.56	2.5
5	189.96	38.29	0.73	33.26	3.0
6	160.77	21.85	0.82	21.02	11.0
7	213.60	25.50	0.88	29.54	12.0
8	243.62	31.10	0.82	30.26	10.0
9	387.25	30.53	0.62	-	3.0
10	403.41	35.65	0.64	-	3.5
11	124.73	29.36	0.02	9.46	27.0
12	225.94	59.29	0.25	-	
13	148.18	54.09	0.02	-	17.5
14	180.50	54.21	0.09	-	12.5
15	405.95	55.49	0.65	-	10.0

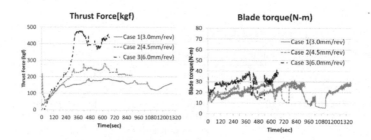

Figure 4. Thrust force and Blode torque with respect to PRev.

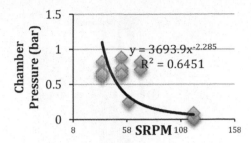

Figure 5.   Screw conveyor RPM versus chamber pressure.

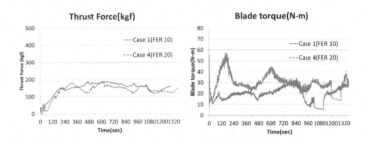

Figure 6.   Thrust force and Blade torque wth respect to FER.

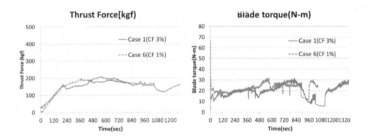

Figure 7.   Thrust force and Blade torque with respect to CF.

increased at the higher FER. One note is that as seen in Figure 6, the foam injection effects are more pronounced for the torque as compared to the thrust. An interesting point for the cases 2 and 3 as compared to the cases 4 and 5 is that in spite of having higher penetration rate values, blade torque shows lower values. This is majorly because of the lower FER value and hence a proper mixed soil product.

In case of CF variation, case 6 (CF=1.0%) showed a slight increase in thrust force, blade torque and vane torque as compared to the case 1(CF 3.0%). The comparison of the effect of CF on the thrust and torque for similar cases (Figure 7) show the change of CF may have some effects on the excavation performance parameters, but not as much as the effects of the change of FER. Additional experiments are underway to optimize the excavation performance, CF, and foam injection.

3.4   Test results of ripper wear

As mentioned above, the rippers were made from replaceable aluminum material (size 20mm×20mm). Figure 8 show the position of each ripper. Outermost ripper (No.1 ripper)

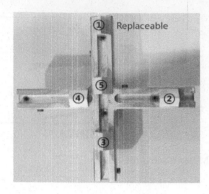

Figure 8.   Position and numbering of each ripper.

showed highest wear because it has the longest rotational distance and largest rotational torque. Wear test results are shown in Table 6 and Figure 9–11.

Table 6 shows the variation of rippers' wear according to the variation of the penetration depth per revolution for the cases of 1 to 3 and 6 to 8. The amount of weight loss decreases as the penetration depth increases (Figure 9). This is because of the shorter rotational distance in the case of higher penetration rate.

The test results about changing FER are shown in Figure 10 for the similar cases of 1–4, 2–5, 6–9, and 7–10. As seen, at higher FER value the wear increased dramatically. This is

Table 6.   SAPT wear results.

Case	Ripper No.	Weight loss (mg)	Weight loss/ travel length (mg/m)	Case	Ripper No.	Weight loss (mg)	Weight loss/ travel length (mg/m)	Case	Ripper No.	Weight loss (mg)	Weight loss/ travel length (mg/m)
1	1	8	0.07	6	1	11	0.10	11	1	19	0.24
	2	2	0.02		2	2	0.02		2	3	0.05
	3	1	0.02		3	0	-		3	3	0.07
	4	1	0.02		4	0	-		4	3	0.10
	5	0	-		5	0	-		5	1	0.09
2	1	6	0.08	7	1	8	0.10	12	1	34	0.43
	2	2	0.03		2	1	0.02		2	9	0.15
	3	1	0.02		3	1	0.02		3	4	0.10
	4	1	0.03		4	1	0.03		4	3	0.10
	5	0	-		5	1	0.09		5	4	0.35
3	1	6	0.10	8	1	7	0.12	13	1	37	0.43
	2	1	0.02		2	2	0.04		2	5	0.08
	3	1	0.03		3	1	0.03		3	4	0.09
	4	0	-		4	1	0.05		4	4	0.13
	5	0	-		5	0	-		5	1	0.08
4	1	20	0.17	9	1	21	0.17	14	1	36	0.52
	2	5	0.05		2	3	0.03		2	5	0.09
	3	2	0.03		3	3	0.05		3	4	0.11
	4	1	0.02		4	1	0.02		4	4	0.16
	5	1	0.06		5	0	-		5	2	0.20
5	1	17	0.21	10	1	9	0.10	15	1	205	0.59
	2	5	0.08		2	6	0.09		2	32	2.92
	3	1	0.02		3	6	0.13		3	18	3.48
	4	1	0.03		4	5	0.16		4	13	3.39
	5	1	0.09		5	6	0.48		5	4	4.34

Note: case number refers to the same case number in Table 4 and 5

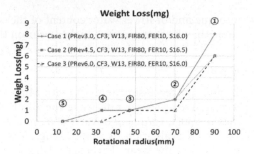

Figure 9.   Weight loss with respect to PRev.

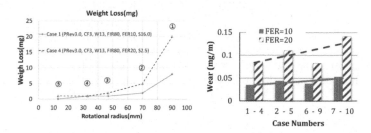

Figure 10.   Weight loss with respect to FER.

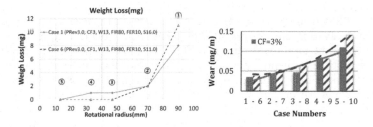

Figure 11.   Weight loss with respect to CF.

because at higher FER value, lower amount of liquid is injected and the soil might become comparably drier.

Finally, the change of the amount of wear caused by the change of CF is shown in Figure 11 for the similar cases of 1–6, 2–7, 3–8 and 4–9. Wear tend to decrease at higher CF value but the difference is very small.

Analysis results show, among the affecting parameters on the wear, torque and water content have the highest effects (Figure 12). These two parameters are directly related to the shear

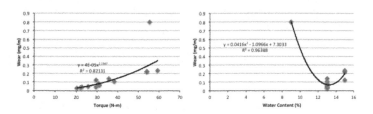

Figure 12.   Torque and water content versus wear.

strength of the soil. Higher torque and/or lower water content of the soil are linked to a stiffer soil, hence a higher shear strength, which can cause a higher wear on the cutting tools.

3.5 *Model for the wear prognosis*

Using the wear data and the multiple regression analysis (with Minitab software version 16), a formula is derived to predict an average wear rate value for all of the rippers on the SAPT device (Eq. 1). As seen in this formula, torque and water content are the two most important influential parameters for the wear evaluation. An interesting outcome from this formula is the reverse effect of water content on the wear. This means when the soil is drier, the wear is higher. As explained previously, this may refer to a higher shear strength conditions.

$$W = 0.0055\,\mathrm{Tq}^{2.21}/\mathrm{WC}^{1.89} \qquad R^2 = 89.4\% \tag{1}$$

where W = average wear in mg/m; Tq = torque in N-m; WC = water content in %..

In order to include more parameters in the formula and to generalize it for a real TBM application, a larger variation of the parameters and the actual TBM cutting tools' wear data are required. These data are going to be collected and analyzed in the next stage of this study.

4 CONCLUSIONS

In this study, for the evaluation of excavation performance and the amount of wear, a new testing machine named as soil abrasion and penetration test (SAPT) with a mini foam generator to inject the foam inside the chamber is developed. This device simulates the working process of an EPB TBM from the point of excavation to the discharge point. Using this testing machine, main excavation parameters such as thrust force, blade torque, chamber pressure and ripper wear can be monitored on a real-time basis. As a result of a parametric study by taking into account the influential parameters, it is found that the foam mixing ratio parameters (both FIR and FER), torque, and water content are the most significant parameters among all other input parameters such as penetration per revolution and foam concentration CF, etc. Consequently, although more foam means more cost, it is worth considering a sufficient amount of foam injection as it can compensate its higher cost by increasing the TBM performance through reduced machine load, amount of wear, and downtime of TBM. Some notable results are listed as follows:

- As foam concentration decreases, the slump value goes down. Results of slump tests show that at 9% water content, optimum excavation performance (slump value between 10 cm and 20 cm) can be obtained when FIR is around 120% and FER is 10 to 15. The results also show at 12% water content, optimum excavation performance can be obtained when FIR is 60% to 100% and FER is around 10 or below.
- The test results show that the thrust force generally increases as the depth of penetration per revolution (PRev) increases. The blade torque and vane torque are slightly increased as PRev increases. In the case of chamber pressure, the results show this pressure is mainly controlled by the screw conveyor speed.
- The test results about varying FER show, both thrust and torque are increased at higher FER values. An interesting point is that the foam injection effects are more pronounced for the torque as compared to the thrust.
- At higher FER values, tool wear increases dramatically.
- Analysis results show, among the affecting parameters on the wear, torque and water content have the highest effects.

In conclusion, it is understood the TBM foam mixing ratio parameters (FIR, FER) play a key role in the TBM performance evaluation. Further experiments on various soil types and site samples are underway to analyze the correlations among the real TBM operational data and lab tests' parameters and to develop prediction models for the TBM performance and wear prognosis.

REFERENCES

Barbero, M., Peila, D., Picchio, A., Chieregato, A., Bozza, F., Mignelli, C., 2012. Procedura sperimentale per la valutazione dell'effetto del condizionamento del terreno sul- l'abrasione degli utensili nello scavo con EPB. Geoingegner. Ambient. Miner. 135, 13–19.

Bosio, F., Bassini, E., Onate Salazar, C.G., Ugues, D., Peila, D., 2018. The influence of microstructure on abrasive wear resistance of selected cemented carbide grades operating as cutting tools in dry and foam conditioned soil. Wear 394–395, 203–216.

Budach, C. & Thewes, M. 2015. Application ranges of EPB shields in coarse ground based on laboratory research, *Tunnelling and Underground Space Technology*, Vol. 50, pp. 296–304.

Gharahbagh, E.A., Rostami, J., Palomino, A.M., 2011. New soil abrasion testing method for soft ground tunneling applications. Tunn. Undergr. Space Technol. 26, 604–613.

Jakobsen, P.D., Langmaack, L., Dahl, F., Breivik, T., 2013. Development of the Soft Ground Abrasion Tester (SGAT) to predict TBM tool wear, torque and thrust. Tunn. Undergr. Space Technol. 38, 398–408.

Jakobsen, P. D. 2014. *Estimation of soft ground tool life in TBM tunneling*, PhD Dissertation. Norwegian University of Science and Technology.

Langmaack L. 2000. Advanced Technology of soil conditioning in EPB shield tunneling, *Proceedings of North American tunneling*, pp. 525–542.

Lovat, P. 2007. TBM Design Considerations: Selection of Earth Pressure Balance or Slurry Pressure Balance Tunnel Boring Machines, *New Advanced Methods for Tunneling Engineering*, TEE, Athens, February 1, 2007.

Peila, D., Picchio, A., Chieregato, A., Barbero, M., Dal Negro, E., Boscaro, A., 2012. Test procedure for assessing the influence of soil conditioning for EPB tunneling on the tool wear. In: Proceedings World Tunnel Congress. Engineering Institute of Thailand, Bangkok.

Peila, D. & Oggeri, C. & Borio, L. 2009. Using the Slump Test to Assess the Behavior of Conditioned Soil for EPB Tunneling, *Environmental & Engineering Geoscience*, Vol. XV, No.3, pp. 167–174.

Rostami, J., Gharahbagh, E.A., Palomino, A.M., Mosleh, M., 2012. Development of soil abrasivity testing for soft ground tunneling using shield machines. Tunn. Undergr. Space Technol. 28, 245–256.

Salazar C.G.O., Todaro C., Bosio F., Bassini E., Ugues D., Peila D. 2018. A new test device for the study of metal wear in conditioned granular soil used in EPB shield tunneling. Tunnelling and Underground Space Technology 73, 212–221.

Thewes, M. & Budach, C. & Galli, M. 2010. Laboratory Tests with various conditioned Soils for Tunnelling with Earth Pressure Balance Shield Machines, *In: Tunnel6/2010, Bauverlag BV GmbH*, Gutersloh, pp 21–30.

Vinai, R. 2006. *A contribution to the study of soil conditioning techniques for EPB TBM applications in cohesionless soils*, PhD Dissertation. Politecnico di Turino.

Vinai, R. & Oggeri, C. & Peila, D. 2008. Soil conditioning of sand for EPB applications: a laboratory research, *Tunnelling and Underground Space Technology*, Vol. 23, pp. 308–318.

Tunnels and Underground Cities: Engineering and Innovation meet Archaeology,
Architecture and Art, Volume 5: Innovation in underground engineering,
materials and equipment - Part 1 – Peila, Viggiani & Celestino (Eds)
© 2020 Taylor & Francis Group, London, ISBN 978-0-367-46870-5

Development of a ground monitoring sensor system for tunnel maintenance using geophysical techniques

J.W. Kim & G.C. Cho
Korea Advanced Institute of Science and Technology, Daejeon, Republic of Korea

ABSTRACT: Recent developments in sensing and monitoring technology allow the application of various sensors and sensor networks for tunnel maintenance. Most maintenance monitoring techniques focus on measuring physical parameters of and within the tunnel structure it-self, with less emphasis placed on the ground conditions surrounding the tunnel structure. Soft soil, weak soil or grouted soil zones are particularly susceptible to long-term changes in ground condition and thus, extensive monitoring is essential to ensure long term tunnel stability and accurate numerical analysis. This study proposes a ground monitoring sensor system for tunnel maintenance using geophysical techniques. The proposed system uses seismic wave velocity and electrical resistivity techniques to estimate the changes in the geomechanical properties of the ground. A low cost prototype sensor was created using open source hardware and software and preliminary tests were conducted on an open area test site. The results show that the system can adequately perform seismic and electrical resistivity tests for tunnel maintenance purposes.

1 INTRODUCTION

Traditional tunnel monitoring techniques primarily rely on manual inspection along with monitoring equipment such as inclinometers, earth pressure sensors, extensometers or pie-zometers to aid in the monitoring of the ground conditions surrounding the tunnel struc-ture. Commonly measured ground parameters using the aforementioned monitoring equipment focus on how changing ground properties affect the tunnel structure itself and less on measuring the actual properties of the ground. Soft soil, weak soil or grouted soil zones along the tunnel span are particularly susceptible to long-term changes in ground condition and thus, extensive monitoring is essential to ensure long term tunnel stability and accurate numerical analysis. However, tunnel ground monitoring is limited by sensor installation and maintenance monitoring costs of the tunnel project.

In this study, a low-cost ground monitoring sensor system for tunnel maintenance using geophysical techniques and open source hardware is proposed. The system uses both seismic wave-based and electrical resistivity-based methods, both of which are commonly conducted for site investigations before construction but seldom used for maintenance monitoring. A low cost prototype sensor and DAQ system is created using open source hardware and software and preliminary tests were conducted on an open area test site. The results show that the system can adequately measure ground vibrations as well as perform seismic and electrical resistivity tests. The sensor data can be stored and sent to a central server via wireless connection.

2 GEOPHYSICAL MONITORING PARAMETERS

2.1 *Seismic wave velocity*

Seismic wave survey tests are frequently conducted prior to underground construction to estimate the ground conditions. Survey techniques vary from borehole-based intrusive tests and non-destructive surface surveys (Stokoe & Santamarina, 2000). The measured wave velocities are correlated with soil properties such as the constrained modulus, bulk modulus, shear modulus, Young's modulus and Poisson's ratio. Non-destructive wave survey techniques have previously been applied to examine the state of tunnel linings, backfill grout and the ground state behind tunnel lining, but limitations exist in the methodologies adopted. Nazarian et al. (1995) conducted impact echo tests using hand-held hammers along the tunnel axis to assess the backfill grout and ground conditions behind the tunnel lining. However, the measurement process is not automated and relies on manpowered impact sources. Gebre (2013) discusses the THEAM^{TM} scheme that uses geophones and electrodynamic shakers to conduct non-invasive tunnel health monitoring using active seismic. In this scheme, the shaker is placed on the bare rock with a concrete base, which limits its application to TBM tunnels with segment linings and backfill grouting. Other methods such as tunnel-ahead prediction or probing techniques may be suitable during construction, but are not applicable for long term tunnel maintenance purposes.

The main problem is related to the placement of seismic wave sources and receivers within the tunnel. Sources and receivers placed on the lining surface limits wave propagation behind the backfill, making it hard to assess the exact conditions of the ground surrounding the tunnel. For exact and accurate measurement of the ground properties however, existing methods for installing tunnel monitoring sensors can be adopted. Tunnel ground monitoring equipment are drilled and subsequently installed behind the tunnel lining. By adopting the same installation process for seismic wave velocity measurements inside tunnels, the monitoring system is similar to that of borehole tests conducted in the field.

2.2 *Title, author and affiliation frame*

Traditional electrical resistivity surveys are conducted to detect subsurface conditions such as archaeological features, anomalies or ground water levels. In sync with piezometers, resistivity surveys within the tunnel can be used to assess water penetration around the tunnel structure. Similarly to the aforementioned seismic survey techniques, electrical resistivity surveys can be carried out on land or using boreholes. When adopting the same borehole installation scheme as the seismic wave velocity mentioned above, accurate analysis techniques are required to convert the measured voltage and resistance data to electrical resistivity. In addition, the sensor casing or borehole casing would have a major effect on the measured resistivity data (Schenkel, 1994).

3 PROPOSED SENSOR SYSTEM

3.1 *Monitoring concept*

The proposed sensor system is derived from traditional borehole techniques used for ground surveying. Each sensor is comprised of a sensor module and non-conducting ex-tension casing. Two pipe-shaped sensors are installed into sensor installation holes drilled along the tunnel axis with backfill grouting. The sensors measure the seismic wave velocity and electrical resistivity of the ground between and around the sensor modules using the aforementioned borehole testing methods. Multiple individual sensor modules can be add-ed for different depths such as the backfill grouted, grouted or ungrouted ground zones. This allows flexible placement of the sensor modules depending on the depth of the drilled installation hole or other site conditions. The module placement within the two sensors should be identical to ensure accurate measurements between sensor installation holes.

3.2 *Sensor module*

Each sensor module consists of a hollow cylindrical non-conducting casing, two ring electrodes placed at each end of the casing, P- and S-wave transmitters and wave receivers. The

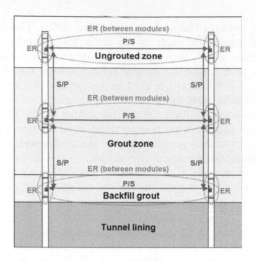

Figure 1. Sensor measurement scheme.

two ends of the sensor module are threaded to allow the installation of end-caps or extension cases. For wave velocity measurements, the P- or S-wave transmitters on one sensor module sends a pulse that is detected by the corresponding module on the opposite sensor or other modules within the same sensor. The wave velocity is calculated by the arrival time and known distance between sensor modules. For electrical resistivity measurements, local resistivity measurements within the same module or between other modules are possible. The electrical resistivity is calculated from the resistance values between ring electrodes using an analytical solution derived from Won (1987). The equation models each electrode as a torus with a major radius r and minor radius t as follows:

$$R = \left| \frac{\rho}{\pi^2 r} \left(\ln\left(\frac{L + \frac{\pi r}{2}}{L} \right) - \ln\left(\frac{t + \frac{\pi r}{2}}{t} \right) \right) \right| \tag{1}$$

where R = measured resistance between two electrodes; ρ = resistivity of the ground; r = major radius of the torus electrode; L = distance between the center of the electrodes; and t = minor radius of the torus electrode. The equation can be used to measure the resistivity of the medium between two electrodes, both between modules and within modules. This allows 6 unique configurations between two sensor modules.

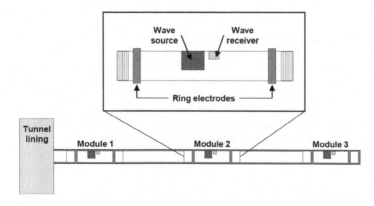

Figure 2. Sensor module design.

Based on the proposed sensor system, a low cost prototype sensor was made. The module and extension bar casings are made of PVC and the ring electrodes are made of copper. The wave source is applied using two solenoid hammers that are oriented 90 degrees from each other to simulate P- and S-waves respectively. The solenoids trigger a steel hammer which applies an impact on the steel impact plate. A tri-axial MEMS accelerometer (Analog ADXL335) is attached to each impact plate and used as the main wave receiver. Compared to traditional geophones used in seismic surveys, MEMS accelerometers are much smaller and have similar measurement accuracy above 5 Hz (Hons et al., 2008).

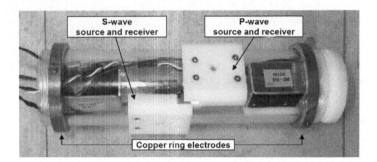

Figure 3. Low cost prototype sensor.

3.3 *Data acquisition*

The data acquisition scheme is shown in Figure 4. The sensor system uses cost-effective open-source hardware devices; Arduino Uno, Arduino Due and Raspberry pi. The Arduino Uno microcontroller is used to measure the voltage and resistance between ring electrodes and to control the solenoid hammers via a 12VDC relay using the AnalogRead() function on Arduino IDE. For accurate voltage-to-resistance readings, a variable reference resistor (10kΩ-10MΩ) was added to the breadboard connecting the wires from the electrodes to the Arduino Uno. The Arduino Due is used to measure the accelerometer signals. Wave velocity measurements require high sampling rates for accurate arrival time measurements and hence, the Arduino Due was selected to pick up the accelerometer signals as it has a higher CPU clock rate compared to the Arduino Uno. The Arduino Due allows sufficient sampling rates up to 1MSPS using the native USB port and direct memory access (DMA) method.

As Arduinos are basic microcontroller units with limited memory and speed, a Raspberry-Pi microcomputer was added to handle the data saving and computation processes and to avoid overloading the Arduinos. The data acquisition program is written in Python and Arduino IDE and the data is saved as a text file. Wifi connection on the Raspberry Pi allows

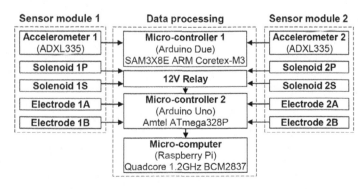

Figure 4. Data acquisition scheme.

Figure 5. Prototype data acquisition device.

remote control of the data acquisition program as well as sending the measurement data to the main server. A prototype DAQ device was made as shown in Figure 5.

4 FIELD APPLICATIONS

4.1 *Test site*

Using the prototype sensors and data acquisition device, field tests were conducted to test the validity of the monitoring system. The test site was located on an undisclosed construction site and the ground consisted of loose gravel and sand. The sensors were installed at a depth of 0.8m with a distance of 2m between the two centers. No backfill grouting was conducted after the installation. Seismic wave velocity and electrical resistivity measurements were taken for a period of 2 days with the data acquisition device.

Figure 6. Field test site.

4.2 *Test results*

A sample electrical resistivity measurement is shown in Figure 7. The site had an average electrical resistivity of 5610.13Ωm, which is within the range of sand and gravel (Gunn et al., 2015), with each data collection taking 5kB. The two lower measured voltage configurations, 1-2 and 3-4, were measured for two electrodes within the same module. When converted to the electrical resistivity value, the analytical solution considers the distance between electrodes and the values were similar to that of the other configurations. Water infiltration tests were conducted by pouring 2L of water from the surface at the center between the sensors. The electrical resistivity showed a 32.8% decrease in resistivity with water penetration with an average value of 3636.5Ωm, showing that the developed system is capable of detecting changes in the environment.

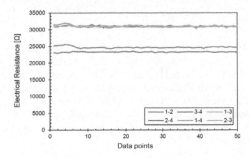

Figure 7. Electrical resistance data sample.

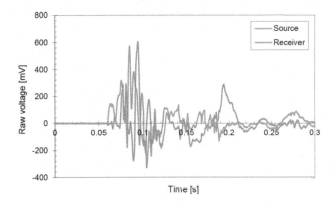

Figure 8. Wave velocity data sample.

A sample seismic wave velocity measurement is shown in Figure 8. The average P- and S-wave velocities of the site were 569m/s and 231m/s respectively with each data collection taking 258kB of memory. The measured values are within the range of sands from previous studies. Assuming a soil density of $1.6g/cm^3$, the mechanical soil properties could be calculated from the measured wave velocities. The ground had an average Young's modulus of 246.4MPa, an average shear modulus of 72.3MPa and average Poisson's ratio of 0.29.

5 CONCLUSIONS

This study proposes a low-cost ground monitoring sensor system for tunnel maintenance. The proposed system uses seismic wave velocity and electrical resistivity techniques to estimate the changes in the geomechanical properties of the ground. A low cost prototype sensor was created using open source hardware and software and preliminary tests were conducted on an open area test site. The field test results show that the system can adequately perform seismic and electrical resistivity tests. Further detailed studies need to be conducted on the feasibility and applicability of the concept for applications to tunnel maintenance.

ACKNOWLEDGEMENTS

This research was supported by the "Development of key subsea tunneling technology" program (13SCIP-B066321-01) funded by the Korea Agency for Infrastructure Technology Advancement (KAIA) and the "U-City Master and Doctor Course Grant (Education) Program" under the Korea Ministry of Land, Infrastructure and Transport (MOLIT).

REFERENCES

Gebre, M.G. 2013. Tunnel Health Monitoring Using Active Seismics. MS Thesis, University of Oslo.

Gunn, D.A., Chambers, J.E., Uhlemann, S., Wilkinson, P.B., Meldrum, P.I., Dijkstra, T.A., Haslam, E., Kirkham, M., Wragg, J., Holyoake, S. & Hughes, P.N. 2015. Moisture monitoring in clay embankments using electrical resistivity tomography. *Construction and Building Materials* 92: 82–74.

Hons, M.S., Stewart, R.R., Lawton, D.C., Bertram, M.B. & Hauer, G. 2008. Field data comparisons of MEMS accelerometers and analog geophones. *The Leading Edge* 27(7): 896–903.

Nazarian, S., Baker, M., & Crain, K. 1995. Use of seismic pavement analyzer in pavement evaluation. *Transportation Research Record* 1505:1–8

Schenkel, C.J. & Morrison, H.F. 1994. Electrical resistivity measurement through metal casing. *Geophysics* 59(7): 1072–1082.

Stokoe, K.H. and Santamarina, J.C. 2000. Seismic-wave-based testing in geotechnical engineering. In GeoEng2000; *ISRM International Symposium*: 1490–1536.

Won, I.J. 1987. The geometrical factor of a marine resistivity probe with four ring electrodes. *IEEE journal of oceanic engineering* 12(1): 301–303.

Tunnels and Underground Cities: Engineering and Innovation meet Archaeology,
Architecture and Art, Volume 5: Innovation in underground engineering,
materials and equipment - Part 1 – Peila, Viggiani & Celestino (Eds)
© 2020 Taylor & Francis Group, London, ISBN 978-0-367-46870-5

Enlargement of in-service evacuation tunnel and automated lining placement

A. Kimura, S. Yamamoto & F. Kusumoto
Shimizu Corporation, Tokyo, Japan

H. Nobunaga & S. Yamanaka
West Nippon Expressway Company Limited, Osaka, Japan

ABSTRACT: The Kawabe Daiichi tunnel has one traffic lane in each direction, with an evacuation tunnel 4.50 meters in diameter nearby. In this project, the current tunnels will be remodeled to a twin tunnel system, by enlarging the evacuation tunnel for use as a new inbound road tunnel, keeping it open 24 hours a day. The new tunnel will be full-face bored using a blasting technique. The cutting face used for enlarging the tunnel's width has a hemispherical shape. For weak ground, the support will be structured in the shape of a ring and the cross section be closed early to maintain tunnel stability mechanically. Lining concrete to be used will be medium fluidity. The concrete is placed automatically with the use of a computer controlled and compacted with a form vibrator, to make the placed concrete uniform and to maximize the automation of lining construction.

1 INTRODUCTION

Yuasa-gobo Road, starting from Gobo City in Wakayama, is a local toll road with a total length of about 19 kilometers (Fig. 1). This project will enlarge the in-service two-traffic-lane Yuasa-gobo Road to four lanes in order to improve traffic safety and traffic flow. In addition, the project will greatly contribute to revitalizing the regions nearby and improving convenience. Moreover, it is expected that the project will reinforce alternative functions for disaster

Figure 1. Location of Yuasa Gobo Road.

relief, eliminate traffic congestion caused by traffic concentration and thereby improve punctuality of public transport.

The Kawabe Daiichi Tunnel project will enlarge the evacuation tunnel which is parallel to the Phase I route which is in service to use as in-bound lanes.

This paper reports on the mountain tunneling method employed in the construction of the Kawabe Daiichi Tunnel, and automated lining concrete with medium fluidity.

2 OVERVIEW OF THE TUNNEL CONSTRUCTION

The Kawabe Daiich Tunnel is located in Wakayama Prefecture. This two-lane tunnel is 2,641 meters long with a standard boring cross-sectional area of 70 to 89 square meters. In this project, a blasting system was employed to enlarge the existing evacuation tunnel, which runs parallel to the in-service Phase I route, into the in-bound lanes while keeping the Phase I tunnel open (Fig. 2). The evacuation tunnel was constructed using TBM 4.5 meters in diameter. Escape passages connected to the evacuation tunnel are placed at intervals of about 400 meters in the tunnel longitudinal direction from the Phase I route to the evacuation tunnel. In emergency situations, people can evacuate through these escape passages and the evacuation tunnel.

For this reason, the prerequisite for the construction of the Phase II route tunnel is to ensure safety of the in-service Phase I tunnel and the functions of the evacuation tunnel as well as reliable construction of the Phase II route tunnel.

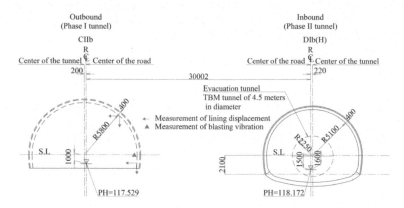

Figure 2. Overview of twin tunnel (enlarging to four lanes) and measurement of the Phase I tunnel.

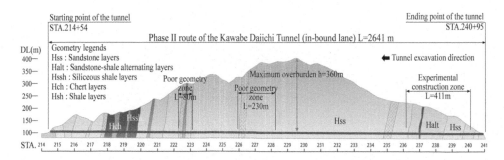

Figure 3. Longitudinal profile of the geometry (in-bound lane).

The geology surrounding the Kawabe Daiichi Tunnel is composed of sandstone, sandstone-shale alternating layers, siliceous shale and chert layers (Fig. 3). The water discharged from the evacuation tunnel is approximately 10 to 15 m³/min.

Table 1. Specifications of the phase II route tunnel.

Tunnel support structure		Arch structure (Standard pattern)			Ring structure (Early closure pattern)	
Support pattern		CIIa	CIIb	DIb	DIIc	DIIcL
Estimated ground competence factor Cf (-)		$2 \leqq$	$2 \leqq$	$2 \leqq$	1.0 ~ 2.0	1.0 ~ 2.0
Round length (m)		1.2	1.2	1.0	1.0	1.0
Deformation allowance (cm)		0	0	0	0	0
Support structure	Thickness of shotcrete (cm)	7	7	10	15	20
	Compression (28 day)	36N/mm²			36N/mm²	
	Arched steel supports	—	HH-100	HH-100	HH-108	HH-154
	Rock bolts	L=3m, 170kN		L=4m, 290kN	L=4m, 290kN	
Early closure structure	Members for early closure	—	—	—	Same as upper half and lower half	
	Ratio of structure radius (r3/r1)	—	—	—	2.0	2.0
	Early closure distance Lf (m)	—	—	—	6	6
Lining thickness (cm)	Arch, side walls	30	30	30	30	40
	Invert	—	—	45	30	40

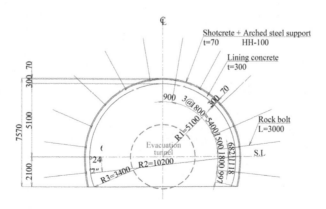

Figure 4. Arch structure CIIb.

3 OVERVIEW OF THE STRUCTURE OF THE PHASE II ROUTE TUNNEL

The supporting system of the Phase II route tunnel is a combination of a ground arch formed in the ground surrounding the tunnel, rock bolts as support members, shotcrete and arched steel supports. The standard support pattern is an arch structure designed according to the class of ground, combining the above mentioned supports (Table 1, Fig. 4). Early closure pattern DIIc of ring structure is employed for the weak ground (Fig. 5).

4 TUNNELING METHOD FOR THE PHASE II ROUTE TUNNEL

In elastic ground, this project plans to use blasting tunneling boring with a spherical cutting face of full cross section which is known to achieve high quality in tunneling, high construction safely and excellent construction performance (Fig. 6) (Kusumoto et al. 2013). Since vibration of this blasting potentially impacts the tunnel route in Phase I, it is decided that the tunnel will be bored by the use of a controlled blasting pattern to guarantee safety on the lining of the tunnel route which is already in service. Although in the ground of plasticity DII, even with a distance of about 3 Ds (D, boring width) provided between the center lines of the two tunnels, there still remains fear that an excessive displacement may adversely affect the Phase I route tunnel already in service, therefore, we will adopt a boring method of closing the full face at an early stage, with a curved cutting face by using an early closing pattern of ring structure (Sato et al. 2013).

Muck produced in the tunneling route will be hauled by a continuous belt conveyer system with a belt width of 610 mm and a conveying capacity of 300 t/h.

Excavation muck will be crushed so that it is less than 200 mm, with the use of a self-propelling crusher installed 50 - 80 meters from the cutting face. Then, the muck will be loaded on the belt via the tail piece of the belt conveyer to be taken out of the tunnel.

5 SUMMARY OF MEASUREMENTS OF THE PHASE II ROUTE TUNNEL

The tunnel will be measured for displacements to determine and evaluate the behavior characteristics and the stability of the tunnel. Reference sections for measurement are provided at intervals of 10 meters in the tunnel boring direction. Displacement will be measured automatically every six to twelve hours with a 3D automatic measurement system (Fig. 5). The

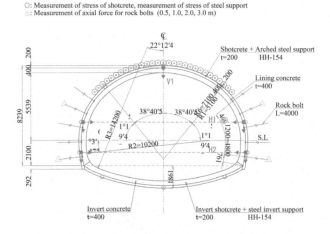

Figure 5. Overview of ring structure DIIcL and measurement monitoring.

Figure 6. Full face excavation with spherical cutting face.

management values of H, crown settlement, and V, convergence displacement, are set at –50 mm and –100 m respectively.

6 CONSTRUCTION IN THE GROUND OF LOW STRENGTH

Plastic ground exhibiting low strength was in a construction segment where the tunneling portion by TBM collapsed and was reinforced when boring an evacuation tunnel to guarantee required safety by placement of ring supports, spraying of mortar in large depths, filling of cavities in order to prevent the collapse of the tunnel. This segment for which collapse measures were taken is located between STA.228 and STA.226. For the emergency parking area in the tunnel, we used an early closing pattern of DIIcL and as an auxiliary method for excavation, used a grouting type forepiling for rock consolidation at the same time.

6.1 Construction method

The construction method is an early closing full face boring to enlarge the width of the TBM tunnel, alternating full face boring and early closure every 3 meters in the progress of boring (Figs. 7 and 8). The basic distance for early closure Lf is 6 m. The construction unit for early closure will be set at Lc = 3 m, considering construction performance and construction speed.

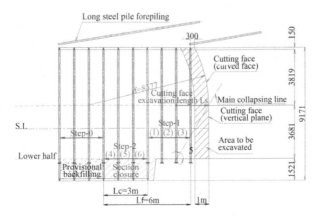

Figure 7. Steps for early closure and geometry of the curved face.

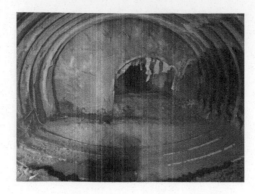

Figure 8. Early closure of the full tunnel face (Lf= 6 m).

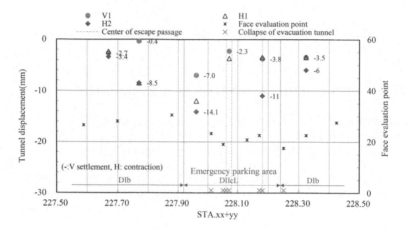

Figure 9. Displacement of the Phase II tunnel and face evaluation points.

The construction step for early closure will be 1.0 meter as unit excavation length, excavating 3 meters for the elements (1), (2) and (3) for full face boring. Then, the inverts (4), (5) and (6), are excavated and dressed at one time for installation of 3 steel inverts, and then shotcreted for three spans. For the invert, the portion of 3 meters is temporally refilled with excavation muck to end up with one cycle of early closure.

6.2 *Characteristics of tunnel behavior*

The geology concerned is composed of brittle shale with no water inflow from the cutting face. V1 of DIb is a maximum settlement of –3.5 mm. DIIcL is a maximum settlement of –7.0 m. H2 of DIb has a maximum contraction of –8.5 mm, and at DIIcL there is a settlement of –14 at maximum (Fig. 9). Convergence at DIIcL is larger at H2 than at H1. The tunnel displacement at DIIcL is slightly larger than at DIb, however sufficiently constrained to assure stability of the tunnel.

7 MEASUREMENT OF PHASE I ROUTE BEHAVIOR

Measurement of the Phase I route was made at three points: one was for blasting vibration at one point for the surface of the lining SL of the Phase I route and two other points were for displacement of the shoulder of the tunnel inner side wall and the bottom of the inner

Figure 10. Automatic measurement system of the lining displacement on the Phase I tunnel in service.

wall (Fig. 2). The former was real time measurement by accelerometer, and the latter two were measurements made at intervals of one hour by a 3D automatic measurement system (Fig. 10).

Measurements were started six months in advance at a time when there was no impact from the Phase I route. The measurement of the low strength plastic ground revealed that the horizontal displacement of the Phase I route lining was in a range of –0.2 to 1.1 mm, while vertical displacement V was +/–0.4 mm, and that there was no impact from the excavation of Phase II. The predominant impacts were from vehicle traffic.

8 LINING CONSTRUCTION

As lining concrete, the project will use a concrete of middle fluidity with a design standard strength of 24 N/mm2, slump flow 35 to 50 cm, with no material segregation. The concrete

Table 2. Mix proportion of middle fluidity lining concrete.

Material age 28 days Compressive strength (N/mm^2)	Coarse aggregate max. size (mm)	Water - binding agent ratio W/C (%)	Fine aggregate ratio s/a (%)	Specific amount (kg/m^3)							Admixture Ad (C × %)
				C	W	S1	S2	G1	G2	Fiber FB	
24	20	50.0	51.0	350	175	612	261	430	430	2.73	1.40

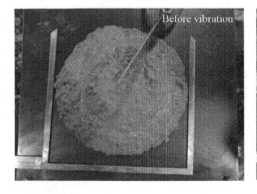

Figure 11. Characteristics of medium fluidity concrete (concrete slump test).

after being placed will be compacted with the vibrator (VB) which is to be installed in the movable form.

8.1 *Concrete mix proportion*

Table 2 shows a mix proportion of the medium fluidity concrete. The cement was an ordinary Portland cement, using an admixture Ad as AE reducing agent of one component type. The fiber (FB) is polypropylene of a specific weight of 0.91, with a diameter of 0.72 × 47. The fresh concrete had good quality, with a slump flow of 43 cm, and after vibration, exhibited no material segregation (Fig. 11).

8.2 *Concrete placement and compaction*

In order to compact medium fluidity lining concrete, an automatic compacting method is used according to the patterns of placing and compaction, and the form vibrator (VB) is controlled with a PC system (Table 3). The form VBs will be provided at intervals of 1.5 meters or less in the vertical direction, and at intervals of 3.0 meters in the longitudinal direction. For concrete placement along the sides of the form, concrete will be placed at 4 locations each on the right and left side of the movable form center. At the right and left of the shoulder, there will be two locations for concrete placement. At the crown, there will be one place where fresh concrete will be upwardly placed from below. The unit placement height is set at 50 cm. Concrete is placed alternatively from right to left, and is compacted at each place of pouring. The type of vibrator (VB) will be changed to the upward placement type when concrete is placed on the upper half. At each stage of vibration, five vibrators at maximum will be put in operation simultaneously, with operation time of 15 seconds per stage.

Table 3. Specifications of vibrators and sensors.

Machine/instrument	Standard and capacity	Number	Use
Form vibrator	550W,100Hz	60	Concrete compaction
Concrete sensors	$1.0N/mm^2$	32	Detection of concrete surface positions
Accelerometer	$max.235m/s^2$	20	Calculation of compaction energy
Pressure/temperature sensors	$1.4N/mm^2,85°C$	8	Lateral pressure, placement stop pressure, form removal strength

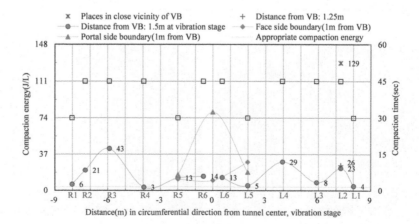

Figure 12.   Compaction energy and compaction time.

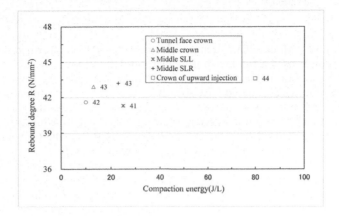

Figure 13. Rebound degree R.

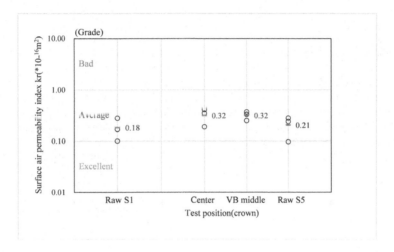

Figure 14. Surface air permeability Kr.

8.3 Construction results

In terms of the quality, the fresh concrete was stable, and after compaction, was able to remain level, with neither bleeding water nor material segregation. The compaction was good. The compaction energy was approximately 30 J/L or less, and the vibration applied was less than 10 times the appropriate compaction energy of 3.7 J/L (Fig. 12).

Three months after construction of the lining, the strength of the lining was measured by the use of a test hammer, and the surface of the lining was tested with the Torrent method for air permeability. The rebounding degree R with the test hammer was in a distribution of 41 - 44 N/mm2. The resulting compressive strength property was estimated to be homogenous (Fig. 13). The indices kr (×10–16 m2) of surface air permeability were distributed in an average range of 0.18 – 0.32 ×10–16 m2. In other words, the permeability obtained is rated "good" in the ordinary grade to achieve required compactness of the lining surface (Fig. 14). The status of the finished lining surface lining was judged excellent because no peeling, air bubbles, uneven colors or construction joint lines were observed visually.

9 RESULTS

In the elastic ground with excellent strength in which the arch was formed and able to stand on its own, the project used a full face boring method with a semi-spherical face and a standard support pattern. For the elastic ground with low strength, we used an early closing method of full section with a curved face of early closure pattern to obtain mechanical stability of the Phase II tunnel, making it possible to check impacts from the Phase I route tunnel and to guarantee function as an emergency passage, which led to a successful completion of the Phase II tunneling route. From this, it was demonstrated that the tunneling project of the Phase II route has a wide range of applicability and is highly effective.

On the other hand, the lining project used a concrete of middle fluidity which was excellent for fluidity and without material segregation. For the compaction of concrete form vibrators were used and controlled by a PC system, enabling pouring and compaction automatically according to the pattern. Therefore, the characteristics of compressive strength were estimated to be homogeneous and suitable for a desired compactness of the lining concrete surface. Consequently, it was demonstrated that the lining concrete increased in quality to be effective in achieving long term endurance.

10 CONCLUSION

The project for converting the Yuasa-gobo local toll road to a road with two lanes in each direction is scheduled to be completed by the end of 2021. On the other hand, for the Kawabe Daiichi tunnel, excavation was started in April 2016 and is currently being continued with expected completion in September 2019 with no accidents and no disasters. The technology used for the construction of this mountain tunnel is considered to greatly contribute to the development of tunneling technology for the Phase II route. We will report the details of this project on a separate occasion.

REFERENCES

Sato, J., Kanematsu, K. & Kusumoto, F. 2013. Behaviour of a tunnel with rapid ring closure and curved face in low-strength ground; *World Tunnel Congress 2013 Geneva.*
Kusumoto, F., Tanimura, K. & Sato, J. 2013. The effect of hemi-spherical tunnel face on the stability of mountain tunnels; *World Tunnel Congress 2013 Geneva.*

*Tunnels and Underground Cities: Engineering and Innovation meet Archaeology,
Architecture and Art, Volume 5: Innovation in underground engineering,
materials and equipment - Part 1 – Peila, Viggiani & Celestino (Eds)*
© 2020 Taylor & Francis Group, London, ISBN 978-0-367-46870-5

Effective water leakage detection by using an innovative optic fiber sensing for aged concrete lining of urban metro lines in Tokyo

S. Konishi, N. Imaizumi & Y. Enokidani
Tokyo Metro Co.,Ltd, Japan

J. Nagaya
Geo-Research Institute,Co., Ltd., Japan

Y. Machijima
LAZOC Corporation, Japan

S. Akutagawa & K. Murakoshi
Kobe University, Japan

ABSTRACT: A trial investigation was undertaken to evaluate the feasibility of applying a uniquely designed optic fiber sensor for detection of water leakage in a metro tunnel in Tokyo. To conduct effective maintenance works for underground infrastructures constructed several decades ago, it is essential to perform a thorough inspection for damage states of concrete structures that may be induced by water leakage from cracks over time. The results of the investigation indicated that the proposed sensor is capable of detecting water leakage both in laboratory and field, suggesting that the method would be readily applicable for practical use in maintenance works for aged concrete structures.

1 INTRODUCTION

Tokyo Metro operates a total of 195km of urban rail networks of which 167km is underground. There have been an increasing number of locations where a maintenance care has to be taken with respect to water leakage through cracks in aged tunnel linings, which might occasionally lead to damaged concrete or rusting of re-bars within the lining concrete if proper treatment is not conducted. It is therefore regarded important that water leakage behavior be appropriately grasped over time such that periodical manual investigation and the following maintenance works can be properly planned and executed.

An innovative optic fiber sensor has been proposed by Akutagawa et al. (2017, 2018) with which presence of water can be clearly identified and recorded. In this method, plastic optic fibers are used as 1-pixel camera in a very different way from other methods of optic fiber sensing, for example, Hill et al.(1978), Meltz et al.(1989) and Funnel et al.(2015). The sensor is made of two plastic optic fibers glued together with both of their tips cut at 45 degrees in such a way that the sensing part has two inclined faces. Akutagawa et al. (2017, 2018) have already confirmed that this sensor can detect presence of water in a very sensitive way so that a small difference in wet state of concrete surface, such as the one shown in Figure 1, could be identified with respect to time elapsed. A series of laboratory and field tests are described in this paper to show how this sensor works and how it can be applied for an important maintenance problem with efficiency and confidence.

Figure 1.   Typical appearance of wet surface of tunnel concrete wall.

2   DETECTION OF WATER BY PLASTIC OPTIC FIBER SENSOR

A new sensor using two Plastic Optic Fibers (in short, POF and its diameter is typically 1mm) (Akutagawa et al., 2017, 2018) is shown in Figure 2. It is called RR sensor, as its measurement concept is derived from the physics of *Reflection* and *Refraction* of light. In this sensor, light is first sent into Fiber 1 as indicated as L1, part of which is refracted as L2 leaving Fiber 1 and the rest is reflected as L3. Then, the light with the flux L3, reaches the side face of Fiber 1; part of it gets reflected, and the rest goes out of Fiber 1. The light that has just left Fiber 1 immediately reaches the side face of Fiber 2, where some of it gets reflected, and the rest finally goes into Fiber 2 as L3′. L3′, again, reaches the inclined face of Fiber 2 and what has happened before on the inclined face of Fiber 1 occurs again in the same manner; the paths of light in Fiber 2 are defined with the fluxes of L4 and L5. The important fact over this sequence of how light travels across two inclined faces is that the magnitude of flux L5 depends primarily on Δn, which is the difference between refractive index of POF, 1.49, and that of the surrounding material. It should also be noted that L5 also depends on how much of the total area of inclined faces are in contact with the surrounding material.

Exact prediction of L5 for a constant and stable L1 is not straightforward as true geometry of fiber faces is 3-dimensional and two fibers must be glued together as shown by the Front view in Figure 2. However, the description of the proposed RR sensor supports the

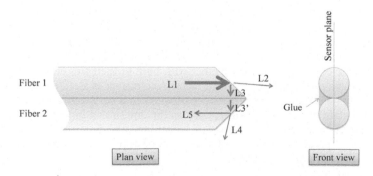

Figure 2.   RR sensor made of plastic optic fibers for water detection.

fundamental strategy for creating a new fiber sensor with inclined front faces. Akutagawa et al (2017, 2018) showed in their previous works that RR sensor could be used to detect presence of water and its water-ice phase change very clearly regardless of whether the water lies solely on table or it is in a soil mass.

3 LABORATORY TESTS

3.1 Behavior of water around RR sensor installed horizontally

Figure 3 shows two possible directions in which an RR sensor could be installed to detect water leakage. The important point not to be missed is that those sensors are installed near a point from which water leaks forming a typical wet triangle zone below.

The first laboratory test was configured in such a way that a tip of an RR sensor, installed in X direction, was in point-contact with vertical surface of a concrete plate, while keeping its sensor plane (shown in Front view of Figure 2) horizontal. A couple of drops of water were given to the contact point and light intensity was recorded as the water disappeared by dropping and also by evaporating over time. Photographs taken in this process are shown in Figure 4. At 20 sec (immediately after the water was given), it is confirmed that much of the surface area of inclined faces are in water because of surface tension of water. As time elapsed, the surface area which is covered by water kept decreasing leading to a gradual increase in the light intensity recorded by the RR sensor as shown in Figure 5. Note that it took around a minute or so for the light intensity comes back to the value recorded before wetted.

3.2 Behavior of water around RR sensor installed vertically

The second laboratory test was configured in such a way that a tip of an RR sensor, installed in Y direction, was in side-face-contact with vertical surface of a concrete plate, while keeping its sensor plane (shown in Front view of Figure 2) vertical. A couple of drops of water were given at around the 80th second only once. Photographs shown in Figure 6 confirm that the water did not stay around the sensor tip for long and dropped quite rapidly. The light

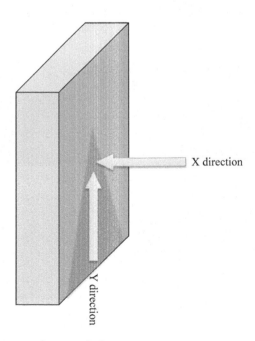

Figure 3. Fundamental strategy for water leakage measurement.

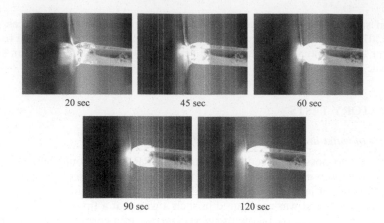

Figure 4.   Behavior of water around RR sensor installed horizontally.

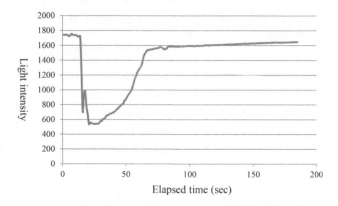

Figure 5.   Light intensity recorded by the RR sensor installed horizontally.

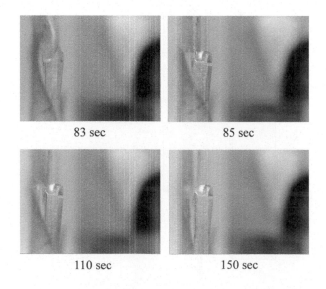

Figure 6.   Behavior of water around RR sensor installed vertically.

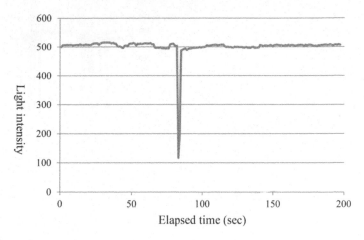

Figure 7. Light intensity recorded by the RR sensor installed vertically.

intensity recorded also supports this phenomenon while showing just one sharp spike in the graph. The difference of water behavior around the sensor tip seems to come from differences in three dimensional geometry of the sensor with respect to flat concrete surface where surface tension of water plays a key role.

3.3 *Simulation of water leakage using a specially designed apparatus*

Having confirmed sensitivities of RR sensors installed in two different directions, an additional test was conducted by using a set of attachment apparatus. As shown in Figure 8, a concrete plate was prepared onto which two RR sensors were installed in X and Y directions, respectively. A small hole was prepared as a water leakage point, which was approximately aligned vertically with the contact points of the RR sensors in X and Y directions.

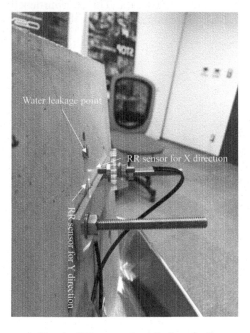

Figure 8. Light intensity recorded by the RR sensor installed vertically.

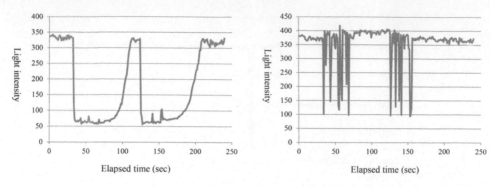

Figure 9. Light intensity recorded by the RR sensors for repeated occurrences of water leakage. (a) X direction (b) Y direction.

Water was supplied through the hole from behind the concrete plate such that it leaked out of the hole streaming downward wetting the RR sensors. As the water supply was given in two main steps, the RR sensors recorded the light intensities for X direction (Fig. 9 (a)) and Y direction (Fig. 9(b)), respectively. It was confirmed that the installation arrangement to detect water leakage from concrete surface was appropriate for use in a field measurement.

4   FIELD TESTS

4.1   *Monitoring site in Tokyo and installation of RR sensors*

In order to perform a trial monitoring of water leakage, a site was chosen along Chiyoda Line of Tokyo Metro as shown in Figure 10. In this monitoring, the main focus was set on the identification of timings of water leakage.

A double deck concrete structure (as shown in Figure 11(a)) was constructed in 1969 and has been in use since then for nearly 50 years. The overburden thickness of soil above the structure is approximately 7m and the underground water table is expected to be somewhere

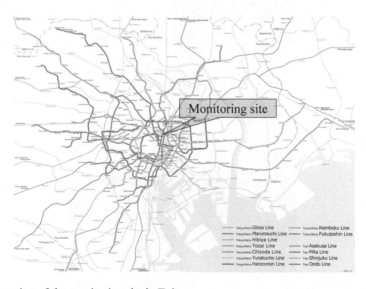

Figure 10.   Location of the monitoring site in Tokyo.

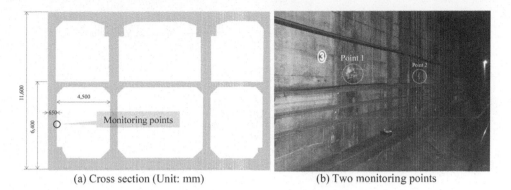

| (a) Cross section (Unit: mm) | (b) Two monitoring points |

Figure 11. Cross section of the monitoring site and monitoring points. a) Cross section (Unit: mm) (b) Two monitoring points.

above the upper end of the concrete structure. Two monitoring points were set up as Points 1 and 2, respectively, as shown in Figure 11(b).

For each monitoring point, accumulated residue material on the surface was removed first and the same attachment apparatus as was used in the laboratory test was used to install two RR sensors, X and Y directions, and a temperature sensor, as shown in Figure 12(a). Once all sensing devices were mounted on the apparatus, they were covered by a black cap box, shown in Figure 12(b), so that light from train would not disturb the measurement process. Data logging was conducted in a room which was approximately 25m away from the monitoring points.

4.2 Results of the monitoring

The results of monitoring are shown in Figures 13, 14 and 15 for the first week. Blue vertical bands indicate time zones when no train was running. Light intensities for X and Y directions were successfully recorded at Point 1, as shown in Figure 13. Whereas the light intensity for X direction showed the smooth and slowly declining trend over a week, that for Y direction showed more variations, which is believed to be related to overall temperature variation. It is also seen that there is no sign of water leakage at Point 1 in the first week.

The light intensity for X direction at Point 2 was not recorded due to a technical problem related to damage to plastic optic fiber. The light intensity for Y direction at Point 2 showed overall variation which is similar to what was observed at Point 1. In addition, the data showed finer variations within a day. Therefore, a detailed look at the data variation in the

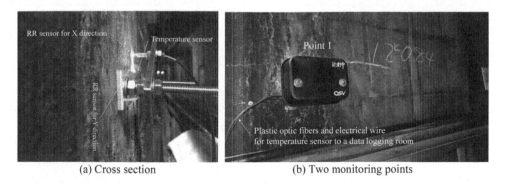

| (a) Cross section | (b) Two monitoring points |

Figure 12. Installation of two RR sensors, X and Y directions, and a temperature sensor at a monitoring point. (a) Cross section (b) Two monitoring points.

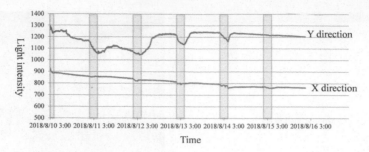

Figure 13.   Light intensity for Point 1 recorded in the first week.

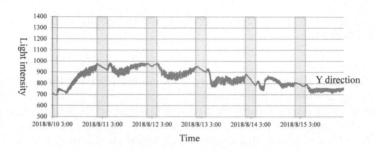

Figure 14.   Light intensity for Point 2 recorded in the first week.

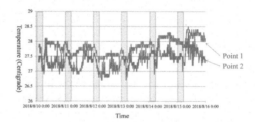

Figure 15.   Temperature profile for the first week with data sampling time interval of 10 minutes.

light intensity for Point 2 is given for a short period, approximately 30 minutes, on August 22nd, 2018, in Figure 16 with the temperature data which was sampled every one minute. Red arrows are expected times of trains passing in front of the monitoring points.

For the first three trains passing in this period, the temperature and also the light intensity tended to rise a little and come back to the values recorded just before the trains. For the fourth train, the temperature rose again as anticipated, whereas the light intensity showed a sharp drop before the timing of the fourth train and came back. A similar sudden drop of the light intensity is also seen in the period between 12:57 and 12:59. These drops in the light intensity are most likely the direct indication of water leakage due to the following reasons:

1. Their timings are irrelevant of passing trains.
2. The light intensity dropped, instead of rising, and its magnitude was much bigger than those variation ranges which seem to have been caused by the effects of passing trains.
3. The shapes of light intensity graphs are in accordance with the shapes recorded for water leakage in the laboratory tests as shown in Figures 7 and 9(b).

Finally, the light intensities recorded from August 20th to 23rd are shown in Figure 17. For Point 1, the recorded light intensities suggest that there was literally no water leakage of considerable degree, whereas a slight and gradual drop of the light intensity is observed for X

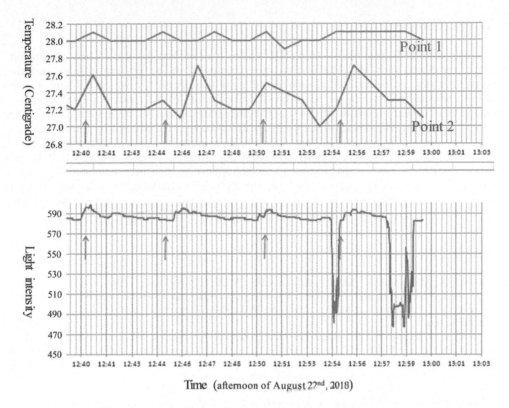

Figure 16. Detailed look at data variation with respect to trains passing.

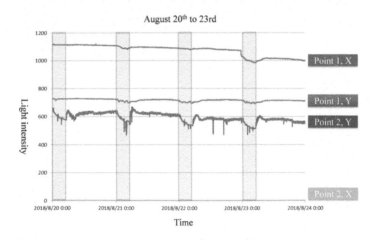

Figure 17. Light intensities recorded from August 20th and 23rd, 2018.

direction. On the other hand, the light intensity recorded in Y direction at Point 2 showed many downward spikes indicating that the water leakage occurred at multiple timings throughout the clock including the periods when no train was running. By correlating these data with respect to temperature environment, underground water level variation and crack width variation with time, a useful set of data would become available for improving maintenance works in future.

5 CONCLUSIONS

A uniquely designed optic fiber sensor, RR sensor, was applied for water leakage detection both in laboratory and field. The investigation conducted in the laboratory suggested that presence of water can be detected clearly regardless of the direction in which RR sensors were used with respect to concrete surface. The trial application of RR sensor for water leakage detection in a 50 year old metro tunnel was conducted in Tokyo. The results obtained showed that light intensity variations recorded by RR sensors were partly affected by trains passing in front of the monitoring points. The exact mechanism on how the light intensity was affected is to be studied in a future study. Finally and most importantly, it was confirmed that the RR sensor captured the light intensity variation which seems to have been caused by the actual water leakage phenomena. Further continued study will be conducted to minimize effect of passing trains on the measurement data and to reveal seasonal characteristics of water leakage so that effective maintenance practices would be possible.

REFERENCES

Akutagawa, S., Machijima, Y., Sato, T. & Takahashi, A. 2017. Experimental characterization of movement of water and air in granular material by using optic fiber sensor with an emphasis on refractive index of light. *Proceedings of the 51st US Rock Mechanics/Geomechanics Symposium. ARMA 17-313.* San Francisco.

Akutagawa, S. & Y. Tanaka. 2018. Experimental observation of hardening process of engineering materials by optic fiber sensor. *Proceedings of the 52st US Rock Mechanics/Geomechanics Symposium. ARMA 18-1127.* Seattle.

Funnell, A., Xu, X., Yan, J., & Soga, K. 2015. Simulation of BOTDA and Rayleigh COTDR systems to study the impact of noise on dynamic sensing. *Int. J. Smart Sens. Intell. Sys.* 8 (3).

Hill, K.O., Y. Fujii, D.C. Johnson & B. S. Kawasaki. 1978. Photosensitivity in optical fiber waveguides: Application to reflection fiber fabrication. *Appl. Phys. Lett.* 32 (10): 647.

Meltz, G. et al. 1989. Formation of Bragg gratings in optical fibers by a transverse holographic method. *Opt. Lett.* 14 (15): 823.

*Tunnels and Underground Cities: Engineering and Innovation meet Archaeology,
Architecture and Art, Volume 5: Innovation in underground engineering,
materials and equipment - Part 1 – Peila, Viggiani & Celestino (Eds)
© 2020 Taylor & Francis Group, London, ISBN 978-0-367-46870-5*

Development and field validation of the Smart Batcher Plant system used for control of as-mixed temperature of concrete

K. Kumagai, T. Tsutsui, A. Hirama, H. Matsuda & M. Takinami
Tobishima Corporation, Tokyo, Japan

S. Kobayashi, N. Endo & H. Hata
Harasho Co., Ltd, Matsue, Japan

ABSTRACT: In conventional tunneling, initial strength, long-term strength and bonding properties of shotcrete are significantly affected by the type and content of setting accelerator. Setting accelerator performance is highly dependent on temperature, and changes in base concrete temperature can affect the quality of shotcrete. The authors have developed "Smart Batcher Plant" that automatically controls the as-mixed temperature of concrete and produces concrete of stable temperature. The plant was field-tested at a road tunnel construction site in a cold region where the lowest temperature in winter can falls to 20°C below the freezing point. This paper briefly describes the Smart Batcher Plant and reports on the field validation of the automatic as-mixed temperature control function, the results of comparison with the conventional construction method, and the verification of quality and workability improvements and cost reduction accomplished by achieving target temperatures.

1 INTRODUCTION

Shotcrete is one of the primary supports used in conventional tunneling. Its initial strength, long-term strength, and bonding properties are greatly affected by the type and content of setting accelerator used, and as shown in Figure 1, the performance of the setting accelerator is highly temperature-dependent. Changes in the base concrete temperature lead directly to changes in shotcrete quality.

Particularly in winter when base concrete temperatures are lower, such measures as increasing setting accelerator content above normal levels are taken to ensure development of satisfactory initial strength, while maintaining the bonding properties of shotcrete. However, exces

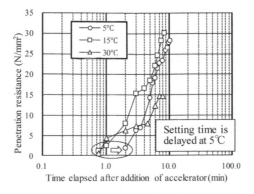

Figure 1. Temperature dependency of setting accelerator.

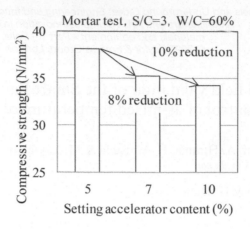

Figure 2.   Effect of setting accelerator content on compressive strength.

sively increasing the amount of setting accelerator added can lead to deterioration of long-terms strength, as shown in Figure 2. Adjustment of the setting accelerator content during construction is normally done while checking the bonding state of the shotcrete. Under certain conditions, fluctuations in the concrete temperature may make it difficult to set the percentage of setting accelerator content for adequate bonding, which can lead to qualitative deterioration. In winter, hot water is used for mixing to raise the concrete temperature. However, there are few examples of constant control of heating or of heat-insulation being used for the mixing water. Usually the as-mixed concrete temperature becomes unstable when the ambient temperature fluctuates greatly.

To overcome these difficulties and achieve stabilization of shotcrete and improve its workability, the authors have developed a "Smart Batcher Plant." This plant is designed to produce concrete at stable temperatures through automatic control of the as-mixed concrete temperature to specified levels. The plant was put into full operation in the Iwai-Area Tunneling Project (undertaken by the Tohoku Regional Development Bureau, MLIT) for the Miyako-Morioka Highway. The highway is being built in an area where winter temperatures can fall 20°C below the freezing point.

This paper provides an outline of the Smart Batcher Plant and describes in detail the results of validating its automatic as-mixed temperature control function when it was put into full operation in the road construction work in the cold region.

2   OUTLINE OF THE SMART BATCHER PLANT

2.1   *Automatic control of the as-mixed temperature*

The Smart Batcher Plant is a concrete production system that provides automatic control of as-mixed concrete temperatures to ensure a stable supply of concrete at optimal temperatures and to enhance the performance of mixing materials. An outline of the plant is shown in Figure 3.

The automatic as-mixed temperature control system consists of three basic functions:

Function I: Heating concrete materials, such as water, fine aggregate (sand), and coarse aggregate (crushed stone)
Function II: Accurate temperature measurement of materials before mixing and continuous temperature measurement during mixing
Function III: A control function for automatic adjustment of the ratio of raw water (cold water) to hot water to achieve targeted as-mixed temperatures

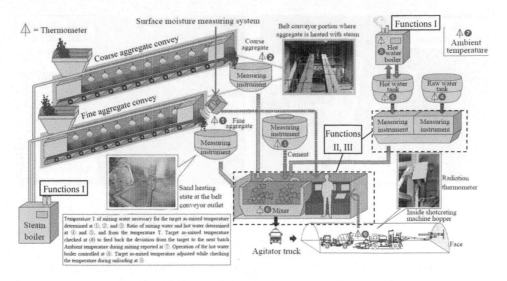

Figure 3. Outline of the Smart Batcher Plant system.

The greatest merit of the Smart Batcher Plant is its two functions, Function II and Function III above, that include accurate measurement of the temperatures of the raw and hot water, the aggregates, the cement and the as-mixed concrete, and the sequential calculation of the heat capacity necessary to obtain the targeted as-mixed temperatures based on the measurements. This enables production of concrete at targeted as-mixed temperatures through automatic adjustment of the amounts of hot and raw water to be added. The plant also measures changes in concrete temperatures from the end of the mixing to the start of the shotcreting, in which a hopper and shotcreting machine near the tunnel face are used. The measurements are reflected in the target for as-mixed temperature in the next cycle, which enables shotcreting to always be executed at the targeted concrete temperature. It should be noted that the temperature of the concrete being mixed is obtained without contact by means of a digital radiation temperature sensor with a shutter for continuous temperature measurement, as shown in Figure 4. This unit is installed on top of the mixer.

In winter in cold regions, automatic control to achieve targeted as-mixed temperatures can be difficult by mixing with hot water alone. In such cases, arrangement of a steam spray system for heating of aggregates, as shown in Figure 5, enables heating of fine and coarse aggregates on the conveyor belts (Function I). With this function, the temperature of the aggregates before mixing can be increased by around 10°C. In addition, heating of aggregates

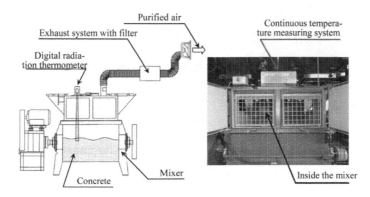

Figure 4. Method used to measure the concrete temperature in the mixer during mixing.

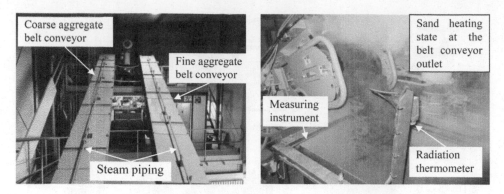

Figure 5. Steam spray on the conveyor belts: Function I.

by the steam spray system enables efficient heating of the materials only in amounts required in the mixing of the concrete for the next batch. A function is also provided for measurement of the percentage of aggregate surface moisture for each batch. This is based on the change in the percentage of surface moisture caused by steam spray, which is fed back for adjustment of the amount of mixing water.

2.2 *Cloud management of actual mixing data*

The function for cloud management of actual mixing data is illustrated in Figure 6.

In conventional batcher plants, the mixing data during tunneling is managed using printed slips. On the other hand, in the Smart Batcher Plant, the actual mixing data is digitalized (as shown in Figure 7) and acquired remotely via the Internet (cloud). The actual data in the mixing cloud includes quality-related data, such as the concrete temperature and the percentage of surface moisture, along with the construction-related data, such as design, and the actual support pattern. These data can be utilized for construction management after automatic editing for each month, and each support pattern by means of a management slip conversion function. The construction manager in charge of plant management can check the entry/exit of mixer trucks and aggregate stock status with images on a PC or smartphone by setting up a Web camera at the entrance or inside the batcher plant.

In this way, the construction manager can deal with quality and material control related to shotcrete in a short period of time while using remotely acquired actual mixing data and images from Web cameras. This is expected to lead to labor saving and greater management efficiency.

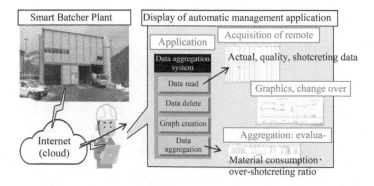

Figure 6. Outline of cloud management function of actual mixing data.

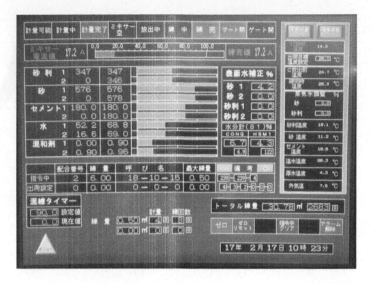

Figure 7. Actual mixing data on the control panel.

3 FIELD VALIDATION IN COLD REGIONS

3.1 *As-mixed temperature of concrete*

Field validation of the plant was implemented in the Iwai Area Tunneling Project (undertaken by the Tohoku Regional Development Bureau, MLIT) to evaluate the functions for automatic control of as-mixed temperature in cold regions.

Figure 8 shows the actual values of ambient and as-mixed temperatures in January 2017. Circle marks in the figure show the average temperature of concrete to be shipped (average temperature of concrete of one truck mixer), and square marks show the as-mixed temperature of each batch. The ambient temperature shown by a solid line represents the observed data at ground observation posts (AMEDAS, name of point: Kuzakai) in the neighborhood of the applied site.

In the field validation, the targeted as-mixed temperature was set to 25°C while taking into account that the temperature drop during the as-mixed period from mixing to shotcreting is around 5°C.

Figure 8 shows that the temperature of shipped concrete was within a range of the targeted 25±3°C. Since the trend of changes in concrete temperature was found to be not related to changes in ambient temperatures, it was confirmed that concrete could be produced at stable

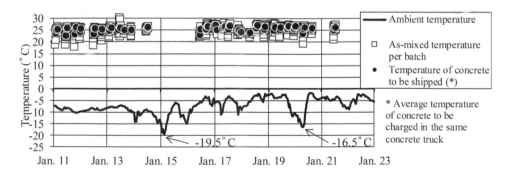

Figure 8. Changes in concrete as-mixed and ambient temperatures over time.

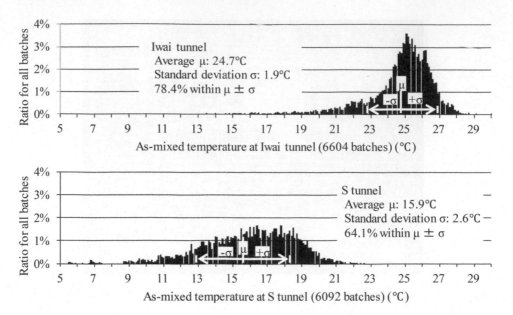

Figure 9.   Frequency distribution of concrete as-mixed temperatures.

as-mixed temperatures without being affected by fluctuation of ambient temperatures even in winter when the ambient temperature can drop to 20°C below the freezing point.

Figure 9 shows the frequency distribution by comparing two types of data. They are actual measurement data on as-mixed temperatures during a period from December 2016 to March 2017 for the S Tunnel Project (road tunnel in Fukui Prefecture), for which concrete was produced with the batcher plant using hot water without automatic temperature control, and the actual data on as-mixed temperatures in this project. As is evident from this figure, in the case of the S tunnel, the as-mixed temperature could not be positively controlled, resulting in an average as-mixed temperature of 15.9°C and a standard deviation of 2.6°C, although the concrete was produced with hot water. In the case of the Smart Batcher Plant, the average as-mixed temperature was 24.7°C and the standard deviation was 1.9°C. This was due to automatic control to the targeted as-mixed temperature of 25°C for all batches. It was confirmed that concrete could be produced and controlled with accurate concrete temperature control.

3.2   *Effects on quality and workability*

The authors hypothesized that strict control of the shotcreting quantity and the percentage of setting accelerator added to the content per unit time during shotcreting achieved by stabilizing the shotcrete temperature would stabilize the bonding properties and workability of shotcrete even in winter when construction conditions are severe with temperatures below the freezing point. It was expected that this would prevent deterioration and variance of quality and contribute to qualitative improvement.

Table 1 shows the shotcrete mix proportion as discussed in this paper.

Table 1.   Mixing of shotcrete.

		Unit weight (kg/m³)				
Water-cement ratio (%)	Fine aggregate ratio (%)	Water	Cement	Fine aggregate	Coarse aggregate	Fine dust reducer (C × 0.5%)
51.1	62.0	184	360	1105	694	1.80

Table 2 shows the shotcreting conditions in the field validation. The shotcrete mix proportion included 360 kg of unit cement with a design standard strength of 18 N/mm². Shotcreting was done using the wet-mix process at an injection rate of about 15m³/hr.

Figure 10 shows the average setting accelerator content before and after activation of the automatic control function during autumn to winter. Figure 11 shows a comparison of the compressive strength achieved during the same period using the conventional method and using the Smart Batcher Plant. Figure 12 shows a similar comparison in terms of the over-shotcreting ratio. The over-shotcreting ratio is the ratio of all quantities of overbreak, rebound, and any surplus quantity to the quantity at the design shotcrete thickness. The discussions below used the over-shotcreting ratio as an index to determine the appropriateness of the bonding properties because improving these properties will lower the over-shotcreting ratio.

Table 2. Shotcreting method.

Shotcreting conditions	Specifications
Shotcreting method	Wet type
Shotcreting machine	Solid spray machine with elevator (FTS., LTD: Hercules)
Discharge	15m³/hr (max. 25m³/hr)

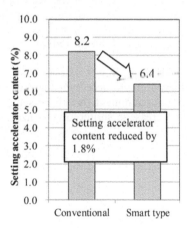

Figure 10. Comparison in terms of setting accelerator content.

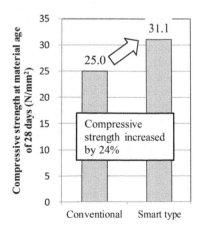

Figure 11. Comparison in terms of compressive strength.

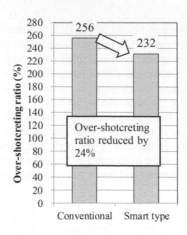

Figure 12.   Comparison in terms of over-shotcreting ratio.

In order to ensure satisfactory bonding properties and development of strength even in winter, the concrete as-mixed temperature was set at 25°C. (The concrete temperature during construction was 20°C.) This enabled reduction of the settling accelerator content from 8.2% to 6.4%, as shown in Fig. 10. As a result, the shotcrete compressive strength at the age of 28 days was 31.1N/mm², which was a 24% increase from the case of the setting accelerator content of 8.2% (Figure 11). This made it possible to reduce the over-shotcreting ratio by 24%, from 256% to 232% (Figure 12).

Installing Functions I, II, and III of the Smart Batcher Plant is an added cost. However, the reduction of material costs achieved due to decreasing the setting accelerator content exceeds the cost of installing the additional equipment. It can be seen that the overall cost reduction becomes even greater when the decrease in shotcreting utilization due to less over-shotcreting is taken into account.

4   CONCLUSIONS

Shotcrete is one of the primary supports used in conventional tunneling. Stabilizing its quality and improving its workability are essential construction management goals. The Smart Batcher Plant developed by the authors enables the provision of a stable supply of concrete at optimal concrete temperatures, which makes it possible to maximize the performance of the materials. The system makes it possible to gain satisfactory workability of shotcrete even in winter periods when the ambient temperature drops sharply. It was confirmed that, when applied to tunneling in cold regions, the use of this technology can lead to qualitative improvement of shotcrete and lower shotcreting costs.

In the future, we will proceed with the addition of a material cooling system to cope with high summer temperatures and rises in concrete temperatures during summer. We also intend to add a mixer heating function to enable automatic control of as-mixed temperatures with greater accuracy and to improve the method used for feedback to the next batch. We will also promote application of the technology in the tunneling project for the Chuo Shinkansen and the Hokkaido Shinkansen. A Smart Batcher Plant has been operating since December 2017 in the Oshima Tunnel (Kitauzura) Specified Construction Project. (This is a Joint Venture for the Sato Mirai Kusawake Tabata Hondo Hokkaido Shinkansen.) A Smart Batcher Plant has also been in operation since June 2018 in the Niseko Tunnel construction project, which is another Specified Construction Project Joint Venture, undertaken for the Tobishima, Daiho, Saito, Shiraki Hokkaido Shinkansen. For these two tunnel projects, a mixer with a capacity of 1.3 m³ has been introduced. This is significantly larger than the normal capacity of 0.5 m³. It increases the concrete supply capacity during split mixing from 18m³/hr to 40m³/hr. We are verifying the effects of automatic as-mixed temperature control along with increased mixer capacity.

In the future, Smart Batcher Plants will be increasingly introduced at construction sites. This will enable further verification of successful experiences and the accumulation of more basic data on the contribution of the automatic as-mixed temperature control function to improvement of shotcrete bonding properties and greater strength. In this way, we will continue our efforts to establish a high-quality shotcreting method without any rebound and with minimum variance in strength.

ACKNOWLEDGEMENTS

We would like to express our deepest gratitude to the Tobishima Construction personnel working at the tunnel site and the staff of Denka Co., Ltd., the manufacturer of the setting accelerator, for their kind cooperation in the development work.

REFERENCES

Hirama, A. & Ando, S. & Araki, A. & Uomoto, T. 2000. Effects of Materials Used on the Strength of Shotcrete. *Annual Collected Papers on Concrete Techznology* 22 (2). Tokyo, Japan: JCI.

Kadota, K. & Marukawa, S. & Nishie, K. & Matsuda, A. & Konishi, M. & Iwamoto, Y. & Teramoto, T. & Takahashi, T. 2000. Seasonal Characteristics of Shotcrete Using Setting Retarder. Japan Society of Civil Engineers. *The 55th Annual Scientific Meeting* (V-222). Tokyo, Japan: JSCE.

Kumagai, K. & Tsutsui, T. & Takinami, M. & Matsuda, H. & Yamada, H. 2016. Development and Field Application of a Smart Batcher Plant for Automatic As-Mixed Temperature Control. *Report of Tunnel Technologies* 26. Tokyo, Japan: JSCE.

Takinami, M. & Tsutsui, T. & Kumagai, K. & Hirama, A. & Matsuda, H. & Yamada, H. & Hata, H. & Endo, N. 2016. Development of a Smart Batcher Plant for Tunnel Shotcrete. *Collected Abstracts of Civil Engineering and Construction Presentations*. Tokyo, Japan: JSCE.

Tunnels and Underground Cities: Engineering and Innovation meet Archaeology, Architecture and Art, Volume 5: Innovation in underground engineering, materials and equipment - Part 1 – Peila, Viggiani & Celestino (Eds)
© 2020 Taylor & Francis Group, London, ISBN 978-0-367-46870-5

Extraction geothermal energy from the rock mass surrounding tunnels and its exploitation. Feasibility study in a sector of the Brenner Base Tunnel

M. Lanconelli, A. Voza & H. Egger
Galleria di Base del Brennero BBT-SE, Bolzano, Italy

ABSTRACT: Based on an experimental study developed in a sector of the Brenner Base Tunnel (BBT) in 2015, this paper summarizes the result of an analysis concerning the exploitation of the geothermal energy from the rock mass around a tunnel. The geothermal heat stored inside the rock mass could be exploited by a closed-loop system composed by heat exchangers in the tunnel lining, so called energy lining. The potential energy of a stretch of the BBT was defined via geothermal characterization, laboratory tests and temperature measurements. The energy lining was designed for the present case. In order to exploit the extracted energy and have a reference to define the convenience of the investment, a road pavement with snow-melting and de-icing systems was supposed as potential energy user. The operating costs were compared with other heating solutions. The study shows the technical and economic feasibility of this solution and represents a step for future research on this subject.

1 INTRODUCTION

The article introduces several topics concerning the low-enthalpy geothermal energy and an innovative technology which can extract heat from the rock mass through deep tunnels. The structural lining of the tunnel can act as a heat exchanger with the surrounding rock mass. Tunnels are particularly good locations for geothermal energy extraction systems, because they are often located deep enough to be affected by geothermal heat. The geothermal energy extracted by this type of underground infrastructure can be used indirectly to heat buildings nearby the tunnel access or for activities where a heat source is required. The energy lining also benefits the tunnel itself by helping to control internal temperatures, reducing ventilation and cooling costs (Zhang et al., 2016a, b). The study focused on the feasibility of a closed-loop system composed by heat exchangers in the access tunnel on the Italian side of the Brenner Base Tunnel. Today energy linings for tunnels are still in a research phase and only few tunnels in the world use this technology, especially in Austria and Germany. The first test field in Italy was inaugurated in 2017 in the Lingotto-Bengasi subway, Turin (Barla, 2018). As described below, the design of the energy lining of the BBT differs from that adopted in Turin comparing technical characteristics, construction method and surrounding environment.

2 GEOTHERMAL ENERGY AND ENERGY LININGS

The temperature underground is influenced by a complex balance between different energy sources. At depths above 100 m, temperatures are influenced by the geothermal flow from the inside of the Earth. Closer to the surface, temperatures depend mainly on solar irradiation.

Except in places with particular geological or volcanic anomalies, the most common geothermal systems, like the one in our study, are low-enthalpy systems in which the liquid heat transfer medium reaches temperatures below 90° C (D.Lgs. n. 22 11/02/2010). The geothermal

Figure 1. Simplified configuration of the energy lining (www.rehau.com).

energy potential at low enthalpy cannot reach temperatures high enough for heating needs, so that frequently the installation of a heat pump is necessary (Tinti, 2008).

In general the most common geothermal systems used to extract the underground heat consist in absorber pipes placed into the ground in which a thermo-vector fluid circulates accumulating heat that is released to the geothermal heat-pump. Among the various types of geothermal system, the most popular is the vertical closed-loop. Energy linings for tunnels have similar characteristics, however in this case the closed-loop circuit ("source circuit") is placed in or near the lining, becoming a heat exchanger with the surrounding rock mass (Lanconelli, 2016) (Figure 1).

The system can be used both for tunnels excavated using drilling and blasting (D&B), as explain later, and for those excavated using a TBM. In the latter case the absorber pipes are preventively placed within the precast elements during their production.

For tunnels excavated by using conventional method (D&B or mechanical equipment) several installation possibilities exist. Absorber pipes can be placed in the final lining, fixed on the primary lining or placed directly into the geotextile of the waterproofing system. The most suitable position for the absorber pipes depends on the temperature domain at the boundaries, the purpose of the energy lining (exploiting energy/cooling the tunnel) and the practical installation constraints of time and cost features in terms of containing timing and costs (Tinti, 2017).

3 GEOTHERMAL ENERGY APPLIED TO THE BRENNER BASE TUNNEL

The Brenner Base Tunnel is a 55-km railway tunnel running north-south and connecting Italy with Austria. It crosses the Eastern Alps under the Brenner Pass. The infrastructure is of primary importance as part of the European TEN Scan-Med Corridor (Scandinavian-Mediterranean), which links La Valletta with Helsinki and is the longest multi-modal link between northern and southern Europe.

The length of the main tunnels between the south portal, Fortezza (Italy), and the north portal, Innsbruck (Austria) is just over 55 km. The tunnel will be directly linked to the existing Innsbruck rail bypass (Figure 2). This means that once the BBT is completed, the underground rail connection will be, at over 64 km, the longest railway tunnel in the world.

Considering that the tunnel portals are strategic points for the transport and exit of extracted geothermal heat, the Mules Access Tunnel represents the most appropriate part of the underground system to install the energy lining.

The Mules Access Tunnel, which was fully excavated in 2009, is the only access tunnel to the Brenner Base Tunnel in Italy. It runs west to east and is 1.780 m long, intersecting the west tunnel (Figure 2). The maximum tunnel bench is about 10 m, while maximum height is around 9 m. The maximum overburden of the Mules Access Tunnel is 1.210 m at chainage 1.540. The excavation of the entire access tunnel took place inside a granitic mass (Brixen Granite). During operational phase of the main tunnels, the Mules Access Tunnel will be used for rescue vehicles in case of emergencies. The selection of this sector of the BBT was made

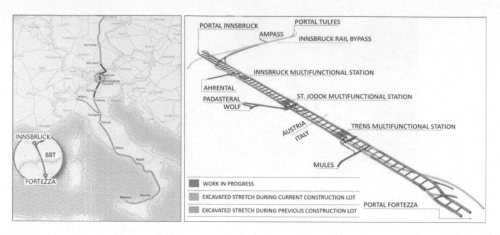

Figure 2.   Brenner Base Tunnel system (www.bbt-se.com).

based on several preliminary considerations. Currently, only the first lining has been applied and the final lining will be realize in later. In this case, the geothermal system could still be easily embedded between first and secondary lining (Lanconelli, 2016).

The Mules Access Tunnel will not have a railway infrastructure nor will it be used for any kind of railway traffic. Therefore, the source circuit and the railway systems could not interfere with each other in any way.

In a 2-km radius around the portal there are several structures that can use the geothermal energy (greenhouses and the residential center of Mules).

3.1   *Measurement of the temperatures*

In the months of July, November and December 2015, temperature measurements were carried out inside the Mules Access Tunnel. Using a laser infra-red thermometer, temperatures on the intrados of the first lining were measured in 6 positions (invert, crown, foot walls, abutments) for 21 sections of the tunnel. Figure 3 is a summary of the temperature trends.

As the chainage increases, and with it the overburden, the temperatures increase. This tendency is more significant in the winter months, in which the difference between the portal (chainage 0) and the end of the tunnel (chainage 1.780) reaches 30 °C in December. Temperatures tend to be uniform, independently from the measurement period, maintaining values between about 16 °C and 20°C. The overburden has a direct effect on the temperature of the lining. The rock mass above acts as a sort of insulating layer for the heat coming from the deeper levels of

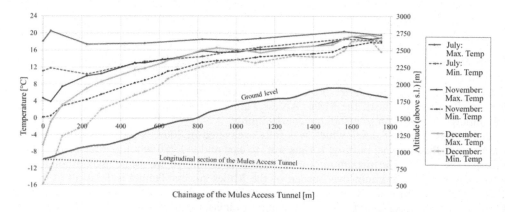

Figure 3.   High and low temperatures measured on the existing first lining.

the Earth's crust (Lanconelli, 2016). Near the portal, on the other hand, the outside climate has a significant effect. For almost all measurements, the highest temperatures were taken near the crown, probably due to the convective movements of the air inside the tunnel.

3.2 *Laboratory tests on sample of rock and shotcrete*

In order to determine the thermal and physical properties of the rock mass around the Mules Access Tunnel and the first lining which has already been applied, thermal tests based on the "Flash Method" were carried out with the NETZSCH LFA 447 NanoFlash® machine of the Technical Physics Laboratory (University of Bologna). The test measures the thermal diffusivity α and the specific heat C_p and, indirectly, thermal the conductivity λ of a material. Brixen Granite and shotcrete were investigated. These parameters are fundamental to calculate the energy potential of the area and therefore to size the entire source circuit inside the tunnel.

The rock samples (diameter 25 mm, thickness 5 mm) were collected during the drilling of borehole near the investigated area inside the Brixen Granite. The shotcrete samples were directly obtained from the primary lining in the Mules Access Tunnel.

The thermal conductivity λ and thermal diffusivity α of the granite are necessary to determine the temperature trends in the rock mass along the longitudinal profile of the tunnel, knowing the ground surface weather conditions. The thermal conductivity λ of the shotcrete samples shows the thermal resistance of the first lining to the flow of heat from the surrounding rock mass.

The calculations describe below were based on the highest value of rock mass thermal conductivity found during the tests (3,44 W/(m·K)), so as to be as conservative as possible. In fact, the temperature inside the rock mass is influenced by the outside environment, so, assuming a high value of λ, the maximum heat dispersion toward the surface is supposed. The thermal conductivity of shotcrete is lower than that of fibre-reinforced concrete, because the steel fibres in the solid concrete increase heat transmission. For the shotcrete of the first lining, already realized during the Aica-Mules costruction lot, and for the concrete of the final lining, to be realized in the Mules 2–3 costruction lot, a thermal conductivity of 1,52 W/(m·K) was assumed, since this was the average value found during the laboratory tests (Lanconelli, 2016).

3.3 *Local geothermal model*

The estimated rock mass temperatures along the Mules Access Tunnel represent the input data for the energy potential calculation of the area of study. A deterministic approach has been considered. By equation (1), the temperatures on the interface between the rock mass and the first lining at the depth of the access tunnel were obtained (Figure 4, dashed line). The

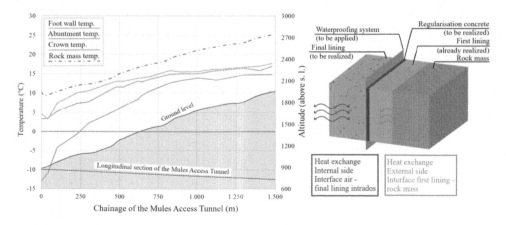

Figure 4. Estimated temperatures on the interface between rock mass and first lining (dashed line) and supposed temperatures on the intrados of the final lining (continuous lines).

ratio between q_{geo} and $\lambda_{granite}$ gives the geothermal gradient, which for the investigated area is about 19 °C/km.

$$T_{rock}(j) = T_{gl}(j) - A(j) \cdot e^{-C(j) \cdot \sqrt{\frac{\pi}{365 \cdot \alpha_{granite}}}} \cdot \cos\left[\frac{2\pi}{365} \cdot \left(\frac{C(j)}{2} \cdot \sqrt{\frac{365}{\pi \cdot \alpha_{granite}}}\right)\right] + \frac{q_{geo} \cdot C(j)}{\lambda_{granite}} \quad (1)$$

T_{rock} (j) = rock temperature at chainage j [°C]; T_{gl} (j) = air temperature at ground level at chainage j [°C]; A (j) = environment air temperature wave amplitude at chainage j [°C]; C (j) = overburden at chainage j [m]; $\alpha_{granite}$ = thermal diffusivity of granite [m^2/d], assumed 0,105 m^2/d based on laboratory tests; q_{geo} = typical geothermal heat flow for the area [W/m^2], assumed 0,065 W/m^2 (Ministero dello Sviluppo Economico, 2010); $\lambda_{granite}$ = thermal conductivity of granite [W/(m·K)], assumed 3,44 W/(m·K) based on laboratory tests.

In according to the temperature measurements in December on the first lining, the temperatures of the intrados of the future final lining were cautiously estimated (Figure 4, continuous line).

Figure 4 shows the estimated temperature domain at the boundaries and the designed lining stratigraphy for the Mules Access Tunnel (BBT-SE, 2015).

Because of natural variability of rock-mass thermal parameters, a probabilistic approach has been developed by the University of Bologna to estimate ground temperature distribution and validate the accuracy of the deterministic approach. Kriging spatial interpolation method is widely used for the estimation of regionalised variables (random variables defined in space) at places where they are not measured, starting from spatially localised measurements and known values (Kasmaee, 2016). This method is included in the class of unbiased estimators, and it is usually the one with minimum uncertainty, as measured by its estimation variance (Chilès, 2012). No substantial differences in terms of results between the deterministic and probabilistic approach was observed.

3.4 Calculation method of the geothermal energy of the Mules Access Tunnel

The temperature of the thermo-vector fluid circulating along the pipes of the source circuit and the geothermal power that can be extracted from the rock mass allow an estimate of the energy resources that can be obtained from the stretch of tunnel under consideration.

To find these two important parameters, heat transmission by conduction in semi-cylindrical geometric model has been used. The system is characterized by two different heat exchanges that take place thanks to two temperature gradients (Figure 4). The first heat exchange involves the working fluid and the rock mass (heat exchange from the rock mass). The second takes place between the working fluid and the intrados of the final lining (heat exchange to the inside of the tunnel) (Lanconelli, 2016). The main assumptions at the base of an energy lining thermal model are all related to the thermal resistances of the materials interposed between the working fluid and the boundary temperatures (rock-mass and air inside the tunnel).

Once the boundary temperature values have been estimated (Figure 4), the tunnel may be separated into fixed length segments and discrete variations in working fluid temperature calculated along the pipelines by cumulating the internal and external contributions for each step. The thermal properties of the fluid, the flow regime, circuit geometry and thermal resistances of materials define the rate and amount of temperature change at the tunnel entrance. Thermal resistances are represented as a series of different strata around the pipes (Tinti, 2017).

The total thermal power Q_i exchanged by circulating fluid along the absorber pipe in each tunnel segment i is expressed by equation (2). The temperature of the outgoing fluid from the sector i $T_{out,i}$ (i.e. $T_{in,\ i+1}$) is then expressed by equation (3).

$$Q_i = \frac{1}{2} \cdot \left(\frac{T_r - T_{in,i}}{R_{ext}}\right) + \frac{1}{2} \cdot \left(\frac{T_a - T_{in,i}}{R_{int}}\right) \quad (2)$$

T_r = temperature on the interface between the rock mass and the first lining (Figure 4); $T_{in,\,i}$ = temperature of the fluid entering on the sector i (for the first segment is assumed $T_{in,\,0} = 0\ °C$); T_a = temperature on the intrados of the final lining (Figure 4); R_{ext} = thermal resistance between the fluid and the rock mass; R_{int} = thermal resistance between the fluid and intrados of the final lining.

$$T_{out,\,i} = T_{in,\,i+1} = T_{in,\,i} + \frac{Q_i}{q_i \cdot c} \tag{3}$$

q_i = mass flow, it depends on working fluid density, working fluid speed and internal area of the absorber pipe; c = fluid heat capacity, assumed 4186 j/(kg·K).

The calculation included the number of probes, the delivery and (insulated) return headers which hydraulically connect the source circuit to the heat pump located at the portal. Corrective coefficients to be applied to the thermal resistances to account for any thermal interferences in multi-pipe systems (UNI 11466, 2012). The calculation was automated to search an optimal system configuration for the Mules Access Tunnel in terms of maximum thermal output accumulation along the source circuit according to surrounding environmental conditions. In other words, the calculation gave important indications for the design of the system, such as the best chainage at which to install the probe systems, the length of the delivery and return pipes, the point in which the return header must be insulated to avoid heat dispersion (Lanconelli, 2016).

3.5 Energy lining layout in the Mules Access Tunnel

In general installing an energy lining system requires detailed calculation of the available space and the additional time needed to install the absorber pipes and main headers. Inserting the energy lining inside the preliminary tunnel design is the simplest option, presenting undoubted simplifications. However, the system alternatively might also be implemented at advanced stages of tunnel realization as it is the case of Mules Access Tunnel (Tinti, 2017).

Various possible configurations were analyzed for the internal source circuit, trying to find the most efficient one in terms of extractable heat output and thermo-vector fluid temperature. The geometric and structural issues of the final lining, the functional requirements of the systems (decrease of load loss, hydraulic balance), the choice of materials to be used, the cost containment and the issues linked to the installation of the energy lining have been considered.

The idea of install the absorber pipes directly on the surface of the primary lining has proven to have a series of advantages in terms of system efficiency and ease of installation. This configuration implies that the absorber pipes are inserted between the primary lining and the waterproofing system, inside the regularization concrete (Figure 5). Whereas the larger heat contribution comes from the rock mass, this configuration allows to reduce the thermal resistance between rock mass and working fluid, in the present case consisting of simple water. The mechanical characteristics of the final lining do not change and the installation of the source circuit does not interfere with other works (Lanconelli, 2016).

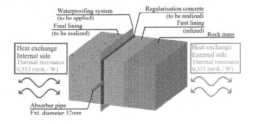

Figure 5.   Absorber pipes positioning and thermal resistances.

The geometrical configuration proposed for the source circuit inside the Access Mules Tunnel is shown in Figure 6 and Figure 7. The parallel line configuration can exploit high quantities of geothermal energy from the tunnel. By a reverse return header pipe (Tichelmann configuration) the water back to the tunnel entrance and close the circuit. Based on the information obtained from the calculation described above (Par. 3.4), a system using 60 delivery absorber pipes reaching longitudinally along the tunnel has been chosen. Different starting points for smaller groups of pipelines are supposed. In this way it is possible to increase the number of tubes, simplify installation, and maintain a system shape that is clean, linear and with no interference between the absorber pipes. Since the source circuit is confined between the primary and the secondary lining counteracting the pressure effects, the absorber pipes – 32 mm external diameter and 2,9 mm thick – are made of PE polyethylene. Absorber pipes are placed at adequate distances from each other to avoid thermal short-circuit and thermal depletion of the underground during short or long-term operation. The total length of the activate stretch is 800 m, from ch. 675 to ch. 1.475. Each absorber pipe develops along the tunnel for 300 m.

Absorber pipes depart, through inlet manifolds, from the main header situated on one side of the tunnel course along the tunnel sections according to the particular design configuration,

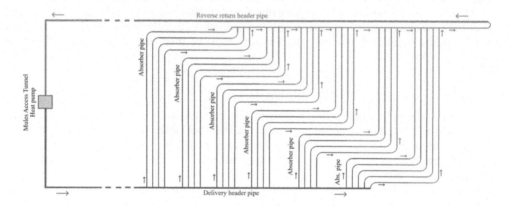

Figure 6.   Simplified pipelines layout of the source circuit inside the Mules Access Tunnel.

Figure 7.   3D reconstruction of the energy lining designed for the Mules Access Tunnel.

returning to the return header (or the reverse return header) on the opposite tunnel side and then connecting to the outlet manifolds. Any short-circuit between two headers is avoided (Tinti, 2017). The main headers are embedded into drainage material (coarse gravel), so there is no confining pressure to contrast the internal water pressure. To counteract the internal radial pressure of the water in the main headers, steel pipes were proposed. Return header insulation is fundamental, preserving the heat accumulated by the source circuit (Lanconelli, 2016).

3.6 *Exploitable power using the energy lining designed for the Mules Access Tunnel*

Figure 8 shows the possible exploitable power and outlet temperature of fluid flow rate using the final configuration describe above. The temperature of the water entering in the source circuit is assumed to be 0 °C. Figure 9 shows the estimated pressure drop at the tunnel entrance of the hydraulic circuit for different flow regimes and the consequent hydraulic pressure.

3.7 *User circuit, energy requirements and system performance*

The user circuit makes the geothermal energy available to the area to be heated. The two circuits interface at a heat pump which transfers heat from a source at a lower temperature (tunnel) to an environment at a higher temperature (end user). It is supposed to give the extracted heat from the tunnel for an automatic snow melting and de-icing road linking the portal of the Mules Access Tunnel and the road SS12. During the operational phase of the BBT the Mules Access Tunnel will be used by rescue vehicles in case of emergency in the main tunnels: in order to support and speed up rescue operations, maintaining the road to the portal in an efficient condition is considered a primary goal. The SIM (Snow and Ice melting System) can remove accumulated snow and ice from an area, such as road pavement, through a system of pipes in which a heated fluid flows (water or a glycol-water mixture). Based on the weather and climate data available for the area, the monthly energy requirements in the

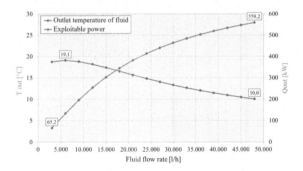

Figure 8. Outlet temperature and thermal power at the tunnel entrance for different flow rates according to the chosen pipeline configuration (inlet temperature equal to 0 °C).

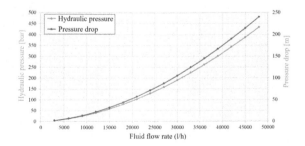

Figure 9. Hydraulic pressure and pressure drops according to the chosen configuration.

coldest months for the 900 m² SIM pavement were obtained (ASHRAE, 2011). A geothermal heat pump (power 600 kW, max outlet temperature 55 °C) was assumed. The main results by using this heat pump with proper circulation pumps were calculated: Figure 10 shows the comparison between thermal energy supply and electric consumption (Tinti, 2016).

The thermal energy produced by the heat pump is markedly greater than electric consumption of heat pump and the circulation pumps. The geothermal system is also perfectly able to handle the energy requirements of the SIM paving, with good efficiency as well.

The monthly COP (Coefficient of Performance), corresponding to the ratio between the available thermal energy produced by the heat pump and the electric consumption, is high (Figure 11). The SPF (Seasonal Performance Factor), corresponding to the average seasonal efficiency of the heat pump, is 4,6 (considering the circulation pumps it is 3,7). This result is higher than the minimum threshold set forth in the standard, which is 2,17 (Decision No. 2013/114/EU).

The PLF (Percentage Load Factor) expresses the rate of utilization of the heat pump in relation to the entire period of availability. The average PLF in winter time is 20,2% (Figure 11). Such a low value implies that the available energy can easily serve other end users as well.

3.8   *Analysis of investment and operating costs*

Table 1 shows an estimate of costs for the source circuit inside the tunnel, for the user circuit (SIM pavement) and the technical equipment (heat pump and circulated pumps). The costs refer to the materials, workforce and rent of the equipment required to install the two circuits was considered. The analysis was elaborated using price lists (Provincia Autonoma di Bolzano, 2015), requests for cost estimates, interviews and websites.

The estimated cost for the entire system is about 482.900 € and includes expenses for materials and workforce for paving the road (39.880 €). For peak thermal power of 560 kW, the cost of the source system is about 340 €/kW. Compared with the cost of a traditional vertical closed loop GHE (Ground Heat Exchangers), around 1250 €/kW, energy lining is considerably more cost-effective, with a cost saving of about 73%.

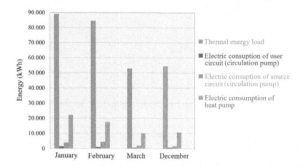

Figure 10.   Thermal energy supplied and electric consumption referred to the case study.

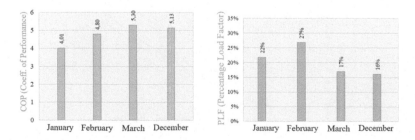

Figure 11.   Monthly COP and PLF referred to the case study.

Table 1. Estimated total investment costs.

Source circuit costs		User circuit costs		Technical equipment costs	
Material	153.985 €	Material	72.236 €	Material	169.500 €
Workforce	40.974 €	Workforce	6.340 €		
		Pavement[*]	39.880 €		
Total	194.959 €	Total	118.456 €	Total	169.500 €

* Cost of materials and workforce for the construction of the pavement.

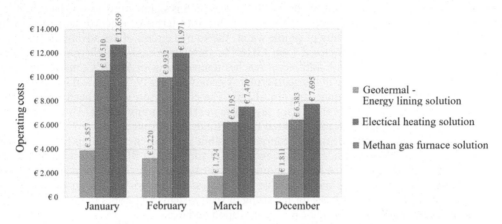

Figure 12. Comparison between the operating costs of the geothermal system and other types of heating.

The monthly operating costs depend essentially on the electricity consumption of the system (heat pump and the circulation pumps). Assuming that electricity costs 0,14 €/kWh, the monthly operating costs for the geothermal system were compared with the operating costs of other types of heating system. Specifically, it assumed that the pavement near the portal of the tunnel is heated with an electrical system or a methane gas furnace. Figure 12 shows the results.

4   CONCLUSION

The article gives an overview of several basic theoretical and practical notions required to study and design energy linings for deep tunnels excavated through rock mass.

The Mules Access Tunnel represents the most appropriate part of the Italian side of the Brenner Base Tunnel to install an energy lining. Using direct measurements of the temperatures inside the Mules Access Tunnel and laboratory tests on rock and shotcrete samples, a picture of the geothermal conditions in the studied case was obtained. The data gathered gave a model of the temperatures domain around the investigated area using both a deterministic and a geostatistic method, which showed no substantial differences in terms of results.

Subsequently, a calculation process was developed by which the geothermal potential of the investigated tunnel was estimated. The calculation shows the optimal system configuration for the source circuit in order to maximize the system's efficiency.

A series of possible configuration and design choices for the geothermal system was considered, preferring the scenarios that maximize system efficiency without altering the resistance or decreasing the functions or the installation of the final lining. The extractable geothermal power and the thermo-vector fluid temperature reached along the source circuit was calculated for each section of the Mules Access Tunnel.

The final proposed configuration consists in a group of 60 PE 32 pipelines, divided in 6 subgroups running parallel to the tunnel length, starting from dedicated niches 100 m apart and terminating at a reverse return header.

At the tunnel entrance the maximum exploitable thermal power is estimated equal to 560 kW, with a maximum outlet temperature of about 19 °C.

The geothermal energy could use for an automatic snow melting and de-icing road (900 m^2) linking the portal of the Mules Access Tunnel and the SS12 road. The climate data of Bressanone were used to identify the most critical energy load for the end-users. In order to ensure an efficient operation of the entire system, a monthly energy consumption analysis was done.

The cost for materials and workforce for the entire system, including the source circuit (194.959 €), user circuit (118.456 €) and technical equipments (169.500 €) is estimated to be 482.900 €. The Mules Access Tunnel energy lining is able to extract about 560 kW of thermal power. This means a cost per unit of about 340 €/kW, which is significantly lower than the 1250 €/kW, usual for a typical geothermal vertical closed-loop. In terms of reached temperatures by the thermo-vector fluid, consisting of simple water, the source circuit offers advantages as compared to a traditional vertical closed loop GHE. In this latter case, the thermo-vector fluid (which is often water mixed with antifreeze) reaches temperatures of 5÷10 °C, markedly lower than the maximum temperature in the geothermal system, which is 19 °C in low flow regime.

Yearly operating costs are quite low (16.612 €). The comparison with other possible solutions for the SIM paving shows significant savings in yearly operating costs (73% as compared to an electrical system and 68% as compared to a methane gas furnace).

This study shows the technical and economic feasibility of the energy lining for the Mules Access Tunnel. However, considering the low value of the average PLF in winter time (about 20 %), additional end-users to be heated should be identify to justify the investment costs.

Since no commonly accepted methods have yet been established for the design of energy linings, this study is a preliminary basis for the sizing of such systems and hopes to inspire the development of further in-depth work that can improve upon and expand the topics discussed here.

REFERENCES

ASHRAE Handbook – HVAC Application, 2011. Snow Melting and Freeze Protection, chapter 51.
Provincia Autonoma di Bolzano – Alto Adige, 2015. Elenco prezzi informativi per opera civili non edili. ACP (Agenzia per i procedimenti e la vigilanza in materia di contratti pubblici di lavori, servizi e forniture).
Barla, M., 2018. Il primo concio termosifone. *Le Strade Aeroporti Autostrade Ferrovie*. 03/2018. 54–57.
BBT-SE, 2015. Opere generali, Relazione tecnica, Disposizioni tecniche di contratto, Lavori in sotterraneo, rivestimenti definitivi. Progetto esecutivo Mules 2–3. 02 H61 DT 990 KTB D0700 11120 21.
Chilès, J.P., Delfiner, P., 2012. Geostatistics, Modeling Spatial Uncertainty. In Wiley (ed.).
Kasmaee S., Tinti F., Ferrari M., Lanconelli M., Boldini D., Bruno R., Egger H., di Bella R., Voza A., Zurlo R., 2016. Use of Universal Kriging as a tool to estimate mountain temperature distribution affected by underground infrastructures: the case of the Brenner Base Tunnel. *European geothermal congress EGEC. Proceedings*. Strasbourg.
Lanconelli, M., 2016. Studio di Fattibilità per lo sfruttamento geotermico delle Gallerie del Brennero. Master Thesis in Civil Engineering, DICAM University of Bologna, Bologna.
Ministero dello Sviluppo Economico, 2010. Dipartimento per l'energia, Direzione generale per le risorse minerarie ed energetiche, Bollettino ufficiale degli idrocarburi e delle georisorse Anno LIV – N.2.
Tinti, F., 2008. Geotermia per la climatizzazione. In Flaccovio Dario (ed).
Tinti F., Boldini D., Ferrari M., Lanconelli M., Kasmaee S., Bruno R., Egger H., Voza A., Zurlo R., 2017. Exploitation of geothermal energy using tunnel lining technology in a mountain environment. A feasibility study for the Brenner Base Tunnel – BBT. *Tunneling and Underground Space Technology*, 182–203.
UNI 11466, 2012. Sistemi geotermici a pompa di calore – Requisiti per il dimensionamento e la progettazione.
Zhang, G., Guo, Y., Zhou, Y., Ye, M., Chen, R., Zhang, H., Yang, J., Chen, J., Zhang, M., Lian, Y., Liu, C., 2016a. Experimental study on the thermal performance of tunnel lining GHE under groundwater flow. *Applied Thermal Engineering*. 106, 784–795.
Zhang, G., Xia, C., Sun, M., Zou, Y., Xiao, S., 2016b. A new model and analytical solution for the heat conduction of tunnel lining ground heat exchangers. *Cold Regions Science Technology*. 88, 59–66.

Tunnels and Underground Cities: Engineering and Innovation meet Archaeology,
Architecture and Art, Volume 5: Innovation in underground engineering,
materials and equipment - Part 1 – Peila, Viggiani & Celestino (Eds)
© 2020 Taylor & Francis Group, London, ISBN 978-0-367-46870-5

Finding the right way underground – The importance of solid control points network

B. Larsson Gruber
The Swedish Transport Administration, Solna, Sweden

T. Jonsson
Sweco Civil, Stockholm, Sweden

ABSTRACT: The project E4 Stockholm Bypass governed by The Swedish transport Administration as client, consists of more than 50 kilometers of tunnel. And the project are now up and running in a highly intensified phase of construction through drilling and blasting as production-method with several different contractors involved.

One of the main issues during the construction-phases of a tunnel, especially as many upcoming tunnel projects tend to be of a very long and deep nature, is to always have a solid control point network up to date, not only above ground, but all the way underground towards the face of the tunnels, all in order to make it possible for the contractors to fulfil all their works inside the required tolerances and achieve breakthroughs in the right tolerances stipulated by the client.

This abstract will describe the clients work in a general view, independently by the different type of long and deep tunnel projects and also as a case-study in detail of the clients work to establish and create a stringent and common set of requirements and conditions in the specific project E4 Stockholm Bypass, governed by The Swedish Transport Administration as client.

1 INTRODUCTION

1.1 *The project E4 Stockholm Bypass*

1.1.1 *Project conditions*

The County of Stockholm accounts for 30% of the national growth and the population is growing by about 20,000 new inhabitants a year. In 2040 there will be about 2.5 million people living in the region. The infrastructure in the region is undersized, the traffic system is very fragile and there is a great need of significant investments in road infrastructure. Therefore, the Government is now investing heavily in the development of roads, rail and subway and has therefore decided to finance the infrastructure in the region with 11 billion Euro during the period 2011–2021. This decision is defined in the so called "Stockholm agreement" where Stockholm Bypass is the absolutely biggest investment. The Stockholm Bypass is a key element in increasing road capacity across the water strait for car and bus traffic and for commercial transport.

The bypass will reduce congestion on the approach roads, fuse the northern and southern parts of the county together and create efficient links between terminals and ports. It will relieve pressure on the inner-city center of the region while linking together and developing the external and regional cores. Stockholm Bypass consists of a relocation of the European highway E4, and as it passes several natural and cultural reserves, the major part of the highway will be located in tunnel. The project consists of approximately 50 kilometers of rock-tunnels and five large interchanges. In the rock-tunnels there are three packages of ramp tunnels for connections to the surface. The tunnel is located as deepest 65 meters below sea level due to fracture zones in the rock in the position of the water passages.

When completed it will take 15 min to travel from north to south, compared to more than one hour today. One of the longest tunnels in the world based on actual traffic intensity, approximately 140 000 cars/day in the year 2035.

Table 1.   Tunnel data (Trafikverket, 2011a).

	Tubes	Lanes	Length [m]	Total Length [m]
Main tunnel 1	2	3	16500	33000
Main tunnel 2	2	3	1800	3600
Surface Connections	-	-	-	17400
Total	-	-	-	54000

One condition that was a real challenge for a major part of the work with the different types of control point networks was the fact that a large part of the areas suitable for marking and sightlines was located inside or close by to areas protected by the laws of nature reserve in Stockholm.

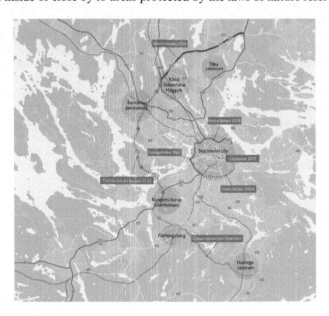

Figure 1.   Layout of the E4 Stockholm Bypass.

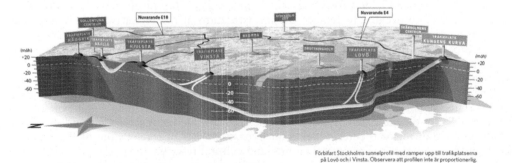

Förbifart Stockholms tunnelprofil med ramper upp till trafikplatserna
på Lovö och i Vinsta. Observera att profilen inte är proportionerlig.

Figure 2.   Vertical layout of E4 Stockholm Bypass.

1.2 Reference systems

The horizontal reference system of the project is SWEREF 99 18 00. It is based on the GRS 80 reference ellipsoid with a transverse Mercator projection with central longitude 18°00" and scale factor 1.

There was an early investigation into the scaling effects of map projection and height above the ellipsoid. Due to the project's north-south-orientation and proximity to the ellipsoid surface the scaling effects are manageable (less than 20 ppm). Had the conditions been less fortunate, an object specific map projection might have been considered.

The vertical reference system is FS RH 00 which is a hybrid of the old RH 1900 system used in the Stockholm region and the new national vertical reference system RH 2000. Heights in FS RH 00 are RH 2000 heights reduced with the mean height difference between RH 2000 and RH 1900 over the project area. This was done because municipal geographical data was not available in RH 2000 at the beginning of the design phase, but the lack of homogeneity of the municipal control point networks in RH 1900 made them unsuitable for tunnel construction.

1.3 Hierarchy and terminology of control point networks

In Sweden, control point networks are categorized in different types depending on the measurement methods used, their purpose and the area they cover.

The national horizontal and vertical control point networks are created and maintained by the Swedish national geodetic survey (Lantmäteriet). They provide national coverage of control points, but with low point density.

Horizontal and Vertical control point networks called connection point networks are created for specific projects or towns by means of static GNSS measurement and precision levelling respectively. They are directly connected to the national control points and their purpose is usually to provide high quality control points approximately 500 m apart, surrounding an area of interest.

A special case of connection point networks are connection point networks for tunnel building, which generally consist of at least three horizontal and two vertical control points outside each tunnel opening. The spacing and accuracy of these control points are especially important for determining the correct direction for tunnel excavation.

The third level networks, after national and connection control point networks, are subdivided into construction control points and maintenance control points.

The construction control point networks are connected to the connection points, and established with such density that they can be used for the everyday measurement needs of the construction project. The markings used only need to be permanent for the duration of the works.

The maintenance control point networks are also connected to the connection points, but are generally created at the end of the project and need to be as permanent as possible, as they are to be used for the long-term maintenance of the construction.

1.4 Standards and regulations relevant to Stockholm Bypass

There are no legal regulations on surveying work in Sweden, but there are several documents specifying what would be considered good practice, and technical specifications intended for use in tender documents for construction contracts.

SIS-TS 21143:2009 Engineering survey for construction works – Surveying and mapping for long range objects, is a technical specification published by the Swedish Standards Institute. Its intended use is as a reference in construction contracts, and it provides specific descriptions of how different measurement tasks should be performed. It has since been revised in 2013 and 2016, but the 2009 edition is the one used in the Stockholm Bypass contracts.

TRVK Mät, Trafikverkets tekniska krav avseende mätningsarbeten för väg is a publication by the Swedish Transport Administration (Trafikverket), detailing their specific technical requirements for geodetic measurement on their roads and railroads. These demands were worked into the Stockholm bypass contracts by references to SIS-TS 21143:2009.

The above documents gave a basis for the contractual texts, but they are severely lacking when it comes to tunneling projects as complex as the Stockholm Bypass, where more than one contractor is involved. Some creativity was therefore needed in creating the surveying texts in the tender documents for the contracts.

In general, Trafikverket does not create its own independent control point networks in the tunnels as they are built, but rely on the contractor following the specifications of the contract and continually reporting their work to the client.

2 DESIGN PHASE

2.1 The general design phase

During the general design phase, the general design of the tunnel system and locations of work tunnels were determined and methods of transporting excavated rock and other logistical and environmental issues were investigated to secure the feasibility of the project and get all the required permits and funding.

During this phase horizontal and vertical connection control point networks were created by the consulting firms hired for the general design phase. The networks consisted of 16 horizontal control points and 40 vertical control points. These control points were not placed with specific regard for the tunnel openings, as their purpose was mainly to serve as a basis for further works after the tunnel designs were finalized, and as control points for checking GNSS equipment and benchmarks for checking the digital terrain models.

The documentation of these connection control point networks was delivered to the client (Trafikverket) and were part of the tender documents for the consulting contract for the detailed design phase.

2.2 The detailed design phase

2.2.1 Surveying tasks

During the detailed design phase there were three main tasks for the surveying departments of the consulting firms that won the contract. Firstly, to do all the detailed measurements required for the design work and control of the work from the previous phase. Secondly, to write the tender texts regarding surveying for the construction contracts. Thirdly, to create the horizontal and vertical connection control point networks for tunnel building at the openings of all the work tunnels.

The first task falls outside the scope of this paper.

The second and third task required some investigation into the accuracies needed, both in control points and measurement methods needed to achieve acceptable tolerances.

2.2.2 Simulation of construction control point networks

It is well known that for the long open horizontal lattice networks necessary for tunnel construction, the expected breakthrough error does not vary linearly with tunnel length, as described in Lewén (2006), pp. 41–43. Because of this, a thorough investigation of expected breakthrough errors using different methods was necessary.

The horizontal construction control point network between the opening of each work tunnel and the point in the bedrock where the main tunnel was supposed to connect to the neighboring contractor's tunnel were simulated in geodetic software to assess what magnitude of errors were to be expected, with different equipment and with different accuracies of the connection control points. Simulations were also carried out with and without gyro theodolite measurements added every kilometer.

The results of simulation of the construction control point network from the opening of one of the work tunnels up to the connection to the neighboring contractor's tunnel are shown

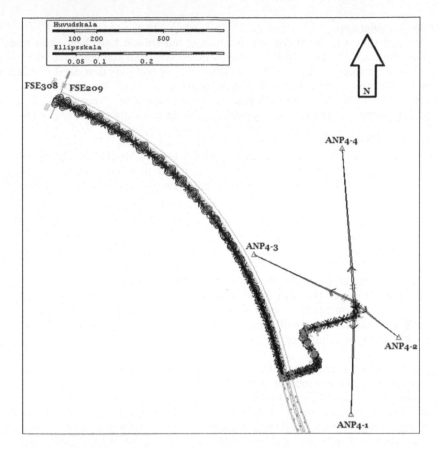

Figure 3. Example of simulated construction control point network with gyro observations implemented presented as error ellipses for each point.

graphically in Figure 1. The total length of the network is approximately 4.5 km. The simulation is showing a worst-case scenario, as in practice both main tunnels will be excavated simultaneously, and more connection measurements between tunnels will be possible. On the other hand, the effects of refraction due to temperature gradients in the air close to the tunnel wall cannot be accurately modeled, so simulated results should be interpreted with some caution.

In the network simulation above, which was for one of the longest tunnel contracts, the resulting error ellipses at the 68% confidence level, using a top of the line total station, had major axes of 69 mm at the points furthest from the work tunnel opening. When gyro theodolite measurements were added to the simulation at every km, the ellipses' major axis dropped in size to15 mm.

Some of the conclusions were that:

- the longer the sight lines between the connection control points could be, the less sensitive the direction of the tunnel would be to small inaccuracies in the connection control points
- measurements with gyro theodolite would be necessary to achieve tolerances
- the use of the highest precision class of total stations was merited for the creation of the construction control point network

Some errors can't be adequately modeled, such as refraction of sight lines because of temperature gradients in the air in the tunnel. Any simulation is therefore a best-case scenario, and great care should be taken to eliminate as many error sources as possible

2.2.3 *Other investigations*

An investigation that wasn't carried out in advance, but should have been, was what effects the deflection of the vertical would have on gyro theodolite measurements over the whole project area. In a later stage of the project calculations using methods described in Featherstone & Rüeger (2000) were carried out to find the approximate size of the effect. Fortunately, the geoid surface close to the Stockholm Bypass has an approximately constant slope, which renders the effects on the gyro observations in relation to the calibration base negligible

2.2.4 *Tender documents*

As there are over ten different rock tunnel contracts for the Stockholm Bypass, the task of coordinating the surveying parts of the tender documents to achieve as great consistency as possible was not easy. With regards to control point networks, the following strategy has been used:

- Total stations with accuracies of 1" for directions and 1 mm + 1 mm/km for lengths should be used by the contractor
- Gyro Theodolite measurements should be performed at every km of tunnel length by the contractor
- The contractor is free to design their construction control point networks as they see fit, but a full simulation of the network should be provided to the client before work starts, where it is clearly shown that no error ellipses larger than 30 mm are expected.
- The vertical construction control point network can be measured trigonometrically by total station, but must be connected to levelled vertical control points every 200 m of tunnel length.
- The client shall have access to all data pertaining to the construction control point networks and gyro theodolite measurements, for control and validation purposes.
- The maintenance control point networks are to be marked with uniform configuration and with specified markings (compliant with the Unikonsol system) across all contracts.
- Measurement data for the maintenance control point networks is to be delivered to the client in specified formats and with specified detailed documentation, so that the client can combine the networks from the different contractors in one network adjustment for the entire tunnel system.

There might be some slight variations in the wording in the different contracts due to the order the tender documents were made in, as the texts kept getting refined after each review, but the general rules above are common to all the rock tunnel contracts.

2.2.5 *The connection control points for tunneling*

The creation of the horizontal connection control point network for tunneling turned out to be a major challenge, in part due to the geographical placement of the work tunnels and in part due to the severe environmental restrictions on cutting down trees or tree limbs for sight lines and better GNSS coverage. To get the long sight lines required in spots were permanent markings could be made and open view to the sky for GNSS measurements, some ingenuity was necessary.

In four cases points had to be placed on buildings: two on water towers and two on residential buildings. Special consoles were made and permanently mounted on the buildings, so that a total station, a reflector or a GNSS antenna could be placed over the exact same point.

Most of the other points were marked with acid resistant steel markers drilled into the bedrock. Twelve of the new points did not have sufficient view to the sky due to vegetation, so the GNSS measurements had to be done with up to 10 m high masts, which had to be laboriously centered over the markers with total stations.

In total the horizontal connection control point network consisted of 49 points, of which 16 were the known connection control points from the general design phase and three would form the calibration base for gyro theodolite measurements. At least 4 new points were placed outside every work tunnel opening.

However, horizontal connection control points were not placed in front of every tunnel opening, as many of the tunnel openings in the finished bypass are not used for tunnel excavation, but will only be opened at the end of the tunnel excavation. Therefore, high directional accuracy in the control point networks outside those tunnel openings is not critical, and the tunneling lattice networks will be connected to the construction control point networks of the above ground works in those areas.

Vertical connection control points for tunneling were established with precision levelling and placed in safe distance from the constructor's work areas.

3   PRODUCTION PHASE

3.1   *Control point network within the Contract works*

3.1.1   *Preconditions for the surveying works*
This chapter will mainly describe the contractors work with the control point networks for the production of tunnels. The preconditions regarding the control point network provided from the client was as described earlier in chapter 1.2, a connection-net consisting of 4 to 5 stable fix points located in relatively close connection to the access-tunnels. The access-tunnels themselves gave a quite difficult preconditions by; only one connection to above ground, quite narrow cross-sections, winding and turning layout both in longitudinal- and vertical alignments and as an extra obstacle a 90 degree connection to the main-tunnels where the main production of the tunnels was planned to be executed.

In this project, and as it's common in Sweden, it's important to highlight the fact that the contractors have the full responsible inside their contract for all the work with the surveying works from start of the contract up to the final deliverance of the contract to the client. All the contractors work with the control point networks are as described further up in chapter 2.2, quite rigorous stipulated in a stringent and common-set of requirements throughout the project, and not divided in different requirements regarding surveying works for the access-tunnels or the main-tunnels.

3.1.2   *The contractors continuous surveying works inside the tunnels*
During the construction phase the surveyors has to make continuous completions of the reference nets according to the speed of progress by the tunnel production, all with new pair of fix points to be established approx. 50 – 100 meters behind the tunnel face. All requirements for the surveying works are stipulated in the specific technical description for surveying that is belonging to the contractual documents. By certain intervals described within the contractual documents, a total re-measurement of the whole control point network established by the contractor shall be performed from the portal-net above ground down to the face of the tunnels. The main principle to perform the survey inside the tunnel is in the way of using free-station establishments with sightlines to targets mounted on the walls, thus minimizing the influence of refraction.

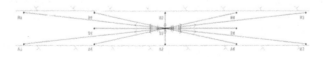

Figure 4.   Principle for measurements method, straight tunnel.

3.1.3   *The contractors independent and continuous control of the control point network*
Besides the contractor's ordinary surveying work with the control point network, a certain amount of extra independent controls of the control point network established by

the contractors shall be done. This control are stipulated to be done by the contractor by the use of high precision surveying gyroscope instruments. Normally control by gyroscope instruments are first performed in the tunnels when they has reached a length of approx. 1 kilometer, but depending of the quite harsh preconditions deriving from the access-tunnels layout an initial control by gyroscope instrument has to be done promptly when the progress of the main-tunnels has reached a sufficient distance for the line of sight, approx. 250 – 350 meters. Further on in each contract, a continuous control by gyroscope instrument is to be done per each produced tunnel length of 1 kilometer. For this special type of measurement the client has established and provide to the contractors a reference base for gyro measurements, located in the middle of Stockholm, close-by to the project, to be used before and after each gyro-campaign.

3.1.4 *Deliveries regarding surveying works from contractor to the client*

As mentioned before, a specific technical description regarding the surveying works is included in the contractual documents. Besides of setting the requirements for all the surveying works, it's also stipulate all different types of documentation that has to be delivered from the contractor to the client. Regarding the surveying works for the control point networks, a certain type of document called Measuring-program (in Swedish Mätprogram), is stipulated to be delivered to the client for examination and to be agreed before actual field-work can take place. This primary document shall consist of actual planning, layout of planned surveying of control point network, simulation of the specific survey work and all other aspects to take care of before the specific field-work.

The follow-up of each specific survey of control point network is to be delivered to the client by the mean of a specific document called Surveying report, (in Swedish Mätteknisk Redovisning), This document consists of the content from the earlier mentioned Measuring-program, and will also contain the actual calculations and final coordinates, this document will as well be delivered to the client for evaluation and final acceptance. All the content in this two types of document are well stipulated in the contractual documents and/or as described in chapter 1.4 above, quoted in the contractual documents.

3.2 *The Clients organization*

The amount of work and responsibility that lies on the contractor sets also requirements on the clients own organization to be able to follow-up the survey work during all the different phases of each contract. The clients' organization for these duties is build-up by the following positions;

• Head of Survey – overall responsibly for a stringent and common follow-up of the contractors' fulfillment of stipulated requirements and deliveries to the client throughout the entire project.
• Surveying Supervisors – responsibly in each separate contract for continuously follow-up of the contractors' survey-work and primary counterpart for questions and orders.
• Expert Support of Survey – back office support that work in close co-operation with each Surveying Supervisor regarding the contractors work with the control point networks.

4 CONTROL POINT NETWORKS IN THE TUNNELS AFTER COMPLETED CONTRACTS

4.1 *Control point network through the entire tunnel*

Once the tunnel-contractors has completed their construction-works, and all the inner works are finalized, each of the tunnel-contractors are stipulated to mark and measure inside their contracts, a final control point network that will be delivered to the client, and further on handled over to the Maintenance- and Service-department, a department that is also belonging to the Swedish Transport Administration, for their use in the upcoming maintenance and service phase.

4.1.1 *Deliveries to the client*
The client will make a final calculation of the control point network throughout the entire tunnel-system. In order to get similar deliveries from each contractor, the requirements are identical in every contract, and by the client a common template for the registration of certain values resulting from the measurements is provided to the contractors.

4.1.2 *Calculation of the entire maintenance control point networks through the entire tunnel*
To avoid having deformations in the maintenance control point networks at the borders between the different contractors, the client will calculate the entire horizontal maintenance control point network in one network adjustment, and deliver new coordinates to the contractors. In some cases this might mean that as-built measurements already performed by the contractors need to be adjusted or commented. To minimize these problems, preliminary adjustments of parts of the network might be necessary.

5 CONCLUSIONS

5.1 *Tender documents*

To let the client achieve good control of the contractors work with the surveying of tunnels, it's important to have personnel involved in early stages of the design-phase, with good knowledge of surveying and different methods of construction. As it is mentioned in several different chapters in this document, to be within the right tolerances will gain economical advances, both for the contractor and the client.

5.2 *Permit issues*

Don't underestimate the need of negotiations and necessary precautions in the contact with other authorities, especially when control point networks are to be established in close vicinity of nature reserve areas.

5.3 *Foresight*

Both theoretical expertise and practical experience are required to anticipate some of the more uncommon problems that might arise because of the general preconditions of the project. What worked perfectly well in the previous project might spell disaster in this one if the preconditions are different enough.

Any geodetic problem that can be identified and mitigated early in the design phases will save a lot of money in the production phase. The earlier the geodetic infrastructure in the form of control point networks can be established, the better and more reliable the design work will be.

5.4 *The contractors survey engineers status and mandate*

It's important that the contractor in an early stage will have an organization ready to deal with survey issues, an organization that will have good knowledge of work in tunnels as well has good knowledge in survey works and the contractual documents. Although deliveries of documents are stipulated in the contracts, and the contractors own internal recordings should be up to date, it is an advantage for the contractor to have an intact organization throughout the entire project.

REFERENCES

Featherstone, W.E. & Rüeger, J.M. 2000. The Importance of Using Deviations of the Vertical for the Reduction of Survey Data to a Geocentric Datum. In *The Australian Surveyor Vol. 45 No. 2*. Brisbane: Queensland Institute of Surveyors.

Lewén, I. 2006. Use of gyroscope in underground control network. *Master of Science Thesis TRITA GIT EX 06-002 Geodesy Report No.3090*. Stockholm: Royal Institute of Technology (KTH).

SIS Förlag AB, 2009. Teknisk Specifikation SIS-TS 21143:2009 Utgåva/edition 3

Trafikverket, TDOK 2014:0571,2016, version 4.0, Krav – Geodetiska mätningsarbeten och geografisk lägesbestämning

Tunnels and Underground Cities: Engineering and Innovation meet Archaeology, Architecture and Art, Volume 5: Innovation in underground engineering, materials and equipment - Part 1 – Peila, Viggiani & Celestino (Eds)
© 2020 Taylor & Francis Group, London, ISBN 978-0-367-46870-5

Implementation of new methods for inspection of tunnels

B. Larsson Gruber
The Swedish Transport Administration, Solna, Sweden

P. Östrand
WSP, Stockholm, Sweden

ABSTRACT: In Sweden, the requirements for inspection of tunnel are stipulated to be performed in hand-close distance, in other means manually and optical by using some sort of lifting device, a monotonously and above all very time-consuming way of inspection.

But as the number of tunnel in Sweden grows, and become longer and longer, a need to implement a different approach to the inspections, are ongoing in The Swedish Transport Administration, governing almost all railway- and road-tunnels in Sweden.

Several evaluations and field tests has been performed by Trafikverket to see the capabilities of different hardware on the market, that has led to the following conclusion - the hardware part of the systems is not the main issue to implement, instead is the main issue what kind of software will the client, in this case The Swedish Transport Administration, rely on?

1 INTRODUCTION

1.1 *Historical*

Looking back in time, until the beginning of the 1900's, we can see that there were not as many requirements as today, and the requirements were not as complex. Environmental and environmental requirements were not in the same extent and safety thinking was not fully developed as today. In return, we did not have as good production aids as today.

Although we did not have the same requirements as today, we apparently built rock tunnels that still have a good function today, and in the 80's, rock tunnels were still built without shotcrete reinforcement and small amount of rock bolts. Today tunnels are still in service, which were built in the 1860s and 1870s.

From 1860 until the middle of the 1950's we only had railway tunnels in Sweden, but then road tunnels began to be built. Not so big and not so long. Until the 1990s, development was modest but when we passed 1990 something happened. Both the number of road tunnels and railway tunnels doubled up to the year of 2000. The development from 2000 until today 2018 has meant that the total tunnel length for rail- and road tunnels increased by almost 400% each.

1.2 *Similarities and differences*

Historically, we'd often talk about railways and roads as different objects in Sweden, and since they had separate administrations, they also had their own specific documents regarding the requirements for construction and inspections of rock tunnels. However since 2010, both administrations have merged into one administration (Trafikverket, The Swedish transport Administration), which meant that requirement documents could also have been merged, as most requirements are common, regardless of whether it concerns rail or road, it's the same

Figure 1. Construction of railway tunnel in the early 1900's.

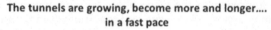

The tunnels are growing, become more and longer....
in a fast pace

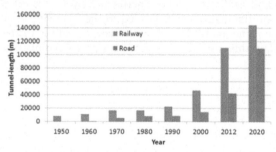

Figure 2. Development of the stock of tunnel meters in Sweden.

rock that traffic should go through. Those new documents, especially TRVK Tunnel 11 and TRVR Tunnel 11 acts as requirements and advices which describes how in detail how service and maintenance should be performed to achieve a product within the requirements regarding tunnels governed by the Swedish Transport Administration.

1.3 *Methods for inspection of tunnels*

Following all these documents and the requirements, the tunnels that are been constructed by The Swedish Transport Administration has in general a stipulated lifespan of 120 years. In order to achieve such a lifespan, a constant run of inspection, at a certain interval, of the structures is needed, together with related maintenance during all those years. Generally in Sweden, the running inspections of rock tunnels are performed in a way what is called "hand-close" distance. It basically means that the inspection are performed in a manual way, very close-by to the surface, e.g. very reliably on what can be monitored visibly, which means that the inspectors must be using lifts, flash lamps etc., all in order to achieve a good inspection. This means that this type of inspection is a method that are very time-consuming and very analogue, which doesn't work well with the tunnel stock becoming more and longer, and in combination with an overall shortened available time frame for inspection – that will be no good solution for the future. An analog way of work that doesn't cope well with the adoption by the Swedish Transport Administration to use digital information as a primary set for documentation.

2 NEEDS AND EVALUATIONS

2.1 Needs

For the mobile inspections of tunnel governed by the Swedish Transport Administration, there is a certain amount of detailed information that is essential to gather for the need for service and maintenance of the tunnels, independently of being a railway or highway tunnels. Since the upcoming procurements of inspections during the upcoming years to come, most certainly will be executed by different consultants, there is also a need of collected information that can be processed without being firmly attached to which specific consultancy that has performed the actual task. The question has bearing on the main fact that the design for the technical lifetime of Swedish tunnels are 120 years, with intervals of inspection of approximately every 5 years. This is leading to a huge amount of upcoming procurements of different consultants and their solutions.

2.2 Procurement document

In order for the Swedish Transport Administration to achieve a way of order to procure upcoming mobile inspections, a specific procurement document was created that specifies all necessary information for both the client and the consultants to fulfill in form of information and specific information for each tunnel facility. The type of information needed as a result from inspections is specified in this procurement document in general consisting of detection of anomalies such as; cracks with a minimum size of 0.5 mm, deformations of minimum size of 1.0 mm, (by the term of deformation means anomalies such as deviations in joints, chipping, protruding corners, anomalies from projected geometry etc.), a wide spectra of information regardless the construction method, for example built by drill and blast or bored by the use of TBM.

2.3 Evaluations

By the needs for the Swedish Transport Administration, to find a method for inspections that are time effective and give information in a digital format was very essential. Several different mobile methods from different consultancies was tested and evaluated and under the management of the Swedish Transport Administration. The evaluations consisted of both knowledge acquisition from existing undergoing inspections throughout Europe, field tests in a the newly built highway tunnel, Norra Länken, in Stockholm, and from two, as

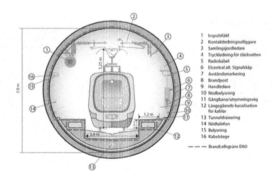

Figure 3. Cross section of railway tunnel Citytunneln, Malmö.

part of the regularly running inspection, mobile inspections in two newly completed, and running with regularly traffic, railway tunnels, the Hallandsåstunnel and Citytunneln in Malmö, which also acted as field tests of the chosen method. The method will be described in detail in the following chapters.

3 METHOD

The methodology used for capturing inspection data for post-processing in tunnels requires high performance sensors that are capable of operating in challenging conditions. Such conditions can be high humidity and low light levels. In addition to these challenges, the data collection process must be able to operate at a relatively high speed due to the limited window of opportunity available for undertaking the tunnel inspection.

The data captured is not intended to be limited to assessing the current state of the infrastructure. It shall also to be used in future applications as an historical database and be a key component for the development of predictive maintenance applications. This places additional demands on data collection and data formats. Accurate positioning is a necessity to ensure that data can be compared over time between different inspections. This allows damage and deterioration to be identified and analyzed, examining how it propagates and changes over time. Standardized data formats should also be used to create a reduced dependency on specific presentation and analysis software. The configuration of survey system can be divided into the following components:

- Positioning and geo referencing of captured data to the current geodetic reference system
- Photography of ceilings and walls with fully coverage
- LiDAR scanning and modelling for detection of deformations

3.1 *Data capture*

The data capture was performed by assembling measurement sensors and lighting on an ATV (All-Terrain Vehicle) adapted to be able to perform collection in both road and rail tunnels. The quality of measurement data gradually deteriorates with increased speed. Therefore, the vehicle speed was optimized so that the highest quality data could be captured in the allocated time window. This resulted in a vehicle speed of approximately 10 km/h.

3.2 *Positioning*

Traditionally, mobile mapping systems use a combination of GNSS (Global Navigation Satellite Systems), Inertial System and an odometer to position the survey system and subsequent

Figure 4.   Data capture in Norra Länken (the northern link) road tunnel.

Figure 5.   Data capture in Citytunneln (The city tunnel), railroad tunnel.

geo-referencing of collected data. These three systems work in unison to deliver an accurate location. GNSS requires a clear line of sight between the survey antenna and the satellites to deliver accurate results. In tunnels this is not possible. Inertial measurement systems can function underground but tend to "drift" over time. This is usually compensated for by utilizing the GNSS input which, as stated, is not an option in a tunnel environment. To avoid a rapid deterioration of positional accuracy, control points with known position were measured in using traditional survey techniques. These points were then utilized to strengthen the inertial measurement system. This reduced measurement uncertainty and allowed for the collection of high quality inspection data.

3.3   Surveying sensors

A mobile mapping system was selected with the ability to use sensors produced by different manufacturers and with different capabilities. This enables the system to be adapted for data collection in different types of infrastructure facilities. It also ensures that the system can evolve as R&D progresses in sensor technology.

3.3.1   LiDAR scanners
For laser scanning, laser scanners less sensitive to moisture and large temperature differences, were used. Six scanners were mounted in different directions to prevent occlusions in the resulting point cloud, which could otherwise be a problem if only one or two scanners were

Figure 6.   Point cloud in Citytunneln.

used on an irregular rock surface. The scanning resulted in a point cloud with a density of approximately 5000 points/m^2.

3.3.2 *Photography*

Two different types of camera solution were used to capture different types of information. Panoramic cameras were used to capture 360° images at regular intervals along the tunnel. These images are then used for a variety of purposes such as data collection and condition assessment. High resolution IR (Infrared) cameras were used for the detection of cracks and water leakage. To account for the poor lighting conditions in tunnels, LED lights were mounted on the vehicle to improve the quality of the panoramic images and IR flashes were used to enable image capture with the IR cameras.

Panoramic images were captured with a 5m interval. The exact location of the camera was provided by the positioning system, which in turn allows the images to be used for measuring coordinates, lengths, area etc.

Nine inspection cameras were mounted on a circular mounting device to cover the entire tunnel. The distance between the exposures was adjusted to cover the entire tunnel structure. The geographic resolution in each image is less than 1mm. This allows minute cracks to be detected.

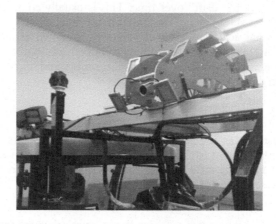

Figure 7. Inspection cameras.

Figure 8. Panoramic image.

Figure 9.   Inspection images of the tunnel roof.

4   IMPLEMENTATION

Deformations in the tunnel walls and ceiling were detected by calculating a difference model based on 2 point clouds that were captured on different occasions. The magnitude of the separation between the two surfaces can be visualized using different colors, making anomalies easy to see. For this method to function correctly it is imperative that the spatial accuracy of the point clouds is high. If the point clouds are incorrectly positioned relative to each other then erroneous anomalies will be detected.

Cracks and leaks were detected visually in inspection images. The possibility of detecting leakage was improved thanks to photography in the IR spectrum. This format and quality-level proved suitable for future automatic image recognition through machine learning. In

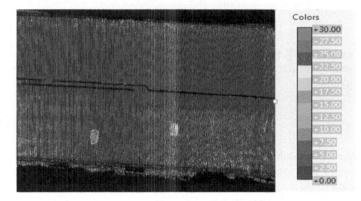

Figure 10.   Deformations identified by plotting separations in point clouds captured on different occasions. Color scale is in millimeters.

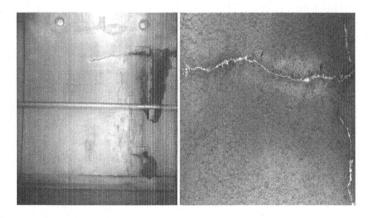

Figure 11.   Cracks and water leaks.

Figure 12. Orbit GIS and tools for measuring and feature extraction.

order to create an automatic image classification system optimized to detect these types of features at an acceptable accuracy level it will be necessary to generate a large reference library of relevant training data.

5 VISUALIZATION

All captured data was georeferenced into the correct geodetic reference frame. That makes it possible to compare data over time and to follow the propagation of damages. Data was adapted to be used in different visualization solutions to distribute and make it available for further analysis and historical monitoring.

Export was carried out to Orbit GIS that are used by the Swedish Road administration to distribute and archive LIDAR point cloud, 360 panoramic- and planar imagery. The software offers tools to measure and perform inventory and condition assessment in geographical data.

Data are provided by a cloud service that streams data, to be shared via web and mobile devices and integrated into CAD-, GIS- and Asset Management software. When fully integrated in Asset Management software, asset databases can be updated at office in desktop solution or traditionally by field inspection.

6 CONCLUSIONS

The Swedish Transport Administration has adopted the use of digital information in all major forthcoming procurements of building contracts. Therefor it is very essential for the administration not to break the chain by lacking the presence of digital information even in the aspect of service and maintenance phases during the tunnels expected life span. The challenge is with the aspect of the quick development in technology to find solutions and visualization tools that can grow in compliance with available systems for the years to come.

REFERENCES

Trafikverket, TDOK 2016:0231,2016, version 1.0, Krav – Tunnelbyggande
Trafikverket, BatMan, Publikation 2006:61
Trafikverket, Rapport 7A, Grundläggande bearbetning av laserdata, publikation 2018:069
SIS Förlag AB, 2009. Teknisk Specifikation SIS-TS 21143:2009 Utgåva/ edition 3
Uppdragsbeskrivning Mät, Inspektion Citytunneln, Malmö publ. 2015-09-01

Tunnels and Underground Cities: Engineering and Innovation meet Archaeology,
Architecture and Art, Volume 5: Innovation in underground engineering,
materials and equipment - Part 1 – Peila, Viggiani & Celestino (Eds)
© 2020 Taylor & Francis Group, London, ISBN 978-0-367-46870-5

Numerical modelling for design of composite concrete lining with sprayed waterproofing membrane in tunnel

C. Lee, S.H. Chang & S.W. Choi
Korea Institute of Civil Engineering and Building Technology, Goyang, Republic of Korea

K. Lee & D. Kim
Incheon National University, Incheon, Republic of Korea

B. Park
University of Science and Technology, Daejeon, Republic of Korea

ABSTRACT: A sprayed waterproofing membrane whose base material is a polymer has a higher initial strength and faster construction time than conventional sheet-type waterproofing materials. Because of its high physical properties, such as adhesion and tensile strength, the shotcrete or concrete reinforced with waterproofing membrane shows as composite materials. In this study, to consider an application to the conventional design method, the numerical method was used. In the numerical analysis, material and contact properties were adopt from previous studies. Because the contact properties between membrane and other supporting materials at the interface is important parameter to understand the supporting mechanism of the sprayed waterproofing membrane, contact properties were mainly considered at the interface. In addition, the conventional design method for the tunnel lining, was reviewed with numerical results.

1 INTRODUCTION

Groundwater is an important consideration in engineering projects because it can weaken and degrade a structure causing subsidence near an excavated section (Nakashima et al. 2015). Excessive leaking can increase construction costs, delay construction, and suspend the operation of structures (ITAtech, 2013). A sprayed waterproof membrane is commonly 3–5 mm thick, with a maximum thickness of less than 10 mm, which is thin compared with shotcrete (EFNARC, 2008). A sprayed waterproof membrane must adhere to its supporting wall by cohesive and tensile strengths, because it is generally constructed between the primary and secondary linings in a tunnel (Su & Bloodworth, 2016). It can thus support and reinforce an underground structure, and may possibly reduce the required thickness of the secondary lining, reinforcement ratio, or concrete grade (Lee et al. 2018).

In this study, the membrane properties and interface properties were obtained from experiments and numerical methods. The properties of the material itself were determined by tensile test and the interface characteristics were determined by LBS (Linear Block-Support, proposed by EFNARC) test. And shear behavior between concrete and membrane were obtained and calibrated from the direct shear test by previous research. And then design criteria for the concrete lining in tunnel was investigated numerically.

2 MATERIAL PROPERTIES OF SPRYED WATERPROOFING MEMBRANE

To evaluate the material properties of the sprayed waterproof membrane, tensile tests were conducted according to ASTM-D638 standards (ASTM, 2010), as shown in Figure 1(a). The

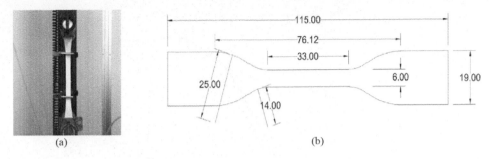

Figure 1. Tensile test for sprayed membrane: (a) photograph during testing and (b) dimensions of the membrane (mm), Type-4 specimen (ASTM, 2010).

dimensions of the Type-4 membrane specimen from the ASTM standards (thickness 3 mm) are shown in Figure 1(b). The specimen is molded in a mold corresponding to Type-4 according to ASTM-D638, and the tensile test was carried out after 28 days of curing age.

The numerical analysis model for the LBS test is shown in Figure 2 (Lee et al. 2018). Because both sides are symmetrical with respect to the center block, only one-half side was modeled. In addition, the center block to which the displacement control is applied can be simulated by controlling its bottom surface displacement (uniform vertical displacement); therefore, modeling the center block of the LBS test is not necessary. The uniform vertical displacement of the center block bottom in the numerical analysis was controlled to match that observed in the linear block-support test. The boundary condition of a side block was completely constrained by the bolt connection, and the maximum relative displacement of the center plane was set to be 7 mm in the vertical (y-axis) direction (the actual failure was observed at a relative displacement of 6.65 mm in the experiment). The cohesive behavior model and interface damage model were implemented to simulate separation of the contacted surfaces of the membrane and concrete block (SIMULIA, 2014).

Figure 3 shows the results of numerical analysis for fitting contact properties and Table 1 presents the best fit parameter values for the damage model. The cohesive stiffness influences the initial slope (initial stiffness) of the force–displacement relationship at the contact, and the maximum stress of damage initiation affects the maximum force at the time of damage occurrence. Finally, fracturing energy is related to the energy required for complete separation of the contact surface.

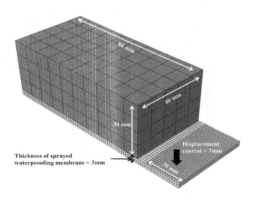

Figure 2. Numerical modeling of the LBS test using ABAQUS (Lee et al. 2018).

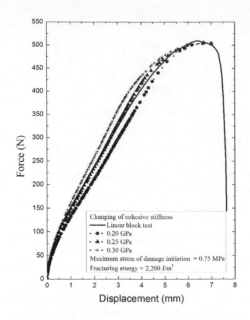

Figure 3. Comparison of LBS test results and numerical analysis in terms of cohesive stiffness (Lee et al. 2018).

Table 1. Contact properties of a sprayed waterproof membrane.

Contact properties	Value
Cohesive stiffness (GPa)	0.25
Maximum nominal stress at damage initiation (MPa)	0.75
Fracturing energy (kJ/m^2)	2.20

3 NUMERICAL MODELING FOR DESIGN CONCRETE LINING

To evaluate the lining thickness in tunnel design, the designing load should be considered. Generally, the designing load is determined by combination of self weight, loosening load of ground and residual hydraulic pressure surrounding tunnel. Figure 4 shows schematics of loosening load and residual hydraulic pressure for tunnel (MOLIT, 2012). And Table 2 shows

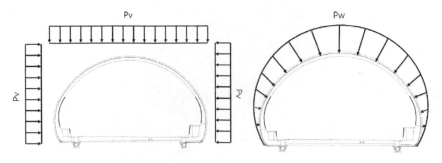

Figure 4. Schematics of (a) loosening load and (b) residual hydraulic pressure for tunnel (MOLIT, 2012).

2434

Table 2. Coefficients of combination for designing load.

Case	Self weight	Rock load	Residual hydraulic pressure	Temperature	Shrinkage
Case 1	1.4				
Case 2	1.2	1.6	1.6	+1.2	
Case 3	1.2	1.6	1.6	-1.2	1.2

* Case 1, 2 & 3 are strength-based and Case 4 is allowable stress design method

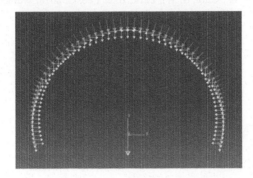

Figure 5. Boundary conditions of tunnel by combination for load (Lee et al. 2017).

coefficient of combination for loads by MOLIT (2012) in South Korea. The designing load was applied in numerical analysis using user subroutines as boundary conditions (Figure 5).

4 RESULTS

From the series of numerical analyses, design thickness for concrete lining can be evaluated regarding the adhesion of sprayed waterproofing membrane. By this method, it is expected that thickness of shotcrete lining with membrane in NATM tunnel also can be evaluated. However, interface parameters for each membrane were not considered in this study. Therefore, various conditions for interface parameters must be taken into account as well as conditions of designing loads in further study.

ACKNOWLEDGEMENT

This research was supported by the Korea Agency for Infrastructure Technology Advancement under the Ministry of Land, Infrastructure and Transport of the Korean government. (Project Number: 18SCIP-B108153-03)

REFERENCES

Nakashima, M., Hammer, A. L., Thewes, M., Elshafie, M. & Soga, K. 2015. Mechanical behavior of a sprayed concrete lining isolated by a sprayed waterproofing membrane. *Tunneling and Underground Space Technology* 47: 143–152.
ITAtech. 2013. *ITAtech design guidance for spray applied waterproofing membranes.* ITAtech Activity Group Lining and Waterproofing. ITAtech Report No. 2.
EFNARC. 2008. *Specification and Guidelines on Thin Spray-on Liners for Mining and Tunneling.* European Federation of National Associations Representing for Concrete.

Su, J. & Bloodworth, A. 2016. Interface parameters of composite sprayed concrete linings in soft ground with spray-applied waterproofing. *Tunnelling and Underground Space Technology* 59: 170–182.

ASTM D638. 2010. *Standard Test Method for Tensile Properties of Plastics*. ASTM International. West Conshohocken. PA.

MOLIT. 2007. *Standard of Tunnel Design*. MOLIT (Ministry of Land, Transport and Maritime Affairs). in Korean.

SIMULIA. 2014. *6.14 Documentation Collection*. ABAQUS/CAE User's Manual.

Lee, K., Kim, D., Chang, S.H., Choi, S. W. & Lee, C. 2017. Analysis of reinforcement effect of TSL (Thin Spray-on Liner) as supports of tunnel by numerical analysis. *J. Korean Geosynthetics Society* 16(4): 151–161.

Lee, K., Kim, D., Chang, S. H., Choi, S. W., Park, B. & Lee, C. 2018. Numerical approach to assessing the contact characteristics of a polymer-based waterproof membrane. *Tunnelling and Underground Space Technology* 79: 242–249.

*Tunnels and Underground Cities: Engineering and Innovation meet Archaeology,
Architecture and Art, Volume 5: Innovation in underground engineering,
materials and equipment - Part 1 – Peila, Viggiani & Celestino (Eds)*
© 2020 Taylor & Francis Group, London, ISBN 978-0-367-46870-5

Effect of machine factors of tunnel boring machines on penetration rates in rocks

G.J. Lee & T.H. Kwon
Department of Civil and Environmental Engineering, Korea Advanced Institute of Science and Technology (KAIST)

H.H. Ryu & K.Y. Kim
Korea Electric Power Research Institute (KEPRI)

T.M. Oh
Department of Civil Engineering, Pusan National University (PNU)

ABSTRACT: Prediction of penetration rate of tunnel boring machine (TBM) is important in tunneling design; however, the effect of machine specifications of the TBM on penetration rate remains poorly identified. This study defined a representative machine factor (MF) composed of thrust force, number of cutters, and TBM diameter, and its correlations with penetration rates, using the previously reported field data. In addition, the correlations between the machine factors and the unconfined compressive strength (UCS) of rocks were examined. The stronger relationship between penetration depth and MF than between penetration rate and MF was found. Finally, the generalized prediction-model to penetration depth per revolution was suggested. The presented findings provide insights into the roles of machine factors in TBM performance.

1 INTRODUCTION

Prediction of penetration rates of the tunnel boring machine (TBM) plays a primary role in estimation of cost and time for tunnel construction. The TBM penetration rate is primarily affected by ground conditions, machine specifications, and engineering judgment of an operator (Torabi et al. 2013; Hassanpour et al. 2010). Recently, several prediction models (Yagiz, 2008; Hamidi et al. 2010; Delisio et al. 2013; Salimi et al. 2016; Naghadehi and Ramezanzadeh, 2017; Jamshidi, 2018) for TBM penetration rate have been presented based on field data; however, mechanical specifications of TBM are hardly considered and included in the models. For instance, for the similar rock properties and ground conditions, the penetration rates differ as the number of cutters increases, even though the TBM diameter and the cutter disc size are the same (Yagiz, 2008; Hamidi et al. 2010; Hassanpour et al. 2010; Delisio et al. 2013; Jain et al. 2016). In this study, we investigated correlations between various combinations of machine factors, such as thrust force, number of cutters, and TBM diameter, and penetration rates, using the previously reported field data (e.g., Yagiz, 2008; Hassanpour et al. 2010; Delisio et al. 2013; Jain et al. 2016; Salimi et al. 2016). Thereby, the penetration rates were attempted to be normalized with proper machine factors, such that the normalized penetration rate can be correlated to the rock properties and ground conditions.

Table 1. Various types of machine factor.

Model	MF type			
	MF$_1$	MF$_2$	MF$_3$	MF$_4$
Model	$\frac{Th}{N}$	$\frac{Th}{N \cdot D}$	$\frac{Th}{N \cdot D^2}$	$\frac{Th \cdot D}{N}$

2 DATA REDUCTION AND DEFINITION OF MACHINE FACTORS

2.1 Data reduction

All regression analyses were conducted using the field data previously published in Hassanpour et al. (2010) and Jain et al. (2016). We extracted the penetration rates, TBM specifications (thrust force, number of cutters, diameter of cutterhead), and UCS values; 27 data from Hassanpour et al. (2010) and 79 data from Jain et al. (2016), respectively. As the parameters representing the penetration rates, the penetration rate (PR) is defined as the penetration depth per time, i.e., length/time (e.g., mm/min); and the penetration depth (Pe) is defined as the penetration depth per revolution (e.g., mm/rev). Thus, PR and Pe can be correlated, i.e., PR (mm/min) = Pe (mm/rev) x RPM (rev/min).

2.2 Definitions of various machine factors of TBM

We chose thrust force, number of cutters, and cutterhead diameters as representative parameters related to machine specifications. It was presumed that the thrust force had a positive relation with the penetration rate. We hypothesized that the number of cutters had a negative effect on the penetration rate, because the force applied per cutter decreases as the cutter number increases for a given thrust force. Meanwhile, the effect of cutterhead diameter on penetration rate is unclear; it can be hypothesized that the larger cutterhead area decreased the pressure by thrust force, on the other hand, it increased the torque at the edge of cutterhead. Thus, we defined four types of machine factors (MF) and examined their correlation with the penetration rate, as shown in Table 1.

Th: Thrust force (kN), N: The number of cutters, D: Diameter of cutterhead (m)

3 RESULTS AND ANALYSIS

3.1 Correlations between machine factors and penetration depth and rate

A regression analysis was conducted for the penetration depth and rate with the machine factors defined in Table 1. Except for MF$_3$, the rest of MFs positively affected the Pe and PR with the positive standardized coefficients. Among the MFs listed in Table 1, MF$_4$ was the most effective for dependent variables to the penetration depth (Pe) and the penetration rate (PR) with the highest correlation coefficient, R. The t- values for MF$_4$ were also larger than 2, thus MF$_4$ was found to be a meaningful independent variable (Table 2). The regression equation for each correlation are as follows:

$$Pe(mm/rev) = 2.760 + 0.048MF_1 \ (kN \cdot m/cutter), (R = 0.69); \tag{1}$$

$$PR(m/h) = 2.566 + 0.007MF_1 \ (kN \cdot m/cutter), (R = 0.29); \tag{2}$$

$$Pe(mm/rev) = 2.140 + 0.224MF_2 \ (kN \cdot m/cutter), (R = 0.62); \tag{3}$$

Table 2. Regression analysis for Pe and PR by machine factors.

| Model | Dependent variables | Independent variables | Unstandardized coefficient | | Standardized coefficients | | | |
			B	Std. error	Beta	t	Sig.	R
1	Pe	(Constant)	2.760	0.460		6.002	0.000	0.687
		MF_1	0.048	0.005	0.687	9.654	0.000	
	PR	(Constant)	2.566	0.200		12.814	0.000	0.289
		MF_1	0.007	0.002	0.289	3.076	0.003	
2	Pe	(Constant)	2.140	0.613		3.491	0.001	0.617
		MF_2	0.224	0.028	0.617	8.001	0.000	
	PR	(Constant)	2.679	0.253		10.573	0.000	0.177
		MF_2	0.021	0.012	0.177	1.837	0.069	
3	Pe	(Constant)	2.222	0.866		2.564	0.012	0.467
		MF_3	0.870	0.162	0.467	5.386	0.000	
	PR	(Constant)	3.145	0.324		9.714	0.000	0.012
		MF_3	-0.007	0.060	-0.012	-0.119	0.905	
4	Pe	(Constant)	3.381	0.373		9.063	0.000	0.721
		MF_4	0.010	0.001	0.721	10.620	0.000	
	PR	(Constant)	2.575	0.166		15.477	0.000	0.355
		MF_4	0.002	0.000	0.355	3.870	0.000	

$$PR(m/h) = 2.679 + 0.021 MF_2 \ (kN \cdot m/cutter), (R = 0.18); \qquad (4)$$

$$Pe(mm/rev) = 2.222 + 0.870 MF_3 \ (kN \cdot m/cutter), (R = 0.47); \qquad (5)$$

$$PR(m/h) = 3.145 - 0.007 \ MF_3 \ (kN \cdot m/cutter), (R = 0.01); \qquad (6)$$

$$Pe(mm/rev) = 3.381 + 0.010 \ MF_4 \ (kN \cdot m/cutter), (R = 0.72); \quad and \qquad (7)$$

$$PR(m/h) = 2.575 + 0.002 \ MF_4 \ (kN \cdot m/cutter), (R = 0.36). \qquad (8)$$

3.2 *Regression of MF and UCS for Pe and PR*

Herein, MF_4 was chosen as a representative MF parameter for multiple regression analysis (Table 2). When comparing the regressions between Model 5 which composed of independent variable of UCS and Model 6 which composed of independent variables of UCS and MF, inclusion of MF as a dependent variable led to the better correlation (Table 3). Pe and PR were correlated to UCS and MF_4 via multivariable regression, and the resulting correlations are as follows:

$$Pe(mm/rev) = 5.374 + 0.024 UCS \ (MPa), (R = 0.19); \qquad (9)$$

$$PR(m/h) = 3.354 - 0.005 UCS \ (MPa), (R = 0.11); \qquad (10)$$

$$Pe(mm/rev) = 5.630 - 0.079 UCS \ (MPa) + 0.061 MF_4 (kN.m/cutter), (R = 0.85); \ and \quad (11)$$

Table 3. Regression analysis for Pe and PR by UCS and MF$_4$.

Model	Dependent Variables	Independent variables	Unstandardized coefficient B	Unstandardized coefficient Std. error	Standardized coefficients Beta	t	Sig.	R
5	Pe	(Constant)	5.374	0.715		7.518	0.000	0.190
		UCS	0.024	0.012	0.190	1.979	0.050	
	PR	(Constant)	3.354	0.239		14.026	0.000	0.110
		UCS	-0.005	0.004	-0.110	-1.131	0.261	
6	Pe	(Constant)	5.630	0.386		14.578	0.000	0.850
		UCS	-0.079	0.009	-0.634	-8.643	0.000	
		MF$_4$	0.016	0.001	1.168	15.937	0.000	
	PR	(Constant)	3.416	0.190		17.995	0.000	0.620
		UCS	-0.030	0.004	-0.717	-6.578	0.000	
		MF$_4$	0.004	0.000	0.861	7.893	0.000	

$$PR(m/h) = 3.416 - 0.030UCS\,(MPa) + 0.004MF_4(kN.m/cutter),\,(R = 0.62). \qquad (12)$$

Figure 1 shows the comparisons between the predictions by Equations 11 and 12 and the measured data. It appears that Pe has a better correlation with UCS and MF$_4$ than PR does. Furthermore, the correlation coefficient R of Model 6 for Pe was 0.85, which means that its determination coefficient, R^2 was 0.72.

3.3 Relationship between UCS and MF

As the mechanical strength of rocks increases, it requires the greater energy to excavate them with disc cutters. Therefore, the greater UCS is expected to require the greater thrust force

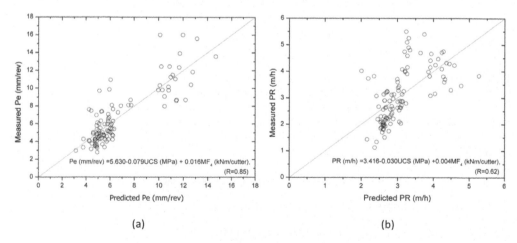

(a)   (b)

Figure 1. (a) Linear relation between measured Pe and predicted Pe (b) Linear relation between measured PR and predicted PR.

and the more energy. In light of this, UCS and machine factors (MF) that are combinations of the thrust force, the number of disc cutters, and the cutterhead diameter are presumed to exhibit a considerable level of correlation.

In this study, we investigated the UCS-MF relationships for different levels of penetration depth Pe or penetration rate PR. Herein, we classified Pe into three classes: (a) 2–5 mm/rev, (b) 5–8 mm/rev, and (c) 8–11 mm/rev. And, PR was classified into four classes: (a) 1–2 m/h, (b) 2–3 m/h, (c) 3–4 m/h, and (d) 4–5 m/h. MF_4 was used as a machine factor.

A positive but non-linear relation between UCS and MF_4 was found, as shown in Figures 2 and 3. For Pe = 2–5 mm/rev, the UCS-MF relation showed the adjusted R^2 of 0.60 (Figure 2a). Likewise, the adjusted R^2 ranged 0.48–0.68 for Pe = 2–11 mm/rev (Figure 2). Meanwhile, the adjusted R^2 of the UCS-MF relation with respect to PR ranged 0.48–0.86 (Figure 3). It appears that as Pe or PR increases, the positive relation between UCS and MF becomes clearer and stronger.

3.4 A simple model to predict Pe using UCS and MF

Pe values were plotted against MF in Figure 4a. As can be seen, two clusters were formed with the threshold MF at 500 kN·m/cutter. Less than 500 kN·m/cutter, majority of Pe ranged 2–8 mm/rev; and above it, Pe ranged 8–16 mm/rev. The plot of PR-MF shown in Figure 4b also shows the similar trend, but with the greater scattering.

Figures 4c and 4d show the relations between RPM and MF and between RPM and MF normalized with UCS. In Figure 4c, a negative RPM-MF relationship when MF less than 500 kN·m/cutter and a positive RPM-MF relationship when MF greater than 500 kN·m/cutter were observed. This complicates the development of a robust prediction model.

Based on the observation in Section 4.3, we attempted to normalize MF with UCS to minimize the effect of UCS on the thrust force. Thereby, we found not strong but meaningful correlations between MF/UCS-RPM, MF/UCS-Pe, and MF/UCS-PR, as shown in Figures 4d-to-4f. Among these correlations, the highest adjusted R^2 was found for the MF/UCS-Pe (i.e., adjusted R^2 = 0.70; Figure 4e). This implies that the normalized machine factor with UCS provides a good prediction to penetration depth per revolution (Pe). The generalized model can be expressed as follows:

$$Pe = \alpha \left[\frac{MF/(1kN \cdot m/cutter)}{UCS/1MPa} \right]^{\beta},$$

(13)

where α and β are empirical parameters. For instance, α and β can be determined by finding the best fitting using the initial advance rate data in a TBM field, then the model with these parameters can be used to predict the penetration depth and advance rate for the rest of tunnel length.

There was no strong correlation not only between RPM and MF (Figure 4c) but also between RPM and MF/UCS, as adjusted R^2 was 0.15 (Figure 4d). In the rock excavation, the RPM is controlled by the TBM operator in consideration of the mechanical condition of the TBM. When Pe is high, the amount of excavation per one rotation of cutterhead increases, and hence the RPM tends to decrease relatively to account for the discharge rate of the excavated rock chips. Furthermore, the higher RPM may cause the less crack transfer at the same thrust force, which in turn leads to the less effective cutter penetration depth (Macias et al. 2014). Therefore, it is reasonable to set RPM as a separate factor from the TBM specification, as RPM is used to compute PR. In light of this, Pe is expected to have better correlation with MF than PR, which is also previously confirmed in Table 2.

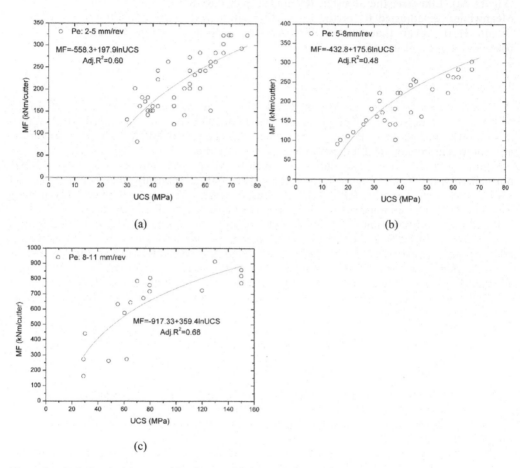

Figure 2. Relation between machine factor of TBM and UCS of rock with Pe.

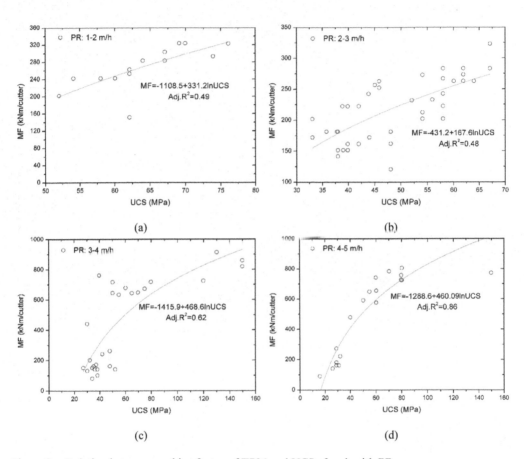

Figure 3. Relation between machine factor of TBM and UCS of rock with PR.

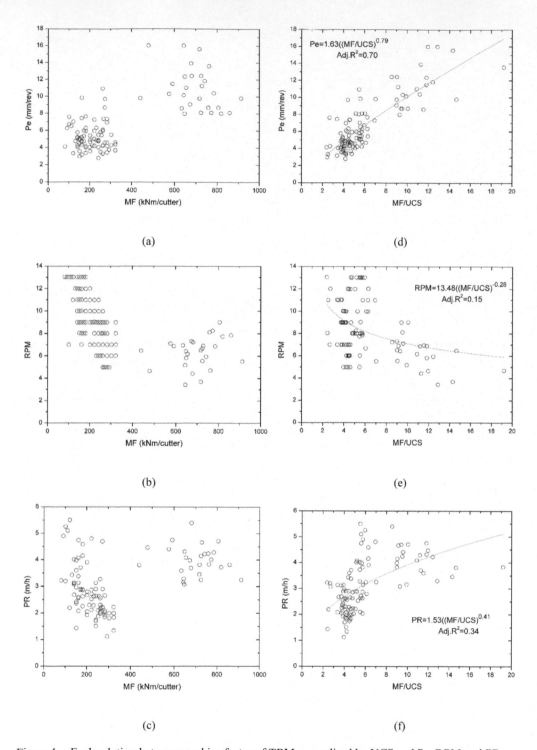

Figure 4.   Each relation between machine factor of TBM normalized by UCS and Pe, RPM and PR.

4 CONCLUSION

In order to study on the effect of TBM specification on penetration rates in rocks, regression analysis was conducted. UCS was used as a representative ground effect factor. Machine factor of a TBM, MF, including cutter head diameter, number of cutters, and thrust force was defined and used as a representative of TBM specification based on the regression analysis results. The stronger relationship between Pe and MF than between PR and MF was found. Through multiple regression analysis, model for Pe having a $R^2 = 0.72$ was drawn. We found not strong but meaningful correlations and the highest adjusted $R^2 = 0.70$, was found for the MF/UCS-Pe and the generalized prediction-model to penetration depth per revolution (Pe) was suggested. It is reasonable to set RPM which is affected by operation's judgment as a separate factor from the TBM specification.

ACKNOWLEDGEMENT

This research was supported by a grant (15SCIP-B105148-01) from the Construction Technology Research Program funded by the Ministry of Land, Infrastructure, and Transport of the Korean government.

REFERENCES

Torabi, S. R., Shirazi, H., Hajali, H., Monjezi, M. 2013. Study of the influence of geotechnical parameters on the TBM performance in Tehran–Shomal highway project using ANN and SPSS, *Arabian Journal of Geosciences*, Vol. 6, No. 4, pp. 1215-1227.

Hassanpour, J., Rostami, J., Khamehchiyan, M., Bruland, A., Tavakoli, H. R. 2010. TBM performance analysis in pyroclastic rocks: a case history of Karaj water conveyance tunnel, *Rock Mechanics and Rock Engineering*, Vol. 43, No. 4, pp. 427-445.

Yagiz, S. 2008. Utilizing rock mass properties for predicting TBM performance in hard rock condition, *Tunnelling and Underground Space Technology*, Vol. 23, No. 3, pp. 326-339.

Hamidi, J. K., Shahriar, K., Rezai, B., Rostami, J. 2010. Performance prediction of hard rock TBM using Rock Mass Rating (RMR) system, *Tunnelling and Underground Space Technology*, Vol. 25, No. 4, pp. 333-345.

Delisio, A., Zhao, J., Einstein, H. H. 2013. Analysis and prediction of TBM performance in blocky rock conditions at the Lötschberg Base Tunnel, *Tunnelling and Underground Space Technology*, Vol. 33, No. 1, pp. 131-142.

Salimi, A., Rostami, J., Moormann, C., Delisio, A. 2016. Application of non-linear regression analysis and artificial intelligence algorithms for performance prediction of hard rock TBMs, *Tunnelling and Underground Space Technology*, Vol. 58, No. 1, pp. 236-246.

Naghadehi, M. Z., Ramezanzadeh, A. 2017, Models for estimation of TBM performance in granitic and mica gneiss hard rocks in a hydropower tunnel, *Bulletin of Engineering Geology and the Environment*, Vol. 76, No. 4, pp. 1627-1641.

Jamshidi, A. 2018, Prediction of TBM penetration rate from brittleness indexes using multiple regression analysis, *Modeling Earth Systems and Environment*, Vol. 4, No. 1, pp. 383-394.

Jain, P., Naithani, A. K., Singh, T. N. 2016. Estimation of the performance of the tunnel boring machine (TBM) using uniaxial compressive strength and rock mass rating classification (RMR)–A case study from the Deccan traps, India, *Journal of the Geological Society of India*, Vol. 87, No. 2, pp. 145-152.

Macias, F. J., Jakobsen, P. D., Bruland, A., Log, S., Grøv, E. 2014. The NTNU prediction model: a tool for planning and risk management in hard rock TBM tunnelling, *In World tunnelling congress 2014*.

Tunnels and Underground Cities: Engineering and Innovation meet Archaeology,
Architecture and Art, Volume 5: Innovation in underground engineering,
materials and equipment - Part 1 – Peila, Viggiani & Celestino (Eds)
© 2020 Taylor & Francis Group, London, ISBN 978-0-367-46870-5

Prediction of EPB shield TBM performance using a lab scale excavation test with different soil conditions

H. Lee, D. Shin & H. Choi
Korea University, Seoul, Republic of Korea

D.Y. Kim & Y.J. Shin
Hyundai Engineering & Construction, Gyeonggi-do, Republic of Korea

ABSTRACT: In the recent decades, the earth pressure balanced tunnel boring machine (EPB TBM) is one of the commonly employed mechanized tunnelling methods, which supports the tunnel face with the excavated soil. During excavation, various additives should be injected into the tunnel face, chamber and screw conveyor in order to improve the mechanical and hydrological properties of the muck. In this paper, a lab-scale simulating apparatus for EPB TBM excavation was devised to study the excavation mechanism of EPB TBMs including the additives injection. Subsequently, a series of excavation tests were conducted with different soil conditions, and the experimental data such as the thrust force, blade torque and abrasion loss of the cutter bits were discussed. In this research, the effect of the fine content and water content of in-situ soil and the injection volume of foam (Foam Injection Ratio, *FIR*) on the mechanical behavior of EPB TBMs were carefully studied in particular.

1 INTRODUCTION

Earth pressure balanced shield tunnel boring machines (EPB shield TBMs) are commonly applied to various tunnel construction projects all over the world since the 1970s after the development of the machine in Japan (Herrenknecht, 2011). As the advance of the EPB TBM under the ground, the excavation chamber is gradually filled with the freshly excavated soil that supports the tunnel face to cope with earth pressure and water pressure ahead of TBM. At this time, the excavated soil should achieve some mechanical and hydrological properties for efficient TBM excavation. For this purpose, additives injection can facilitate to attain target properties of soil bodies at the tunnel face, and inside of the chamber and screw conveyor.

In order to verify the behavior of conditioned soil inside of EPB TBMs, Maidl (1995) experimentally studied the conditioning characteristics including foam penetration, permeability and compressibility, and suggested application ranges of EPB TBM with conditioned soil. In addition, various reproducible test methods were suggested in order to determine the characteristics of the conditioned soil (Budach, 2012). To investigate the behavior of conditioned soil under pressure, Lisa Mori (2016) examined the effect of total stress, effective stress and void ratio. In the laboratory, Viani et al. (2008) used a slump test to analyze the global characteristics of conditioned soil, and conducted a series of screw conveyor tests for quantitatively evaluating conditioned soils. Jakobsen et al. (2013) developed Soft Ground Abrasion Tester (SGAT) to predict TBM tool wear, torque and thrust on soft ground. Gharahbagh et al. (2011, 2014) reviewed the influence of soil conditioning on tool wear and torque requirement by devising Penn State Soil Abrasion testing device. Also, in Italy, wear of TBM drilling tools with conditioned soil was intensively examined in micro point of view (Bosio, 2018; Salazar, 2018).

Although many researchers have carried out numerous studies as documented above, no specific criteria have not been suggested yet, which can define the exact state of the conditioned soil during the EPB TBM operation. Besides, no study has not been performed with a simulation of EPB TBM excavation mode including the simultaneous injection of two additives.

Accordingly, in this paper, a new lab-scale test apparatus was devised in order to simulate and estimate the performance of EPB TBMs, which is focused on thrust force, torque and abrasion excavating with different soil conditions. The developed equipment is designed to excavate the soil in a desired condition in the vertical direction with constant penetration rate, and at the same time, additives such as water, foam and polymer can be injected in front of the excavation face with the rotating blade.

As mentioned above, the purpose of the present study is to estimate the influence of the different soil conditions on TBM operation in the laboratory. Therefore, a series of lab tests were carried out in the lab-scale simulating equipment and the representative TBM advance performance data (e.g. thrust force, blade torque and cutter bit abrasion) was obained during the test. From the test results, the effect of fine content in soil, water sprinkling and foam injection in EPB TBMs excavation was examined.

2 SOIL CONDITIONING IN EPB SHIELD TBM

Figure 1 shows the scheme of EPB TBMs with soil conditioning system (Budach, 2012). The chamber is filled with soil, conditioning additives are injected, and cutterhead with cutting tools rotates in order to create a pulpy soil paste. During this operation, the most frequently used conditioners are foam, polymers, and bentonite slurry. In this study, since all experiments performed with foam injection as a conditioning agent, the general details of foam are described in the following paragraph.

The primary objective of mechanical and hydrological muck properties through foam injection in EPB TBMs can be summarized as follows: 1) Lowering permeability of the excavated soil in the chamber could prevent an inducing seepage force in front of the tunnel face and reduce effective face support. 2) Creation of a homogeneous and compressible soil paste could maintain pressure and reduce pressure fluctuations in the working chamber. 3) Reduction of internal friction between soil particle could lower the torque of the cutterhead, screw conveyor and abrasion of tools. 4) Reduction of soil stickiness could decrease clogging risks in case of excavating cohesive grounds. However, the injection of the required foam to achieve optimum efficiency is different from a stiff clay to sandy gravel. Correspondingly, the application of foam which will be used for a particular site has to be determined by laboratory tests with the in-situ soil (Langmaack, 2000; Maidl et al., 2013; Budach & Thewes, 2015).

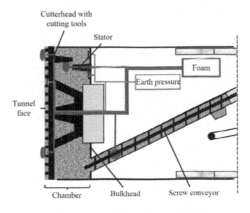

Figure 1.   The composition of EPB shield TBM with soil conditioning system (Budach, 2012).

Representative soil conditioning terminologies frequently using in practice and research area when applying foam as a conditioning agent are shown in Equation 1~3. Surfactant concentration (C_f) is the concentration of surfactant in the surfactant solution (Eq. 1). Foam Injection Ratio (*FIR*) is the ratio of injected foam volume to the in-situ soil to be excavated (Eq. 2). Foam Expansion Ratio (*FER*) is the ratio of foam volume to surfactant solution volume (Eq. 3). The typical application range of C_f is 0.5% to 5%, *FIR* is 30% to 60%, and *FER* is 5 to 30 (EFNARC, 2005). In this study, V_{es} was calculated by considering penetration rate (*PR*) and the diameter of the excavation chamber (200 mm).

$$C_f = \frac{V_{sf}}{V_{sf} + V_w} = \frac{V_{sf}}{V_l} \times 100\% \tag{1}$$

$$FIR = \frac{V_f}{V_{es}} \times 100\% = \frac{V_l + V_a}{V_{es}} \times 100\% \tag{2}$$

$$FER = \frac{V_f}{V_l} = \frac{V_l + V_a}{V_l} \tag{3}$$

where V_{sf} = volume of the surfactant; V_w = volume of water in the solution; V_l = volume of the surfactant liquid; V_a = volume of air; and V_{es} = the in-situ volume of soil to be excavated.

3 LAB SCALE EPB SHIELD TBM EXCAVATION TEST

3.1 Experimental apparatus

In this paper, lab scale test apparatus in order to simulate advance of EPB shield TBM was devised. The equipment is mainly composed of the excavation chamber, the screw conveyor, blade with five cutter bits, three-floor molds for the preparation of soil samples, and the foam generator and the water pump system for supplying foam and water. It can simulate the excavation mode of EPB TBM advance: excavation with the rotation of the blade, soil conditioning with foam or polymer, and extracting excavated soil through the screw conveyor. The overall configuration of the experimental apparatus is illustrated in Figure 2.

In the excavation stage, the machine penetrates the prepared soil with rotating of the blade that replaceable five aluminum cutter bits installed with different distance from the axis of the rotation. While excavation, load cell and torque meter installed upper part of the rotating shaft gather thrust force and blade torque data respectively. Also, foam generator and water pump system supply constant quality of foam and water with a set amount for increase soil

Figure 2. Configuration of the EPB shield TBM excavation lab test apparatus.

consistency and reducing overload in the excavation tasks. In order to achieve this, two pipes are embedded inside the blade rotary shaft. Therefore, two different type of additives can be injected simultaneously during the excavation test. Finally, aluminum cutter bits mounted on the blade are designed to be separable and allow weight change measurements of cutter bits before and after the experiment.

3.2 Specimens for experiment

Three different soils for the experiment which have different fines (d < 0.075 mm) were prepared by the sieve analysis. These three samples were mainly composed of artificial silica sand with 3%, 15% and 30% of fine contents, respectively, and illite was selected as representing fine particle fraction. Due to the scale of the test equipment, especially diameter of the screw conveyor (40 mm), the samples were formed with artificial silica sand of which have a particle size less than 2 mm. The grain-size distribution curves of three soils used in this study are shown in Figure 3. All the specimens were uniformly compacted with five-floors and stacked in three-floors of specimen molds (Figure 2). All of the specimens in this research were placed with 10% of water content after a sufficient mixing. On the other hand, in some cases, a constant amount of water was instilled during the experiment to increase water content from 10% to 15%.

3.3 Test conditions

As the focal point of the current study was the investigation on the EPB shield TBM excavation performance according to soil conditions, operating parameters of the EPB TBM experiment were fixed such as penetration rate (PR, 45 mm/min), blade rotation speed (10 rpm), and screw conveyor speed (120 rpm). The experiment was performed in a total of five cases with different soil conditions as documented in table 1 and parameters were determined to evaluate the effect of soil type with different fine contents, sprinkling water, and foam injection ratio (FIR) in TBM driving. The soil-foam mixing ratio of all cases was selected by conducting a

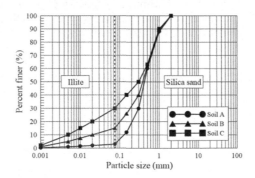

Figure 3.   Grain-size distribution curves of test specimens.

Table 1.   Soil conditions of the EPB TBM excavation test.

Case	Soil type	C_f %	FIR %	FER	Water content %
Case 1	Soil A	3	40	15	10
Case 2	Soil B	3	60	15	10
Case 3	Soil C	3	80	15	10
Case 4	Soil C	3	30	15	from 10 to 15
Case 5	Soil C	3	40	15	from 10 to 15

preliminary slump test (ASTM C143, 2015), which is a convenient and rapid test for determining the optimum workability of the conditioned soil. In case of the test with water sprinkling, a constant amount of water was simultaneously injected with foam considering the penetration rate (PR), the area of the excavation chamber, and the wet unit weight of the specimen.

4 EXPERIMENT RESULT AND DISCUSSION

4.1 *Effect of fine contents (d < 0.075 mm, Illite)*

Figure 4 shows the acquired thrust force and blade torque data from conducting experiments with a different type of soils which have different fine contents (Case 1 & 2). In the case of experiment Case 1 performed with soil containing low fine contents (3%), the test was ceased due to the limitation of load cell measuring the thrust force. In other words, the test was abnormally discontinued even though sufficient workability was achieved by preliminary slump test. However, the specimen was replaced after the fines were added up to 15% (Case 2), the experiment proceeded without any problems as reported in Figure 4.

Based on the experimental results, it can be concluded that achieving enough consistency of the coarse-grained soil with low fines is relatively challenging when using only foams as additives. Because of insufficient consistency, the conditioned soil in the chamber forming arch and extraction through the screw conveyor became nearly impossible. Consequently, the machine came across the thrust force limitation. Experiment results imply that when managing EPB TBM in the coarse-grained sand with scant fine contents, thrust force could be excessively loaded to the machine due to the difficulty in muck discharging. Accordingly, optimal TBM drive can be accomplished by injecting not only foam but also high-density slurries or polymers as additives for satisfying enough consistency of the soil paste.

4.2 *Effect of sprinkling water*

Figure 5 illustrates thrust force and blade torque data during excavation tests performed with or without sprinkling water (Case 3 & 4). Without injecting water (Case 3), the test was halted due to the limitation of the torque meter which is a different type of trouble in Case 1. On the other hand, when experiment conducted with the addition of water (Case 4), blade torque limit of the machine did not occur even though *FIR* was dramatically decreased from 80% to 30%.

In the experiment of Case 3, after the equipment was shut down by blade torque limit, inside of the excavation chamber was filled with very stiff soils. Subsequently, it can be

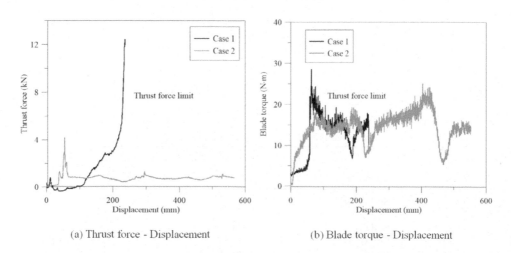

(a) Thrust force - Displacement   (b) Blade torque - Displacement

Figure 4.   Thrust force and blade torque difference between case 1 and 2 (effect of fine content).

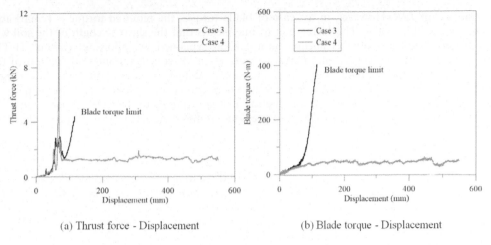

(a) Thrust force - Displacement (b) Blade torque - Displacement

Figure 5. Thrust force and blade torque difference between case 3 and 4 (effect of watering).

inferred that the overload of the blade torque was caused by the formation of the excessive shear strength of the soil ahead of the excavation face and inside of the bulk chamber. Therefore, this result suggests that when operating EPB TBMs in soil with plenty of fines and low water contents as Soil C in this research, the addition of water during excavation is indispensable. In this case, the only foam injection does not work as a function of additives because fines absorb water which is an essential component of the foam. In other words, the water absorbing by fines leads to destroy foam and deprive its function as reducing internal friction of soil paste. As a result, the shear strength of the soil cannot be reduced by the injection of the foam alone when the soil contains a certain amount of fine content (above 30%). Experimental results indicate that EPB TBMs driving in the ground with fines more than 30% can be more efficiently and economically carried out by injecting water and employing a small amount of foam as additives to prevent some clogging risks.

4.3 Effect of foam injection ratio (FIR)

Experiments Case 4 and 5 were conducted with different *FIR* and the test results is shown in Figure 6. In the case of thrust force, the results did not show significant difference with

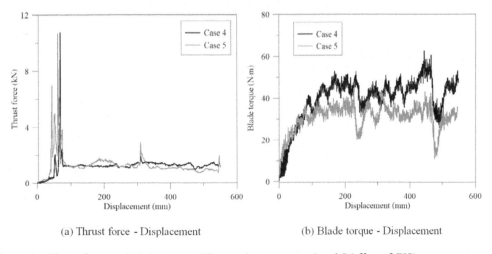

(a) Thrust force - Displacement (b) Blade torque - Displacement

Figure 6. Thrust force and blade torque difference between case 4 and 5 (effect of *FIR*).

increase an of *FIR*. However, in the case of blade torque, the required torque is remarkably decreased by about 30%. The decrease in torque means that the shear strength of the soil was significantly reduced and this phenomenon can be explained by following reasons: 1) The foam which is mainly consist of air fills the porosity and reduces the internal friction of the soil. 2) More surfactant reduces the surface tension of the pores water and reduce the strength of the water bond between each soil particles. Therefore, it is considered that the increase of the *FIR* will have a great effect on reduction in the required torque when operation EPM TBM and these results are consistent with previous studies conducted by many researchers.

4.4 *Cutter bits abrasion*

In all experiments, the amount of weight loss on the cutter bits was calculated by measuring the weight of cutter bits before and after the test. As shown in Figure 7, five cutter bits were attached to the blade to have a different distance from the axis of the blade rotation. Therefore, all the cutter bits have different total travel distances during penetration so that different weight loss of cutter bits could be measured. Due to the different travel distance, the weight of cutter bits shows negligible difference before and after the test except for the cutter bit number one and two. The results of the cutter bit wear measurements of all experimental cases are documented in Table 2. Through the measured weight loss of cutter bit number one which has the longest travel distance, the result confirms that TBM operation with sufficient water contents and adequate *FIR*, which depends on different soil type could definitely reduce the wear of excavation tools as well as steel body of EPB TBMs.

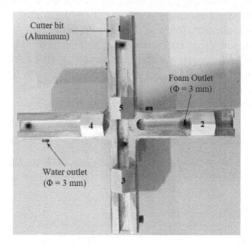

Figure 7.   Blade and disposition of cutter bits for measuring abrasion.

Table 2.   Weight loss of cutter bits after experiment.

Cutter bit number	1	2	3	4	5
	Weight loss				
Case	mg				
Case 1	Immeasurable				
Case 2	11	3	2	2	1
Case 3	Immeasurable				
Case 4	37	5	4	1	
Case 5	19	3	3	2	1

5 CONCLUSIONS

The results of this study can be summarized as follows:

1) In this study, the experimental equipment that can simulate the excavation mode of EPB TBMs including additives injection was suggested, and a series of experiments conducted with different soil conditions. During experiments, foam and water were simultaneously injected with excavation and thrust force, blade torque, and cutter bits abrasion data were gathered.
2) Depending on the fine content of the soil, the tendency of the TBM operation data was considerably changed. When the amount of the fine contents is insufficient (3%), the consistency of the muck was hardly achieved despite enough amount of injected foam and thrust force was excessively loaded. As a result, for sound EPB TBM driving in coarse-grained soil without fines, high-density slurries or polymers have to be additionally injected.
3) When the fine contents occupying a relatively large proportion (above 30%), a different type of trouble in operating TBM could occur. Because high contents of fines impair foams with absorbing liquid, it causes the TBM torque to be easily overloaded. In this case, the excessive torque can be significantly decreased by sprinkling appropriate amount of water into the soil.
4) As a result of the experiment conducted with different *FIR*, the increase of *FIR* accompanied by a proper sprinkling of water is expected to reduce the mechanical loading of TBM significantly. In particular, the decrease of the required torque was more prominent than the reduction of the required thrust force.
5) Based on the measurements of the cutter bits abrasion after the tests, it is confirmed that appropriate foam and water injection, which is depending on the different type of soil could prominently reduce the abrasion of the cutter bits as well as the wear of the machine itself.

REFERENCES

ASTM C145, 2015. Standard Test Method for Slump of Hydraulic-Cement Concrete, *ASTM Standard*

Bosio, F., Bassini, E., Salazar, C. G. O., Ugues, D., & Peila, D. 2018. The influence of microstructure on abrasive wear resistance of selected cemented carbide grades operating as cutting tools in dry and foam conditioned soil. *Wear*, 394, 203–216.

Budach, C., & Thewes, M. 2015. Application ranges of EPB shields in coarse ground based on laboratory research. *Tunnelling and Underground Space Technology*, 50, 296–304.

Budach, C. 2012. Untersuchungen zum erweiterten Einsatz von Erddruckschilden in grobkörnigem Lockergestein *(Doctoral dissertation, Institut für Konstruktiven Ingenieurbau, Ruhr-Universität Bochum)*.

Efnarc, A. 2005. Specification and guidelines for the use of specialist products for mechanized tunnelling (TBM) in soft ground and hard rock. *Recommendation of European Federation of Producers and Contractors of Specialist Products for Structures*.

Gharahbagh, E. A., Rostami, J., & Palomino, A. M. 2011. New soil abrasion testing method for soft ground tunneling applications. *Tunnelling and Underground Space Technology*, 26(5),604–613.

Gharahbagh, E. A., Rostami, J., & Talebi, K. 2014. Experimental study of the effect of conditioning on abrasive wear and torque requirement of full face tunneling machines. *Tunnelling and Underground Space Technology*, 41, 127–136.

Herrenknecht, M., Thewes, M., & Budach, C. 2011. The development of earth pressure shields: from the beginning to the present/Entwicklung der Erddruckschilde: Von den Anfängen bis zur Gegenwart. *Geomechanics and Tunnelling*, 4(1),11–35.

Jakobsen, P. D., Langmaack, L., Dahl, F., & Breivik, T. 2013. Development of the Soft Ground Abrasion Tester (SGAT) to predict TBM tool wear, torque and thrust. *Tunnelling and underground space technology*, 38, 398–408.

Langmaack, L. 2000. Advanced technology of soil conditioning in EPB shield tunnelling. *Proceedings of North American tunneling, 2000*, 525–542.

Maidl, B., Herrenknecht, M., Maidl, U., & Wehrmeyer, G. 2013. *Mechanised shield tunnelling*. John Wiley & Sons.

Maidl, U. 1995. Erweiterung der Einsatzbereiche der Erddruckschilde durch bodenkonditionierung mit Schaum *(Doctoral dissertation, Institut für Konstruktiven Ingenieurbau, Ruhr-Universität Bochum)*.

Mori, L. 2016. Advancing understanding of the relationship between soil conditioning and earth pressure balance tunnel boring machine chamber and shield annulus behavior *(Doctoral dissertation, Colorado School of Mines. Arthur Lakes Library)*.

Salazar, C. G. O., Todaro, C., Bosio, F., Bassini, E., Ugues, D., & Peila, D. 2018. A new test device for the study of metal wear in conditioned granular soil used in EPB shield tunneling. *Tunnelling and Underground Space Technology*, 73, 212–221.

Vinai, R., Oggeri, C., & Peila, D. 2008. Soil conditioning of sand for EPB applications: A laboratory research. *Tunnelling and underground space technology*, 23(3),308–317.

Tunnels and Underground Cities: Engineering and Innovation meet Archaeology,
Architecture and Art, Volume 5: Innovation in underground engineering,
materials and equipment - Part 1 – Peila, Viggiani & Celestino (Eds)
© 2020 Taylor & Francis Group, London, ISBN 978-0-367-46870-5

A study on the efficient crack detection algorithms for tunnel lining based on deep learning approach

J.S. Lee, S.H. Hwang, M. Sagong & I.Y. Choi
Korea Railroad Research Institute, Kyunggi Province, Republic of Korea

ABSTRACT: Efficient detection of damage such as cracks and proper management of the lining are essential during the life cycle of a tunnel structure. For these purposes, various crack detection algorithms have been suggested. In order to better understand the State of the Art of crack detection technologies, convolutional neural network models based on Deep Learning (DL) techniques were compared with each other and the optimal crack detection algorithm in terms of accuracy was further investigated in this study. Specifically, a semantic segmentation model showed much better results compared with those by conventional bounding-box models, due to the intrinsic nature of the cracks. Field application of the model was tested under various conditions. Future work on model improvement will also be discussed in this paper.

1 INTRODUCTION

Since most railway tunnels are in service during the daytime, inspection and maintenance work is usually carried out during the night time; in some cases, only 4 to 5 hours are available, especially in the case of high-speed rail tunnels. To increase the efficiency of inspection and maintenance work, a Track Measurement Vehicle (TMV) together with a tunnel profiler has been introduced to facilitate detection of obstacles within the architectural limit of a tunnel. Although overall inspection based on TMV is efficient in terms of the time required to measure and interpret results, other information on tunnel defects such as cracks and groundwater leakage cannot be obtained in that way.

Recently, new concepts of crack detection devices such as an autonomous robotic system (Menendez, 2018) have been deployed to measure the width and depth of detected cracks inside tunnels. For this, a vision camera was used to detect cracks based on a Deep Convolutional Neural Network (DCNN) model with real crack images for training, and an ultrasonic sensor was utilized to estimate crack depth. A good summary of the recent developments in tunnel inspection techniques can be found in Attard, et al. (2018).

Crack detection of concrete lining is mostly made based on the image processing, and quite a few Deep Learning (DL) techniques such as DCNN have been widely employed in recent days. Additionally, classical image processing models such as the Gabor filter with rotation invariant characteristics (Medina, 2017) and a Scale Invariant Feature Transform (SIFT; Hassaballah, 2016) have been tested for modeling of crack patterns in terms of a gradient concept. However, it has been noted that due to crack width typically being quite small relative to crack length, conventional region-based (i.e., bounding-box based) DCNN models yield erroneous results in many cases (Chen, 2018a). Furthermore, the boundary, i.e., edge of the lining crack, cannot be retrieved by means of bounding-boxes. In the following, various Region-based DCNN (RCNN) models will be introduced to evaluate the accuracy of the test results, while a rather new concept of pixel-based Semantic Segmentation (SS) will also be considered for comparison of the results with those of RCNN.

Although DCNN can be directly applied in maintenance work on tunnel defects, severity and priority should also be included in the Decision Support System (DSS) for tunnel maintenance. For this purpose, DL including Machine Learning (ML) techniques can be combined to find the optimum schedule and repair methods in DSS. In this regard, a decision support system of the tunnel structure based on ML technologies (Gatsoulis, 2016) has been proposed for quantitative analysis of tunnel conditions. For this, a Support Vector Machine (SVM) or an Artificial Neural Network (ANN) model (Lee, 2018a) can be introduced as a ML tool. Meanwhile, DL approaches to crack detection such as that of Thakker (2015) can be imported to the DSS solution so that overall maintenance scheduling in terms of Regions of Interest (ROIs) can be visualized in a systematic way.

In the following, various DL-based crack detection models are introduced, and the efficient ones for detection of cracks and severity classification according to thickness are subsequently highlighted. Recent developments in region-based DCNN models as well as pixel-based SS models are compared with each other. The accuracy of DCNN models for various crack widths is investigated with field data, and test results on lining cracks are also explained. Future work on crack-edge detection for development of improved crack detection algorithms will also be considered.

2 RELATED WORK

2.1 Introduction

Apart from the classical image processing models mentioned above, DCNN models are widely used to detect cracks and, in some cases, to classify them according to their widths. In the following, various DCNN models that have been used to detect concrete cracks as well as possible future improvements will be discussed.

2.2 Region-based Deep Convolutional Neural Network (RCNN) models for crack detection

Ever since the concept of RCNN for image classification was introduced (Krizhevsky, 2012), region-based bounding-boxes (BB) have been normally used to capture and train objects, because RCNN results were faster and more accurate than those of the classical pattern recognition methods. Subsequently, a region proposal based on the selective search method was used to extract the features of images by employing a CNN having 5 convolutional layers and 2 fully connected layers. Finally, SVM was employed to classify the objects (RCNN, Girshick, 2014).

RCNN was improved by introducing a special convolutional network that covered the entire image and multiple ROIs simultaneously (Girshick, 2015). By extracting the feature vector for each ROI, the final object detection in terms of bounding-box coordinates was processed by means of special softmax probabilities. Thereby, approximately 9 times faster output compared with RCNN was realized (FCNN), and the application to crack detection is rather straight-forward.

Another RCNN improvement was implemented by Ren (2016), who used a Region Proposal Network (RPN) to share the convolutional features with the detection network so that faster object detection could be realized by combining FCNN and RPN simultaneously (FFCNN). The so-called convolution and anchor boxes were employed to facilitate the prediction of the region proposal at multiple scales and aspect ratios. Recently, a few more upgraded models based on several bounding-boxes have been proposed for reduction of the time needed to extract the feature map. Application of RCNN to a concrete structure was made in Cha (2017), who used a total of 8 layers of convolutions, poolings and softmax. The outside of the concrete building was taken for use as a training dataset, and a sliding window concept was used to locate the cracks within an image. Accurate test results were obtained by employing more than 10K images during the training session, although a thickness model was not included. Meanwhile, Xue (2018) applied RCNN to segment lining structures and trained images using 3 classifying labels: crack, leakage and scratch. The RPN and ROI concepts were used during feature extraction and estimation of the bounding-box location, respectively, and a total of 22 layers of convolutions, max poolings and inceptions in Szegedy (2015) were utilized in the network. Since the slenderness of the crack was very narrow, the aspect ratio of the anchor box was also elongated accordingly. About 10K images were used during the training session, and the accuracy was apparently better than that of AlexNex (Krizhevsky, 2012). Although much better results for crack detection were foreseen, the crack thickness was not involved in the modeling process, and either the modification of CNN or dataset increment will be necessary to classify cracks according to their thicknesses.

2.3 Pixel-based Semantic Segmentation (SS) model for crack detection

So far, a typical bounding-box concept has been used to detect objects by using convolutional, pooling and fully connected layers. A different approach whereby pixel information can be included in CNN and subsequently in the feature map was attempted, and a fully convolutional network (FCN) concept was established for detection of objects on the basis of pixel-wise information (Long, 2015). The so-called Semantic Segmentation (SS) of the objects was based on the pixel information labelled with the class of its enclosing object or region.

Further development of SS, called SegNet, was made by Badrinarayanan (2017), whereby a convolutional encoder-decoder layers were fully utilized to generate the feature maps and, subsequently, to obtain pixel-wise classification with the decoder network. The main difference between FCN and SegNet lies in the processes of upsampling steps: whereas a deconvolution process of the input feature map was introduced into the FCN model, max pooling indices were used in the decoding process of SegNet. Therefore, slightly better results were obtained by adopting SegNet in the example cases (Badrinarayanan, 2017). Like RCNN, a few enhanced models were recently proposed in terms of SS, DeepLab (Chen, 2018b) being one of the efficient modeling schemes in this regard. Since max pooling and the striding were normally used in DCNN to reduce the number of parameters, the resolution of the output result was not high enough to distinguish the boundaries or edges of objects. To recover proper resolution, a special upsampling layer called atrous convolution filter as well as a conditional random field model were included in DeepLab (Chen, 2018b).

Application of the SS model to crack detection can be found in Huang (2018), where the FCN model was applied to the detection of cracks in a shield tunnel lining. As mentioned above, cracks in the lining structure were retrieved by employing a deconvolution process to upsample the feature maps, and two stream defect-detection schemes were used to model cases of overlapping crack and leakage. Meanwhile, Lee, et al. (2018b) tried to distinguish cracks from remaining slender objects by using SegNet model (Badrinarayanan, 2017), and classification according to thickness was successful even if the thicknesses of the objects were roughly 2mm and 5mm.

2.4 Edge detection with DCNN

Even though both RCNN and SS models have been successfully applied to detect cracks within tunnel lining or concrete structures, better results can be foreseen if the boundary or edge of the crack is clearly defined at the output stage of the DL. In this regard, several possibilities within the DCNN framework can be considered. A simple remedy is to skip the pooling layers in DCNN, which will yield a feature map of better resolution, though another disadvantage such as prolonged CPU time will be unavoidable. A more feasible model would include a special layer related to boundary detection. In fact, Marmanis (2018) has proposed an edge-detection layer in SS for refinement of satellite images. For this, in order to obtain smooth boundaries, a normalized digital surface model and a regression loss function were added to SegNet. Another possibility is to adjust the input training data using refinement techniques for improved accuracy.

In the present study, the emphasis was placed on crack classification according to thickness for detection of very fine cracks within a lining structure. The optimal edge-detection algorithm will be subsequently incorporated into DCNN, and the results will be published elsewhere.

3 CLASSIFICATION OF CRACKS WITH DCNN

3.1 *Thickness classification with DCNN models*

In the beginning, the various RCNN models mentioned above were considered in order to evaluate classification accuracy between thin and thick cracks. Unfortunately, the training image sets of the various crack thicknesses were not easy to prepare and, therefore, artificial cracks were generated as shown in Figure 1. Images of short leaves with relatively small thicknesses and thick branches were collected and trained with 2 classes of bounding-boxes as shown in Figures 1(a) & (b).

Training and testing of the images were mainly performed with MATLAB (MATLAB, 2018), and an Intel i7-7820X CPU having 64G RAM memory as well as a GTX 1080Ti GPU were utilized for the image processing. The precision, recall and intersection of union (IoU) of the test results could be calculated as follows:

$$P_r = \frac{t_p}{t_p + f_p}; \ R_r = \frac{t_p}{t_p + f_n}; \ IoU = \frac{t_p}{t_p + f_p + f_n} \tag{1}$$

Figure 1.   Crack detection with various DCNN models according to crack thickness.
(a) Sample image of leaves (b) Sample image of branches (c) Result of RCNN
(c) Result of FCNN   (d) Result of FFCNN   (f) Result of SS DCNN

2458

Table 1. Average precisions of thickness test with various DCNN models. (Lee, 2018b).

Model	Leaf Avg. P_r	Branch Avg. P_r
RCNN	0.04	0.03
FCNN	0.04	0.00
FFCNN	0.35	0.34
SS (R_r)	0.96	0.81

where, P_r and R_r represent the precision and recall, respectively, while t_p, f_p and f_n mean the true positive, false positive and false negative data, respectively. A total of 70 images were collected, and 70% of them were used in the training session. Since the total number of training images was quite small, the expected test result was not satisfactory in terms of precision. However, the test result of SS was not bad at all, as shown in Table 1, and this was partly due to the fact that the precision calculation of the bounding-box in eq. (1) included location and score data such that the estimated bounding-boxes of the narrow objects were different from those of the ground truth data. Meanwhile, the accuracy, i.e., the recall rate of SS in our case, was calculated on the pixel level so that the overall estimated values were very similar to those of the ground truth data in the case of the test results of Lee (2018b), see Table 1.

It is clear from Figure 1 and Table 1 that the detection of narrow objects with different thicknesses is not easy to accomplish in terms of bounding-box model, partly because the training images were rather small and partly because the bounding boxes contained meaningless background compared with the object under consideration. Therefore, the precision of RCNN models can be improved if the total number of images is increased or if the bounding-box is divided into smaller ones so that the parts of a crack can fit into the small boxes. It was also noted from Figure 1(f) that the boundary of the leaves was rather blurred when the SS model was employed, and so further refinement will be necessary to accurately classify cracks according to their thicknesses.

3.2 Crack and noise classification with DCNN models

The crack detection capability of DCNN has been demonstrated with objects having different thicknesses. The possibility of crack detection under the condition of similar objects was subsequently tested with another image set. For this, 83 images of cracks including tree branches were collected under the daylight condition. Figures 2(a) & (b) show the ground truth images with the bounding-box and semantic segmentation model, respectively. The detection results are shown in Table 2, and, regardless of the detection model, the overall detection accuracy was far lower than that shown in Table 1. This was mainly owed to the fact that the object under consideration, i.e., a real crack, was not easy to detect even with the naked eye, and the object was again very slender in this case. As mentioned above, if the number of images is increased and the size of bounding-box is reduced, a better result can be foreseen. Some of the good results are shown in Figures 2(c) & (d); the predicted thickness of the crack in Figure 2(d) needed to be improved in order to be able to model the fine details of the object.

So far, the applicability and accuracy of the various DCNN models have been investigated with various image sets, and it was found that an improvement of the existing models was needed to detect and classify the lining cracks according to their thicknesses. Although marginal improvement can be expected if the number of training image sets is increased and fine tuning of the bounding-box model is imposed, a special tool such as an additional layer to specify the boundary of the crack along with minimization of resolution change will be considered in order to improve the detection accuracies of the models.

Table 2. Average precisions of DCNN models with crack-like objects. (Lee, 2018b).

Model	Crack Avg. P_r	Branch Avg. P_r
FFCNN	0.16	0.10
SS (R_r)	0.76	0.81

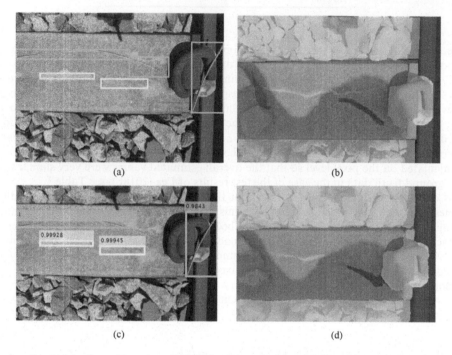

(a)   (b)

(c)   (d)

Figure 2. Crack detection with various DCNN models and with crack-like objects.
(a) Sample image of bounding-box model   (b) Sample image of SS model
(c) Result of FFCNN   (d) Result of SS

4 APPLICATION OF DCNN MODELS TO LINING CRACKS

Finally, application of DCNN models to the case of lining cracks including leakage was carried out to find the advantages and disadvantages of the models, and to enable future improvement thereby. For this, another training set of 82 images containing lining cracks, groundwater leakage and efflorescence were prepared and resized to 320×480 pixels to speed up the training process. Meanwhile, 70% of the images were used in the process of training. Additional lighting equipment was used to clearly capture the crack images, and 4 categories of lining wall were classified in the process of segmentations.

Figures 3(a) & (b) illustrate the training image of the crack and leakage together with efflorescence at the same time, and similar digitization was also done to obtain segmentation information. Figure 3(c) shows one of the FFCNN results, and a noisy output was predicted partly because the training set was rather small and partly because the background image, i.e. lining, was not homogeneous in terms of texture. Meanwhile, the SS prediction shown in Figure 3(d) was in good agreement with the ground truth data, though a small patch of miscalculation, i.e., leakage, was detected. Again, the thickness of the crack was not realistic, and a substantial

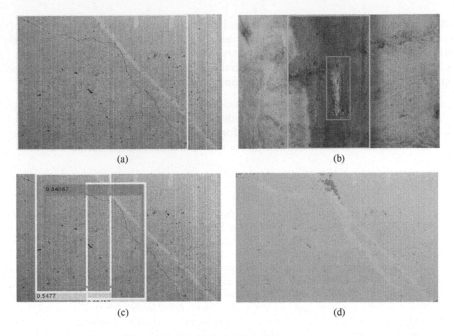

Figure 3.   Crack detection of lining with FFCNN & SS models.
  (a) Sample image with bounding-box model   (b) Sample image with SS model
  (c) Result of FFCNN   (d) Result of SS

improvement needs to be achieved in this regard. The overall accuracy of crack, leakage and efflorescence in terms of average precision was found to be 0.3/0.1/0.1 in FFCNN, while the overall precision of SS was 0.6/0.9/0.9. The test results for RCNN and FCNN are not listed here, as the same tendency shown above could be predicted.

Crack detection improvement can be thought to proceed in two directions: first, an additional layer having an edge detection algorithm involving either gray-color classification or an orientation gradient can be included in the model; second, another classification module based on SVM can be included so that the predicted cracks are classified according to the minimum or average crack thickness of SS results.

5   CONCLUSIONS

Cracks and groundwater leaks through the tunnel lining are not easy to visualize by tunnel inspection car or TMV. Image processing techniques can, therefore, be employed to better investigate tunnel conditions. Various deep convolutional neural networks including RCNN, FCNN, FFCNN and semantic segmentation models have been introduced to evaluate the detectability of slender objects such as cracks and leakage within tunnel lining. The following conclusions were drawn based on the application of DCNN models under various field conditions:

– RCNN based on bounding-boxes inherently include shortcomings when applied to slender objects since the bounding-box for a specific object contains meaningless background information as well. To remove unnecessary information, either very small boxes or a huge number of training images have to be used to model cracks. Classification of cracks according to thickness can be another challenge when using RCNN-based models. A possible combination of RCNN and SS can be an option for future study.
– Semantic segmentation based on pixel information shows better results for modeling of slender objects including cracks. This is partly due to the fact that the profile of meaningful objects is precisely digitized, and partly owed to the SS framework being comprised of very deep layers

including encoding and decoding processes. However, digitization of narrow cracks possibly incurs errors during the preparation of the training data. Further enhancement to enable automatic detection of the boundary of ground truth data will be necessary. Furthermore, for precise classification of crack thickness, a special DL layer to distinguish the edge of the object from the remaining background needs to be included in the SS model.

Future work on crack detection within tunnel lining will seek to improve lighting equipment, which currently is not easy to setup or operate during image collection. Unmanned aerial vehicles will be an option to consider, although GPS information would not be available inside the tunnel. Also, a DCNN model with enhanced equipment will be investigated in a future study.

REFERENCES

Attard, L., et al. 2018. Tunnel inspection using photogrammetric techniques and image processing: A review. *ISPRS J. Photo. Remote Sens.*, 144: 180–188.

Badrinarayanan, V., Kendall, A. & Cipolla, R. 2017. SegNet: A deep convolutional encoder-decoder architecture for image segmentation, *IEEE Trans. Pattern Anal. Mach. Learn.*, 39(12): 2481–2495.

Cha, Y., Choi, W. & Buyukozturk, O. 2017. Deep learning-based crack damage detection using convolutional neural networks, *Comp. Aided Civil Infra. Eng.*, 32: 361–378.

Chen, F. & Jahanshahi, M. 2018a. NB-CNN: Deep learning-based crack detection using convolutional neural network and naïve Bayes data fusion, *IEEE Trans. Ind. Electro.*, 65(5): 4392–4400.

Chen, L., et al. 2018b. DeepLab: Semantic image segmentation with deep convolutional nets, atrous convolution, and fully connected CRFs, *IEEE Trans. Pattern Anal. Mach. Learn.*, 40(4): 834–847.

Gatsoulis, Y., et al. 2016. Learning the repair urgency for a decision support system for tunnel maintenance. In G.A. Kaminka (ed.) *Frontiers in artificial intelligence and application*; *Proc. ECAI 2016*, Hague.

Girshick, R., et al. 2014. Rich feature hierarchies for accurate object detection and semantic segmentation. *Proc. CVPR*, 580–587.

Girshick, R. 2015. Fast R-CNN. *Proc. IEEE Int. Conf. Computer Vision.*

Hassaballah, M., Abdelmgeid, A. & Alshazly, H. 2016. Image features detection, description and matching. In A. Awad & M. Hassaballah (eds), Image Feature Detectors and Descriptors, *Proc.*, 11–45, Springer.

Huang, H.W., Li, Q.T. & Zhang, D.M. 2018. Deep learning based image recognition for crack and leakage defects of metro shield tunnel. *Tunnel. Under. Space Tech.*, 77: 166–176.

Krizhevsky, A., Sutskever, I. & Hinton, G. 2012. Image net classification with deep convolutional neural network. *Proc.* Advances in neural information processing systems, 25.

Lee, J.S., et al. 2018a. Prediction of track deterioration using maintenance data and machine learning schemes. *J. Transp. Eng., A.*, 144(9): 04018045–1:9.

Lee, J.S., et al. 2018b. A comparative study on the crack detection of sleepers with region- and semantic segmentation-based models of deep learning, *J. Korean Soc. Railway*, submitted, in Korean

Long, J., Shelhamer, E. & Darrell, T. 2015. Fully convolutional networks for semantic segmentation, *Proc.* IEEE CVPR, 3431–3440.

Marmanis, D., et al. 2018. Classification with an edge: Improving semantic image segmentation with boundary detection, *ISPRS J. Photo. Rem. Sens.*, 135: 158–172.

MATLAB, 2018. MATLAB 2018a, Mathworks.

Medina, R., et al. 2017. Crack detection in concrete tunnels using a Gabor filter invariant to rotation, *Sensors*, 17, 1670: 1–16.

Menendez, E., et al. 2018. Tunnel structural inspection and assessment using an autonomous robotic system, *Auto. Constr.*, 87: 117–126.

Ning, Q., Zhu, J. & Chen, C. 2018. Very fast semantic image segmentation using hierarchical dilation and feature refining, *Cog. Comp.*, 10: 62–72.

Oliveira, H. & Correia, P. 2017. Road surface crack detection: Improved segmentation with pixel-based refinement, *Proc.* 25th Eur. Sig. Process. Conf. (EUSIPCO), 2080–2084.

Ren S., et al. 2016. Faster R-CNN: Towards real-time object detection with region proposal networks, *arXiv*:1506.01497v3.

Szegedy, C., et al. 2015. Going deeper with convolutions, *Proc.* CVPR 2015.

Thakker, D., et al. 2015. PADTUN – Using semantic technologies in tunnel diagnosis and maintenance domain. In F. Gandon, et al. (eds), The semantic web. Latest advances and new domains. *Proc. ESWC*, 683–698, Springer.

Xue, Y. & Li, Y. 2018. A fast detection method via region-based fully convolutional neural networks for shield tunnel lining defects. *Comp. Aided Civil Infra. Eng.*, 33: 638–654.

Tunnels and Underground Cities: Engineering and Innovation meet Archaeology,
Architecture and Art, Volume 5: Innovation in underground engineering,
materials and equipment - Part 1 – Peila, Viggiani & Celestino (Eds)
© 2020 Taylor & Francis Group, London, ISBN 978-0-367-46870-5

New design method for energy slabs in underground space using artificial neural network

S. Lee, D. Kim & H. Choi
Korea University, Seoul, Republic of Korea

K. Oh
Incheon Transit Corporation, Incheon, Republic of Korea

S. Park
Korea Military Academy, Seoul, Republic of Korea

ABSTRACT: The energy slab is a ground-coupled heat exchanger consisting of the horizontal ground heat exchanger (HGHE). The energy slab is installed as one component of the floor slab layers to utilize the underground structure as a hybrid energy structure. In this study, a design method for energy slabs using the artificial neural network (ANN) is proposed. Before establishing the design method, it is necessary to accurately identify system parameters that may affect the thermal performance of energy slabs. Therefore, the effect of system parameters on the thermal performance of energy slabs was investigated by numerical analysis for ideal conditions. Then, a set of databases constructed through the numerical analysis was implemented by the ANN. From this approach, the thermal performance of energy slabs for a typical condition can be quickly predicted without time-consuming numerical simulations.

1 INSTRUCTION

The Ground Source Heat Pump (GSHP) system utilizes the geothermal energy in the heating and cooling of buildings by constructing Ground Heat Exchangers (GHEXs) in underground. The GHEXs allow circulating a working fluid through heat exchange pipes to induce heat exchange between the working fluid, the wall of heat exchange pipe and the surrounding medium. While a closed-loop vertical GHEX is mostly used in practice, it demands a high initial investment cost because of the additional drilling of the borehole and the need for the construction area (Boënnec, 2008). Especially, the drilling cost occupies more than 50% of the total construction cost. In this respect, novel types of GHEX have been developed to reduce the construction cost, which can be installed in underground structure elements, e.g., energy textile (Lee et al., 2012), energy pile (Brandl, 2006; De Moel et al., 2004; Morino and Oka, 1994), etc. Among them, the energy slab can be a tactical way to use the underground space for GHEXs effectively. The energy slab is installed as one component of the floor slab layers to utilize the building structure as a hybrid energy structure (Choi, 2012; Choi and Sohn, 2012). Constructing an energy slab is also an effective approach to reducing the construction cost without additional drilling, because it is buried underground with a horizontal layout. If the energy slab is applicated in the underground structures, its thermal performance would be improved than results of previous studies (Choi, 2012; Choi and Sohn, 2012) because of high thermal conductivity of underground formations and less influence of ambient air to the energy slab. Despite these advantages, the design methods or algorithms for predicting the thermal performance of energy slabs have not been established yet.

Two methods have been used to predict the thermal performance of various types of GHEXs. First, Park et al. (2018) proposed engineering charts to predict the thermal performance of cast-in-place energy piles. In this method, the thermal performance of energy piles after three months of operation can be predicted. Furthermore, the designer can predict the thermal performance of energy piles for long-term operation (over three months) with consideration of the temperature difference between the ground formation and inlet fluid. Second, a machine learning such as Artificial Neuron Network (ANN) or Adaptive Neuro-Fuzzy Interference System (ANFIS) were adopted. Esen and Mustafa (2010), Sun et al. (2015) developed the ANN and ANFIS models to predict the COPs of GHEXs. Moreover, Makasis et al. (2018) conducted a machine learning approach to energy pile design, by inputting annual thermal loads.

In this paper, a new design method for predicting the thermal performance of energy slabs, which are applicated in the underground structure, was proposed. First, a Computational Fluid Dynamic (CFD) model was developed for ideal conditions. Then, a family of engineering charts was constructed along with a series of CFD simulations considering key influential parameters that affect the thermal performance of energy slabs (i.e., thermal conductivity of the concrete slab, ground formation and thermal insulation material, the flow rate of working fluid, the initial temperature of surrounding mediums and the interval of heat exchange pipes). Finally, the prediction model for thermal performance of energy slabs was developed by adopting an ANN model based on CFD simulation data, and the applicability of the developed ANN was evaluated.

2 DEVELOPMENT OF CFD MODEL AND NUMERICAL DATABASE CONSTRUCTION

2.1 Development of CFD model

A CFD model was developed using a commercial 3D FE analysis program, COMSOL Multiphysics. In this numerical model, the heat transfer in the medium by convection and conduction can be accurately simulated (Park et al., 2016, Park et al., 2018). The overview of entire geometry and mesh configuration of the model are shown in Figure 1. The model was implemented with the energy slab installed in the underground structure. The number of meshes is set to about 80,000 to improve the accuracy of the model, and the 3-D tetrahedral element is used for consisting the geometry. The size of floor and wall concrete slabs working as an energy slab is 5 m X 5 m. The internal space of the concrete slab was modeled with air, and the external surface was surrounded by the underground formation. The thermal properties of each component input into the CFD model are shown in Table 1.

Table 2 shows the applied operation condition of the CFD model. The initial temperatures of ground formation and concrete slabs are assumed to be constant temperature because the structure is installed in an underground formation. The intermittent operation of 8 hours operation - 16 hours pause was considered to simulate a typical commercial building. The inlet fluid temperature was kept at 30°C during operation to simulate the cooling condition. The internal temperature of the concrete slab is assumed to be 24°C. In order to prevent the degradation of the thermal performance of the energy slab due to the interactions between the energy slab and air inside the underground structure, the thermal insulation layer (i.e., Phenol Foam board, k = 0.018 W/mK) is installed in the floor concrete slab of the CFD model.

2.2 Construction of database on thermal performance

In this section, the key influential parameters that affect the thermal performance of energy slabs were selected. First, the thermal conductivity of concrete slab and ground formation was considered. As the thermal conductivity of surrounding medium increases, the thermal performance of GHEXs increase (Lee et al., 2010, Park et al., 2011, Kim et al., 2017,

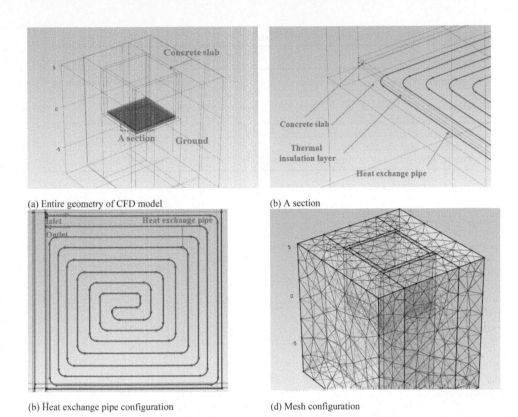

(a) Entire geometry of CFD model   (b) A section

(b) Heat exchange pipe configuration   (d) Mesh configuration

Figure 1. Overview of developed CFD model.

Table 1. Thermal properties of CFD model.

Type	Ground formation	Concrete slab	Heat exchange pipe (High-density polyethylene)	Working fluid
Density (kg/m^3)	1,820	3,640	950	998.2
Specific heat capacity (J/kgK)	1,480	840	2,302	4,182
Thermal conductivity (W/mK)	1.3	2.05	0.4	0.6
Viscosity (kg/ms)	-	-	-	0.001

Park et al., 2016, Chen et al., 2018). Second, the flow rate of a working fluid is regarded as a significant influential parameter on the thermal performance according to the previous studies (Park, 2016, Misra et al., 2013, Carotenuto et al., 2017). Third, the initial temperature of surrounding medium was considered. The temperature difference between the surrounding medium and inlet fluid shows a linear relation with the thermal performance (Park et al., 2018). As the temperature difference increases, the thermal performance of GHEXs increases. Thus, various temperature conditions were applied to surrounding medium in the course of simulation while the temperature of inlet fluid was maintained to be 30°. As mentioned in section 2.1, the thermal insulation layer was installed in the energy slabs to relieve thermal interaction between the energy slab and air inside the underground structure. If the thermal conductivity of thermal insulation material is not low

Table 2. Operation condition of CFD model.

Type	Condition
Operation type	Activating 8 hours–Deactivating 16 hours (intermittent operation)
Flow rate	11.35 liter/min
Inlet temperature	30 °C (cooling operation)
Initial temperature of surrounding medium	15 °C
Internal temperature of concrete slab	24 °C
Thermal conductivity of thermal insulation layer	0.018 W/mK (PF board)

Table 3. Total simulation cases.

Type	Applied value					
Thermal conductivity of concrete slab (W/mK)	1.75	1.90	**2.05**		2.30	2.45
Thermal conductivity of ground formation (W/mK)	1.1	1.2	**1.3**		1.4	1.5
Flow rate of working fluid (liter/min)	7.57	9.46	**11.36**		13.25	
Initial temperature of surrounding medium (°C)	13	14	**15**		16	
Thermal conductivity of thermal insulation material (W/mK)	**0.018**	0.05	0.1	0.3	0.5	2.05
Interval of heat exchange pipe (length) (cm)	**30 (76.66 m)**	40 (59.11 m)	50 (49.32 m)	60 (41.47 m)		

*Bold: base condition

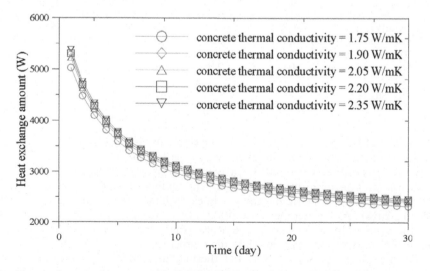

Figure 2. Heat exchange amount according to thermal conductivity of concrete slab.

enough to prevent the thermal interaction, the thermal performance of energy slabs would be reduced. Therefore, the thermal conductivity of a thermal insulation layer also should be considered as a key influential parameter. Finally, an interval between each heat exchange pipe also affects the thermal performance due to the existence of thermal interference. As the total pipe length installed in an energy slab increases, the heat exchange amount can also increase while the efficiency and long-term performance significantly decrease. As a result, the total simulation cases were chosen as shown in Table 3. Figures 2–7 show engineering charts for designing energy slabs with consideration of the mentioned key influential parameters, and comparative evaluations were summarized in Table 4.

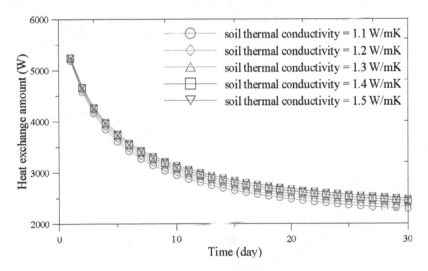

Figure 3.   Heat exchange amount according to thermal conductivity of ground formation.

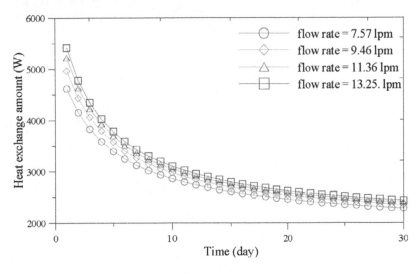

Figure 4.   Heat exchange amount according to flow rate of working fluid.

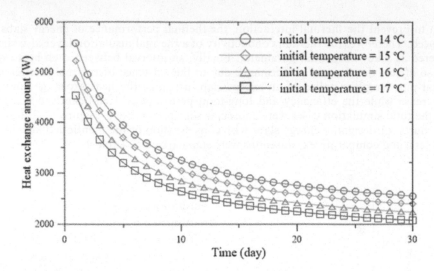

Figure 5.   Heat exchange amount according to initial temperature of surrounding medium.

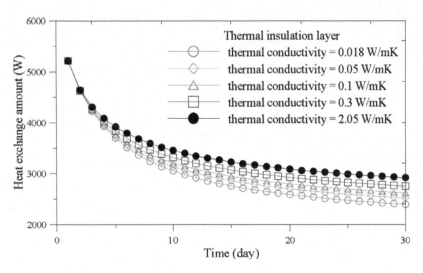

Figure 6.   Heat exchange amount according to thermal conductivity of thermal insulation layer.

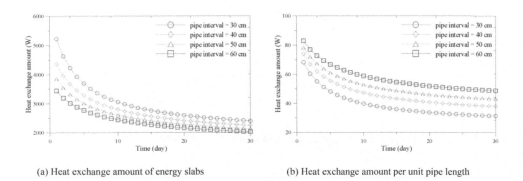

(a) Heat exchange amount of energy slabs   (b) Heat exchange amount per unit pipe length

Figure 7.   Heat exchange amount according to interval of heat exchange pipe.

Table 4. Thermal performance of energy slab according to various key influential parameters.

Key influential parameter	Conclusion
Thermal conductivity of concrete slab	No significant effect on thermal performance of energy slab
Thermal conductivity of ground formation	As the thermal conductivity of ground formation increases, the thermal performance of energy slab increases.
Flow rate of working fluid	As the flow rate of a working fluid increases, the thermal performance of energy slab increases (non-linear relationship).
Initial temperature of surrounding medium	As the initial temperature of surrounding medium increases, the thermal performance of energy slab decreases
Thermal conductivity of thermal insulation material	As the thermal conductivity of thermal insulation material decreases, the thermal performance of energy slab decreases.
Interval of heat exchange pipe (length)	As the heat exchange pipe interval decreases, the thermal performance of energy slab increase (non-linear relationship).

3 DESIGN METHOD FOR ENERGY SLAB USING ANN

The Artificial Neuron Network (ANN) is an information processing model, which consists of several neurons. In this model, in the course of data training, the neurons learn the relationship between output and input data. Figure 8 shows a schematic of ANN.

In the training section, various learning algorithms can be adopted, such as Levenberg-Marquardt (LM), Bayesian Regularization (BR), Scaled Conjugate Gradient (SCG), etc. Each learning algorithm possesses unique properties, and the optimal learning algorithm differs according to the characteristics of training data. Additionally, the ANN has a hidden layer, which can extract and store the information of data set. If the number of neurons in the hidden layer is too small to get sufficient information, the accuracy of the ANN model may be signficantly reduced. On the contrary, too many neurons in the hidden layer may lead to getting irregular information and requiring additional learning time. Consequently, it is necessary to define the acceptable learning algorithm and the optimum number of neurons in the hidden layer. In this paper, the LM and BR algorithms were compared to each other according to the number of neurons in the hidden layer. A total of 89 datasets (63 for the training section, 13 for the validation section and 13 for the testing section) was used in the ANN model. As shown in Figure 8, the ANN model was developed to predict the outlet temperature with a total of 7 input data. The predicted outlet temperature can be used to calculate the heat exchange amount of energy slabs (Equation 1).

$$Q = C \dot{m} \, \Delta T = C \dot{m} \left(T_{in} - T_{out} \right) \tag{1}$$

where, Q = heat exchange amount of energy slab (W); $\dot{m}$= flow rate of working fluid (liter/min); C = specific heat of circulating fluid (J/kgK); ΔT = temperature difference (°C); T_{in} = Inlet temperature (°C); T_{out} = Outlet temperature (°C).

As a result, the BR algorithm with 8 neurons in the hidden layer has the lowest Mean Square Error (MSE) of 0.00142 corresponding to R of 0.997673 (referred to Table 5).

To verify the proposed ANN model, the outlet temperature of energy slabs was evaluated with the randomly selected input data between the maximum and minimum input value. Then, the result was compared with the result of CFD numerical analysis that was simulated by applying the same input data in the model. As shown in Table 6, the trained ANN has a good agreement with the numerical simulation in the CFD model. Consequently, the ANN model allows the designers to effectively estimate the thermal performance of energy slabs without time-consuming numerical simulation.

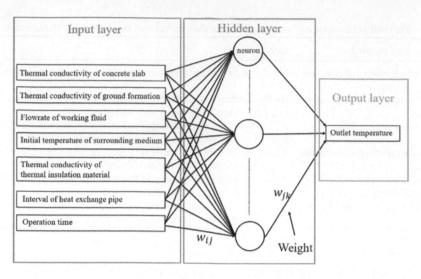

Figure 8. Schematics of ANN structure.

Table 5. MSE and R according to the number of neurons in hidden layer.

Levenberg-Marquardt algorithm			Bayesian Regularization		
The number of neurons	MSE	R	The number of neurons	MSE	R
1	0.02906	0.97720	1	0.025694	0.969834
2	0.02211	0.97591	2	0.031009	0.985444
3	0.01382	0.98799	3	0.003295	0.99734
4	0.01096	0.98480	4	0.007478	0.993684
5	0.01588	0.98700	5	0.004626	0.994469
6	0.01072	0.99004	6	0.006116	0.99418
7	0.00741	0.99097	7	0.011153	0.981538
8	0.01004	0.98793	8	0.00142	0.997673
9	0.01755	0.98792	9	0.003609	0.997268
10	0.01442	0.98625	10	0.001512	0.998285

Table 6. Verification of proposed ANN model.

Case	Thermal conductivity of concrete slab (W/mK)	Thermal conductivity of ground formation (W/mK)	Flow rate of working fluid (liter/min)	Initial temperature of surrounding medium (°C)	Thermal conductivity of thermal insulation material (W/mK)	Interval of heat exchange pipe (cm)	MSE
1	2.00	1.35	10	15.5	0.018	30	0.53
2	1.80	1.23	8	15.2	0.018	30	0.61

4 CONCLUSION

In this study, a comprehensive design method for energy slabs which are applicated in the underground structures was discussed. A CFD numerical model was developed to construct a family of numerical databases. Then, a ANN model was trained using the database obtained from a series of CFD simulation. Finally, the ANN model was verified with the developed CFD model. Some important findings are summarized as follows:

1) The thermal conductivity of ground formation, concrete slab and thermal insulation layer, the flowrate of working fluid, the Initial temperature of surrounding medium and the Interval of heat exchange pipe were selected as the key influential parameters on the thermal performance of energy slabs. Among the parameters, the thermal conductivity of concrete slab has the least effect on the thermal performance of energy slabs.

2) The ANN model was trained using the data selected from the constructed engineering charts. The BR algorithm with 8 hidden layer neurons was found to be an optimum set of the ANN model. Then, the predicted values from the ANN model and the CFD model with randomly selected inputs were compared for verification of the developed ANN model. Consequently, the developed ANN model shows a good agreement with the CFD numerical analysis. This result indicates that the designers can predict the thermal performance of energy slabs accurately with the aid of the developed ANN model without time-consuming and recursive numerical simulations.

ACKNOWLEDGEMENT

This research was supported by National Research Foundation of Korea Government (NRF-2014R1A2A2A01007883) and by Korea Institute of Energy Technology Evaluation and Planning (KETEP), Ministry of Knowledge Economy (No. 20153030111110).

REFERENCES

Brandl, H. 2006. Energy foundations and other thermo-active ground structures. *Geotechnique* 56(2): 81–122.

BoëNnec, O. 2008. Shallow ground energy systems. *Proceedings of the Institution of Civil Engineers-Energy* 161(2): 57–61.

Carotenuto, A., Marotta, P., Massarotti, N., Mauro, A. & Normino, G. 2017. Energy piles for ground source heat pump applications: Comparison of heat transfer performance for different design and operating parameters. *Applied Thermal Engineering* 124: 1492–1504.

Chen, F., Mao, J., Chen, S., Li, C., Hou, P. & Liao, L. 2018. Efficiency analysis of utilizing phase change materials as grout for a vertical U-tube heat exchanger coupled ground source heat pump system. *Applied Thermal Engineering* 130: 698–709.

Choi, J. M. 2012. Heating and cooling performance of a ground coupled heat pump system with energy-slab, *Korean Journal of Air-Conditioning and Refrigeration Engineering* 24(2): 196–203.

Choi, J. M. & Sohn, B. H. 2012. Performance analysis of energy-slab ground-coupled heat exchanger, *Korean Journal of Air-Conditioning and Refrigeration Engineering* 24(6): 487–496.

De Moel, M., Bach, P. M., Bouazza, A., Singh, R. M. & Sun, J. O. 2010. Technological advances and applications of geothermal energy pile foundations and their feasibility in Australia. *Renewable and Sustainable Energy Reviews* 14(9): 2683–2696.

Esen, H. & Inalli, M. 2010. ANN and ANFIS models for performance evaluation of a vertical ground source heat pump system. *Expert Systems with Applications* 37(12): 8134–8147.

Kim, D., Kim, G., Kim, D. & Baek, H. 2017. Experimental and numerical investigation of thermal properties of cement-based grouts used for vertical ground heat exchanger. *Renewable Energy* 112: 260–267.

Lee, C., Lee, K., Choi, H. & Choi, H. P. 2010. Characteristics of thermally-enhanced bentonite grouts for geothermal heat exchanger in South Korea. *Science in China Series E: Technological Sciences* 53 (1): 123–128.

Lee, C., Park, S., Won, J., Jeoung, J., Sohn, B. & Choi, H. 2012. Evaluation of thermal performance of energy textile installed in Tunnel. *Renewable Energy* 42: 11–22.

Makasis, N., Narsilio, G. A. & Bidarmaghz, A. 2018. A machine learning approach to energy pile design. *Computers and Geotechnics* 97: 189–203.

Misra, M., Bansal, V., Agrawal, G. D., Mathur, J. & Aseri, T. K. 2013. CFD analysis based parametric study of derating factor for Earth Air Tunnel Heat Exchanger. *Applied Energy* 103: 266–277.

Morino, K. & Oka, T. 1994. Study on heat exchanged in soil by circulating water in a steel pile, *Energy and Buildings* 21(1): 65–78

Park, M., Min, S., Lim, J., Choi, J. M. & Choi, H. 2011. Applicability of cement-based grout for ground heat exchanger considering heating-cooling cycles. *Science China Technological Sciences* 54(7): 1661–1667.

Park, S., Lee, S., Lee, H., Pham, K. & Choi, H. 2016. Effect of borehole material on analytical solutions of the heat transfer model of ground heat exchangers considering groundwater flow. *Energies* 9 (5): 318.

Park, S., Lee, S., Oh, K., Kim, D. & Choi, H. 2018. Engineering chart for thermal performance of cast-in-place energy pile considering thermal resistance, *Applied Thermal Engineering* 130: 899–921.

Sun, W., Hu, P., Lei, F., Zhu, N. & Jiang, Z. 2015. Case study of performance evaluation of ground source heat pump system based on ANN and ANFIS models. *Applied Thermal Engineering* 87: 586–594.

*Tunnels and Underground Cities: Engineering and Innovation meet Archaeology,
Architecture and Art, Volume 5: Innovation in underground engineering,
materials and equipment - Part 1 – Peila, Viggiani & Celestino (Eds)*
© 2020 Taylor & Francis Group, London, ISBN 978-0-367-46870-5

The research on longitudinal deformation features of shield tunnel when considering the variation of grout viscosity: A case study

Y. Liang & L.C. Huang
School of Aeronautics and Astronautics, Sun Yat-Sen University, Guangzhou, China

J.J. Ma
School of Civil Engineering, Sun Yat-Sen University, Guangzhou, China

ABSTRACT: The variation of grouting pressure has a significant impact on the internal force and deformation of adjacent segments, during synchronous grouting process of shield tunnel construction. To ensure the accurate control of grouting pressure and the safety for shield driving, the grout pressure space-time distribution along longitudinal direction of tunnel had been studied. Based on the study of grout viscosity time-variation, the process of grout diffusion and dissipation had been united. Considering the grout movement, the deformation formulas of tunnels along longitudinal direction had been obtained, with Winkler elastic foundation beam model. The result of theoretical calculation had been compared with field test data, and it shows that: the increase of grout viscosity and decrease of permeability of soil layers will affect the extent of grouting pressure dissipation, and then affect the deformation of tunnel segments. The result calculated from theoretical models are closer to field monitoring data when considering the time-varying effect of grout viscosity. The research provides a guideline for safety control of segment during construction.

1 INTRODUCTION

The shield tunnel technology is widely used in the projects such as river tunnel and urban subway. With the advance of the tunnel shield, a void between the lining segments and the surrounding stratum at the tail part of the tunnel shield (rock-segment void) is created, after the segments are pushed out of the shield tail. The grouting process must be precisely controlled and monitored. Any inappropriate settings in the grouting process, i.e. the injecting pressure or the injecting locations, would increase the risk of segments instability, such as floating, dislocation, cracking, crushing and many others. It is important to study the floating amount of shield tunnel caused by the grouting pressure.

Many scholars have studied the uplift of shield tunnels with different methods, establishing theoretical or numerical models. However, the effects of pressure dissipation and viscosity variation with time have not been considered during these studies. Based on the Winkler elastic foundation beam model, this paper studied the uplift rule of shield tunnel during construction, considering the time-varying characteristics of grouting pressure.

2 THE SPATIAL AND TEMPORAL DISTRIBUTION OF GROUTING PRESSURE ALONG THE LONGITUDINAL DIRECTION OF TUNNEL

The grout diffusion along the longitudinal direction of tunnel usually lasts for several hours. In order to formulate the grout pressure distribution along the longitudinal direction, the following assumptions are made:

- The grout fills up 'annular cake' rapidly when diffusing along the circumferential direction. Then the 'annular cake' diffuses, as a whole, along the longitudinal direction.
- During longitudinal diffusion process, the grout remains Bingham fluid.
- The grout flow will not be affected or blocked by the grout filled previously.

The mechanical model is shown in Figure. 1. x direction means longitudinal and y direction means transverse direction along the tunnel.

The microelement of grout is adopted to carry the force analysis. p_l means the longitudinal grouting pressure, dl means the length of microelement of x direction, dy means that of y direction. τ is the shear force of grout, h is the fill width of grout. The mechanical equilibrium equation can be obtained as follow:

$$p_l dy - (p_l + dp_l)dy - \tau dl + (\tau + d\tau)dl = 0 \qquad (1)$$

Or:

$$\frac{d\tau}{dy} = \frac{dp_l}{dl} \qquad (2)$$

Set $dp_l/dl = d\tau/dy = B$, integrate it along x direction with the boundary conditions: $l = 0$, $p_l = p_{G0}$ (initial grouting pressure), the grout pressure distribution along x direction can be given as

$$p_l = p_{G0} + Bl \qquad (3)$$

Then Equation (2) can also be integrated along y direction with the boundary conditions: $y = 0$, $\tau = 0$. The grout shear force distribution along y direction can be given as

$$\tau = By \qquad (4)$$

As the Bingham fluid, the grout shear force can also be expressed as

$$\tau = \tau_0 + \mu\gamma = \tau_0 - \mu\frac{dv}{dy} \qquad (5)$$

Where τ_0 is the static shear force, μ is the plastic viscosity coefficient, γ is the shearing rate and v is the fluid velocity

Substitute Equation (4) into Equation (5)

$$dv = \frac{1}{\mu}(\tau_0 - By)dy \qquad (6)$$

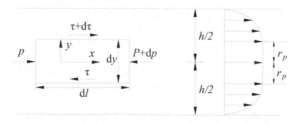

Figure 1.   Grout diffusion along longitudinal direction of tunnel.

It can be seen from Figure 1 that: in the range of $|y| \leq r_p$, the Bingham fluid don't suffer the shear action, which means the adjacent layers of fluid remain still. While in the range of $r_p < |y| \leq h/2$, the fluid considered is in motion relative to adjacent layers. The nuclear radius (r_p) of Bingham fluid can be given as

$$r_p = \frac{\tau_0}{B} \tag{7}$$

Then Equation (6) can also be integrated along y direction with the boundary conditions: $y = h/2$, $v = 0$. The grout velocity (v) within the range of $r_p < |y| \leq h/2$ can be obtained

$$v = \frac{1}{\mu}\left[\frac{B}{2}\left(\frac{h^2}{4} - y^2 \right) - \tau_0\left(\frac{h}{2} - y \right) \right] \tag{8}$$

The grout velocity (v_p) within the range of $|y| \leq r_p$ can also be obtained when substitute y in Equation (8) with r_p

$$v_p = \frac{1}{\mu}\left[\frac{B}{2}\left(\frac{h^2}{4} - r_p^2 \right) - \tau_0\left(\frac{h}{2} - r_p \right) \right] \tag{9}$$

The total grout velocity along y direction can be given as

$$v = \begin{cases} \frac{1}{\mu}\left[\frac{B}{2}\left(\frac{h^2}{4} - r_p^2 \right) - \tau_0\left(\frac{h}{2} - r_p \right) \right] & |y| \leq r_p \\ \frac{1}{\mu}\left[\frac{B}{2}\left(\frac{h^2}{4} - y^2 \right) - \tau_0\left(\frac{h}{2} - y \right) \right] & r_p < |y| \leq h/2 \end{cases} \tag{10}$$

The grout flow q can be solved as

$$q = b\int_{-h/2}^{h/2} v\,dy = \frac{b}{\mu}\left[\frac{2B}{3}\left(\frac{h^3}{8} - r_p^3 \right) - \tau_0\left(\frac{h^2}{4} - r_p^2 \right) \right] \tag{11}$$

Substitute Equation (7) into Equation (11), the parameter B can be solved by Equation (12)

$$B^3 - \left(\frac{3\tau_0}{h} + \frac{12\mu q}{bh^3} \right)B^2 + \frac{4\tau_0^3}{h^3} = 0 \tag{12}$$

Substitute parameter B into Equation (3), the grout pressure distribution along longitudinal direction can be obtained, not considering the grouting pressure dissipation.

As the grout diffuses along the longitudinal direction, the liquid in the grout penetrates into the surrounding stratum, the particles in grout will fill the pores of surrounding stratum, and the permeability coefficient of it gradually decreases. The consolidation occurs, and the grouting pressure gradually dissipates. Based on previous research result, when considering pressure dissipation effect, the grouting pressure at time t ($P_d(t)$) during the longitudinal diffusion process can be expressed as

$$P_d(t) = p_{G0} - \frac{2G}{r}\frac{1}{1-v}\frac{n_i - n_e}{1 - n_i}x(t) \tag{13}$$

Where G is the shear modulus of surrounding stratum, r is the thickness of the disturbed stratum, n_i is the initial porosity of the grout, n_e is the porosity of the consolidated grout layer. $x(t)$ is the consolidated thickness of grout at time t, which can be obtained by follows

$$\left(f_g + f_s(t)\right)\frac{dx}{dt} + \alpha x = \eta \tag{14}$$

Where α, η are parameters about formation condition and grout properties. f_g and $f_s(t)$ are seepage resistance of liquid penetrates from grout and into the surrounding stratum. The grout used for synchronous grouting is generally cement based. Therefore, the seepage resistance $f_s(t)$ will increase with the grout viscosity increase with time. In this work, an exponential evolution of the grout viscosity $\mu(t)$ is adopted

$$\mu(t) = \mu_0 e^{\xi t} \tag{15}$$

Where the μ_0 is the initial grout viscosity, ξ is a material constant, which can be obtained from experiment.

The process of grout diffusion and dissipation occur simultaneously. Combing Equations (3) and (13), The distribution of grouting pressure along the longitudinal direction of tunnel can be obtained, considering the effect of grout dissipation and the variation of grout viscosity.

$$\begin{cases} p_l = p_{G0} + Bl; \\ P_d(t) = p_l - \dfrac{2G}{r}\dfrac{1}{1-\upsilon}\dfrac{n_i - n_e}{1 - n_i}x(t) \end{cases} \tag{16}$$

3 WINKLER ELASTIC FOUNDATION BEAM MODEL

The theory of elastic foundation beam is based on Winkler hypothesis. Winkler elastic foundation beam model, semi-infinite elastic foundation model and double-parameter elastic foundation model are commonly used. The basic assumption of the Winkler elastic foundation model is as follow

$$p(x) = k_g s \tag{17}$$

Where $p(x)$ is the pressure on the foundation at any point. s is the foundation settlement at that point. k_g is the elastic foundation coefficient.

As shown in Figure 2, a rectangular coordinate system is established with the axis of elastic beam as x axis. The load $q(x) = nr + z$ is distributed linearly between any position of a and b along the tunnel structure.

According to the Winkler elastic foundation beam theory, the deflection of the infinite length of elastic foundation beam is as follows when $q(x) = nr + z$ is applied between any position of a and b

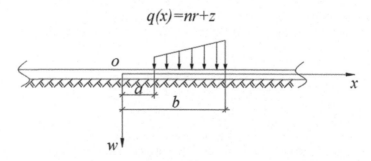

Figure 2.   The linear load distribution on the Winkler elastic foundation beam.

$$w_0 = \int_a^b dw_0 = \int_b^b \frac{(nr+z)\lambda}{2K} e^{-\lambda|x-r|}(\cos(\lambda|x-r|) + \sin(\lambda|x-r|))dr \qquad (18)$$

Where a and b are the load range, λ is the foundation flexibility coefficient, which is equal to: $\lambda = (K/4EI)^{0.25}$, K is the concentrated foundation coefficient, $K=k_gB$, and B is the width of foundation beam.

For a semi-infinite foundation beam, supposing that the beam is extended to the left side as the infinite foundation beam, we can obtain the internal forces M_0 and Q_0 at the point $x = 0$, and the deflection line when $x > 0$, according to equation (18) and its symmetry. Since there is no shear force or bending moment at the free end of a semi-infinite beam, it is necessary to add a concentrated couple M'_0 and concentrated force Q'_0 at the free end of an infinite beam, which is equal in size and opposite in direction to M_0 and Q_0. Based on the superposition principle, the deflection curve w of the half infinite beam can be obtained

$$w = w_0 + \frac{e^{-\lambda x}}{2\lambda^3 EI}[Q'_0 \cos \lambda x - \lambda M'_0(\cos \lambda x - \sin \lambda x)] \qquad (19)$$

4 ENGINEERING VERIFICATION OF THE PROPOSED MODEL

Based on Nanhu Road cross-river tunnel project of China, a typical cross section has been chosen as an example for calculation. The cross section of the tunnel concerned in this work lies under the river, and the cover depth is 10.79m (from the vault of tunnel to river bottom). The permeability coefficient of surrounding stratum is very large. The geological section map is show in Figure 3. The grout is injected from the shield tail into the rock-segment gap, through 6 grouting holes, and the grouting pressure is set from 0.1MPa to 0.18MPa.

The other parameters of stratum and grout need for calculation are listed in Table 1.

4.1 *The diffusion and dissipation of grouting pressure*

Grout viscosity rheological test is conducted to obtain the temporal variation of the grout viscosity. The grout viscosity at different time is plotted in Figure 4. The exponential evolution of the grout viscosity with time are given to fit the experiment data. The fitting result is $\mu(t) = 0.907e^{0.0107t}$.

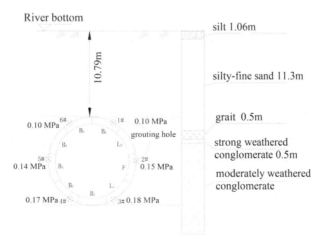

Figure 3.　The cross section of the shield tunnel and typical geological layer.

Table 1. Grouting and stratum parameters used in Nanhu road shield tunnel.

Shield advance rate s/(m/s)	Segment radius R/m	tail void thickness T/m	initial grout pressure p_{G0}/MPa	grout flow q/m³/s	initial stratum permeability coefficient ks/(m·s⁻¹)
0.00035	5.65	0.165	0.18	3.9×10⁻⁴	1.85×10⁻⁶

grout density ρ/(kg/m³)	Plastic viscosity μ/Pa·s	static shear force τ_0/Pa	initial grout porosity n_i	grout cake porosity n_e	initial grout permeability coefficient k_0/(m·s⁻¹)
1900	1	90	0.425	0.417	4.7×10⁻⁸

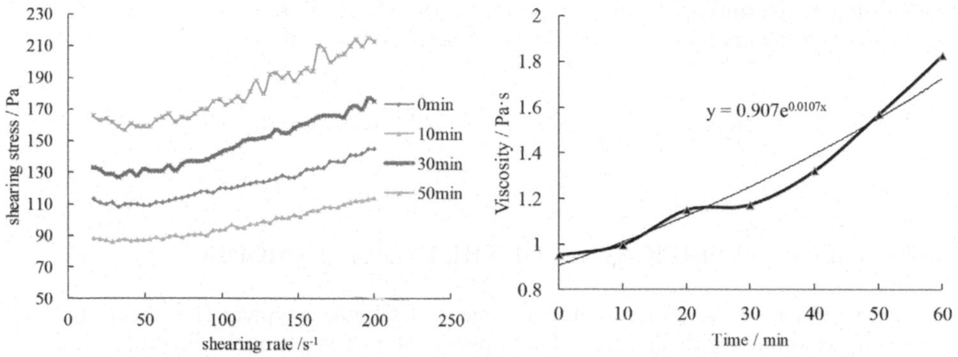

(a) Shear stress vs. shear rate   (b) Fitting results

Figure 4. Temporal variation of the grout viscosity evaluated from laboratory: (a) The recorded shear stress vs. shear rate at different time, and (b) the fitting results of the temporal variation of the grout viscosity.

Based on the parameters collected above, the grouting pressure dissipation curve with time is shown in Figure 5, according to equation (13).

It can be seen from the Figure 5 that during the synchronous grouting process, the grouting pressure will decrease significantly. If the variation of grouting viscosity is taken into account, the dissipation rate of grouting pressure is relatively smaller, and the final dissipation amplitude of grouting pressure is also smaller, compared to the condition not considering the

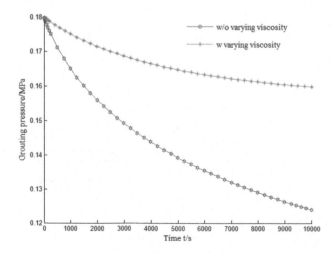

Figure 5. The dissipation of grouting pressure with time.

variation of grouting viscosity. Since the grouting process generally lasts for several hours, the grouting pressure at time 7500s is compared with that at the initial time. Set the initial grouting pressure (p_{G0}) = 0.18 MPa. If not considering the variation of grout viscosity, the grouting pressure ($P_d(t)$) at time 7500s is 0.131 MPa, if considering varying viscosity, the grouting pressure ($P_d(t)$) at time 7500s is 0.162 MPa.

The distribution of grouting pressure along the tunnel longitudinal direction can be deduced from equation (16), as shown in Figure 6. The pressure decreases linearly with distance away from the grouting cross section. The grout pressure at different locations along the longitudinal direction all decreases by the same rate with time going on, i.e. the pressure-distance line moves downward parallel. It can be observed that, if not considering varying viscosity, the grout pressure decreases by 27.2% and propagates about 8.85 m at the end of grouting process; if considering varying viscosity, the grout pressure decreases by 10.0% and propagates about 10.95 m.

By fitting the data in Figure. 6, the distribution of longitudinal grouting pressure at the time $t = 0$ s and $t = 7,500$ s can be obtained as follows (unit: MPa):

$$\begin{cases} P_l(0) & = 0.180 - 0.0148l & 0 \le l \le 12.16\,m \\ P_{l1}(7500) = 0.131 - 0.0148l & 0 \le l \le 8.85\,m \\ P_{l2}(7500) = 0.162 - 0.0148l & 0 \le l \le 10.95\,m \end{cases} \tag{20}$$

Where $P_l(0)$ presents the longitudinal distribution of grouting pressure at time $t = 0$ s. P_{l1} (7500) and $P_{l2}(7500)$ present the longitudinal distribution of grouting pressure at time $t = 7500$ s, without or with considering the variation of grout viscosity. In addition to the influence of grouting pressure, the influence of tunnel weight and slurry buoyancy should also be considered, when carrying the longitudinal tunnel floating calculation. The slurry buoyancy and tunnel gravity are applied to the foundation beam as uniformly distributed load, and the longitudinal force distribution formula of the foundation beam is obtained after superimposing with the grouting pressure as follows (unit: kN/m)

$$\begin{cases} F_l(0) & = 1714.98 - 58.38l & 0 \le l \le 12.16\,m \\ F_{l1}(7500) = 1521.70 - 58.38l & 0 \le l \le 8.85\,m \\ F_{l2}(7500) = 1643.98 - 58.38l & 0 \le l \le 10.95\,m \end{cases} \tag{21}$$

Where $F_l(0)$ presents the longitudinal force distribution at time $t = 0$ s. $F_{l1}(7500)$ and $F_{l2}(7500)$ present the longitudinal force distribution at time $t = 7500$ s, without or with considering the variation of grout viscosity.

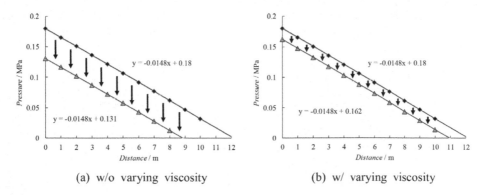

(a) w/o varying viscosity   (b) w/ varying viscosity

Figure 6. Grout pressure distribution along longitudinal direction versus the diffusion distance, in the case of t = 0 and t = 7500 s.

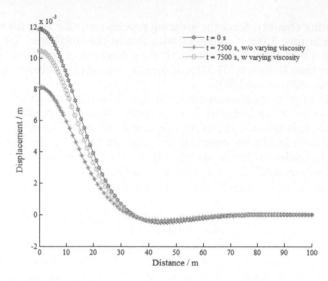

Figure 7. The predicted value of floating on the tunnel segment based on Winkler elastic foundation beam model.

4.2 *Prediction of longitudinal floating based on Winkler elastic foundation model*

Assuming that the tunnel structure is completely homogeneous, the longitudinal equivalent stiffness is calculated as $(EI)_{eq}=9.06\times10^8\text{kN·m}^2$. In addition, considering the existence of shield tail gap, the ground reaction coefficient k_g needs to be deducted, taking $1.0\times10^4\text{kN·m}$. Combined with equations (16), (19) and (21), the vertical displacement curve of the floating segment in this project can be obtained, as shown in Figure 7.

It can be concluded from the Figure 7 that after the tunnel segments are detached from the shield tail under the combined force, the segments adjacent to the shield tail are highly displaced. The amount of uplift decreases dramatically with the increase of longitudinal distance, even negative displacement occurs. And then the amount of displacement gradually goes to zero. The grouting pressure is not attenuated at the initial moment, the maximum displacement reaches 11.9mm (blue line). If considering the dissipation of grouting pressure and the variation of grout viscosity at the same time, the vertical maximum displacement is 10.5mm (green line), compared to that (8.1mm, red line) not considering the variation of grout viscosity.

In the actual construction process, the tunnel segments floated upward immediately after it came out of the shield tail, and the floating speed was relatively fast, the floating amount reached nearly 9.9mm. After that, the grout gradually hardened, and the tunnel gradually became stable after 6 days. The final uplift of the tunnel segment is about 15.0mm

5 CONCLUSION

The grouting process is of great importance to the stability of lining segments in shield tunnelling. This paper proposed a novel model to characterize the temporal and spatial distribution of the grouting pressure, and a prediction of longitudinal tunnel floating feature. And the results evaluated from the model are benchmarked against field monitored data. Summarizing the results in this work, the following conclusions are made.

(1) The temporal and spatial distribution of grout pressure along tunnel longitudinal direction can be well characterized by the proposed model, which is derived based on the diffusion and consolidation theory, as well as considering the temporal variation of grout viscosity.

(2) If the variation of grout viscosity is taken into account, the dissipation rate of grouting pressure is relatively smaller, and the final dissipation amplitude of that is also smaller. The reason is that: as the grout viscosity increases, the grout liquidity and permeability of surrounding soil decrease, which will affect the extent of grouting pressure dissipation, and then affect the distribution of deformation of tunnel segments.

(3) The predicted value of floating on the tunnel segment based on Winkler elastic foundation beam model is closed to that monitored in the field, if considering the dissipation of grouting pressure and the variation of grout viscosity at the same time.

ACKNOWLEDGEMENT

This work has been supported by the National Natural Science Foundation of China (No. 51708564; No. 51678578), China Postdoctoral Science Foundation (No. 2018M633223), the Guangdong Natural Science Foundation of China (No. 2016A030313233), the Guangzhou Science & Technology Program of China (No. 201804010107; No. 201704020139), the Fundamental Research Funds of Sun Yat-Sen University(18lgpy31). These supports are gratefully acknowledged.

REFERENCES

Liang, Y. Huang, L.C. 2018. Grout pressure distribution characteristics in space-time domain of shield tunnels during synchronous grouting. *Journal of Harbin Institute of Technology* 50(3):2-7

Liang, Y. Yang, J.S., Wang, S.Y., Zeng, X.Y. 2015. A study on grout consolidation and dissipation mechanism during shield backfilled grouting with considering time effect. *Rock and Soil Mechanic* 36 (12): 3374-3380.

Lin, J.X., Yang, F.O., Shang, T.P., Xie, B. 2008. Study on tunnel stability against uplift of super-large diameter shield tunneling. In The Shanghai Yangtze River Tunnel. Theory, Design and Construction, Huang R. (ed.). *Taylor & Francis group: London*;267–274.

Talmon, A.M., Bezuijen, A. 2008. Simulating the consolidation of TBM grout at Noordplaspolder. *Tunnelling and Underground Space Technology incorporating Trenchless Technology Research* 24(5): 493-499.

Talmon, A.M., Bezuijen, A. 2013. Analytical model for the beam action of a tunnel lining during construction. *International Journal for Numerical and Analytical Methods in Geomechanics* 37(2): 181-200.

Talmon, A.M., Bezuijen, A. 2013. Calculation of longitudinal bending moment and shear force for Shanghai Yangtze River Tunnel Application of lessons from Dutch research[J]. *Tunnelling and Underground Space Technology* 35: 161-171.

Wang, D.Y., Yuan, J.X., Zhu, Z.G., Zhu, Y.Q. 2014. Theoretical solution of longitudinal upward movement of underwater shield tunnel and its application. *Rock and Soil Mechanics* 35(11):3079-3085.

Ye, F. Zhu, H.Z., Ding, W.Q. 2008. Longitudinal Upward Movement Analysis of Shield Tunnel Based on Elastic Foundation Beam. *China Railway Science* 29(4):65-69

Zhou, J. 2014. Tunnel segment uplift model of earth pressure balance shield in soft soils during subway tunnel construction. *International Journal of Rail Transportation* 2(4): 221-238.

Tunnels and Underground Cities: Engineering and Innovation meet Archaeology,
Architecture and Art, Volume 5: Innovation in underground engineering,
materials and equipment - Part 1 – Peila, Viggiani & Celestino (Eds)
© 2020 Taylor & Francis Group, London, ISBN 978-0-367-46870-5

Safety approach for tunnel lining calculations in 3D-continuum models

Y.A.B.F. Liem, J.T.S. Vervoort & M.H.A. Brugman
Arthe Civil & Structure, Houten, The Netherlands

M. Partovi
DIANA BV, Delft, The Netherlands

ABSTRACT: The state-of-the-art method for calculations on bored tunnel lining is considered to be a 3D-continuum model using Finite Element Software, in which both the soil and concrete lining and their interaction are modelled. According to Eurocode 0 it is preferred to perform a ULS verification using partial factors on individual loads and resistances as described in Eurocode 1, 2 and 7. Eurocode 7 provides 3 design approaches which may be applied. A common approach is to perform SLS calculations and apply an overall safety factor on the resulting forces in order to obtain ULS results. This is similar to Design Approach 2 in Eurocode 7 (EC7-DA2). However, a more economic design may be reached by applying partial factors on the loads and soil parameters rather than the calculation results. In this paper an approach is presented for ULS tunnel lining calculations in a finite element environment, following Design Approach 3 as described in EN-1997 and applying partial factors within the finite element model to achieve the desired safety level (EC7-DA3). Furthermore, a case study is presented, illustrating the difference between the results for EC7-DA2 and EC7-DA3.

1 INTRODUCTION

Traditionally, the prefab concrete lining for shield driven tunnels may be calculated using framework models in which the segmented lining is represented by beams. The support of the lining by the surrounding soil can be modelled as pressure only springs, who's stiffness is generally determined using formulae as presented by Schulze & Duddeck (1964). The loads acting on the tunnel lining are modelled explicitly. This method is commonly referred to as the "Duddeck method", and the Design Approaches described in the Eurocodes can be applied to perform ULS calculations.

However, the structure soil interaction and its effect on the loads cannot be modelled accurately with this approach. Continuum models in which both tunnel lining and soil can be modelled have become more widely used as these allow for a more accurate assessment of the soil-lining interaction. These models are usually made in Finite Element Software. As geotechnical actions and support are not modelled explicitly, applying partial factors is not as straightforward compared to framework models (Boxheimer et al 2008).

Since the introduction of Eurocode 7, tunnels are classified as a geotechnical construction (European Committee for Standardization 2004). Therefore, ULS verifications of the structure should be made using one of the design approaches as prescribed by EC7. This paper describes the application of the Design Approaches from Eurocode 7 for the design of a shield driven tunnel lining in a continuum model. Section 2 describes the available safety approaches from Eurocode 7 with their application of partial factors, and how they may be applied to a continuum model. Section 3 provides the proposed approach of ULS modelling for continuum models. Finally, section 4 provides a case study based on the design calculations for the RijnlandRoute bored tunnel and compares the results for the various design approaches.

2 DESIGN APPROACHES

Eurocode 7 specicfies three possible Design approaches for ULS calculations. The national annex to the Eurocode may specify which Design Approach is to be used. Otherwise, the designer may choose one based on preference or practicability. All three approaches are expected to result in a sufficiently safe design.

The design approaches prescribe one or more combinations of partial factors to be applied either on the loads and soil parameters directly, or on the forces resulting from the calculations without partial factors.

2.1 Partial factors

In the Eurocode, partial factors are divided into four categories:

- A1, for structural actions
- A2, for geotechnical actions
- M1/M2, for soil parameters
- R1/R2/R3/R4, for foundations and (slope) stability

For a shield driven tunnel lining, factors for structural and geotechnical actions and soil parameters are relevant. Depending on the chosen Design Approach, a selection of these may be applied. For all soil parameters an analysis should be performed whether decreasing the value is favourable or unfavourable. When a decrease proves favourable, a partial factor of 1,00 should be applied instead. The following tables present the partial factors for reliability class 3.

2.2 Design Approaches

The Design Approaches prescribe combinations of partial factors for which a "limit state of rupture or excessive deformation" should be verified. As factor set R1 to R4 are not relevant for shield driven tunnels, this leaves the following combinations for the design approaches:

Design Approach 1: Combination 1: A1"+" M1
Combination 2: A2 "+" M2

Table 1. Partial factors for loads in A1 and A2.

Action	Factor	
	A1	A2
Permanent unfavourable	1,50	1,00
Permanent favourable	0,90	1,00
Variable unfavourable	1,65	1,45
Variable favourable	0,00	0,00

Table 2. Partial factors for soil parameters.

Soil parameter	Symbol	Factor	
		M1	M2
Friction angle	$\gamma_{\varphi'}$	1,00	1,25
Effective cohesion	$\gamma_{c'}$	1,00	1,25
Volume weight	γ_γ	1,00	1,00
Stiffness*	γ_e	1,00	1,30

* partial factor on stiffness from Dutch National Annex

Design Approach 2: Combination 1: A1"+" M1

Design Approach 3: Combination 1: (A1 or A2) "+" M2

In Design Approach 1 (EC7-DA1), both Combination 1 and Combination 2 should be checked, in which the partial factors are applied on the actions and ground strength parameters. However, as the loads from for example the soil are not modelled explicitly in a continuum model, it is not possible to apply a partial factor 1,5 to the soil load. Therefore, EC7-DA1 cannot be applied properly when using a finite element model.

In EC7-DA2, partial factors may be applied to either actions or the effects of actions and to ground resistances. Again, it is not possible to apply the partial factors from set A1 onto the soil loads. Therefore, the only viable approach for EC7-DA2 is to apply partial factors on the effects of actions, and ground resistances. For structural linear calculations, it is possible to calculate each load case separately, and combine them afterwards with their corresponding partial factors. For nonlinear calculations however, this is not necessarily possible. Therefore, a more practical approach is to apply overall factors to the calculated forces.

For the main reinforcement of a segmented tunnel lining, normal forces are generally favourable while bending moment is unfavourable. Consequently, a factor 0,9 should be applied on normal force, while bending moment is factored by a value somewhere between 1,50 and 1,65 depending on the relative effect of the permanent and variable unfavourable loads. It is not straightforward to accurately assess this relation. A sensible choice is to apply a factor of 1,65.

EC7-DA3 finally, partial factors may again be applied to either actions or the effects of actions and to ground resistances. However, EC7-DA3 makes a distinction between structural actions (A1) and geotechnical actions (A2). Unfavourable geotechnical actions and structural actions are modelled explicitly, and the partial factors from set A1 or A2 can be applied within the model. All implicit actions, such as soil loads, are generally permanent geotechnical actions, which receive a partial factor of 1,00 from set A2. As a result, it is possible to include all partial factors within the model, rather than applying them to the calculation results.

Concluding this chapter, EC7-DA1 cannot be applied in a continuum model as partial factors on soil loads other than 1,00 cannot be modelled properly. EC7-DA2 can be used by applying partial factors to the effects of actions. In EC7-DA3 all partial factors can be included within the model, as it prescribes a partial factor of 1,00 on permanent geotechnical actions.

3 CASE: ULS DESIGN OF A TUNNEL

3.1 Introduction

This case presents a calculation of the Structural Limit State and Ground Limit State for the segmental lining of a shield driven tunnel. The conditions are based on the RijnlandRoute tunnel in the Netherlands. For sake of clarity, not all load combinations are presented. Also, only one tunnel tube is modelled, whereas in reality two tubes are constructed.

3.2 Tunnel geometry

The tunnel is built up with 2 m long rings, each divided into 7 roughly equal segments. The rings are constructed in a staggered configuration. As the keystone is similar in dimensions to the other segments, there will be no X-joints and the location of the keystone is not relevant for the calculations. Rings are connected using dowels, of which 28 are applied per ring. Guiding rods are applied in the longitudinal joints.

3.3 Soil conditions

The soil profile consists of three layers: a relatively strong Pleistocene sand layer below a more clayey sand layer, covered by soft clay. The tunnel axis is located at a depth of 26 meters below surface, and lies completely in the Pleistocene sand.

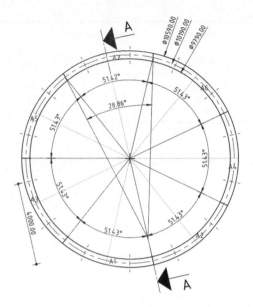

Figure 1. Ring geometry.

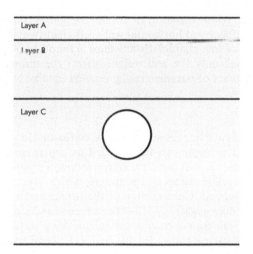

Figure 2. Soil stratigraphy.

Soil parameters were determined by a statistical analysis of the results of laboratory tests performed on actual samples from the field. The statistical analysis results in average, high and low characteristic values for every parameter. These are presented in table 3.

According to Eurocode 7, the most unfavourable combination of low and high values for parameters should be used for independent parameters. Laboratoy tests have shown that in this project, some soil parameters are co-dependent. For the Pleistocene sand, a higher volume weight is correlated with a higher stiffness and friction angle. Therefore it is allowed to use either high, medium or low characteristic values for all parameters, whichever is unfavourable. Doing this will result in a more economic design, however additional analyses are needed in order to determine which values need to be used, as this greatly depends on the range between the upper and lower boundary for the individual parameters. Whether the high average or low values should be used can change between projects or even soil types.

Table 3. Soil parameters.

| characteristic value | volume weight | | strength params. | | modulus of elasticity | | |
	γ_{bulk} [kN/m³]	γ_{sat} [kN/m³]	φ' [°]	c' [kN/m²]	$E_{50;ref}$ [MPa]	$E_{ur;ref}$ [MPa]	$E_{oed;ref}$ [MPa]
$X_{gem;k;low}$	18,4	19,4	35,5	1,7	26	102	26
$X_{gem;k}$	18,9	19,6	37,5	2,0	32	128	32
$X_{gem;k;high}$	19,3	19,9	40,0	2,4	38	154	38

Table 4. Results sensitivity analysis soil parameters Layer C.

Parameter values	M_{min} [kNm]	corr. N [kN]	M_{max} [kNm]	corr. N [kN]
low	-135	-1637	150	-1998
average	-130	-1621	140	-1988
high	-122	-1597	129	-1975

Whether the set of low, average and high characteristic values are unfavourable is determined using simplified 2D finite element analyses, in which the tunnel is modelled as a continuous ring in a homogenous soil body. Table 4 shows the extreme bending moments and their accompanying normal forces.

For Layer C, a relatively low variation in volume weight is found. Bending moments are affected mostly by soil stiffness and horizontal soil coefficient (which is dependent on angle of internal friction). The set of low characteristic values is found to be unfavourable. For layers A and B, above the tunnel, only the unit weight affects the tunnel. The high characteristic value is used. The complete set of parameters is presented in table 5.

3.4 Hydraulic head

While groundwater may always be present, there can be fluctuations in the hydraulic head. As a result, part of the water pressure can be regarded as a permanent load, and part can be regarded as a variable load. Initial pore pressures follow a hydrostatic gradient. A higher water pressure increases normal forces in the tunnel, which generally is a favourable effect. Drained conditions are assumed. Construction of the tunnel and external loads do not cause excess pore pressures that may negatively affect the effective soil stresses.

When sufficient data from measurements are available, a statistical analysis can be performed to determine the extreme values for the hydraulic head with a chance of occurrence

Table 5. Soil parameters.

Parameter		Layer A	Layer B	Layer C	Unit
Soil model		Hard. Soil	Hard. Soil	Hard. Soil	-
Unit weight	$\gamma_{unsat}/\gamma_{sat}$	12,7/12,7	17,6/18,4	18,4/19,4	kN/m³
Triaxial stiffness	$E_{50;char}$	1100	11000	26000	kN/m²
Oedometer stiffness	$E_{oed;char}$	600	11000	26000	kN/m² kN/m²
Unloading stiffness	$E_{ur;char}$	4800	45000	102000	kN/m²
Reference stress	p_{ref}	100	100	100	kN/m²
Power	M	0,90	0,60	0,55	-
Poisson ratio	v	0,17	0,20	0,17	-
Cohesion	c'	6,0	0	0	kN/m²
Friction angle	φ'	22,6	34,0	35,5	°
Dilatancy angle	d_f	0	0	0	°
Initial stress ratio	K_0	0,62	0,44	0,42	-

that fits the required safety level. The lower bound value determines the permanent part of the load, while the variance is the variable load.

3.5 *Modelling*

The tunnel and soil continuum were modelled using the Finite Element software DIANA FEA. Two half widths of tunnel rings (1,0 m) have been modelled as plate elements. The connecting dowels are simulated by linear elastic zero thickness interface elements. The longitudinal joints are modelled with a zero thickness interface using the Janssen material model included in the DIANA software (Manie & Kikstra 2018). This material simulates the deformation in the longitudinal joints based on the formulae presented by Janssen (1983). The soil body is modelled as 3D volume elements using the Modified Mohr-Coulomb material model . This soil model uses Hardening Soil design parameters (Manie & Kikstra 2018). For the soil lining interaction, a zero thickness interface is modelled with Mohr-Coulomb properties. As the friction between lining and soil is uncertain, the friction angle is reduced to near zero. Figures 2 and 3 illustrate the Finite Element mesh.

A three dimensional model is required in order to correctly assess the effect of the staggered configuration of the lining rings. As the contact area in the longitudinal joint has a height of only 220 mm, compared to the 400 mm thickness of the segments, the longitudinal joints do not behave as stiff as the concrete segments. Due to the staggered configuration, forces will be transferred to adjacent rings depending on the locations of the joints.

While soil deformations may be approximated in a two dimensional model by reducing the stiffness of a continuous ring, the redistribution of load and the resulting forces can only be assessed properly by actually modelling multiple rings and joints.

3.6 *Construction sequence*

Due to the conical shape of the TBM shield, the surrounding soil will relax to a certain degree. Subsequent application of grout under high pressure in the tail void gap will then put additional stress on the soil. The combined effect of 'constructing' the tunnel is sometimes accounted for by the convergence confinement method which applies to 2D calculations (Eisenstein & Branco 1991).

In three dimensional models, a sequenced calculation may be made, in which tunnel rings are excavated and installed one by one. A face pressure, relaxation of soil, and grout pressure may be modelled explicitly. The effect of the sequenced construction, including soil relaxation and grout pressures is found to have a favourable effect when structural design of

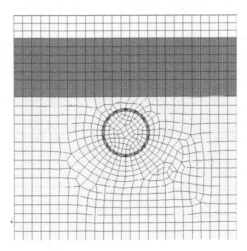

Figure 3.   Finite Element Mesh.

Figure 4. Tunnel Segments.

the lining is concerned. However, these effects are greatly dependent on the execution of the construction works, where especially grout pressures are difficult to control during execution. A certain amount of safety is incorporated in the model by disregarding these favourable but highly uncertain effects. Instead, the construction is modelled as if the excavation of soil and installation of the lining for the complete tunnel occurs instantaneously. The temporary situation in which a tunnel ring is floating inside the grout should be assessed in a separate calculation model.

3.7 Model phasing

The calculation starts by determining the initial soil stresses in the virgin soil. Then, the soil is excavated and the lining activated. After activation of the tunnel lining, additional loads are applied in separate phases. First, a load from the tunnel inlay is activated, representing the cable duct, backfill and asphalt structure. Then, a surface surcharge of 20 kN/m^2 is activated. Other loads, such as passage of the TBM backup train, temperature variations, and traffic inside the tunnel, are excluded from this example calculation.

The calculation scheme is derived from Brinkgreve & Post (2015). The initial stresses in the virgin soil are calculated in Phase 0. The tunnel tube is then excavated and activated at once in Phase 1. In phase 2, the inlay is modelled as a distributed load on the tunnel invert. A surface surcharge load of 20 kN/m^2 is then applied on the full extent of the model in phase 3.

The calculation scheme for the case study is shown below.

Phases 1 to 3 are SLS calculations for different load combinations, in which the characteristic values for soil parameters and loads are applied. Phases 4 to 6 are similar to phases 1 to 3 respectively, however partial safety factors are applied to the relevant parameters and loads in order to obtain ULS values. For EC7-DA3, all partial factors are incorporated in the model. For EC-DA2, partial factors on effects of actions are applied as an overall factor on the calculation results.

Table 6. Calculation scheme.

Phase	State	Start from phase
0. Initial phase	SLS	-
1. Activate tunnel tube 1	SLS	0
2. Activate tunnel inlay	SLS	1
3. Variable surcharge 20 kN/m^2	SLS	2
4. ULS of phase 1	ULS	0
5. ULS of phase 2	ULS	1
6. ULS of phase 3	ULS	2

Table 7. Calculation results SLS.

Phase	M_{max} [kNm/m]	N_{Mmax} [kN/m]	M_{min} [kNm/m]	N_{Mmin} [kN/m]
0. Initial phase	-	-	-	-
1. Activate tunnel tube 1	150	2012	-135	1637
2. Activate tunnel inlay	143	1979	-131	1654
3. Variable surcharge 20 kN/m^2	261	2454	-266	1822

Table 8. Calculation results ULS for EC7-DA3.

Phase	M_{max} [kNm/m]	N_{Mmax} [kN/m]	M_{min} [kNm/m]	N_{Mmin} [kN/m]
4. ULS of phase 1	183	2025	-163	1635
5. ULS of phase 2	168	1980	-152	1657
6. ULS of phase 3	392	2739	-379	1934

Table 9. Calculation results ULS for EC7-DA2.

Phase	M_{max} [kNm/m]	N_{Mmax} [kN/m]	M_{min} [kNm/m]	N_{Mmin} [kN/m]
4. ULS of phase 1	248	1811	-223	1473
5. ULS of phase 2	236	1781	-216	1489
6. ULS of phase 3	431	2209	-439	1640

3.8 *Results*

The SLS results for extreme bending moments and the normal forces corresponding to these bending moments are shown in table 7. Table 8 presents the ULS results for EC7-DA3. In table 8, the SLS results from the Finite Element calculations are multiplied by a partial factor of 1,65 for bending moment (unfavourable) and 0,9 for normal forces (favourable) to obtain ULS results for EC7-DA2.

Figure 5 shows the calculation results plotted against the capacity of a 1 m concrete tunnel section with a 400 mm thickness and 10–100 mm main reinforcement. For EC7-DA3, the results fall within the capacity of the section. The normative results for EC7-DA2 however, fall outside the section capacity. Additional reinforcement would be required when using EC7-DA2.

Figure 5. Distributed bending moments.

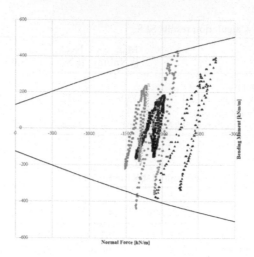

Figure 6.   Bending moments versus capacity EC7-DA2 (grey), EC7-DA3 (black).

4   DISCUSSION OF THE RESULTS

From the SLS calculations we can make a number of observations. First is the effect of the tunnel inlay, which has a mild, but favourable effect on the normative forces in the tunnel. This is explained by how the soil load presses on the tunnel lining. As the vertical soil stresses are significantly greater than the horizontal soil stresses, the tunnel will initially deform into an flat horizontal oval shape. The bending moments at the top and invert act inwards, while the bending moments at the sides act outwards. Any action inside the tunnel on the tunnel invert then acts counter to the initial load from the soil. This can be extrapolated to a degree for all structural actions on the tunnel invert, as long as these are significantly smaller than the soil pressure on the tunnel. For very shallow tunnels, the deformation of the tunnel may actually resemble more of a standing oval due to uplift. Actions on the tunnel invert are then expected to be unfavourable. As a whole, the effect of the inlay is relatively small.

Phase 3 shows the effect of a surface load after construction of the tunnel. Bending moments more than double, while normal forces increase as well, but to a much smaller extent. This shows how surface loads can significantly influence the reinforcement design of the tunnel, and why it is of paramount importance not to underestimate these future loads during design, as well as not to exceed the loads that are used for the design by future surface works. Conversely, being very conservative in the estimation of surcharge loads can lead to an uneconomic design.

From the results of the EC7-DA3 calculations it can be observed that between the SLS and ULS calculations of phase 1, only a 20% increase is found in the bending moments, while normal forces remain practically unchanged. This increase can be almost entirely attributed to the partial factor of 1,3 on the stiffness of the soil, which is added by the Dutch National Annex. The partial factor on permanent geotechnical actions is 1,00, there are no substantial structural actions, and a partial factor of the internal friction angle would actually be favourable and is therefore set to 1,00. Without this partial factor on stiffness, no significant differences would be found.

Phase 2 shows the effect of structural actions on the tunnel invert. In comparison to the results for phase 1, the effect is limited.

Between the SLS and ULS calculation of phase 3, a 40% increase is found in the bending moments compared to a 10% increase in normal forces. This is explained by the partial factor of 1,35 on the surcharge load.

From the results it is clear that the ULS bending moments directly obtained from the Finite Element model by EC7-DA3 are more favourable compared to the EC7-DA2 results. This is due to the distinction made by EC7-DA3, where the partial factor for permanent geotechnical actions is 1,00, while in EC7-DA2 it is 1,50. Also, the normal forces are reduced with a partial factor 0,9 in EC7-DA2, where in EC7-DA3 they are similar between SLS and ULS, or increase when the surface load is also increased by its partial factor.

Figure 5 confirms our findings, in that more reinforcement would be needed when EC7-DA2 is used compared to EC7-DA3.

In general we can state that EC7-DA3 provides a more economic design than EC7-DA2. We have also observed that most of the safety for the ULS calculations comes from the partial factor on stiffness parameters, which is specific to the Dutch national annex only. Without it, only a small difference may be found between SLS and ULS calculations, and one may argue whether this provides enough safety in the design.

5 CONCLUSIONS AND REMARKS

This paper presents a comparison between the different Design Approaches for geotechnical structures as described in EN-1997-1 within a finite element environment, with regards to the structural design of the lining for a shield driven tunnel. A case study is performed to illustrate the integration of the Design Approaches in the finite element environment, and show the difference in the results.

Both Design Approach 2 and 3 may be applied to perform Ultimate Limit State (ULS) calculations in finite element models. As all actions that are transferred through the soil are modelled implicitly, partial factors for EC7-DA2 should be applied on the effects of actions. In EC7-DA3, partial factors for permanent geotechnical actions are 1,00, which allows all partial factors to be modelled in the finite element model.

Ultimately, EC7-DA3 can lead to a more economic design, at the expense of additional time and analyses during the design phase.

When applying EC7-DA3 as described, the engineer should be confident that he or she provides a sufficient level of safety, regardless of what the Eurocodes allow for. This may be achieved partly by incorporating a margin of safety in the construction model, for which we propose to disregard favourable effects of the tunnelling process which are difficult to predict and quantify, partly by incorporating a partial factor on the soil stiffness as prescribed in the Dutch national annex, and partly by taking a somewhat conservative attitude towards the estimation of future surface loads.

REFERENCES

Brinkgreve, R.B.J. & Post, M. 2015. Geotechnical Ultimate Limit State Design Using Finite Elements. In T. Schweckendiek et al. (eds.), *Geotechnical Risk and Safety V; 5th International Symposium on Geotechnical Safety and Risk; Proc. symp., Rotterdam, 13-16 October 2015*. Amsterdam: IOS Press

Boxheimer, S. et al 2008. E & C-contract voor tunnellining. *Cement* 2008(2): 54–57.

Eisenstein, Z. & Branco, P. 1991. Convergence-Confinement Method in Shallow Tunnels. *Tunneling and Underground Space Technology* 6(3):343–346.

European Committee for Standardization 2004. *Eurocode 7: Geotechnical Design – Part 1: General Rules (EN 1997-1)*. Brussels: European Committee for Standardization.

Janssen, P. 1983. *Tragverhalten von Tunnelbauten mit Gelenktübbings*. Braunschweig: University of Braunschweig.

Manie, J. & Kikstra, W. 2018. *DIANA – Finite Element Analysis User's Manual release 10.2 Material Library*, Delft: DIANA FEA BV.

Schulze, H. & Duddeck H. 1964. Spannungen in schildvorgetriebenen Tunneln. *Beton Stahlbetonbau* 59: 169–175.

Tunnels and Underground Cities: Engineering and Innovation meet Archaeology,
Architecture and Art, Volume 5: Innovation in underground engineering,
materials and equipment - Part 1 – Peila, Viggiani & Celestino (Eds)
© 2020 Taylor & Francis Group, London, ISBN 978-0-367-46870-5

Plug-in crosscut element – How standardized pre-manufactured crosscut wall systems improve overall tunneling effectiveness

M. Lierau
Elkuch Bator AG, Herzogenbuchsee, Switzerland

R. Binder
Biprotec GmbH, Weistrach, Austria

ABSTRACT: The Plug-in Crosscut Element (PCE) is a highly standardized, high precision pre-manufactured and fully assembled ready-to-install concrete wall system. It is designed to close cross connecting galleries in a tunnel in one step. Its main advantages consists of (1) 15% less manufacturing cost, (2) up to 90% installation and commissioning time saving, (3) max. access to the cross connecting gallery until the last moment, (4) increased safety due to less manpower required inside the tunnel, (5) high system reliability due to pre-verified modules. These advantages result from standardized and highly repetitive production steps outside the tunnel and a clever logistics concept. The PCE, including its fully operational sub-systems is pre-installed, tested and accepted before entering the tunnel. Its logistics concept is based on a versatile applicable manipulator, which can be fixed to any type of transportation system. The transportation, installation and commissioning time can be accurately planned, optimized and executed.

1 INTRODUCTION

1.1 *The Outset - Conclusions from a monumental project*

With the opening of the Gotthard Base Tunnel on June 1st 2016, a technical milestone in the history of mankind was successfully handed over to the tunnel operator. One of the key aspects of the successful completion of the project was, and still is, logistics. With respect to the installation of cross connecting gallery escape doors, transportation to and from the installation site located every approx. 350m throughout the entire tunnel, is a significant cost factor. In the 57km long Gotthard Base Tunnel, travel time to a door could last up to 2.5h - one way!

Logistics become an even more significant cost driver during maintenance and refurbishment. The impact of the associated logistic costs is often not well addressed in long tunnel projects during the planning and procurement stage, as experience today is still somewhat limited and long term aspects of tunnel operations (e.g. concept of refurbishment after 30 years) quite unclear. Furthermore, the economics of logistics are often impacted by prerequisites and requirements, which result of updated maintenance concepts and schedules long after the product has been designed. It was concluded that for a next long tunnel project, much more focus must be laid on the economics of logistics and the overall timeline.

1.2 *PCE – an innovative logistics and installation concept*

The impulse however for this innovation was given by the installation situation in the Koralm Base Tunnel in Austria end of 2016. The small diameter of the crosscut as well as the concrete buildings inside the cross connecting gallery prevent the installation of larger systems such as doors, fans, etc. to the inside of the crosscut wall. A concept of pre-fabricated crosscut wall elements with previously fitted subsystems was drafted in order to take account of the challenging installation environment and an optimized logistics concept. The idea of a Plug-in Crosscut Element (PCE) was born.

Although the idea had to be eventually abandoned for the Koralm Base Tunnel due to the too far progressed project plan, the idea was nevertheless further explored. The overwhelming feedback clearly indicated: the concept was reasonable, technically and economically feasible and bore an innovative logistics and installation approach with had the potential of setting a new standard in the tunneling industry.

1.3 SCAUT – the global tunneling competence center

Key challenges of this project were the many technical aspects of tunneling which had to fill the team's knowhow gap and which needed to be combined simultaneously. It was clear from the outset to the project core team, that only an interdisciplinary approach comprising among other tunnel planning an engineering, optical measurement, pre-cast element production, fire protection and safety system engineering, etc. would lead to a success. It was the Swiss Centre of Applied Underground Technologies (SCAUT), which provided the platform for such a venture and hence would play a key role throughout the subsequent stages of the project.

Through the support of SCAUT, Elkuch Bator AG and biprotec GmbH teamed up with several leading industry partners, filling the technology gaps with their respective knowhow.

2 THE CONCEPT OF THE PCE

2.1 Conventional escape door installation procedure

A typical installation procedure for crosscut gallery escape doors in tunnels consists of:

1. measurement of the free opening in the tunnel
2. installation of the top rail incl. drive (core hole drilling)
3. grouting of the bottom rail
4. installation of the left and right door frame (core hole drilling)
5. mounting of the door leaf/leaves
6. installation of the control cabinets
7. electrical connection
8. commissioning

For the installation of the escape doors in the Loetschberg, Gotthard and Ceneri Base tunnel, additional special tools such as mounting robots have been used. Not including the delivery of standardized transport boxes to the corresponding crosscut, the standard installation time a door would take an experienced mounting team of 2 mechanical and 1 electrical technician between 2.5 to 3 days depending on the space and access available (Figure 1). The installation time also accounted for the travelling time with the construction train to the installation site, which in case of the 57km long Gotthard Base Tunnel would take up to 2.5 h. This left only 3-4 h for the installation of the doors per day.

In fact, the installation of one door required a minimum of 5 trips into the tunnel, not including delivery, quality checks, repairs, several commissioning and service trips etc. The associated costs comprised the hourly salary including underground and risk surcharge as well as the cost for the safety training, equipment and heat test (i.e. tunnel fitness test).

2.2 Objectives of the PCE project

The constraints and experiences from long railway tunnels and the requirements from the Koralm Base Tunnel, eventually lead to the PCE's market requirements: The Plug-in Crosscut Element (PCE) is a highly standardized, high precision pre-manufactured and fully assembled

Figure 1.   Typical installation site for tunnel escape doors.

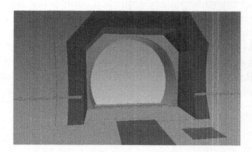
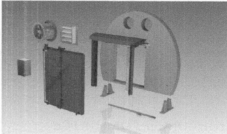

Figure 2.   Crosscut view from tunnel and pre-assembly of fully functional sub-systems.

Figure 3.   Crosscut view from tunnel and pre-assembly of fully functional sub-systems.

ready-to-install concrete wall system. It is designed to close cross connecting galleries in a tunnel within 90 minutes. Its main advantages consists of:

- High precision wall element with workshop production process safety level
- 15% less manufacturing cost
- Up to 90% installation and commissioning time saving
- Pre-installation and testing of all sub-assembly and systems such as fans, control cabinets, fire shutters, doors, lighting, signs, etc.
- High system reliability due to pre-verified modules
- Increased safety due to less manpower required inside the tunnel, safer working conditions and healthier environment
- Full accessibility of crosscut for further installations until the very last moment

These advantages result from standardized and highly repetitive production steps outside the tunnel. The entire PCE, including its readily operational sub-systems such as lighting, doors etc. are fully pre-installed, tested and accepted before entering the tunnel (Figure 2).

The PCE's logistics concept is based on a versatile applicable manipulator, which can be fixed to any type of ISO normed transportation system such as rail wagon, flatbed trailer or any other special vehicle. The manipulator accurately moves the PCE into position where it is bolted to the crosscut and sealed (Figure 3).

3 IMPLEMENTATION OF THE PROTOTYPE

3.1 *Planning Stage*

As the PCE merges activities from the structural work and the equipment installation phase, the concept of standardized plug-in crosscut elements has to be considered at an early stage of the general project planning. In particular, a special attention has to be given to interferences during transportation, weights of elements, electro-mechanical sub-system selection and an adapted interface at the crosscut ends which allows an optimal installation of the PCE.

Once integrated in the planning, the concrete element is designed based upon a 3D laser scan of the crosscut gallery entrance profile. Available measurement technologies on the market are sufficient to provide the required accuracy.

In case of the PCE prototype, the profile of the Brenner Base Tunnel stood model. Its profile was rebuilt in the Hagerbach Test Gallery (VSH) in Switzerland, where the entire PCE concept was tested on 1:1 scale in late 2017 and early 2018 (Figure 4).

3.2 *Standardized production outside the tunnel*

From this CAD project, the PCE is manufactured in a common concrete workshop outside the tunnel, ensuring high process reliability and security (Figure 5). The adjustable mold is aligned with the crosscut's geometry to reduce tolerances down to 20 mm +/-10 mm. After the installation of the concrete reinforcement, all necessary interfaces such as bolts, threaded sleeves, RFID tags etc. are precisely prepositioned. The mold is then filled with concrete to produce wall elements of a thickness between 25 and 30 cm, depending on the static and dynamic loads inside the tunnel.

Once cured, the pre-fabricated concrete wall element is set and mounted on jigs that provide fixation for the concrete element during assembly, transport and installation. All additional sub-assemblies and systems such as fan, aeration tube interface, door, control cabinet, lighting etc. are then mounted to the Plug-in Crosscut Element. As all sub-systems are fully functional, the PCE can be verified, pre-commissioned and pre-accepted. All associated documents are

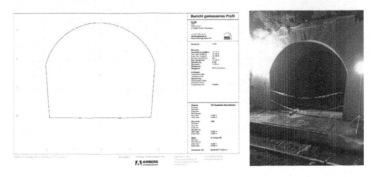

Figure 4. Scan of crosscut profile and Test Gallery crosscut preparation.

Figure 5.   Production of PCE. Positioning of core wire and inlets.

logged on to the system and remain trackable with the RFID tag, which is cast into the Element.

3.3   *Installation procedure inside the tunnel*

Once the tunnel is ready to receive the PCE's, a rail car or a flatbed trailer loads the elements weighing up to 8 t apiece. In the case of a rail car, up to six PCEs for a workman shift - given 90 minutes installation time per PCE - are loaded. The PCEs are transported to their specific installation site. The rail car's final position in front of the crosscut will be defined according to a laser based marking.

Once the car is in position, guide rails are unfolded. Of the recommended 3 installation engineers, one is positioned inside the crosscut, one outside in the vicinity of the PCE and one at the controls of the manipulator. The manipulator then lifts the PCE into its final position. In order to avoid interference with the catenary wires, the PCEs has to be slightly tilted when moved into position in the crosscut. The PCE is set into the groove of the crosscut. Once in place, it is secured with a bolt to the crosscut. At this stage, the PCE is electrically connected and the sub-systems fully operational. While part of the team is filling the groove with a common grout for tunnel applications, the other part of the team disconnects and pulls back the manipulator on the rail car, where it receives a new PCE for the next installation site.

Figure 6.   Installation of the PCE in the Hagerbach Test Gallery.

In case of the prototype testing the Hagerbach Test Gallery, the rail car was simulated by means of a 2.1 t steel structure to which a manipulator of 1.1 t was mounted (Figure 6). The manipulator's adjustment capability in case of the prototype allows for 160mm lateral displacement, 500mm lift and 45° tilt. The PCE is moved into position with a speed of 0.03 m/sec.

Once the grout has cured and a recommended minimum set of 6 – 8 PCEs are installed, the remaining opening around the PCE, will be sealed. In order to withstand the dynamic loads and to fulfill the demanding requirements in high-speed railway tunnels, a special sealing technology is applied in two steps.

Step 1: The gap is filled with a pre-treated back rod, meeting the common fire resistance requirements. This step is part of the previously described PCE installation procedure.

Step 2: Once cured, a composite resin is inject through the 2 inch pipe connections built into the PCE. Excess resin protruding from the relieve ducts on top of the PCE indicate the full spread of the resin and a high quality bonding.

This last operational step concludes a 90 minute installation procedure.

4 RESULTS

4.1 Standardized crosscut closure

On 31st November 2017, the first prototype of the Plug-in Crosscut Element was successfully tested in the Hagerbach Test Gallery in a real railway tunnel environment. During the premiere, the PCE equipped with a high-speed railway tunnel escape door including its control cabinet and a total weight of over 7 t was set into position in just under 5 minutes with a team of 3 (unexperienced) technicians.

Today the PCE has been installed and backed out over 20 times. During this time, the installation team consisted of mostly 2 technicians. The fastest installation time was just under 4 minutes and averages around 4.5 minutes with a team of 3 technicians. The operation in a real tunnel environment is simple and reproducible, easily controllable and reversible. No special technology other than a manipulator is needed for a fast positioning of the PCE.

4.2 Sealing trials

In order to verify the injection, a test rig has been developed to simulate the interface between the crosscut wall and the PCE. The test rig consisted on the lower part of a 1:1 scale model (i.e. concrete block) of the crosscut wall and the PCE joint. The 2 concrete elements were grouted into a foundation while preserving a 20 +/-10mm gap as expected in real environments. Back rods were bonded into the ends of the gap, forming a firm block (Figure 7). This block was successfully water pressure tested up to 5 bar.

On top of this block a 3 m tube was mounted to simulate the remaining bonding joint. A crude pressure vessel contained the resin, which was injected with up to 5 bar through the pipe connections into the test assembly. The assembly was completely filled in less than 10 minutes. Given its potting and curing time, the filler hardens within 24 hours and bears the loads. The trial was carried out two times.

While a full scale sealing test has not been carried out at the time this paper was completed, the test clearly demonstrated the operational reliability of the sealing technology. It can be applied on single units or in a continuous application process (i.e. several PCEs).

Figure 7. Fire resistant filler injection trials.

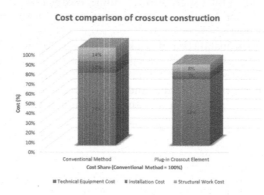

Figure 8. Construction cost comparison between conventional method and PCE concept.

4.3 *Cost assessment*

Main cost driver of the PCE derive from the technical equipment mounted to the PCE. However, the major concept differences between the conventional method and the PCE technology are the fabrication of standardized crosscut wall modules and the increased logistical efficiency due to cost incurrence outside the tunnel. Due to the concept of prefabrication, the PCE's cost is reduced up to 15 % compared to the conventional method (Figure 8). The savings account for:

- Concrete building costs are lower as finished concrete is used instead of site concrete.
- Installation cost is calculated on workshop level during the day and not at tunnel level. All works inside the tunnels are considered, e.g. drilling and sealing of the PCE is included in the cost calculation.
- Overall installation efficiency is higher as free access to all parts of the PCE is given at all time outside the tunnel.

Figure 9.   Installation cost comparison between conventional method and PCE concept.

- Mounting racks for the door are avoided as they are simpler and integrated in the PCE.

The main advantages though of the PCE system are the benefits from a much faster installation and commissioning time. Currently, a well experienced installation team of three members need about three days for the installation and electrical connection of one cross connecting gallery door in the Loetschberg, Gotthard and Ceneri Base Tunnel. All trials and tests indicated, that overall objective to install one PCE within 90 minutes can be achieved. This accounts for cost and time savings up to 90% (Figure 9).

Not included in this calculation are other project benefits such as:

- increased tunnel safety due to less personnel in the tunnel
- more efficient crosscuts completion due to unconstraint access until the very last moment
- less track occupancy and increased supply chain efficiency as installation is reduced to a very short period of time
- less errors and downtimes due to pre-commissioning of the sub-systems installed on the PCE
- higher system reliability
- etc.

4.4   Development time

The highly interdisciplinary and cross-functional project faced many challenges. Starting with the unknown success potential and common skepticism at the outset of the project, various knowhow gaps had to be filled and technical as well as economical questions to be answered. The anticipated R&D cost analysis indicated an immediate termination of the project. With the background of a niche supplier to the tunneling industry a task that under normal circumstances would have been abandoned immediately.

In this context, it is remarkable that the development time from the first sketch until the project's proof of concept (i.e. first live demonstration of the prototype) was accomplished in only 12 months. Key to this fast development was the support of SCAUT.

SCAUT supported the project team and later on the collaboration partners in keeping an out-of-the-box thinking. It assisted the team with professional insights and access to an extremely broad and knowledgeable industry network, some of which became partner in this venture. SCAUT significantly contributed to the fact, that the project increased its credibility as all the tests were performed in a real tunnel environment at the Hagerbach Test Gallery and that the R&D cost burden was spread among the participating partners. Today, the project still collaborates with SCAUT with respect to the global marketing of the concept.

5 CONCLUSION

Tunnel crosscuts connecting the running tubes are important infrastructural measures to achieve the safety goals for long tunnels. They serve as primary self-rescue measure in case of a forced stop in the tunnel. At both end, walls terminate crosscuts, which are among other equipped with emergency escape doors. With conventional methods, the installation and electrical connection of one door in the tunnel takes a well experienced team of 3 workers up to 3 days. Since the wall construction is mainly done by the main contractor, accessibility to the crosscut for further installations is limited at an early stage.

The Plug-in Crosscut Element (PCE) is an integral prefabricated push-in wall element for terminating the tunnel crosscut that is immediately ready for use (plug-and-play). The simple but ingenious idea: Prefabricated in the workshop, these fully functional and pre-tested wall systems are delivered to the crosscut in the tunnel an installed with special transport aids. Elaborate on-site work, such as for formwork, concrete work, drilling and system installations and assembly, is eliminated.

Trials indicated, that an overall installation time is reduced to less than 90 minutes. This means up to 90% cost savings due to a faster installation. The production costs of the PCE are 15% lower than the conventional method given a continuous and standardized production outside the tunnel. Further cost saving potential result from a transfer of work from the tunnel into a workshop.

Overwhelming feedback from the tunneling industry is, that this may set a standard and will be standard technology in less than 10 years.

Tunnels and Underground Cities: Engineering and Innovation meet Archaeology,
Architecture and Art, Volume 5: Innovation in underground engineering,
materials and equipment - Part 1 – Peila, Viggiani & Celestino (Eds)
© 2020 Taylor & Francis Group, London, ISBN 978-0-367-46870-5

Soil conditioning for EPB shield tunneling with the auxiliary air pressure balance mode

P. Liu & S. Wang
Central South University, Changsha, Hunan Province, People's Republic of China

T. Qu
Swansea University, Swansea, Wales, UK

ABSTRACT: Soil clogging frequently happens when earth pressure balance (EPB) shield passes through clayey or clay-rich rock ground and tends to be responsible for abrasion of disk cutter, excessive thrust and torque, low advancing speed, and even blocked cutterhead. To avoid the clogging of shield, the auxiliary air pressure balance mode is widely applied in Guangzhou, China. Based a construction case in Guangzhou Metro Line 21, the soil conditioning for the auxiliary air pressure balance mode was studied in the slightly weathered conglomerate formation with an average uniaxial compressive strength of 29.7 MPa. This research shows that the effect of foam was better than that of dispersant. The ideal value of slump test for the muck was in the range of 2-6 cm to avoid muck spewing and hold the pressure in the excavation chamber. This study provides a good reference for EPB shield tunneling in conglomerate formation.

1 INTRODUCTION

Earth Pressure Balance (EPB) shields have been widely adopted for tunnel construction due to their high engineering efficiency and safety. The excavated material is used as the supporting medium in the excavation chamber to prevent the water spewing and balance the ground pressure at excavation face. As stated by many researchers, in clayey ground, the muck easily sticks to the working parts such as cutter head, chamber bulkhead and screw conveyer, resulting in low tunneling rate and high energy consumption. The usual approach to avoid clay clogging is to decrease the consistency by adding water and/or injecting soil conditioners into the clayey muck.

Several researchers have worked on soil conditioning for EPB shield tunneling in clayey ground. Milligan (2000) stated that dispersants, which increase the overall negative surface charge on solid particles, should be applied in clayey soils. Messerklinger et al. (2011) and Zumsteg & Puzrin (2012) determined the shear strength and adhesion of clay paste conditioned with foam and dispersant through a pressurized vane shear apparatus, Hobart mortar mixer and shear plate apparatus, where an enhanced interaction mechanism was proposed by Zumsteg et al. (2013) later on. Peila et al. (2015) determined the conditioning parameters by slump tests, and the dynamic and static lateral adhesion tests were carried out to evaluate the adhesion of the conditioned soil. Heuser et al. (2012) proposed that the electro-osmosis method could be applied to reduce the adhesion of clays on the tunnel boring machines. Using a large vane device, Merritt (2005) measured the undrained shear strength of clayey paste before and after being conditioned, presenting that they were not different significantly with those obtained from the normal shear vane apparatus. Qiao (2009) studied the permeability, plastic flow, shear strength and compression of red clay conditioned with foam and analyzed the micro-mechanism of foam improving clay. Liu et al. (2018) studies changes of the Atterberg limits of clays due to addition of dispersants and their electrochemical mechanisms.

However, soil clogging of EPB shields also may occur in sedimentary rock ground, such as slightly weathered conglomerate formation. The solid rock shows no consistency and cannot therefore undergo a change of consistency due to its mineral grain bonding. However, the solid rock can be transformed into soft muck with the effect of water in combination with mechanical action during excavation (Thewes & Hollmann, 2016). Few studies on the soil conditioning for EPB shield tunneling with the auxiliary air pressure balance mode were conducted in the past.

This paper investigates the soil conditioning for EPB shield tunneling with the auxiliary air pressure balance mode based on a construction case of shield tunnel in Guangzhou Metro Line 21. The type of soil conditioners was determined first with Atterberg limits tests. The slump tests were conducted to achieve the consistency of the ideal muck. Then the parameters of soil conditioning were applied and adjusted in the site according to the state of soil discharged from the screw conveyor. Finally, the tunneling parameters of EPB were analyzed.

2 PROJECT OVERVIEW

The project is a part of the Guangzhou Metro Line 21, and the length of one of shield tunnels is 1023.2 m. The buried depth of the tunnel excavated with EPB shield ranges from 12.1 m to 20.6 m. The tunnel was excavated through firstly argillaceous siltstone and then conglomerate, and above the rock layer five types of soil are present, including plain fill, silt-fine sand, medium-coarse sand, mucky soil and silt clay. Therein, the length of tunnel in slightly weathered conglomerate formation is about 584 m. Table 1 lists the main parameters of the slightly weathered conglomerate.

The gradation curve of the conglomerate muck discharged from the EPB shield, which is being adopted in Guangzhou, China, is presented in Figure 1. The soil has a similar nature to

Table 1. The main parameters of the slightly weathered conglomerate.

Items	Natural compressive strength MPa	Saturated compressive strength MPa	Dried compressive strength MPa	Softening coefficient	Description
Parameters	29.7	22.6	33.2	0.68	Brownish red, gravel-like texture, bedded structure and argillaceous-alcareous cement

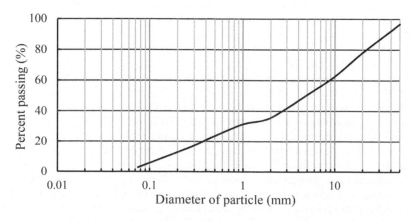

Figure 1. Grain size distribution curve of the investigated conglomerate muck.

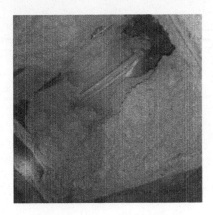

Figure 2. Muck sticking to the disc cutter.

soft ground. The muck can also be changed to a plastic consistency and cause clogging (Figure 2). In order to avoid the clogging of shield in the formation with high sealing performance, there is about one-third or half of chamber to be filled with soil muck for the EPB shield tunneling in Guangzhou, China (Zhu et al., 2017). This excavation mode of the EPB operation is called as auxiliary air pressure balance one. Soil conditioning is also applied to decrease muck stickiness and avoid clogging tendency of the muck.

3 LABORATORY TESTS IN SOIL CONDITIONING

Dispersant and foam agent are commonly adopted as soil conditioners to prevent clogging. The Atterberg limits can be used to evaluate the effects of soil conditioners on the clayey soil (Liu et al., 2018). The type of soil conditioners was determined through testing the changes of Atterberg limits induced by soil conditioners. The slump tests also were conducted in the laboratory to assess the fluidity of muck. The muck was taken from the conveyor belt with the EPB tunneling in the slightly weathered conglomerate formation and then dried as the tested soil.

3.1 Atterberg limits tests

The main reason to induce clogging is the high content of clay and fine particles. The clay and fine content should be fully conditioned to achieve the ideal effect of soil conditioning. Thus, the soil particles of less than 0.5 mm in diameter were selected to determine the Atterberg limits to evaluate the effects of soil conditioners. The foam agent and dispersant were provided by a manufacturer. The main components of foam agent and dispersant are surfactant and sodium polyacrylate, respectively.

The fall cone approach was adopted herein to determine the Atterberg limits of the soil. The soil was mixed with a certain amount of distilled water and then stored in a sealed container for 24 h. The Atterberg limits were determined after the wet soil specimens were mixed respectively with foam and dispersant. Figure 3 shows the results of Atterberg limits for different conditioning cases. The "PL" and "LL" are the abbreviations of plastic limit and liquid limit, respectively. The "D" and "F" represent the dispersant and foam, respectively. The content of foam or dispersant is the volume ratio of foam agent or dispersant to that of dry soil. The Atterberg limits kept almost constant with an increase in the dispersant content. The liquid limit decreased while the plastic limit did not change with an increase in the foam content. It indicated the conditioning effect of the foam was better than that of the dispersant. Thus the foam agent was applied at the construction site to condition the soil for the EPB shield tunneling.

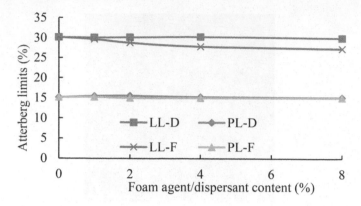

Figure 3.   Changes of Atterberg limits of the soil.

Table 2.   Results of laboratory slump tests

w	FIR	Slump value	Picture	w	FIR	Slump value	Picture
%	%	cm		%	%	cm	
10	0	0		15	20	18.9	
10	10	1.7		15	40	22.6	
10	20	2.8		20	0	19.7	
10	40	7.1		20	10	21.2	
15	0	11.8		20	20	23.0	
15	10	15.8		20	40	26.5	

3.2 *Slump tests*

The slump tests were conducted in the laboratory to evaluate the fluidity of muck and obtain the parameters of soil conditioning. The water contents (w) for testing were selected as 10%, 15% and 20%, respectively. The foam injection ratios (FIRs) were 0, 10%, 20% and 40%, respectively. Table 2 lists the results of the slump tests. The value of slump test was affected by water content and FIR. An increase in either water content or FIR led to a higher slump value. When the water content was 10%, the soil was stiff and the foam easily dissipated. The value of slump test was relatively low and increased slowly (0~7.1cm) with an increase in FIR (0~40%). The consistency of muck increased obviously after the water content increased to 15%. The slump value ranged from 15.8 cm to 22.6 cm with an increase in the FIR. When the water content reached 20%, the slum value was almost higher than 20 cm and the muck is too soft to suitable for EPB shield tunneling.

4 SOIL CONDITIONING IN FIELD AND ANALYSIS OF TUNNELING PARAMETERS

4.1 *Soil conditioning in field*

The EPB shield tunneled with auxiliary air pressure mode in the slightly weathered conglomerate formation. At the start of shield tunneling section, the foam injection ratio was set as 10% and certain amount of water was injected to ensure the muck being discharged from the screw conveyor with a slump value of 15 cm. However, the soil spewed at the outlet of the screw conveyor, i.e., the conditioned soil was squeezed out directly by the chamber pressure, which ranged from 2.2 to 2.6 bar. It was nearly impossible with the excessive fluidity of conditioned soil to control the soil pressure in chamber, and then and the soil conditioning was not suitable for this tunneling mode. The injection volumes of the foam and water were reduced to

| (a) Slump value of 5.4 cm | (b) Slump value of 2.2cm | (c) Slump value of 1.6cm |

Figure 4.   Results of in-situ slump tests.

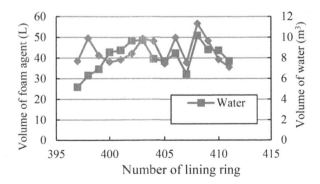

Figure 5.   Volume of injected foam agent and water.

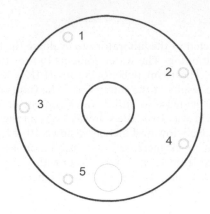

Figure 6.   Distribution of the pressure sensors in the chamber.

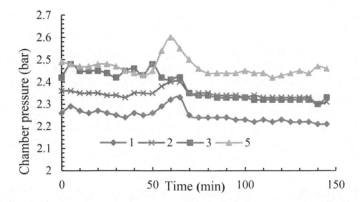

Figure 7.   Changes of chamber pressure during excavation of No. 411 ring.

degenerate the fluidity of conditioned soil for the EPB tunneling later on. The slump tests were conducted in field to explore the suitable consistency of the conditioned soil, and their results are presented in Figure 4. The soil with a slump test of 2–6 cm was smoothly discharged from the screw conveyor, and it will be indicated in the following section that the chamber pressure kept steady. However, the water content of the soil was relatively low (about 11~14 %) and the foam with an injection ratio of 8% was completely dissipated. Therefore, the foam cannot improve the fluidity of the soil at the low water content. However, the foam agent can effectively reduce the friction between the cutting tools and the stratum before the excavation face, and so the foam solution with a concentration of 0.5% was injected into the front of the cutter head to reduce cutter wear.

4.2   Analysis of tunneling parameters

Figure 5 shows the volume of the injected foam agent and water for each tunneling cycle equal to 1.5 m in length. Due to the groundwater in the slightly weathered conglomerate formation, the injection rates of water and foam agent needed to be adjusted to achieve the suitable state of the muck being discharged from the shield screw conveyor and the control of chamber pressure. As a result, the injected volumes of foam agent and foam were not steady in Figure 5. The injected volume of foam solution per lining ring was about 8.6 m^3 in average, and the injection ratio of foam solution (the volume ratio of foam solution to excavated soil for each lining ring) was 18.4%.

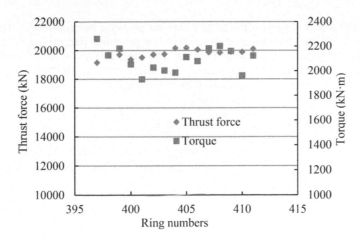

Figure 8. Thrust force and torque of the shield.

Figure 6 shows the distribution of the pressure sensors in the chamber. Five pressure sensors were installed on the chamber wall to monitor the pressure. The sensor in the bottom right corner (No. 4) was damaged during the excavation and no data was recorded. Figure 7 shows the changes of chamber pressure against time during the excavation of the No. 411 line ring. The pressure in the bottom was higher than that in the top under the deadweight of muck in the chamber. The chamber pressure, ranging from 2.2 to 2.6 bar, was almost steady. Thus, the consistency of the muck in shield chamber was suitable, and the field indicated that the muck was discharged smoothly and no muck spewing happened.

Figure 8 shows the total thrust and torque of the shield during the excavation of No. 397–411 lining rings. The thrust force was in the range of 18000~20000kN while the torque was in the range of 1900~2200kN•m. The thrust and torque of the shield were generally stable and their fluctuations were small. The tendency of clogging was not observed in field.

Therefore, when the EPB shield tunneled with the auxiliary air pressure mode in the slightly weathered conglomerate formation, the slump value of the muck should be controlled within the range of 2~6cm. The muck can be smoothly discharged with the screw conveyor. Through the analysis of the tunneling parameters of the shield, it can be found that the chamber pressure did not vary significantly. The fluctuations of the total thrust and torque were also small. However, the parameters of the soil conditioning were greatly affected by the water content in the stratum. It was necessary to timely adjust the parameters of soil conditioning according to the state of the muck.

5   CONCLUSIONS AND PROSPECTS

The Atterberg limits of the fine soil broken from the conglomerate kept almost constant with an increase in the dispersant content. The liquid limit decreased while the plastic limit did not change with an increase in the foam content. It is indicated the conditioning effect of the foam was better than that of dispersant to avoid soil conditioning during the EPB shield tunneling in the conglomerate.

The ideal value of slump test for the muck was 2~6 cm to avoid muck spewing when the EPB shield tunnels in the slightly conglomerate formation with the auxiliary air pressure balance mode. The suitable muck was achieved with a foam solution (a concentration of 0.5%) injection ratio of 18.4% in mass. The chamber pressure, thrust force and torque of the shield were basically steady and their fluctuations were small. With the conditioning approach, the shield tunnels smoothly.

When the EPB shield tunnels with the auxiliary air pressure balance mode, the ideal state of muck is affected by the stratum condition, chamber pressure, parameters of the screw conveyor and even the volume of muck in chamber. The ideal softness state of the muck is different to that proposed by previous studies, such as a slump value of 10–20 cm and a consistency index of 0.4–0.75. It needs to be studied further in the future.

ACKNOWLEDGEMENT

The financial support from the National Natural Science Foundation of China (No. 51778637) and the National Key R&D Program of China (No. 2017YFB1201204) are acknowledged and appreciated.

REFERENCES

Heuser, M., Spagnoli, G., Leroy, P., Klitzsch, N. & Stanjek, H. 2012. Electro-osmotic flow in clays and its potential for reducing clogging in mechanical tunnel driving. Bulletin of Engineering Geology & the Environment 71: 721–733.

Liu P, Wang S, Ge L, Thewes M, Yang J & Xia Y. 2018. Changes of Atterberg limits and electrochemical behaviors of clays with dispersants as conditioning agents for EPB shield tunnelling. Tunnelling & Underground Space Technology 73:244–251.

Liu P, Wang S, Yang J & Hu Q. 2018. Effect of soil conditioner on Atterberg limits of clays and its mechanism. Journal of Harbin Institute of Technology 50(6): 91–96 (in Chinese).

Merritt, A.S. 2005. Conditioning of Clay Soils for Tunnelling Machine Screw Conveyors. University of Cambridge.UK

Messerklinger, S., Zumsteg, R. & Puzrin, A.M. 2011. A new pressurized vane shear apparatus. Geotechnical Testing Journal 34 (2): 112–121.

Milligan, G., 2000. Lubrication and Soil Conditioning in Tunnelling, Pipe Jacking and Microtunnelling: A State-of-the-Art Review. Geotechnical Consulting Group,London, UK

Peila, D., Picchio, A. & Martinelli, D., 2015. Laboratory tests on soil conditioning of clayey soil. Acta G eotechnica 11 (5): 1–14.

Qiao, G., 2009. Development of New Foam Agent for EPB Shield Machine and Foam-soil Modification (Ph.D. Dissertation). China University of Mining & Technology, China (in Chinese).

Thewes M. & Hollmann F. 2016. Assessment of clay soils and clay-rich rock for clogging of TBMs[J]. Tunnelling & Underground Space Technology 57:122–128.

Zhu W, Zhong C, Huang W, He T, Zhang B & Zhu S. 2017. Key techniques for the auxiliary air pressure balance mode for shield tunneling. Modern Tunneling Technique 54(1):1–8 (in Chinese).

Zumsteg, R., Plötze, M. & Puzrin, A.M. 2013. Reduction of the clogging potential of clays: new chemical applications and novel quantification approaches. Géotechnique 63 (4): 276–286.

Zumsteg, R. & Puzrin, A.M. 2012. Stickiness and adhesion of conditioned clay pastes. Tunnelling & Underground Space Technology 31 (5): 86–96.

Tunnels and Underground Cities: Engineering and Innovation meet Archaeology,
Architecture and Art, Volume 5: Innovation in underground engineering,
materials and equipment - Part 1 – Peila, Viggiani & Celestino (Eds)
© 2020 Taylor & Francis Group, London, ISBN 978-0-367-46870-5

Catania metro tunnel: Advancement and wear predictive models suitability in EPB excavation through variable mixed-face conditions

V. Lo Faro, E. Nuzzo, E. Chimenti & G. Brino
SWS Engineering, Trento, Italy.

A. Alvarez-Merayo
Cooperativa muratori e cementisti, Ravenna, Italy.

ABSTRACT: "Nesima-Misterbianco" lot I, with its 1.76km-length tunnel excavated with a Ø10.6m EPB is part of expansion project of Catania metro-line. The tunnel is the first application of an EPB through the complex geological context of Catania, characterized by the presence of numerous volcanic formations. The EPB advancement faced and led to a great deal of critical aspects during the excavation: presence of high variability of formations at tunnel faces, abrasiveness, high ground permeability and a high level of cutter wear. An unceasing manual set-up of excavation parameters and ring-by-ring data processing is needed to improve the understanding and knowhow on EPB possible performance in this context. This paper proposes a comparison between principal prediction models in literature and EPB parameters recorded during tunnel excavation, the main focus is understand how the geological context and mixed-face conditions have influenced EPB performance during the excavation.

1 INTRODUCTION

Nowadays mechanized excavation to realize underground infrastructures is continuously increasing. Greater knowledge and experience in mechanized tunneling are leading to the design of more ambitious and complex projects with ever-new challenge in term of length and excavation diameter.

In these new challenging projects, it became even more important an accurate estimation of work duration, a correct organization of the ordinary maintenance of the TBM and a reliable prediction of cutter disks and tools consumption, caused by wear.

This paper will analyze the return of experience of the first mechanized tunneling application in Catania; the tunnel has been realized through the formation of lava rock with deal of several critical geomorphological aspects, like mixed-face conditions. The back-analysis of TBM data, recorded during tunnel realization, they have been compared with most widespread performance predictive models known in literature, in order to evaluate the reliability of theoretical estimation on tunneling projects in difficult geomorphological conditions like Catania ones.

2 CATANIA METRO TUNNEL PROJECT

Catania metro extension project includes the realization of a17.4km-length underground infrastructure that will connect the urban district of Misterbianco with Catania city center and "Fontanarossa" airport. This paper will analyze TBM advancement during the excavation of Nesima-Misterbianco-lot I tunnel. It has an overall length of about 1.76km (ch 5 + 283.7 – ch 7 + 031.78) and it has been realized by TBM-EPBs with an excavation diameter of 10.6m

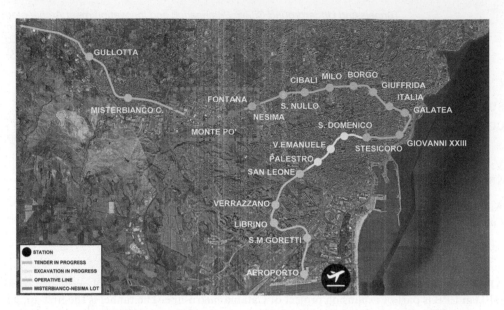

Figure 1. Catania metro alignment.

Table 1. Herrenknecht S454.1 Main Features.

Characteristic	Description
Type	TBM-EPB shield
Diameter of excavation	10.60m
Overall length	≈ 95m (shield + 6 back up gantries)
Maximum advancing thrust	110370 kN
Cutterhead weight	≈ 200 tons
Cutter disks	n° 55 single disk + n° 4 double disk
Cutter knife	n° 182
Bucket	n° 42

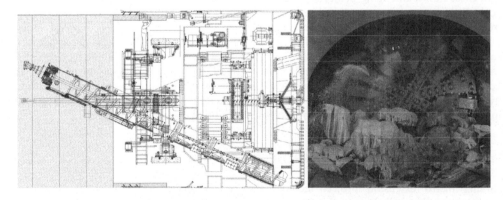

Figure 2. S454.1 Herrenknecht section & Monte Pò Station breakthrough.

3 GEOLOGICAL FEATURES

The presence of active Etna volcano is the most relevant geological aspect in Sicily and among all Sicilian cities, Catania is probably the most affected by its influence, due to its close proximity to the volcano.

Catania geomorphology is indeed the result of the overlapping of different lava rock structures, formed in different eras and often coming from different eruptive axes. The succession of several lava flows in ages (prehistoric and historic), has determined the formation of a complex, chaotic and highly heterogeneous geology with alternation of volcanic formations to clay, alluvial deposits and sandy silts. Along tunnel alignment, the excavation in lava rock formations has constituted about the 47% of the entire length (830m), while the other 53% (930m) has characterized by the presence of volcanic sands (with a variable degree of compaction), breccia and clay and silty clay deposits.

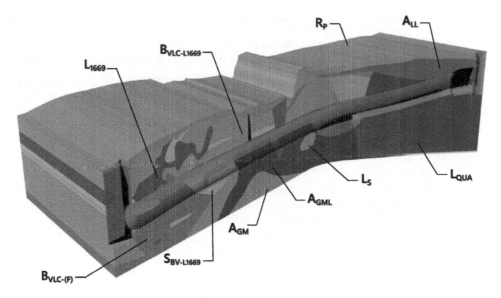

Figure 3.   Geological profile of the tunnel project.

Table 2.   Geological formations crossed by the excavation.

Formation	ID	Description	γ ton/mc	φ °	c' kPa	UCS Mpa
R$_P$ - Superficial Deposits		Superficial deposits sandy or silty-sand poorly cemented and heterogeneous	1.85	32.5	5	--
A$_{LL}$ - Alluvial Deposit		Loose soils silty-clay, poorly cemented with incorporated boulders of lava rock.	1.95	21.5	5	--
L$_{QUA}$ - Lava Rock*		Basaltic lava rock highly tough with a variable persistency of fracturing.	2.50	57	800	140
L$_{1669}$ - Lava Rock*		Blocky and disjointed lava rock with possible presence of cavity or voids	2.40	57	800	140
S$_{BV-L1669}$ - Volcanic Sand*		Volcanic Sands with a highly variable cementation and poor granulometryc assortment.	1.95	36	10	--
B$_{VLC-L1669}$ - breccia*		Volcanic Breccia with a highly variable cementation and poor granulometryc assortment.	1.95	41.5	10	--
A$_{GM}$-A$_{GML}$ - Silty clay/clay		Marine-origins deposits, normal-consolidated with good mechanical characteristics.	2.05	22.5	17.5	--

* the abrasiveness of lava rock has been considered taking into account the follow values:
   S$_J$ = 5.15 1/10mm, S$_{20}$ = 53.2% , CAI = 3 1/10mm and 20% of content in quartz (%qz)

4 TBM PERFORMANCE PREDICTIVE MODELS: LITERATURE REVIEWS

Since the first phases of a tunneling project, TBM performance prediction is fundamental to define time and cost of the excavation process. Many aspects influence the industrial production process of a mechanized tunnel e.g. ring installation time, maximum penetration, maximum TBM thrust and torque, cutter disks wear and geology plays always a fundamental role in estimation of all these parameters.

Theoretical predictive models can be used to have a preliminary estimation of TBM performances and cutter consumption using analytical or empirical equations. Based on back analysis of several excavation data, empirical models are constantly updated to spread the field of possible application and to adapt to technological improvement of excavation machines.

4.1 Cutter wear predictive models

Cutter wear predictive models can mainly be classified in:

Analytical models, based on theoretical assumptions;
Empirical models, based on the back-analysis of the data-set of several realized excavation.

Frenzel (2010) extended the field of the application of the CSM model (Colorado School of Mines) developed by Rostami and Ozdemir (2005) and updated by Rostami himself (2008).

The model returns the net volume of rock excavated per cutter in mc/cutter:

$$\Delta V_C = \Delta L_C \cdot \left(\frac{\pi \cdot \Phi_{TBM}^2}{4} \right) \tag{1}$$

where $\varnothing_{TBM}$ is TBM excavation diameter (10.6m) and the average tunnel length covered by TBM, until the consumption of a cutter occurs [m]:

$$\Delta L_c = \frac{U_C \cdot p}{n_{cuttters}} \tag{2}$$

where U_C is the average number of rotation that cutter can sustain in its life (≈ 25700), p is cutter penetration (_8mm/rev -average value of TBM advancement in lava_) and $n_{cutters}$ is the total number of tools on the cutter-head (_63_).

Maidl et al. (2001), states that tools wear is connected to the abrasiveness of the rock (_CAI, in this application is equal to 3_) and the uniaxial compressive strength (_140 MPa_). From _Maidl et al._ diagram, it is possible to observe the following series of equations for the Specific Disk Cutter Wear Rate (SDCWR), expressed in mc/cutter:

$$\text{For CAI} : 2 \rightarrow SDCWR = 10^{\left(2 \cdot 10^{-5} \cdot \sigma_c^2 - 8,8 \cdot 10^{-3} \cdot \sigma_c + 3,9944\right)}$$

$$\text{For CAI} : 3 \rightarrow SDCWR = 10^{\left(7 \cdot 10^{-6} \cdot \sigma_c^2 - 5,6 \cdot 10^{-3} \cdot \sigma_c + 3,1889\right)}$$

$$\text{For CAI} : 4 \rightarrow SDCWR = 10^{\left(10^{-5} \cdot \sigma_c^2 - 3,7 \cdot 10^{-3} \cdot \sigma_c + 2,8387\right)} \tag{3}$$

$$\text{For CAI} : 5 \rightarrow SDCWR = 10^{\left(10^{-5} \cdot \sigma_c^2 - 3,7 \cdot 10^{-3} \cdot \sigma_c + 2,4669\right)}$$

$$\text{For CAI} : 6 \rightarrow SDCWR = 10^{\left(5 \cdot 10^{-6} \cdot \sigma_c^2 - 3,7 \cdot 10^{-3} \cdot \sigma_c + 2,2371\right)}$$

NTNU model is a prediction model developed in Trondheim Norwegian University of Science and Technology. The version used for this paper comes from Bruland (1998). The average life of a cutter ring expressed in mc/cutter is Cutter Ring Life CRL:

$$CRL = \frac{\Phi_{TBM}^2 \cdot \pi}{4} \cdot i_0 \cdot \frac{v_{rot} \cdot 60}{1000} \cdot H_H \qquad (4)$$

where v_{rot} is the disks cutter maximum rotation speed (*5 rpm*), H_H is the average cutter ring life (*equal to 2.12 and function of: excavation diameter, CLI, n° cutter of the cutter-head, quartz content into the rock and cutter disk diameter*) and i_0 is the basic penetration rate (*8mm/rev-average value of TBM advancement in lava*).

4.2 TBM performances predictive models

Also in these cases, it is possible to classify the model in <u>analytical models</u> and <u>empirical models</u> *(the value of 1.5rpm for cutter-head rotation, it has been used into the models to obtain an estimation of daily advancement)*.

Innaurato et al. (1991) integrated the study performed by Casinelli et al. (1992), developing a predictive model based on the Rock Structure Rating (*RSR, in this application equal to 62.45*) and uniaxial compressive strength (*UCS 140 MPa*):

$$p = 40,41 \cdot \sigma_C^{-0.437} - 0,047 \cdot RSR + 3,15 \qquad (5)$$

Farrokh et al. (2012) based his model on 17 different excavation projects, realized with a great variability of geology condition and TBM parameters. The author developed a predictive model of the penetration, expressed:

$$p = e\big(0.41 + 0.404 \cdot \Phi_{TBM} - 0.27 \cdot \Phi_{TBM}^2 + 0.0691 \cdot RT_C - 0.00431 \cdot \sigma_c +$$

$$0.0902 \cdot RDQ_C + 8.93 \cdot 10^{-4} \cdot F_N\big) \qquad (6)$$

where RQD_C (*equal to 2*) and RT_C (*equal to 2*) are from numerical code of the model and F_N is the thrust force for a single disk (*equal to 208 kN and mainly function of width of the bit and cutter disk diameter*).

Vergara et al. (2017) proposed a predictive model for the excavation realized in mixed-face ground conditions.

The author introduced a new penetration index MFPI related to a weighted rock mass rating RMR_m, function of the percentage of rock or soil on the excavation faces. For the predictive model is been used a value 15 MN for advancing thrust on the face of the excavation (*average value obtain from the excavation data*)

$$MFPI = 2.12 \cdot e^{(0.02\,RMR_m)} \quad \left[MN \cdot rev/_{mm}\right] \qquad (7)$$

where

$$RMR_m = \frac{\sum_{n=1}^{n=5} RMR_n \cdot P_n\%}{100} \qquad (8)$$

where RMR_n is the RMR of n-rock and $P_n\%$ is a percentage of n-rock on the excavation face (for this application has been used a percentage of 10%, 20% and 30%).

Maidl et al. (2011) proposed a prediction model for the daily rate of advancement of the EPB shielded machines. The model considers the complex system of factors that occurs during excavation cycle, e.g. ring installation time, stoppage and real advancement time etc.

The daily advancement P_D, can be calculated:

$$P_D = N\, L_{SEG} \tag{9}$$

where L_{SEG} is the longitudinal length of the segmental lining (*1.5m*) and $N = T_D / T_{CYCLE}$ is the ratio between daily worked hours (*24h*) and total cycle time (*≈1.8h*).

5 THEORETICAL AND EXCAVATION DATA COMPARISON

The 1.76km-lenght tunnel "Nesima-Misterbianco lot I" is the first application of a mechanized excavation in Catania geology. The excavation along the entire tunnel alignment has overcame several criticalities, e.g. high water inflow, presence of mix-ground excavation face, presence of exceptionally tough lava rock, etc.

In the following chapters is shown the comparison between excavation data acquired (postprocessed for the TBM advancement parameters analysis) and the predictive models estimation, in terms of cutter wear and advancement rate.

The main goal of this paper is analyse all cases of bad data correlations and understanding the principal causes associated with. To achieve this focus it has been combined the geomorphological features of the excavation with the leading mechanisms that can occur on the excavation face. As is shown in the follow, in all those cases where the cutter replacements are due not only for abrasive effects of the ground, but also to the presence of dynamic interactions between disks and excavation face, the theoretical models estimations does not find a good matching with the excavation data. For this reason, for the geomorphological context with mixed-face conditions, it is proposed an empirical correction of the predictive models in order to take into account the presence of dynamics effects.

5.1 *Excavation data comparison with predictive models estimations*

The comparison between wear TBM data-set with the theoretical estimations (predictive models), it shows a good matching along the stretches of tunnel excavated through lava rock formation: theoretical range for wearing cutter tools is equal to 280÷450 mc/cutter, very similar to the average values of CLI recorded during the excavation advancement in lava, 300÷400 mc/cutter. Differently, along the stretches of excavation characterized by the presence of breccia (usually with a massive presence of boulders) and the mixed-face condition, the recorded values of wearing during TBM advancing do not show this correspondence. The average values of CLI during the excavation advancement in breccia (rings 350–490) is equal to 330 mc/cutter, similar to the lower limit of the theoretical range obtained with the predictive models. Along the excavation in mixed-face condition, the average recorded value of CLI is equal to 200÷250 mc/cutter even smaller than the theoretical lower limit equal to 280 mc/cutter (Figure 4).

The comparison between real advancement rate (AR) with theoretical values, also show remarkable differences along the stretches of excavation realized in breccia formation and in mixface conditions. The average recorded values of 9 m/day are lower than the minimum theoretical advancement value obtained from the predictive models, equal to 10 m/day (Figure 5).

It is also necessary to highlight, that the average AR in the first stretch of the tunnel in lava is lower than the theoretical production values obtained from the predictive models, about 5.5 m/day compared to 10 m/day. These differences have principally due to the massive presence of water inflow on the excavation face encountered during the TBM advancement in this stretch. The issues related to this phenomena, have negatively affected the excavation because of the normal phases of advancement have been alternating with those connected with the water disposal and management of the water flow.

5.2 *Empirical correction of the predictive model estimations*

The comparison shown in the previous chapter has highlighted that theoretical values of rate of advancement and cutter disk life do not constitute a reliable estimation for the excavations

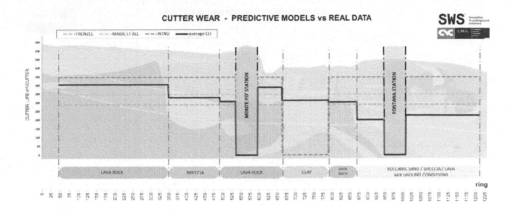

Figure 4.   Cutter wear – Theoretical CLI vs real wear comparison.

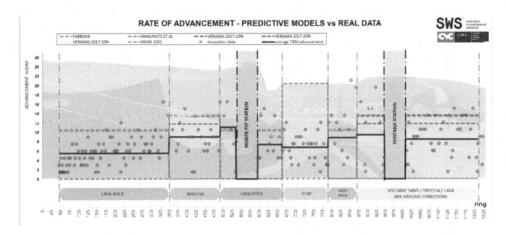

Figure 5.   Daily advancement rate – Theoretical vs real data excavation comparison.

in complex geomorphological condition, e.g. breccia (with boulders) or in mixed-face conditions. It has been analyzed the possible excavation aspects in common for these two geomorphological conditions, in order to understanding and study the origin of these remarkable differences from theoretical to real values. One common aspect analyzed, that can have negatively influenced the excavation advancement is certainly the presence of heterogeneity on the excavation face (boulders in breccia and lava with sand and breccia in mixed-face condition), due to the presence of material with different mechanic characteristics. The presence of this heterogeneity on the excavation face can lead the arise of dynamics effects on the cutter disks, that are very different from the wear mechanism induced by abrasion, on which are mainly based the predictive models known in literature. So, in these geomorphological contexts of excavation, the need of cutter tool replacements and TBM stoppages not depending only on the abrasiveness of the excavated ground, but also to the presence or not of dynamics effects on the cutter tools.

In order to study the possible influence of dynamics effects on excavation advancement and on cutter disks replacements, it has been plotted the graph of occurrence of cutter disk replacement due to breakage of the tool (percentage measured on the total cutter replacement %BR, Figure 6).

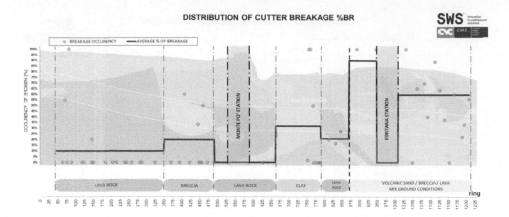

Figure 6. Cutter replacement due to breakage of the bit (percentage).

The stretches of tunnel excavated along breccia and mix-face conditions, like it shown on the graph, have a high percentage of cutter replacement due to the phenomena of cutter tool breakage. In the mixed-face excavation, the breakage of the cutter disk is even the predominant causes for tools replacement. It is reasonable said that for these cases, the cutter life estimation with theoretical predictive models is overcstimated, because it is not been taking into account the massive presence of dynamic effects along the excavation face.

In order to obtain a reliable estimation of wearing, for the excavation stretches with average values of cutter breakage higher than 20%, it has been proposed the introduction an empirical corrective factor to apply to the theoretical estimation. This corrective factor has been introduced in order to take into account the presence of the dynamic effects before descripted and it is related to the percentage of cutter replacement induced by cutter bit breakage (%BR.) The cutter life index modified, is obtain from follow equation:

$$CLI_m = CLI_{MODEL} \cdot k_1 \qquad (10)$$

Where CLI_{MODEL} is the cutter life index calculated with the theoretical models and k_1 is the empirical corrective factor, equal to:

$$k_1 = 1 - 0.5 \cdot \%BR \qquad (11)$$

%BR is the breakage occurrence expressed in percentage.

A similar approach it has been proposed also for the theoretical estimation of daily advancement rate. Also in this case, the results obtained with the predictive models tend to overestimated the TBM advancement, because it is not taking into account the presence of the dynamic effects on the excavation face. It has been introduced a second corrective factor to apply to the theoretical advancement of the predictive models, in order to achieve a reliable estimation of the TBM advancement along these geomorphological conditions. The modified AR is:

$$AR_m = AR_{MODEL} \cdot k_2 \qquad (12)$$

Where AR_{MODEL} is the rate of advancement calculated with the theoretical models and k_2 is the empirical corrective factor, that is equal to:

$$k_2 = 1 - 0.5 \cdot \left(\frac{V_{excavation}}{CLI_m}\right) \qquad (13)$$

$V_{excavation}$ is the volume excavated for single ring advancement in mc.

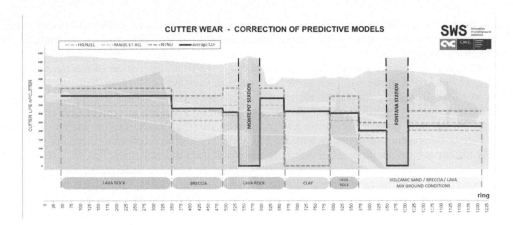

Figure 7. Cutter wear – Theoretical modified CLI vs real wear comparison.

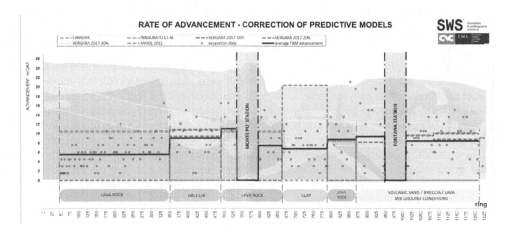

Figure 8. Daily rate of advancement – Theoretical modified AR vs TBM advancement comparison.

In the follow graphs are shown the comparison of theoretical estimation of the modified models with excavation data of wearing and TBM daily advancement.

The introduction of corrective factors, for the estimation of TBM advancement and cutter disks wearing, it has been involved a better compatibility between theoretical predictive model results and real excavation data.

Taking into account the consequences due to the presence of dynamic effects on the excavation face, not negligible effects in these geomorphological contexts (presence of grounds with different mechanic characteristics), it allow to achieve a better advancement & wear data correlations with theoretical ranges obtained from the predictive models.

6 CONCLUSION

The actual scenario of world tunneling is moving to ever more challenging projects, with realization of underground infrastructures in geological contexts where it was been impossible, until now, designing any possible feasibility solution. In this prospective, it is necessary find innovative solutions in order to transfer the nowadays-acquired knowledge, to these new complex contexts of mechanized tunneling. Therefore, the truly purpose is find planning and constructive tools able to anticipate and analyze specific features or eventual critical issues connected to the excavation, with estimations as reliable as possible and contextualized to ever new excavation conditions.

"Nesima-Misterbianco Lot I" project is for example the first application of mechanized tunneling with an EPB in the complex geological volcanic-context of Catania. Starting from the data acquired from this excavation experience, it has been possible evaluate the reliability of the more common theoretical models for the prediction of TBM advancement rate and cutter disks wearing. The comparisons have showed that in cases of mixed-face condition (excavation face with presence of grounds with different mechanic characteristics) there are remarkable differences between theoretical estimation and real data of excavation.

The same excavation stretches that have shown these remarkable differences, also shown an high percentage of cutter replacements due to the bit breakage (%BR). These tools breakage are connected to the elevated presence of dynamics effects on the excavation face due to the mixed-face excavation condition. The mechanism induced by the presence of dynamics effects it is not considered into the predictive models estimations and therefore the negative influence of these effects on CLI and TBM advancement rate is not taking into account: in cases of mixed-face excavation, thus the theoretical estimation tend to overestimate the CLI and AR of the excavation.

For these geomorphological contexts, it has been proposed therefore the introduction of empirical corrective factors, based on the presence of these dynamics effects and related on the cutter disks breakage occurrence (%BR equal to 20÷40% in breccia with boulders and until 100% in excavation faces characterized by the presence of breccia and lava rock or sand and lava rock). Applying these corrective factors to the theoretical estimations of CLI and AR, it has been achieved a better compatibility between theoretical predictive model results and real excavation data also for that cases of excavation in mixed-face conditions.

However it is important to highlight that the proposed approach and the introduction of these empirical corrective factors, it is based on a limited number of data available and related to a single case of excavation. For this reason, in order to obtain ever more reliable estimations of excavation advancements and cutters disk replacements in these difficult geomorphological contexts, it would be interesting and also necessary a better calibration of the corrective factors also from a statistical point of view, obtainable only with the processing of new excavation-data in similar contexts.

REFERENCES

Brino, G. & Peila, D. et all. 2015. Prediction of performance and cutter wear in rock TBM: application to Koralm tunnel project. *Geoingegneria Ambientale e Mineraria*. vol.52 (2): 41–54. Turin

Bruland, A. 2014. The NTNU Prediction Model for TBM Performance. *NFF Norwegian Tunnelling Society & Norwegian TBM Tunnelling*. vol. 23: 179–189

Cassinelli, F. & Cina, S. et all. 1982. Power consumption and metal wear in tunnel-boring machines: analysis of tunnel-boring operation in hard rock. *Tunnelling '82*. pp.73–81. London

Farrokh, E. & Rostami, J. et all. 2012. Study of various models for estimation of penetration rate of hard rock TBMs. *Tunnelling and Underground Space Technology*. 30:110–123.

Innaurato, N. & Mancini R. et all. 1991. Forecasting and effective TBM performances in a rapid excavation of a tunnel in Italy. *Proceedings of the Seventh International Congress ISRM*. pp.1009–1014. Aanchen

Maidl, U. & Comulada, M. 2011. Prediction of EPB Shield Performance in soils. *Proceedings of Rapid excavation and tunneling conference*. pp 1083.

Maidl, U. & Schmid, L. et all. 2001. Tunnelbohrmaschinen im Hartgestein. *Verlag Ernst and Sohn*. ISBN-10:3433014531 pp.350. Berlin.

Oggeri, C. & Oreste, P. 2012. The wear of tunnel boring machine excavation tools in rock. *American Journal of Applied Sciences*. 9(10):1606–1617.

Rostami, J. & Ozdemir, L. 1993. A new model for performance prediction of hard rock TBM. *Rapid Excavation and Tunneling Conference*. pp 793–809.

Vergara, M.I. & Saroglou, C. 2017. Prediction of TBM performance in mixed-face ground conditions. *Tunneling and underground space technology*.

Tunnels and Underground Cities: Engineering and Innovation meet Archaeology,
Architecture and Art, Volume 5: Innovation in underground engineering,
materials and equipment - Part 1 – Peila, Viggiani & Celestino (Eds)
© 2020 Taylor & Francis Group, London, ISBN 978-0-367-46870-5

Explosive spalling in reinforced concrete tunnels exposed to fire: Experimental assessment and numerical modelling

F. Lo Monte & R. Felicetti
Politecnico di Milano, Milan, Italy

A. Meda
Università degli Studi di Roma Tor Vergata, Rome, Italy

A. Bortolussi
Bekaert Maccaferri Underground Solutions, Aalst - Erembodegem, Belgium

ABSTRACT: Fire is a severe scenario for tunnels, since high temperature induces both the decay of mechanical properties and the rise of indirect actions. Furthermore, explosive spalling can take place in concrete (namely, the violent expulsion of shards from the exposed face), with a relevant impact in fire resistance and in the retrofitting phase after the fire. Hence, preliminary screening of concretes before construction plays a key role to reduce spalling risk with the optimal compromise between costs and efficacy. In this context, three test setups were adopted for the assessment of spalling sensitivity in concrete: hot spot, biaxial loading and full-scale tests. The experimental campaign is presented with the aim of explaining the different results. Afterwards, on the basis of the experimental evidence, it is briefly described how to implement a finite element model of a tunnel lining with precast R/C segments, in order to assess the fire resistance taking spalling phenomenon into account.

1 INTRODUCTION

Fire is a very severe scenario for strategic infrastructures such as tunnels, due to some particular conditions. Firstly, the structural geometry and the fire load can lead to very high maximum temperatures, even higher than 1000°C for the most severe scenarios (FIT, 2005). Furthermore, the inherent structural redundancy of this kind of structures leads to high indirect actions, which can trigger local collapse. This makes of primary importance the proper evaluation of fire performance.

Thanks to inherent material properties such as incombustibility and thermal insulation capability, concrete members generally show good fire performance, provided that little or no spalling takes place. Spalling phenomenon, in fact, is the violent expulsion of concrete chunks from the hot face, which reduces the thermal protection of the reinforcing bars to the flames, hence increasing the overall loss of bearing capacity.

The two main actors behind spalling are (see Figure 1): (a) stress caused by applied forces, indirect action and thermal gradients, and (b) pore vapour pressure because of water vaporization. Fracture tests in concrete specimens under sustained pore pressure during heating showed that the combination of pressure and stress is necessary for triggering spalling (Felicetti et al., 2017).

When concrete is heated, thermal gradients induce thermal stresses characterized by compression next to the exposed face and tension in the colder core. Compressive stress and the subsequent cracking may locally reduce the mechanical stability.

On the other hand, pore pressure rises due to water evaporation in the pores and it is responsible for the violent propagation of cracks. Due to pressure gradients, moisture is pushed toward the inner core, where vapour can condensate due to the lower temperature so

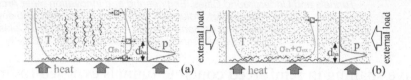

Figure 1.   Cracking pattern in a slab due to (a) thermal stress only and (b) both thermal and load-induced stress. σ_{th} = thermal stress, σ_{ex} = load-induced stress, p = pore pressure and d_{ha}= hygrally-active region.

locally increasing the water content. In dense concretes, such as High-Performance Concretes (HPC), pores saturation can be attained (the so-called moisture clog). This makes HPC more prone to explosive spalling because of the higher values of pore pressure, even higher than 4 MPa (Kalifa et al., 2000 and 2001, Lo Monte et al., 2019).

The most effective way to reduce the propensity to spalling of R/C structures is to add poly-propylene fibre (Kalifa et al., 2000; Khoury, 2008), also in combination with other kind of fibres. Cement paste microcracking accompanying polypropylene fibre melting (Khoury, 2008; Rossino et al., 2013; Pistol et al., 2014) allows to reduce pore pressure and, possibly, thermal stress (Huismann et al., 2012 and Lo Monte and Felicetti, 2017).

So far, the most used way for the assessment of spalling sensitivity is experimental testing, which however is influenced by many aspects such as concrete specimen geometry, loading and heating conditions (all factors governing thermal gradients and stress, concrete permeability evolution and pore pressure development). Hence, the test setup must be carefully designed in order to obtain results representative of the case at issue.

As regards the heating curve, the most known fire curves are related to different fire loads and ventilation factors. The fire curve should be chosen according to the structural context (buildings, industrial plants, tunnels).

Specimen geometry and loading conditions are strongly interconnected, due to the combination of thermal and applied stresses. According to the dimensions of the setup, tests can be subdivided in small-, medium- and full-scale tests (Krzemień and Hager, 2015). In all cases, an external load can be applied during heating.

Examples of small-scale investigations can be found in in the literature, such as Kalifa (2000), Mindeguia et al. (2010), Toropovs et al. (2015), Lo Monte and Gambarova (2015), Felicetti et al. (2017) and Lo Monte et al. (2018), in which temperature, and pressure in some cases, were measured in unloaded small prismatic specimens. Sizeable spalling was observed in small-scale tests when samples were loaded during heating (Hertz and Sørensen, 2005; Tanibe et al., 2011; Connolly, 1995).

On the other hand, in medium-scale tests, slabs or prismatic specimens are generally heated on one side, either in loaded or unloaded conditions. External load can be applied in one or two directions by using passive restraint or active actuators. Examples of one direction loading can be found in Heel and Kusterle (2004), Sjöström et al. (2012), Boström et al. (2007), Jansson and Boström (2008), where pre- or post-tensioning systems have been adopted, or in Carré et al. (2013) and Rickard et al. (2016), where active actuators have been used. Loading in two directions is generally more complex to apply, via either passive restraining (as in Connolly,1995; Heel and Kusterle, 2004) or hydraulic jacks (as in Lo Monte and Felicetti, 2017). Although it is difficult to draw some general conclusions due to the high number of influencing parameters, it seems clear that spalling propensity and severity are higher in loaded sample, especially in biaxial loading conditions (Miah at al., 2016).

Finally, full-scale investigations consist in testing structural members or substructures under restraining and loading conditions almost identical to the real ones.

The different levels of investigations, obviously entail an increasing burden in terms of time and cost. This should be kept in mind in the preliminary screening of concrete mixes, in order to reach the optimal compromise between cost and efficacy.

Full-scale tests, for example, are the most representative of the real conditions, but are costly and time demanding, hence unsuitable for a first categorization of several concrete mixes. On the contrary, the test setup should take into account the effective working conditions of the structures. For instance, small-scale tests on unloaded samples can be very useful in studying particular mechanisms behind spalling, but can be too far from the service conditions of the structural member at issue.

In this context, three different test setups have been compared in this study for the evaluation of concrete spalling sensitivity. In the following the results are briefly discussed in order to highlight the main influencing factors and a possible common key of interpretation.

Once spalling behaviour has been characterized experimentally, the results can be generalized via numerical modelling and the structural analysis of a tunnel lining exposed to fire can be performed via a non-linear finite element software.

2 EXPERIMENTAL CAMPAIGN

Listed for increasing effort in terms of cost and time, the three test setups adopted are (see also Lo Monte et al., 2017a): (1) hot spot test, (2) biaxial loading test and (3) full-scale test.

Hot spot test (Figure 2a) consists in heating a small part (approximatively 250 mm in diameter) of one face of a 1 m-side concrete cube. The target temperature of 800°C is reached instantaneously at the beginning of the test and then kept constant. Although no external load is applied, significant stresses are introduced due to the effective constrain provided by the surrounding cold concrete against the thermal dilation of the heated region.

In biaxial loading test (Figure 2b), unreinforced concrete slabs (800x800x100 mm) are subjected to biaxial membrane compression, while heated at the intrados according to the Standard Fire curve (Lo Monte and Felicetti, 2017). Biaxial loading allows to more easily highlight spalling phenomena thanks to the almost axisymmetric in-plane compression. Vertical deflection has been monitored via 6 displacement transducers. Although not implemented in the present case, temperature and pore pressure could be monitored within the specimen by means of special sensors (Lo Monte and Felicetti, 2017). In previous experiments, moisture front survey has been implemented via Ground-Penetrating Radar technique (Lo Monte et al., 2017b). The described setup allows to control the main parameters involved, such as heating curve and external load, thus easing the interpretation of the results.

Full-scale test has been performed at the Leipzig Institute for Materials Research and Testing – MFPA (Work Group 3.2 - Fire Behaviour of Building Components and Special Constructions). A tunnel lining segment has been loaded vertically in three points and horizontally in longitudinal direction via hydraulic jacks (see Figure 3). Load has been applied before heating and then kept constant, while heating has been applied at the intrados according to the RWS fire curve. During the test, horizontal and vertical displacements at the actuators have been monitored, together with the temperature within the thickness.

Two concretes have been investigated (f_{cm} = 45 and 50 MPa) containing various amount of polypropylene (L = 6 mm, Ø = 18 μm and melting point T = 165°C) and steel (hook end, L = 60 mm and Ø = 750 μm) fibres, for a total number of 8 mixes (Table 1).

(a)   (b)

Figure 2. (a) hot spot test: concrete cube exposed to a jet flame; (b) biaxial loading test: view of the concrete slab within the loading system placed over the horizontal furnace.

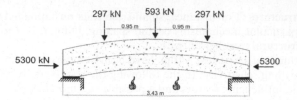

Figure 3. Full-scale test: scheme of loading and heating of the tunnel lining segment. (P_1 = 593 kN, P_2 = 297 kN and H = 5300 kN).

Table 1. Concrete mixes.

mix	f_{cm}[MPa]	polypropylene fibre [kg/m^3]	steel fibre [kg/m^3]
45_Plain	45		
45_1.5	45	1.5	
45_2.0	45	2.0	
45_1.5_4D	45	1.5	40
50_Plain	50		
50_1.5	50	1.5	
50_2.0	50	2.0	
50_1.5_4D	50	1.5	40

Hot spot test was performed on all concretes. Afterwards, mixes with 2.0 kg/m^3 of polypropylene fibre were not considered for biaxial loading tests, since hot spot test proved as 1.5 kg/m^3 of polypropylene fibre was already effective in avoiding spalling phenomena. Finally, full-scale test was performed on one plain mix only for final validation.

In hot spot tests, fire duration was set to 15 min, being enough to observe possible concrete detachments. Spalling was observed only in the two plain mixes, 45_Plain and 50_Plain, already at 45 and 60 s, with final maximum spalling depths of 100 and 120 mm, respectively.

Also in biaxial loading test, spalling occurred only in plain mixes, 45_Plain and 50_Plain, after 17 and 15 min, respectively. Spalling progression involved the whole exposed face with a final average depth of 55 and 62 mm for 45_Plain and 50_Plain, respectively. In both hot spot and biaxial loading tests, no spalling at all was observed in concretes with polypropylene fibre.

In real scale test, only mix 45_Plain was tested. Spalling started after 3 min of heating with a final average depth of 154 mm.

In order to understand the differences among the test setups, non-linear thermo-mechanical analyses of the different experimental procedures have been performed, so to compute thermal field and stress state in the specimens in a significant time range (Lo Monte et al., 2017a). In particular, numerical simulations showed that:

1. Although fire curves and temperature evolution are rather different among the test setups, spalling takes place when the temperature at the heated face is within a narrow range (300–350°C).
2. The stress normal to the exposed face is negligible in biaxial loading and full-scale tests, since heating flux is essentially unidimensional, while tension even approaching concrete tensile strength arise in the hot spot test.
3. At the onset of spalling, the maximum compression at the hot face is within a rather small range of values among the different setups, namely between 17 and 23 MPa, albeit they are reached for different fire durations (1, 3, 15 min for hot spot, full-scale and biaxial loading test, respectively).

Compressive stress in the order of 20 MPa is expected to be enough for triggering spalling thanks to the contribution of pore pressure (usually in the range of 0.5–2.5 MPa), which induces an equivalent tensile hydrostatic stress acting on the solid skeleton.

In conclusion, in spite of the differences among experimental procedures and final results in terms of spalling onset time and final depth, common triggering conditions can be detected. Such differences, however, can be explained only once the effects of specimen geometry, loading and heating conditions are carefully analysed.

3 NUMERICAL STRUCTURAL ANALYSES OF A PRECAST SEGMENTAL TUNNEL

Thanks to the first screening investigations via hot spot test, and the following assessment of the effect of external compression via biaxial loading test, the final choice of the mix can be properly done. Afterwards, full-scale test can be used to calibrate the numerical model of the structure in terms of detachment progression during heating. In order to assess the overall fire performance of a tunnel lining, a non-linear finite element thermo-mechanical analysis should be performed.

Concrete mechanical behaviour can be modelled according to a few models from the literature, as for example *concrete damaged plasticity* model (implemented in Abaqus), or the *total strain rotating crack* non-linear approach (implemented in DIANA). As regards the uniaxial constitutive laws, the curves in compression suggested by EC2 (EN 1992-1-2:2004) are in rather good agreement with the experimental results (Bamonte and Lo Monte, 2015), while in tension an elasto-brittle model can be adopted. The variation of all the properties with temperature are provided as functions of temperature in EC2 (Figure 4).

In order to perform a plastic structural analysis, the role of reinforcements is fundamental, this making necessary modelling the rebars. Also the variation with temperature of steel mechanical behaviour is well described in EC2.

Soil should also be properly introduced in the model, since it governs (a) the initial state of stress in the tunnel and (b) the constrain provided to the dilation of the lining during heating and the subsequent indirect actions. The effect of soil can be taken into account by modelling the soil itself, as in Lilliu and Meda (2013), or using a combination of elastic springs and initial pressure to the lining as implemented in Colombo et al. (2015).

Furthermore, in precast segmental tunnel lining, a key role is played by the joints between consecutive segments, since they lead to a relaxation of the internal actions thanks to partial rotation. Such joints should be modelled as compression-only interfaces which do not allow any tension. As regards the transverse behaviour, perfect adhesion can be implemented, since the friction force is hardly activated during heating. Also the constrain between lining and soil should be implemented as compression-only interface.

A simplified approach for describing the joints between consecutive segments consists in introducing particular elastic beam elements (Colombo et al., 2015), this decreasing the computational burden.

The thermo-mechanical analysis is sequentially coupled, since firstly the thermal analysis is performed to work out the thermal field, and then the mechanical analysis is carried out to assess the structural behaviour. This is still valid when spalling is considered, only if detachment progression

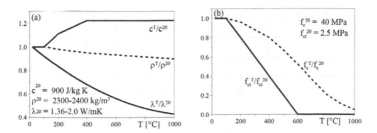

Figure 4. Normalized variation of (a) density ρ^T, specific heat c^T and thermal conductivity λ^T, and (b) of compressive and tensile strengths $f_c{}^T$ and $f_{ct}{}^T$, for concrete as suggested by EC2.

is given a priori. On the contrary, if spalling depends on thermal damage and stress field, thermal and mechanical problems are fully coupled (see also Gawin et al., 2011).

Heating should be applied by introducing the heat flux between concrete and hot gasses via convection and radiation at the exposed faces Convection coefficient and concrete emissivity at the exposed face are reported in EC2 depending on the fire curve adopted.

Spalling can be taken into account in a simplified way by progressively deactivating the external layer supposed to spall and by applying convection and radiation to the new exposed (inner) faces (Lilliu and Meda, 2013). Once known the spalling rate (for example between 100 and 200 mm/h, according to the type of concrete and to the fire curve used), a suitable thickness of the layers to be deactivated during time can be chosen. It is worth noting that when plastic numerical analyses are carried out, no further strength check is needed. If the analysis reaches convergence at any time step of fire duration, it means that equilibrium can be find and collapse do not occur. In Figure 6 typical profiles of temperature and stress are shown in a generic section of the lining at different time steps, in absence of spalling.

A quite repeatable result is that the whole section is under compression, since the soil surrounding the lining constrains its thermal dilation, this introducing compressive axial force. This axial force, however, is concentrated in the inner part of the lining (where thermal dilation is higher), hence introducing a negative moment, which can significantly change the initial stress distribution. In this regard, parametric numerical analyses are in progress, investigating the effect of lining thickness. On the basis of preliminary results, it has been observed that for increasing values of thickness, the compressive axial force introduced by heating decreases (since the average temperature of the lining is lower), while the negative bending moment may increase (because of the higher lever arm of the dilating hot layer with respect to the lining axis). This suggests that increasing tunnel lining thickness does not necessary increase fire resistance.

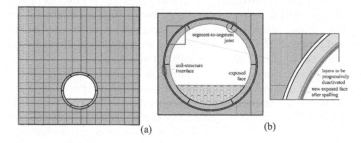

Figure 5. (a) Finite element model of a tunnel lining and surrounding soil, and (b) detail of the lining.

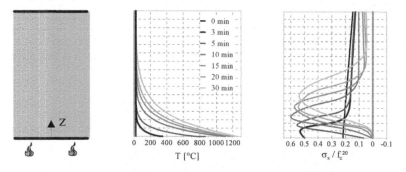

Figure 6. Temperature and stress profiles in a generic section of the tunnel lining at different time steps, in absence of spalling.

Finally, it should be noted that as an alternative to plastic structural analysis, an elastic approach could be adopted. In such case, concrete and still are considered linear elastic, but with elastic modulus and thermal expansion coefficient depending on temperature. Once reached the design fire duration, the sectional strength checks are necessary to assess that in any point of the lining (considering also the segment-to-segment joints), the bearing capacity is higher than the action developed during heating.

A further critical aspect is represented by the behaviour of R/C structures under natural fires, since the overall damage can increase during cooling and collapse in such phase cannot be excluded a priori. This problem is still to an open issue in the literature (Bamonte et al., 2018).

4 CONCLUSIONS

The problem of explosive spalling in R/C tunnel exposed to fire has been examined from the experimental assessment to the numerical structural modelling.

Firstly, the critical issues in experimental testing have been discussed on the bases of the results obtained on the same concrete mixes from three different setups. It is shown as the initial screening of different mixes should be performed via simple testing, as compromise among time, cost and representativeness of the effective working conditions of the structural members. Once the mix is chosen, full-scale test is very useful in calibrating numerical models, which allow to generalize experimental results and to assess the fire performance of the real structure.

In the present study, two test setups have been used in the screening phase, the first one (namely the hot spot test) very simple and able to exclude some mixes, and the second one (biaxial loading test) which allowed to study the effect of the load. Finally, full-scale test has been carried out with the final mix on a tunnel lining segment. Afterward it is discussed how non-linear numerical analyses can be performed enforcing spalling progression (observed experimentally), so assessing the structural stability for the design fire duration.

ACKNOWLEDGEMENTS

The Authors are grateful to Leipzig Institute for Materials Research and Testing – MFPA (Work Group 3.2 - Fire Behaviour of Building Components and Special Constructions) for providing the results regarding the full-scale test on the tunnel lining segment.

REFERENCES

Bamonte, P., Kalaba, N. & Felicetti, R. 2018. Computational study on prestressed concrete members exposed to natural fires. *Fire Safety Journal* 97: 54–65. DOI: 10.1016/j.firesaf.2018.02.006

Bamont, P. & Lo Monte, F. 2015. Reinforced Concrete Columns Exposed to Standard Fire: Comparison among Different Constitutive Models for Concrete at High Temperature. *Fire Safety Journal* 71: 310–323. DOI: http://dx.doi.org/10.1016/j.firesaf.2014.11.014.

Boström, L., Wickström, U., & Adl-Zarrabi, B. 2007. Effect of Specimen Size and Loading Conditions on Spalling of Concrete. *Fire and Materials* 31: 173–186.

Carré, H., Pimienta, P., La Borderie, C., Pereira, F., & Mindeguia, J. C. 2013. Effect of Compressive Loading on the Risk of Spalling. *Proceedings of the 3rd International Workshop on Concrete Spalling due to Fire Exposure, pp. 01007, Paris (France), September 25–27.*

Colombo, M., Martinelli, P. & di Prisco, M. 2015. A design approach for tunnels exposed to blast and fire. *Structural Concrete* 2.

Connolly, R. 1995. *The Spalling of Concrete in Fires.* PhD Thesis, University of Aston in Birmingham.

EN 1992-1-2:2004. *Eurocode 2 - Design of concrete structures - Part 1–2: General rules - Structural fire design.* European Committee for Standardization (CEN), Brussels (Belgium).

Felicetti, R., Lo Monte, F. & Pimienta, P. 2017. A New Test Method to Study the Influence of Pore Pres-sure on Fracture Behaviour of Concrete during Heating. *Cement and Concrete Research* 94: 13–23.

FIT. 2005. *Technical Report – Part 1: Design Fire Scenarios, European Thematic Network FIT – Fire in Tunnels,* Rapporteur A. Haack, supported by the European Community under the 5th Framework Programme 'Competitive and Sustainable Growth' Contract n° G1RT-CT-2001-05017.

Gawin, D., Pesavento, P. & Schrefler, B.A. 2011. What physical phenomena can be neglected when modelling concrete at high temperature? A comparative study. Part 2: Comparison between models. *International Journal of Solids and Structures* 48: 1945–1961.

Heel, A. & Kusterle, W. 2004. Die Brandbeständigkeit von Faser-, Stahl- und Spannbeton [Fire resistance of fiber-reinforced, reinforced, and prestressed concrete]" (in German). *Tech. Rep. 544, Bundesminis-teriumfür Verkehr, Innovation und Technologie, Vienna (Austria).*

Hertz, K. & Sørensen, L. 2005. Test method for spalling of fire exposed concrete. *Fire Safety Journal* 40: pp. 466–476.

Huismann, S., Weise, F., Meng, B., Schneider, U. 2012. Transient strain of high strength concrete at elevat-ed temperatures and the impact of polypropylene fibers. *Materials and Structures* 45: 793–801.

Jansson, R. & Boström, L. 2008. Spalling of concrete exposed to fire. *SP Technical Research Institute of Sweden, Borås.*

Kalifa, P., Chéné, G. & Gallé, C. 2001. High-temperature behaviour of HPC with polypropylene fibres. From spalling to microstructure. *Cement and Concrete Research* 31: 1487–1499.

Kalifa, P., Menneteau, F. D. & Quenard, D. 2000. Spalling and pore pressure in HPC at high temperatures. *Cement and Concrete Research* 30: 1915–1927.

Khoury, G. A. 2000. Effect of fire on concrete and concrete structures. *Progress in Structural Engineering and Materials* 2: 429–447.

Khoury, G. A. 2008. Polypropylene Fibres in Heated Concrete. Part 2: Pressure Relief Mechanisms and Modelling Criteria. *Magazine of Concrete Research* 60 (3): 189–204.

Krzemieńa, K. & Hager, I. 2015. Assessment of concrete susceptibility to fire spalling: A report on the state-of-the-art in testing procedures. *Procedia Engineering* 108: 285–292.

Lilliu, G. & Meda, A. 2013. Nonlinear Phased Analysis of Reinforced Concrete Tunnels Under Fire Exposure. *Journal of Structural Fire Engineering* 4 (3).

Lo Monte, F. & Felicetti, R. 2017. Heated slabs under biaxial compressive loading: a test set-up for the assessment of concrete sensitivity to spalling. *Materials and Structures* 50 (192).

Lo Monte, F., Felicetti, R. & Miah, Md. J. 2019. The Influence of Pore Pressure on Fracture Behaviour of Normal-Strength and High-Performance Concretes at High Temperature. *Cement and Concrete Composites.*

Lo Monte, F., Felicetti, R., Meda, A. & Bortolussi, A. 2017a. Influence of the Test Method in the Assessment of Concrete Sensitivity to Explosive Spalling. *Proceedings of the 5th International Workshop on "Concrete Spalling due to Fire Exposure", Boras (Sweden), October 12–13.*

Lo Monte, F. & Gambarova, P. G. 2015. Corner spalling and tension stiffening in heat-damaged R/C mem-bers: a preliminary investigation. *Materials and Structures* 48: 3657–3673.

Lo Monte, F., Lombardi, F., Felicetti, R. & Lualdi, M. 2017b. Ground-Penetrating Radar Monitoring of Con-crete at High Temperature. *Construction and Building Materials* 151 (1): 881–888.

Miah, Md J., Lo Monte, F., Felicetti, R., Carré, H., Pimienta, P. & Borderie, C. L. 2016. Fire spalling behaviour of concrete: Role of mechanical loading (uniaxial and biaxial) and cement type. *Key Engineering Materials* 711: 549–555.

Mindeguia, J. C., Pimienta, P., Noumowé, A. & Kanema, M. 2010. Temperature, Pore Pressure and Mass Variation of Concrete Subjected to High Temperature – Experimental and Numerical Discussion on Spalling Risk. *Cement and Concrete Research* 40: 477–487.

Pistol, K., Weise, F., Meng, B. & Diederichs, U. 2014. Polypropylene fibres and micro cracking in fire exposed concrete. *Advanced Materials Research* 897: 284–289.

Rickard, I., Bisby, L., Deeny, S. & Maluk, C. 2016. Predictive Testing for Heat Induced Spalling of Concrete Tunnels - the Influence of Mechanical Loading. *Proceedings of the 9th international Conference "Structures in Fire 2016 – SIF'16), 217–224, Princeton (USA), June 8–10.*

Rossino, C., Lo Monte, F., Cangiano, S., Felicetti, R. & Gambarova, P.G. 2013. Concrete Spalling Sensitivi-ty versus Microstructure: Preliminary Results on the Effect of Polypropylene Fibers. *Proceedings of the 3rd International Workshop on Concrete Spalling due to Fire Exposure, September 25–27, Paris (France),* DOI: 10.1051/matecconf/20130602002.

Sjöström, J., Lange, D., Jansson, R. & Boström, L. 2012. Directional Dependence of Deflections and Dam-ages during Fire Tests of Post-Tensioned Concrete Slabs. *Proceedings of the 7th Int.Conf. on Structures in Fire – SIF'12, June 6–8, 2012, Zurich (Switzerland):* 589–598.

Tanibe, T., Ozawa, M., Lustoza, R.L., Kikuchi, K. & Morimoto, H. 2011. Explosive spalling behaviour in re-strained concrete in the event of fire. *Proceedings of the 2nd International RILEM Workshop on "Concrete Spalling due to Fire Exposure", Delft (the Netherlands), October 5–7:* 319–326.

Toropovs, N., Lo Monte, F., Wyrzykowski, M., Weber, B., Sahmenko, G., Vontobel, P., Felicetti, R. & Lura, P. 2015. Real-time measurements of temperature, pressure and moisture profiles in High-Performance Concrete exposed to high temperatures during neutron radiography imaging. *Cement and Concrete Research* 68: 166–173.

Tunnels and Underground Cities: Engineering and Innovation meet Archaeology,
Architecture and Art, Volume 5: Innovation in underground engineering,
materials and equipment - Part 1 – Peila, Viggiani & Celestino (Eds)
© 2020 Taylor & Francis Group, London, ISBN 978-0-367-46870-5

Alternative umbrella arches: The use of composite pile roofs

F. Lopez, F. von Havranek & G. Severi
Friedr. Ischebeck GmbH, Ennepetal, Germany

ABSTRACT: Umbrella arches belong to the most widely-used reinforcement technics in conventional tunneling, especially for low-depth tunnels (H/D=1 to 5) in unfavorable ground conditions (soils with poor mechanical properties, fractured rocks, etc.) and are installed primarily to increase the safety of the excavation front and to reduce settlements. Due to several constructive advantages, self-drilling hollow bars have been used in tunneling to provide a cost-effective alternative type of umbrella arch: the composite pile roof (CPR). While the conventional steel pipe roof (PR) undergoes essentially bending moments, the CPR relies on the ability of the self-drilling hollow bars to transfer axial forces. In terms of safety, the final result for the tunnel excavation remains the same. The following paper presents an overview of the structural behavior of the above mentioned constructive procedures. The relevant considerations for the design of a CPR will be discussed.

1 INTRODUCTION

1.1 *Umbrella arches*

Umbrella arches are used as temporary longitudinal tunnel reinforcement (pre-support) of the excavation front, in the time when the top heading (roof) is unsupported (between the excavation and the application of the structural shotcrete lining). The arch is materialized by a set-up of forepoles above and around the crown of the tunnel face (Figure 1 and Figure 2).

Tubular steel sections are used regularly as the load bearing reinforcement elements (forepoles) of umbrella arches, with lengths greater than the excavation height (H_e in Figure 2) and a shallow installation angle, typically between 3° and 8° from the horizontal in the longitudinal direction of the tunnel alignment. The design length of the forepoles is so defined, that the elements are long enough to be embedded past the empirically derived Rankine active line, in front of the excavation face.

The umbrella arch configurations correspond to the most widely spread (conventional) pipe roof (PR) and the less known composite pile roof (CPR).

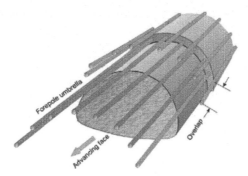

Figure 1.   Umbrella arch as longitudinal tunnel reinforcement after Aksoy & Onagan (2010).

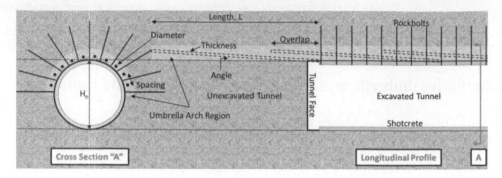

Figure 2. Umbrella arch after Oek et al.(2014).

1.2 Pipe roof (PR)

In the conventional pipe roof, steel pipes (usually S235 or S355) are installed as a forepoling system (Figure 3). The diameter of the steel pipes ranges between 60 mm and 200 mm, with a wall thickness of 4 mm to 8 mm. The length of one umbrella is commonly 12 m or 15 m. The excavated length underneath (pipe roof field length) ranges from 6 m to 12 m (Volkmann & Schubert 2007).

The forepoles are regularly installed as a cased-drilling system, where the steel pipes follow directly the drill bit supporting immediately the borehole (Figure 4). The center to center distance lays in the range of 30 cm – 60 cm.

1.3 Composite pile roof (CPR)

Composite pile roofs (CPR) have been used since 2001 as an alternative to the conventional pile roof (PR), to materialize forepoled umbrella archs. The load bearing elements are continuous threaded self-drilling hollow bars, also known in the literature as IBO-anchors and/or tubular injected micropiles, made out of seamless fine-grained steel pipes, installed via rotary percussive drilling.

During the drilling process, the hollow bars are continuously grouted (dynamic injection), building a rough interlocking at the interface grout-soil, increasing the skin friction (Lopez & Severi, 2018). The components, the installation process and a typical cross-section of the grouted body of a self-drilling micropile are presented in Figure 5.

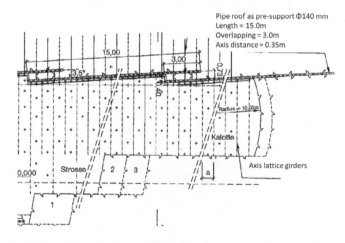

Figure 3. Longitudinal section of a PR, modified after Jodl et al. (2005).

Figure 4.   Installation and components of a PR, modified after Atlas Copco (2011).

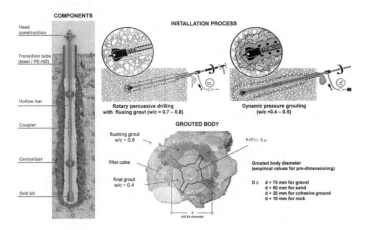

Figure 5.   Self-drilling micropiles: components, installation and grouted body after Lopez & Severi (2018).

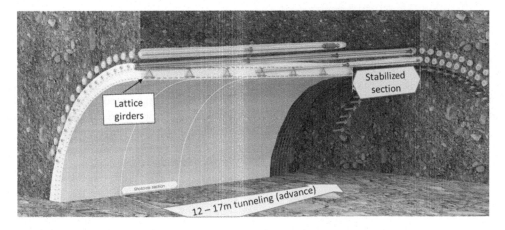

Figure 6.   Schematic presentation of a CPR, modified after von Havranek (2017).

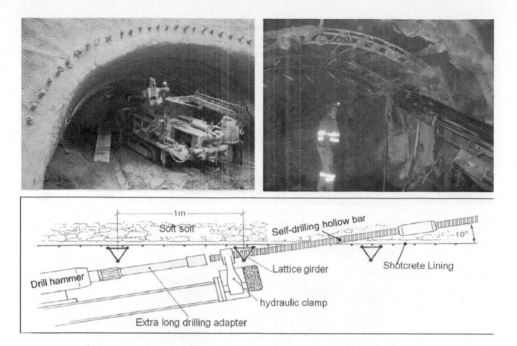

Figure 7. Installation of a CPR after Friedr. Ischebeck GmbH (2018).

Self-drilling hollow bars are regularly used in tunneling as radial bolts and spiles. Although similar in their configuration, spiles and forepoles differ from the way their support is needed, which ultimately defines their installation length: spiles are short longitudinal support elements, installed when the geological structure itself controls a possible local failure of the portion of the tunnel (Volkmann et al. 2006) with a length measuring less than the excavation height (H_e in Figure 2). Self-drilling IBO-anchors or tubular injected micropiles used as forepoles with lengths up to 18 m with center to center distance in the range of 30 cm – 50 cm, as presented schematically in Figure 6 (von Havranek 2017).

For the installation of CPR a drilling rig provided with a drifter (hydraulic hammer) is required (Figure 7).

2 LOAD BEARING BEHAVIOR

2.1 Pipe roof (PR)

It is generally accepted that the pipes of the arch undergo essentially bending moments. The load bearing behavior of a PR can be considered analogous to supported beam: each pipe is founded in the ground (ahead of the face) as well as on the lining (behind the face) in the longitudinal- and radial direction (Volkmann & Schubert 2007). Two simplified static models are presented in Figure 8.

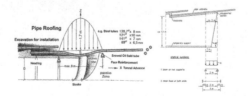

Figure 8. Simplified static models of a PR. Left: after Ischebeck (2005). Right: after Wittke et al. (2002).

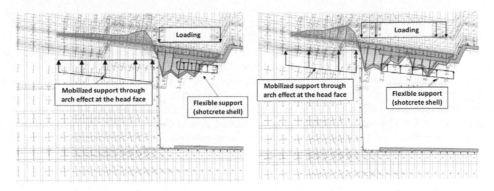

Figure 9. Bending moments acting on the umbrella pipes after two subsequent strokes, modified after Eckl (2012).

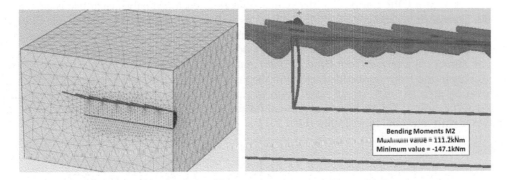

Figure 10. Bending moments acting on the umbrella pipes, modified after Monnet & Jahangir (2014).

This pipe roof support effect is affected by the bending of the pipes so the second moment of the area defines the activation speed of the support effect by bending. Pipes with a larger diameter activate a more powerful support effect at similar displacements compared to smaller diameters. The pipe wall thickness, on the other hand, defines the critical moment and/or the maximum bending (Volkmann & Schubert 2007).

Several numerical investigations, i.e. Janin (2012), Eckl (2012), Monnet & Jahangir (2014), identify the occurrence of bending, however the shape of the bending envelope varies considerably, mainly influenced by the adopted connections to the supports (shotcrete/lattice girders and the ground ahead of the face) (Figure 9 and Figure 10). The analysis of Eckl (2012) shows

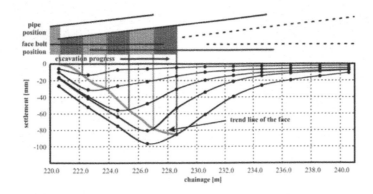

Figure 11. In-situ deflection curve of an umbrella arch in rock after Volkmann & Schubert (2007).

that under similar conditions (ground, tunnel geometry, etc.), the magnitude of the bending moments depends on the bending stiffness (E*I) of the pipe, being directly proportional to it.

The bending effect can be also derived from in-situ deflection measurements, carried out with in-place inclinometer chains, as presented in Figure 11 (Volkmann & Schubert 2007).

However, the results of numerical investigations (i.e. Volkmann 2004, Mämpel & Faber 2005, Eckl 2012) identify also the occurrence of normal forces acting on the pipes, which more likely result from a restraint of the axial displacements (reactions). The analysis of Eckl (2012) shows that under similar conditions (ground, tunnel geometry, etc.), for the commonly used pipe sizes, the magnitude of the normal forces increases slightly with the increasing axial stiffness (E*A) of the pipes.

2.2 Composite Pile roof (CPR)

The results of numerical analysis, i.e. Mämpel & Faber (2005), Brandl (2005) and Eckl (2012) show that the composite piles of the arch undergo essentially axial (normal) forces. The load bearing behavior of a CPR can be considered analogous to a tensioned cable: each pile is embedded in the internal lining/shotcrete (behind the face) as well as in the ground (ahead of the face), as shown in Figure 12. The self-weight of the soil (funnel) acts on the umbrella arch, activating axial (normal) forces, which are transferred to the supports over skin friction (von Havranek, 2017).

The numerical analysis from Brandl (2005) and Eckl (2012) confirm the occurrence of tensile forces in the front (embedded length in the ground) but present compressive forces in the excavation area (stroke), which result from the arch effects occurring at the face (front) and top (roof) heading (Figure 13). As stablished for the PR, CPR also undergo bending moments, however with marginal magnitudes (Eckl 2012).

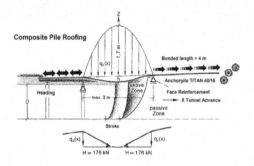

Figure 12.   Simplified static model of a CPR after Ischebeck (2005).

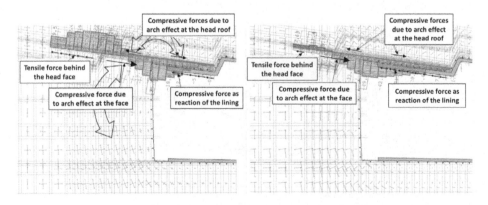

Figure 13.   Axial forces acting on the umbrella piles after two subsequent strokes, modified after Eckl (2012).

3 COMPARISON BETWEEN PR AND CPR

3.1 *Settlements at the surface and at the roof ridge*

The results of several numerical investigations (i.e. Brandl 2005, Janin 2012, Eckl 2012 and Monnet & Jahangir 2014) show that umbrella arches do not seem to have significant effects on the surface settlements. To this effect, to use a small hollow bar (CPR) or a large diameter steel pipe (PR) does not seem to modify significantly the results, either (Figure 14). The latter applies also to the evaluated roof ridge settlements, as presented in Figure 15.

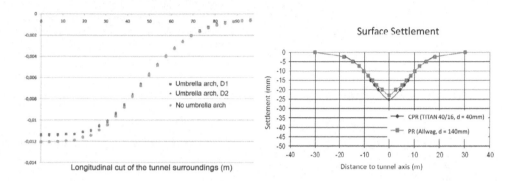

Figure 14. Surface settlements. Left: longitudinal cut after Monnet & Jahangir (2014). Right: cross cut modified after Brandl (2005, unpubl.).

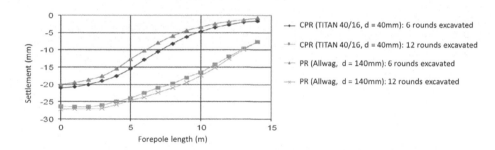

Figure 15. Roof ridge settlements (center forepole), modified after Brandl (2005, unpubl.).

3.2 *Normal forces acting on the forepoles*

The results of the parametric numerical investigation from Eckl (2012) show a dependency between the normal (axial) forces acting on a forepole and their axial stiffness (E*A), given only by the mechanical and geometrical properties of the steel tubular section, according to the parameter definition of the structural beam elements.

The normal forces acting on the forepoles increase with the corresponding increment of the axial stiffness. The increment is rather appreciable when going from the regular sizes of composite piles (diameter = 30mm - 73mm) to the regularly used steel pipes (diameter = 140mm).

3.3 *Bending moments acting on the forepoles*

As previously mentioned, the results of the parametric analysis carried out by Eckl (2012) show also a dependency between the bending moments acting on a forepole and their bending

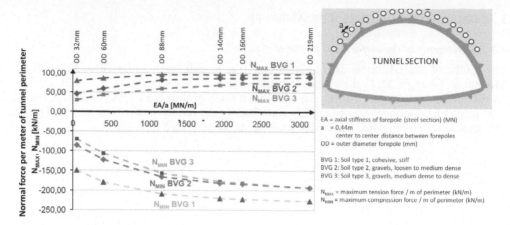

Figure 16. Normal forces acting on the forepoles in dependence of the axial stiffness, modified after Eckl (2012).

stiffness (E*I), defined again by the mechanical and geometrical properties of the steel tubular section only.

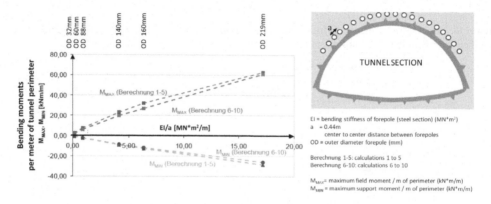

Figure 17. Bending moments acting on the forepoles in dependence of the bending stiffness, modified after Eckl (2012).

The bending moments acting on the forepoles increase considerably with the corresponding increment of the bending stiffness. This proportionality is evident for the regular sizes of steel pipes (diameter = 140mm – 219mm), confirming that the load bearing behavior of PR is primarily governed by bending moments, and secondarily by normal (axial) forces.

Under the same conditions, for the regular sizes of composite piles (with diameters between 30mm - 73mm), the bending moments are approx. 30times lower than the obtained values for the regular used steel pipes (diameter = 140mm). This confirms, that the development of normal (axial) forces defines the load bearing behavior of CPR, thus these values are relevant for their design. Bending moments, on the other hand, are so marginal that play rather a subordinate role.

4 CONCLUSIONS

Based on the review of different technical documents, this paper presented an overview of the load bearing behavior of the commonly used umbrella arch configuration, consistent of a pipe roof (PR), and the less known composite pile roof (CPR).The forepoles of a CPR are materialized by continuous threaded self-drilling hollow bars with small diameters, commonly between 30mm – 73mm, also known in the literature as IBO-anchors and/or tubular injected micropiles.

The load bearing behavior of the CPR was described in detail in this paper. Due to their high axial stiffness (E*A), CPR undergo mainly axial forces, transferring the loads to the ground (ahead of the excavation face) and to the lining (behind the excavation face) via friction, while their reduced bending stiffness (E*I) is enough to resist the marginal acting bending moments.

The findings presented in this document allow to affirm that, although having a different load bearing behavior, both PR and CPR can provide the same safety level to the longitudinal tunnel reinforcement, equally fulfilling the structural requirements of an umbrella arch.

In this context, CPR can be considered as a time and cost-effective alternative to the conventional PR, since self-drilling elements with small diameters can be installed using very flexible equipment, obtaining high drilling performances even under restricted space conditions.

Finally, the experience acquired from the use of self-drilling hollow bars in other geotechnical applications, such as micropiling and soil nailing, shows that the load transfer via friction is influenced by the soundness of the grouted body, which is largely affected by the thread geometry of the bars and the stiffness of the coupled connections.

REFERENCES

Aksoy, C.O. & Onargan, T. 2010. The role of umbrella arch and face bolt as deformation preventing support system in preventing building damages. In *Tunneling and underground space technology* 25: 553–559

Atlas Copco 2011. Pipe roofing solutions (Technical brochure, unpubl.)

Brandl, J. 2005 (Geoconsult Holding). Metro de Santiago Tunel de Acceso Pique Estación El Parrón, Design of Ischebeck Pipe roof system (unpubl.).

Janin, J-P. 2012. *Tunnels en milieu urbain: Prévisions des tassements avec prise en compte des présoutènements (renforcement du front de taille et voûte- parapluie)*. Doctoral Thesis, Institut national des sciences appliquées de Lyon.

Friedr. Ischebeck GmbH 2018. TITAN for tunnels and mining (Technical brochure, unpubl.)

Ischebeck, E. F. 2005 (unpubl.). New approaches in Tunneling with Composite Canopies TITAN Installation-Design-Monitoring. *Summer School on Rational Tunneling, Innsbruck, 3–6 October 2005*.

Eckl, M. 2012. *Tragverhalten von Rohrschirmdecken beim Tunnelbau im Lockergestein*. Doctoral Thesis, Technische Universität München.

Jodl, H. G., Altinger, G., Bichler, M., Kriebaum, W. & Schlosser, W. 2005. Vortriebsmethoden und Ausbau von Tunnels. In Ernst & Sohn (eds), *Betonkalender 2005 – Band 1 Tunnelbauwerke*: 21–118

Lopez, F. & Severi, G. 2018. Micropiling in Urban Infrastructure: Advantages, Experience and Challenges. *Proc. DFI-EFFC Intern. Conf. on Deep Foundations and Ground Improvement, Rome, 5–8 June 2018*.

Mämpel, H. & Faber, U. 2005 (Ingenieurbüro Prof. Maidl, unpubl.). Ischebeck GmbH, Verbundschirm aus Injektionsbohrankern Typ TITAN 40/16 – WIBOREX 40/16, Gutachten zum Baustellenversuch.

Monnet, A. & Jahangir, E. 2014. Preliminary 3D Modelling of Structural behaviour of Face Bolting and Umbrella Arch in Tunneling, *Plaxis Bulletin – Autumn issue 2014 (unpubl.)*

Oek, J., Vlachopoulus, N. & Marinos, V. 2014. Umbrella Arch Nomenclature and Selection Methodology for Temporary Support Systems for the Design and Construction of Tunnels. In Springer (eds), *Geotechnical and Geological Engineering* 32(1): 97–130.

Volkmann, G.M. 2004. A contribution to the Effect and Behavior of Pipe Roof Supports. In Schubert (eds). *Proc. Eurock 2004 & 53rd Geomechanics Colloquy, Salzburg, 6–8 October 2004*.

Volkmann, G.M., Schubert, W. & Button, E. 2006. A contribution to the Design of Tunnels Supported by a Pipe Roof. In Yale (eds), *Proc. 41st U.S. Rock Mechanics Symp., Golden, 17–21 June 2006*.

Volkmann, G.M. & Schubert, W. 2007. Geotechnical Model for Pipe Supports in Tunneling. In Taylor & Francis Group (eds), *Proc. ITA-AITES World Tunnel Congress, Prague, 5–10 May 2007*.

von Havranek, F. 2017. Verbundschirm TITAN – Eine Alternative zu herkömmlichen Rohrschirmen im Tunnelbau. *Proc. 3rd Int. Symp. FreiBERGbau 2017, Freiberg, 5–6 October 2017*.

Wittke, W., Pierau, B. & Erichsen, C. 2002. *New Austrian Tunneling Method (NATM) – Stability Analysis and Design*. Essen: VGE-Verlag Glückauf GmbH

Tunnels and Underground Cities: Engineering and Innovation meet Archaeology,
Architecture and Art, Volume 5: Innovation in underground engineering,
materials and equipment - Part 1 – Peila, Viggiani & Celestino (Eds)
© 2020 Taylor & Francis Group, London, ISBN 978-0-367-46870-5

Long-term durability analysis and lifetime prediction of PVC waterproofing membranes

A. Luciani, C. Todaro, D. Martinelli, A. Carigi & D. Peila
Politecnico di Torino, Turin, Italy

M. Leotta
Mapei SpA, Milan, Italy

E. Dal Negro
UTT Mapei, Milan, Italy

ABSTRACT: Water inflow is an important issues in tunnels because it causes damage to structures and plants. Waterproofing systems are a key aspect for an effective design. Risk analysis can be useful to design the waterproofing, considering long-term efficiency and costs. Moreover, durability is important due to the lifetime requirements of modern tunnels. Nevertheless, there is a lack of knowledge on this aspect for waterproofing membranes. PVC, which is one of the most used materials for this application, is degrades due to loss of plasticizers and polymer alteration. Accelerated ageing tests have been developed to study its long-term durability, but none of those is simulating the real conditions of the membrane in a tunnel. Therefore, an accelerated aging test simulating the tunnel conditions, i.e. water flow on the PVC between concrete layers, has been developed. The mechanical parameters of aged specimens are studied to develop a long-term extrapolation of the results to the non-accelerated jobsite conditions.

1 INTRODUCTION

The interaction of underground structures with water is of overwhelming importance for a good project, technically and economically speaking (Luciani et al., In press; Peila et al., 2016). Indeed, water influences construction technologies and creates serious issues during operation, leading to reductions of efficiency, disruption and expensive refurbishment works. Water inflows have been recognized by several authors among the most important causes of damages to underground structures (ITA, 1999; Sandrone & Labiouse, 2011). Nowadays, several technologies are available for tunnel waterproofing, depending on the geological conditions and construction techniques (Luciani & Peila, 2018; Peila et al., 2015; Thewes & Budach, 2009). The best choice of the waterproofing technology is a function of boundary conditions, of possible hazards and of the lifetime cost of the waterproofing, i.e. installation, maintenance and refurbishment costs.

In these evaluations, risk analysis can be a useful tool to evaluate all possible cases and analyze the long-term cost of the solution (Luciani et al., 2018). Risk analysis is widely used in geotechnical and tunnel engineering (Bianchi et al., 2012; Eskesen et al., 2004; Guglielmetti et al., 2007; ITIG, 2012) to define from the very beginning of a project all possible interaction and consequences, to minimize risk and reduce life-long costs of the structure. However, nowadays this approach is not applied for waterproofing systems.

Nevertheless, risk analysis can hardly consider the durability of the waterproofing solutions. For road and train tunnels, a dry or almost dry surface is usually required. Moreover, tunnels are often designed with a life span of more than 100 years. For all the previous reasons, durability is a key parameter for tunnel waterproofing. However, few information are available on this aspect.

The materials used for waterproofing membranes are relatively young (60 years since they are used in the market) compared to the tunnel requested lifetime. Many studies have been developed on the durability of PVC and TPO but they are all focusing on environmental conditions that are not applicable to underground (e.g. high temperatures, UV rays). Seldom studies have analyzed the problem of degradation in the tunnel conditions. In few cases the analysis of real site aged materials have been performed (Mahner et al., 2018; Usman & Galler, 2014), but there are few data, only related to maximum 40 years and it is not possible to compare the properties of the aged materials to the original ones, because they were no more available. Once the time-dependent properties of the materials are analyzed, they can be compared to the properties that are required from the waterproofing during the whole structure life (e.g. watertightness, tensile strength, drainage capacity) project-related. From this comparison the time to failure of the waterproofing can be evaluated (i.e. the time when available properties are lower than the requested ones) and compared to the design life of the structure (ISO/TS 13434).

In this paper, an accelerated ageing test simulating the tunnel condition is proposed for PVC membranes. The mechanical properties of the original and aged membranes are compared. This comparison, done for samples aged at 3 different temperatures, permits a long-term extrapolation of the results to the jobsite conditions in order to define the durability of the membranes.

2 RISK ANALYSIS APPROACH

A risk analysis approach has been developed to help the choice of the most suitable waterproofing system to be used in a project. This method allows to consider the long-term cost of the waterproofing, keeping in mind not only the initial cost of different solutions, but also the long-term efficiency, the consequences of possible damages to the structure and maintenance costs.

With this aim, a risk register has been developed containing all possible interactions between underground structures and water. The hazardous events have been linked one to the other in a cause-consequence relationship through an event tree starting from water inflow probability. The consequences are evaluated as a cost, based on 5 different categories: injury to third parties, damage or economic loss to third parties, harm to the environment, disruption, economic loss to the owner. These categories are derived from those proposed by ITA (Eskesen et al., 2004). The risk is then evaluated as a cost, using the event tree to compute the probability of each hazardous event. In order to allow a statistical evaluation, the Monte Carlo method is used, iterating the computation. This permits the evaluation not only of the average or maximum value of the risk, but of the whole probability distribution of the risk and of the probability of failure of a waterproofing system.

The effect of each waterproofing system can be estimated repeating the computational procedure considering each technology. The technical solutions available for water management in underground excavations do not change the consequences of the hazards, but reduce the probability of water inflow. Therefore, the evaluation of the effect of a technical solution implies the evaluation of the ability of that solution to reduce the probability of water inflow. Once this value is known, the event tree can be re-computed using the new water inflow probability value and the residual risk evaluated. Their efficiency in reducing the probability of water inflow can be estimated through a fault tree analyzing the possible cause-consequence connections before water inflow occurs (Figure 1).

Giving a likelihood value to the possible causes of ineffectiveness, the combined probability can be assessed and used as input in the event tree. In this evaluation, it is important to consider also the possibility, availability and efficiency of repair of damage during the life of the work. The cost of execution of each technical solution is also computed, considering possible repair costs.

The result of this approach is both a comparison between initial and residual risk and a cost-benefit comparison among different technical solutions. Figure 2 reports the cost-benefit comparison between the no-waterproofing solution and two possible membrane solutions for a metro tunnel in urbanized area: system 1 is a single layer PVC waterproofing membrane, while system 2 is a double layer PVC membrane with re-injectable valves and testable with vacuum system.

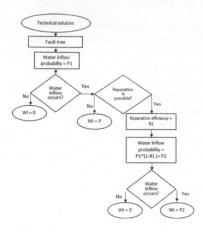

Figure 1. Flow chart of the procedure for the evaluation of mitigation measure effect (modified from Luciani & Peila, 2018).

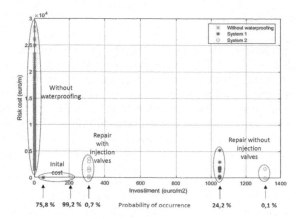

Figure 2. Cost-benefit comparison between two different types of waterproofing for a metro tunnel in urbanized area (modified from Luciani & Peila, 2018).

The obtained results permit to asses that the case without waterproofing has, as expected, a very high risk without any initial cost. The two waterproofed cases have lower risk, and the difference in mean risk between the two is lower than the difference in initial investment. However, the probability of water inflow is bigger for System 1 than for System 2 and, therefore, the related risk is higher. Moreover, the reparability of System 2 permits low costs of repair in case of leakage. The case of repair without foreseen system for System 2 is the case when the repair with re-injectable valves has been performed but it has not been efficient. In the simulation this has occurred only 78 time on 100000 iteration (0,078 %), while the repair without foreseen system for System 1 has occurred 12318 times (12,32 %).

3 PVC DURABILITY

Even if risk analysis is a powerful tool for the evaluation of long-term efficiency and cost of waterproofing systems, it is not able to consider the ageing and durability of the membranes.

Plasticized Poly Vinyl Chloride (P-PVC) is nowadays the most used material for waterproofing membranes. It is composed of PVC, plasticizers, fillers, stabilizers, pigments and other elements. The content of plasticizer is usually ranging between 20 and 35% of the total weight. The P-PVC is subjected to several possible degradations actions that can reduce its properties (Wypych, 2015):

- thermal degradation: for high temperatures HCl is released, the PVC chains break and thus the properties of the material are reduced. The effect of thermal degradation can be reduced with the use of stabilizers;
- UV degradation and weathering: UV rays and weathering can induce fast degradation of polymer chains, damaging the material, causing a reduction of mechanical properties and dimensions;
- microbiological degradation: P-PVC can be subjected to the attack of bacteria and fungi that feed themselves with polymers and plasticizers, inducing a reduction of the plasticizer content, and a stiffening of the material. The effect of microbiological attack can be reduced with antifungal and antibacterial added in the material;
- loss of plasticizer: the plasticizers used for getting an elastic and workable material are usually not chemically linked to the PVC chains, therefore, with time, they tend to migrate from PVC to the surrounding environment. This phenomenon induces the stiffening of the material, a reduction in size and an increasing of the permeability.

Several standardized ageing tests have been developed to study the durability of P-PVC in various conditions. These tests differ for the environment and the conditions of ageing, considering different degradation phenomena. The tests can be summarized in oxidation test, in an air environment accelerated by heat (EN 14575, 2005), immersed test, in water environment accelerated by heat (EN 14415, 2004; SIA V280, 2009) or in different chemical solutions environment accelerated by emphasizing the chemical condition (EN 1847, 2009), micro-organism test (EN 12225, 2000) and UV weathering test (EN 12224, 2000).

None of these tests represents the real conditions of the membrane once it is installed in the tunnel.

For P-PVC membrane oxidation is negligible below 150°C (Titov, 1984), i.e. in all real tunnel applications, UV and weathering are only limited to the storage phase of the material before the installation and are therefore negligible if storage is done in accordance to correct management procedures. The presence of fungi or bacteria is inhibited by the absence of air circulation and the chemical conditions (i.e. pH>8.5) due to the presence of concrete.

Finally, immersed tests seem to be the most representative, simulating mostly the loss of plasticizer from the membrane accelerated using hot water. It is important to consider that, if the temperature is too high, dehydrochlorination can occur due to the higher activation energy. The two phenomena will cooperate in degrading the material. This is not a realistic condition on the jobsite, where the temperature is not enough to start dehydrochlorination.

Nevertheless, also immersed tests, where small specimens are immersed in water, are not simulating exactly the tunnel conditions. In the tunnel, only one side of the membrane is exposed to the water, while the intrados side is in direct contact with the final lining concrete. This means that the path of diffusion of the plasticizer can be different due to a non-uniform flow of the plasticizer and to the different surface conditions. Moreover, due to the small size the diffusion can be almost three-dimensional, while the membrane can be considered as a uniaxial condition due to the very small thickness compared to length and high (some mm compared to 2 to 10 m). Furthermore, the worst condition for the degradation of a real tunnel is when there is a constant flow of incoming water through the geotextile on the extrados side of the membrane, accelerating the removal of the plasticizer from the surface, and consequently the diffusion of plasticizer from the membrane to the surface. Finally, the presence of spritz-beton and concrete creates a particular environment in terms of pH that can affect the membrane degradation and that is not considered in the immersion tests.

Therefore, there was the need for the correct choice to have a test device able to better simulate the tunnel real conditions and to obtain a more realistic estimation of the long-term behavior of P-PVC waterproofing membrane.

4 PROPOSED TUNNEL WATERPROOFING AGEING TEST

4.1 *Test layout*

In order to better simulate the real conditions of the waterproofing membrane in the tunnel the specimens has been created (Figure 3) with 4 layers:

- a 5 cm slab of concrete to simulate the first phase lining;
- a layer of polypropylene (PP) geotextile simulating the protection layer;
- the P-PVC waterproofing membrane;
- a 15 cm slab of concrete to simulate the final lining.

In the test a constant flow of water has been maintained through the PP geotextile to reproduce the flow of the drained water. Water is only flowing on the geotextile side and not on the surface between membrane and the final lining. The PVC specimens dimensions are 150x150 mm; this dimension has been chosen to avoid boundary interference, to simulate the uniaxial behavior of plasticizer flow and to allow mechanical tests on the aged specimens. Bigger specimens would result in difficulties due to the size of the device and to the water flow, and should not have increased significantly the results. On the top of the samples a rigid PVC pipe is installed. This pipe has 150x2mm hoses in the bottom where the PP geotextile is inserted. Therefore, the water flowing in the pipe flows along the geotextile through the sample and then in a U-shaped rigid PVC pipe positioned below the samples. This U-shaped pipe collects the water to the storage tank. A pump keeps the flow pumping water from the storage tank to the inlet pipe. The temperature of the water is kept constant by an electric device installed between the pump and the inlet pipe.

The tested membrane is a commercial standard two-colour signal layer 2 mm co-extruded P-PVC membrane. This membrane has 30% of plasticizer. The protection layer is a commercial Polypropylene non-woven geotextile 500g/m^2. This is the minimum request for tunnel protection layer. In this case, no drainage layer has been considered and the drainage function has been performed by the protection layer, as is often in tunnels without big water flows.

The ageing tests have been performed at three different temperatures: 45°C, 60°C and 75°C for a duration of 3 months.

4.2 *Tests on the aged material*

The following properties have been measured on the original and aged specimens:

- size of the sides of the specimens, measured with a caliper with a precision of 0.02 mm, before and after the ageing;
- weight of the specimens. Before ageing the PVC specimens have been kept in a desiccator and weighted to the nearest 0.1 mg. The weight has been repeated until a constant weight has been reached. This value has been used as initial weight (m_1). After the ageing, the specimens have been cleaned with alcohol to remove all impurities, dried in an oven at 60°C

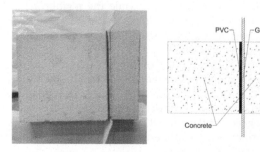

Figure 3.   Sample lateral view.

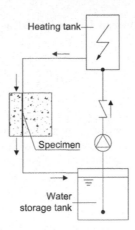

Figure 4. Scheme of the ageing device

for 48 h, then cooled to room temperature in a desiccator and weighted to the nearest 0.1 mg. This value has been used as final weight (m_2).

From calibration tests, the 48 h in the oven at 60°C have result to influence the weight for less than 0.1% of the initial weight, it is therefore negligible in comparison to the weight losses.

- tensile strength and elongation at break in accordance with EN ISO 527 (2012);
- hardness of the specimens have been evaluated with a Shore A durometer in accordance to ISO 7619 (2010);
- foldability at low temperatures has been evaluated following EN 495 (2013).

4.3 Interpretation of the data

In order to asses a long-term durability, an extrapolation to the jobsite temperature can be performed based on the values measured in the accelerated tests. In tests accelerated increasing temperature, the extrapolation through Arrhenius' formula is widely applied. It correlates the rate at which a phenomenon occurs with the temperature following the relationship

$$A = A_0 \cdot e^{\left(\frac{-E}{RT}\right)} \tag{1}$$

where A is the rate of degradation, A_0 a constant, E the activation energy, R the gas constant and T the absolute temperature.

However, the plasticizer loss seems not to be a constant process, as shown by the trend of the test at 60°C (Figure 5).

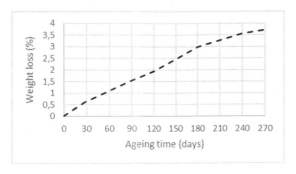

Figure 5. Weight loss at 60°C for 9 month of ageing.

This is because the phenomenon can be considered as a diffusion phenomenon (Neogi, 1996), following Fick's law

$$\frac{\partial c}{\partial t} = D\frac{\partial^2 c}{\partial x^2} \tag{2}$$

where c is the concentration, t the time, x the position in the thickness and D the coefficient of diffusivity. In this case, the concentration of the plasticizer is function of time and of the thickness. Therefore, the process tends to slow down with time. To consider the effect of temperature Fick's law and Arrhenius' equation can be coupled using

$$D = D_0 \cdot e^{\left(\frac{-E}{RT}\right)} \tag{3}$$

In this way the loss of plasticizer evaluated with accelerated tests can be extrapolated to job-site temperature and the long-term durability of the membrane can be evaluated.

5 PRELIMINARY RESULTS

5.1 Size

Due to plasticizer loss the specimens shrink. As an example, Figure 6 shows the average deformation in the transversal and longitudinal direction for 75°C test. Since P-PVC membranes are made by extrusion, some residual stresses are present in the membrane. Therefore, when subjected to relatively high temperatures, such as those of the accelerated ageing tests, relaxation of those stresses leads to dimensional variation: the dimensions in the transversal direction increase and the longitudinal ones decrease. Calibration tests have been performed in order to evaluate this effect. Relaxation release does not occur at jobsite temperature. Therefore, the measured values have been scaled with the calibration values of relaxation. Figure 7 shows the corresponding residual area of the specimens compared to the original one. It is clear that with ageing, and consequently with plasticizer loss, the membrane volume reduces (the thickness deformation has not been analyzed due to the very low values). This can lead to tensile forces in the membrane due to the shrinkage between shotcrete and the final lining.

Nevertheless, this phenomenon can be offset by the high elongation of the membrane and by the long-term stress relaxation of P-PVC.

5.2 Weight

P-PVC weight loss in the designed test can be considered due only to plasticizer loss from the membrane. Therefore, to analyze the plasticizer loss the weight variation has been studied, and the results are reported in Figure 8.

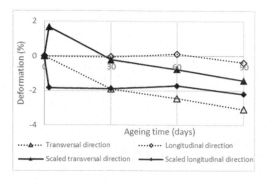

Figure 6.   Deformation of the tested specimens at 75°C.

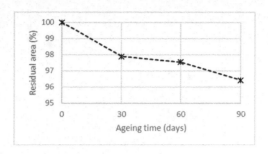

Figure 7.   Residual area of the tested specimens at 75°C.

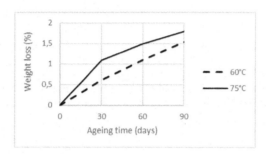

Figure 8.   Weight loss of the specimens.

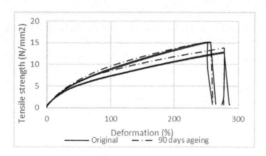

Figure 9.   Stress-strain graph of the original and 90 days aged samples in the transversal and longitudinal direction.

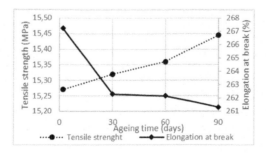

Figure 10.   Tensile strength and elongation at break of specimens aged at 60°C.

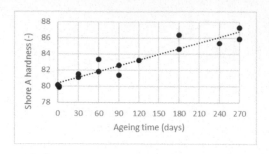

Figure 11.   Shore A hardness of specimens aged at 60°C.

5.3   *Mechanical properties*

The tensile strength and elongation at break are reported in Figure 9 and Figure 10. With ageing and plasticizer loss, the material becomes stiffer, with higher tensile strength and lower elongations at break. All specimens passed the foldability at low temperatures test at -25°C without cracks or failure. The hardness of the membrane increase with the ageing, due to the plasticizer loss, as shown in Figure 11 for specimens aged at 60°C.

6   CONCLUSIONS

The requirements for underground structures becomes always more demanding and the life-time of the project is nowadays more than 100 years. In this frame it is always more important to focus on the long-term efficiency of the components of the tunnel.

Even if waterproofing is only a small amount of the whole cost of the project, it is a key component of the structure for the high impact of possible damages and inefficiencies. There-fore, the long-term efficiency of this component is of great importance. In this paper, two important aspects have been developed to correctly consider this issue.

At first, in the design phase, the correct choice of the waterproofing system through a risk analysis. This leads to the best solution in terms of cost-benefit on the total lifetime of the project. Without the risk analysis approach, unforeseen events can produce unexpected costs and inefficiency of the project.

On the other hand, a careful analysis of the durability of the used material is essential. Now-adays, there is a lack of information on this aspect and the use of accelerated ageing test is unavoidable to analyze the long-term behavior. From the developed test, with the new devel-oped ageing device, the ageing of the material is characterized by reduction of the elastic behavior and stiffening of the membrane. The accelerated ageing tests performed highlight the behavior of membranes during their life. While plasticizer gets lost, the material shrinks in all directions and becomes slowly stiffer, with higher hardness and lower elongation at breaks. Plasticizer loss seems to be the easier parameter to follow in order to verify the ageing. Even though it is not a parameter directly useful in the design, it is linked to the mechanical proper-ties of the material.

The preliminary results show the feasibility of the proposed test procedure and the great importance of duration of these membranes. The results of the aged tests can be compared to the job-site expected values using theoretical relationships: these relationships permit to extrapolate to the job site temperature results obtained in accelerated test and, therefore, to foresee the long-term behavior in the real structure. These results have to be compared to the actual loads and required properties of the project, in order to define the suitability of the waterproofing system for the design life of the structure. Additional studies are ongoing on these aspects to define a correlation among the required properties and the time-dependent behavior of waterproofing systems.

REFERENCES

Bianchi, G. W., Piraud, J., Robert, A., & Egal, E. 2012. Recommendation on the characterisation of geological, hydrogeological and geotechnical uncertainties and risks. *Tunnels et Espace Souterrain* 232: 315–355.

EN 12224. 2000. *Geotextiles and geotextile-related products - Determination of the resistance to weathering.*

EN 12225. 2000. *Geotextiles and geotextile-related products - Method for determining the microbiological resistance by a soil burial test.*

EN 14415. 2004. *Geosynthetic barriers - Test method for determining the resistance to leaching.*

EN 14575. 2005. *Geosynthetic barriers - Screening test method for determining the resistance to oxidation.*

EN 1847. 2009. *Flexible sheets for waterproofing - Plastics and rubber sheets for roof waterproofing - Methods for exposure to liquid chemicals, including water.*

EN 495. 2013. *Flexible sheets for waterproofing - Determination of foldability at low temperature.*

EN ISO 527. 2012. *Plastics - Determination of tensile properties.*

Eskesen, S. D., Tengbor, P., Kampmann, J., & Veicherts, T. H. 2004. Guidelines for tunnelling risk management: International Tunnelling Association, working group no. 2. *Tunnelling and Underground Space Technology* 19: 217–237.

Guglielmetti, V., Grasso, P., Mahtab, A., & Xu, S. 2007. *Mechanized Tunnelling in Urban Areas: design methodology and construction control.* London: Taylor & Francis.

ISO/TS 13434. 2008. Geosynthetics — Guidelines for the assessment of durability.

ISO 7619-1. 2010. *Rubber, vulcanized or thermoplastic - Determination of indentation hardness - Part 1: Durometer method (Shore hardness).*

ITA. 1999. Study of methods for repair of tunnel linings.

ITIG. 2012. A code of practice for risk management of tunnel works. *Tunnels et Espace Souterrain* 232: 315–355.

Luciani, A., Onate Salazar, C. G., & Peila, D. Impermeabilizzazione di gallerie: problematiche ed aspetti realizzativi. *Geoingegneria Ambientale e Mineraria.* (In Press)

Luciani, A., & Peila, D. 2018. Tunnel Waterproofing: Available Technologies and Evaluation Through Risk Analysis. *International Journal of Civil Engineering.* Advance online publication. DOI: 10.1007/s40999-018-0328-6.

Luciani, A., Peila, D., Pavese, E., & Leotta, M. 2018. Risk Assessment approach for waterproofing design. In *The role of underground space in future sustainable citie, World Tunnel Congress 2018.* Dubai.

Mahner, D., Peter, C., & Sauerlander, B. 2018. Long-term behaviour of plastic sheet membranes. *Tunnel* (1): 12–19.

Neogi, P. 1996. Transport phenomena in polymer membranes. In P. Neogi (Ed.), *Diffusion in polymers.* New York, USA: Marcel Decker, Inc.

Peila, D., Chieregato, A., Martinelli, D., Salazar, C. O., Shah, R., Boscaro, A., Dal Negro, E., Picchio, A. 2015. Long term behavior of two component back-fill grout mix used in full face mechanized tunneling. *Geoingegneria Ambientale e Mineraria* 144(1).

Peila, D., Martinelli, D., & Luciani, A. 2016. Use of tunnels for landslide stabilization. *Geoingegneria Ambientale e Mineraria* 148(2).

Sandrone, F., & Labiouse, V. 2011. Identification and analysis of Swiss National Road tunnels pathologies. *Tunnelling and Underground Space Technology* 26(2): 374–390.

SIA V280. 2009. *Kunststoffdichtungsbahnen (Kunststoff- und Elastomerbahnen, Geosynthetische Kunststoffdichtungsbahnen) - Produkte- und Baustoffprüfung, Anwendungsgebiete.*

Thewes, M., & Budach, C. 2009. Grouting of the annular gap in shield tunnelling. An important factor for minimization of settlements and production performance. In *Proceedings of ITA-AITES World Tunnel Congress 2009, Budapest, 23-28 May 2009.*

Titov, W. V. 1984. *PVC Technology.* New York, USA: Elsevier Science Publishing Co.

Usman, M., & Galler, R. 2014. Ageing and degradation of PVC geomembrane liners in tunnels. In *International Conference on Chemical, Civil and Environmental Engineering*: 58–65. Singapore.

Wypych, G. 2015. *PVC Degradation & Stabilization.* Toronto, Canada: ChemTec Publishing.

Tunnels and Underground Cities: Engineering and Innovation meet Archaeology, Architecture and Art, Volume 5: Innovation in underground engineering, materials and equipment - Part 1 – Peila, Viggiani & Celestino (Eds)
© 2020 Taylor & Francis Group, London, ISBN 978-0-367-46870-5

An innovative approach in tunnel planning and construction through 30 years of experience

P. Lunardi
Lunardi Geo Engineering, Milan, Italy

G. Cassani, A. Bellocchio, M. Gatti & F. Pennino
Rocksoil S.p.A., Milan, Italy

ABSTRACT: Analysis of controlled deformation in rocks and soils is a universal approach in planning and construction of underground works (named A.DE.CO), which stands apart from all previous methods for having always pursued to attain the respect of times and costs of completing works, independent of the excavation system used, whether it be mechanized or traditional. Additionally, respecting times and costs of production works is in the interests of Clients and the Contractors alike, and can be achieved only by means of industrialization of the excavation. The starting point of the approach is the reinforcement -in poor geotechnical conditions- of the tunnel core-face of the excavation by means of fiber glass elements. The practice, which started almost 30 years ago in Italy, has proven to be suitable in the most complex geological contexts, in clays, in soft grounds and in rocks and for whatever stress state conditions and tunnel dimensions.

1 INTRODUCTION

Analysis of controlled deformation in rocks and soils approach represents a universal approach to planning and construction of underground works. This method was developed in the '80s in Italy by Professor Pietro Lunardi to deal with heterogeneity of soils and difficult ground conditions. Its introduction to tunneling practice can be dated to 1985, almost thirty years ago, when it was introduced for the construction of some tunnels of the new High Speed railway line from Florence to Rome (the Talleto, Caprenne, Tasso, Terranova, Le Ville, Crepacuore and Poggio Rolando tunnels) for a total length of about 11,0 km.

Italian country is characterized by a very complicated geological and geomechanical contexts: difficult ground conditions like landslide, heterogeneity of soils like flysch, clays and claystones, low strength ground parameters. At the same time Italy presents the longest network of tunnels worldwide (more than 6000 Km). A high percentage of them is in urban area (more than 800 Km in last 15 years). Over 1000 Km of tunnels have been constructed in urban and extra-urban areas. Even though being constructed in very complex and difficult ground conditions, generally they were completed (from the design stage to the construction and actual opening of activity) with industrial methods and respect to the foreseen construction times and costs. Today we know that full-face advancement is all the more necessary the more difficult the excavation conditions are.

The paper summarizes the innovative approach for the design and construction of tunnels through the important experience gained during the design and construction of the biggest infrastructural projects developed in Italy during the last 30 years.

The significant knowhow was fundamental to outline the criteria that helped to define modern design and construction concepts through the experience of:

- High Speed Railway System - Milan to Naples Railway Line: Bologna-Florence Section (Total underground excavation 104,3 km)
- High Speed Railway System - Milan to Naples Railway Line: Rome-Naples Section (Total underground excavation 28,3 km)
- High Speed Railway System - Milan to Genova Railway Line (Total underground excavation 106 km)
- Highway A1 Milan-Naples: Bologna - Florence Section (Total underground excavation 62,2 km)

The method has been traded in Europe, too. Tartaguille, Visnove and Sochi Tunnels are the best known examples.

2 ANALYSIS OF CONTROLLED DEFORMATION IN ROCKS AND SOILS APPROACH

The idea began to make headway in the authors' mind as part of the effort to develop the Analysis of controlled deformation in rocks and soils approach to tunnel design and construction at the beginning of the '80s.

This kind of approach is based on the following two main concepts:

- the significance of the deformation response of the ground during excavation, which the tunnel engineer has to be able to fully analyse and then control;
- the use of the advance – core of the tunnel as a stabilization tool to control deformation response and to get a prompt stability of the excavation.

After more than 30 years it has been widely demonstrated that it is possible to safely and efficiently drive tunnel in whatever geomechanical and in situ stress context, even in the most critical ground conditions, working full face and reinforcing in different ways the core - face of advance. Main advantages of this approach are the industrialization of the works: one of the great innovation of Analysis of controlled deformation in rocks and soils approach is the simplification and cleanliness of the construction operations into the tunnel. With comparison to the past tunnel practice, there are always few workers moving into the tunnel (not more than six persons in each stage of construction), with few and powerful machines and equipment to be used), with a clear sequence of operations. There is never a superposition of different kind of interventions or operations. All these aspects lead to an increase in safety conditions and to an industrialization of the work into the tunnel:

- there is a direct relationship between the number of workers and of the machinery and safety: the less they are, the greater is the safety;
- excellent production rates - in whatever ground and overburden it is possible to ensure 1- 4 m/day of advance rate in tunnels with a cross section of about 160 sq.m.

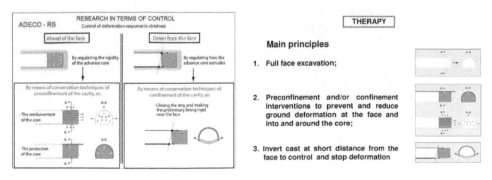

Figure 1.   The analysis of controlled deformation in rocks and soils approach for tunnelling – main principles.

– certainty of costs - the deep understanding of the deformation response to be expected during excavation and the correct design of the necessary interventions to be adopted, are the key point to have a good design, ensuring sure time and costs of construction.

To deal with all possible situations it is necessary, especially in the worst geomechanical conditions, to act as soon as possible to prevent deformation around the cavity, this experimental evidence brought to the concept to reinforce and strengthen the core – face to counteract deformation at its real onset, that is usual at 1 to 2 diameter ahead of the face of excavation, deep inside the core of advance. This is, at the end, the great innovation introduced by the new approach with respect to other tunnel design approaches: to control efficiently deformation in bad soils (clays, sands, soft and weak rock) to act just from inside the cavity may be ineffective, a preconfinement action has to be applied. Preconfinement of the cavity can be achieved using different techniques depending on the type of ground, the in situ stress and the presence of water. Sub horizontal jet grouting, fiber glass elements, injections by TAM (Tubes A Manchettes), etc. may be used to get an efficient reinforcement of the core and face.

There are two different kind of interventions:

– protective conservation is the one set in advance, working around the perimeter of the cavity to form a protective shell around the core able to reduce deformation on the core-face system;
– reinforcement conservation is the one set directly into the core of advance, to improve its natural strength and deformation parameters.

Systematic fiber glass reinforcement of the advance core is a typical conservative measure introduced by the approach in tunneling practice. This intervention is not to be seen simply as a traditional soil nailing dimensioned at equilibrium limit. It means not just the stabilization of the face as in traditional tunnel face bolting, but also the reduction of deformation of the ground by improving the characteristics of strength and elasticity of the ground. This assumption is very important: face grouting is not to be dimensioned just as a stabilizing element but as a tool to reduce ground deformation as close as possible to the elastic limit. Very often this concept has been the less understood point of the approach. The technology mainly consists of dry drilling a series of holes into the tunnel face; fiber glass elements are then inserted in the holes and injected by cement mortar. It may be applied in cohesive o semi cohesive soils and in soft rocks and combines great strength properties of the material with high fragility, never becoming an obstacle for excavation. Length, number, overlap, cross – section area and geometrical distribution of the reinforcement constitute the parameters that characterize this reinforcement technique. Starting from its first application in 1985 there was a sort of evolution with respect to these parameters: this evolution came out first of all from design and construction experience and then from the evolution of the fiber glass elements themselves, and of the machines used to drill and to install the reinforcement.

Figure 2.   Core face reinforcement in clayey materials, cemented sand, soft rocks.

In 1985 the usual length of these elements was 15 m and the overlap between one grouting intervention and the other could vary between 5 and 6 m. Nowadays, thanks to the improvement of the equipment to be used for fast horizontally drilling, it is possible to reach 24 m of length for the fiber glass elements and the overlapping may vary between 6 and 12 m. This innovation allowed a best control of deformation of the core – face, especially in tunnels with a great diameter, and a further increase in the rate of advance of excavation: the longest the fiber elements are, the less it is necessary to stop to make a new grouting intervention. This way it has been possible to reach a production rate of 35 – 45 m/month for a section of about 160 sq.m. in clays and soft rocks. This production rate also includes the final lining casting and may be intended as the rate to get the completion of the tunnel (grouting, excavation and final lining). In the years several fiberglass profiles were tested and adopted. The first element to be used was a 40/60 mm pipe; this showed to be very good but had sometimes problem of transportation with length greater than 15 m.

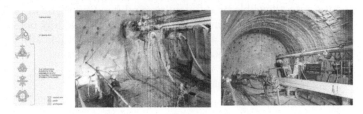

Figure 3.   Fiber glass element profile Core face reinforcement execution insertion and execution drilling of the fiber glass element.

To solve this problem and to reduce the final cost on site of the fiber glass elements, special profiles were introduced using band of fiber glass differently composed together. These elements are more flexible than pipes and can be rolled and transported in length up to 24 m. They are then assembled on site; it must be considered that tubular elements always have a considerable shear resistance that the above mentioned elements do not have; this aspect must be considered in the design of the intervention. Another important issue to be treated talking about the execution of a core-face reinforcement using fiber glass elements is cementation. When inserted into the hole, the element is then cemented to fill the drilling hole. The system works thanks to the adherence between the grout itself and the ground. The greatest is the adherence, the most effective is the core-face reinforcement. Several efforts were made in these almost thirty years to improve adherence: one of the most important was the use of expanding mortars. The expansion is obtained using several agents to be mixed together with the cement in the injected grout. These new mortars had a great success in the middle of the "90s because of the increase of adherence they are capable of. The first systematic application was in some tunnels of the Bologna – Florence High Speed railway line in 1997. These mortars are constituted mixing on site cement and an alluminate expanding agent.

Approximately in 2010 a new ground element has been introduced in the Italian market, in order to improve adherence characteristics: it is a fiber glass element constituted by a corrugated pipe. It is proven that these new elements give very good results about adherence characteristic.

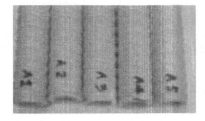

Figure 4.   Corrugated ground elements and Fiber glass extraction test.

Numerous in situ tests and measurements were performed during tunnel advance for in – depth study of both the nature of the interaction between the fiber glass elements and the surrounding ground (extraction tests, strength and deformation tests) and the effect of the reinforcement of the core-face in these three decades (extrusion and convergence measures).

Different methods of measuring extrusion were developed and are now widely used in tunneling along with the more traditional convergence measurements.

Extrusion measures were done inserting incremental extensometers into the core. The results of extrusion, preconvergence and convergence measurements allowed to increase the theoretical knowledge of the stress – strain behavior of a tunnel at the face and they confirmed the effectiveness of the new technology in controlling deformation. Monitoring data represents an important tool to fine tuning the face reinforcement intervention, increasing or diminishing the overlapping and increasing or diminishing the number of elements to be applied into the face. The approach is in this way very flexible and guarantees to work always at the maximum production rate allowable by different ground conditions. In worst geomechanical conditions this tool is absolutely indispensable and effective to avoid any possible trouble into the tunnel, including collapses. Fitting all this information it is practically impossible to be found unprepared to unforeseen geological and geotechnical variations.

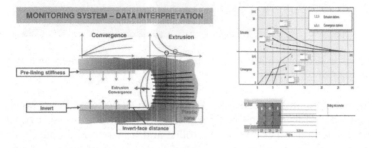

Figure 5.   Monitoring data – combined extrusion and convergence measurements.

The results obtained directly from the tunnels advance are the ones that better give the opportunity to support the success of the ADECO – RS approach when using fiber glass elements as reinforcement of the core – face of advance.

Different parameters control the core – face deformation when driving a tunnel:

a. Ground parameters (cohesion, friction angle, modulus of elasticity, pore pressure, water flow, overburden, stress state, ground constitutive model, stratigraphy and strata inclination)
b. Tunnel parameters (dimension and geometry, construction stages and sequences)
c. Reinforcement parameters (number of elements into the face, position/distribution, inclination, length, overlapping, area, tensile strength, shear strength, Young modulus, diameter of the drillings).

A good 3D FEM or DEM model, able to produce results comparable with the ones collected on site during excavation, may help the tunnel engineer to make several sensitive analyses and to guide the final choose of the distribution and length of the fiberglass elements at the face and into the core. 3D models are useful to give evidence of the behavior of the reinforced core; if we compare the analyses of a tunnel at low overburden with those of a tunnel at high overburden, in absence and presence of core-face reinforcement, it is proven that the use of core-face reinforcement produces the effect to create a similar behavior of the tunnel at low overburden with the one at high. The stress redistribution induced by the reinforcement during excavation is generally more homogeneous on the height of the tunnel and the displacements of the core-face are reduced. This is very important to control the risk of potential collapses starting from the crown in tunnels at shallow overburden. This is a crucial issue in urban areas, where important interferences with existing buildings and underground utilities have to be considered and preserved. But it is important, too, to have not significant stops of the tunnel during construction, to avoid any kind of collapse even in less sensible areas. These stops should cause in fact major time and cost.

In the most recent Italian codes and prescriptions for public works (D.M. 20 august 2012 n. 161), fiberglass element is finally described as an inert material. This has definitively solved the environmental issue of muck transportation to final landfill when using this kind of reinforcement.

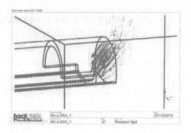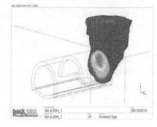

Figure 6. Numerical analysis of a tunnel at low overburden (20 m).

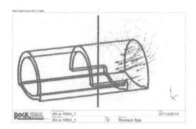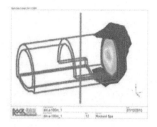

Figure 7. Numerical analysis of a tunnel at high overburden (100 m).

3 CASE HISTORIES

ADECO-RS approach has been adopted in the most complicated geological and geomechanical contexts. A tough geological study of the area interested by the future infrastructure is the starting point of an efficient and successful underground project. The following road and rail tunnels represent the consolidation of the of ADECO-RS approach.

3.1 *High Speed Railwayy system – Milan to naples Railway Line-Florence Section*

The new railway line between Bologna and Florence is part of the Italian Milan-Rome-Naples route of the High Speed/Capacity Train which constitutes the southern terminal of the European rail network. The total length of the alignment is more than 78.5 km, constituted by twin track underground tunnel. This was a pilot experience for the whole of the major infrastructure sector and not just because the project was awarded to a single general contractor for the first time in Italy, but also because of the technical difficulties that had to be overcome. The geological and geotechnical context appeared to be, and so it was, one of the most difficult and complex in the world. A wide and heterogeneous variety of grounds had to be tunnelled, affected by groundwater and gas at times and under overburdens ranging from nil to very deep. A new design and construction approach was selected to tackle the heterogeneity of the materials and conditions, which is based on the "Analysis of the COntrolled DEformation in Rocks and Soils" (ADECO-RS). By using this approach, which is based on a clear distinction between the design stage and the construction stage of tunnels, it was possible to reliably estimate construction costs and times for the project in advance and this in turn made it possible to award contracts for the work on a lump sum "turnkey" basis for the first time in the history of underground works for projects of these dimensions and difficulty. Tunnel construction on the Apennines section between Bologna and Florence of the new Milan-Rome-Naples high speed/capacity railway line finished after less than six years since the first excavation of the running tunnels started. Today the route from Bolognas to Florence with more than 70 km of twin track tunnel, which has a cross section of 140 sq. m. is under traffic. The exceptional complexity of the grounds to be crossed was well known. It had already been tackled with great difficulty for the construction of the "Direttissima" railway line inaugurated in 1934 and currently still in service. It was then decided to invest a sum of € 84 million, 2% of the total cost of the project in the geological survey campaigns required for detailed design of the new high speed/capacity railway line. This provided a geological-geomechanical characterisation of the ground to be tunnelled that was very detailed and above all accurate. The formations are primarily of flyschoid, clay and argillite formations and loose grounds, at times with substantial water tables, accounting for more than 70% of the underground alignment, with overburdens varying between 0 and 600 m.

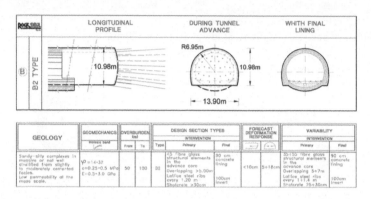

Figure 8.   Section type variability as a function of the actual response of the ground to excavation.

The contract for the entire section of the railway between Bologna and Florence was awarded on a rigorous lump sum basis (€ 4,209 million) by FIAT S.p.A., the general contractor, which accepted responsibility for all unforeseen events, including geological risks on the basis of the final design as illustrated above. It subcontracted all the various activities out to the CAVET consortium (land expropriation, design, construction, testing, etc.). Immediately after the contract was awarded, the construction design of works began at the same time as excavation work (July 1996). Additional survey data and direct observation in the field generally confirmed the validity of the detailed design specifications, while minor refinements were made in the construction design phase.

3.2   High Speed Railway System – Milan to Naples Railway Line-Naples Section

The new Rome-Naples railway is part of the High Speed Train Milan-Naples line which in turn represents a southern terminal of the European High Speed network. The total length of the line is 204 km and 28.3 km of this (equal to 13% of the total length) runs underground. Rocksoil designed 22 tunnels for a total length of 21.8 km. The underground route runs through ground which can basically be classified as having two different types of origin: pyroclastic ground and lava flows, generated by eruptions of the volcanic complexes of Latium, of Valle del Sacco and of Campania; sedimentary rocks of the flyschoid and carbonatic type (marly and limy argillites) belonging to the Apennine system.

The overburdens vary greatly but never exceed 110 m., while they are often very shallow at the portals. The final and construction design of the tunnels was performed using the Analysis of Controlled Deformation in Rocks and Soils approach (ADECO-RS).

3.3   High Speed Railway System – Milan to Genova Railway Line

The Terzo Valico is the new high-speed, high capacity railway line that will improve connections between the Liguria port system and the main railway lines of Northern Italy and the rest of Europe. The Terzo Valico represents one of the strategic projects of national interest and is part of the Rhine-Alpine Corridor, one of the main corridors of the trans-European strategic transport network (TEN-T core network) connecting Europe's most densely populated and most important industrial regions. The total length of the alignment will be approximately 53 km.

The project requires the construction of 36 km of tunnels running through the Apennine Mountains between Piedmont and Liguria. The full scope of underground works, including dual tube single-track running tunnels, accesses and interconnections, for a total underground excavation of about 106 Km. About 60% of the design alignment crosses the lithological unit defined as Argille a Palombini – claystone schist with limestone lens – (hereinafter aP), which therefore is particularly important for the implementation of the project.

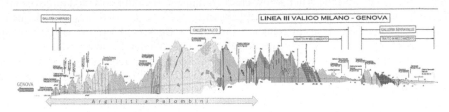

Figure 9. Longitudinal geological profile.

The excavation intersects the aP at the southern part of the alignment, along the stretch Genoa-Voltaggio (AL), thus embracing the excavation of mainline tunnels, access tunnels and interconnections with the old Voltri railway. Excavations in aP are mainly designed and executed with conventional method by full face tunnel advance and cross sections area between 75 and 400 sq.m. The only exception is the Polcevera access, which was carried out by an EPB Tunnel Boring Machine, with an excavation diameter of approximately 10 m.

Figure 10. "Argille a Palombini" Formation – claystone schist with limestone lens.

Last but not least, the project and construction for the new high speed railway line faced one of the most relevant environmental issue: safe excavation and managing of rocks, that may contain asbestos in relevant amounts. Asbestos is a group of fibrous minerals that mainly occurs in mafic and ultramafic rocks (ophiolitic sequences). This work focuses the criteria to better evaluate the amount of asbestos fibres in the metaophiolites belonging to Sestri-Voltaggio Zone (Liguria, Northern Italy), the criteria for the environmental monitoring system and management of disposal materials used by Cociv for checking asbestos risk along tunnel alignment. The Cravasco Tunnel adit section, one of the four accesses tunnels to the main tunnel of the Terzo Valico, of approx. 1,260m in length has represented the pilot experience to face tunnel excavation issues when asbestos is present. Cravasco Tunnel crosses a part of the complex geological context within the Sestri-Voltaggio Zone. This area represents the core zone of Sestri – Voltaggio area and is constituted by a large number of different lithologies: from dolomites to chalk stones, from argilloschists to metabasalts to serpentinites and ophicalcites.

3.4 *Motorway A1 Milan-Naples_ Bologna – Florence Section*

The accurate geological, geotechnical and hydrogeological model for Tunnel construction on the Apennines section between Bologna and Florence of the new Motorway A1 (total of underground excavation of about 62 Km) detected the presence of a very complex geological and geomorphological context in the area between Bologna and Florence. Complexity derived both for the presence of a heterogeneity of soil conditions (flysch, clays and claystones) and landslide soil conditions. When tunneling in presence of a landslide, the main design target should be the reduction of the disturbance caused in the ground by the excavation. The "ADECO – RS approach" to tunneling design and construction is well capable to manage this kind of difficult and delicate situations.. A detailed monitoring system should be designed in detail, highlighting the reading frequency and the criteria/flow-chart to manage the data collected and to enable corrective actions in the presence of unexpected behaviour or exceeding the thresholds defined. Some recent experiences, collected on different job sites of the Bologna–Florence Motorway, are presented in the following. Geological-geotechnical context is mainly affected by flysch consisting of argillites, siltites and clayey marls. The interaction between the underground excavations and the overlying slopes, affected by a complex system of landslides, will be focused through the exam of movement data, derived from an extensive monitoring program. Both conventional and mechanized tunneling will be analyzed to compare the results and to evaluate similitudes and differences.

Val di Sambro Tunnels are located at the base of a slope degrading to the Setta River, mainly consisting of deposits (silty clays and sandy silts) laying on the rock-mass (flysch of Monghidoro Formation); the thickness of the deposits is ranging between few meters up to 35-40 m, and represents a big landslide, quiescent in nature, with local active area of lower thickness (max.10-15 m). During the design stage, the alignment of the tunnels was located deep into the slope, so to present a big distance from the topographic surface and from the base of the landslide, with overburden into the range of 60-90 m. Further investigations and a new geological model were necessary. The area interested by the movement is very large, even if within the same it was possible to point out several components: the main one is connected to the deep landslide. It is located about ten meters over the tunnels alignment and with a planimetric angle of 45°. The new information about geological model and the data collected about interferences gave the possibility to redefine the design data and revalue the tunnel construction process. Once known the real geometry of the landslide, it was not possible to avoid interference with the tunnels constructions, so that it became very important to adopt a construction system able to minimize the impact with the pre-existences (the buildings and the slope itself!), minimizing the deformation inside the ground during excavation. Before the tunnels excavation was stopped, conventional system was adopted, without face support, just confining the cavity by steel ribs and shotcrete. This system was upgraded, in the spirit of ADECO Approach, according to Lunardi (2006). New pre-confinement and confinement actions were defined, in order to minimize face extrusion, convergence and settlements into the tunnel, which had reached 10-15 cm in the first stage of excavation. Several support sections were prepared according to the distance from the landslide surface, the landslide nature (active or deep landslide) and the face position with respect to the buildings location; specific "guidelines" were prepared too, in order to govern the support section application and to fine tuning the number of intervention.

Sparvo Tunnels were located in the opposite bank of the Setta River with respect to Valdisambro Tunnels; a bridge between the two tunnels crosses the river. But the slope of this bank is, at the same time, interested by several landslides, because of the poor condition of the ground, mainly represented by clays of the "Argillite a Palombini" formation, and considering the water level, which is very high. To face the tunnels excavation, a TBM-EPB was adopted; it was a very challenger project, considering the excavation diameter (15.56 m) and the TBM working parameters, such as thrust (up tp 390 MN) and chamber pressure (3.5-4.0 bar in crown, up to 7-8 bar at the base of the machine). The geotechnical dimensioning of the TBM and the construction process are described in Gatti et al. (2011) and Lunardi, Cassani and Gatti (2013). With respect to the interference with the landslides located along the tunnels alignment, similar criteria with Valdisambro Tunnels were used: the aim of the construction method was to minimize the deformation response after excavation, mainly maintaining the clays' strength parameters closed to the peak value (elastic domain), neglecting, as much as possible, plasticization in the ground. Also this goal was achieved applying the ADECO Approach: several detailed analyses were performed to define the correct chamber pressure to properly confine the core-face. For the tunnel stretch under discussion, with overburden equal to 110-120 m, chamber pressure equal to 3.5 bar (at the top of the excavation chamber) was defined, to confine external ground pressure with low stress-strain redistribution. In this way, as assessed in Gatti (2011), it was possible to reduce the plasticization ring around the excavation profile, and consequently on one hand the friction forces on the TBM shield ("jamming risk control") and on the other the propagation of the movements towards the ground surface.

4 ADECO-RS APPROACH IN THE WORLD

ADECO-RS approach was successfully implemented not only in Italy. The first case history is represented by the Tartuaguille Tunnel (1997), new High Speed Rail line between Marseille and Lyon in France. More recently it was applied in Russia (Sochi Motorway, T8 and T8A tunnels) and in Slovak Republic (D1 Motorway, Visnove Tunnel). One of the most important and recent case history of application of ADECO-RS approach all over the world is represented by the Sochi Tunnels construction in Russia. During the preparation for the XXII Winter Olympic Games in Sochi (Russia) from 7 to 23 February 2014, the Russian Federation allocated important investments in fixing the city's lack of infrastructure and in strengthening its transportation network. One of these projects is the new Sochi by-pass motorway, also known as "Dubler Kurortnogo Prospekta", which runs parallel to the black sea and makes it possible to reach the Olympic sites and the Adler airport without having to cross the city. The new Sochi by-pass motorway presents two separate carriageways, each with 2 lanes per direction, adjusted for a design speed of 120 km/h. Construction required the excavation of a series of tunnels. The T8 and T8A tunnels crossed the most difficult geological context because of the presence of clays, claystones and sandstones, which are flysch conditions. Extra difficulties were represented by the reduced times available to design, construct and open the works, considering the length of the underground layout (1,550 m for the T8 tunnel and 1,523 for the T8A tunnel) and the size of the excavation faces, ranging from 120 sq.m up to 220 sq.m. The "Dubler Kurortnogo Prospekta" tunnel, is made up of 6 double-bore tunnels for a total of: 10 tunnels bored using the NATM approach (New Austrian Tunnelling Method) and 2 tunnels using the ADECO-RS approach (Tunnels 8 and 8a). It was the first time that these two approaches were directly and reliably compared to each other, in relatively similar conditions (despite the difficulty regarding those tunnels constructed using the ADECO-RS). The two approaches are compared both in terms of average production and in terms of "bored volume/month" and in terms of "meters of tunnel finished/month". The geological conditions affecting the 8-8a tunnel excavation were clearly worse than those of the other tunnels, and the excavation sizes were also greater. The tunnels bored using the NATM passed through better contexts with excavation sizes just equal to 113-115 sq.m. The comparison between the two approaches clearly exhibits extreme efficiency of the ADECO-RS system over the NATM: in terms of volume of excavation/month for single face the ADECO-RS approach produced results 2.4 times higher than the NATM; in terms of linear metres of completed tunnel, productivity was 40% greater than the NATM approach. The ADECO-RS approach is naturally fast in closing the lining, in the case of SOCHI this was completed under 3 weeks. Instead, the NATM approach showed much greater times: from a minimum of 10 weeks to a maximum of 43; such a time frame has clear repercussions on safety. From this point of view, the ADECO.RS method can complete all necessary safety measures for the tunnel in a short time frame, thus greatly limiting the deformation phenomena linked to excavation.

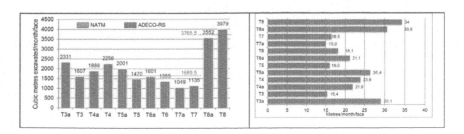

Figure 11. Comparison of ADECO-RS and NATM in the Sochi Tunnels, Russia.

5 CONCLUSIONS

The important knowhow and experience described above produced the standards for full face tunnel excavation. The first idea began to make headway in the authors' mind as part of the effort to develop the ADECO–RS approach to tunnel design and construction at the beginning of the '80s. The new railway High Speed/Capacity line between Bologna and Florence was a pilot experience for the whole of the major infrastructure sector because of the technical difficulties that had to be overcome. One of the most difficult and complex geological context in the world has been faced successfully with ADECO-RS method. Today we know that full-face advancement is all the more necessary the more difficult the excavation conditions are. To deal with all possible situations it is necessary, especially in the worst geomechanical conditions, to act as soon as possible to prevent deformation around the cavity, this experimental evidence brought to concept to work with a reinforcement and strengthening of the core – face to counteract deformation at its real onset, that is usual from 1 to 2 diameter ahead of the face of excavation, deep inside the core of advance. This is at the end the great innovation introduced by ADECO-RS with respect to other tunnel design approaches: to control efficiently deformation in bad soils (clays, sands, soft and weak rock) to act just from inside the cavity may be ineffective, a preconfinement action has to be applied. These important results have been made possible since the public administrations have adopted modern design and construction approaches in their contract specifications, based on strictly scientific criteria.

The memory of previous experiences contributed to develop and improve the approach, that it is still an ongoing process, looking for further improvements of equipment, solutions, safety and environmental aspects.

REFERENCES

Lunardi, P. 2000. Design & constructing tunnels – ADECO-RS approach. In *Tunnels & Tunnelling International*, Special Issue, May.

Lunardi, P. 2001. The ADECO-RS approach in the design and construction of the underground works of Rome to Naples High Speed Railway Line: a comparison between final design specifications, construction design and "as built". In *AITES-ITA World Tunnel Congress "Progress in tunnelling after 2000", Milan, 10th-13th June 2001*, Vol. 3, 329–340

Lunardi, P. 2008. *Design and construction of tunnels: Analysis of Controlled Deformation in Rock and Soils (ADECO-RS)*. Berlin: Springer.

Lunardi P., Bindi R. & Cassani G. 2007. From the ADECO-RS approach to the tunnelling industrialisation. In *Proceedings of the International Conference on "Tunnels, drivers of change"*, Madrid, 5-7 November.

Lunardi P., Bindi R. & Cassani G. 2014. The reinforcement of the core face: history and state of the art of the Italian technology that has revolutionized the world of tunneling. Some reflections. In *Proceedings of the World Tunnel Congress 2014-Tunnels for a better life*. Foz de Iguaçu, Brazil

Lunardi G., Cassani G. & Bellocchio A. 2014. Construction of the T8 and T8A tunnels in the "Dubler Kurortnogo prospekta" in Sochi (Russia). First implementation of the ADECO-RS approach in the Russian Federation. In *Gallerie e grandi opere in Sotterraneo n. 110*

Lunardi P., Cassani G. & Gatti M. 2017. Planning of tunnels in landslides situation: the experience in the Italian Apennines. In *Proceedings of the World Tunnel Congress 2017-Surface challenges-Underground solutions*. Bergen, Norway

Lunardi G., Cassani G., Bellocchio A., Frandino M. & Baccolini L. 2017. High Speed Railway Milan – Genoa, Parametric analysis of rock stress-strain control during tunnel excavation in the "Argille a Palombini" Formation. In *Proceedings of the World Tunnel Congress 2017-Surface challenges-Underground solutions*. Bergen, Norway

Lunardi P., Cassani G., Bellocchio A., Pennino F. & Poma F. 2017. Naturally occurring asbestos in the Rocks belonging to Sestri –Voltaggio Zone (Liguria, Northern Italy), Excavation railway tunnels management. *In Proceedings of the World Tunnel Congress 2017-Surface challenges-Underground solutions*. Bergen, Norway

Author Index